Indian Institute of Metals Series

About the Book Series:

The study of metallurgy and materials science is vital for developing advanced materials for diverse applications. In the last decade, the progress in this field has been rapid and extensive, giving us a new array of materials, with a wide range of applications, and a variety of possibilities for design of new materials, processing and characterizing the materials. In order to make this growing volume of knowledge available, an initiative to publish a series of books in Metallurgy and Materials Science was taken during the Diamond Jubilee year of the Indian Institute of Metals (IIM) in the year 2006, and has been published in partnership with Springer since 2016.

This book series publishes different categories of publications: textbooks to satisfy the requirements of students and beginners in the field, monographs on select topics by experts in the field, professional books to cater to the needs of practising engineers, and proceedings of select international conferences organized by IIM after mandatory peer review. The series publishes across all areas of materials sciences and metallurgy. An panel of eminent international and national experts serves as the advisory body in overseeing the selection of topics, important areas to be covered, and the selection of contributing authors.

Raghumani S. Ningthoujam · A. K. Tyagi
Editors

Handbook of Materials Science, Volume 3

Catalytic Materials

THE INDIAN INSTITUTE OF METALS
IIM
Metallurgy
Materials Engineering

Springer

Editors
Raghumani S. Ningthoujam
Chemistry Division
Bhabha Atomic Research Centre
Mumbai, Maharashtra, India

A. K. Tyagi
Homi Bhabha National Institute
Mumbai, Maharashtra, India

ISSN 2509-6400 ISSN 2509-6419 (electronic)
Indian Institute of Metals Series
ISBN 978-981-96-8228-7 ISBN 978-981-96-8229-4 (eBook)
https://doi.org/10.1007/978-981-96-8229-4

This Springer imprint is published by the registered company Springer Nature Singapore Pte Ltd.
The registered company address is: 152 Beach Road, #21-01/04 Gateway East, Singapore 189721, Singapore

Preface

The growth of human civilization has always depended on materials. Thus, evolution of materials has been an integral part of human beings right from the primitive times. The importance of materials can be gauged by the fact that various historic ages have been named based on the materials used by the people belonging to a particular era, such as stone age, bronze age, iron age, concrete age, steel age, and silicon age. Some of the earlier processes like forging and smelting have played a key role in growth of materials and their large-scale deployment for the benefit of human-kind. However, during the initial phase, the materials were used without much understanding of the factors responsible for their functioning. For example, Romans who were great innovators of glasses and invented the glass window could not predict how important this material would be in future. It is believed that the first metal tool appeared perhaps only six thousand years ago as a result of knowledge of melting and casting processes. Materials that have immensely changed our lives are wood, clay, bronze, iron, steel, cement, polymers, aluminum, glasses, and semiconductors. It shall not be wrong to claim that the twenty-first century is going to witness a tremendous progress in the field of materials science as the need for advanced materials in various areas, such as engineering, electronics, communications, healthcare, societal sectors, and natural resources management; generation of wealth from waste is going to exhibit rapid growth. Even the demand for materials for high-tech sectors shall witness exponential growth. Net-zero target will also be intricately linked with the advancement in the field of materials science. The quest for developing better, efficient, and cost-effective materials has been a constant endeavor and shall continue forever. It is generally believed that materials are going to play most important role to achieve the sustainable development goals, in particular for clean energy, clean water, climate change, health care, and industries, innovation, and infrastructures. In this context, there is an urgent need to develop efficient materials consisting of earth-abundant elements. The solution for some of the challenges, such as production of green hydrogen and green synthesis of ammonia, will also come from advanced materials. The rational design of newer and efficient materials by interplay of synthesis methodologies, crystallographic structures, morphologies, and dimensionality is going to be of paramount importance. In coming years, a diverse range of materials will have to be developed to meet these goals. In view of this, it was felt appropriate to publish a *Handbook of Materials Science* comprising of ten volumes on optical materials, magnetic materials, catalytic materials, photocatalytic materials, nuclear materials, sensor materials, materials for health care, energy materials, materials for environmental applications, and soft materials.

Volume 3 (Catalytic Materials): Materials used to efficiently enhance chemical reactions are referred as catalytic materials without any change in its properties after the reaction. These materials can be in the form of ions, molecules, and cluster of atoms or particles. Particles are of metallic, semiconducting, or insulating types. By changing size and shape of particles, their catalytic properties may vary. Even, if dimension of particles changes from 0D, 1D, and 2D to 3D, its catalytic property varies considerably. Nowadays, catalytic materials in the form of heterogeneous system or core–shell model or hybrid model are coming up. Their catalytic property can vary with temperature, light, electric field, magnetic field, solvent, or surrounding environment. Catalyses can be divided into three categories, namely homogeneous, heterogeneous, and enzymatic catalyses. In this volume, the basic concepts of catalyst, adsorption, absorption, desorption, adsorption isotherm, frustrated Lewis pairs (FLPs) and its importance in cleaving the H–H bond, practical applications of FLPs in reactions, such as hydroboration, hydroarylation, hydrosilylation, polymerization, transfer hydrogenation, amination, and N–H activation, clusters (aggregates of a finite number of atoms/molecules held together by interactions ranging from van der Waals forces to ionic or covalent or metallic bonds) and its applications in oxidation of CO and hydrogenation reaction, synthesis of rare earth (RE) and mixed metal nanoparticles and their catalytic activities toward CO oxidation, metal–organic frameworks (MOFs) materials and its applications in organic transformation, especially oxidation reaction, reduction, hydrogenation, metal-catalyzed cross-coupling reaction, and acid–base catalyzed reaction, the critical role of photocatalysis in addressing environmental and energy challenges within organic synthesis, enzyme catalysis, a biological molecule, biocatalysts for clean and sustainable routes for production of pharmaceuticals, agrochemicals, polymers, food additives, transition metal oxides and hydroxide-based catalysts, graphene oxide (GO), reduced graphene oxide (r-GO), and g-C_3N_4 for degradation of organic pollutants under solar radiation, the ionic liquid in the catalytic applications, MOFs for nitroarene reduction, heterogeneous electrophilic catalysis, synthesis gas (syngas), conversion from CO_2 to dimethyl ether (DME), the C–H bond activation, single-atom catalysts (SACs), the application of artificial intelligence (AI) and machine learning in designing catalysts, metal–metal-oxide interfaces on CO oxidation and ammonia borane hydrolysis, turnover numbers, noble metal catalysts, metal complexes, layered double hydroxide (LDH)-based electrocatalyst, density functional theory (DFT) to understand adsorption and desorption, and MOF-supported transition metal nanocatalysts for the generation of hydrogen from different hydrogen storage materials, such as NH_3BH_3, $NaBH_4$, and $N_2H_4BH_3$, are provided.

We are immensely thankful to all the authors for their rich contributions toward this handbook series. Although due efforts have been taken to make the book as error-free as possible, some may have crept-in unnoticed. We shall be thankful to the readers for bringing such unintentional errors to our notice. Finally, we sincerely hope that our efforts will be of use to a wide range of researchers working in the field of materials science.

Mumbai, India

Raghumani S. Ningthoujam
A. K. Tyagi

Contents

About the Editors

Dr. Raghumani S. Ningthoujam obtained M.Sc. in Chemistry from Manipur University in 1994. He was awarded Ph.D. in Chemistry from IIT Kanpur in area of superconducting, electron transport and magnetic properties of nanostructured transition metal nitrides in 2004. He joined BARC (Bhabha Atomic Research Centre), Mumbai as Scientific Officer (D) in 2006 after completion of Dr. K. S. Krishnan Fellowship. Presently, he is working in area of Luminescent and Magnetic Nanomaterials and their applications in sensors, imaging, diagnosis and therapy. He did Post-doctoral Fellowship at University of Victoria, Canada in the area of quantum dots. Presently, he is Scientific Officer (G) at Chemistry Division, BARC, Mumbai. He is recognized for Guideship of Ph.D. at Homi Bhabha National Institute (HBNI), Mumbai and Mumbai University, Mumbai. Many research scholars and students finished their projects under his guidance. He has published about 175 Papers in the refereed Journals, six Review Articles and 22 Book Chapters and edited five Books. In recognition of his significant contributions to the chemical science, he has been awarded DAE-Scientific and Technical Excellence Award (2012) Young Achiever Award, SSPS (2010), Chemical Research Society of India's Bronze Medal (2022), Materials Research Society of India's Medal (2022) Society for Materials Chemistry's Silver Medal (2023) and P. N. Sazhin, Russian Academia, State Atomic Energy Corporation Rosatom, Moscow, Russia (2024).

He has been elected as a Fellow, The National Academy of Sciences, India (FNASc) in 2016, a Fellow,

Maharashtra Academy of Sciences (FMASc) in 2013 and a Fellow, Physics Academy of the North East (2024).

Dr. A. K. Tyagi obtained his M.Sc. (Chemistry) degree in 1985 from Meerut University, Meerut and joined BARC Training School, Mumbai in the same year. After completing one year orientation course in Nuclear Science and Technology, he joined Chemistry Division, Bhabha Atomic Research Centre (BARC), Mumbai in 1986. During his career spanning close to four decades, he occupied various positions at BARC such as Head, Chemistry Division, Head, Analytical Chemistry Division, Head, Radiation Medicine Centre, Director, Chemistry Group and Director, Bio-Science Group. He was a Distinguished Scientist, DAE. Presently, he is Dean and Senior Professor at Homi Bhabha National Institute (HBNI), Mumbai. His research interests are in the field chemistry of materials, which includes functional materials, nanomaterials, nuclear materials, energy materials, metastable materials, hybrid materials and structure-property correlation. He has published more than 700 papers in journals, several books and has supervised 38 Ph.D. students.

He was awarded Ph.D. by Mumbai University, Mumbai in 1992. He did postdoctoral research at Max-Planck Institute for Solid State Research (MPI-FKF), Stuttgart, Germany during 1995–1996. In addition, he has also visited Institute of Superior Technology, Portugal; Institute for Chemical Process and Environmental Technology, Ottawa, Canada; Dalhousie University, Halifax, Canada; Moscow State University, Moscow, Russia; Institute for Materials, Nantes, France; University of Malay, Malaysia; National Institute of Materials Science, Tsukuba, Japan; National University of Singapore, Singapore; Royal Institute of Technology, Stockholm, Sweden; Rice University, Houston, USA; Shanghai, China; University of Valencia, Valencia, Spain; Weizmann Institute of Science, Israel; University of Queensland, Brisbane, Australia; US-Air Force Research Lab, Dayton, USA, Institute for Studies of Nanostructured Materials, Palermo, Italy and iThemba Labs, Cape Town, South Africa.

He has been conferred with a number of awards such as such as DAE-Homi Bhabha Science and Technology Award, DAE-SRC Outstanding Researcher

Award, DAE-Group Achievement Award (four times); MRSI Medal; MRSI-ICSC Materials Science Senior Award; MRSI-CNR Rao Prize in Advanced Materials; MRSI Distinguished Materials Scientist of the year award; CRSI Bronze Medal; CRSI-CNR Rao National Prize in Chemical Sciences; CRSI-Silver Medal; Rajib Goyal Prize in Chemical Sciences; Metallurgist of the Year award from Ministry of Steel, GoI; National Prize in Solid State and Materials Chemistry from JNCASR; ISCA Acharya PC Ray Memorial Award; NETZSCH—ITAS Award by Indian Thermal Analysis Society, DN Agarwal Memorial Award from Indian Ceramics Society and Gold medal of Chirantan Rasayan Sanstha. Recently, he was conferred with the prestigious **Vigyan Shri Award** of Government of India. He is a recipient of J. C. Bose National Fellowship.

He is an elected Fellow of the Indian National Science Academy (FNA), Indian Academy of Sciences (FASc); National Academy of Sciences, India (FNASc); Indian National Academy of Engineering (FNAE); Maharashtra Academy of Sciences; Royal Society of Chemistry, UK (FRSC); Asia Pacific Academy of Materials; World Academy of Ceramics; African Academy of Sciences and The World Academy of Sciences (FTWAS).

Chapter 1
Basic Concepts of Catalytic Materials and Summary of All the Chapters

Raghumani S. Ningthoujam and A. K. Tyagi

1.1 Introduction

Catalysts are materials, which are added intentionally in a reaction and increase the rate of reaction (Ningthoujam and Gajbhiye 2015; Santen et al. 1999; Wishiak and "The history of catalysis. From the Beginning Nobel Prizes", 2021; Chorkendorff et al. 2007; Whittall 2021; Mohammed 2020; Jens 2014; Bond 1972). The reaction follows an alternative pathway in presence of catalyst. The activation energy (E_a) required for a chemical reaction decreases when a catalyst is employed. Usually, catalyst forms intermediate(s) with reactants, which makes facile path way for formation of the final product(s). Catalysts remain unchanged after the completion of reaction. Catalytic property of a material depends on temperature, pressure, surface area, and porosity and reaction vessel in some cases. Catalysis is a process of the increase in rate of a chemical reaction due to an added substance called a catalyst. A catalyst may enhance rates of both forward and backward reactions, thus attaining equilibrium faster.

Let us take a chemical reaction of "A" and "B" reactants to form the products "C" and "D".

$$\text{A} + \text{B (reactants)} \rightarrow \text{C} + \text{D (products)} \tag{1.1}$$

The reaction can be faster when a foreign substance (E) is added to reaction (1.1).

The term "Catalysis" was coined in 1835 by Jons Jakob Berzelius (1779–1848), which was derived from the Greek words "kata" meaning down and "lyien" meaning

R. S. Ningthoujam (✉)
Chemistry Division, Bhabha Atomic Research Centre, Mumbai, India
e-mail: rsn@barc.gov.in

R. S. Ningthoujam · A. K. Tyagi
Homi Bhabha National Institute, Mumbai, India
e-mail: aktyagi@barc.gov.in

R. S. Ningthoujam and A. K. Tyagi (eds.), *Handbook of Materials Science, Volume 3*,
Indian Institute of Metals Series, https://doi.org/10.1007/978-981-96-8229-4_1

loosen. When human civilization started, role of catalysts was known. This improves yield of products in a reaction by addition of external material, which does not change its property after completion of reaction. However, this term was not used for so many years. In chemical industries, catalysts had been used for large scale production of NH_3, urea, and drugs leading to industrial and green revolution in the world.

The reaction path to convert reactants to the products can be expressed in term of the Gibbs free energy change (ΔG) in a closed system:

$$\Delta G = \Delta H - T\Delta S \tag{1.2}$$

H, T, and S are the enthalpy, absolute temperature, and entropy, respectively.

Without catalyst, reaction path needs activation energy (E_a), which is shown in Fig. 1.1. Here, enthalpy of products is less than that of reactants, ΔH becomes negative (exothermic). In a closed system, ΔG is negative and the reaction becomes spontaneous. In some cases, ΔH can be positive (endothermic). Depending on T and ΔS, ΔG may be negative or positive. ΔG of a chemical reaction is unchanged with or without catalyst.

In some reactions, one or more intermediates (I_1, I_2, I_3,...) may be formed kinetically before reaching thermodynamically stable phases, such as C and D from the reactants, such as A and B as represented in Fig. 1.2.

Catalytic materials can be of solid, liquid, or gas or mixture phase. Even, it can be atoms, molecules, clusters, or particles. Solid phases commonly include oxides, nitrides, carbides, phosphate, phosphides, halides, sulfides, alloys, composites, and hetero-structures (Ghedini et al. 2021; Ningthoujam and Kulshreshtha 2009; Xu et al.

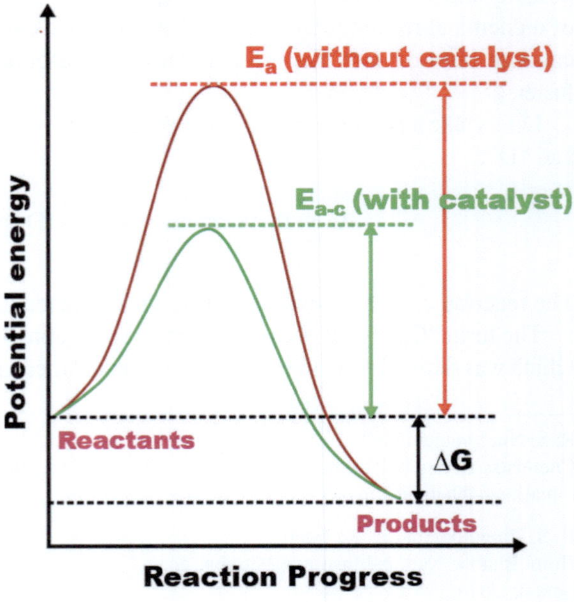

Fig. 1.1 Energy versus reaction progress of a chemical reaction. Here, reactants (say **A** and **B**) interact with each other in a medium to produce products (say **C** and **D**). An activation energy (E_a) required in a reaction is more than that with catalyst (E_{a-c})

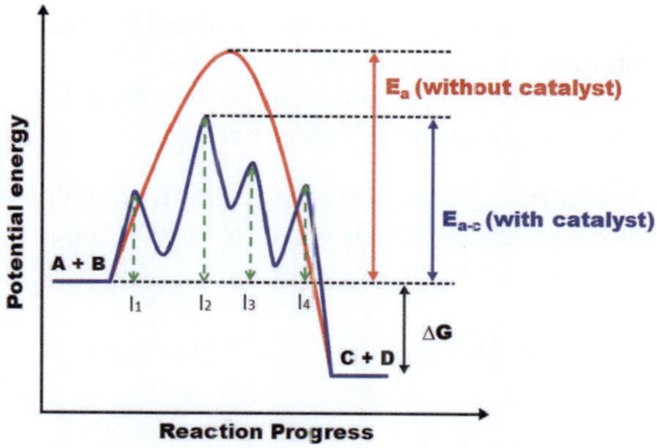

Fig. 1.2 Energy versus reaction progress of a chemical reaction. Here, reactants (say **A** and **B**) interact with each other in a medium to produce products (say **C** and **D**). In between reactants (**A** and **B**) and products (**C** and **D**), there are intermediate metastable phases or intermediates (I_1, I_2, I_3, —), which are kinetically stable. An activation energy (E_a) required in a reaction is more than that with catalyst ($E_{a\text{-}c}$)

2025; Yang et al. 2025; Li et al. 2023; Parchur and Ningthoujam 2012; Umare et al. 2008; Peng et al. 2023; Dongil 2019; Prats and Stamatakis 2024; Czaplicka et al. 2021; Liu et al. 2024, 2023; Buttard and Guinchard 2023; Ren et al. 2023; Zhang and Findlater 2023; Ningthoujam and Gajbhiye 2008; Gajbhiye et al. 2008; Gajbhiye and Ningthoujam 2004; Ningthoujam et al. 2009; Arandiyan et al. 2023; Zhu and Xiao 2023; Singh et al. 2010, 2008, 2024; Shete et al. 2013; Nikam et al. 2014; Phaomei et al. 2011; Thorat et al. 2014; Rawool et al. 2018). Noble metals (such as Pt) were used as catalysis in many applications. However, it is very expensive. Nowadays, such noble metals are replaced by transition metal nitrides, carbides having similar electronic structures (Singh et al. 2010, 2008; Shete et al. 2013; Nikam et al. 2014; Phaomei et al. 2011).

1.2 Classification

Catalysts can be classified into three: 1. Homogenous, 2. Heterogeneous, and 3. Enzymes and other biocatalysts (Santen et al. 1999; Wishiak 2021; Chorkendorff et al. 2007; Whittall 2021; Mohammed 2020; Jens 2014; Bond 1972; Ghedini et al. 2021).

The catalysis is known as a homogeneous catalysis when components, such as catalyst and reactants, are in the same phase (usually gaseous or liquid), whereas the catalysis is known as a heterogeneous, whose components such as catalyst and reactants are not in the same phase (solid and liquid, solid and gas, gas and liquid or solid, liquid and gas). In addition, enzymatic reaction and other biocatalysis reaction are often considered as a third category. This is shown in Fig. 1.3. Further, homogeneous or heterogeneous catalysis is grouped into two: acid and base catalyses depending on type of catalysts (acid, H^+ and base, OH^-).

Difference between homogeneous and heterogeneous catalyses is represented in Fig. 1.4. The following examples provide the difference.

1. Homogeneous catalysis: 2 molecules of SO_2 (gas) react with 1 molecule of O_2 (gas) to produce 2 molecules of SO_3 (gas). Here, NO (gas) is used as catalyst. Reactants and catalyst are in gaseous phase.

$$2SO_2(gas) + O_2(gas) \overset{NO(Gas)}{\rightarrow} 2SO_3(gas) \tag{1.3}$$

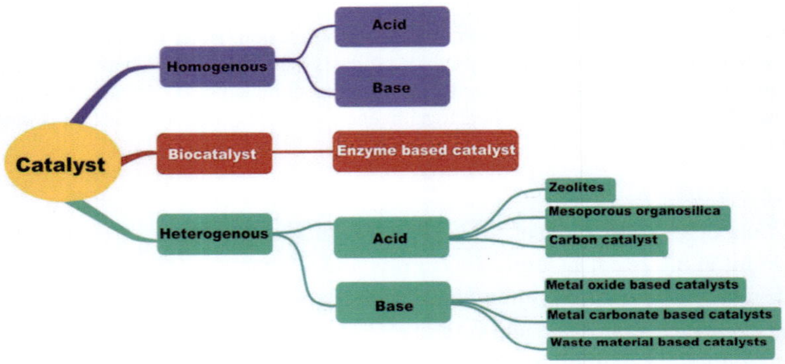

Fig. 1.3 Classification of catalysts: homogeneous, heterogeneous, and biocatalysts

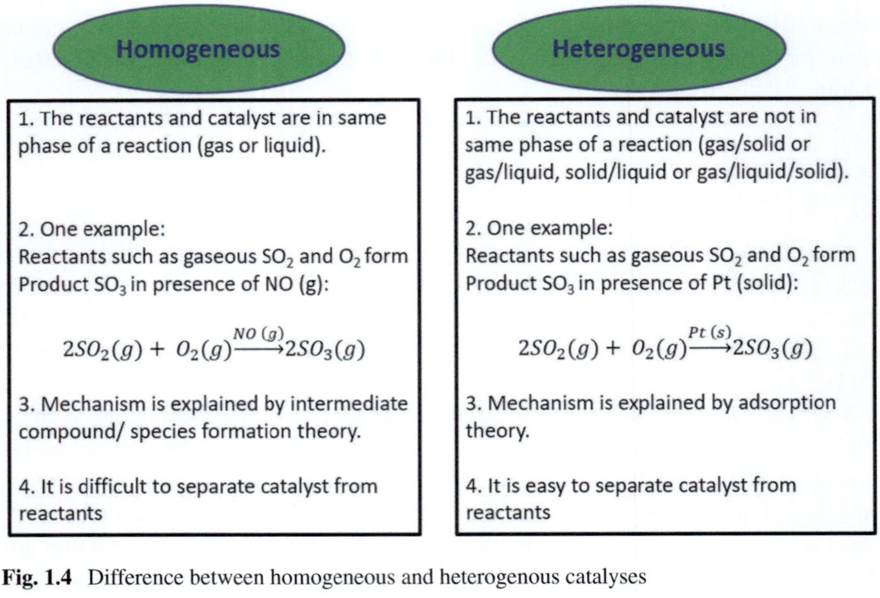

Fig. 1.4 Difference between homogeneous and heterogenous catalyses

2. Heterogeneous catalysis: 2 molecules of SO_2 (gas) react with 1 molecule of O_2 (gas) to produce 2 molecules of SO_3 (gas). Here, Pt (solid) is used as catalyst. Reactants are in gaseous phase, whereas catalyst is in solid phase.

$$2SO_2(gas) + O_2(gas) \overset{Pt(Solid)}{\rightarrow} 2SO_3(gas) \quad (1.4)$$

In homogeneous catalysis, intermediate compound or species is formed by interaction of catalyst with one of reactants (Santen et al. 1999; Wishiak 2021; Chorkendorff et al. 2007; Whittall 2021; Mohammed 2020; Jens 2014; Bond 1972; Ghedini et al. 2021; Puri et al. 1989). Subsequently, intermediate compound reacts with another reactant to form product. Simultaneously, catalyst is generated in its original form. For example, the formation of SO_3 product from reactants SO_2 and O_2 takes place in the presence of nitric oxide (NO) as catalyst. Here, SO_2, O_2, and NO are in gaseous phase. The formation of product is explained by the formation of intermediate compound NO_2 from NO and O_2 and further interacts with another reactant SO_2.

$$2NO(gas, catalyst) + O_2(gas, reactant) \rightarrow 2NO_2(gas, intermediate\ compound) \quad (1.5)$$

$$2NO_2(gas) + 2SO_2(gas, reactant) \rightarrow 2SO_3(gas, product) + 2NO(catalyst) \quad (1.6)$$

Without catalyst, reaction would have taken place as

$$2SO_2(gas) + O_2(gas) \rightarrow 2SO_3(gas) \quad (1.7)$$

The reaction (1.6) is faster than that of reaction (1.7) as NO_2 gas is a stronger oxidizing agent than O_2.

In heterogenous catalysis (gas on solid), the different steps are associated in specific sequence (Fig. 1.5) (Santen et al. 1999; Wishiak 2021; Chorkendorff et al. 2007; Whittall 2021; Mohammed 2020; Jens 2014; Bond 1972; Ghedini et al. 2021; Puri et al. 1989; Zhang and Zhang 2024): (i) diffusion of reactants (gases) on surface of solid, (ii) adsorption of reactants on surface of solid, (iii) chemical reaction on surface of solid for the formation of products, (iv) desorption of products, (v) diffusion of products away from the surface of solid.

In case of enzymatic reaction, substrate (S, say molecule or reactants) comes to specific site of enzyme (E). Here, enzyme acts as catalyst. It forms complex of enzyme–substrate that finally converted to products through enzyme–products complex (Fig. 1.6). This is represented by the following (Whittall 2021; Mohammed 2020; Puri et al. 1989; http: 2025):

$$E + S \rightarrow [ES] \rightarrow [EP] \rightarrow E + P \quad (1.8)$$

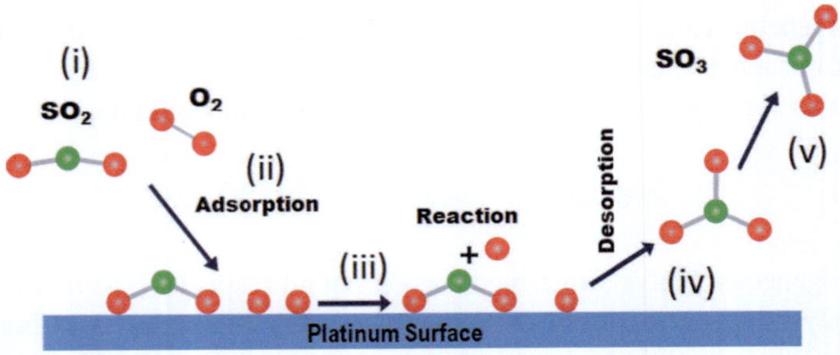

Fig. 1.5 Heterogenous catalysis (gas on solid): The formation of SO_3 from reactants SO_2 and O_2 in the presence of platinum catalyst

Fig. 1.6 Function of enzyme in a biological system

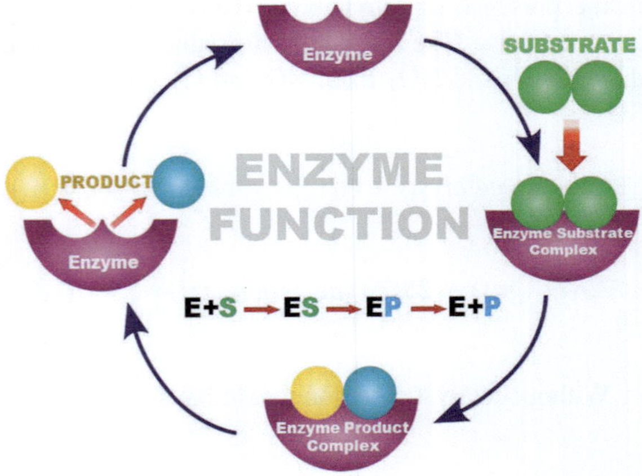

Enzymes are the proteins with high relative molar mass of 10 kDa or more. Each enzyme has a specific catalytic reaction. For examples, enzyme diastase can convert starch to maltose. Enzyme maltase can convert maltose to glucose. Enzyme zymase can convert glucose to ethanol. Enzyme urease can convert urea to ammonia.

$$2(C_6H_{10}O_5)_n(Starch) + nH_2O \overset{Diastase}{\rightarrow} nC_{12}H_{22}O_{11}(Maltose) \tag{1.9}$$

$$C_{12}H_{22}O_{11}(Maltose) + nH_2O \overset{Maltase}{\rightarrow} 2C_6H_{12}O_6(Glucose) \tag{1.10}$$

$$C_6H_{12}O_6(Glucose) + nH_2O \overset{Zymase}{\rightarrow} 2C_2H_5OH(Ethanol) + 2CO_2 \tag{1.11}$$

1.3 Characteristics of Catalytic Reactions

In general, catalytic reactions have the following characteristics (Santen et al. 1999; Wishiak 2021; Chorkendorff et al. 2007; Whittall 2021; Mohammed 2020; Jens 2014; Bond 1972; Puri et al. 1989):

i. The amount and chemical composition of catalyst are almost unchanged.
ii. Small amount of catalyst is required. However, in some cases of homogeneous reaction, a large amount of catalysts is required.
iii. In reversible reaction, catalyst does not alter the position of equilibrium.
iv. Catalyst does not initiate the reaction.
v. In general, every catalyst has specific activity for a particular reaction.
vi. Catalyst does not change the thermodynamic yield of products.

1.4 Particle Size Effects in Heterogeneous Reaction

In heterogeneous catalysis, solid materials in general have been used as catalysts (Santen et al. 1999; Wishiak 2021; Chorkendorff et al. 2007; Whittall 2021; Mohammed 2020; Jens 2014; Bond 1972). There is a difference in catalytic activity upon decreasing particle size from micron to nano range. With decrease in particle size, surface area per particle increases. Higher the surface area, higher is the number of active sites. Reactivity between gas or liquid and surface sites available on particle of catalyst is important. Only when interaction between them is more, the formation of products is enhanced. In general, as particle size decreases, catalytic activity tends to increase.

1.5 Role of Porosity and Surface Area of Adsorbent in Heterogeneous Reaction

There is a relationship between catalytic activity and porosity of host in heterogeneous catalysts (Santen et al. 1999; Wishiak 2021; Chorkendorff et al. 2007; Whittall 2021; Mohammed 2020; Jens 2014; Bond 1972). Some of the solids have large surface area as well as large porosity. For example, SiO_2 can have different surface areas as well as different porosities. Its reactivity is dependent on types of gases or environment available.

1.6 Types of Gases on Adsorption in Heterogeneous Reaction

Those gases, which can be liquefied easily, will be more adsorbed on surface of adsorbent as compared to those, which are difficult to liquefy. CO_2 (bp ~ −78 °C), NH_3 (bp ~ −33 °C), SO_2(bp ~ −10 °C) gases can be adsorbed on surface of adsorbent as compared to those of H_2 (bp ~ −253 °C) and He (bp ~ −269 °C) (Santen et al. 1999; Wishiak 2021; Chorkendorff et al. 2007; Whittall 2021; Mohammed 2020; Jens 2014; Bond 1972; Puri et al. 1989).

1.7 Selectivity in Heterogeneous Reaction

In heterogenous catalysis, some catalysts can be used for a particular reaction, but not all reactions universally. Selectivity depends on sites available on solid, porosity, temperature, AC frequency of experiment, environment/medium, pH, pressure, electron density of states, band gap, etc. (Santen et al. 1999; Wishiak 2021; Chorkendorff et al. 2007; Whittall 2021; Mohammed 2020; Jens 2014; Bond 1972; Puri et al. 1989).

1.8 Pressure and Temperature in Heterogeneous Reaction

Pressure is an important parameter in a gaseous system. With increases of pressure in a gaseous system, contact between reactants and catalyst increases (Fig. 1.7) (Santen et al. 1999; Wishiak 2021; Chorkendorff et al. 2007; Whittall 2021; Mohammed 2020; Jens 2014; Bond 1972; Puri et al. 1989). In general, reaction will be faster in presence of pressure. After a certain pressure, no more adsorption of gases on solid surface takes place as there is no more space on surface. In case of temperature, some reactions may be faster or slow depending on types of chemistry between reactants and catalyst with increase of temperature. Adsorption of gas on the surface of adsorbent can be of two types with increase of temperature: (i) decrease in amount of gas adsorbed on surface of adsorbent with increase of temperature and (ii) initial increase in amount of gas adsorbed on surface of adsorbent with increase of temperature and followed by decrease with further increase of temperature (Fig. 1.8).

1.9 Precatalysts

Precatalysts are the precursors to catalysts (Santen et al. 1999; Wishiak 2021; Chorkendorff et al. 2007; Whittall 2021; Mohammed 2020; Jens 2014; Bond 1972; Ghedini et al. 2021; Puri et al. 1989; https: 2025; Eastwood et al. 2024). The conversion

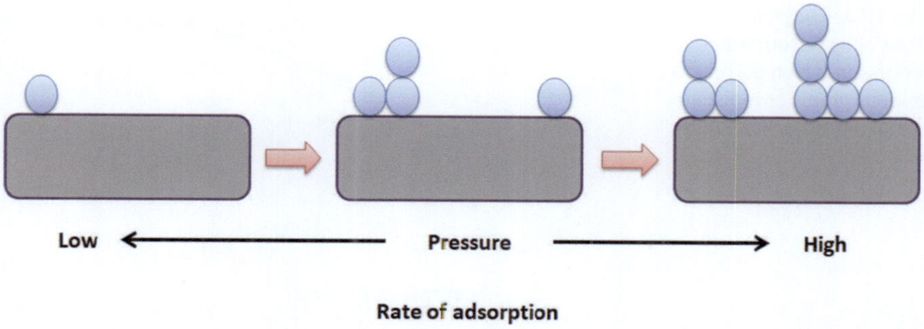

Fig. 1.7 Interaction between gas molecules or atoms and solid surface (catalyst) at different pressures (low to high)

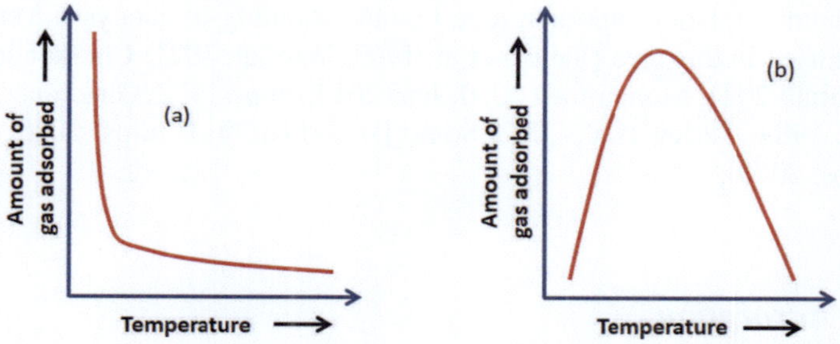

Fig. 1.8 Adsorbed gas versus temperature on gas/solid interface. Here, gas molecules or atoms act as adsorbate, whereas solid acts as adsorbent. In **a** amount of adsorbed gases on solid surface decreases with temperature. In **b** adsorbed gases on the solid surface increases with temperature and decreases with further increase of temperature

process of a precatalyst to a catalyst is known catalyst activation (Fig. 1.9). Metal halides are used as precatalysts for alkene polymerization in Ziegler–Natta catalysis. Mn(0) (pre)catalyst [$Mn_2(CO)_{10}$] has been used in C–H bond functionalization reactions.

Fig. 1.9 Roles of pro-catalyst and catalyst in a chemical reaction

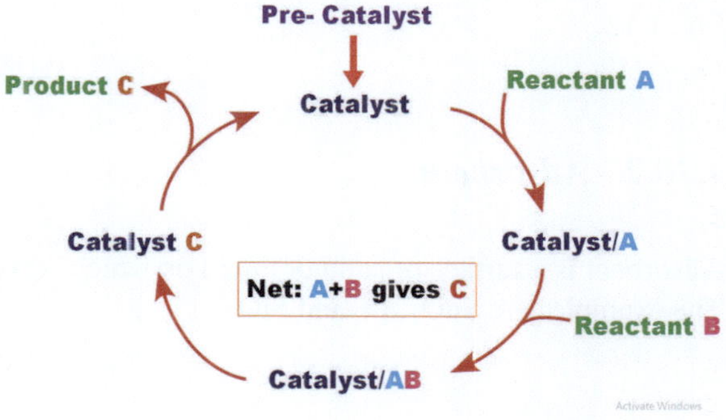

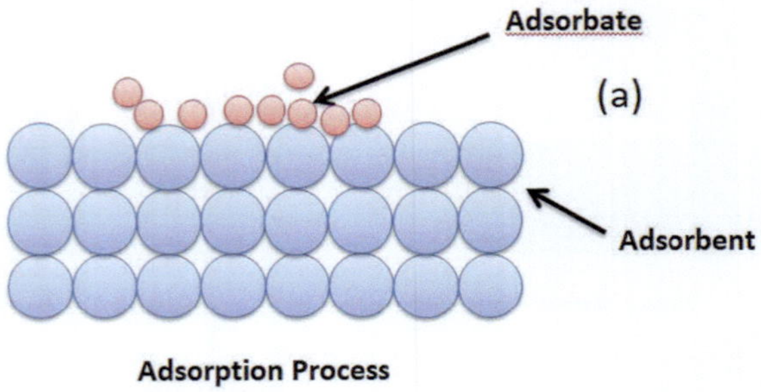

Fig. 1.10 Diagram illustrating adsorption process in which adsorbate gaseous atoms or molecules are on the surface of solid adsorbent

1.10 Terms Used in Surface Chemistry

Surface chemistry is an important area in understanding of catalysis. Many terms have been used in this area (Santen et al. 1999; Wishiak 2021; Chorkendorff et al. 2007; Whittall 2021; Mohammed 2020; Jens 2014; Bond 1972; Ghedini et al. 2021; Puri et al. 1989; Carley et al. 2002; Somorjai and Li 2011; https: 2025; Aljamali 2021; https: 2025).

1.10.1 Adsorption

Adsorption is a process of more interaction of gas or liquid on the surface of solid or liquid due to weak forces of attraction (Figs. 1.10 and 1.11), for example, adsorption of NH_3 gas on SnO_2. Two types of adsorption on the basis of chemical bonding between adsorbate and adsorbent take place: Physical and chemical adsorptions. In physical adsorption, the bonding between adsorbate and adsorbent is weak. The heat of adsorption is less than 40 kJ mol^{-1}. Multilayer formation takes place. The adsorption is appreciable at temperature below the boiling point of gas. In chemical adsorption, the bonding between adsorbate and adsorbent is strong. Monolayer formation takes place. The adsorption can occur in different temperature ranges. The heat of adsorption in general is greater than 200 kJ mol^{-1}. This is illustrated in Fig. 1.12.

1.10.2 Adsorbent

Adsorbent is a surface of solid or liquid on which adsorption takes place (Fig. 1.13). The examples are SnO_2, Pt, and SiO_2.

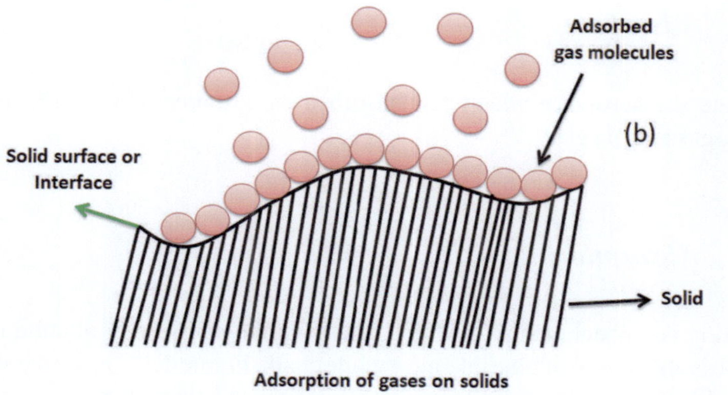

Fig. 1.11 Diagram illustrating adsorption process in which adsorbate gaseous atoms or molecules are on the surface of solid adsorbent. The solid surface acts as interface

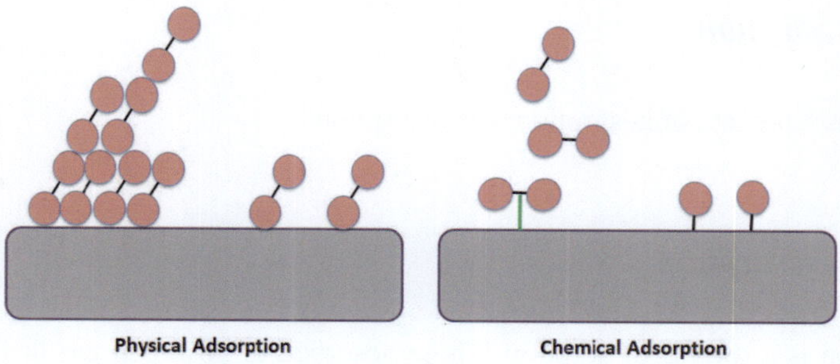

Fig. 1.12 Two types of adsorption take place: Physical and chemical adsorptions

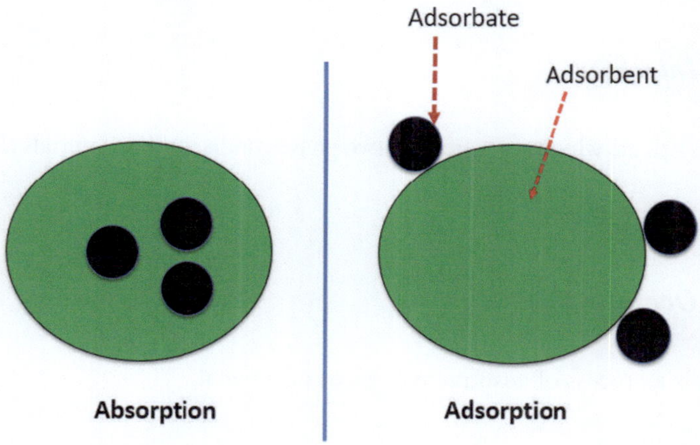

Fig. 1.13 Schematic diagram illustrating difference between adsorption and absorption. Solid black ball represents adsorbate (gas molecules), and elliptical green shape represents absorbent (solid)

1.10.3 Adsorbate

Adsorbate is a substance which gets adsorbed on adsorbent (Fig. 1.13). Examples are NH_3 gas and H_2 gas.

1.10.4 Absorption

Absorption is a phenomenon in which gas or liquid enters body of solid or liquid. Example is absorption of blue ink into a while chalk. Figure 1.13 shows the schematic diagram illustrating difference between adsorption and absorption.

1.10.5 Sorption

Sorption is a process of adsorption and absorption.

1.10.6 Desorption

Desorption is a reverse of sorption. Those adsorbed or absorbed gas or liquid are coming out from medium (solid or liquid).

1.10.7 Interface

Interface is a plane which separates any two phases or dissimilar materials (Figs. 1.11 and 1.14).

1.10.8 Occlusion

Occlusion is a process of adsorption of gases on a metal.

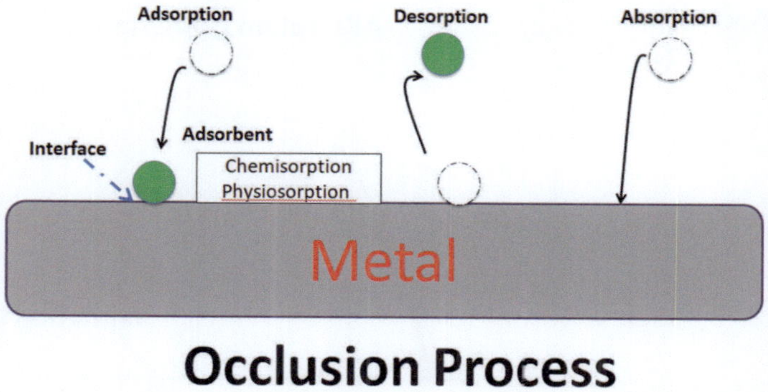

Fig. 1.14 Occlusion: gases (molecules or atoms) adsorbed on metal. Interface can be seen between gas molecules and metal solid

1.10.9 *Types of Adsorption Isotherm*

Adsorption can be of positive and negative types (Fig. 1.15). In case of positive adsorption, concentration of adsorbate on the surface of adsorbent is more than that in the bulk of adsorbate at a particular temperature. Here, bulk means original concentration of adsorbate. In case of negative adsorption, concentration of adsorbate on the surface of adsorbent is less than that in the bulk of adsorbate at a particular temperature. Both are expressed in mass of gases (X, adsorbate) adsorbed per mass of a substance (m, adsorbent). The term adsorption isotherm refers to phenomenon of adsorption of adsorbate (gas) on the surface of adsorbent (solid) at a particular temperature.

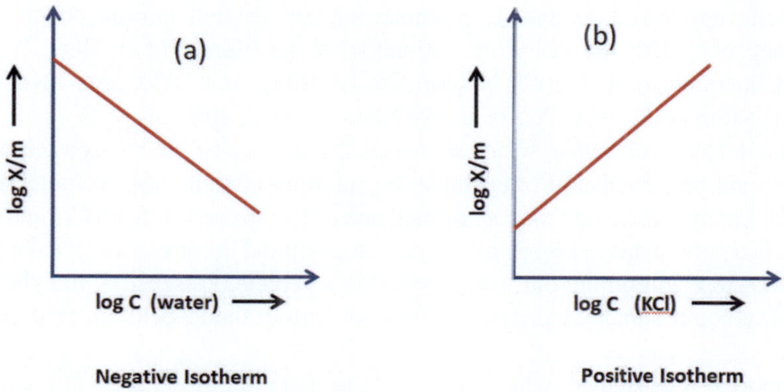

Fig. 1.15 Types of adsorption: **a** negative isotherm and **b** positive isotherm

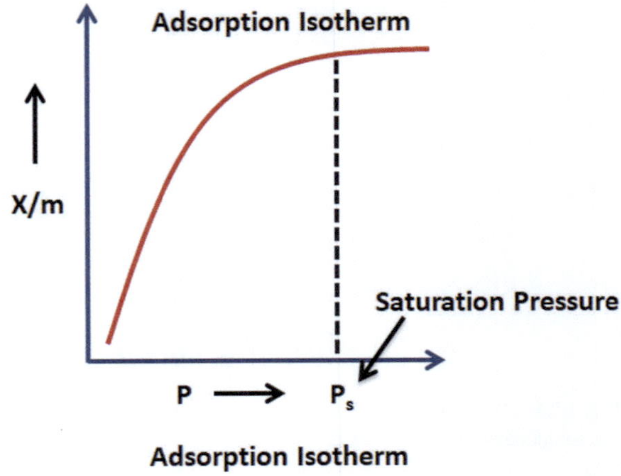

Fig. 1.16 Type I adsorption isotherm

Adsorption Isotherms of Gas Over Solid Can Be Grouped into Five Types

Type I adsorption isotherm: Here, monolayer of gas molecules is adsorbed on surface of solid (Fig. 1.16). Pore size of solid is almost equal to the diameter of gas molecules. At a particular temperature, it shows the increase of absorption with pressure, and it becomes saturated above a particular pressure as there is no longer space of pore of solid for gas molecules. Example is adsorption of N_2 over charcoal at −183 °C. This is a kind of chemisorption. The Langmuir isotherm belongs to Type I.

Type II adsorption isotherm: Here, multilayer of gas molecules are adsorbed on the surface of solid after the completion of monolayer. Here, pore sizes are in macron range, which is larger than that of the diameter of gas molecules. Example is adsorption of N_2 over nonporous silica gel at −195 °C. This is a kind of physisorption.

Type III, IV, and V adsorption isotherms are associated in other specific systems.

Specific terms used in catalysis: promoter, support, catalytic poison, positive catalysts, negative catalysts, inhibitors, and autocatalysis (Santen et al. 1999; Wishiak 2021; Chorkendorff et al. 2007; Whittall 2021; Mohammed 2020; Jens 2014; Bond 1972; Ghedini et al. 2021; Puri et al. 1989; Siregar et al. 2019).

Promoter is a substance, which increases the activity of the catalyst. However, this may not be a catalyst. For example, the addition of vanadium promoter on the Rh/SiO_2 catalyst increases the rate of methane (CH_4) formation from CO and H_2 at 293 K. Promoter helps in bringing catalyst in the original in conversion of toxic gases (CO, NO) to CO_2 coming out from exhaust in a vehicle. Presence of molybdenum in small amount enhances catalytic activity of iron in the Haber's process of NH_3 synthesis.

Support is a substance, which reduces agglomeration of particles of a catalyst. Gold nanoparticles supported on TiO_2 have different properties as compared to unsupported gold nanoparticles.

Catalytic poison occurs when the foreign substance is present in a reaction, which retards or stops the activity of the catalyst. The substance may be the impurity associated, defects or cracks developed during reaction. For example, oxygen acts as poison to the Fe-based ammonia synthesis catalyst.

In case of positive catalysts, reaction becomes faster due to the presence of substance. For example, decomposition of $KClO_3$ becomes faster after the addition of MnO_2 (solid) at 370 °C.

In case of negative catalyst, reaction becomes slower due to presence of substance. For example, oxidation of $NaSO_3$ to Na_2SO_4 becomes slow in the presence of alcohol. Another example is addition of tetra ethyl lead (TEL) to petrol retard the ignition of petrol vapors on compression in an internal combustion engine. This decreases the knocking effect.

Inhibitors are the substances which induce the retardation of a chemical reaction. It is similar to negative catalyst. Inhibitors are found in chemical reaction and biological systems.

Autocatalysis is a reaction in which a product of a chemical reaction acts as catalysis. Examples are below:

(1) Hydrolysis of an Ester: During the hydrolysis of ethyl acetate, acetic acid (CH_3COOH) and ethanol are formed. Here, the product acetic acid acts as a catalyst for the reaction.

$$CH_3COOCH_2CH_3 + H_2O \rightarrow CH_3COOH(Catalyst) + CH_3CH_2OH \quad (1.12)$$

(2) Oxidation of Oxalic acid: During oxidation of oxalic acid with acidified potassium permanganate, manganous sulfate ($MnSO_4$) is produced and acts as a catalyst for the reaction.

$$2KMnO_4 + 5H_2C_2O_4 + 3H_2SO_4 \rightarrow 1MnSO_4(Catalyst) + K_2SO_4 + 8H_2O + 10CO_2 \quad (1.13)$$

1.11 Applications of Catalyst

Applications of catalysts include the following (Santen et al. 1999; Wishiak 2021; Chorkendorff et al. 2007; Whittall 2021; Mohammed 2020; Jens 2014; Bond 1972; Ghedini et al. 2021; Puri et al. 1989):

1. Exponential increase of industrial products such as ammonia, urea, and drugs.
2. Preparation of fine chemicals.
3. Preparation of large-scale production of carbon nanotubes.
4. Selectivity of a chemical reaction and yield.
5. A stereoselective reaction (formation of one stereoisomer selectively over another)
6. Cracking of hydrocarbons and petroleum products.

7. Reduction of harmful products.
8. Hydrogenation of oils.
9. Evolution of green chemistry.
10. Conversion of starch into maltose, conversion of sugar into glucose and fructose, and conversion of glucose into ethanol.
11. Large-scale production of H_2 as source of energy.

1.12 A List of Scientists, Who Got Nobel Prizes in Catalysis

A few scientists got The Nobel Prize in chemistry related to catalysis (Wishiak 2021; https: 2025). These include the following scientists: Wilhelm Ostwald got The Nobel Prize in chemistry 1909 "in recognition of his work on catalysis and for his investigations into the fundamental principles governing chemical equilibria and rates of reaction". In 1912, Nobel Prize in chemistry has been awarded to both scientist: Victor Grignard "for the discovery of the so-called Grignard reagent, which in recent years has greatly advanced the progress of organic chemistry" and Paul Sabatier "for his method of hydrogenating organic compounds in the presence of finely disintegrated metals whereby the progress of organic chemistry has been greatly advanced in recent years".

Fritz Haber got Nobel Prize in chemistry in 1918 for synthesis of ammonia from N_2 and H_2 in presence of catalyst. Carl Bosch also got Nobel Prize in chemistry in 1931 for synthesis of materials in high pressure. The members of the Institution of Chemical Engineers voted Carl Bosch along with Fritz Haber "the world's most influential chemical engineers of all time" during their times. Their partnership established "the Haber–Bosch process" in the world in the synthesis of ammonia from nitrogen captured from the air. This brought the Green Revolution in cultivation of crops.

$$N_2 + 3H_2 \overset{Catlyst}{\rightarrow} 2NH_3 \tag{1.14}$$

The Nobel Prize in chemistry 1963 was awarded jointly to Karl Ziegler and Giulio Natta "for their discoveries in the field of the chemistry and technology of high polymers". Here, they used catalysts to improve yield. It is useful in industries. Ziegler–Natta catalysts are used to polymerize terminal alkenes.

The 2021 Nobel Prize in chemistry has been awarded to Benjamin List of the Max Planck Institute for Kohlenforschung and David MacMillan of Princeton University for their work in establishing asymmetric organocatalysis as a powerful tool in organic synthesis.

1.13 Summary of All the Chapters

Chapter 2: Frustrated Lewis Pairs (FLPs) have revolutionized main-group chemistry, particularly in recent decades. The discovery that sterically hindered, unquenched Lewis acid–base pairs can activate molecular hydrogen by cleaving the H–H bond was a groundbreaking advancement, sparking extensive research and new directions in the field. This discovery has driven the rapid diversification of FLP systems, extending beyond traditional boron- and phosphorus-based frameworks to include a broader spectrum of donors and acceptors. These advancements have significantly impacted both homogeneous and heterogeneous catalyses. This chapter provides a comprehensive introduction to FLP chemistry, guiding researchers through its foundational principles, synthetic methodologies, and wide-ranging applications. It begins by highlighting the role of catalysis in small molecule activation and bond formation, laying the foundation for the emergence of the FLP concept. Key historical milestones and the fundamental principles of FLPs are discussed to establish their significance in the broader chemical landscape. The chapter further delves into the synthetic strategies employed to design FLPs, covering classical systems as well as advanced frameworks involving phosphorus, nitrogen, and carbenes. Practical applications of FLPs are reviewed in depth, highlighting their catalytic versatility in reactions such as hydroboration, hydroarylation, hydrosilylation, polymerization, transfer hydrogenation, amination, and N–H activation. Finally, the chapter expands on the evolving potential of FLPs beyond traditional main-group chemistry. It explores their integration with transition metals, rare earth elements, and metal–organic frameworks (MOFs), showcasing how these hybrid systems open new frontiers in catalysis and mechanistic understanding. This dynamic interplay between FLPs and other domains hints at broader implications for future catalytic technologies.

Chapter 3: Clusters are aggregates of a finite number of atoms/molecules held together by interactions ranging from van der Waals forces to ionic or covalent or metallic bonds. Size may vary from 5 to 20 Å. Their properties are different from constituent atoms or molecules or bulk particle. They have following characteristics: (i) higher surface-to-volume ratio of clusters, as large number of atoms resides on the surface, (ii) due to reduced size, quantum confinement also contributes to unusual properties of clusters, and (iii) unique geometric and electronic structures of clusters, which vary as a function of cluster size and composition. Depending on the nature of constituent atoms and binding property, clusters can be broadly classified into following categories: molecular cluster, rare gas clusters, ionic clusters, semiconductor clusters, and metal clusters. They have been used in many applications including catalysis. In this chapter, synthesis of gas-phase free clusters and stabilization of these free clusters by different methods have been mentioned. The property of free clusters is different from supported clusters. Many characterization techniques are provided. They have been used as catalyst in oxidation of CO, hydrogenation reaction, etc. Also, they have been used as biosensor and as bio-imaging agent. Interestingly, clusters can interact with intense laser pulses (ns to ps) and sometimes, energetic multiply charged cations and electrons can be generated. Thus, cluster qualifies for new class of materials with exotic properties which has immense applications in the fields of physics, chemistry, and biology.

Chapter 4: There is an incredible interest and investment about rare earth (RE) and mixed metal nanoparticles (NPs), as they have found multipurpose applications, stretching from catalysis, sensing, biomedicine, photonics, to quantum memory and signal processing. They are also important in recent trends of energy storage application as electro catalysts in fuel cells, perovskite oxide bifunctional catalysts for batteries, electrode materials in supercapacitors, etc. The synthesis and application of RE and mixed metal oxide NPs have been reviewed in this chapter. In particular, the authors have reviewed some of the exemplary literature reports on synthesis and applications of RE and mixed metal oxide NPs in diverse fields. The field of nanoscience and technology has shown an exponential growth in the past few decades due to their huge applications in diverse fields. A variety of synthetic techniques have been reported so far and new synthetic methods are still being explored. It is expected that this compendium will be helpful to get an overall idea on current trend in the synthesis and application of NPs.

Chapter 5: Metal–organic frameworks (MOFs) materials have attracted great attention for applications in organic transformation in the last two decades. It exhibited excellent activity as a catalyst in organic transformation due to its unique properties like easily synthesizable, surface functional groups, different size cavities with various functional groups, high surface area which governs the high absorption capacity, various functionalities that provide the platform where host–guest interaction; further, the MOF materials permit modification in the catalyst for organic transformation reactions. In this chapter, the main focus is given to understanding the synthesis, and modification in MOFs design to superior potential applications such as catalysis in the organic synthesis reaction. This provides the strategy for controlling the design of highly active MOFs and MOF-based material which catalyzes the chemical reactions and adjusts the optimized condition to acquire highly active MOFs material. Understanding the structure–reactivity relation is a major concern in organic transformation reactions such as coupling reactions, oxidations, and reductions of various substrates, acid- and base-catalysis in which MOFs material are used as catalysts. The special structural properties of MOFs material, such as cavity size, porosity, and surface area, porosity, play a crucial role in organic transformation. The structural properties of the MOFs material can be tailored according to the desired chemical transformation. An overview of the synthesis and applications of metal–organic frameworks noble material in organic transformation, especially oxidation reaction, reduction, hydrogenation, metal-catalyzed cross-coupling reaction, and acid–base-catalyzed reaction, is presented and discussed.

Chapter 6: This chapter delves into the critical role of photocatalysis in addressing environmental and energy challenges within organic synthesis. The field has witnessed significant advancements, with the development and application of both homogeneous and heterogeneous photocatalysts for visible-light-driven organic reactions. Among these, semiconductor nanostructured materials stand out due to their unique properties, including tunable band gaps, high photon absorption, exposed active sites, charge carrier generation, affordability, and ease of preparation. This provides an extensive overview of various nano-semiconductor photocatalysts utilized in organic transformations, such as CdS, Cu: ZnS, InP/ZnS quantum dots, CdSe, ZnSe/CdS, carbon quantum dots, perovskites, carbon nano-onions, and graphitic carbon nitride (g-C_3N_4). It emphasizes strategies to enhance their photocatalytic efficiency, including morphology control, bandgap engineering, modification of crystal facets, incorporation of functional groups, creation of surface vacancies, and the addition of metal or nonmetal nanoparticles as co-catalysts. By summarizing these advancements and strategies, this offers valuable insights into the development of innovative photocatalysts for visible-light-driven organic synthesis, aiming to inspire further research and progress in the field.

Chapter 7: The chapter titled "Enzyme Catalysis: A Green Chemistry Perspective" documents the incredible journey of biocatalysts from "a topic of academic interest" to "a few technological applications" to become 'indispensable tools for innovative organic process chemistry/biotechnology'. Since the activity, selectivity, and stability of enzymes can be manipulated in a predictive manner to suit our needs, enzymes are used by scientists/biotechnologists to meet the ever-increasing demands of mankind for fine and bulk chemicals. Biocatalysts help to design clean and sustainable routes for production of pharmaceuticals, agrochemicals, polymers, food additives, etc. with minimal or no waste production, apart from taking care of the economic, environmental, and safety aspects of the processes. Besides, they offer potential solutions to grave problems related to depletion of natural sources and energy, by minimizing heat generation and carbon emission. This begins with establishing biocatalysts as flag-bearers of green chemistry, giving a brief account on the numerous benefits associated with them. It gives an overview of the *unique enzyme structure* (the primary, secondary, and tertiary structures which give the enzymes and their final folded, globular, compact, and robust forms), the inherent chirality in the enzymes (along with their catalytic and binding sites leading to enantiomeric recognition of substrate in a 3-D space), the various mechanisms, thermodynamic and kinetic driving forces leading to enzyme–substrate bindings, and hence explains the science behind the performance of enzymes as supercatalysts. An idea about the various sources and types of enzymes along with the pros and cons of using whole cell-based or isolated enzymes are also delineated in this. This is followed by a discussion on new paradigms and emerging technologies (directed evolution, recombinant DNA technology, enzyme immobilization, solvent engineering, etc.) to highlight their contributions in "process chemistry" largely protecting the objectives of "green chemistry". Finally, representative examples of biocatalytic syntheses of a few drugs/drug intermediates, some representative few food additives, biofuels, polymers, peptides, and novel biocatalysts-based technologies are mentioned to demonstrate the wide application, relevance and acceptance of enzymatic protocols globally. Also, this provides a quick insight of the evolutionary journey of enzymes and how the champion catalysts have integrated the natural world with that of academia, research, and industry.

Chapter 8: Recently, transition metal oxides and hydroxides-based catalysts are adding new paradigms in the usage of catalysts nanoparticles. These catalysts show much higher redox activity, structural versatility, abundance and allow low-cost fabrication, which makes them commercially viable. Tailored synthesis protocols lead to such materials with increased specific surface area, enhanced electron transfer and higher mechanical stability. As a result, transition metal oxide-based catalysts are now being utilized in newer areas, which range from energy, electrocatalysis, environmental remediation, dye degradation, CO_2 capture, etc. This chapter presents the synthesis processes, which are now utilized to obtain novel catalyst nanoparticles. Some of the synthesis processes are: hydrothermal, sol–gel, co-precipitation, and electrodeposition techniques. The parameters that need to be considered to obtain the desired size, shape, and crystallinity are also discussed. As the morphology of nanoparticles changes from 0D, 1D to 2D, an interesting modulation in their catalytic properties are observed. The reaction rates and order also change as a function of particle size.

Chapter 9: This chapter includes a brief discussion on development of sustainable and affordable methods for the never-ending challenge in the photocatalytic degradation of toxic organic pollutants. This elaborates on the basics of photocatalysis as well as advancements achieved in photocatalyst developments for the effective remediation of organic pollutants from water. Advantages of graphene oxide (GO) in particular for enhancing the catalytic performance of photocatalyst based on semiconductor and plasmonic nanomaterials. Due to its large interior surface area and quick electron conductivity, and high optical transmittance, fostering GO as a fundamental foundation of an efficient hybrid photocatalyst. The efficient degradation of organic toxicants is made possible by quick photo-charge separation and delayed charge recombination. This chapter provides an overview of current progress in GO-, r-GO-, and g-C_3N_4-hybridized plasmonic and semiconductor materials employed in efficient remediation of organic pollutants under solar radiation. It is also supposed that this chapter will provoke higher-level exploration and will definitely attain new possibilities that improve novel hybrid GO-, r-GO-, and g-C_3N_4-based plasmonic and semiconductor photocatalysts.

Chapter 10: In the last 20 years, ionic liquids have attracted considerable attention for applications in various advanced scientific fields including catalysis for organic synthesis. The revolution of the use of the ionic liquid for organic catalytic synthesis is started from the recognition that ionic liquid can be used as reaction media or solvent. Later ionic liquids are recognized as effective versatile catalysts for organic synthesis. Various physico-chemical features, such as wide liquid range, designing of ionic liquid, solvation ability, dissolution ability, supporting tendency with better surface area of the ionic liquid, are accountable for the catalytic applications. However, for the large-scale practical applications from the laboratory to industrial scale-up synthesis, there is a need to address several challenging issues such as synthesizing of biodegradable, nontoxic ionic liquid, industrial infrastructural updating with corrosion-free large-scale equipment/reactors at the industrial large scale. Here, this summarizes various types of reaction examples in the organic synthesis that are performed by catalysis through the use of ionic liquids. Moreover, the role of physico-chemical features of ionic liquid in catalysis with plausible reaction mechanism, justification, and future outlook in the field is also summarized.

Chapter 11: Nitro-containing compounds are recognized as toxic due to the high reactivity of nitro functional groups, which can initiate various undesirable reactions. As a result, these compounds present a serious hazard to both the environment and living organisms. In particular, nitroarenes, common by-products of numerous industrial processes, are classified as major water pollutants. Traditional strategies to mitigate their harmful effects, such as catalytic reduction into less toxic products like amines, have been widely studied, emphasizing their benefits in terms of energy efficiency and minimal solvent use. Recent research highlights the catalytic potential of metal nanoparticles, their limitations, and the growing interest in Metal–Organic Frameworks (MOFs) as a promising alternative. MOFs offer unique advantages, including tunable structures, stable surfaces, and well-dispersed active sites. This chapter reviews various MOFs and their applications as catalysts for nitroarene reduction, demonstrating their high efficiency and reusability in aqueous systems. It also delves into MOF design strategies for enhanced catalytic activity, kinetic analyses, catalytic mechanisms, and the incorporation of MOF composites to improve performance.

Chapter 12: Efforts to bridge the gap between heterogeneous and homogeneous catalysis have been widely reported, aiming to combine the advantages of both systems. Specifically, poly-oxometallates (POMs) of hexavalent molybdenum (Mo) and tungsten (W), after incorporating zirconium (Zr IV) and hafnium (Hf IV) to form butterfly-type cores, have demonstrated excellent catalytic activity in oxidation reactions. Another promising approach involves the use of stabilized metal nanoparticles (NPs) as catalysts, often employing tetra N-alkylammonium cations for stabilization via steric, electrosteric, and electrostatic effects. The stabilization of metal NPs has also been effectively achieved using polymers and dendrimers. The dendrimer approach enables the stabilization of smaller-sized metal NPs, which can then be supported on inorganic materials such as SBA-15, silica, and ceria. Although the heterogeneity of these systems initially limited their catalytic activity compared to homogeneous systems, treating dendrimer-derived supported NPs with oxidizing agents like $PhICl_2$ and Cl_2 resulted in metal–chlorine bond formation. This, in turn, enhanced the electrophilicity of the active metal NP sites, enabling catalytic performance comparable to or better than homogeneous protocols. This chapter outlines the systematic development of heterogeneous electrophilic catalysis and related methodologies that have contributed to bridging the gap between heterogeneous and homogeneous systems.

Chapter 13: Dimethyl ether (DME) is an important chemical with applications as a clean alternative fuel, aerosol propellant, and chemical intermediate. Its production can be achieved through several synthesis routes, each offering distinct advantages and challenges. The most established method is the methanol-to-DME (MTD) process, which involves dehydrating methanol using a solid acid catalyst. This method is efficient and scalable but faces challenges like catalyst deactivation and high energy demands for water removal. Recent advancements aim to improve catalyst stability, activity, and process integration to reduce energy consumption. Another route is direct synthesis from synthesis gas (syngas), where syngas is converted to methanol via Fischer–Tropsch synthesis, followed by dehydration to form DME. This method offers flexibility in feedstock use but is more complex and energy-intensive compared to MTD. Improvements in catalysts for syngas conversion and integration with renewable feedstock are ongoing. Biomass-to-DME involves gasifying biomass to produce syngas, which is then converted into methanol and DME. This method is renewable and carbon–neutral but faces high costs related to biomass collection and gasification. Technological developments in biomass gasification and catalyst efficiency are improving its viability. The use of CO_2 as a feedstock for DME synthesis is a promising emerging area. CO_2 can be converted into methanol or directly into DME using renewable energy sources. While still in early stages, this process holds potential for reducing greenhouse gas emissions, with advancements in electrochemical and photo-catalytic technologies exploring more efficient CO_2 reduction. Therefore, while traditional methanol-to-DME synthesis remains dominant, ongoing research into syngas, biomass, CO_2 utilization, and renewable energy-driven processes is advancing the production of DME in more sustainable and efficient ways. Future breakthroughs in catalyst design, process integration, and energy management could significantly improve the economic and environmental feasibility of DME production.

Chapter 14: The C–H bond activation is a pivotal process in modern catalysis, enabling the functionalization of hydrocarbons and the creation of complex chemical structures. Traditionally, C–H bonds are strong and relatively inert, making their activation a significant challenge in synthetic chemistry. Heterogeneous catalysis, where the catalyst is in a different phase from the reactants, has emerged as a powerful tool for enabling C–H activation, offering benefits such as high selectivity, ease of separation, and reusability. The most common heterogeneous catalysts for C–H activation are metal-based, utilizing transition metals such as platinum (Pt), palladium (Pd), rhodium (Rh), and copper (Cu). These metals can interact with C–H bonds, breaking them and forming reactive intermediates such as metal–hydride or metal–alkyl complexes, which can undergo further transformations. These reactions are essential in processes like cross-coupling, oxidation, and functionalization, where the activation of C–H bonds allows for the introduction of new functional groups or the formation of new C–C bonds. An exciting development in this area is the use of single-atom catalysts (SACs), where metal atoms are dispersed at the atomic level on a support material. SACs offer unique advantages in terms of reactivity and selectivity, allowing for more efficient activation of C–H bonds with fewer by-products. These catalysts are especially useful in selective oxidation reactions, where high precision is required to target specific C–H bonds. In addition to metal-based catalysts, nonmetallic catalysts, including heteroatom-doped carbons (e.g., nitrogen-doped carbons), are gaining attention. These catalysts activate C–H bonds through electron donation or withdrawal facilitated by the heteroatoms, providing an alternative to traditional metal catalysts. They have shown promise in dehydrogenation and functionalization reactions. Despite their potential, several challenges remain in heterogeneous C–H activation. Achieving high selectivity remains difficult, especially in complex molecules with multiple C–H bonds. Catalyst deactivation due to sintering, leaching, or poisoning is another significant hurdle, reducing the lifespan and efficiency of metal catalysts. Additionally, many C–H activation reactions require high temperatures and pressures, leading to high energy consumption. Efforts to improve catalyst stability through better support materials and more robust metals have been successful. New strategies, such as dual-metal catalysts and composite materials, have been developed to enhance both reactivity and selectivity. Moreover, the application of artificial intelligence (AI) and machine learning is helping to design catalysts with optimized properties for specific C–H activation reactions.

Chapter 15: Metal–metal-oxide interfaces are becoming important subject of study in catalysis. A study of the interfaces in terms of chemical composition and microstructure/nanostructure helps in understanding the chemical processes, reaction, and mechanism at the surfaces. From a practical point of views, such as the reaction completion, conversion and selectivity, endurance of the catalysts can be enhanced. Such an approach brings out innovations in the synthetic strategies of new catalysts for industrial reactions. Here, the thermodynamics behind the stability of metal supported catalysts is discussed and followed by general synthesis strategies for obtaining finely dispersed metal supported oxide systems. The characterization of metal–metal-oxide interfaces has been discussed followed by a discussion on the effect of metal–metal-oxide interfaces on CO oxidation and ammonia borane hydrolysis. At the end, the importance of single atom and single cluster catalysis and operando techniques has been highlighted as the future prospects.

Chapter 16: Modern synthetic chemistry has changed the course of human life by contributing immensely toward essential commodities, such as pharmaceuticals, polymers, natural products, and the list is not exhaustive. Coupling reactions are at the heart of the synthetic processes for these industries. Homogeneous catalysis involving group 10 metals has become indispensable in various coupling reactions, such as Heck, Suzuki, Sonogashira, etc. Over the years, these catalysts have provided superior reaction rates, high turnover numbers (TON), and high selectivity and yields. This chapter furnishes a comprehensive summery to group 10 molecular materials, especially Ni- and Pd-based complexes as efficient heterogeneous catalysts for important coupling reactions. The chapter begins with a general introduction pertaining to the importance of coupling reactions with an eventual discussion on the catalysts facilitating the process. The chapter then expands to nickel and palladium complexes which have been widely explored by several research groups as catalysts for coupling reaction. Various synthesis methods of several Ni- and Pd-based catalysts have been categorically presented according to the variety of ligand systems utilized for their synthesis. These ligand systems range from thiosemicarbazones, NSN pincer ligands to thione systems. Special attention has been given into palladium catalysts as their chemistry is comparatively more versatile. Several interesting systems, such as palladacycles, triazole core, have been concisely presented. Each of these sections is accompanied with a flavor of the various coupling reactions; these catalysts have proved to be useful. Finally, the chapter is concluded with a brief discussion and future perspective on how to improve the catalytic activity of these complexes and how to design new complexes which may open new horizons in the field.

Chapter 17: Renewable energy, such as solar and wind-driven electrocatalytic water splitting, is an economically achievable approach to produce the green and clean hydrogen energy. Electrocatalytic water splitting plays a major role for the development of sustainable and eco-friendly energy conversion to green fuel generation that eventually addresses the world's fossil fuel demand. Hydrogen evolution reaction (HER) and oxygen evolution reaction (OER) are considered as the two major half reactions of complete water electrolysis process that decides the fruitful beginning of one-pot overall water splitting (OWS). To optimize and achieve superior activity in these reactions, it is vital to design highly active electrocatalysts. Layered double-hydroxide (LDH)-based electrocatalysts have taken very much interest in current research and challenge in energy conversion towards water splitting process. The layered structured, multiple composition, unique electronic structure, exchangeable anion property makes the LDH-based materials are explored as efficient electrocatalyst for water splitting. Nevertheless, various limitations, including low thermal stability, small surface area, less intrinsic activity, and lower conductivity of LDHs, obstruct their electrocatalytic application. Therefore, LDH-derived materials are being worked as the high-efficient electrocatalysts for water reduction and oxidation due to their impressive electronic and structural traits. The LDH derived heterostructures show unbelievable electrocatalytic performance along with fascinating characteristics owing to the synergistic benefits of LDHs themselves (intrinsic layered structure and adjustable metal cations composition) and the exceptional characteristic features of their derivatives. Also, the uniform dispersion of multivalent metals present in the laminar structure of LDHs can create a close hetero-interface in the derived compound and augment the synergistic interaction among the two metal components in the resultant material. Thus, the derived material offers an improved surface area, a greater number of electroactive sites, and unified charge-transfer, demonstrating in outstanding activity towards water splitting. In this book chapter, the recent advances and efforts in synthetic strategy of LDH-derived nanomaterials (mixed metal oxides, sulfides, selenides, oxyhydroxides, nitrides, and phosphides), discussing the basic principle of electrocatalytic water splitting, physico-chemical characterization techniques, and catalytic performance are provided. The perceptions into the improved catalytic activities and future prospective are also discussed.

Chapter 18: Carbon monoxide (CO) represents one of the toxic gaseous pollutants that act as a chemical asphyxiant. It is released in to the atmosphere by various anthropogenic activities, with the incomplete combustion of fossil fuel as the major cause. Its affinity to reacts with hemoglobin and thereby affecting the oxygen transportation ability of red blood cells makes it harmful to both humans and animals. One way to abate this toxic pollutant molecule is through oxidative conversion to carbon dioxide using metals catalysts. The surfaces of transition metal atoms and/or their oxides are known to be a potential candidate for CO oxidation. Extensive research has been conducted and reported by various research communities to understand the potential use of noble metals, such as platinum, palladium, gold, silver, for CO oxidation. Moreover, non-noble metal/ metal oxide catalysts with diverse morphological features are also reported for their efficacy toward the oxidative removal of CO. The researchers are focusing more on the development of base metal catalysts due to several advantages such as low cost, environmentally friendly, and easy availability. In this regard, this chapter describes the recent advances on the theoretical and experimental studies of transition metal catalysts including noble metals towards the oxidative removal of carbon monoxide. Different approaches have been devised for controlling the catalyst structures precisely by controlling the metal sizes from nano to sub-nano or even single metal atoms, thereby enhancing the catalytic activity. The recent advances in characterization techniques have helped in the development of novel catalysts and reaction routes for CO oxidation. Computational methods are capable of providing invaluable insights into the mechanistic part of the reaction, and it is possible to calculate various parameters for complex catalytic systems. Density functional theory (DFT) is used as a tool to understand the reaction intermediates, adsorption and desorption characteristics, charge transfer interactions, and the change in electronic properties of the catalyst before and after the reaction.

Chapter 19: The use of less abundant fossil fuel as energy sources leads to various environmental problems, including global warming, air pollution, and climate change. In addition, the escalating demand for energy necessitates to find out an alternative energy source that is clean, sustainable, and environmentally benign. One such alternate is hydrogen as it possesses properties such as high calorific value (~282 kJ mol^{-1}), zero pollutant emissions, and renewable nature. By using a suitable catalyst, H_2 can be generated from various chemical hydrogen storage materials in which hydrogen is bonded chemically with other elements. The efficacy of a catalytic system is dependent on various parameters including morphology, nanostructure, and the dispersion. This can be achieved by supporting the metal nanocatalysts on porous support materials such as zeolites, metal–organic frameworks (MOFs), porous silica, activated carbon, porous organic cages (POC) as well as porous organic polymers (POP). Metal–organic frameworks (MOFs), porous organic–inorganic hybrid crystalline materials, possess high surface area which ensures easy access to active sites, thereby facilitating the reactions. Unlike other traditional support materials, MOFs offer the advantage of tunable porosity in terms of size and shape through the selection of appropriate organic linkers and metal ions or clusters (the secondary building unit of MOFs) which in turn allows for the uniform confinement of metal nanocatalysts over it. Additionally, the metal–support interaction in MOFs can be easily adjusted through controlled pyrolysis of nodes and linkers. Literature reports the MOF-supported noble and non-noble metal catalysts (MIL-101, ZIF-8, ZIF-67, and MIL-96) for the production of hydrogen from various solid hydrogen storage materials. This chapter covers the recent advances in synthesis and properties of MOFs-supported transition metal nanocatalysts for the generation of hydrogen from different hydrogen storage materials, such as NH_3BH_3, $NaBH_4$, and $N_2H_4BH_3$. The functionalization of MOFs with groups like -NH_2 increases the metal anchoring sites there by ensuring the uniform dispersion of the mental nanocatalysts and hence the activity. Various noble metals like Ru, Rh, Ag, and Pd have been reported for the hydrolysis of NH_3BH_3, $NaBH_{4,}$ and $N_2H_4BH_{3,}$ with the highest TOF of 1080 min^{-1}. Non-noble metals like Cu, Co, Ni, Fe, etc. are also reported for their considerable activity toward hydrogen production with TOF varying from 0.8 min^{-1} to 2935 min^{-1}. The recyclability studies reported ensure that MOFs are highly stable and can be re-used effectively even after 30 cycles. Thus, it can be summarized from the reported literature that MOFs can be an appropriate support for the nanocatalyst materials and superior hydrogen generation activity can be achieved by fine-tuning the property of the support MOF as well as by possibly alloying the nanometal catalyst with other metal ions.

1.14 Conclusion

In conclusion, basic concepts of catalytic materials and summaries of all chapters are provided. This will be useful in basic understanding of terms associated in surface chemistry and catalysis. Also, before going to all chapters, readers can find appropriate chapter.

Acknowledgements We thank authors of all chapters (2–19) for providing summary of their chapters. We also thank Ruchi Agrawal and Kedukhro Khupfu in this chapter for their help in drafting figures.

References

Aljamali, N.M., Aldujaili, R.A.B., Alfatlawi, I.O.: Physical and chemical adsorption and its applications. Int. J. Thermodyn. Chem. Kinet. **7** (2021). https://doi.org/10.37628/IJTCK

Arandiyan, H., Sudarsanam, P., Bhargava, S.K., Lee, A.F., Wilson, K.: Perovskite catalysts for biomass valorization. ACS Catal. **13**(12), 7879–7916 (2023)

Bond, G.C.: Principles of Catalysis. Published by The Chemical Society, Burliungton House, London (1972)

Buttard, F., Guinchard, X.: Spiroindoles as intermediates/products in transition metal-catalyzed dearomatization of indoles. ACS Catal. **13**(14), 9442–9475 (2023)

Carley, A.F., Davies, P.R., Hutchings, G.J., Spencer, M.S. (eds.): Surface Chemistry and Catalysis. Publisher Springer, New York, NY (2002)

Chorkendorff, I., Niemantsverdriet, J.W.: Concepts of modern catalysis and kinetics. Second Edition. Copyright ©: WILEY-VCH Verlag GmbH & Co. KGaA, Weinheim (2007)

Czaplicka, N., Rogala, A., Wysocka, I.: Metal (Mo, W, Ti) carbide catalysts: synthesis and application as alternative catalysts for dry reforming of hydrocarbons—a review. Int. J. Mol. Sci. **22**, 1233 (2021)

Dongil, A.B.: Recent progress on transition metal nitrides nanoparticles as heterogeneous catalysts. Nanomater. **9**, 1111 (2019)

Eastwood, J.B., Burden, T.J., Hammarback, L.A., Horbaczewskyj, C., Tanner, T.F.N., Clark, I.P., Greetham, G., Towrie, M., Fairlamb, I.J.S., Lynam, J.M.: The importance of understanding (pre)catalyst activation in versatile C-H bond functionalisations catalysed by $[Mn_2(CO)_{10}]$. Chem. Sci. **15**, 9183–9191 (2024)

Gajbhiye, N.S., Sharma, S., Ningthoujam, R.S.: Synthesis of self-assembled monodisperse 3 nm FePd nanoparticles: Phase transition, magnetic study, and surface effect. J. Appl. Phys. **104**, 123906 (2008)

Gajbhiye, N.S., Ningthoujam, R.S.: Structural, electrical and magnetic studies of nanocrystalline δ-MoN and γ-Mo_2N. phys. stat. sol. (c) 1, 12 3449−3454 (2004)

Ghedini, E., Taghavi, S., Menegazzo, F., Signoretto, M.: A Review on the Efficient Catalysts for Algae Transesterification to Biodiesel. Sustainability **13**, 10479 (2021)

https://chemistry-desk.blogspot.com/2012/10/adsorption.html

https://en.wikipedia.org/wiki/Catalytic_cycle

https://laboratoryinfo.com/absorption-vs-adsorption/

https://www.grassrootshealth.net/blog/poor-health-condition-pain-fatigue-due-enzyme-nutrient-deficiency/

https://www.nobelprize.org/prizes/lists/all-nobel-prizes-in-chemistry/

John Whittall.: Applied biocatalysis: the chemist's enzyme toolbox. This edition first published 2021 © 2021 John Wiley & Sons Ltd. The Atrium, Southern Gate, Chichester, West Sussex, PO19 8SQ, UK

Jaime Wishiak.: The history of catalysis. From the Beginning Nobel Prizes. J.: Educ. Quim. **21** 60–69 (2021)

Jens, K.: Nørskov, Felix Studt, Frank Abild-Pedersen, Thomas Bligaard, Fundamental Concepts in Heterogeneous Catalysis. Published by John Wiley & Sons Inc., Hoboken, New Jersey (2014)

Li, J., Zhang, C., Li, Y., Pan, Y., Liu, Y.: Rational Design and Structural Regulation of Robust Catalysts for Electrocatalytic Hydrodechlorination: From Nanostructures to Single Atoms. ACS Catal. **13**(14), 9633–9655 (2023)

Liu, J., Ren, H., Tang, T., Wang, J., Fang, J., Huang, C., Zheng, Z., Qin, B.: The biocatalysis in cancer therapy. ACS Catal. **13**(12), 7730–7755 (2023)

Liu, T., Chen, C., Liu, S., Chen, Z., Pu, Z., Huang, Q., Zhang, L., Al-Enizi, A.M., Nafady, A., Sun, S., Zhang, G.: Transition metal phosphides as noble-metal-alternative co-catalysts for solar hydrogen production. Coord. Chem. Rev. **521**, 216145 (2024)

Mohammed, A.A.: Khalid, "Biophysical Chemistry—Advance Applications." IntechOpen (2020). https://doi.org/10.5772/intechopen.73426ISBN978-1-78984-048-3

Nikam, D.S., Jadhav, S.V., Khot, V.M., Ningthoujam, R.S., Hong, C.K., Malic, S.S., Pawar, S.H.: Colloidal stability of polyethylene glycol functionalized $Co_{0.5}Zn_{0.5}Fe_2O_4$ nanoparticles: effect of pH, sample and salt concentration for hyperthermia application, RSC Adv., **4** 12662–12671 (2014)

Ningthoujam, R.S., Kulshreshtha, S.K.: Nanocrystalline SnO_2 from thermal decomposition of tin citrate crystal: Luminescence and Raman studies. Mater. Res. Bull. **44**, 1, 8 57–62 (2009)

Ningthoujam, R.S., Gajbhiye, N.S., Sachil Sharma.: Reduction mechanism of Ni^{2+} into Ni nanoparticles prepared from different precursors: Magnetic studies. Pramana Vol. 72, Journal of Physics 577–586 (2009)

Ningthoujam, R.S., Gajbhiye, N.S.: Magnetic study of single domain ε-Fe_3N nanoparticles synthesized by precursor technique. Mater. Res. Bull. **43**(5), 1079–1085 (2008)

Ningthoujam, R.S., Gajbhiye, N.S.: Synthesis, electron transport properties of transition metal nitrides and applications. Prog. Mater. Sci. **70**, 50–151 (2015)

Parchur, A.K., Ningthoujam, R.S.: Preparation, microstructure and crystal structure studies of Li^{+} co-doped $YPO_4:Eu^{3+}$. RSC Adv. **2**, 10854–10858 (2012)

Peng, J., Giner-Sanz, J.J., Giordano, L., Mounfield, W.P., III., Leverick, G.M., Yu, Y., Roman-Leshkov, Y., Shao-Horn, Y.: Design principles for transition metal nitride stability and ammonia generation in acid. Joule **7**, 150–167 (2023)

Phaomei, G., Ningthoujam, R.S., Singh, W.R., Loitongbam, R.S., Singh, N.S., Rath, A., Juluri, R.R., Vatsa, R.K.: Luminescence switching behavior through redox reaction in Ce^{3+} co-doped $LaPO_4:Tb^{3+}$ nanorods: Re-dispersible and polymer film. Dalton Trans. **40**, 11571–11580 (2011)

Prats, H., Stamatakis, M.: Transition metal carbides as supports for catalytic metal particles: recent progress and opportunities. J. Phys. Chem. Lett. **15**, 3450–3460 (2024)

Puri, B.R., Sharma, L.R., Pathania, M.S.: Principles of Physical Chemistry, 30th edn. Shoban Lal Nagin Chand & Co Publications, Jalandhar City, Punjab, India (1989)

Rawool, S.A., Pai, M.R., Banerjee, A.M., Arya, A., Ningthoujam, R.S., Tewari, R., Rao, R., Chalke, B., Ayyub, P., Tripathi, A.K., Bharadwaj, S.R.: Pn Heterojunctions in $NiO:TiO_2$ composites with type-II band alignment assisting sunlight driven photocatalytic H_2 generation. Appl. Catal. B **221**, 443–458 (2018)

Ren, Y., Yang, Y., Wei, M.: Recent advances on heterogeneous non-noble metal catalysts toward selective hydrogenation reactions. ACS Catal. **13**(13), 8902–8924 (2023)

Santen, R.A., Moulijn, J.A., van Leeuwen Piet W.N., Averill, B.A.: Catalysis: An Integrated Approach. 2nd Edition, 1999, Elsevier Sci. Technol. Books Publ.

Shete, P.B., Patil, R.M., Ningthoujam, R.S., Ghosh, S.J., Pawar, S.H.: Magnetic core–shell structures for magnetic fluid hyperthermia therapy application. New J. Chem. **37**, 3784–3792 (2013)

Singh, L.R., Ningthoujam, R.S., Sudarsan, V., Singh, S.D., Kulshreshtha, S.K.: Probing of surface Eu^{3+} ions present in ZnO: Eu nanoparticles by covering ZnO: Eu core with Y_2O3 shell: Luminescence study. J. Lumin. **128**(9), 1544–1550 (2008)

Singh, N.S., Ningthoujam, R.S., Singh, S.D., Viswanadh, B., Manoj, N., Vatsa, R.K.: Preparation of highly crystalline blue emitting $MVO_4:Tm^{3+}$ (M=Gd, Y) spherical nanoparticles: Effects of activator concentration and annealing temperature on luminescence, lifetime and quantum yield. J. Lumin. **130**(12), 2452–2459 (2010)

Singh, R.V., Pai, M.R., Banerjee, A.M., Shrivastava, A., Sinha, U.K., Dutta, B., Hassan, P.A., Ningthoujam, R.S., Ghosh, R., Nath, S., Sharma, R.K., Jagannath, Bapat, R.D.: Interfacial Engineering over Pt–Calcium Ferrite/2D Carbon Nitride Nanosheet p–n Heterojunctions for Superior Photocatalytic Properties, ACS Omega 9, **38** 40182–40203 (2024)

Siregar, M.A., Siregar, C.A., Muharnif, Siregar, AM., Maulana, I.: Application of catalytic converter copper catalyst with honeycomb surfaces to reduce emissions of flue gas in motorcycles, IOP Conf. Ser.: Mater. Sci. Eng. **674** 012060 (2019)

Somorjai, G.A., Li, Y.: Introduction to Surface Chemistry and Catalysis, 2nd edn. Published by Wiley, Kindle Edition (2011)

Thorat, N.D., Otari, S.V., Bohara, R.A., Yadav, H.M., Khot, V.M., Salunkhec, A.B., Phadatare, M.R., Prasad, A.I., Ningthoujam, R.S., Pawar, S.H.: Structured superparamagnetic nanoparticles for high performance mediator of magnetic fluid hyperthermia: Synthesis, colloidal stability and biocompatibility evaluation. Mater. Sci. Eng. C **42**, 637–646 (2014)

Umare, S.S., Ningthoujam, R.S., Sharma, S.J., Shrivastava, S., Kurian, S., Gajbhiye, N.S.: Mössbauer and magnetic studies on nanocrystalline $NiFe_2O_4$ particles prepared by ethylene glycol route. In: Gajbhiye, N.S., Date, S.K. (eds.) ICAME (2007). Springer, Berlin, Heidelberg. (2008). https://doi.org/10.1007/978-3-540-78697-9_88.

Xu, H., Zhao, W., Li, D., Ding, S., Xiao, C., Zeng, L.: Emerging supported metal atomic clusters for electrocatalytic. ACS Catal. **15**, 2434–2458 (2025)

Yang, X., Deng, Y., Ling, D., Li, T., Chen, L., Jin, Z.: Recent progress in chiral quaternary ammonium salt-promoted asymmetric nucleophilic additions, ACS Catal. 15, XXX 1973–2001 (2025)

Zhang, S., Findlater, M.: Electrochemically driven hydrogen atom transfer catalysis: a tool for C(sp_3)/Si–H functionalization and hydrofunctionalization of Alkenes. ACS Catal. **13**(13), 8731–8751 (2023)

Zhang, W., Zhang, S.: Catalysts: materials. Encycl. Condens. Matter Phys. (Second. Ed.) **5**, 738–749 (2024)

Zhu, S.-C., Xiao, F.-X.: Transition metal chalcogenides quantum dots: emerging building blocks toward solar-to-hydrogen conversion. ACS Catal. **13**(11), 7269–7309 (2023)

Chapter 2
Frustrated Lewis Pairs (FLPs): Concept, Synthesis and Their Implications in Catalysis

Adish Tyagi, Gourab Karmakar, and Rohit Singh Chauhan

2.1 Introduction

> To be effective, the two functional groups must be so disposed that they can interact simultaneously with a hydrogen molecule, but at the same time are prevented from interacting with (neutralizing) each other.
>
> J. Halpern

Catalysis and activation of chemical bonds have made enormous contribution in the advancement of chemistry and widened its application from food security, to pharmaceuticals, materials for energy conversion and storage, environmental remediation and so on. In addition, catalysis played phenomenal part in aiding eco-friendliness, feasibility and economics of around 90% of the chemical processes and was considered as one of the chief factors for sustainability. Until recently, the field of catalysis was heavily dominated by the transition metals and their compounds, owing to the presence of number of unpaired d-orbital electrons, ability to acquire variable oxidation state, ability to accept electron density from a ligand (include complex molecules to as simple as hydrogen molecule) as well as donate electron density in anti-bonding orbitals of ligands through metal–ligand back bonding and flexible coordination modes. Such was their dominance that they seem to be indispensable in the chemical industry, particularly petroleum, polymer, methanol production, CO_2 sequestration, etc. However, concerning issues with transition metal catalysis, such as cost effectiveness, toxicity and availability, have initiated the quest for organic

A. Tyagi (✉) · G. Karmakar
Chemistry Division, Bhabha Atomic Research Centre, Mumbai, India
e-mail: tyagia@barc.gov.in

Homi Bhabha National Institute (HBNI), Mumbai, India

R. S. Chauhan (✉)
Department of Chemistry, K. J. Somaiya College of Science and Commerce, Mumbai, India
e-mail: rohit.chauhan@somaiya.edu

R. S. Ningthoujam and A. K. Tyagi (eds.), *Handbook of Materials Science, Volume 3*, Indian Institute of Metals Series, https://doi.org/10.1007/978-981-96-8229-4_2

alternatives which is capable of activating small important molecular systems such as dihydrogen (H_2), carbon monoxide (CO), carbon dioxide (CO_2) and other covalent bonds (e.g., C–H) bonds. One such alternative is found in frustrated Lewis pair (FLP) that provides a wonderful opportunity in metal free activation of indolent chemical bonds and discarding the long-standing conviction that presence of metals is essential for bond or small molecule activation.

2.1.1 Basic Concept of FLP

FLP is a chemistry marvel, emerged mainly from the pioneering work of Prof. Douglas W. Stephan's group (Stephan 2016). FLPs are defined as any Lewis acid-Lewis base combination, which are sterically prevented from forming dative bonds, unlike in classical Lewis acid–base adducts as illustrated in Fig. 2.1 (Stephan 2016). Quite interestingly, in FLPs the Lewis acid and basic sites are available for interaction with the third molecule simultaneously. This unquenched Lewis acidity and basicity in FLPs which are available of further reactivity has ignited much attention and initiated numerous investigations into the activation of small molecules and catalysis. However, before initiating the discussion on examples of metal-free activation of H_2 molecules and other small molecules by FLPs or arriving at the synthetic protocols to access FLPs, it will be important and interesting to understand how the FLP concept was conceived, nurtured and matured for important applications.

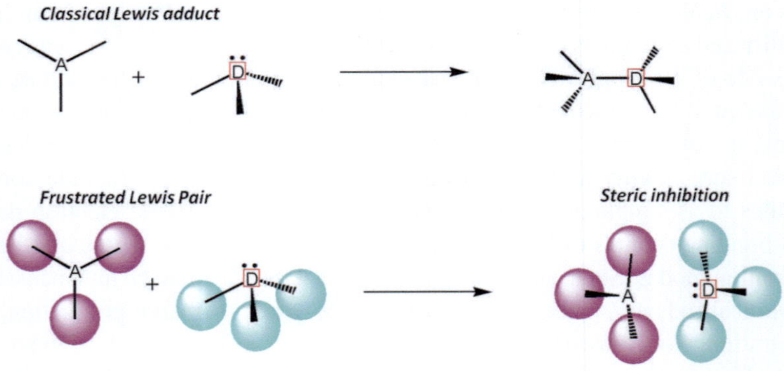

Fig. 2.1 Illustration to explain the difference between the classical Lewis acid–base adduct and frustrated Lewis pair. Redrawn from Stephan (2016)

2.1.2 *Historical Overview*

In 1923, G. N. Lewis classified the acid and base behavior depends upon electron pair acceptance and donation, termed as Lewis acid and base respectively. Lewis base donates electron pairs through its highest occupied molecular orbitals (HOMO) to the lowest unoccupied molecular orbitals (LUMO) of Lewis acids, resulting in neutralization through adduct formation. An adduct $NH_3 \rightarrow BH_3$ is a very common text book example to epitomize this concept. However, way back it was observed that not all Lewis acids and bases form adducts. For instance, Brown et al. in 1942 have reported an unusual example of Lewis acid and base interaction without having a dative bonding interaction between them. They showed that while 2,6-lutidine and BF_3 readily reacts and forms adduct, 2,6-lutidine and BMe_3 failed to do so. This contradictory behavior in case of 2,6-lutidine and BMe_3 is due to steric congestion of methyl group attached to acid (boron atom) as well as base (lutidine carbon) (Brown et al. 1942). Inspired by this, Stephan and coworkers showed 2,6-lutidine and $B(C_6F_5)_3$ form a reversible adduct with a very weak B–N bond capable of activating the hydrogen molecule as shown in Scheme 2.1.

Wittig and Benz isolated zwitterionic addition product stemming out from a reaction between BPh_3 and PPh_3 with benzyne. Further progressive domain of research work shows same evidence of combination effect rather than expected dative bond formation, e.g., addition of BPh_3 and trityl cation to butadiene or triphenyl phosphine.

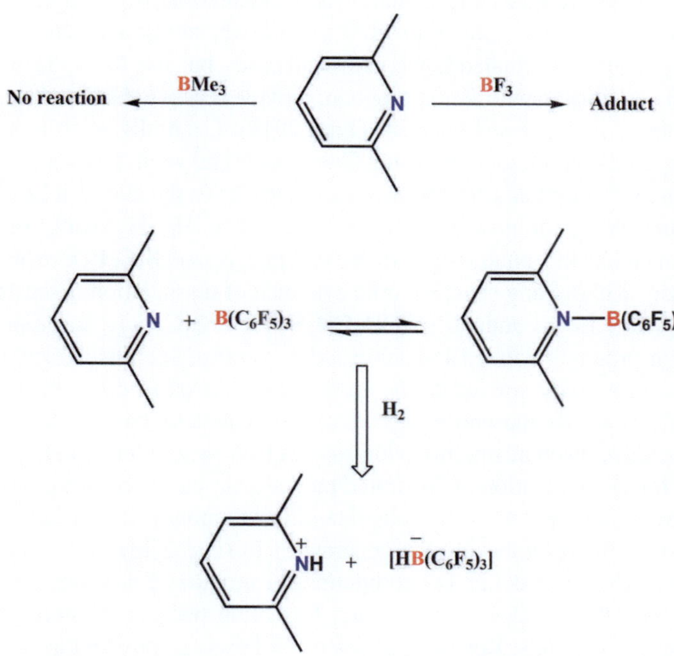

Scheme 2.1 Reactivity of lutidine with different substituted boron

Even though to have notable outcomes from these reactions, chemistry of such sterically encumbered Lewis acid base pair or "antagonistic pairs" as commonly referred in those days, remained dormant for a long time as a manifest from the poor number of reported articles during the tenure of 1960 till 1990.

Toward the end of twentieth century, the $B(C_6F_5)_3$-catalyzed hydrosilylation of ketones by Piers and coworkers marks the true beginning of FLP chemistry. In this example, they have demonstrated that Si–H bond which was attacked by ketone was previously activated by $B(C_6F_5)_3$. Action of $B(C_6F_5)_3$ and ketone was found to be concerted in nature. This was further confirmed by the systematic work done by Oestreich and coworkers (Fontaine and Stephan 2017). The insightful explanation of $B(C_6H_5)_3$-catalyzed hydrosilylation of ketones is considered as the beginning of the chemistry of frustrated Lewis pairs encompassing diverse area which include (i) application in activation in small molecule, (ii) alternate for metal-free catalytic process, (iii) reductions of amines, aniline and aromatic compounds, (iv) enough room for various coordination possibility and (v) "tailor-made" structure depend upon substrate.

2.1.3 *Understanding the Nature of FLP*

Elementary understanding of FLPs suggested that steric inhibition between Lewis acid and base for coordinate bond formation is an essential step to grant FLPs the great potential as shown in Fig. 2.2a (Jupp 2019). However, subsequent studies revealed that not only the un-coordinated Lewis acids and bases, but also the acid base adducts which are in equilibrium with free species can provide access to FLPs and can be used to activate small molecules (Fig. 2.2b) (Jupp 2019). Generally, such type of equilibriums required combination of strong Lewis acids and weak Lewis bases. Recent studies from Prof. Stephan's groups have shown that even the classical Lewis adduct of a particular type, for instance Verkade's base $[N(CH_2CH_2NMe)_3\ PB(C_6F_5)_3]$ derived from proazaphosphatrane $[P(MeNCH_2CH_2)_3N]$ and $B(C_6F_5)_3$ exhibit typical FLP reactivity without any spectroscopic evidence of dissociation (Johnstone et al. 2018). The stable Lewis adduct: $[N(CH_2CH_2NMe)_3\ PB(C_6F_5)_3]$ surprisingly gave FLP addition products with FLP established substrates, such as azides, CO_2, etc. (Fig. 2.2c). This study proved to be instrumental in expanding the horizon of accessing FLPs across the entire spectrum of Lewis acid–base systems. Besides this, an alternative mechanism, involving single electron transfer from Lewis base to Lewis acid for the generation of frustrated radical pair, has also been coined. These frustrated pairs can react with the substrate either homolytically or heterolytically depending upon the reactants used to generate FLPs (Fig. 2.2d).

Broadly, FLPs can be defined as compounds or mixtures containing a Lewis acid and Lewis base that evades strong adduct formation because of steric hindrance and geometric constraints. Retentive character of Lewis acidity and basicity within the same molecular system can render a cooperative effect of the functionalities and makes the incoming reagent/molecule polarizable, which results activation

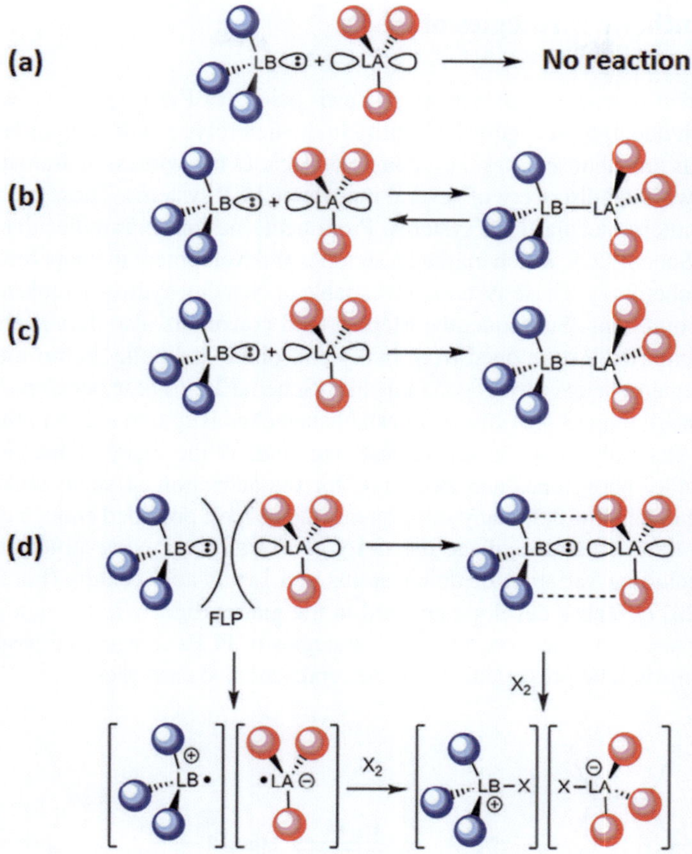

Fig. 2.2 Illustration of the possible ways of accessing the reactivity of FLPs. **a** Sterically congested Lewis base and acid pair. **b** Lewis acid–base equilibrium leading to FLP. **c** Inducing FLP reactivity in classical Lewis acid base adducts. (d) Homolytic and heterolytic reaction pathways for FLPs. Reprinted with permission from Elsevier Jupp (2019)

of incoming molecule much more facile. Based on these characteristics, renewed interest in the FLP-derived chemistry was stimulated by the work of Prof. Douglas W. Stephan, who showed various reactions of sterically crowded tertiary phosphines with $B(C_6F_5)_3$. These reactions afforded a series of zwitter ions $[R_3P(C_6F_4)BF(C_6F_5)_2]$ depending upon the nature of substituent (R = Ph, Et, Bu, $C_6H_2Me_3$, etc.) which has shown their utility as a catalyst for hydrogenation of imines, nitriles and aziridines. In the same prospect, various Lewis bases, viz. phosphine/amine/ferrocene-derived FLPs have been synthesized and studied for their further utility. In short, it is not wrong to say that "the articulation of the frustrated Lewis pair (FLP) concept has helped to foster a resurgence of main group small-molecule activation chemistry in the last decade" (Johnstone et al. 2018).

2.2 Synthetic Strategies of FLPs

The noteworthy reactivity of "frustrated Lewis pairs" (FLPs) is a progressive domain of research due to their established utility in the heterolytic activation of H_2, alkene and various greenhouse gases. Past couple of decades has witnessed dramatic rise of this field with the discovery of several innovative FLP systems. These include both inter and intermolecular FLP systems. Prominent ones are intramolecular FLPs as shown in Scheme 2.1, which marked a significant development in cooperative Lewis acid/base chemistry. These systems are capable of cleaving hydrogen molecule under ambient conditions. Subsequently, Stephan and coworkers also demonstrated that intermolecular FLP developed from bulky phosphine and bulky boron moiety can also conveniently cleave hydrogen molecule (Scheme 2.2). These benchmark studies mark the beginning of previously not established chemistry derived from the combination of sterically crowded Lewis base and acid. While many of the FLPs have utilized bulky phosphines and $B(C_6H_5)_3$ for the activation of small molecules as illustrated in Scheme 2.3, many subsequent studies have provided enough evidences regarding the modulation of the reactivity of FLPs by applying structure modifications including variation of donor atoms and Lewis acid centers (Fontaine and Stephan 2017). These developments led to the giant progress in the field of metal free catalysis. In this section, synthetic strategies of FLPs categorized based on the donor atoms will be presented with some representative examples.

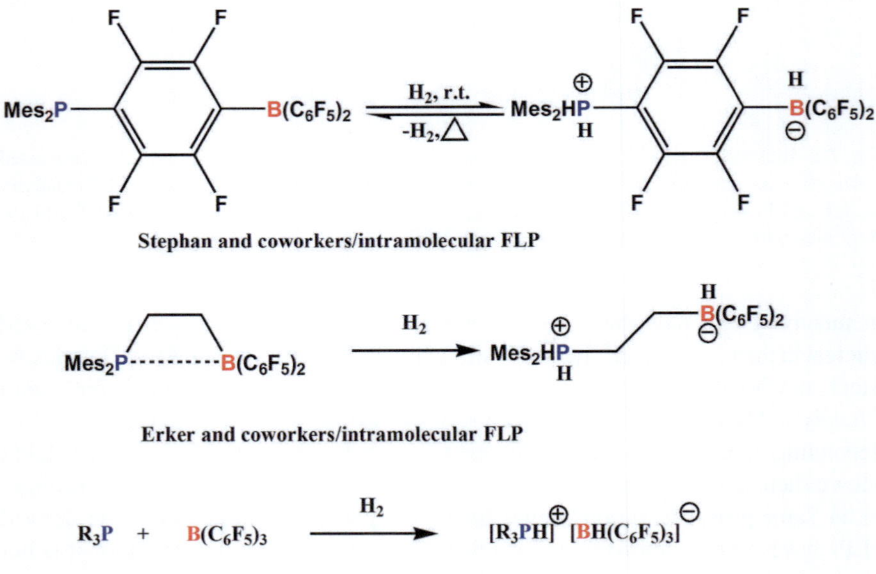

Scheme 2.2 Examples of intra- and intermolecular FLP. Redrawn from Stephan and Erker (2015)

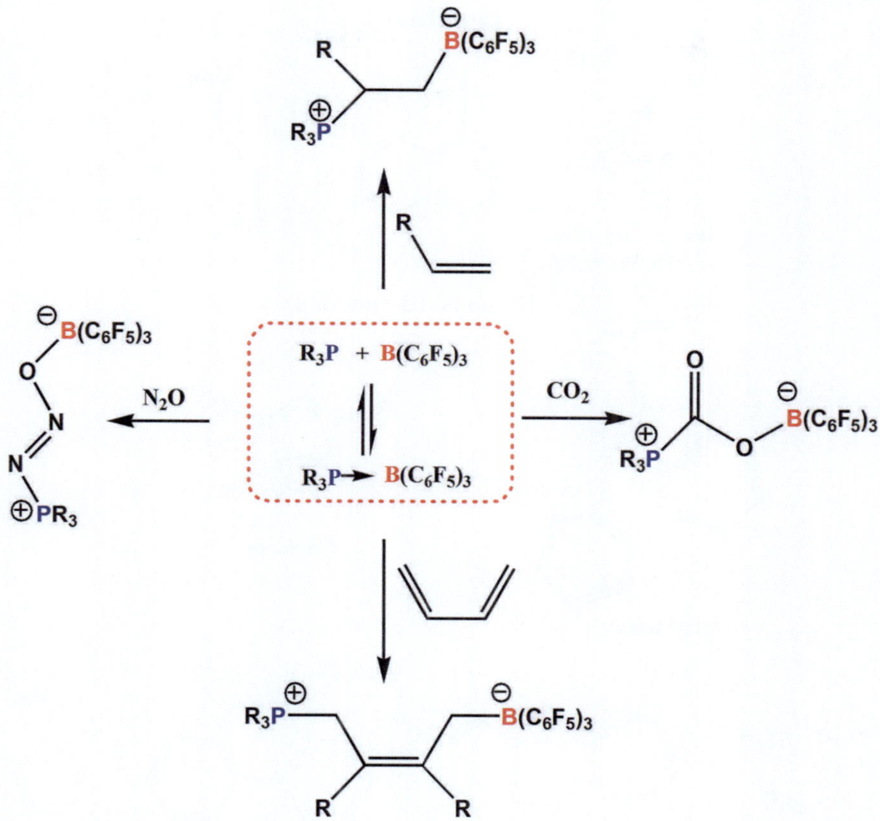

Scheme 2.3 Small molecule activation catalysis achieved by the combination of bulkyphosphine/borane combinations

A representative class of various donors P, N, C centered atoms with their sterically encumbrance (Mes, cy, etc.) as shown in Scheme 2.4 and frequently utilized acceptors like $B(C_6F_5)_3$, triaryl and trialkyl boranes, as described herein. For the sake of convenience, the discussion has been categorized as traditional FLPs and advance FLPs. On the demarcation line lies the discovery of *p*-$(Mes_2PH)C_6F_4(BH(C_6F_5)_2)$(where Mes sands for mesityl group) in 2006 by Stephan and coworkers, which could reversibly liberate hydrogen (Scheme 2.3), probably the first example of metal free hydrogen activation. After the discovery of this unusual compound generated from the special combination of Lewis acid and base which are sterically frustrated and differ from classical adduct, Stephan group termed this class of compounds as frustrated Lewis pairs or FLPs. The zwitterionic species/FLPs discovered prior to this will be discussed under traditional FLPs, while FLPs synthesized after this will describe the advancement in this field.

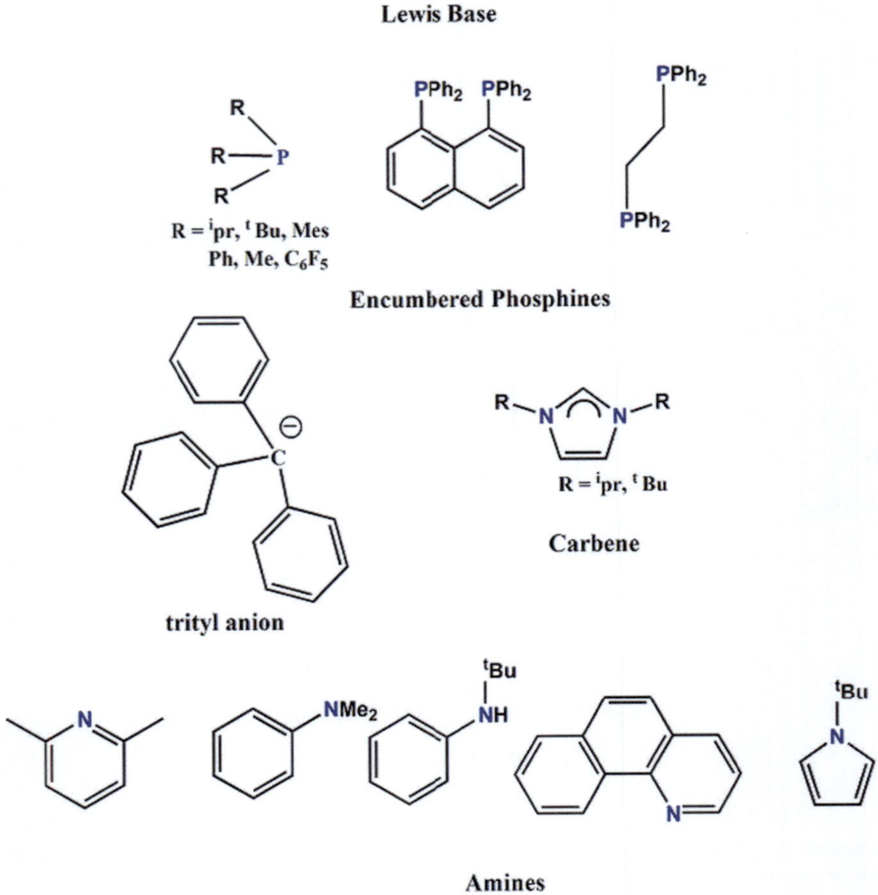

Scheme 2.4 Series of Lewis bases applied in various FLPs

2.3 Traditional FLPs

The very first zwitterionic structure was synthesized by Wittig and Ruckert from the reaction between trityl anion [$NaCPh_3$] and (C_4H_4O)BPh_3 (Wittig 1950). Finally, the reaction results ring opening of borane-bounded THF and isolates the sodium salt of [$Ph_3CC(CH_2)_4OBPh_3$] (Scheme 2.5). Wittig and Benz (1959) described that o-fluorobromobenzene on treatment with active magnesium turnings generates benzyne (in situ mode) which on further simultaneous addition with triphenylphosphine (Lewis donor) and triphenylborane (Lewis acceptor) yielded phenylene bridged zwitterionic phosphonium borate. The latter prospective product has also been obtained by the same path where 2,2'-biphenylenephenyl phosphine as a Lewis base (Scheme 2.5) was used (Wittig and Benz 1959; Stone 1958). These examples could now consider as beginner "frustrated Lewis pair".

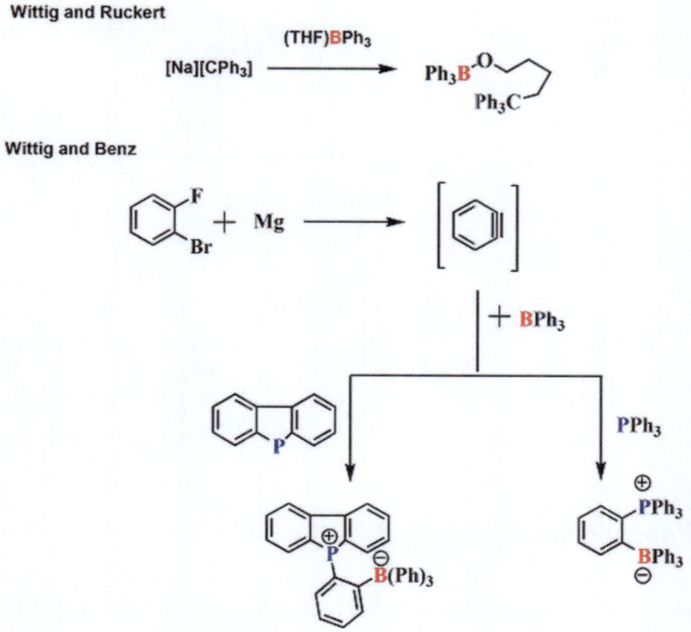

Scheme 2.5 Reaction strategy for the very first "frustrated Lewis pairs"

Subsequently, Tochtermann*et al.* have reported that reaction of trityl anion [$NaCPh_3$] and BPh_3 to butadiene and acetaphthylene leads the product resulted (Tochtermann 1966) by 1, 2 and 1, 6 addition respectively (Scheme 2.6) which is notwithstanding with desired prospect of the formation of either polymerization or Lewis acid–base adducts. This remarkable reactivity was pointed as "Antagonistisches Paar" (Tochtermann 1966). Similarly, the sterically demanded amines on reaction with trityl sodium afforded an iminium cation (Lankamp 1968) which is formed due to hydride abstraction from alpha carbon. However, a reaction with pyridine yielded a carbocation due to attach at *para* position carbon (Lankamp 1968).

2.4 Advance FLPs

In this article, the discussion on the advance FLPs will be presented on the basis donor atom present. Accordingly, FLPs have been categorized as phosphorus-based FLP, nitrogen-based FLP- and carbene-derived FLP. Toward the end, recent advances in this field like transition element-based FLPs, lanthanide element-based FLPs and FLP-MOF (metal organic framework) composite will be discussed.

Scheme 2.6 Synthesis of carbon-based classified frustrated Lewis pairs

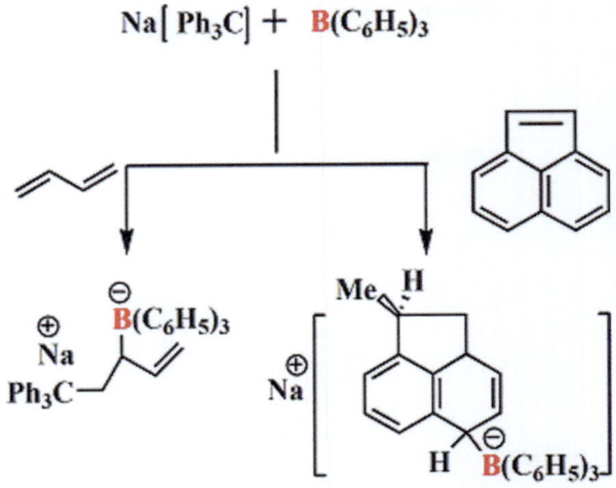

2.4.1 Phosphorus-Based FLP

The chemistry of phosphorus-based FLP is much explored with respect to various other FLP. In 2006, Stephan et al. have reported that the intramolecular Lewis acid/base pair *p*-$(Mes_2P)C_6F_4[B(C_6F_5)_2]$ can reversibly activate H_2 (Welch et al. 2006). This study dislodges the 100 years of central dogma that for the activation of H_2 molecule, presence of metal is essential. Subsequent reports from the same group demonstrated that even the simple combinations of Lewis acids like tris(pentafluorophenyl)borane and bases such as tertiary phosphine where steric congestion precludes classical Lewis adduct formation, exhibit the reactivity of FLP (Welch and Stephan 2007). Resulted adducts automatically undergone to rearranged zwitterionic complex of the composition $[R_3P(C_6F_4)BF(C_6F_5)]$ (R = ^{i}pr, cy, Ph, Et, n-Bu, p-$CF_3C_6H_4$), by the attack at *para* carbon of phenyl ring with synchronized relocation of fluoride to boron atom. These simple combinations of sterically congested phosphines and boranes can activate dihydrogen molecule. For instance, 1:1 molar mixture of R_3P and $B(C_6F_5)_3$ did not showed any experimental evidence of formation of adduct, however in presence of dihydrogen molecule, zwitterionic compound $[R_3PH][HB(C_6F_5)_3]$ is formed. Several combinations stem out from the variation of organic group attached to phosphine and borane were seen to activate H_2 to different extent. Additionally, there are some combinations like $(C_6H_2Me_3)_3P$ and BPh_3, $(C_6F_5)_3P$ and $B(C_6F_5)_3$ or $^tBu_3P$ and $BMes_3$ which showed negligible hydrogen activation. From these studies it is inferred that for the activation H_2 molecule, not only a sterically frustrated Lewis pair but favorable electronic factors (Lewis acidic and basic character) are equally needed.

The reactivity of FLPs originated from the combination of R_3P and $B(C_6F_5)_3$ has also been explored with olefins. Indeed, the addition of olefin to the mixture of R_3P and $B(C_6F_5)_3$ resulted a zwitterionic compound (Scheme 2.7a) (McCahill and

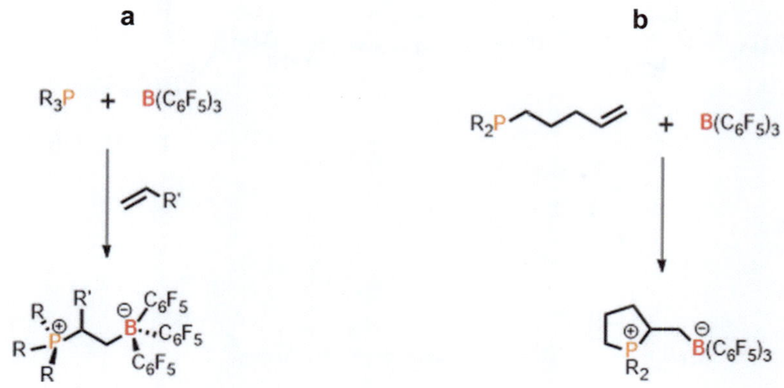

Scheme 2.7 Reaction of phosphine–borane system with olefins, **a** when olefin is added to the mixture of phosphine and borane **b** when olefin is the part of phosphine. Redrawn from Farrell (2012), Farrell (2015)

Stephan 2007). These studies were extended to include the olefinic unit in the phosphine itself. This olefinic decorated phosphine when react with $B(C_6F_5)_3$ resulted in the formation of cyclized phosphonium borate (Scheme 2.7b).

Alternative investigations have reported various new phosphorus-based FLPs derived from different alkyl linked phosphino-boranes systems. In this case, dimesityl alkenylphosphane reacted with Pier's borane $[HB(C_6H_5)_2]$ forms a Lewis acid–base adduct at low temperature (Scheme 2.8). The coexistence of dynamic equilibrium between adduct and reactant shifts an anti-Markovnikov product yielded through hydroboration reaction at room temperature (Spies et al. 2007). The spectroscopic studies of latter complexes have established weak interaction between P and B atom within the molecule. The ^{31}P NMR and ^{11}B NMR showed resonances corresponding to intermediate of phosphane–phosphonium and borane–borate character (Spies et al. 2007). The resultant compound showed FLP reactivity by activating dihydrogen to yield the respective ethylene bridged zwitterionic phosphonium–hydridoborate system, confirming through single-crystal X-ray analysis (Fig. 2.3). Further, the resultant zwitterionic molecule can do hydride transfer as inferred from its reaction with benzaldehyde. Even cyclcohexenyl bridged phosphane FLP is characterized by single-crystal X-ray analysis also. Cyclohexane ring adopts chair confirmation.

Grobe et al. have claimed the notable formation of reactive tricyclic intermediate of borane alkyne (Scheme 2.9) (Grobe et al. 2006). Parallelly, Kehr's group members opted new approach for the isolation of unsaturated FLPs in which dimesityl phosphinoacetylene reacts with $B(C_6F_5)_3$ at room temperature (Ekkert et al. 2011) (Scheme 2.10). This reaction proceeds initially by the addition of $B(C_6F_5)$ to alkyne which followed by the migration of Mes_2P toward acetylenic carbon.

In another study, $B(C_6F_5)_3$ and proazaphosphatrane: $[P(MeNCH_2CH_2)_3N]$ were combined to form the compound $[(C_6F_5)_3BP(MeNCH_2CH_2)_3N]$. The latter was explored for its FLP kind of reactivity (Johnstone et al. 2018). Interestingly, this compound does not show any spectroscopic evidence of dissociation into

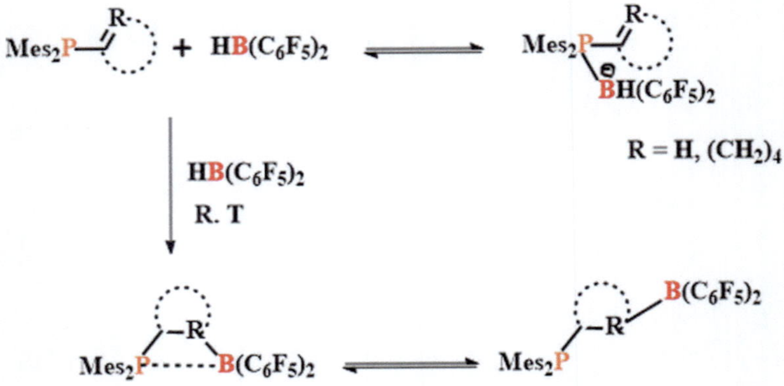

Scheme 2.8 Generalized scheme for the synthesis of various vinyl phosphane ligand-derived FLPs

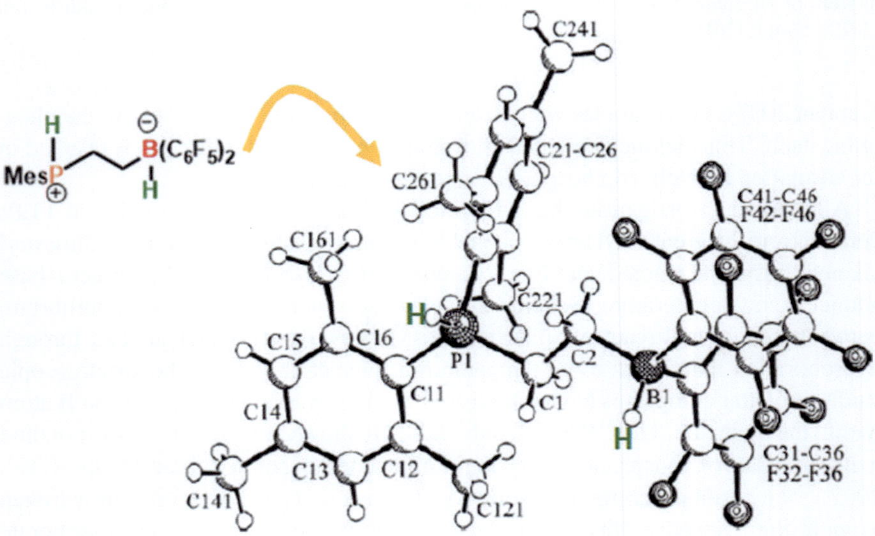

Fig. 2.3 Structural evidence of dihydrogen activation by alkenylphosphane-derived FLP. Adapted and Redrawn with permission from RSC, Spies et al. (2007)

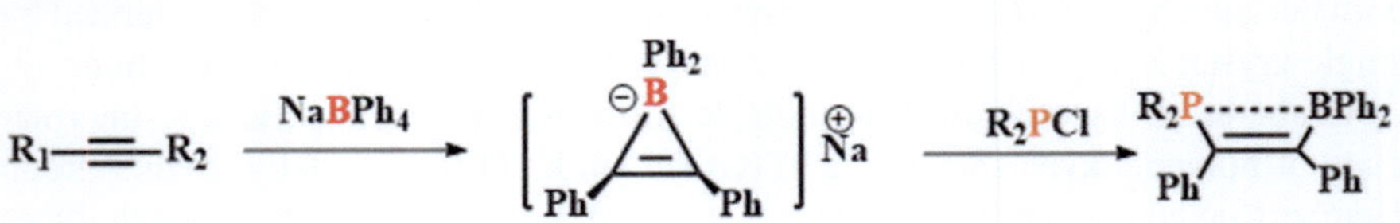

Scheme 2.9 Preparation of unsaturated FLPs by substitution through R_2PCl

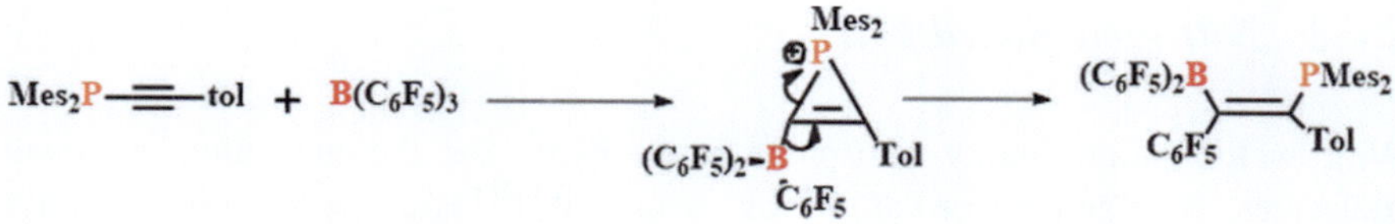

Scheme 2.10 Preparation of unsaturated FLPs by applying $B(C_6F_5)_3$

its constituent Lewis acid and base. However, this compound exhibit FLP reactivity and form FLP addition product with wide variety of substrates as shown in Scheme 2.11. From computational studies, it was inferred that the dissociation of $[(C_6F_5)_3BP(MeNCH_2CH_2)_3N]$ is thermodynamically favored under ambient conditions. This particular example revealed that FLP nature can be experienced even when steric congestion and electronic frustrations are not sufficient to produce experimentally observable formation of the dissociated Lewis acid/base.

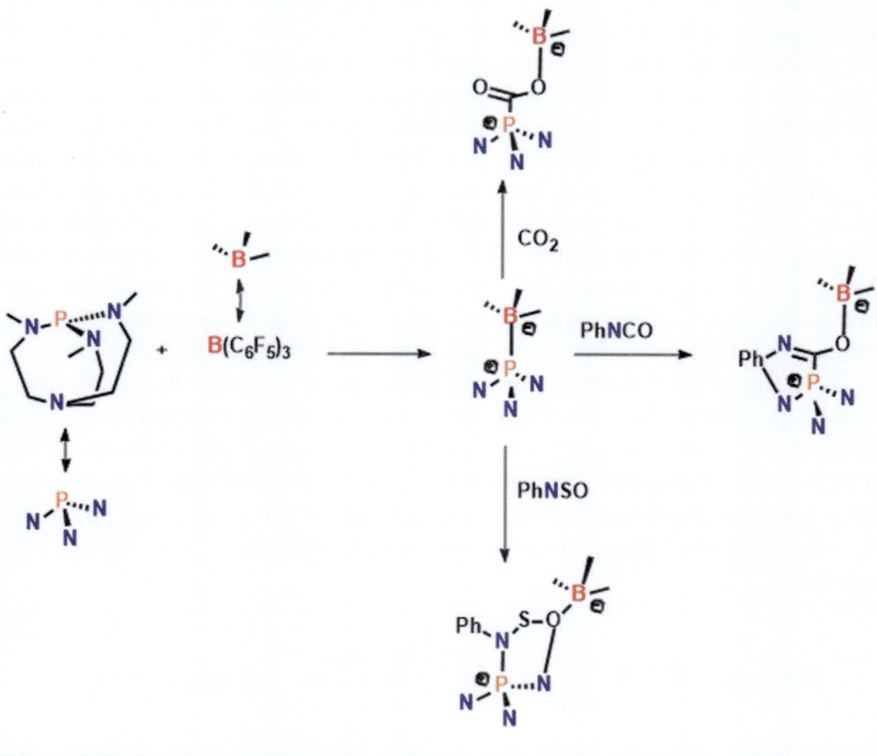

Scheme 2.11 Synthesis and applications of FLP derived from derived from $B(C_6F_5)_3$ and proazaphosphatrane

2.4.2 Nitrogen-Based FLP

With well-explored bulky phosphine-based FLPs for the activation of small molecules, Nitrogen donor-based FLP is matter of logical extension nowadays (Sumerin et al. 2008a). Especially shorter bond length of carbon nitrogen as compared to carbon-phosphorus makes it a lively area of further investigation to synthesize new variety of sterically encumbered amines. First time in the year of 1942, Brown et al. (Cabrera et al. 2006) reported that reaction in between lutidine and boranes favors the isolation of expected classical Lewis acid–base adduct. However, the reaction of lutidine and $B(C_6F_5)_3$ notably finds an attention due to an equilibrium exists between Lewis acid and base and resulted adduct as product (Fontaine and Stephan 2017). However, performing the reaction of $B(C_6F_5)_3$ with alkyl amines produced mixture of salts and zwitterions as shown in Scheme 2.12 (Sumerin et al. 2011). The mixture of bulky amines and borane is capable of cleaving H–H bond, though the temperature of cleavage largely depends upon substituent present on amine N. Furthermore, it is also observed that mixture of bulky amines and borane can also catalyze the transfer hydrogenation of N-heterocycles, mines and enamines. Interestingly, bulky amines like ($^i Pr_2NH$) act as a source of hydrogen (Farrell et al. 2011).

In most of the above-mentioned FLPs with phosphines, amines or carbenes, commonly used Lewis acid is tris(pentafluorophenyl)borane [$B(C_6F_5)_3$]. However, due to hard Lewis acidity of this borane, it can be rendered inactive by common oxygen and nitrogen containing molecules leading to classical Lewis adduct. Thus, careful design of Lewis base is essential for efficient catalytic hydrogenation reactions in case of [$B(C_6F_5)_3$]-derived FLPs. To circumvent this limitation and increase the functional group tolerance, Erös et al. (2010) designed a strategy wherein they increase the steric hindrance around the boron center by involving mesityl group and decrease the congestion around the Lewis base as depicted in Fig. 2.4.

The increase steric hindrance around boron center decreases its complexation ability with normal Lewis base. Furthermore, this shielding around boron also averts its addition to olefins, thus offering a wonderful chance to study the chemoselectivity of FLP catalyzed reactions. This innovatively developed FLP system could selectively hydrogenate enamine and imines in presence of other functional group like methoxy, allyl, etc. This development was of high significance in the field of FLP chemistry.

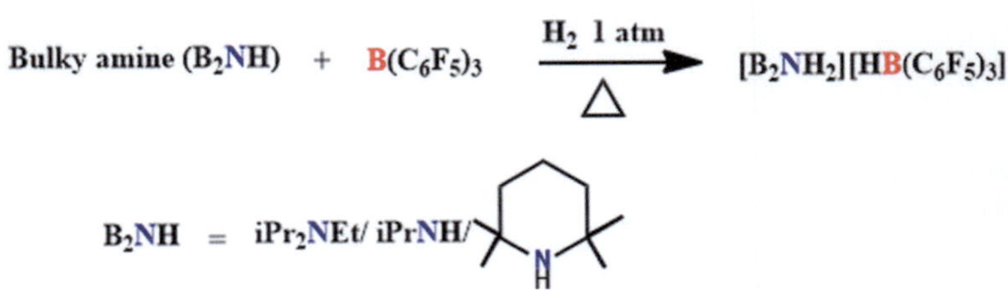

Scheme 2.12 Heterolytic activation of hydrogen by the synergistic action of bulky amine and boranes. (Redrawn from reference (Sumerin et al. 2011))

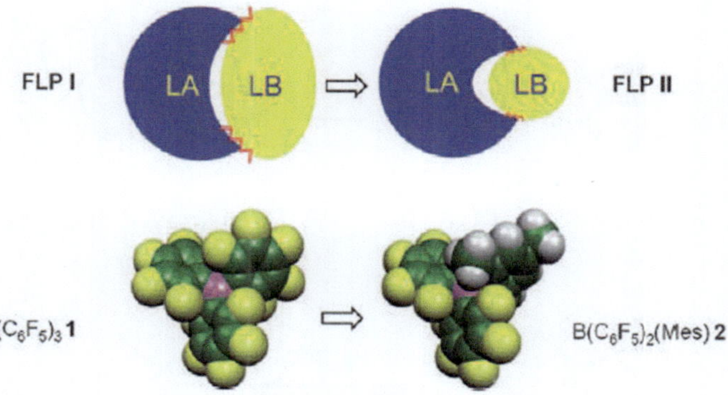

Fig. 2.4 Strategy to improve functional group tolerance of $B(C_5F_5)_3$. Reproduced with permission under Creative common license from Wiley, Ref. (Erös et al. 2010)

Meanwhile, various new sterically hindered amine-based FLPs have been synthesized and get studied for their utility for H_2 activation. One such FLP system is *ansa*-aminoborane which falls under category of intramolecular FLP system (Sumerin et al. 2008b; Schulz et al. 2011). These FLPs have been derived by the initial alkylation of encumbered amine with bromobenzylbromide followed by lithiation acquire substitution through cooled solution of $(C_6F_5)_2BCl$ as shown in Scheme 2.13. This advantageous scheme has been applied on various amines like 2,2,6,6-Tetramethylpiperidine and enamines derivatives, etc. The *ansa*-aminoborane class of FLP could reversibly activate hydrogen through an intramolecular mechanism and form a molecular hydrogen tweezers having partial covalent interaction between dihydrogen. These molecular tweezers could be used as an efficient catalyst for the preparation amines from enamines and imines.

Another prominent path for the N-based intramolecular FLP is the reaction of enamine and Pier's borane. The complex comprises four-membered cyclic ring formed due to weak nitrogen boron bond. Phenyl substituted carbon is diastereotopic in nature which is characterized by ^1H and ^{13}C NMR resonances. However, any attempts to activate hydrogen were unsuccessful with this system. An analogous compound was also obtained by reaction of $[HB(C_6F_5)_2]$ and piperdino-cyclohexene or dimethyl amino (Scheme 2.14). The coordination of N and B (N-B = 1.824 Å) was established through single-crystal X ray analysis (Chernichenko et al. 2012). Range of N-B bond length indicates weaker nature of interaction which was also been evident by nature of ^{19}F NMR spectra at various temperatures. Spectra showed pairs of resonances at low temperature; however, coalescence of resonances appeared at higher temperature, indicates reversibility of N-B bond (Haenel et al. 2006). Various combinations of this class exhibited reactivity similar/slightly lower to P/B FLP analogues. Nevertheless, this class adds to the library of molecules that can reversibly activate dihydrogen (Stephan and Erker 2015).

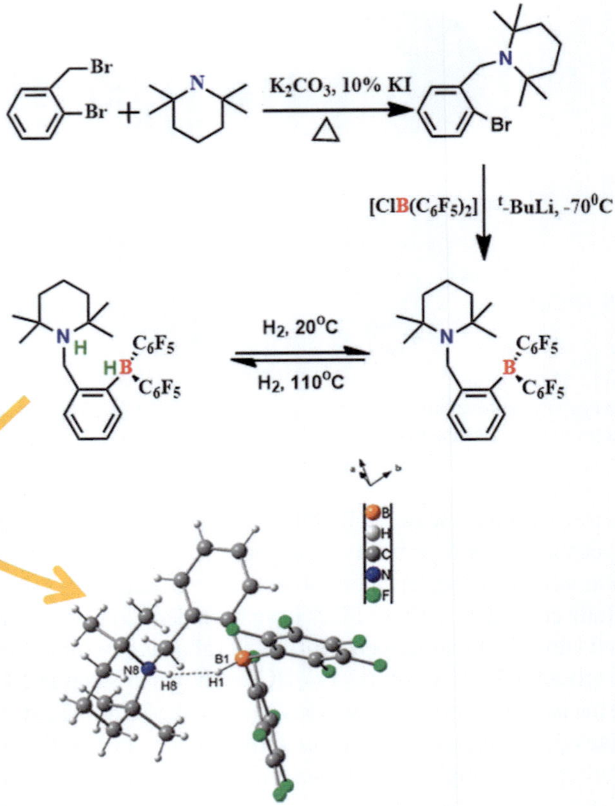

Scheme 2.13 Preparation of *ansa*-aminoborane FLPs and their application in reversible hydrogen activation. Adopted with permission from ACS, Ref. (Sumerin et al. 2008b)

An alternative strategy to induce steric frustration is to encumbered amine molecules such as secondary amine within a cavity like rotaxanes. This mechanically interlocked secondary amine exhibited the reactivity of FLP by activating H_2 molecule (Scheme 2.15) (Caputo et al. 2013). Whereas reaction of secondary amine with Lewis acid molecule results in the formation of classical Lewis adduct, secondary amine incorporated into rotaxanes exhibited FLP kind of activity by heterolytic activation of hydrogen molecule. More importantly, such systems where Lewis donors are under the protection of macrocyclic rings showed FLP activity without the need to increase the bulk around Lewis base.

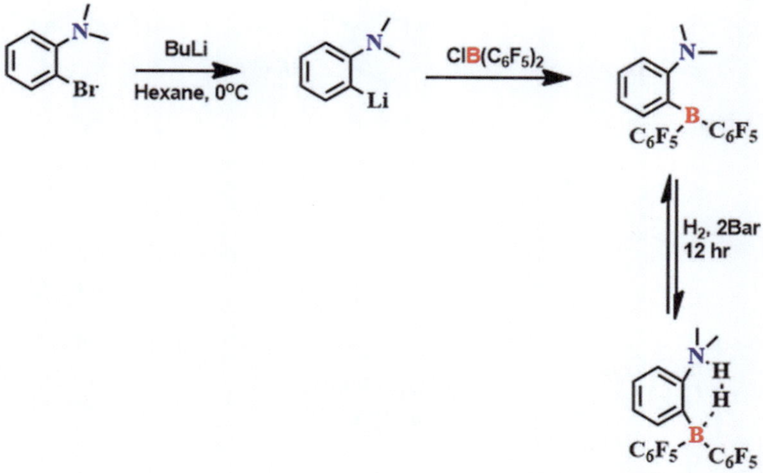

Scheme 2.14 Activation of hydrogen by bis(pentafluorophenyl)borane and dimethyl amino constituted FLP (Chernichenko et al. 2012)

2.4.3 *Carbene-Derived FLP*

FLP field is largely dominated by bulky phosphines as Lewis base and fluorinated borane $B(C_6F_5)_3$ systems which could efficiently activate hydrogen molecule leading to the formation of phosphonium borate salts $[R_3PH][HB(C_6F_5)_3]$. To widen the scope of Lewis acid and base that exhibit such reactivity, several studies on various molecules have been conducted. One such class of compound is NHCs (N-Heterocyclic carbene). Due to the close nature between electron-rich oganophosphines and N-heterocyclic carbenes, similar order of reactivity was anticipated for borane–carbene FLP systems. This understanding paved the way for the development of next-generation FLP systems with sterically encumbered carbon-based Lewis donors NHC. The sterically bulky NHCs like 1,3-di-tert-butylimidazolin-2-ylidene and 1,3-bis(2,6-diisopropylphenyl)imidazol-2-ylidene (IDipp), on reaction with $B(C_6F_5)_3$, afforded a workable molecular system for the activation of H–H, C-O and N–H bonds as shown in Scheme 2.16. (Holschumacher et al. 2008; Chase and Stephan 2008). In particular, reaction between IDipp and $B(C_6F_5)_3$ resulted light brown product of composition $[(IDipp)B(C_6F_5)_3]$ which showed single resonance at −15.6 ppm in the $^{11}B\{^1H\}$ NMR and up fielded resonances at 6.20 and 6.32 ppm in ^1H NMR categorically defend the existing bond between B-C (Holschumacher et al. 2008). The bond length of former bond is quite longer as earlier reported values which may be due to steric hindrance of isopropyl group. The reaction of I^tBu with $B(C_6F_5)_3$ results into $[I^tBuH][HB(C_6F_5)_3]$ leading to the activation of H_2; structure is presented in Fig. 2.3. The compound is thermally stable up to 150 °C and quite interesting as neither of the reactant alone can react with hydrogen (Chase et al. 2009).

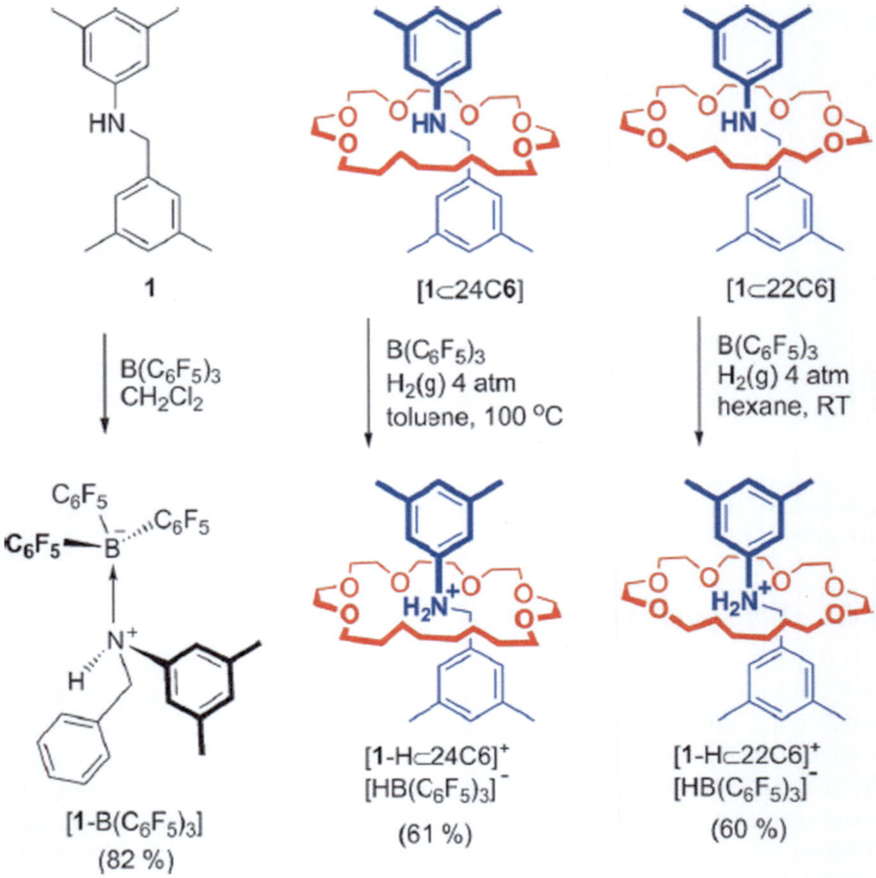

Scheme 2.15 Reactivity of unencumbered and rotaxane-encumbered secondary amines-Borane systems. Reproduced with permission under Creative common license from Wiley, Ref. (Caputo et al. 2013)

Similarly, NHC carbene on treatment with amine-derived Lewis acid–base product like $PhH_2NB(C_6F_5)_3$, $PhH_2NB(C_6F_5)_3$ and $H_3NB(C_6F_5)_3$ yields serendipitous isolation of products with composition like $[I^tBuH][Ph_2NB(C_6F_5)_3]$, $[I^tBuH][PhHNB(C_6F_5)_3]$ and $[I^tBuH][H_2NB(C_6F_5)_3]$, respectively (Chase et al. 2009). These imidazolium salts have shown shielded peaks at −10.0, −9.7 and −7.4 ppm. These upfield resonances indicate the formation of amido-borate anions. The structural studies showed that these FLPs could conveniently activate N–H bond shown in Fig. 2.5. Furthermore, the N–H bond activation was found to be highly dependent on the amine substituent. Stephan group have also explored the $[CPh_3]^+$ Lewis pair chemistry, as a substitute of boranes. They have reported that reaction between $[CPh_3][B(C_6F_5)_4]$ and bulky NHC in toluene solution afforded formulation of as $[C_3H_2N_2{}^tBu_2(C_6H_5)CPh_2][B(C_6F_5)_4]$. The latter complex has formed due to

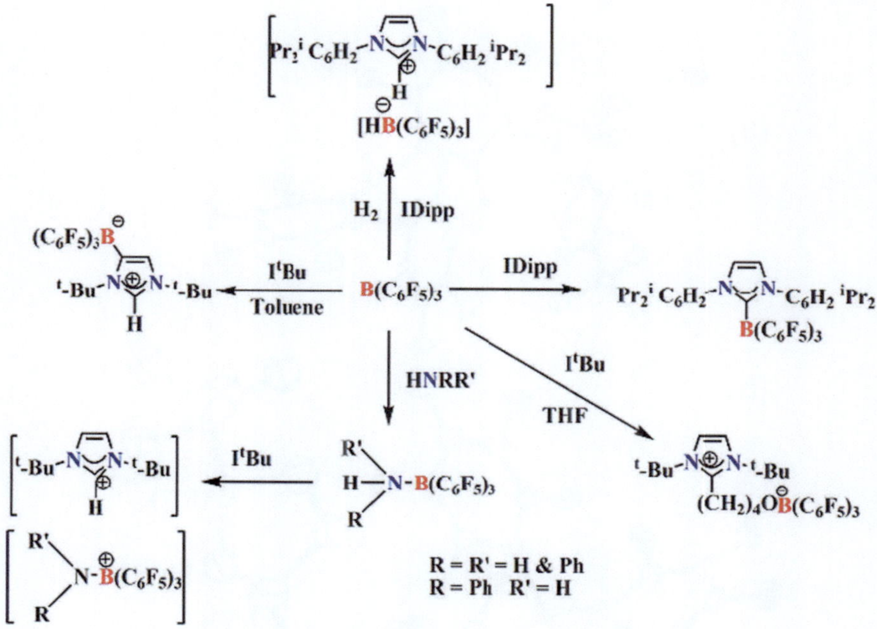

Scheme 2.16 Reactions involved in various NHC-derived FLPs

nucleophilic substitution at *para* position of phenyl ring of trityl cation which is differed from the cases in which phosphonium cation is generated by the migration of *para* hydrogen to cationic carbon (Cabrera et al. 2006) (Scheme 2.16). In these cases, the B-N distance is in the range of 1.53 Å which is significant shorter than other reported complexes.

Tamm et al. have reported that unsaturated N-heterocyclic carbine and its dihydro congener on reaction with $B(C_6F_3)_3$ in THF solution afforded white zwitterionic structure of imidazolidinium–borane (Holschumacher et al. 2008, 2009) (Scheme 2.15). Incorporation of THF as bridging component is verified by the characteristic resonances of THF in the ^1H NMR and $^{13}C\{^1H\}$ NMR spectrum. Structures of the complexes have been verified through single-crystal X-ray, and it elucidates the shortening of C–N bond distances. An anomalous carbene and borane adduct has also been isolated on stirring the same above reaction mixture for longer time. Adduct has a characteristic resonance at −15.7 ppm in ^{11}B NMR and indicates similarity with composition of $C_3H_2N_2{}^tBu\text{-}B(C_6F_5)_3$ (Scheme 2.15). Isolation of such product indicates the self-dehydrogenation of NHC systems.

Later, carbenes have also been utilized to stabilize the borenium salt which derives its Lewis acidity from the positive charge. These carbene-borenium systems could be conveniently employed as metal-free catalyst for the hydrogenation of imines and enamines (Farrell et al. 2012). In presence of imine, borenium cation and imine constitute FLP which heterolytically cleave H_2 followed by subsequent hydride transfer resulting in the reduction of transient iminium cation to amine (Scheme 2.17).

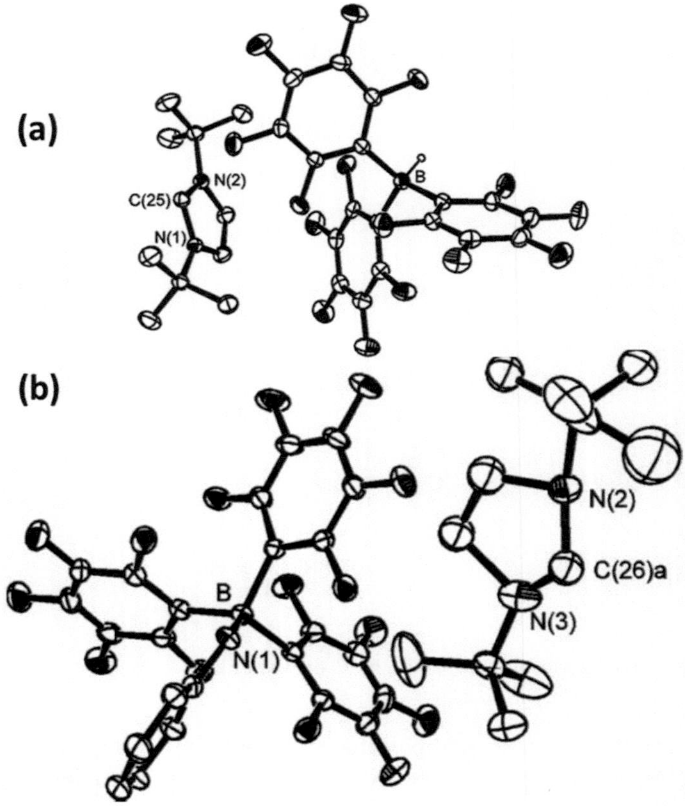

Fig. 2.5 ORTEP diagram of **a** [I^{t}BuH][HB(C_6F_5)$_3$] and **b** [I^{t}BuH][PhHNB(C_6F_5)$_3$] showing activation of H–H and N–H bond. Adopted with permission from Elsevier, Ref. (Cabrera et al. 2006)

Subsequent studies unravel the effect of substituent on carbene and the activity of borenium catalyst. It was inferred that the increase in the bulkiness of substituent causes decrease in the catalytic activity. Further, the presence of electron withdrawing group on NHC reduces carbene electron donation ability and thereby increases the Lewis acidity at boron which leads to enhancement in the catalytic activity of borenium cation-derived FLP. On such FLP structure is given in Fig. 2.6 (Farrell et al. 2015).

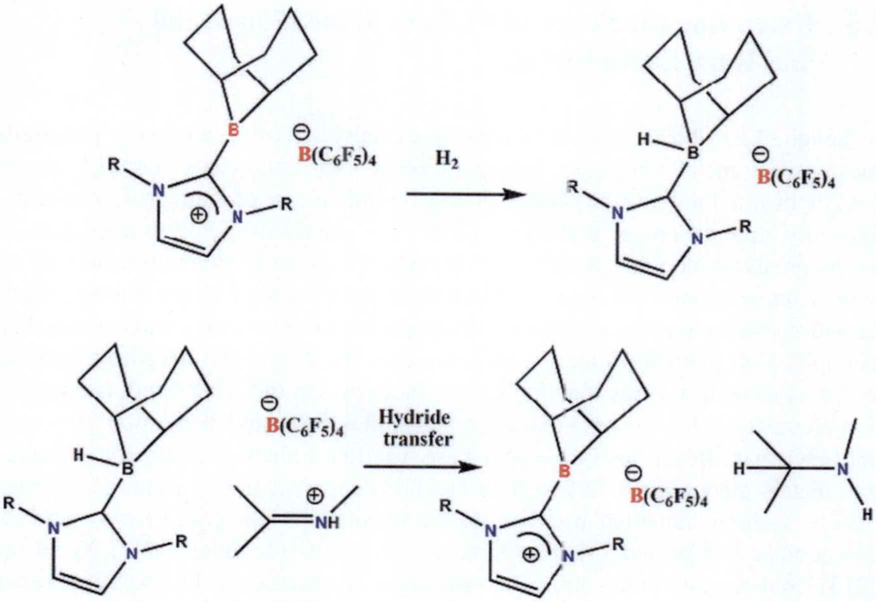

Scheme 2.17 NHC-stabilized borenium ion causing hydrogenation of iminium cation. Redrawn with permission from Ref. (Farrell et al. 2012)

Fig. 2.6 Electron withdrawing group decorated carbene borenium FLP system with high hydrogenation catalytic activity

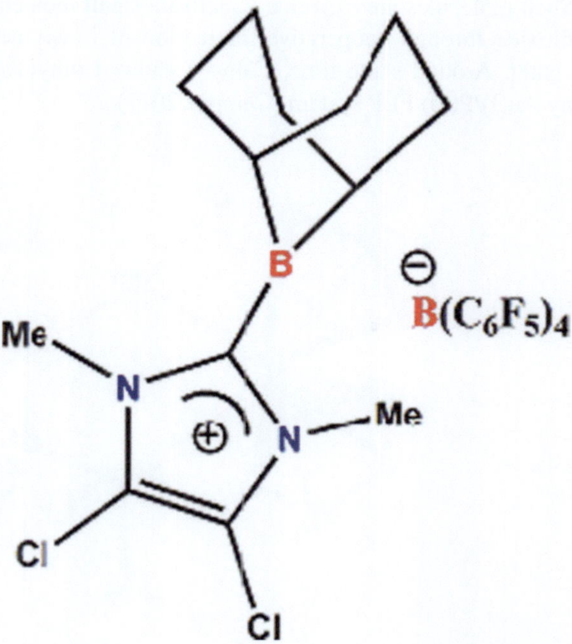

2.5 Extending the Scope of FLPs to Transition Metal and Rare Earth Metals

Although FLPs were developed as metal-free catalysts, it offers a valuable paradigm for transition metal complexes. Incorporation of transition metals in to FLP brings with them structural diversity which is very limited in case of main group elements. Secondly, applications of main group FLPs are mainly restricted to activation of hydrogen and some other small molecules which is possibly due to the inability of main group elements to involve in electron transfer reactions. Furthermore, many transition metals have been well understood for their Lewis acidity and basicity. For example, high oxidation state group 4 elements (Ti, Zr and Hf) have been utilized as Lewis acids in various transformation reactions. On the other hand, late transition elements in low or zero oxidation states like Pt(0) have been used as potential Lewis base in various Lewis acid–base reaction. Enhanced reactivity of transition metals incorporated FLP system has been apparent from various pioneering studies wherein transition metal replacing either the main group Lewis acid or Lewis basic component (MettersOJ et al. 2016a, 2016b; Erker 2011; Xu et al. 2013). Subsequent studies aimed at creating a heterometallic FLP system where both Lewis acid as well as Lewis base features transition metals. In this direction, Wass and co-workers (Chapman et al. 2016) isolated early late transition metal FLP $[Cp_2Zr\{Pt(Me)\}(OC(CF_3)_2CH_2P^tBu_2)]$, crystal structure shown in Fig. 2.7. Such molecules are expected to activate small molecules like hydrogen and carbon dioxide through cooperative interaction of Lewis acid and base transition metal center. Around same time, Campos showed dihydrogen and acetylene activation by Au(I)/Pt(0) FLP system (Campos 2017).

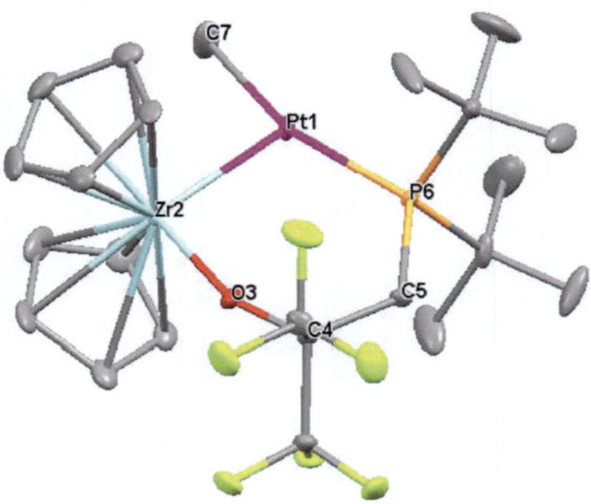

Fig. 2.7 Crystal structure of unexpected $[Cp_2Zr\{Pt(Me)\}(OC(CF_3)_2CH_2P^tBu_2)]$ heterobimetallic complex. Adopted with permission from Ref. (Chapman et al. 2016)

Further progress in this field was witnessed when rare earth metal ion-based FLPs were developed (Xu et al. 2017; Xu and Xu 2018). These FLPs were found to efficient in causing polymerization of conjugated polar alkenes. From mechanistic studies, it is inferred that polymerization of conjugate polar alkene is instigated by FLP type 1,4 addition rather than orthodox single electron transfer initiation or RE-E (E = H, C, N, etc.) bond insertion. Several combinations of rare earth metal ion-based Lewis acids and Lewis bases including sterically congested phosphines and N-heterocyclic carbenes have been surveyed. Interestingly, it was found that the extent of polymerization is not only a function of the ionic radii of the rare earth metals, rather it also depends on the steric and electronic aspects of Lewis bases.

2.6 FLP Connect to Metal–Organic Framework (MOF)

Metal–organic frameworks or coordination polymers are high-stable porous crystalline framework materials synthesized from secondary building units (metal cations or clusters) and polydentate organic linkers. In last two decades, since their discovery, this field has rose to prominence owing to their widespread use in energy conversion and storage, environment and healthcare applications. Furthermore, these wonder materials have been explored for their applications in organic transformation reaction. On the other hand, from our discussion, it is quite apparent that FLP systems are highly effective in activating small molecules, including hydrogen and carbon dioxide. However, that being claimed, actual implementation of these systems faces grave concerns pertaining to stability. There are other concerns also, like catalyst-product separation and recyclability. So, in order to address these rising concerns on FLPs, utilizing MOFs to support FLPs becomes an attractive or rather inevitable option. In this direction, Niu et al. have demonstrated a clever strategy to incorporate FLPs into MOFs (Niu et al. 2018). The group took activated MIL-101(Cr) MOF and sequentially exposed it to a solution of 1,4-diazabicyclo[2.2.2]octane (DABCO) followed by $B(C_6F_5)_3$. The nitrogen atoms of ligands present in the pores coordinate with $B(C_6F_5)_3$ leading to the formation of MOF encapsulated FLP DABCO/$B(C_6F_5)_3$ (Scheme 2.18). This catalyst has been conveniently employed for the reduction of imines by hydrogen as well as hydroboration of imines.

2.7 Conclusion and Future Scope

The FLP concept constitutes a set of model reaction, in which a sterically bulky Lewis acid and base act on a substrate molecule. This apparently simple but powerful tool uncovered the application of main-group elements in hydrogenation catalysis including reduction of imines, nitriles and aziridines and small-molecule activation. Majority of the earlier studies have been carried out on B- or P-based FLP systems. However, diversification in this field was swiftly achieved as numerous other donors

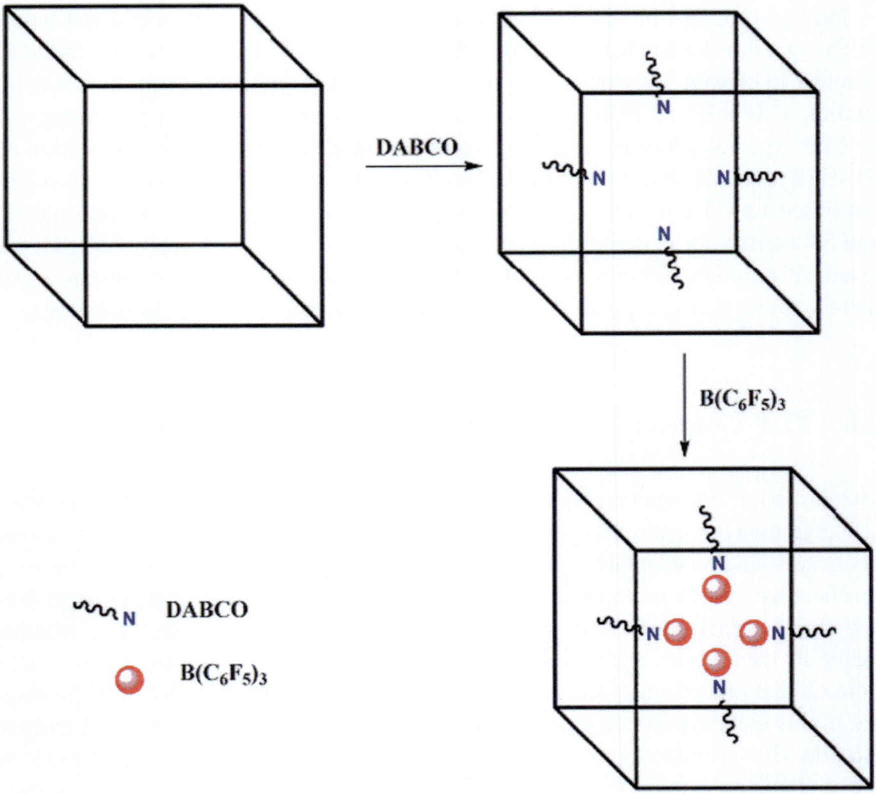

Scheme 2.18 Strategy to decorate MOF with FLP

and acceptors capable of specific small molecule and bond activation begun to be explored in recent years. Such developments are expected to provide new insight toward the discovery of a plethora of new homogeneous and heterogeneous catalytic systems. With the obvious evolution of the idea, new mechanistic recognition will definitely follow. Apart from the main group-based FLPs, quest for systems devoid of main group elements is also accelerating as an obvious development. Practical applications of these molecular materials in synthetic inorganic, organic and polymer chemistry are also emerging parallelly. Thus, while FLP chemistry has been confined so far on applications of synthetic main group chemistry, the concept is definitely destined for broader impact. Witnessing how the creativity of the scientific fraternity addresses the existing challenges associated and how the FLP concept is exploited for unorthodox area of chemistry will be a thrilling experience.

References

Brown, H.C., Schlesinger, H.I., Cardon, S.Z.: J. Am. Chem. Soc. **64**, 325–329 (1942)
Cabrera, L., Welch, G.C., Masuda, J.D., Wei, P., Stephan, D.W.: Inorg. Chim. Acta **359**, 3066–3071 (2006)
Campos, J.: J. Am. Chem. Soc. **139**, 2944–2947 (2017)
Caputo, C.B., Zhu, K.L., Vukotic, V.N., Loeb, S.J., Stephan, D.W.: Angew. Chem. Int. Ed. **52**, 960–963 (2013)
Chapman, A.M., Flynn, S.R., Wass, D.F.: Inorg. Chem. **55**:1017–1021 (2016)
Chase, P.A., Stephan, D.W.: Angew. Chem. Int Ed. **47**:7433–7437 (2008)
Chase, P.A.,Gille, A.L., Gilbert, T.M., Stephan, D.W.: Dalton. Trans. 7179–7188 (2009)
Chernichenko, K., Nieger, M., Leskela, M., Repo, T.: Dalton Trans. **41**, 9029–9032 (2012)
Ekkert, O., Kehr, G., Frohlich, R., Erker, G.: J. Am. Chem. Soc. **133**, 4610–4616 (2011)
Erker, G.: Dalton Trans. **40**, 7475–7483 (2011)
Erös, G., Mehdi, H., Pápai, I., Rokob, T.A., Király, P., Tárkányi, G., Soós, T.: Angew. Chem. Int. Ed. **49**, 6559–6563 (2010)
Farrell, J.M., Heiden, Z.M., Stephan, D.W.: Organometallics **30**, 4497–4500 (2011)
Farrell, J.M., Hatnean, J.A., Stephan, D.W.: J. Am. Chem. Soc. **134**, 15728–15731 (2012)
Farrell, J.M., Posaratnanathan, R.T., Stephan, D.W.: Chem. Sci. **6**, 2010–2015 (2015)
Fontaine, F.G., Stephan, D.W.: Phil. Trans. r. Soc. A **375**, 20170004 (2017)
Grobe. J., Lütke-Brochtrup, K., Krebs, B., Läge, M., Niemeyer, H.H., Würthwein, E.U.: Z Naturforsch **61b**:882–895 (2006)
Haenel, M.W., Narangerel, J., Richter, U.B., Rufinska, A.: Angew. Chem. Int. Ed. **45**, 1061–1066 (2006)
Holschumacher, D., Bannenberg, T., Hrib, C.G., Jones, P.G., Tamm, M.: Angew. Chem. Int. Ed. **47**, 7428–7432 (2008)
Holschumacher, D.,Taouss, C., Bannenberg, T., Hrib, C.G., Daniliuc, Jones, P.G., Tamm, M.: Dalton. Trans. 6927–6929 (2009)
Johnstone, T.C., Wee, G.N.J.H., Stephan, D.W.: Angew. Chem. Int. Ed. **130**, 5983–5986 (2018)
Jupp, A.R., Stephan, D.W.: Trends Chem **1**: 35–48 (2019)
LankampH, NautaWT.: Maclean C. Tetrahedron Lett. **9**, 249–254 (1968)
McCahill, J.S.J., Welch, G.C., Stephan, D.W.: Angew. Chem. Int. Ed **46**:4968–4971 (2007)
Metters, O.J., Forrest, S.J.K., Sparkes, H.A., MannersI, Wass, D.F.: J. Am. Chem. Soc. **138**:1994–2003 (2016)
Metters, O.J., Flynn, S.R., Dowds, C.K., Sparkes, H.A., Manners, I., Wass, D.F.: ACS Catal **6**:6601–6611 (2016)
Niu, Z., Gunatilleke, W.D.C.B., Sun, Q., Lan, P.C., Perman, J., Ma, J.G., Cheng, Y., Aguila, B., Ma, S.: Chem **4**, 2587–2599 (2018)
Schulz, F., Sumerin, V., Heikkinen, S., Pedersen, B., Wang, C., Atsumi, M., Leskelä, M., Repo, T., Pyykkö, P., Petry, W., Rieger, B.: J. Am. Chem. Soc. **133**, 20245–20257 (2011)
Spies, P., Erker, G., Kehr, G., Bergander, K., Fröhlich, R., Grimme, S., Stephan, D.W.: Chem. Commun. **43**, 5072–5074 (2007)
Stephan, D.W.: Sci. **354** (2016). aaf7229
Stephan, D.W., Erker, G.: Angew. Chem. Int. Ed. **54**, 6400–6441 (2015)
Stone, F.G.A.: Chem. Rev. **58**, 101–129 (1958)
Sumerin, V., Schulz, F., Nieger, M., Leskela, M., Repo, T., Rieger, B.: Angew. Chem. Int. Ed. **47**, 6001–6003 (2008a)
Sumerin, V., Schulz, F., Atsumi, M., Wang, C., Nieger, M., Leskelä, M., Repo, T., Pyykkö, P. Rieger, B.: J. Am. Chem. Soc. **130**, 14117–14119 (2008b)
Sumerin, V., Chernichenko, K., Nieger, M., Leskela, M., Rieger, B., Repo, T.: Adv. Synth. Catal **353**, 2093–2110 (2011)
Tochtermann, W.: Angew. Chem. Int. Ed. **5**:351–371 (1966)
Welch, G.C., Stephan, D.W.: Jam. Chem. Soc. **129**, 1880–1881 (2007)

Welch, G.C., Juan, R.R.S., Masuda, J.D., Stephan, D.W.: Sci. **314**, 1124–1126 (2006)
Wittig, G.: RückertA. Justus Liebigs Ann. Chem. **566**, 101–113 (1950)
Wittig, G., Benz, E.: Chem. Ber-Recl **92**, 1999–2013 (1959)
Xu, P., Xu, X.: ACS Catal. **8**, 198–202 (2018)
Xu, X., Kehr, G., Daniliuc, C.G., Erker, G.: Angew. Chem. Int. Ed. **52**, 13629–13632 (2013)
Xu, P., Yao, Y., Xu, X.: Chem. Eur. J. **23**, 1263–1267 (2017)

Chapter 3
Free and Supported Clusters: Unique Materials Influencing Chemistry at Sub-Nanoscale

Soumitra Das, Pramod Sharma, and Chiranjib Majumder

3.1 Introduction

Clusters are aggregates of finite number of atoms/molecules held together by interactions ranging from weak van der Waals forces to strong ionic bonds. These clusters may consist of aggregates made up of as low as two atoms/molecules and can extend up to few hundred/thousand atoms (Heer 1993). The term "cluster' was first coined by F. A. Cotton in the year 1964 to describe compounds with metal–metal bonds. The forces holding these clusters are such that they do not saturate, as a result, one can alter the number of atoms/molecules within the clusters deliberately, causing change in the local structure and properties of the clusters. This characteristic is regarded as primary property of clusters and is referred to as stack-ability. As a result, clusters exhibit the physical and chemical properties, which are strongly influenced by size, geometry, composition and topology of clusters.

In the last few decades, the field of clusters has been of interest to the scientific community, due to potential of clusters to offer ways to make altogether new kinds of material, as their structural and electronic properties bear resemblance to neither atom/molecules they are composed of, nor to their corresponding bulk matter. Also, clusters aid in synthesis of multicomponent materials of sub-nanoscale dimensions with well-defined structure, where composition can be tweaked at atomic scale, leading to synthesis of species/mixture, non-existent under normal conditions. Though these clusters are intermediated between atoms/molecules and bulk-matter, their properties cannot be simply inferred by linear fitting/interpolation of the properties from atoms to the corresponding bulk matter. This unusual disparity in properties has been ascribed to several factors, like—

S. Das (✉) · P. Sharma · C. Majumder
Chemistry Division, Bhabha Atomic Research Centre, Mumbai, India
e-mail: soumitra@barc.gov.in

R. S. Ningthoujam and A. K. Tyagi (eds.), *Handbook of Materials Science, Volume 3*, Indian Institute of Metals Series, https://doi.org/10.1007/978-981-96-8229-4_3

- Higher surface-to-volume ratio of clusters, as large number of atoms resides on the surface.
- Due to reduced size, quantum confinement also contributes to unusual properties of clusters.
- Unique geometric and electronic structures of clusters, which vary as a function of cluster size and composition.

Understanding influence of these factors at atomic level is essential for fundamental understanding of the cluster properties and for designing novel cluster materials for technological applications. Cluster materials have received enormous attention due to their potential applications in area of energy, environment and chemistry. Besides, because of their small size, low toxicity and biocompatibility they also find wide use in biological, medical and sensing applications (Mathew and Pradeep 2014). Below we will first discuss about unique properties of clusters followed by their influence on chemical and physical properties at sub-nanoscale.

3.1.1 Unique Characteristics of Clusters

Clusters exhibit physical and chemical properties, which vary significantly upon addition/removal of a single atom/molecule. They exhibit unique thermodynamic, structural, energetic, optical, electronic, magnetic and chemical properties. These properties differ quite drastically compared to bulk material and arise due to finite size of the cluster system (Kaukonen et al. 1991). For example, gold is considered to be the noblest metal and exhibits poor reactivity owing to its low affinity towards molecules, due to its higher ionisation potential in bulk form. While gold clusters are known to be highly chemically active. These distinct properties exhibited by clusters as a function of cluster size have been ascribed to several factors, i.e. higher surface to bulk ratio of atoms, superatomic character arising due to sharing of electrons within cluster atoms (as compared to bulk), electronic shell closing (Jellium model), geometric shell closing and also due to quantum confinement. As the size of clusters is much smaller than the de Broglie wavelength, they exhibit quantum size effect. Consequently, clusters are often designated as quantum clusters (Woodruff 2007). A cluster of given size can acquire several stable cluster structure arrangements, which are thermodynamically feasible. Besides, they can also easily transform from one structure to another via rearrangement, upon perturbation. Clusters also exhibit interesting electronic properties. As a function of cluster size, clusters exhibit systematic emergence of discrete electronic states, thus depicting molecule-like properties (Mathew and Pradeep 2014). Therefore, smaller clusters are often classified as insulators, due to higher band gap. With increase in cluster size the band-gap decrease. For larger clusters the mean-free path of electron (in case of metal clusters) is in the range of 5–15 nm, thus permitting un-scattered electron flow (Somorjai et al. 2006). These unique properties of clusters are exploited for their potential applications in various fields ranging from microelectronics to information storage and sensors to catalysis. Thus, properties exhibited by clusters depend on their geometric structure, electronic structure and size, which are interrelated.

3.1.2 Cluster Properties: Role of Atomicity and Cluster Structure

Clusters exhibit characteristic properties as a function of cluster size and shape. The structure, stability, physical and chemical properties of clusters strongly depend on the type of interactions, which hold the constituent species within the cluster. Depending on the nature of constituent atoms and binding property, clusters can be broadly classified into following categories.

(a) **Molecular Cluster:** Clusters formed by aggregation of molecules, without disturbing their composition are called molecular clusters. These clusters exhibit weak interactions, i.e. van der Waals interaction, dipole–dipole interactions, higher-order multipolar interactions and hydrogen bonding.
(b) **Rare Gas Clusters:** Rare gases like argon, krypton and xenon can form clusters at lower temperatures. These clusters are held together by weak van der Waals type of interactions. The magnitude of these interaction increases with increasing atomic mass.
(c) **Ionic Clusters:** Ionic clusters or hetero-atomic clusters are formed by atoms with large difference in the electronegativity. i.e. cations and anions. The overall charge of the cluster is determined by the stoichiometry of the cluster.
(d) **Semiconductor Clusters:** These clusters are aggregate of elements which have semiconducting property, i.e. Si, Ge and other compounds like GaAs, CdTe. These clusters exhibit covalent bonding, which are highly directional and strong.
(e) **Metal Clusters:** Clusters composed of elements, which exhibit metallic properties in bulk phase, i.e. Na, Al, Cu, Ag, Au, etc. are called metal clusters. The interatomic forces in metallic clusters are a blend of several types of interaction, which depends on size of cluster.

In the past, several studies have demonstrated that chemical/physical properties of matter can be tuned by varying their size, shape and composition, at nano-metric and sub-nanometric level. Based on dimensions, properties of matter can be broadly classified into three size regimes. In size regime of 100 nm or higher, the properties of matter correspond to those of their bulk counterpart and are independent of size. While in the size regime of 1–100 nm, compared to bulk matter, properties of the material scale as a function of size. This second size regime corresponds to nanoparticles. In this regime, different properties of nanoparticle have been found to vary with size. For example, in gold nanoparticles, melting point was found to increase with particle size (Qiao et al. 2014). In the third size regime, which corresponds to matter with size below 1 nm the properties are non-scalable as a function of size. This size regime corresponds to clusters, which are aggregate of finite number of atoms (2–1000's of atoms). In this size regime, properties of material vary significantly upon addition or removal of an atom, as well as on charge (positive, neutral or negative), shape and composition. In this size regime quantum size effects determine properties of matter. Thus, understanding the fundamental principles governing this non-scalable size regime will steer research beyond nanotechnology. An atom-by-atom variation in cluster size causes severe changes in the chemical and physical properties of clusters. Often, change in physical/chemical properties of clusters over an order of magnitude occurs by simply addition or removal of an atom. These effects can be easily envisioned by carrying out studies on mass-selected gas-phase clusters. Though properties of small cluster vary non-linear with size, for larger clusters the variation in properties is smooth and can be scaled as a function of cluster size (Fig. 3.1). Also, electrons in smaller clusters are not delocalised like bulk metal and hence these clusters are called non-metallic. Compared with the transition-metal cluster, their post-transition element counterpart exhibits distinct properties. For example, often the bare transition-metal clusters are unstable and stabilised by ligands while the post-transition element bare clusters are usually stable and are often destabilised upon interaction with ligands. Also, transition-metal clusters form highly symmetrical metal frame structure.

The geometry of transition metal clusters is determined by their partially filled (fully filled in the case of noble metals) *d* shell, which favours compact structures in order to maximise interaction among localised *d* orbitals due to hybridisation of the *ns*-(*n* − 1)*d* orbitals, facilitated by relativistic contraction of the inner *s* and *p* shells. Due to this effect which is different for different elements, they exhibit varying cluster size limit for transition from 2 to 3D structure. As a result, while ground state geometries of Ni_4 and Pd_4 are tetrahedral, lowest energy isomers of Cu, Ag and Pt composed of six atoms form planar structures (Fernández and Boronat 2018). Here, it is worth mentioning that properties and bonding in metal clusters vary significantly

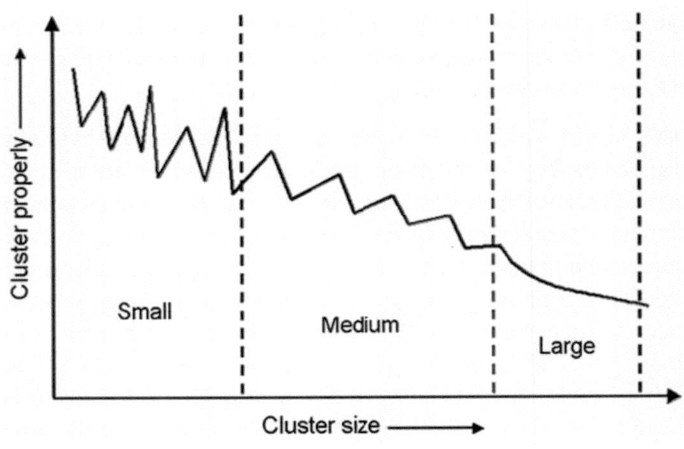

Fig. 3.1 Variation of cluster properties as function of size

for metals of different blocks in the periodic table. For example, clusters of iron, cobalt and nickel exhibit superparamagnetic behaviour with strong size-dependent magnetic moments. For smaller clusters (< 30 atoms) these clusters display atom-like magnetic moment, while for larger cluster (~700 atoms) the magnetic moment is found to approach bulk limit (Billas et al. 1994). Magnetism has also been detected in clusters of some elements which are nonmagnetic in bulk phase (Cox et al. 1993). Such as, though ferromagnetic behaviour has not been observed for pure 4d transition metals, but clusters of rhodium, $Rh_n(n = 12–32)$ exhibit giant magnetic moments. Similarly, although bulk Pd metal is diamagnetic in nature, its clusters of smaller size are found to show significant magnetic moment. Thus, varying cluster size provides a convenient handle for tuning/engineering properties of clusters for various applications.

3.1.3 Free Clusters vs Surface Deposited Clusters

The field of cluster science is interdisciplinary in nature, owing to distinctive approaches required for their synthesis and characterisation. Clusters have been produced and characterised in gaseous, liquid and solid phase. These clusters have been investigated in various forms, i.e. bare clusters, clusters protected with ligands, clusters deposited on various supports-like organic/inorganic framework and surfaces. Several techniques have been devised for in-situ and ex-situ characterisation of these bare and supported sub-nanoscale clusters. Mass spectrometry has been widely used for characterising clusters based on their mass and charge. Using matrix-assisted laser desorption ionisation (MALDI) and electron spray ionisation (ESI) methods, precise formula of ligand-protected cluster has been determined. Extended X-ray absorption fine structure (EXAFS) spectroscopy has also been used extensively to determine structure of supported and non-supported metal clusters (Sudheeshkumar et al. 2020). Besides, metal clusters exhibit unique absorption features in the visible region, which is ascribed to their discrete electronic structures. As a result, different clusters reveal variable features in the optical absorption region, based on the structure of their core. Hence, UV–Vis spectroscopy is a facile technique to carry out cluster speciation in solution. Several characterisation techniques are also available for characterisation of cluster supported materials, i.e. X-ray photoelectron spectroscopy (XPS), (scanning) transmission electron microscopy ((S)TEM), scanning tunnelling microscopy (STM), atomic force microscopy (AFM), and synchrotron, for assessing influence of support on clusters.

A large number of studies have been carried out in gas-phase, to investigate size and structure dependent variation in physical and chemical properties of cluster. Though, generation, isolation and characterisation of atomically precise clusters are quite feasible in gas-phase, for practical applications it is essential to synthesis/transfer these atomically precise clusters in solution or solid substrate. In solution, clusters have been synthesised in-situ from their mononuclear precursors by using various solution-based methods. Being susceptible to sintering, they are often protected by ligands (i.e. phosphenes, thiolates, selenolates, carbenes, etc. (Sudheeshkumar et al. 2020) or supported by organic/inorganic framework (Fernández and Boronat 2018). The ligand shell stabilises the under-coordinated cluster species, thus preventing their aggregation. Ligand-protected metal clusters with atom-precise composition (especially comprising of Au and Ag atoms), owing to their excellent physicochemical properties, have received significant attention, over last few decades. These clusters have potential applications in bio-sensing, drug delivery, bio-imaging and catalysis, etc.

Surface deposited clusters represent a unique and complex class of chemical system, as compared to free clusters and ideal crystal surfaces. Interaction with surface confers clusters with unique electronic and geometric properties, different from those of free cluster. As a result, support materials have a significant influence on cluster properties via charge transfer and by influencing cluster mobility. Though clusters of specific size can be generated and characterised in gas-phase, deposition of size-selected clusters on surface is an arduous task. Since the size-selected clusters are prone to undergo modification during the process of deposition or after deposition on surface. Hence to prevent fragmentation of the clusters and alteration of substrate surface, it is necessary that the kinetic energy of clusters is strictly regulated (<1 eV/atom) to carry out soft landing of clusters (soft-landing condition) (Woodruff 2007). For cluster deposition, upon soft landing the ionised size-selected clusters lose their charge to the surface and transform to neutral species following deposition. Electronic and structural properties of the clusters are altered upon deposition. With advances in synthesis techniques and instrumentation, size-selected clusters produced in gas-phase are mass-selected and deposited on substrate with well-defined coverage (Halder et al. 2018). Thus, one can meticulously control size and composition of the cluster deposited on substrate at microscopic level.

Cluster has important implications on various branches of chemistry ranging from solid-state chemistry/physics, coordination chemistry, organometallic chemistry and catalysis (homogeneous and heterogeneous). Since reactions on clusters are analogous to reaction on surfaces/bulk, they are employed as model systems for fundamental understanding of processes like etching, adsorption and catalysis, at molecular level. Several complex reactions like Fischer–Tropsch, polymerisation, partial oxidation and dehydrogenation reactions have been investigated using clusters and cluster-supported materials.

Studies on clusters have also been of interest for several decades due to their fascinating size-dependent chemical reactivity/properties. Cluster (mainly metal clusters) exhibits a unique chemical (catalytic) property depending on atomicity, structure/morphology, oxidation state and electronic structure. In the field of catalysis, studies on clusters have been carried out to understand effects of various cluster parameters, i.e. cluster size, cluster composition, cluster structure and reaction conditions, on activity and selectivity of cluster for developing and fine-tuning efficient catalysts for various applications. Studies have been carried out on clusters deposited on different supports/surfaces, to understand influence of support on catalytic activity of clusters. Several studies suggest that catalytic activity of free-clusters and supported clusters are strongly influenced by their electronic structure and geometry. These properties in turn are primarily influenced by interaction between cluster and support. Besides, the electronic structure can be fine-tuned by precisely varying the number of cluster atoms.

Clusters deposited on support are considered to be ideal system, for their utilisation as catalysts (heterogeneous catalysis). Here, metal clusters are deposited on solid supports such as metal oxides, silica, carbon and zeolites. (Dong et al. 2021) This kind of clusters is called supported metal cluster (SMCs) which has immense use in research and industrial application. These clusters are used in catalytic reactions under various extreme conditions and subsequently regenerated/recovered after reaction which is not possible in case of ligand protected monodispersed clusters (homogeneous catalysis). Also, for catalyst utilising precious metals, clusters can have a significant economic impact as they offer 100% atom utilisation efficiency, thus significantly reducing the usage of costly precious metals (Tyo and Vajda 2015). Also, these supported clusters have demonstrated outstanding performance in several catalytic transformation reactions. This is facilitated by presence of metal–metal bonds among supported atomic clusters, leading to synergistic effects among every constituent metal atom, thus aiding in achieving unique catalytic properties. Additionally, catalytic activity can be enhanced by varying/tuning cluster size, as well as by altering/engineering interaction between the clusters and their support. Support material also induces support effects on deposited clusters, which are important for various catalytic processes. The support surface can lead to modification of their structural and electronic properties. Besides, it provides features like—lowering of activation barrier for a catalytic reaction, by providing alternate reaction pathways, smaller size of cluster increases the number of under-coordinated atoms (due to higher surface-to-volume ratio)—resulting in higher catalytic activity. During the course of a catalytic reaction upon interaction with reagents, the geometry of SMCs is facilely affected by harsh conditions of a reaction (high temperature, pH and the exterior environment) due to limited number of atoms present within a cluster. As a result, SMC often lose their rigidity and get transformed into fluxional species having metastable geometries, which are conducive for higher reactivity (Dong et al. 2021).

At this point, it is pertinent to compare catalytic behaviour of nanoparticles and clusters. As compared to nanoparticles, most of the constituent atoms of the cluster are exposed to maximum extent and are available for the reactant molecules, which is unique. Also, clusters exhibit discrete electronic structures, which vary with size. Their electronic properties differ significantly from their analogues nanoparticle systems, which often exhibit plasmon band due to metallic nature. Though nanoparticles have also been promoted as efficient catalyst for several reactions, as compared to clusters, nanoparticles offer lower catalytic selectivity due to their large size and complicated structure. On a nanoparticle as the reactant molecules can get absorbed on different sites of particle surface, i.e. planes, corner sites and edge sites, which exhibit varying degree of selectivity and activity. Besides large nanoparticles might acquire different geometric structures, which may also contribute in varying interaction of reactant molecule with nanoparticles (Rong et al. 2020).

In this chapter, we will discuss various methods employed for synthesis and characterisation of atomically precise clusters in gas-phase and deposited on different supports. Also, various studies carried out to illustrate unique properties and reactivity of clusters will be discussed. It includes studies carried out under vacuum conditions as well as under ambient conditions. In addition, computational studies carried out to understand cluster characteristics and to design novel cluster materials with desired properties and reactivity will also be highlighted. Besides potential applications of size-specific clusters, in catalysis and various other applications will be highlighted.

3.2 Different Strategies for Synthesis of Clusters

Synthesis of unique material requires novel approach for their generation and characterisation due to higher reactivity and lower stability. Often synthesis of isolated clusters is a herculean task. In addition, clusters produced by various methods exhibit log-normal size distribution. Thus, isolation of size-selected clusters is always a tedious work for carrying out various studies. With advancements in instrumentation, synthesis and characterisation of clusters have been carried out in gaseous, solution and solid phases. Often synthesis and characterisation of cluster is carried out in-situ. For practical application, clusters are often stabilised by protecting ligands in solution or by deposition on a solid substrate. In the following section, some of the popular methods used for cluster generation are discussed below.

3.2.1 Gas-Phase Synthesis and Deposition

Gas-phase synthesis of free clusters and deposition of these clusters on suitable substrate (supported clusters) have been carried out in a number of ways. Preparation of surface supported clusters consists of three steps, i.e. preparation of gas-phase clusters followed by selection of clusters, based on their size. Finally, size-selected clusters are deposited on appropriate substrate and have been tested for various applications ranging from catalysis, sensing, bio-imaging, etc. Bare clusters in gas-phase are often generated by magnetron sputtering, laser vaporisation, hot and cold discharge ions source, etc. Using these techniques, clusters of desired material are produced in gas-phase with varying size distribution (log-normal distribution). To isolate clusters of well-defined composition and size, selection is mandatory. For size selection, there are a number of ways such as by using quadrupole mass filter (electrostatic and radiofrequency), time-of-flight method, etc. which can separate clusters of desired size from the stream of cluster beam. The size-selected clusters are subsequently deposited on suitable substrate by soft landing technique, so that the fragmentation of the cluster is prevented and structure and composition of the cluster remains unchanged. For soft landing, kinetic energy of the cluster should be decelerated to <1 eV/atom which is achieved using different electro-optics (Winans et al. 2006). Moreover, to avoid the possibility of coalescence on surface, cluster

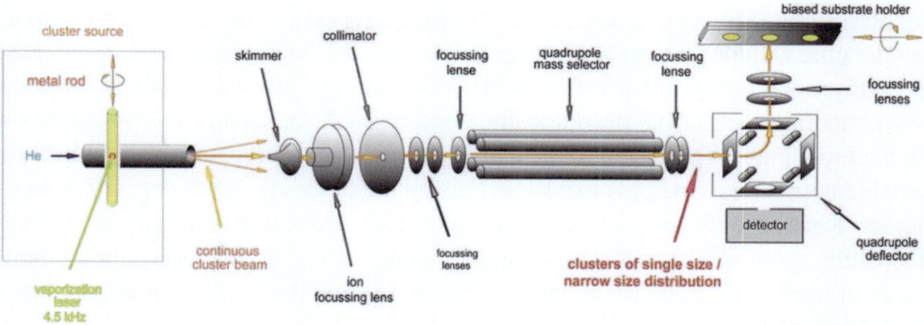

Fig. 3.2 Schematics of cluster deposition set up under ultra-high vacuum conditions, reproduced with permission from reference (Winans et al. 2006)

deposition is often carried out at lower temperature. Schematics of cluster deposition set up are shown in Fig. 3.2.

Gas-phase synthesis of clusters and deposition on substrate requires ultrahigh-vacuum conditions. However, achieving ultrahigh-vacuum condition is a tedious task and it is still a challenge to use the deposited clusters under standard operating conditions for various applications. Studies have been carried out on deposited clusters under in-situ conditions, so that effect of oxidation state of cluster constituents, structural changes in support and cluster, contamination from external environment can be minimised. Thus, size-selected method is very effective method for synthesis of free and supported clusters which can be prepared precisely with desired composition. However, it requires sophisticated instrument and ultrahigh vacuum conditions, which limits its application under normal conditions.

3.2.2 Atomic Layer Deposition Method

Atomic layer deposition (ALD) is one of the important techniques where supported clusters can be synthesised with exceptional conformality of the structures, controlling the thickness in atomic precision and enability of the composition. During this process, gaseous precursors are undergoing surface reaction with the substrate sequentially. Each precursor molecule reacts with the substrate in a self-limiting way so that reaction stops after occupying all the reactive sites of the substrate. This process is distinctively different from chemical vapour deposition, as precursors never present simultaneously in the reactor; rather, the reactants are inserted as a series of sequential, non-overlapping way. During a complete cycle of atomic layer deposition, amount of materials deposited on the surface depends on the nature of interaction between the surface and the precursor. Atomic layer deposition cycle can be performed multiple times to increase the thickness of the layer depending upon the experimental requirement. During this procedure, temperature of the substrate is generally kept low for fragile, thermally unstable precursor. The temperature range which is suitable for saturated growth of the precursor on the substrate is called 'ALD temperature window'. Deposition of precursor outside the temperature range will lead to poor growth on the substrate, slow reaction kinetics, precursor condensation (at low temperature), thermal decomposition and rapid desorption of the precursor (at high temperature). Therefore, for desirable growth of the precursor on the substrate, suitable temperature range is expected. Advantage of atomic layer deposition techniques over conventional chemical/physical vapour deposition technique is uniform deposition with high aspect ratio of the structure. Also, the thickness and composition of the layer can be easily tailored by controlling the number of atomic layer deposition cycles.

3.2.3 Wet Chemical Method

Wet chemical method is widely used for synthesising supported clusters. Wet chemical method provides better approach for facile and feasible synthesis of supported clusters, as compared to gas-phase synthesis. During this process, supported clusters are synthesised via adsorption of ligand-protected clusters on the solid surface. Using this method, various metal carbonyl clusters have been immobilised with thiolate and/or phosphine protected metal clusters on solid substrate in solvent. In this method, supported clusters can be prepared with uniformity; however, efficient co-ordination between ligands with cluster weakens the overall catalytic reactivity.

Impregnation method is one common example of wet chemistry method where supported clusters are prepared. In this method, inorganic metal salt solution is incorporated on a solid support. Subsequently, the support is dried and subjected to reduction (Munnik et al. 2015). However, during this process, the interaction of clusters with their support was found to be very weak. Therefore, strong electrostatic adsorption (SEA) method has been applied which is a special type of impregnation method. In SEA, stronger interaction between cluster precursor and support is assisted by adjusting proper pH which leads to more dispersed cluster atoms after reduction. (Fig. 3.3) Using this method, both monometallic and bimetallic clusters have been prepared on suitable substrate in basic solution ($pH = 11–12$). By this process homogeneously alloyed bimetallic clusters of Pt, Pd, Cu, Ni and Co were synthesised on silica surface (Zhu et al. 2008).

Using wet chemical method, supported cluster can be synthesised at industrial scale. However, this method has also few drawbacks, as the supported clusters are produced with broad size distribution. The capping ligand plays a very important role in controlling the morphology and aggregation of the clusters on the surface. Sometimes, these ligands interact with the cluster and change the oxidation states of the constituent species thus affecting activity of the supported clusters. As a result, it is difficult to synthesise supported clusters with desired chemical composition.

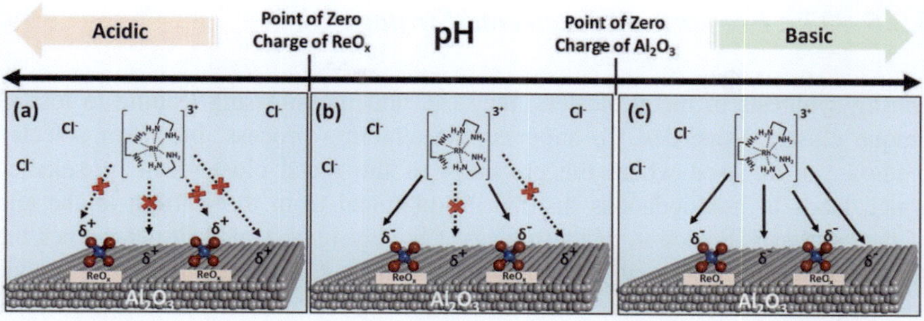

Fig. 3.3 Schematics of strong electrostatic adsorption (SEA)-based strategy to prepare Rh-ReO_x clusters supported on Al_2O_3 substrate, reproduced with permission from reference (Ro et al. 2019)

3.2.4 Host–Guest Strategy

Another method to prepare supported clusters is host–guest strategy where cluster synthesis is carried out in channels and cavities of molecular framework. Metal organic frameworks (MOFs) are a type of porous materials in which small atomic clusters can be confined in a controlled way. Pardo and co-workers synthesised Pd_4 clusters stabilised on MOF via a three-step process (Fortea-Pérez et al. 2017). In the first step is the transmetallation process, Mg^{2+} and Cu^{2+}ions present in the MOF are exchanged with Ni^{2+} and Cu^{2+}. In the second step, Ni^{2+} ions were exchanged by $[Pd(NH_3)_4]^{2+}$ and finally, $NaBH_4$ was used as reducing agent to produce Pd_4-MOF. Using single crystal XRD, quasi linear structure of Pd_4 was established within the MOF. MOF-supported Pd_4 cluster shows better catalytic efficiency as compared to metallic Pd_4 in carbine mediated reactions of diazoacetates. Other Pd salts and Pd nanoparticle show lower catalytic activity under similar experimental conditions. Using host–guest strategy, Wu et al. *had* co-doped Fe^{3+} and Co^{2+} in zeolitic imidazolate frameworks (ZIF) matrix (Wang et al. 2017). Fe-Co bonds were formed within the support matrix, resulting in higher efficiency and more stability as compared to single Fe or Co doped cluster system. Fe-Co dual metal catalysts exhibited 100% conversion of CO oxidation at low temperature, while single atom catalyst of Co and Fe showed very low conversion of CO. Similar type of dual metal site catalyst has been synthesised using Ni–Fe metals in ZIF matrix where Fe–Ni co-ordination has been found in the supported cluster assembly. These dual centred metal catalysts were found to be more efficient as compared to their single atom counterpart. Theoretical study reveals that presence of several types of metal atoms provides more avenues for manipulation of electronic structure than one type of atom. The enhanced catalytic activity of bimetallic supported cluster system is attributed to inter-cluster polarisation and atomic interface effects. Using different reagent molecules absorbed in different type of metal sites, the reaction barrier can be decreased which leads to enhanced catalytic activity. Therefore, host–guest strategy is very effective method to synthesise supported metal clusters.

3.2.5 *The Precursor-Preselected Strategy*

During synthesis of metal clusters, they can undergo sintering leading to loss of unique cluster properties. To suppress the sintering process, precursor selected method can be used where the precursor of the metal cluster can be selected which leads to homogeneous distribution of metal from the starting to the end of the synthesis. Selection of the precursor is very important as it determines the supported metal clusters without significant sintering. Generally, metal carbonyls are synthesised using carbonyl ligand ($M_n(CO)_m$) as removal of CO ligand is comparatively easy. Though use of metal carbonyl precursor produces reasonably dispersed metal species in the cluster however cluster aggregation can't be prevented completely during decarbonylation process. To confine the metal complex, different porous MOFs, ZIFs of comparable pore size can be used. Moreover, mesoporous carbon nitride (C_3N_4) was used as a support to stabilise Fe_2 metal clusters where $Fe_2O_4C_{14}H_{10}$ was used as precursor (Tian et al. 2018). Vorobyeva *et. al* synthesised Pd_1, Pd_2 and Pd_3 clusters in C_3N_4 matrix where $Pd(NH_3)_4(NO_3)_2$, $[PdCl(C_3H_5)]_2$ and $[Pd(OAc)_2]_3$ were used as precursors, respectively (Vorobyeva et al. 2019). Pd_3/C_3N_4 was found to be more effective catalyst as compared to Pd_1/C_3N_4, Pd_2/C_3N_4 in hydrogenation reaction of different functionalised alkynes as it provides reduced hydrogen activation barrier. However, Pd_1/C_3N_4 shows higher catalytic activity in Suzuki coupling reaction leading to better chemoselectivity and higher stability. Researchers also used CeO_2 as matrix for metal cluster of palladium. Pd_n/CeO_2 (n = 1 or 6) was synthesised using 1,3,5-tris(4-pyridyl)-2,4,6-triazine precursor. Pd_1/CeO_2 was found to be highly active for oxidation reaction of benzoyl alcohol while Pd_6/CeO_2 was found to be inactive. Therefore, using various porous materials such as MOFs, C_3N_4 and CeO_2 supported atomic clusters can be synthesised by this method.

3.3 Characterisation Methods for Free and Surface Deposited Clusters

To understand the properties of cluster and cluster supported material and effect of cluster size on their unique properties, cluster characterisation plays an important role. For characterisation of cluster, different microscopic and spectroscopic techniques have been widely used. In the subsequent section, important techniques applied for characterisation of isolated gas-phase clusters and supported clusters have been discussed.

3.3.1 *Isolated Gas-Phase Clusters*

For characterisation of isolated gas-phase clusters, mass spectrometry is the universal tool which has been widely used over the years. Different kinds of mass spectrometer such as quadrupole, magnetic sector or time-of-flight are used, depending on the experimental requirement. As mass spectrometry technique directly provides the mass/charge of the ions, it is easier to characterise the cluster ions using this technique. Quadrupole mass spectrometer can be used for characterisation as well as isolation of the desired ions for further application. Major advantage of mass spectrometry technique is that it is very sensitive technique though it is unable to distinguish isomers of cluster ions. For differentiation of different isomers of cluster with identical cluster composition, ion mobility technique has been coupled with mass spectrometry which is very useful to characterise the cluster ions.

3.3.1.1 Ion Mobility Mass Spectrometry

Ion mobility technique is a well-established analytical technique which has been used to characterise cluster ions. The ions are separated according to their mobility through a buffer gas under the influence of an electric field (electrophoretic mobility). The mobility of cluster ion through buffer gas under the influence of electric field depends on the shape of the cluster ion. Typical time required to separate the ions using the ion mobility technique is of the order of milliseconds. When ion mobility spectrometer (IM) is coupled with mass spectrometer (MS), it becomes a very powerful technique for separation of isomers of a particular ion. Using IM-MS technique, chemical composition and mass distribution of stable anion, cation of sodium fluoride cluster has been determined. Also, the technique is very useful for understanding the stable composition of various transition metal oxide clusters (Ohshimo et al. 2014). Further, ion mobility mass spectrometry has been used to distinguish different isomers of disaccharide cluster ions which is useful to understand the isomeric heterogeneity of carbohydrate compounds (Li et al. 2013).

3.3.1.2 Selected Ion Flow Tube Mass Spectrometry

Selected ion flow tube (SIFT) is a technique which was developed to understand the ion–molecule reaction kinetics. When mass spectrometer is coupled with selected ion flow tube, the method becomes very sensitive and is widely used to determine the trace gas composition quantitatively. As the technique (SIFT-MS) is very sensitive having detection limit below ppt level, it can be used conveniently for characterisation of free clusters both qualitatively and quantitatively. In this method, samples are ionised via chemical ionisation method where selected positive precursor ions are used for well-defined time period along the flow tube. In general, H_3O^+, NO^+, O_2^+ ions were chosen as precursor ion which react with sample to produce different kinds secondary ions based on proton affinity or ionisation energy of the sample. The secondary ions are directed to the second quadrupole mass filter and separated according to their mass to charge ratio. Using this technique, two competitive processes, i.e. electron detachment and thermal dissociation of water clusters, have been studied (Viggiano et al. 1998).

3.3.2 *Surface Deposited Clusters*

For characterisation of deposited clusters, different sensitive techniques have been used which includes both microscopic and spectroscopic technique. Some of the important microscopic and spectroscopic techniques are briefly described below-

3.3.2.1 Transmission Electron Microscopy

Transmission electron microscopy (TEM) is a sensitive microscopic technique which has been developed by Max Knoll and Ernst Ruska in 1931. In TEM, electron gun is produced by heating tungsten filament and accelerated by applying high voltage. The accelerated electron gun guided to the sample via magnetic lenses and transmitted through the sample. The electron beams are projected to phosphorescent screen which is able to create an image of the sample. The image is captured using charge-coupled device (CCD) camera, which is placed below the screen. The sample is kept 20–100 nm thick so that electron beam can pass through it. Using TEM, one can get detailed information of the structure and morphology of supported cluster sample. Using TEM, Freund and co-workers studied melting process of platinum cluster deposited on γ-Al_2O_3 as a function of electron beam current (Nepijko et al. 1999).

Scanning transmission electron microscope (STEM) is a sophisticated version of TEM where electron beam is focussed to a fine spot of ~ 0.2 nm which provides atomic resolution image of the sample. STEM is widely used for characterisation of supported clusters used for different application. Palmer et al. used STEM for determination of morphology of hexane thiolate monolayer protected gold cluster (Au_{38}) (Wang et al. 2010).

3.3.2.2 Scanning Tunnelling Microscopy

Scanning tunnelling microscope (STM) is another useful microscopic technique to characterise supported clusters developed by Gerd Binning and Heinrich Rohrer in 1981. The characterisation technique relies on the concept to quantum tunnelling of the electrons. A very sharp and fine tip has been scanned above the surface of the sample. When bias voltage is applied between the sample and the tip, electrons tunnel through the vacuum between them. The resulting tunnel current is recorded

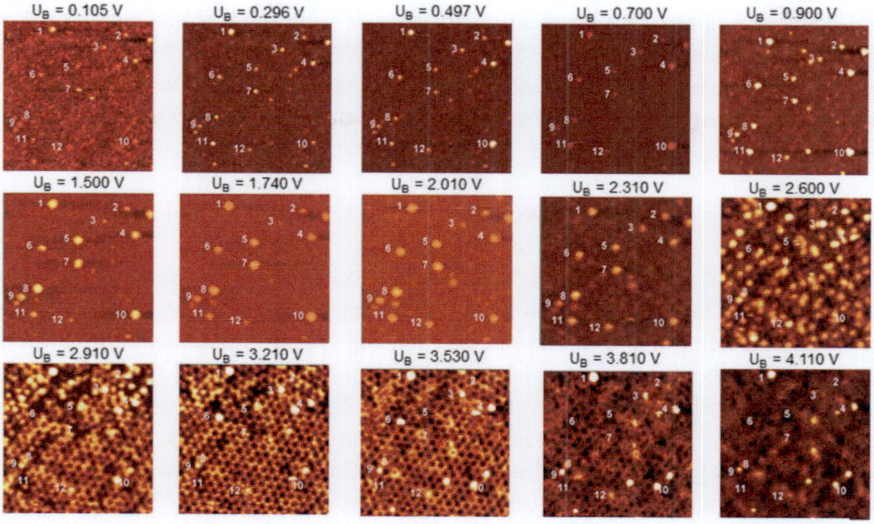

Fig. 3.4 STM images of Fe clusters supported on $Al_2O_3/Ni_3Al(111)$ surface at different bias voltage at 160 K, reproduced with permission form reference (Krupski 2014)

as a function of tip position, applied voltage, etc. and finally presented in image form. The distance between the sample and the tip is kept ~ 4–7 Å where the tip will experience repulsive attraction. The tunnelling current is generally kept in sub-nanoampere range and is amplified as close to the scanner as possible. Using STM, growth of iron cluster was studied on $Al_2O_3/Ni_3Al(111)$ surface where cluster shape was found to depend on temperature. As evident from Fig. 3.4, Fe clusters form regular hexagonal lattice on $Al_2O_3/Ni_3Al(111)$ surface (Krupski 2014).

3.3.2.3 Atomic Force Microscopy

Atomic force microscopy (AFM) is a very high-resolution scanning probe type microscopy developed by Greg Binnig and Calvin Quatein 1986. AFM is sophisticated version of STM where working principles are similar however it can work in different mode demonstrating the versatility of the microscope for surface analysis. In AFM, a nanoscale tip is attached to a small cantilever which interacts with the sample surface through a raster scanning motion. By up-down and side to side motion of the AFM tip, surface has been scanned using a laser beam. The reflected laser beam is monitored by a position-sensitive photodetector (PSPD) that indicates the vertical and lateral probe of the tip. Based on the deflection of the tip as function of voltage, surface morphology can be determined. Depending on the experimental requirement, AFM can be operated in three different modes, i.e. contact mode (static mode), dynamic or tapping mode and non-contact mode depending on the requirement of the experiment. Using non-contact mode of AFM, mechanical properties of gold cluster supported on atomically flat substrate were investigated and the value was compared with bulk sample (Schaefer et al. 1993). Further, AFM was also used to investigate nanometre size water cluster where various mechanical properties like elasticity, viscosity and dissipation energy were determined (Lee et al. 2009).

3.3.2.4 X-Ray Absorption Spectroscopy

X-ray absorption spectroscopy (XAS) is another powerful characterisation technique which provides information about the local geometry and electronic structure of the supported clusters. XAS has been carried out using synchrotron radiation which acts as tuneable high intense X-ray source. When X-ray is absorbed by the constituents of cluster, it undergoes excitation and emits either fluorescence X-ray or Auger electrons. Generally, XAS spectrum is generated in three main regions. First one deals with absorption threshold determination by the transition to the lowest unoccupied states. Second range includes X-ray absorption near edge structure (XANES) where transitions occur between core and the quasi-bound states and photoelectron energies are in the range of 10–150 eV. In this range, multi-electron excitations and many bodies final states in strongly correlated systems are most relevant. XANES spectrum reflects the valence state and co-ordination geometry of the atom which can be analysed using standard spectra. The third range includes high kinetic energy of the photoelectrons where absorption spectrum is dominated by extended X-ray absorption fine structure (EXAFS). EXAFS spectrum provides abundant local structural information on the absorbing atoms. By fitting the spectrum and series of data transformation of both XANES and EXAFS spectrum, one can determine the bond distances, co-ordination numbers, species of the neighbouring atoms, etc. Using XANES spectra and oxidation state, composition stoichiometry of copper oxide clusters (Cu_4O_x and Cu_5O_y) were determined where partial reduction of copper oxide was observed at higher temperature (Zandkarimi et al. 2020). Also, size, shape and interatomic distance of silver clusters were extracted using in-situ XANES data coupled with theoretical XANES simulation and machine learning methods (Timoshenkoa et al. 2020).

3.3.2.5 X-Ray Photoelectron Spectroscopy

X-ray photoelectron spectroscopy (XPS) is a powerful spectroscopic technique used for characterisation of surfaces deposited with clusters. XPS technique relies on photoelectric effect and can identify the elements present in the surface, chemical state of the elements, overall electronic structure and density of electronic states of the sample. In XPS spectrum, the number of electrons emitted is plotted at a specific binding energy as a result each element produces a set of characteristic peaks which corresponds to electron configuration within the atoms. The peak intensity for a particular energy indicates the amount of element present in the sample. Apart from quantitative analysis, this technique is able to produce information about the chemical

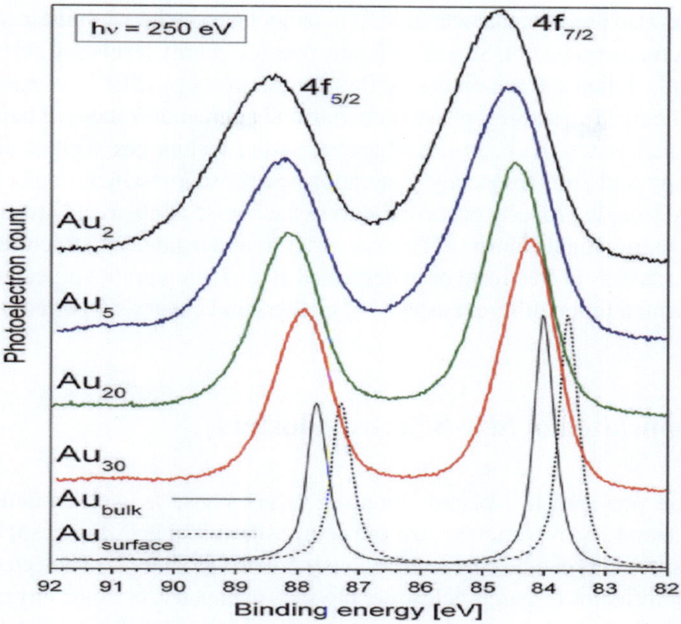

Fig. 3.5 X-ray photoelectron spectra of deposited Au-clusters (on SiO_2/Si) in comparison with the bulk peak of a thick Au-film, reproduced with permission from reference (Mao et al. 2014)

state of the element and their binding environment in few nano-meter length scale. Using XPS spectra of mass-selected gold cluster (n = 1–35) supported on silica layer, binding energy and chemisorption interaction between cluster and support have been investigated (Peters et al. 2013). Further, ambient pressure XPS was used to understand the electronic structure of silver cluster deposited on alumina and titania. Silver cluster showed better stability against sintering on titania surface as compared to alumina (Fig. 3.5) (Mao et al. 2014).

3.3.3 Miscellaneous Techniques

Apart from above mentioned well-established characterisation techniques, there are also a few other techniques which have been used to understand the property of supported clusters. Single crystal XRD has been utilised for understanding the crystal structure of the clusters prepared by organometallic route and stabilised using suitable precursor. Scanning Electron Microscope (SEM) coupled with Energy Dispersive X-ray Analysis (EDX) is another common technique which provides both image of the sample surface for determination of morphology (via SEM) and elemental identification and quantitative compositional information (via EDX). Photoluminescence technique is also used for characterisation of nanoclusters where luminescence intensity in the emission spectra, Stokes shift, etc. provide energy levels and surface states of the sample. Often surface-enhanced Raman spectroscopy (SERS) is used to determine the chemical structures, phase transition and aggregation states of bare clusters and deposited states. Also, in-situ characterisation techniques such as IR-Visible spectroscopy and sum frequency generation spectroscopy which depict real-time variation of sample property occurring during the course of chemical processes have been used to investigate clusters. Thus, in order to understand the structure property relation of clusters in free form or in deposited state, large number of techniques are available which probe different aspects of clusters and cluster supported materials.

3.4 Application of Size-Selected Clusters

Clusters are prepared in isolated condition in gas-phase as well as deposited on suitable substrate with particular size and composition. Using different sophisticated techniques, clusters can be characterised very precisely. These size-selected clusters are used in different research fields, i.e. physics, chemistry, biology, environmental science, etc. In recent years, clusters are most widely used for catalytic reaction where they enhance the rate of reaction by lowering the activation energy. Moreover, clusters are used for bio-sensing and bio-imaging applications where early detection of various diseases is possible. Further, various environmentally important reaction was studied in laboratory where cluster acts as a nano-reactor. Also, isolated gas-phase clusters are used for ionisation studies in physics where high energy ions and electrons are produced. In the subsequent section, application of clusters in different fields has been summarised briefly.

3.4.1 Catalysis

Using clusters with definite numbers of atoms, catalytic reactions can be studied at atomic level and it helps to understand the complex mechanism in depth. Size-selected clusters are considered to be better candidate for catalysis as compared to single atom catalyst or nanoparticle. In case of single atom catalyst, reaction leads to higher activation barrier while application of nanoparticle leads to too strong interaction which blocks the catalytic sites. Therefore, to achieve intermediate interaction between catalytic sites with the atomic species, supported metal clusters are ideal where size of the clusters is sufficient to manipulate the catalytic efficiency.

3.4.1.1 Clusters for Elucidating Mechanism of Catalytic Reactions

As described earlier, size-selected clusters are widely used as catalysts where reaction rates have been enhanced manifold. Apart from catalyst, cluster acts as a medium of various catalytic reactions which help to understand the mechanism of reaction in molecular level. With the knowledge of the reaction mechanism, the yield and efficiency of the reactions can be enhanced by several orders. Therefore, cluster acts as a very useful medium for mimicking catalytic reaction in the laboratory.

Acetylene (C_2H_2) upon cycloaddition (polymerisation) leads to unsaturated hydrocarbons, in presence of catalyst. For studying this polymerisation reaction of acetylene, gas-phase pure C_2H_2 clusters were prepared and subsequently, various transition metal like Fe^+, V^+, Co^+, Ni^+ were interacted for catalytic activation of polymerisation reaction. Ion signals corresponding to adduct of metal ions with acetylene molecule was observed in the mass spectra along with fragment ions $C_6H_4^+$, $C_6H_5^+$, $C_8H_6^+$ and $C_8H_7^+$. These fragment ions are covalently bonded hydrocarbons which are produced upon metal ion induced polymerisation reaction. During this catalytic polymerisation reaction, the C-H bond in acetylene has been activated and subsequently, acetylene molecules are added sequentially to form these fragment ions (Sharma et al. 2011). Thus, using metal doped molecular clusters in laboratory, one can elucidate the mechanism of catalytic reaction which helpful to develop efficient catalyst for industrial scale application.

Clusters as Catalyst for CO Oxidation and CO_2 Conversion

Among various reactions, CO oxidation reaction is one of the important and thoroughly studied reaction in catalysis due to its environmental importance (Zhou et al. 2015). Cluster supported on suitable substrates (like Au, Pt, Pd, Rh, etc.) have been screened for catalysing oxidation reaction of CO to CO_2. Among these, Au metal was extensively used with different supports such as Co_3O_4, TiO_2, TiO_x, FeO_x, CeO_2 and SiO_2. Haruta et al. first reported the application of Au catalyst prepared by co-precipitation method and CO oxidation was observed at -70 ^{0}C (Haruta et al. 1987). Catalytic activity of gold particle further enhanced upon dispersion with transition metal oxides, i.e. Fe_2O_3, Co_3O_4 and NiO (Haruta et al. 1989). When gold clusters were deposited on TiO_2/Mo support, catalytic property was found to vary with particle shape, thickness, oxidation state of the metal, etc. The bilayer Au clusters showed turnover frequency (TOF) to be ~ 3.7 s^{-1}, which is ~ 10 times higher as compared to that of monolayer support Au clusters. As turnover frequency indicates the number of molecular reaction or the catalytic cycles per unit time, higher TOF signifies efficiency of bilayer Au clusters (Chen and Goodman 2004). Anderson and co-workers reported catalytic activity of gold clusters supported on rutile TiO_2 surface as a function of cluster size (n = 1−7) (Lee et al. 2004). Au_1 and Au_2 sites strong adsorption of CO molecule thus they are found to be catalytically inactive. On the contrary, CO molecule is found to be weakly adsorbed on slightly larger clusters (n = 3, 4, 7) and Au_7 was catalytically most active as evident from Fig. 3.6. Further,

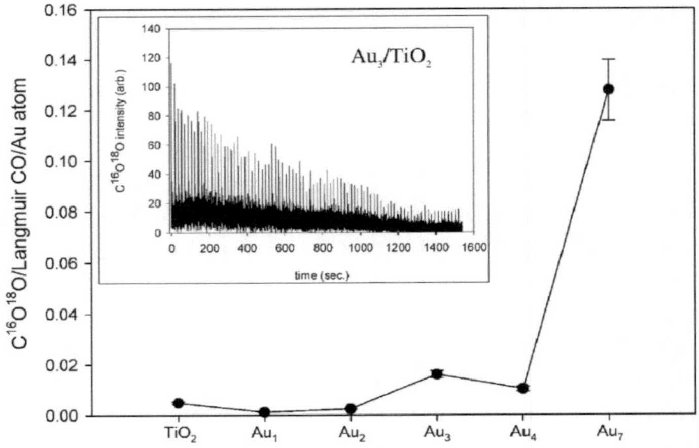

Fig. 3.6 Activity of CO oxidation reaction for clusters of different size, inset shows effect of 100 CO pulses on Au_3/TiO_2, reproduced with permission from reference (Lee et al. 2004)

Herzing *et.al.* studied catalytic activity of gold nanocrystals on iron oxide support and identified the active site using scanning transmission electron microscope (STEM) (Herzing et al. 2008). Higher catalytic activity was observed for samples having both isolated Au atoms and mono/bilayer Au clusters as compared to catalysts having only isolated Au atoms. Among various configurations of catalysts, bilayer Au clusters (n ~ 10) having diameter ~ 0.5 nm are found to be the dominant active species in CO oxidation reaction.

Apart from supported gold clusters, platinum group metal (PGM) (mainly Pt, Pd, Ru, Rh) catalysts are also used extensively for conversion of CO to CO_2. Different forms of Pt on anatase TiO_2 surface was studied for CO oxidation reaction. Three different types of Pt, i.e. isolated Pt atoms (Pt_{iso}), pre-oxidised Pt cluster (Pt_{ox}) and pre-reduced Pt cluster (Pt_{metal}) were embedded on TiO_2 surface. CO adsorption energy was found in the order—$Pt_{iso} \ll Pt_{metal} < Pt_{ox}$ and Pt_{iso} exhibited twofold greater turnover frequency (TOF) for CO oxidation as compared to Pt_{metal} clusters. Higher catalytic activity of Pt_{iso} is explained for the reaction as all the atoms are exposed for catalysis (DeRita et al. 2017). Further, single atom Pt catalyst on FeO_x surface was studied for CO oxidation where the catalyst exhibits very high efficiency and excellent stability. According to density functional theory calculation, the high catalytic activity of single Pt atom is attributed to partially vacant 5d orbitals of high valent Pt atoms (Qiao et al. 2011). Ding et al. studied CO oxidation reaction using Pt catalyst on mesoporopus zeolite HZSM-5 and SiO_2 support where Pt clusters show enhanced activity as compared to single Pt atom (Ding et al. 2015). This is because the smaller clusters have lower d-band centres which results in less electron back donation to the antibonding orbital of O_2 molecule and subsequently, O_2 activation will be less efficient.

Anderson and co-workers studied temperature programmed reaction of CO and O_2 catalysed by size-selected Pd clusters (n = 1, 2, 4, 7, 10, 16, 20, 25) supported on rutile TiO_2 (110) surface (Kaden et al. 2009). X-ray photoemission spectroscopy revealed that Pd 3d bonding energy varied gradually with cluster size which in turn influenced to CO oxidation activity. Li et al. also studied supported Pd for oxidation of CO over various substrates such as CeO_2 and CeO_2-ZrO_2 (Li et al. 2020). For Pd/CeO_2 catalyst, more surface oxygen species were found to interact with Pd to form Pd–O bond as a result more Pd^{2+} ions were present for interaction as compared to Pd/CeO_2-ZrO_2. As Pd^{2+} ion interacts strongly with CO thus Pd/CeO_2 is catalytically less active as compared to Pd/CeO_2-ZrO_2.

Apart from Au, Pd and Pt catalyst, rhodium (Rh) and ruthenium (Ru) clusters are also explored as catalyst in CO oxidation reaction (Dey and Dhal 2016). Guan *et.al.* studied Rh sub-nanoclusters deposited over TiO_2 support which catalyses CO at cryogenic temperature (235 K) (Guan et al. 2016). For supported nanocluster based on Rh, turnover frequency was three orders of magnitude higher as compared to Rh-based catalyst or comparable to Au/TiO_2 catalyst for CO oxidation. The optimised cluster size was found to be 0.4 to 0.8 nm for O_2 activation where Rh-O–O-Ti species play an important role in CO oxidation at such low temperature. Apart from supported metal clusters, core–shell type nanoparticles of platinum group metals are used for CO oxidation reaction. Nilekar et al. synthesised Ru, Rh, Ir, Pd, Au nanoparticle covered with 1–2 monolayer thick Pt atoms (Nilekar et al. 2010). Among these catalysts, Ru@Pt core–shell nanoparticle was found to be most active followed by Rh@Pt, Ir@Pt, Pd@Pt, pure Pt and Au@Pt nanoparticles. Density functional theory calculation of these catalyst shows binding energy of CO as follows—Ru@Pt < Rh@Pt < Ir@Pt < Pd@Pt < Pt < Au@Pt. Thus, weakening of the meta-CO bond leads to enhanced catalytic activity of the core–shell nanoparticle.

3.4.1.2 Hydrogenation and Dehydrogenation Reactions Mediated by Clusters

Hydrogenation/dehydrogenation reaction is one of the important reactions for industrial applications as well as basic science understanding. In this section, we will discuss application of free and supported clusters for various important hydrogenation and dehydrogenation reactions. Hydrogenation of acetylene is one of the important reactions in polymer industry where acetylene is converted to ethylene. However, over hydrogenation of acetylene can lead to formation of ethane which is undesirable in polymerisation process. Therefore, studies have been carried out to design catalyst which can selectively transform acetylene to ethylene. Wang et al. synthesised Pd nanocluster on sodalite (SOD) zeolite for semi-hydrogenation of acetylene (Wang et al. 2019). The small pore size of the SOD support facilitates interaction of H_2 molecule with the confined Pd clusters. H_2 dissociates on the surface of SOD and OH is produced which react acetylene. DFT calculation suggests higher desorption energy for acetylene on Pd(111) as compared to acetylene on SOD which effectively suppressed the over-hydrogenation of ethylene. Further, Pd-based catalyst supported on exfoliated carbon nitride (ECN) was synthesised for semi-hydrogenation reaction of alkynes. Pd_3/ECN catalyst showed 3.8 times higher conversion rate as compared to Pd_1/ECN and Pd_2/ECN for hydrogenation reaction for 2-methyl-3-butyn-2-ol. Higher catalytic activity of Pd_3/ECN was attributed due to barrier less haemolytic dissociation of H_2 while in Pd_1/ECN and Pd_2/ECN, H_2 dissociation occurs via heterolytic path having higher barrier (Vorobyeva 2019).

Also, catalytic hydrogenation reaction of alkenes was studied to understand structure sensitivity of the catalyst. Crampton et al. prepared Pt clusters (n = 8–15) on MgO support for hydrogenation of ethylene (Crampton et al. 2016a). At lower reaction temperature (~ 150 K), catalytic hydrogenation starts for clusters of Pt_{10} and exhibit maximum reactivity for Pt_{13}. At higher reaction temperature (at 400 K), Pt_{13} clusters are passivated by the formation of poisonous ethylidene. Further, dendrimer encapsulated Pt_{12} clusters on porous carbon support was used for hydrogenation reaction of styrene. Catalytic activity of Pt clusters was found to be much higher compare to supported Pt nanoparticle establishing the role of structure–property relationship in catalysis reaction (Takahashi et al. 2013). Apart from the structure, atomicity of the cluster material and type of supports also play a decisive role in determining the catalytic activity of the supported clusters. Gates et al. synthesised Ir_4 clusters on γ-Al_2O_3 and MgO support for catalytic hydrogenation reaction of propene (Argo et al. 2002). Γ-Al_2O_3 supported Ir_4 cluster showed 10 times more activity for hydrogenation reaction as compared to MgO support catalyst. Lower catalytic activity of MgO supported cluster is attributed to higher electron releasing property of MgO to Ir_4. Higher electron density on Ir_4 leads to weaken the adsorption process of reactant and intermediate species and thus exhibit low catalytic activity. In order to enhance the catalytic activity, size, structure and ligand, support of the cluster have been fine tuned. For hydrogenation reaction of ethylene, highest catalytic activity has been observed for Ir_4 clusters supported over zeolite support followed by Ir complex on zeolite and Ir_4 cluster on MgO. Higher activity of the acidic zeolite support catalyst is ascribed to its electron deficient nature which enables Ir atoms to bond with C_2H_4 and H_2 (Lu et al. 2011). In another work, role of support was further established where Pt_{13} clusters were studied on different types of support, i.e. amorphous silica thin films. O-rich silica films supported Pt_{13} catalyst showed threefold higher rate for ethylene hydrogenation reaction as compared to silicon rich and stoichiometric thin films. Due to lower electron density in O-rich silica, the activation barrier of hydrogenation of ethylene decreases while Si-rich silica provides more electron to Pt_{13} cluster thus lowering the activation barrier of reverse ethylene hydrogenation (Crampton et al. 2016b).

Hydrogenation of carbon dioxide (CO_2) is another important reaction which has been extensively studied by researchers using various catalysts as CO_2 is a major component of greenhouse gas. Upon hydrogenation of CO_2, various products such as CH_4, CH_3OH and CO can be formed though appropriate tuning of the catalyst. Yan et al. studied hydrogenation of CO_2 using Rh clusters on $SrTiO_3$ supports where CO was formed with 95% selectivity (Yan et al. 2018). The higher selectivity is attributed to the lower binding strength of CO which arises due to strong interaction between the Rh atom and the oxide support of the surface defects. Matsubu et al. studied Rh metal on TiO_2 support for catalytic hydrogenation of CO_2 where isolated atoms and Rh nanoparticle on same support exhibited different reaction pathways

(Matsubu et al. 2015). Rh nanoparticle on TiO_2 support leads to formation of CH_4 upon hydrogenation of CO_2 while disintegration of the nanoparticle to Rh_{iso} diverts the reaction pathways to form CO as the hydrogenation product. Thus, structure of the catalyst was found to be decisive in controlling the end product of CO_2 reduction reaction. Kwak et al. used different concentrations of Ru metal catalyst supported on Al_2O_3 surface for reduction of CO_2 where products obtained were found to be dependent on the content of Ru metal in the catalyst (Kwak et al. 2013). At low metal loading, the active metal phase was atomically dispersed leading to formation of CO. On the contrary, increased metal loading leads to selectively form CH_4 upon hydrogenation of CO_2. STEM images and activation energy of these two reactions suggest that the reaction pathways and the active centres vary with Rh content. Atomically dispersed Pd metal catalyst supported on Al_2O_3 and multi-wall carbon nanotube (MWCNT) have also been studied for hydrogenation reaction of CO_2. Pd/Al_2O_3 was found to be a superior catalyst due to presence of oxide support which is critical for catalytic activity of the metal catalyst. Also, Ir on CeO_2 support was studied for hydrogenation of CO_2 reaction where strong metal support interaction was found to be critical for observed reaction pathways. In case of metallic Ir particles, methane was obtained as the end product of the reaction while in case of partially oxidised Ir, CO was found to be major product. Thus, modulation of chemical state of metal species via metal support interaction plays an important role in governing the hydrogenation reaction of CO_2 (Li et al. 2017). Yao et al. prepared Co cluster on γ-Mo_2N support which showed high selectivity for CO generation over CH_4 in CO_2 hydrogenation reaction. Due to electronic perturbation, nitride forms strong bond with Co, as a result it is unable to dissociate CO_2 completely to form CH_x. Thus, nitride supports behaves differently as compared to oxide support and is able to control the reaction mechanism in a selective way (Yao et al. 2019).

Another important hydrogenation reaction is conversion of nitro aromatic compounds to aniline which is one of the important intermediates for production of dyes, pharmaceuticals, etc. When nitro compounds are present with other reducible groups like hydroxyl, vinyl or acetyl group, selective hydrogenation of the nitro compounds is challenging task. Beller and co-workers synthesised Fe_2O_3-based catalyst for pyrolysis of iron-phenanthroline complex on carbon producing nitroarenes in very high yield (Jagadeesh et al. 2013). Shimizu et al. prepared highly selective silver cluster catalyst on θ-Al_2O_3 support for hydrogenation of nitrostyrene (Shimizu et al. 2010). The catalytic activity was found to increase with decrease in size of silver cluster. Pt/TiO_2 catalyst was also for hydrogenation reaction of 4-nitrostyrene where ~100% conversion to 4-amonistyrene was observed (Makosch et al. 2012). Gold nanoparticles on TiO_2 support are explored as catalyst for hydrogenation reaction of nitrostyrene. High chemoselectivity of this catalyst is attributed to cooperative effect of gold nanoparticle with the support (Boronat et al. 2007). Wei et al. used FeO_x supported Pt single atom catalyst for chemoselective hydrogenation of different substituted nitroarenes. Its chemoselectivity is ascribed to presence of positive charge on Pt atom and absence of Pt–Pt metallic bonds (Wei et al. 2014). Pd clusters on CeO_2 support were used for hydrogenation of 4-nitrophenol where 99.9% chemoselectivity was obtained due to high dispersion of Pd metal on the support (Zhang et al. 2016).

Thus, using various metal clusters over suitable support, higher chemoselectivity of hydrogenation reaction of nitroarenes can be obtained.

Apart from hydrogenation of various compounds, dehydrogenation of various alkanes, acids, etc. are very important for industrial applications. Studies have been devoted to design efficient catalyst for dehydrogenation reaction. Bimetallic Pt–Sn clusters on Al_2O_3 support are used for dehydrogenation reaction of propane. Due to presence of Sn in the catalyst, electron density on Pt increases, as a result desorption temperature of propylene was lowered (Shi et al. 2015). Isolated Pt on CeO_2 supports was also used for propane dehydrogenation reaction at 680 ^{0}C. Due to strong binding for propylene and carbon–carbon cleavage, selectivity of the catalyst was found too poor. When Sn was added to Pt, the catalyst showed enhanced selectivity for propylene (Xiong et al. 2017). Zhang et al. synthesised Sn assisted Pt clusters over the core shell nanodiamond @graphene hybrid support (ND@G) for dehydrogenation reaction of n-butane (Zhang et al. 2019). Pt/ND@G and Pt3/ND@G showed fast deactivation while PtSn/ND@G exhibited higher performance as well as better stability.

Now days, main focus is on synthesise of materials which can store hydrogen for future use. For recovery of the stored hydrogen from these materials, dehydrogenation reaction is very essential and application of suitable catalyst can moderate the required reaction condition. Formic acid is a potential material for hydrogen generation as it is less toxic, less flammability and high capacity for hydrogen storage. For decomposition of formic acid, Au clusters over ZrO_2 support were used which leads to formation of H_2 and CO_2. The catalyst showed high activity and stability even at very low pressure of H_2, indicating its potential for hydrogen generation (Shekhar et al. 2012). Ammonia borane is another important compound which has very high capacity for hydrogen storage. For dehydrogenation reaction of ammonia borane, PtNi/NiO clusters supported on porous hollow silica support showed higher selectivity as compared to Ni/NiO clusters for hydrogen generation (Ge et al. 2017). Dimer of Pt on graphene support was also explored for hydrogenation reaction of ammonia borane which showed 34 times higher TOF as compared to single Pt catalyst. DFT calculation showed higher adsorption energy of H_2 for Pt_1 site as compared to Pt_2 sites thus enhanced catalytic activity was observed in Pt_2 (Yan et al. 2017).

3.4.2 Environment

Recent studies have shown that clusters can be successfully applied to various environmentally relevant problems such as water and air pollution where pollutants can be controlled by chemical sensor based on cluster or supported cluster materials.

3.4.2.1 Chemical Sensing

Presence of various heavy metals like Hg, Mn, Cu, Fe, Zn, etc. in drinking water beyond safe limit, cause various toxic and carcinogenic effects. These metals are non-biodegradable and make complexes with biomolecules and proteins, thus creating serious health hazards for living organism. In order to detect the heavy metals in the drinking water, protein protected gold and silver clusters are extensively used where luminescence property of the clusters has been explored. When the heavy metal such as Hg^{2+} interacts with Au^{+} or Ag^{+} (metallophilic interaction), the fluorescence of the metal cluster gets quenched and the intensity of florescence decreases with concentration of heavy metals (Xie et al. 2010). Bovine serum albumin (BSA) protected gold and silver cluster have been used for detection of mercury ions (Yang et al. 2017). 11-mercaptoundecanoic acid protected Au clusters show good sensitivity for detection of Hg^{2+} ion. Complexation with chelating agent 2,6-pyridinedicarboxylic acid with gold cluster induces selectivity for Hg^{2+} ions as compared to other ions (Huang et al. 2007). Mercury ions also form chelating complexes with various carboxylic acids such as dihydrolipoic acid, glutathione which leads to fluorescence quenching of clusters. Further, Ag_{15} cluster supported over silica coated gold mesoflowers have been successfully utilised for detection of Hg^{2+} ions at concentration as low as sub-zeptomole level (Mathew et al. 2012). For detection of other heavy metal ions in aqueous medium, along with gold and silver, clusters of other elements have also been explored. For example, Bovine serum albumin (BSA) protected copper clusters has been utilised for detection of Pb^{2+} ions in aqueous solution at ppm level. Florescence quenching of the copper cluster was ascribed to complexation of BSA with lead ions through carboxylate groups (Goswami et al. 2011).

Apart from various heavy metals, anions such as cyanides, halides, sulphides and nitrites also create serious environment pollution and health hazard. Therefore, efforts have been directed to detect these anions in various forms using ligand-protected clusters. Lu et al. explored gold cluster-based sensor for detection of cyanides (Liu et al. 2010). As a result of reaction between cyanide ion and gold cluster, OH^{-} was produced and this causes changes the pH of the aqueous solution. Fluorescence yield was found to vary with the pH of the solution. Using the fluorescence yield, cyanide anions were detected in aqueous solution using gold cluster-based sensor (Fig. 3.7). For detection of various halide ions, hyperbranched polyethyleneimine functionalised silver cluster has been utilised. Halide ions were detected based on the reaction of halide ions with silver atoms which causes change in solubility product of silver (Qu et al. 2012).

Halocarbons are another important class of pollutant which are present in the air and water due to their extensive use as refrigerants, industrial solvents, lubricants, etc. In order to degrade these compounds at room temperature, mercaptosuccinic acid (MSA) protected silver clusters have been used. In the aqueous solution, Cl^{-} ions are formed due to dissociation of halocarbon which subsequently replaces the thiolates group on the surface of the silver cluster producing AgCl. Based on the stability of the cluster, halocarbons are detected in the aqueous medium, and cluster

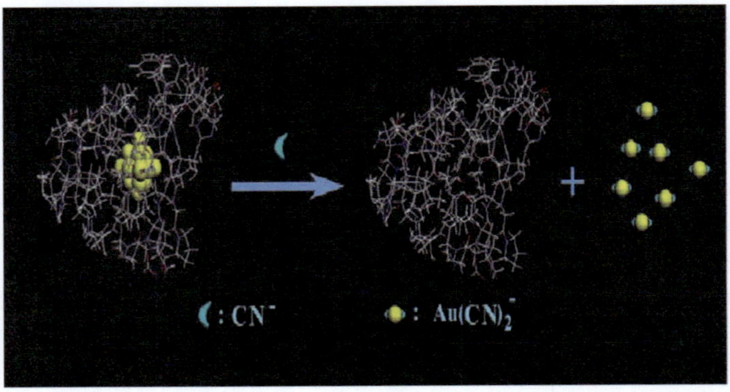

Fig. 3.7 Schematic illustration of the detection of cyanide ions from aqueous solution using Au clusters, reproduced with permission from reference (Liu et al. 2010)

shows more efficiency as compared to corresponding nanoparticle (Bootharaju et al. 2013).

3.4.2.2 Understanding/Mimicking Atmospheric Process Under Laboratory Conditions

In the rarified atmosphere and interstellar space, due to presence of different molecular clusters, dust particles, various chemical reactions are initiated or altered. As cluster provides heterogeneous surface for adsorption of chemical species, they facilitate various chemical reactions at much lower temperature. For example, moisture present in the atmosphere forms complexes with various pollutants such as O_3, SO_2, NO_2, NH_3, HNO_3, H_2SO_4, HCl, O_2 and N_2. Ozone-water complex ($O_3.H_2O$) undergoes photolysis to produce OH radical. These atmospheric chemical reactions are mimicked in laboratory by generating water cluster doped with different chemical species. Water cluster doped with various acids exhibit suppression in rate of hydrolysis at water–air interface (Vaida 2011). On the contrary, hydrolysis of acid in water is a spontaneous process in bulk phase. Further, fragmentation pattern of pure halocarbons (CF_2Cl_2) and halocarbons mixed with water cluster has also been investigated. Pure CF_2Cl_2 exhibits Cl fragmentation upon interaction with 193 nm laser excitation, while for mixed cluster the fragmentation channel is suppressed under similar experimental conditions. The difference in the fragmentation pattern is attributed due to surface orientation of CF_2Cl_2 molecule in the water cluster (Poterya et al. 2014). In another study, stability and lifetime of pyruvic acid have been investigated as pyruvic acid is an important organic compound present in the interstellar medium. Upon clustering with water, degree of fragmentation of pyruvic acid has been found to decrease significantly. Suppression of fragmentation of pyruvic acid in presence of water cluster is ascribed to non-dissociative attachment of electrons (Pysanenko et al. 2021). Thus, in laboratory utilising clusters, various atmospherically important chemical reactions can be studied.

3.4.3 Biology

In early days, various nanoparticles were used for medical applications (diagnosis and treatment) as they exhibit good optical properties and due to their high surface-to-volume ratio. However, due to lower quantum yield of plasmonic nanoparticles, they are not useful for bio-imaging and bio-labelling applications. Clusters with smaller sizes and higher quantum yield were found to be better candidate for medical application as compared to nanoparticle. Due to smaller size, clusters show lower cytotoxicity and cause less burden on excretory system. Moreover, smaller cluster size helps in target specific drug delivery quite easily and luminescence property of the cluster assists in tracking the drug release and subsequent modifications in the cell after release of the drug. Apart from drug delivery applications, clusters act as a nano-reactor for studying various biological reactions in laboratory which helps to understand the mechanism of biological reactions. Therefore, in recent years, cluster science has been widely exploited in biological applications and medicinal science.

3.4.3.1 Understanding Biological Processes at Molecular Level

Hydrogen bonding plays an important role in biological system as DNA strands are held together by this weak intermolecular interaction. Also, hydrogen bonding helps to maintain the three-dimensional structure of various biomolecules which are crucial for their activity in the living organism. To understand various biological processes, studies have often been carried out in weakly bound intermolecular clusters. As 70–80% of human body is made up of water, a large number of studies have been carried out on micro-hydrated clusters of biomolecules for understanding the role of solvation on structure, activity and radiation stability. During radiation therapy, living cells are exposed to high energy radiation, leading to low energy electrons. Studies were carried out upon exposure of low energy electron on micro-hydrated biomolecules to evaluate the radiation stability and identify potential radiosensitiser. For example, nimorazole is used as radiosensitiser in oxygen deficient tumour cell. When low energy electron interacts with nimorazole cluster, parent radical anions are formed via associative electron attachment. Generally, anions are formed via dissociative electron attachment, while in nimorazole cluster, anions are formed by associative electron attachment. This mechanism was further supported by experiments with nimorazole doped water cluster where anions were stabilised by energy dissipation to the surrounding environment (Meibner et al. 2019). In another study, non-hydrated and hydrated clusters of pyridine and pyrimidine have been investigated where binding energy was found to be smaller for protonated cluster (11.0 kcal/mol) as compared to non-hydrated counterpart (15.6 kcal/mol). Lowering of binding energy was explained based on formation of weak $CH^{\delta+}$ OH_2 hydrogen bond and stronger NH^+ OH_2 in the respective complexes (Hamid et al. 2014). Moreover, proton transfer reaction which is important in biological process, has been extensively studied by researchers in different aromatic compounds. For example, studies on pyrazine, pyrimidine, tetrahydrofuran, etc. which are the building blocks for proteins and nucleic acids have been carried out. When pyrimidine doped water clusters were ionised with different wavelengths, series of protonated $C_4H_4N_2$–$(H_2O)H^+$ was observed. The protonated cluster ions were produced as a result of intra-cluster proton transfer reaction and where initial excitation or ionisation of water molecule triggers the ionisation process in the doped cluster (Zhang et al. 2002). When tetrahydrofuran clusters were ionised by 355 nm laser excitation, protonated cluster ions were observed without formation of parent ions (Sharma and Das 2015). Upon interaction with electron beam, vertical detachment energies of the solvated electron in tetrahydrofuran (THF) were found to increase from 1.96 to 2.71 eV with increase in cluster size. This result indicates stabilisation of solvated electrons in larger clusters.

3.4.3.2 Bio-Sensing and Bio-Imaging

Ligand-protected clusters are often used as sensor for detection of various biomolecules (biothiols, proteins, DNA, ATP, etc.) and reactive oxygen species (ROS) (like H_2O_2). Sensing of these biologically important species is important for diagnosis of various diseases. Biothiols such as cysteine (Cys) and glutathione (GSH) are produced in cellular metabolism by redox reactions. Based on deficiency of Cys in body, various medical problems such as liver damage and edema can be diagnosed. GSH protected silver clusters were used for detection of Cys level in human body where silver cluster shows better selectivity towards Cys as compared to other non-thiol containing compounds. Higher selectivity for Cys was explained based on specific Ag-thiol interaction (Yuan et al. 2012). Similarly, horseradish peroxidase (HRP) enzyme protected gold clusters were used for detection of hydrogen peroxide (H_2O_2) in the solution. The structure and conformation of HRP gets altered in presence of H_2O_2, thus leading to fluorescence quenching of the gold cluster (Fig. 3.8) (Wen et al. 2011).

For sensing of very low concentration of proteins like human Immunoglobulin G in aqueous solution, Poly(amidoamine)(PAMAM) dendrimer encapsulated gold clusters have been used. Sensing of the protein molecules was carried out based on luminescent property of gold cluster (Triulzi et al. 2006). Ligand-protected clusters were also explored for sensing of various other proteins like thrombin, thyroglobulin, concanavalin, etc. For sensing ATP molecule, DNA-protected silver clusters were used where Ag clusters act as fluorescent molecular beckons (Zhang et al. 2012).

Apart from bio-sensing, various ligand-protected metal clusters are used for bio-imaging application. Clusters are suitable for bio-imaging application as they have smaller size, having biocompatibility, provide high fluorescence quantum yield and are stable against photo bleaching. Due to luminescent property of metal clusters, they are excellent candidate for biological marker in NIR region. Fluorophore and

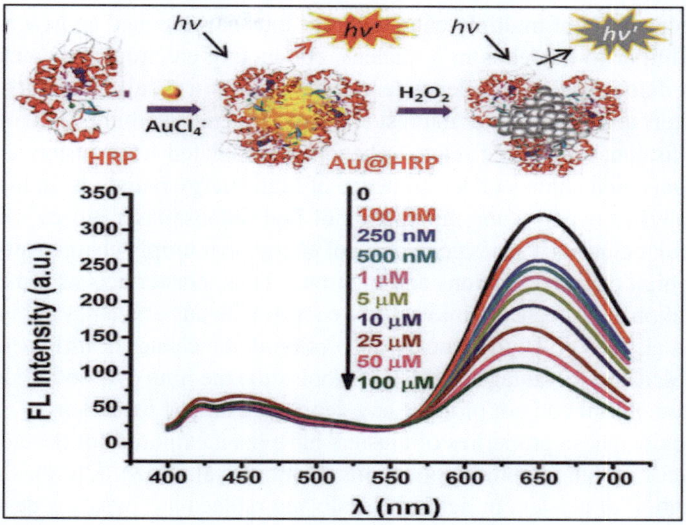

Fig. 3.8 Change in emission spectra of Au@HRP clusters with increase in concentration of H_2O_2. Inset shows the schematic of the formation of Au@HRP clusters and it's quenching with H_2O_2, reproduced with permission from reference (Wen et al. 2011)

thioflavin T were attached to silver cluster for bio-imaging application of amyloid fibrils. Bright green fluorescence was observed from the cluster in aqueous solution without photo bleaching (Makarava et al. 2005). Argyrophilic protein protected silver clusters were also used for imaging of living cells (Yu et al. 2007). DNA encapsulated silver cluster conjugated with avidin protein was explored for surface labelling of live cells (Yu et al. 2008). Dihydrolipoic acid protected silver clusters were conjugated with various biomolecules like polyethylene glycol, avidin, BSA, etc. and subsequently used for imaging application. Apart from silver cluster, gold clusters have been utilised for imaging application. Pradeep and co-workers have successfully utilised streptavidin functionalised gold cluster (Au_{23}) and folic acid conjugated Au@BSA for imaging human hepatoma cell and oral carcinoma cell, respectively (Muhammed et al. 2010, 2009). Luminescence property of the functionalised gold cluster has been explored in imaging the biologically important cells in both the cases.

3.4.4 Physics

The interdisciplinary field of cluster science is also of interest to astrophysicists, atomic and molecular physicists, solid-state physicists, nuclear physicists, plasma physicists and technologists. Observation of electronic shell structure and plasmon resonances in metal clusters, investigation of magic numbers for different types of clusters, generation of multiply charge cluster ions have opened up new avenues in physics. Similar to nucleons in a nucleus, conducting electrons in a metal cluster behave as delocalised and independent fermions. As a result, analogies between metal clusters and nuclei are of interest to nuclear physicists, for understanding shell structure, fission, fusion and related phenomena. Isolated free clusters are widely used for ionisation studies under influence of high energy radiations in high energy and astrophysics experiments. Interaction of high intense laser pulses with atomic and molecular clusters, causes generation of energetic multiply charged atomic ions, electrons, high energy radiations and neutrons. Thus, cluster acts as a medium for transformation of laser energy into particle energy, with any efficiency as high as 95% (Ditmire et al. 1997a). Higher energy absorption in the cluster is attributed to solid/liquid like density. Advantage of cluster as ionisation medium over solid/liquid materials is those clusters do not produce any debris and having local density equivalent to solid/liquid. These properties of clusters have been explored for development of XUV sources for lithographic applications (Sharma et al. 2018). A promising potential application of clusters in field of atomic and molecular physics is development of table-top accelerators.

3.4.4.1 Multiply Charged Cations Generation: Development of Table-Top Accelerator

Interaction of high intense laser pulses with clusters leads to generation of highly charged cations and electrons, with MeV kinetic energy. Efforts are underway to further optimise the yield and kinetic energy of ions/electrons and design table-top accelerators, based on efficient laser-cluster interaction. The efficiency of laser-cluster interaction depends on various experimental parameters such as laser intensity, pulse duration, cluster size and cluster constituents. At low and moderate intensity regime (~10^9–10^{12} W/cm^2), studies have been carried out using nanosecond and picosecond laser pulses for xenon clusters. From that, multiply charged cations of xenon (up to Xe^{10+}) were observed with keV kinetic energy (Fig. 3.9) (Das et al. 2021). Generation of multiply charged cations under low laser intensity condition shows cluster size dependence and kinetic energy of multiply cations was found to enhance with cluster size (Das et al. 2012). Energetic multiply charged cations of various elements, such as Sn^{13+}, Si^{4+} and Fe^{6+}, have also been reported to be produced, upon subjecting various organometallic clusters to intense laser pulses (Badani et al. 2011, 2012). Interaction of laser pulses with molecular clusters leads to generation of multiply charged cations. Multiply charged Cl^{11+}, I^{10+}, Br^{5+} were observed upon ionisation of alkyl halide clusters using nanosecond and picosecond laser pulses at different laser wavelengths (Das et al. 2011, 2013, 2015). When molecular clusters were interacted with high intense laser pulses (> 10^{14} W/cm^2), even higher charge states were observed (I^{15+}) where kinetic energy of the ions are hundreds of eV (Ford et al. 1999). Atomic cluster interacts even more efficiently with intense laser pulses leading to formation of higher charge state as well as higher kinetic energy (Ditmire et al. 1997b). The kinetic energy of the cations obtained

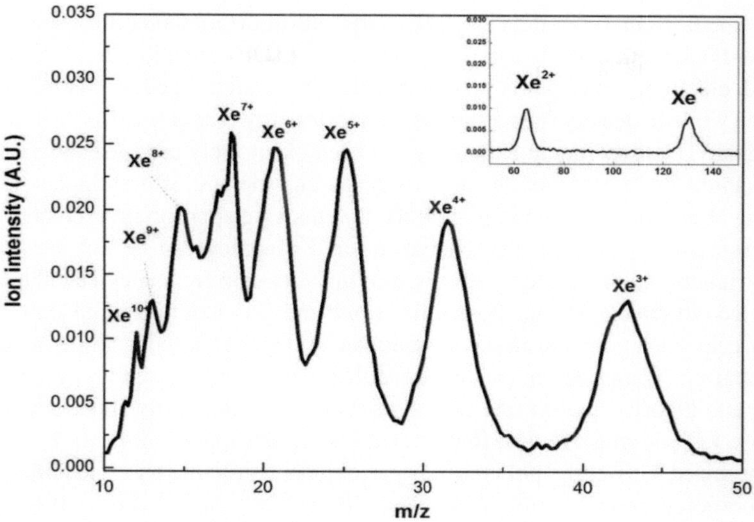

Fig. 3.9 Time-of-flight mass spectrum of xenon clusters subjected to photo ionisation by 532 nm laser pulse of ~ 10^9 W/cm^2, reproduced with permission from reference (Das et al. 2012)

in atomic cluster is four orders of magnitude higher as compared to molecular clusters. Thus, atomic clusters are potential candidate to produce energetic cations which can be used for table-top accelerator. Mathur and co-workers observed asymmetric distribution of the energetic ions produced upon interaction of argon clusters with femtosecond laser pulses of intensity ~ 10^{15} W/cm^2 (Kumarappan et al. 2001). This observation further promotes the perception of table-top accelerator by altering the experimental conditions, direction of energetic multiply charged cations can be controlled. Therefore, cluster is an important class of materials which can be utilised for the various applications as they show unique features.

3.5 Theoretical Methods/Strategies for Gaining Insight into Structure and Reactivity of Clusters

With the development of various theoretical models and computer simulations, field of cluster chemistry got additional drive as these tools predict the structure of the cluster, its evolution with size, thermodynamic behaviour, physico-chemical properties of cluster, etc. Moreover, theoretical calculations aid in understanding the experiments in a deeper way. Thus, theoretical calculations are complementary technique for any experimental observation. Here, we will discuss the commonly used theoretical methodology which has been applied for free clusters as well as clusters supported on surface.

For smaller cluster system (~ 100 atoms) ab initio methods are used such as Hartree–Fock method or density functional theory (DFT) method. In Hartree–Fock method, electronic energy of a system has been calculated based on electronic wave function while in density functional theory, total energy is expressed as the total electron density. Density functional theory has been extensively used for understanding the structure and properties of the cluster, as this method affords higher level of accuracy with lower computational effort. By choosing appropriate exchange correlation function for a particular system, desired accuracy can be achieved. Using density functional theory, local energy minimisation and frequency calculation can be carried out which provide the geometry and energetics of the equilibrium cluster structure. Further, potential energy surface can be explored based on the local minima configuration. Thus, one can locate the saddle point and subsequently using the transition state theory, reaction rate can be predicted. Using density functional theory, structure of gold clusters of different size in gas-phase and on α-Al_2O_3 substrate was optimised. Gas-phase gold clusters prefer planer structure and also retain planer configuration upon deposition on substrate (Rajesh et al. 2014) (Fig. 3.10).

Apart from theoretical calculations, simulation studies have also been carried out on clusters. Two approaches most commonly adapted for simulation are molecular dynamic and Monte Carlo simulation. In molecular dynamic simulation, time dependent molecular behaviour has been examined such as vibrational motion or Brownian motion. Initially, the atoms are assigned with a particular velocity which indicates the total kinetic energy of the system or the simulation temperature. When the temperature of the system has been increased and allows the energy to equilibrate among the atoms. Melting behaviour of metal cluster of different size was studied

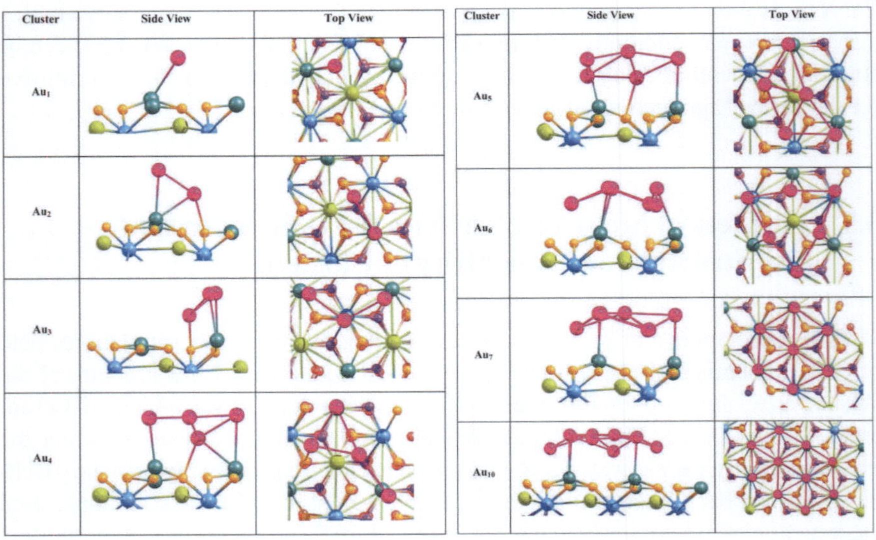

Fig. 3.10 Lowest energy configuration of gold clusters Ag_n (n = 1–7, 10) deposited on α-Al_2O_3 surface, reproduced with permission from reference (Rajesh et al. 2014)

using molecular dynamic simulation. Melting point was found to vary with atomic number for small size clusters while for larger clusters, initial structure (Icosahedra or octahedra) plays an important role for different metal clusters (Hamid et al. 2015).

In Monte Carlo (MC) simulation, numerical technique has been used where random numbers are simulated employing stochastic model of a physical system or a phenomenon. A large number of particle configurations which are in thermodynamic equilibrium, are created on a computer using a random number generator. MC is also an excellent alternative to molecular dynamics (MD) when it is difficult to extract the intermolecular force law from the potential function. Using MC simulation, electronic and optical properties of thiolate protected gold cluster [Au_{38} $(SR)_{24}$] were calculated as a function of temperature which showed excellent match with experimentally observed result (Pihlajamaki et al. 2020).

3.6 Future Direction of Cluster Research in Material Chemistry

In this book chapter, we have discussed synthesis, characterisation and application of free and supported clusters in catalysis, environmental studies, biology as well as physics. Using various modern synthesis methodology and sophisticated characterisation techniques, clusters of desired size and composition can be prepared. High selectivity of a particular reaction channel and efficient conversion of reactant to the desired product can be achieved using suitable clusters. However, there are still a few aspects that have not been explored yet and can improve the efficiency of the clusters to a greater extent.

First of all, there are still no supported clusters which exhibit very high metal coverage and stability simultaneously. Generally, metal coverage is kept low to avoid the possibility of aggregation which in turn lowers the active site and make it unsuitable for large scale application. Moreover, there are different types of support having various kinds of surface defects and functional group which leads to structural non-uniformity. Thus, clusters supported on surface suffer from non-uniformity of the metal content on the surface of the support. As a result, it is very difficult to determine the active site of the supported metal clusters which participate in the reaction. Also, as supported metal clusters are synthesised using wet chemistry method, cause inhomogeneous distribution of the metal contents. To understand the structure–property relationship of the supported metal clusters, exact active site elucidation is necessary. Therefore, it is quite important to synthesise clusters with high structural uniformity so that they can be used more efficiently. Also, it is important to characterise the clusters under working condition (in-situ) as the structure of the cluster as well as the active site often modified at elevated temperature which significantly alters the activity of the cluster. Therefore, future experimental setup should be acquainted with in-situ structural characterisation technique which can throw more light on comprehensive mechanism of the reaction. Finally, to support the experimental result, theoretical modelling should provide the details of energetics and structure of the intermediates of the reaction. Theoretically, one can calculate the thermodynamically most stable cluster structure however most active structure of cluster configuration could be metastable isomer with different coordination geometry. Therefore, accurate modelling of the cluster will be helpful to understand the detailed reaction pathways with structural information of the intermediates.

3.7 Conclusions/Summary

Clusters are intermediate in size between individual atoms/molecules and the bulk material. Due to quantum size effect, cluster shows distinctively different property from atoms, molecules or bulk counterpart and therefore, they have been widely explored in different area of research. In this chapter, synthesis, characterisation and application of clusters in chemical and biological reactions have been summarised. Clusters prepared in gas-phase, can be size-selected clusters and used for desired application. However, application of clusters prepared by gas-phase has several practical limitations which can be overcome by depositing them on suitable substrate. Clusters deposited on various substrates show distinct features which can be utilised judiciously for various reactions. Surface deposited clusters are widely used in catalytic reactions such as oxidation of CO, selective hydrogenation of alkynes, alkenes, hydrogenation of CO_2 and nitro compounds. Apart from deposition on substrate, clusters are also being stabilised using various types of ligands such as organic moiety, proteins, DNA and biomolecules. These ligand-protected clusters are very useful for environmental, biological and medical applications. Fluorescence property of the ligand-protected clusters can be used for detection of pollutants at ppm/ppb level in the environment. Ligand-protected clusters are very useful candidate for sensing biologically important molecule in medical diagnosis. For example, fluorescence property of clusters is exploited in bio-imaging. Thus, clusters both in free form and stabilised on substrate, constitute a class of compounds which have immense potential to alter various chemical and biological reactions. Moreover, free clusters are used as target for laser-cluster interaction study which can produce highly energetic multiply charged cations. These energetic multiply charged cations can be potentially applied for ion implantation, lithography and table-top accelerator. Theoretical study aids in understanding the mechanism of cluster-mediated reactions at the molecular level and designing cluster systems for futuristic applications ranging from energy, environmental chemistry, biology, medical science, etc.

References

Argo, A.M., Odzak, J.F., Lai, F.S., Gates, B.C.: Nature **415**, 623–626 (2002)

Badani, P.M., Das, S., Sharma, P., Vatsa, R.K.: Rapid Commun. Mass Spectrom. **26**, 2204–2210 (2012)

Badani, P.M., Das, S., Rao, M.V., Sharma, P., Vatsa, R.K.: AIP Adv **1**: 042164(1–13) (2011)
Billas, I.M.L., Chatelain, A., de Heer, W.A.: Sci. **265**, 1682–1684 (1994)
Bootharaju, M.S., Deepesh, G.K., Udayabhaskararao, T., Pradeep, T.: J. Mat. Chem. A **1**, 611–620 (2013)
Boronat, M., Concepcion, P., Corma, A., Gonzalez, S., Illas, F., Serna, P.A.: J. Am. Chem. Soc. **129**, 16230–16237 (2007)
Chen, M.S., Goodman, D.W.: Sci. **306**, 252–255 (2004)
Cox, A.J., Loaderback, J.G., Bloomfield, L.A.: Phys. Rev. Lett. **71**, 923–926 (1993)
Crampton, A.S., Rotzer, M.D., Schweinberger, F.F., Yoon, B., Landman, U., Heiz, U.: Angew. Chem. **55**, 8953–8957 (2016b)
Crampton, A.S., Rotzer, M.D., Ridge, C.J., Schweinberger, F.F., Heiz, U., Yoon, B., Landman, U.: Nat. Commun **7**: 10389 (1–12) (2016)
Das, S., Badani, P.M., Sharma, P., Vatsa, R.K., Das, D., Majumder, A., Das, A.K.: Rapid Commun. Mass Spectrom. **25**, 1028–1036 (2011)
Das, S., Badani, P.M., Sharma, P., Vatsa, R.K.: Chem. Phys. Lett. **552**, 13–19 (2012)
Das, S., Badani, P.M., Sharma, P., Vatsa, R.K.: RSC Adv. **3**, 12867–12873 (2013)
Das, S., Sharma, P., Badani, P.M., Vatsa, R.K.: RSC Adv. **5**, 8887–8894 (2015)
Das, S., Sharma, P., Singh, R., Vatsa, R.K., Tripathi, V.K.: J. Photochemphotobiol A **404**: 112884 (1–9) (2021)
de Heer, W.A.: Rev. Mod. Phys. **65**, 611–676 (1993)
DeRita, L., Dai, S., Lopez-Zepeda, K., Pham, N., Graham, G.W., Pan, X., Christopher, P.: J. Am. Chem. Soc. **139**, 14150–14165 (2017)
Dey, S., Dhal, G.C.: Polytech. **3**, 26–42 (2016)
Ding, K., Gulec, A., Johnson, A.M., Schweitzer, N.M., Stucky, G.D., Marks, L.D., Stair, P.C.: Sci. **350**, 189–192 (2015)
Ditmire, T., Smith, R.A., Tisch, J.W.G., Hutchinson, M.H.R.: Phys. Rev. Lett. **78**, 3121–3124 (1997a)
Ditmire, T., Tisch, J.W.G., Springate, E., Mason, M.B., Hay, N., Smith, R.A., Marangos, J., Hutchinson, M.H.R.: Nature **386**, 54–56 (1997b)
Dong, C., Li, Y., Cheng, D., Zhang, M., Liu, J., Wang, Y.G., Xiao, D., Ma, D.: ACS Catal. **10**, 11011–11045 (2021)
Fernández, E., Boronat, M.: J. Phys. Condens. Matter 31: 013002 (pp 24) (2018)
Ford, J.V., Zhong, Q., Poth, L., Castleman, A.W., Jr.: J ChemPhys **110**, 6257–6267 (1999)
Fortea-Pérez, F.R., Mon, M., Ferrando-Soria, J., Boronat, M., Leyva-Pérez, A., Corma, A., Herrera, J.M., Osadchii, D., Gascon, J., Armentano, D., Pardo, E.: Nat. Mater. **16**, 760–766 (2017)
Ge, Y., Ye, W., Shah, Z.H., Lin, X., Lu, R., Zhang, S.: ACS Appl. Mat. Inter. **9**, 3749–3756 (2017)
Goswami, N., Giri, A., Bootharaju, M.S., Xavier, P.L., Pradeep, T., Pal, S.K.: Anal. Chem. **83**, 9676–9680 (2011)
Guan, H., Lin, J., Qiao, B., Yang, X., Li, L., Miao, S., Liu, J., Wang, A., Wang, X., Zhang, T.: Angew. Chem. **55**, 2820–2824 (2016)
Halder, A., Curtiss, L. A., Fortunelli, A., Vajda, S.: J. Chem. Phys. **148**:110901 (pp. 15) (2018)
Hamid, I., Fang, M., Duana, H.: AIP Adv. **5**: 047129 (1–10) (2015)
Hamid, A.M,, Samy, El-Shall, M., Hilal, R., Elroby, S., Aziz, S.G.: J. Chem. Phys. **141**: 054305 (1–11) (2014)
Haruta, M., Kobayashi, T., Sano, H., Yamada, N.: Chem. Lett. **16**, 405–408 (1987)
Haruta, M., Yamada, N., Kobayashi, T., Iijima, S.: J. Catal. **115**, 301–309 (1989)
Herzing, A.A., Kiely, C.J., Carley, A.F., Landon, P., Hutchings, G.J.: Sci. **321**, 1331–1335 (2008)
Huang, C.C., Yang, Z., Lee, K.H., Chang, H.T.: Angew. Chem. **46**, 6824–6828 (2007)
Jagadeesh, R.V., Surkus, A.E., Junge, H., Pohl, M.M., Radnik, J., Rabeah, J., Huan, H., Schuunemann, V., Bruckner, A., Beller, M.: Sci. **342**, 1073–1076 (2013)
Kaden, W.E., Wu, T., Kunkel, W.A., Anderson, S.L.: Sci. **326**, 826–829 (2009)
Kaukonen, H.P., Landman, U., Cleveland, C.L.: J ChemPhys **95**, 4997–5013 (1991)
Krupski, A.: J. Phys. Condens. Matter **26**: 053001 (1–26) (2014)

Kumarappan, V., Krishnamurthy, M., Mathur, D.: Phys. Rev. Lett. **87**: 085005 (1–4) (2001)
Kwak, J.H., Kovarik, L., Szanyi, J.: ACS Catal. **3**, 2449–2455 (2013)
Lee, M., Sung, B., Hashemi, N., Jhe, W.: Faraday Discuss. **141**, 415–421 (2009)
Lee, S., Fan, C., Wu, T., Anderson, S.L.: J. Am. Chem. Soc. 126: 5682−5683 (2004)
Li, H., Bendiak, B., Siems, W.F., Gang, D.R.: Hill Jr HH. Rapid Commun. Mass Spectrom. **27**, 2699–2709 (2013)
Li, S., Xu, Y., Chen, Y., Li, W., Lin, L., Li, M., Deng, Y., Wang, X., Ge, B., Yang, C., Yao, S., Xie, J., Li, Y., Liu, X., Ma, D.: Angew. Chem. **56**, 10761–10765 (2017)
Li, H., Shen, M., Wang, J., Wang, H., Wang, J.: Ind. Eng. Chem. Res. **59**, 1467–1476 (2020)
Liu, Y., Ai, K., Cheng, X., Huo, L., Lu, L.: Adv. Funct. Mater. **20**, 951–956 (2010)
Lu, J., Serna, P., Aydin, C., Browning, N.D., Gates, B.C.: J. Am. Chem. Soc. **133**, 16186–16195 (2011)
Makarava, N., Parfenov, A., Baskakov, I.V.: Biophys. J. **89**, 572–580 (2005)
Makosch, M., Lin, W.I., Bumbalek, V., Sa, J., Medlin, J.W., Hungerbuhler, K., van Bokhoven, J.A.: ACS Catal. **2**, 2079–2081 (2012)
Mao, B., Chang, R., Shi, L., Zhuo, Q., Rani, S., Liu, X., Tyo, E.C., Vajda, S., Wang, S., Liu, Z.: Phys. Chem. Chem. Phys. **16**, 26645–26652 (2014)
Mathew, A., Pradeep, T.: Part PartSyst Char **31**, 1017–1053 (2014)
Mathew, A., Sajanlal, P.R., Pradeep, T.: Angew. Chem. **51**, 9596–9600 (2012)
Matsubu, J.C., Yang, V.N., Christopher, P.: J. Am. Chem. Soc. **137**, 3076–3084 (2015)
Meibner, R., Kocisek, J., Feketeova, L., Fedor, J., Farnik, M., Limão-Vieira, P., Illenberger, E., Denifl, S.: Nat. Commu 10: 2388 (1–7) (2019)
Muhammed, M.A.H., Verma, P.K., Pal, S.K., Kumar, R.C.A., Paul, S., Omkumar, R.V., Pradeep, T.: Chem. Eur. **J15**, 10110–10120 (2009)
Muhammed, M.A.H., Verma, P.K., Pal, S.K., Retnakumari, A., Koyakutty, M., Nair, S., Pradeep, T.: Chem. Eur. J. **16**: 10103−10112 (2010)
Munnik, P., de Jongh, P.E., de Jong, K.P.: Chem. Rev. **115**, 6687–6718 (2015)
Nepijko, S.A., Klimenkov, M., Kuhlenbeck, H., Freund, H.J.: J. Vac. Sci. Technol. **17**, 577–583 (1999)
Nilekar, A.U., Alayoglu, S., Eichhorn, B., Mavrikakis, M.: J. Am. Chem. Soc. **132**, 7418–7428 (2010)
Ohshimo, K., Komukai, T., Takahashi, T., Norimasa, N., Wu, J.W.J., Moriyama, R., Koyasu, K., Misaizu, F.: Mass Spectrom **3**: S0043 (1–7) (2014)
Peters, S., Peredkov, S., Neeb, M., Eberhardt, W., Al-Hada, M.: Surf. Sci. **608**, 129–134 (2013)
Pihlajamaki, A., Hamalainen, J., Linja, J., Nieminen, P., Malola, S., Karkkainen, T., Hakkinen Hannu.: J. Phys. Chem. A. **124**: 4827−4836 (2020)
Poterya, V., Kocisek, J., Lengyel, J., Svrckova, P., Pysanenko, A., Hollas, D., Slavíček, P., Fárník, M.: J. Phys. Chem. **A118**, 4740–4749 (2014)
Pysanenko, A., Grygoryeva, K., Kocisek, J., Ragesh, Kumar, T.P., Fedor, J., Oncak, M., Farnik, M.: Phys. Chem. Chem. Phys, **23**: 4317–4325 (2021)
Qiao, B., Wang, A., Yang, X., Allard, L.F., Jiang, Z., Cui, Y., Liu, J., Li, J., Zhang, T.: Nat. Chem. **3**, 634–641 (2011)
Qiao, Z., Feng, H., Zhou, J.: Ph Transit **87**, 59–70 (2014)
Qu, F., Li, N.B., Luo, H.Q.: Anal. Chem. **84**, 10373–10379 (2012)
Rajesh, C., Nigam, S., Majumder, C.: Phys. Chem. Chem. Phys. 16:26561−26569 (2014)
Ro, I., Xu, M., Graham, G.W., Pan, X., Christopher, P.: ACS Catal. **9**, 10899–10912 (2019)
Rong, H., Ji, S., Zhang, J., Wang, D., Li, Y.: Nat. Commun. **11**: 5884 (1–14) (2020)
Schaefer, D.M., Patil, A., Andres, R.P., Reifenberger, R.: Appl. Phys. Lett. **63**, 1492–1494 (1993)
Sharma, P., Attah, I., Momoh, P., El-Shall, M.S.: Int. J. Mass Spectrom. **300**, 81–90 (2011)
Sharma, P., Das, S., Vatsa, R.K.: Rapid Commun. Mass Spectrom. **32**, 1135–1140 (2018)
Sharma, P., Das, S., Vatsa, R.K.: Rapid Commun. Mass. Spectrom **29**: 45−53 (2015)
Shekhar, M., Wang, J., Lee, W.S., Williams, W.D., Kim, S.M., Stach, E.A., Miller, J.T., Delgass, W.N., Ribeiro, F.H.: J. Am. Chem. Soc. **134**, 4700–4708 (2012)

Shi, L., Deng, G.M., Li, W.C., Miao, S., Wang, Q.N., Zhang, W.P., Lu, A.H.: Angew. Chem. **54**, 13994–13998 (2015)

Shimizu, K., Miyamoto, Y., Satsuma, A.: J. Catal. **270**, 86–94 (2010)

Somorjai, G.A., Contreras, A.M., Montano, M., Rioux, R.M.: PNAS **103**, 10577–10583 (2006)

Sudheeshkumar, V., Sulaimana, K.O., Scott, R.W.J.: Nanoscale Adv. **2**, 55–69 (2020)

Takahashi, M., Imaoka, T., Hongo, Y., Yamamoto, K.: Angew. Chem. **52**, 7419–7421 (2013)

Tian, S., Fu, Q., Chen, W., Feng, Q., Chen, Z., Zhang, J., Cheong, W., Yu, R., Gu, L., Dong, J., Luo, J., Chen, C., Peng, Q., Drax, C., Wang, D., Li, Y.: Nat. Commun. **9**: 2353 (1–7) (2018)

Timoshenkoa, J., Roese, S., Hövel, H., Frenkela, A.I.: Radiat. Phys. Chem **175**: 108049 (1–4) (2020)

Triulzi, R.C., Micic, M., Giordani, S., Serry, M., Chiou, W., Leblanc, R.M.: Chem. Commun. **48**, 5068–5070 (2006)

Tyo, E., Vajda, S.: Nature Nanotech **10**, 577–588 (2015)

Vaida, V.: J. Chem. Phys. **135**: 020901 (1–9) (2011)

Viggiano, A.A., Arnold, S.T., Morris, R.A.: Int. Rev. Phys. Chem. **17**, 147–184 (1998)

Vorobyeva, E.: FakoE, Chen Z, Collins SM, Johnstone D, Midgley PA, Hauert R, Safonova OV, Vile G, Lopez N, Mitchell S, Porez-Ramírez. J. Angew. Chem. **58**, 8724–8729 (2019)

Vorobyeva, E., Fako, E., Chen, Z., Collins, S.M., Johnstone, D., Midgley, P.A., Hauert, R., Safonova, O.V., Vilé, G., López, N., Mitchell, S., Pérez-Ramírez, J.: Angew. Chem. **58**, 8724–8729 (2019)

Wang, Z.W., Toikkanen, O., Yin, F., Li, Z.Y., Quinn, B.M., Palmer, R.E.: J. Am. Chem. Soc. **132**, 2854–2855 (2010)

Wang, J., Huang, Z., Liu, W., Chang, C., Tang, H., Li, Z., Chen, W., Jia, C., Yao, T., Wei, S., Wu, Y., Li, Y.: J. Am. Chem. Soc. **139**, 17281–17284 (2017)

Wang, S., Zhao, Z., Chang, X., Zhao, J., Tian, H., Yang, C., Li, M., Fu, Q., Mu, R., Gong, J.: Angew. Chem. **58**, 7668–7672 (2019)

Wei, H., Liu, X., Wang, A., Zhang, L., Qiao, B., Yang, X., Huang, Y., Miao, S., Liu, J., Zhang, T.: Nat. Commu. 5: 5634 (1−8) (2014)

Wen, F., Dong, Y., Feng, L., Wang, S., Zhang, S., Zhang, X.: Anal. Chem. **83**:1193−1196 (2011)

Winans, R.E., Vajda, S., Ballentine, G.E., Elam, J.W., Lee, B., Pellin, M.J., Seifert, S., Tikhonov, G.Y., Tomczyk, N.A.: Top. Catal. **39**, 145–149 (2006)

Woodruff, D.P.: Atomic Clusters: From Gas-phase to Deposited: Volume 12, Elsevier Science; 1st edition (2007)

Xie, J., Zheng, Y., Ying, J.Y.: ChemCommun46: 961−963 (2010)

Xiong, H., Lin, S., Goetze, J., Pletcher, P., Guo, H., Kovarik, L., Artyushkova, K., Weckhuysen, B.M., Datye, A.K.: Angew. Chem. **56**, 8986–8991 (2017)

Yan, B., Wu, Q., Cen, J., Timoshenko, J., Frenkel, A.I., Su, D., Chen, X., Parise, J.B., Stach, E., Orlov, A., Chen, J.G.: Appl. Catal. B **237**, 1003–1011 (2018)

Yan, H., Lin, Y., Wu, H., Zhang, W., Sun, Z., Cheng, H., Liu, W., Wang, C., Li, J., Huang, X., Yao, T., Yang, J., Wei, S., Lu, J.: Nat. Commu. 8: 1070 (1–11) (2017)

Yang, G., Zhang, H., Wang, Y., Liu, X., Luo, Z., Yao, J.: Sens. Actuators b. Chem. **251**, 773–780 (2017)

Yao, S., Lin, L., Liao, W., Rui, N., Li, N., Liu, Z., Cen, J., Zhang, F., Li, X., Song, L.: Betancourt De Leon L, SuD, Senanayake SD, Liu P, Ma D, Chen JG, Rodriguez JA. ACS Catal. **9**, 9087–9097 (2019)

Yu, J., Patel, S.A., Dickson, R.M.: Angew. Chem. **46**, 2028–2030 (2007)

Yu, J., Choi, S., Richards, C.I., Antoku, Y., Dickson, R.M.: Photochem. Photobiol. **84**, 1435–1439 (2008)

Yuan, X., Tay, Y., Dou, X., Luo, Z., Leong, D.T., Xie, J.: Anal. Chem. **85**, 1913–1919 (2012)

Zandkarimi, B., Sun, G., Halder, A., Seifert, S., Vajda, S., Sautet, P., Alexandrova, A.N.: J. Phys. Chem. C **124**, 10057–10066 (2020)

Zhang, B., Cai, Y., Mu, X., Lou, N., Wang, X.: J. Chem. Phys. **117**, 3701–3710 (2002)

Zhang, M., Guo, S., Li, Y., Zuoa, P., Ye, B.: Chem. Commun. **48**, 5488–5490 (2012)

Zhang, S., Chang, C.R., Huang, Z.Q., Li, J., Wu, Z., Ma, Y., Zhang, Z., Wang, Y., Qu, Y.: J. Am. Chem. Soc. **138**, 2629–2637 (2016)

Zhang, J., Deng, Y., Cai, X., Chen, Y., Peng, M., Jia, Z., Jiang, Z., Ren, P., Yao, S., Xie, J., Xiao, D., Wen, X., Wang, N., Liu, H., Ma, D.: ACS Catal. **9**, 5998–6005 (2019)
Zhou, Y., Wanga, Z., Liu, C.: Catal. Sci. Technol. **5**, 69–81 (2015)
Zhu, W., Wang, X., Li, C., Chen, X., Li, W., Liu, Z., Liang, C.: J. Catal. **260**, 329–341 (2008)

Chapter 4
Synthesis Techniques of Nanostructured Rare Earth Metal Oxides and Mixed Metal Oxides for Catalytic and Related Applications

Deboshree Mukherjee, Debasis Das, and Benjaram M. Reddy

4.1 Introduction

'There's Plenty of Room at the Bottom: An Invitation to Enter a New Field of Physics', the famous presentation by Richard P. Feynman in 1959, which highlighted the futuristic vision of technology towards the atomic scale, is considered to be the beginning of nanotechnology (Feynman 2012). Surprisingly, at the time when, a computer used to occupy an entire room or several, the Nobel laureate physicist put forward the idea of reducing the integrated circuits of a computer to diameters between 10 and 100 atoms. While Feynman apprehended the concept of miniaturization of machines, the credit of introducing the term 'nano' goes to Norio Taniguchi, for his presentation 'On the basic concept of nanotechnology' in 1974 (Taniguchi 1974). Here, it is also important to emphasize that the term nanoparticles can be considered as the renaming of the metallic colloids discovered by Michael Faraday in 1857 (Thompson 2007), while conducting research on the optical property of gold. Richard Zsigmondy won the Nobel Prize in Chemistry in 1925 for proving that colloidal dispersions are heterogeneous. His methodological innovations have been crucial for the study of contemporary colloidal chemistry and nanotechnology. The characteristic properties of nanomaterials compared to bulk occur primarily

D. Mukherjee (✉) · D. Das
The Department of Chemistry, The University of Burdwan, Burdwan, West Bengal, India
e-mail: deb.mu.82@gmail.com

D. Das
e-mail: ddas@chem.buruniv.ac.in

B. M. Reddy (✉)
Department of Chemistry and Materials Center for Sustainable Energy and Environment, Birla Institute of Technology and Science (BITS) Pilani, Hyderabad Campus, Hyderabad, India
e-mail: bmreddy@hyderabad.bits-pilani.ac.in; bmreddy.iict@gmail.com

R. S. Ningthoujam and A. K. Tyagi (eds.), *Handbook of Materials Science, Volume 3*,
Indian Institute of Metals Series, https://doi.org/10.1007/978-981-96-8229-4_4

due to the large surface-area-to-volume ratio and dominance of quantum effects at nanoscale. These effects result in the net improvement of the properties like chemical reactivity, electrical characteristics, optical behaviour, and magnetic properties in nanomaterials. Metal and metal oxide NPs have become indispensable in modern science and technology.

Metal oxide catalysis came into prominence in the mid-1950s after it was discovered that they can catalyse different classes of chemical reactions including oxidation–reduction, hydrotreating, and acid-–base reactions, especially in petroleum chemistry (Ma et al. 2015). In modern pharmaceutical and fine-chemical industries, metal oxide nanoparticle catalysts are absolutely essential (López and Padrón 2022). More recently, metal oxide nanoparticle catalysts found application in environmental remediation, artificial photosynthesis, etc. (Jesus et al. 2022; Zhang and Sun 2019). Nanostructured heterogeneous materials are used in modern electrochemical energy conversion systems as electrode catalysts, electrolyte additives, and polymer membranes. (Zhang et al. 2014; Melchionna and Fornasiero 2014). The NPs with fluorescence property have served as a foundation for constructing powerful light-emitting devices and also for photodynamic therapy (Yu et al. 2022a).

Uses of metal oxide NPs depend on their various characteristics, including crystallinity, structural and chemical stability, size and shape, surface area, corrosion resistance, conductivity, etc. Hence, for the commercial application of metal oxide NPs, it is highly desirable to have good control over their properties during synthesis. A good example of the influence of the particle shape and size on the NP properties was demonstrated in an investigation by Haruta et al. which suggest that catalyst-support surface contact play a key role in determining the CO oxidation catalytic activity of gold NPs. The NPs of hemispherical shape outperformed the spherical NPs (Ishida et al. 2020). In a recent study, enhanced benzene oxidation catalytic activity of ultrafine Pt NPs was attributed to their confinement into the hollow ZSM-5 zeolite cages (Tian et al. 2022). Face selective catalytic activity of NPs was reported by several authors in literature (Choudary et al. 2003; Yoshihara and Campbell 1996; Zhou et al. 2005; Xu et al. 2006; Trovarelli and Llorca 2017). These literature reports demonstrate the importance of the size and shape-controlled synthesis of NPs. However, there still remains challenges in controlling the size, size distribution, shape, and dispersion of NPs.

The synthesis techniques for NPs can generally be split into two categories: (i) physical techniques include electro spraying, ball milling, laser ablation, sputtering, electron beam evaporation, etc. and (ii) chemical processes such as co-precipitation method, sol–gel method, hydrothermal method, polyol method, chemical vapour deposition, micro emulsion technique, etc. While chemical techniques adopt bottom to up approach, which involves the association of atoms or molecules towards the formation of different sizes of NPs, physical methods apply top to down strategy, where the NP synthesis takes place via depletion of the bulk of that. However, top to down approach has several disadvantages including reduced scope for controlling the shape and size distribution of the NPs, higher expense of the synthesis technique, etc. Whereas the chemical methods allow the synthesis of NPs with greater control over size, shape, composition, etc. In the following section chemical methods of NP synthesis techniques and their application will be discussed with special emphasis on the RE metal and metal oxides.

4.2 Synthetic Methods and Applications of Metal Oxide NPs

4.2.1 Co-Precipitation Method

Co-precipitation method of metal oxide NP synthesis in aqueous medium is a convenient and green approach. This synthetic method is eco-benign as it proceeds in aqueous medium and avoids use of harsh chemicals, uses less expensive starting materials and regents, takes place under gentle conditions, and can be performed in a larger scale with precision. Reddy et al. have reported the co-precipitation synthesis of several RE (La, Pr, Tb, Sm, Eu) (Reddy et al. 2009, 2017; Thrimurthulu et al. 2012; Vinodkumar et al. 2013) and transition metal (Zr, Ti, Fe, Mn, Co, Ni, Cu) (Venkataswamy et al. 2015; Katta et al. 2010; Reddy et al. 2007, 2008; Sudarsanam et al. 2014; Mukherjee et al. 2018) doped CeO_2-based nanomaterials for the CO oxidation and diesel soot oxidation towards automobile exhaust purification catalysis. Partial burn of fossil fuels in the automobile engines generates noxious elements like CO, NO_x, volatile organic compounds, soot particles, etc. 'Three-way-catalysis' technology (Schematic presentation in Fig. 4.1), which is applied for automobile exhaust cleaning, has huge application of CeO_2 NPs as the redox component of the catalyst.

The RE oxide CeO_2 gained importance due to its oxygen storage capacity (OSC) and redox catalysis. Ce can reversibly switch between the + 3 and + 4 oxidation states, resulting in reversible release and uptake of O_2 producing oxygen vacancies in the CeO_2 crystal lattice at room temperature, depending on the surrounding partial

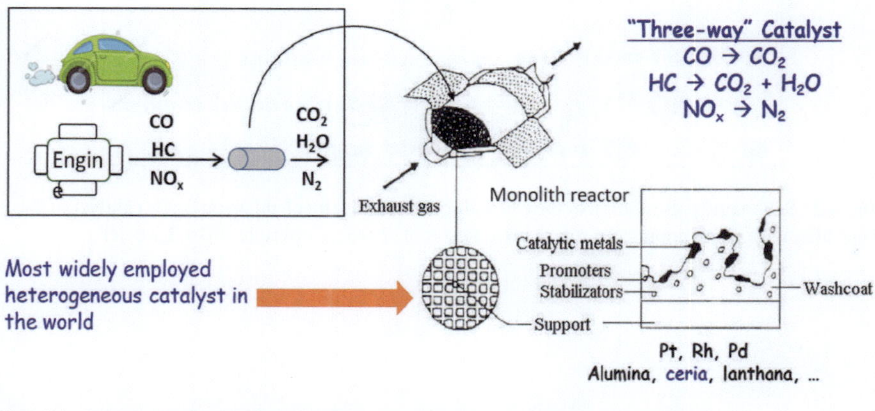

Fig. 4.1 Schematic presentation of three-way-catalysis for automobile exhaust

pressure of O_2 (Trovarelli 1996; Senanayake et al. 2013). Nano-CeO_2 shows excellent redox catalytic activity compared to bulk. This is attributed to higher oxygen vacancy formation and specific surface area. However, OSC of pure ceria is low for its commercial applications. Loss of catalytically active surface area at exposure to higher temperature is another reason for decline in catalytic activity. Doping of foreign metal ion in CeO_2 significantly enhances its OSC and thermal stability. In an experiment, a number of CeO_2-based catalysts were prepared with transition and RE metal dopants of different reducibility and sizes, namely, Zr, Hf, Fe, Mn, Pr, La, etc. by a facile co-precipitation approach (Mukherjee et al. 2016). Characterization of the samples indicated that compared to pure CeO_2, all of the doped samples had smaller crystallite sizes, higher BET surface area, and more oxygen vacancies. The CO and soot oxidation catalysis of Mn doped CeO_2 was considerably higher. The high CO oxidation catalysis of Mn doped CeO_2 was correlated with the multiple valence states of Mn ions promoting the redox catalytic activity of CeO_2, the existence of weakly bound oxygen atoms at lattice sites, and the surface attached active oxygen species present on the Mn-CeO_2 surface (Fig. 4.2).

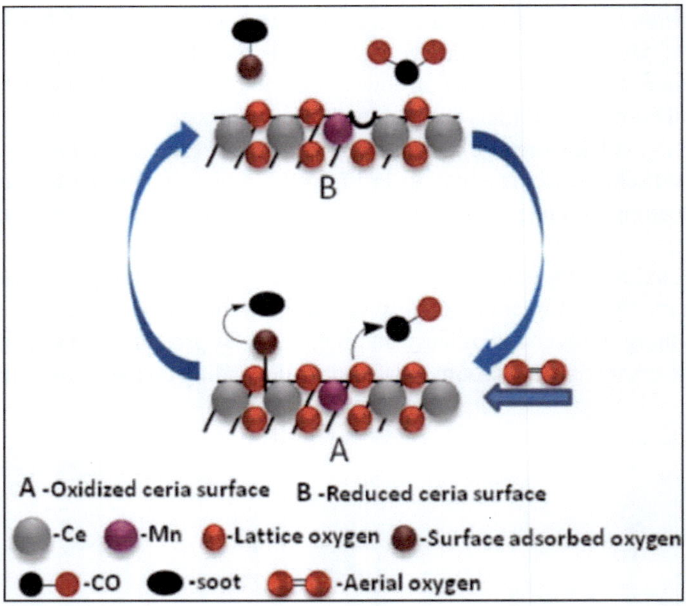

Fig. 4.2 Schematic presentation of the CO and soot oxidation mechanism on CeO_2 catalyst surface. Reprinted with permission from ref. (Mukherjee et al. 2016). Copyright 2016, Elsevier

4.2.2 Thermal Decomposition Method

This is a facile single step method for fabrication of NPs with controlled morphology, size, and particle size distribution. It is a well-known fact that, by adjusting the nucleation step, the thermal decomposition approach allows for fine-tuning of the nanoparticle's size, form, and particle growth by using appropriate surfactants (Wei et al. 2019). Superior semiconductor nanocrystals can be fabricated by thermalizing organometallic precursors, such as acetylacetonate, carbonate, alkoxide, and carbonyl derivatives of the precursor metal (Murray et al. 1993; Peng et al. 1998, 2000). Kim et al. (2013) demonstrated the synthesis of hierarchically nanoporous magnesia and ceria by using thermally less stable and labile aliphatic carboxylate ligand-based MOF precursors (*aph*-MOF). During the thermolysis process, the ligands were converted into carbogenic moieties through chemical breakdown, and contained as vesicles in the solids (Fig. 4.3). The vesicles also stopped the metal oxide nanocrystals from aggregating. Ultimately, the trapped organic moieties evaporated after thermolysis at higher temperatures, creating extremely porous nanostructures made of nanocrystalline metal oxides. The porous materials showed excellent adsorption properties applicable for CO_2 absorption from flue gas.

There are several other literature reports of RE NP synthesis by thermal decomposition. The RE NPs have very interesting optical properties. Upconversion (UC) is the process of producing visible or UV light by the longer wavelength excitation of trivalent RE ions supported by a solid-state host. This process found potential applications in lasers, displays and clinical diagnostic assays. Nanocrystalline $Y_2O_3:Eu^{3+}$ is one of the researched materials. The microcrystalline form of the material is frequently applied as the red element in cathode ray tubes and fluorescent lighting. Both $Y_2O_3:Eu^{3+}$ and $Y_2O_2S:Eu^{3+}$ are commercially available. In the red region, these materials exhibit an external quantum efficiency higher than 45% (López-Luke et al. 2014; Herrera-Rodríguez et al. 2022). Ibarra–Sánchez et al. (Jesús et al. 2018) reported Yb^{3+} and Er^{3+} co-doped Fe_3O_4 NPs by thermal decomposition method. The superparamagnetic behaviour of Fe_3O_4 and Yb^{3+} and Er^{3+} co-doped

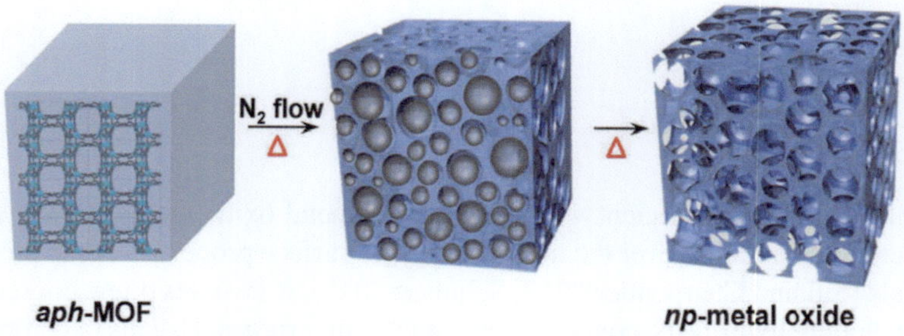

Fig. 4.3 Schematic view of the direct conversion from *aph*-MOF to *np*-Metal oxide by heating under nitrogen atmosphere. Reprinted with permission from ref. (Kim et al. 2013). Copyright 2013, American Chemical Society

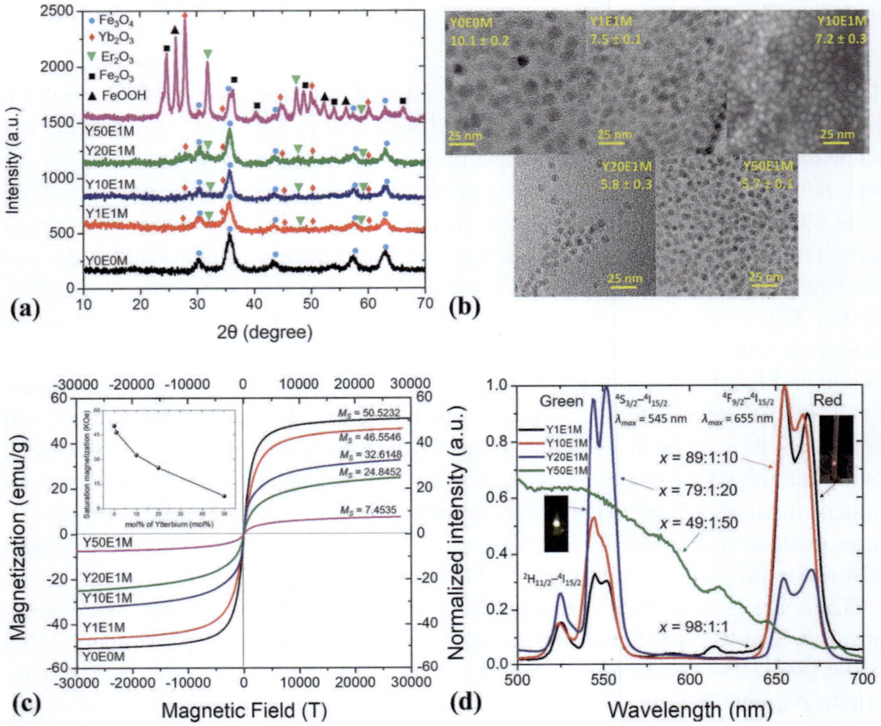

Fig. 4.4 **a** X–ray diffraction patterns, **b** HRTEM image, **c** room temperature hysteresis loops. **d** The UC emission spectra of Fe_3O_4:Yb^{3+}:Er^{3+} synthesized at different concentration of rare earth ions: 0–0 mol% (Y0E0M), 1–1 mol% (Y1E1M), 10–1 mol% (Y10E1M), 20–1 mol% (Y20E1M), 50–1 mol% (Y50E1M). Reprinted with permission from ref. (Jesús et al. 2018). Copyright 2018, Elsevier

Fe_3O_4 NPs was demonstrated from the hysteresis loop (Fig. 4.4). The RE co-doped Fe_3O_4 NPs also showed improved optical property and hyperthermia behaviour that are applicable in optical imaging and hyperthermia therapy.

4.2.3 Sol–Gel Method

Sol–gel method is an efficient way of preparing metal oxide nanoparticles, which offers crucial regulation over the morphology and surface properties of the prepared materials (Sui and Charpentier 2012; Schubert 2015). It involves a few consecutive steps including hydrolysis, condensation, and drying process. Usually, a component of high chemical activity is used as the precursor. The raw material is uniformly mixed in the solvent, which hydrolyse and condense to form a stable transparent sol system. Aged colloidal particles in the sol slowly polymerize to create a gel that contains a three-dimensional network. The solvent contained in the gel network

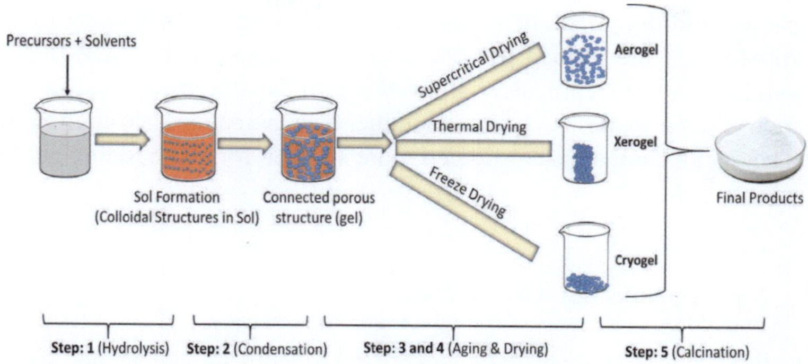

Fig. 4.5 Schematic presentation of the reaction pathway for the production of metal oxide nanostructures in the sol–gel method. Reprinted with permission from ref. (Parashar et al. 2020). Copyright 2020, Springer Nature

loses fluidity to form a gel. Finally, the gel so formed is dried, sintered and solidified to obtain the nanomaterial. The drying process is challenging due to the disruption of the gel's structure caused by the separation of organic component from water to create gel. Several drying techniques, such as atmospheric/thermal drying, freeze-drying, and supercritical drying, each have an impact on the gel network's structure differently, as shown in Fig. 4.5. High temperature heating of the porous gel results in densification and the pores may be eliminated to create xerogel, which has a high gel shrinkage and a relatively low surface area and pore volume. However, aerogels created by supercritical drying preserve significant pore volume and surface area while virtually completely preserving the original network present in the gel. The third drying method involves freezing of the solvents resulting in the formation of cryogel; compared to xerogel, cryogel's gel network shrinkage is substantially less. The sample is next subjected to heat treatment or calcination to remove any remaining materials and water molecules. The calcination temperature is a crucial variable in influencing the material's density and pore size (Parashar et al. 2020).

The sol–gel method can be broadly classified into two categories based on the solvent used in process. When water is used as reaction medium, it is named aqueous sol–gel, and non-aqueous sol–gel method, if organic solvent is used as the reaction medium.

4.2.3.1 Aqueous Sol–Gel Method

In this process, the required oxygen for metal oxide formation is provided by the water solvent. The metal precursors for this process such as metal acetates, nitrates, sulphates, chlorides, and metal alkoxides are typically used. Nonetheless, due to the high reaction affinity of metal alkoxides for water, they are frequently utilized as the precursors for the creation of metal NPs. The aqueous sol–gel approach does come with certain challenges, though. In many instances, the crucial phases of hydrolysis, condensation, and drying occur simultaneously, making it challenging to control the morphology of the formed particles and ensure repeatability. The bulk metal oxides synthesis is not significantly affected due to the aforementioned factors, but the creation of nano-oxides is. As a result, rather than their nanoscale equivalents, the aqueous sol–gel technique is considered more suitable for synthesizing bulk metal oxides.

4.2.3.2 Non-Aqueous Sol–Gel Method

In this method, the fundamental drawbacks of the aqueous sol–gel method are absent. In the non-aqueous sol–gel method, the metal precursors or solvents such as alcohols, ketones, and aldehydes provide the oxygen needed for the creation of metal oxide. These organic solvents also deliver oxygen while providing a flexible tool for adjusting a number of important factors, including the final oxide material's composition, particle size, surface characteristics, and morphology. Non-aqueous sol–gel approaches have demonstrated excellent influence on the creation of nano-oxides compared to that of aqueous sol–gel approach, while not being as popular as aqueous sol–gel approaches. You can divide the non-aqueous sol–gel method into two categories: surfactant controlled and solvent controlled approaches for the production of metal oxide NPs. In a surfactant-controlled technique, hot injection involves the direct transformation of a metal precursor into the appropriate metal oxide at a higher temperature range. This technique enables exceptional control over the shape, development, and prevention of particle aggregation.

For example, the fabrication of Eu_2O_3 nanotubes by porous anodic alumina (PAA) template assisted sol–gel template method was reported by Wu et al. (Wu et al. 2004). The formation of uniform Eu_2O_3 nanotubes was observed. The outer diameters of the Eu_2O_3 nanotubes were about 70 nm, which were demonstrated as the clear replication of the pore structure of PAA templates. The thickness of the nanotube wall was estimated to be around 5 nm. The open end of single nanotube was also clearly visible in TEM image as shown in Fig. 4.6.

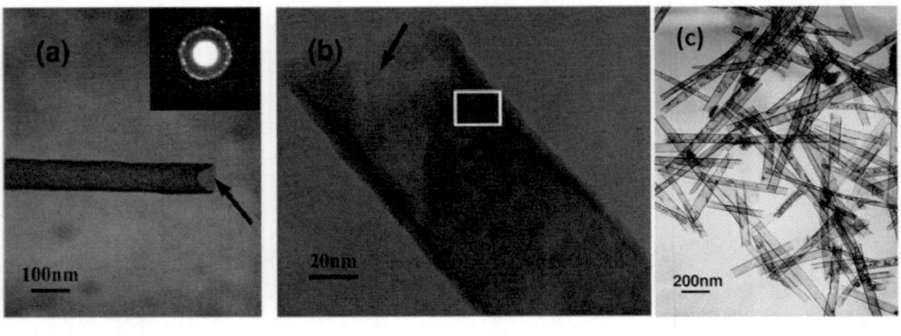

Fig. 4.6 **a** TEM image of single Eu_2O_3 nanotube and its SAED pattern (inset). **b** HREM image of a nanotube. **c** Typical SEM image of Eu_2O_3 nanotube arrays. Reprinted with permission from ref. (Wu et al. 2004). Copyright 2004, American Chemical Society

4.2.4 Hydrothermal and Solvothermal Method

Hydrothermal or solvothermal is another popular method of nanomaterial synthesis which offers critical control over the morphology of the synthesized particle. In this method, the metal precursors and the reagent's solution in water or organic solvent are filled in a stainless-steel autoclave reactor and placed into the chamber to carry out the reaction under high temperature and pressure. When non-aqueous solvents are used for the material fabrication, the term 'solvothermal' is used, whereas in case of water as solvent, it is termed as hydrothermal method. Advantage of this method is that the properties, morphology, particle size and shape of nano materials can be easily manipulated by changing the reaction's time, temperature, medium, pressure, pH, and reactant concentration, as well as the autoclave's filled capacity. For example, the synthesis of one-dimensional CeO_2 nanorod (Fig. 4.7) doped with different concentration of rare earth Y, $Ce_{1-x}Y_xO_{2-x/2}$ ($0 \leq x \leq 1$) was carried out by a facile hydrothermal method (Yang et al. 2010). The investigation demonstrated high doping concentration of 90% and the oxygen vacancies were also could be tuned in a wide range. Interestingly, the role of oxygen vacancies on the catalytic properties had been systematically investigated. It was found that the amounts of the dopant (Y^{3+}) determined the concentration, type, and distribution of oxygen vacancies of these one-dimensional solid solutions. This also controlled the reducibility and catalytic activity of the materials.

There are a number of interesting literature reports of obtaining well-defined crystalline structure of RE metal oxide NPs including Eu doped calcium hydroxyapatite NPs, Eu:$GdPO_4$ nanocubes, hexagonal-phase $NaYF_4$:Yb, Er/Tm nanocrystals, $LiYF_4$:Nd^{3+} NPs by conventional or microwave assisted hydrothermal procedure for bio imaging applications (Escudero et al. 2013; Li and Zhang 2008; Rodriguez-Liviano et al. 2013). The TEM images of the synthesized materials are presented in Fig. 4.8.

4.2.5 Polyol Method

The polyol method is a liquid phase synthesis route of producing NPs utilizing high boiling and multi-valent alcohols. The term polyol came from the usage of polyols including ethylene glycol, di-ethylene glycol, propylene glycol, butylene glycol, etc. in this process. The polyols influence particle development during synthesis by acting as both a solvent and a reducing agent. The high boiling point of polyols makes it possible to synthesize substances at temperatures between 200 and 300 °C without using an autoclave or high pressure. The polyols are good solvents to solubilize a range of starting materials allowing the usage of simple and cheap metal precursors. The advantage of high chelating ability of polyols is the key factor in controlling the nucleation, growth, and agglomeration of the particles prepared by this process. The boiling point and reduction potential of the polyol must be taken into consideration

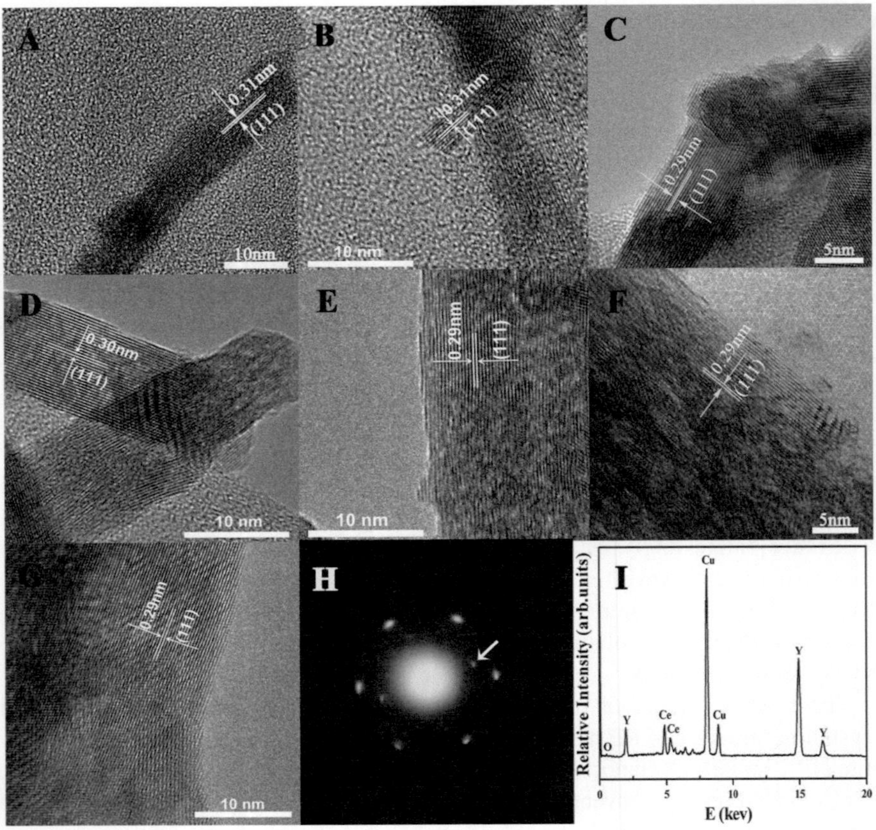

Fig. 4.7 HRTEM images of the oxide solid solutions: (A) $Ce_{1-x}Y_xO_{2-x/2}$ x = 0, (B) $Ce_{1-x}Y_xO_{2-x/2}$ x = 0.1, (C) $Ce_{1-x}Y_xO_{2-x/2}$ x = 0.3, (D) $Ce_{1-x}Y_xO_{2-x/2}$ x = 0.5, (E) $Ce_{1-x}Y_xO_{2-x/2}$ x = 0.9, (F) $Ce_{1-x}Y_xO_{2-x/2}$ x = 0.9, and (G) Y_2O_3. (H) SAED pattern for a single nanorod of $Ce_{1-x}Y_xO_{2-x/2}$ x = 0.3. (I) EDS spectra for one of the typical nanorods. Reprinted with permission from ref. (Yang et al. 2010). Copyright 2010, American Chemical Society

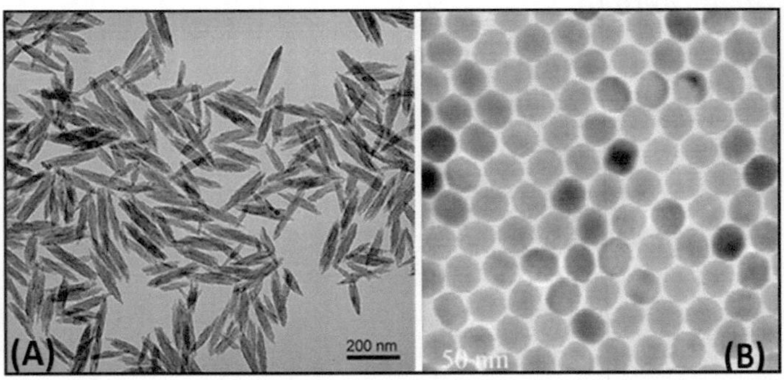

Fig. 4.8 TEM micrograph of **a** Europium-doped calcium hydroxyapatite nanophosphors functionalized with PAA (Escudero et al. 2013). **b** NaYF4:Yb, Er nanospheres (Li and Zhang 2008) Copyright, American Chemical Society & Royal Society of Chemistry

when choosing a polyol for the synthesis of NPs. Zheng et al. (Zheng et al. 2015) reported the synthesis of $CeO_2/CoSe_2$ nanobelt composite (Fig. 4.9) by a simple polyol reduction route. CeO_2 NPs were grown in-situ on $CoSe_2$/diethylenetriamine nanobelt structures. The catalyst exhibited excellent oxygen evolution reaction (OER).

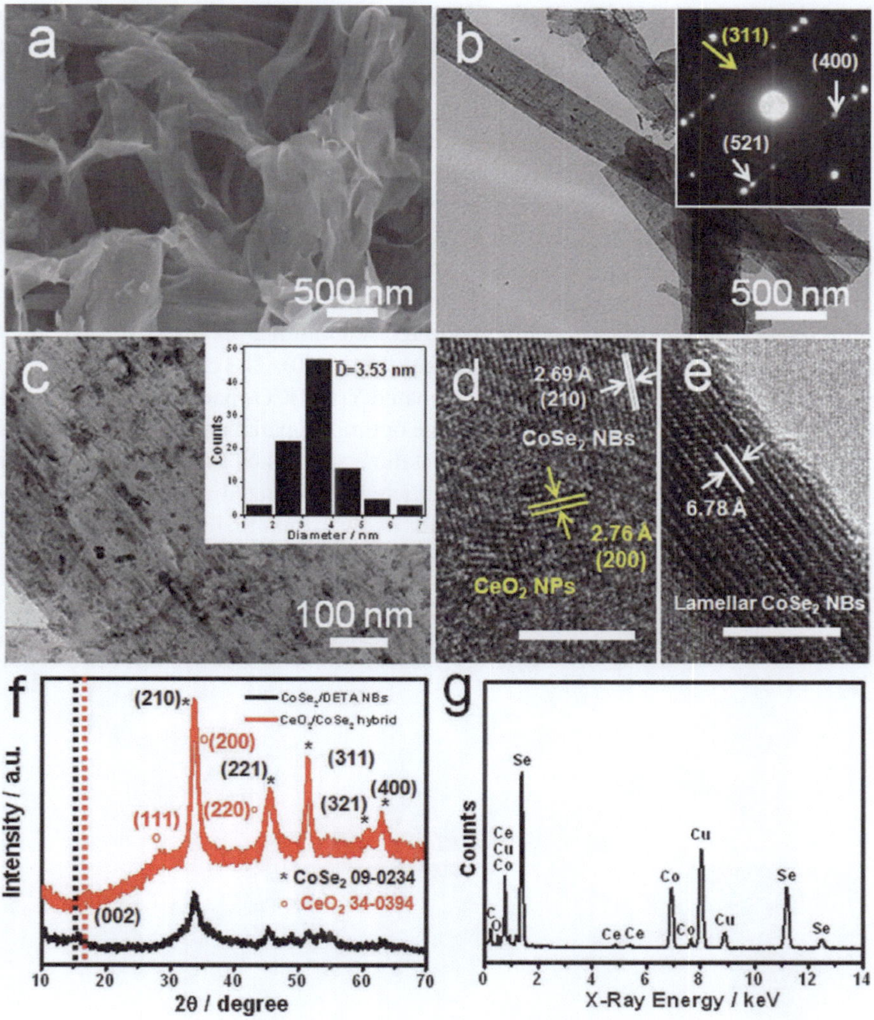

Fig. 4.9 **a** SEM image of $CeO_2/CoSe_2$ nanobelt composite. **b, c** TEM images with different magnifications of $CeO_2/CoSe_2$ nanobelt composite prepared at 278 °C for 1 h. The inset in **b** shows the corresponding SAED pattern, and the inset in **c** shows the corresponding particle size histogram. **d, e** HRTEM images of the $CeO_2/CoSe_2$ nanobelt composite. Scale bar: 5 nm. **f** XRD patterns and **g** EDX spectrum of the nanocomposite. Reprinted with permission from ref. (Zheng et al. 2015). Copyright 2015, John Wiley and Sons

4.2.6 Microwave-Assisted Method

The liquid-phase synthetic routes of nanoparticle synthesis usually apply conventional heating for driving the chemical reactions. However, the drawbacks like high thermal gradient effects, slow reaction kinetics, undesirable side-reaction are associated with conventional heating, particularly, for large-scale production of NPs (Priecel and Lopez-Sanchez 2019). Microwave method is rather advantageous in addressing these issues as microwave utilize electromagnetic energy with right frequency to carry out the chemical reaction. Ionic conduction and dipole interactions are the main mechanisms by which the materials and microwave interact during the reaction. Yet, when there is coupling between the target compound's constituent parts and the oscillating microwave electric field, the interactions can be successful. Microwave assisted method was effectively utilized in preparing RE metal oxide NPs. One such literature report used microwave assisted urea precipitation method to obtain spherical monodispersed NPs of $Y(OH)CO_3$ (Malek Khachatourian et al. 2015). The reaction parameters such as, metal ions concentration, urea/metal ions ratio, temperature and reaction time had precise control over the size, shape and size distribution of the obtained nanospheres (Fig. 4.10). The cubic Y_2O_3 was also obtained by the heat treatment of the as prepared $Y(OH)CO_3$ particles at the temperatures above 600 °C. Furthermore, using the optimized parameters for the synthesis of Y_2O_3 particles, the $Y_2O_3:Eu^{3+}$ nanocrystalline spheres of 100–350 nm diameter were also synthesized. The $Y_2O_3:Eu^{3+}$ particle demonstrated good photoluminescence property. A strong red emission peak of Eu^{3+} ion at 613 nm was observed, due to the $^5D_0 \rightarrow {}^7F_2$ electronic dipole transition of Eu^{3+} ions, making it a suitable candidate for bio imaging application and display devices.

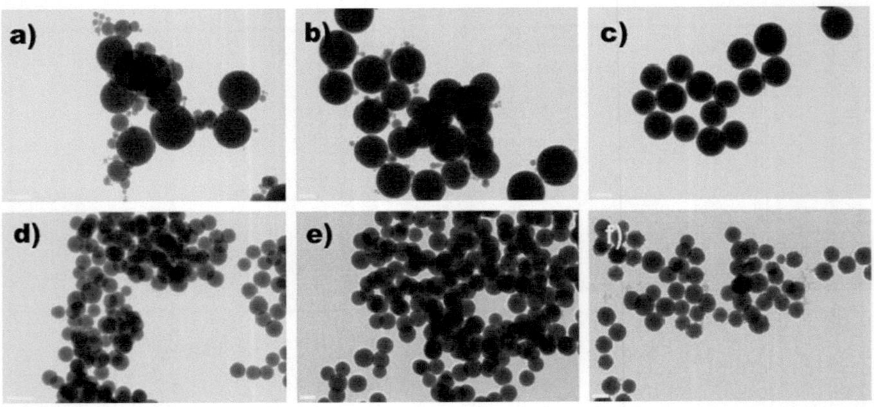

Fig. 4.10 TEM micrographs of as prepared particles with $[Y^{3+}] = 0.005$ M and reaction time of 15 min for different $[urea]/[Y^{3+}]$ (u) and reaction temperatures of **a** u = 10, 80 °C, **b** u = 20, 80 °C, **c** u = 20, 90 °C, **d** u = 50, 90 °C, **e** u = 60, 90 °C and **f** u = 80, 100 °C (the length of the scale bar represents 200 nm). Reprinted with permission from ref. (Malek Khachatourian et al. 2015). Copyright 2015, Elsevier

4.2.7 Sonochemical Method

One of the more recently developed methods that offers a flexible way to obtain nanostructures is the sonochemical process. The basic difference between the other conventional synthetic routes from sonochemical method is that it applies high intensity ultra sounds to induce transient conditions. Generally, in this method high intensity ultra-sound creates bubbles in the reagent solutions. The high temperature and pressure created in the centre of the bubbles formed during the acoustic cavitation helps in the generation of the transient states. The required conditions in sonochemical method include: (i) speed of sound in liquid should be in the range 1000 to 1500 ms^{-1}, (ii) ultra-sonic wavelength is required between 10 cm to 100 μm, and (iii) the frequency range of 20 kHz to 15 MHz is necessary. Alammar et al. (2015) reported a one-step ultrasound synthesis of CeO_2 NPs in different ionic liquids. Nano-CeO_2 was produced with a variety of morphologies, including nanospheres, nanorods, nanoribbons, nanoflowers, etc. depending on the synthetic parameters of the ionic liquid, Ce(IV) precursor, heating method, and precipitator. (Fig. 4.11). The CeO_2 nanospheres prepared in 1-butyl-3-methylimidazolium bis(trifluoromethanesulfonylamide ([C4mim][Tf2N]) by sonochemically, exhibited the best performance for low-temperature CO oxidation.

4.2.8 Micro-Emulsion Method

The synthesis of several RE NPs by using micro-emulsion method has been reported in the literature. The foundation of the micro-emulsion method is the development of reverse micelles as a result of the surfactant's thermodynamically stable structure, which enclose a nanoscale water core in contact with their hydrophilic head groups. While the water core contains the water-soluble polar and ionic substances, the solvents help in solvating the hydrophobic tails of the surfactant and remain unaltered. Shape-controlled synthesis of CeO_2 nanopowder by microemulsion method was reported by Bai et al. (Bai et al. 2006). Interestingly, different reaction temperatures resulted in the difference in confinement effect of microemulsion droplets. This effect prominently influenced the formation morphology structures as shown in Figs. 4.12 and 4.13.

4.2.9 Synthesis of Two-Dimensional NPs

Because of the high aspect ratio, distinctive surface chemistry, and quantum-size effect, two-dimensional (2D) nanomaterials frequently have exceptional physiochemical properties. The enormous collection of 2D materials that has been developed after the discovery of graphene, includes transition metal dichalcogenides,

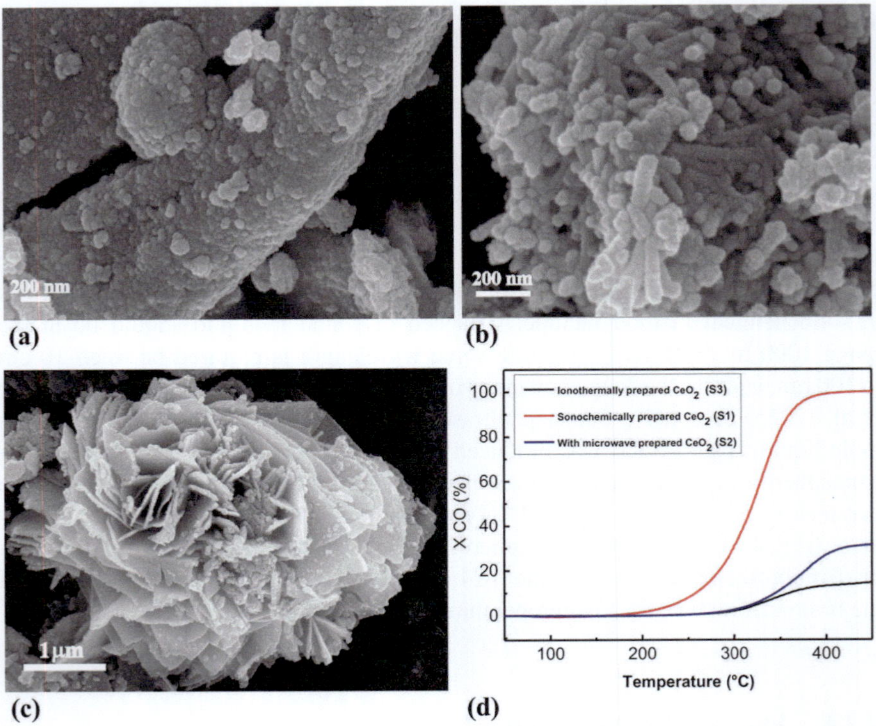

Fig. 4.11 SEM images of CeO_2 prepared in [C4mim][Tf2N] using **a** ultrasound (as-prepared, S1), **b** microwave (425 °C calcined, S2), **c** ionothermal route (425 °C calcined, S3), and **d** CO oxidation over CeO_2 obtained by different preparation methods. Reprinted with permission from ref. (Alammar et al. 2015). Copyright 2015, American Chemical Society

hexagonal boron nitride, graphitic carbon nitride, black phosphorus, etc. Because of their intriguing characteristics, fascinating anisotropic structure, and potential applications in sensing, bio imaging, photo thermal treatment, solar cells, and other fields, 2D metallic nanomaterials have recently attracted a lot of scientific interest (Tan et al. 2017). Based on the HSAB theory, RE ions are hard acids and prefer to have affinities towards bases. In a contemporary work (Li et al. 2022), NH_4X (X^- = Cl^-, B^-, I^-) is employed as the facet-controlling assistor (FCA), and halide ions (X^-) belonging to the base act as the active assistor. The introduction of an FCA suppressed the isotropic growth of NPs along the three-dimensional (3D) direction and controlled the final exposing facet (Fig. 4.14). The strategy was generalized for the facet-controllable synthesis of a series of 2D RE oxide NPs, including CeO_2, Nd_2O_3, Sm_2O_3, Eu_2O_3, Dy_2O_3, Ho_2O_3, Y_2O_3, etc.

Fig. 4.12 TEM **a**, **b** and SEM **c**, **d** micrographs of CeO_2 obtained by microemulsion at low temperature of 80 °C and calcination at 450 °C (Bai et al. 2006). Copyright 2006, Elsevier

Fig. 4.13 TEM **a**, **b** and SEM **c**, **d** micrographs of CeO_2 prepared by increasing the microemulsion temperature to 180 °C and calcination at 450 °C (Bai et al. 2006). Copyright 2006, Elsevier

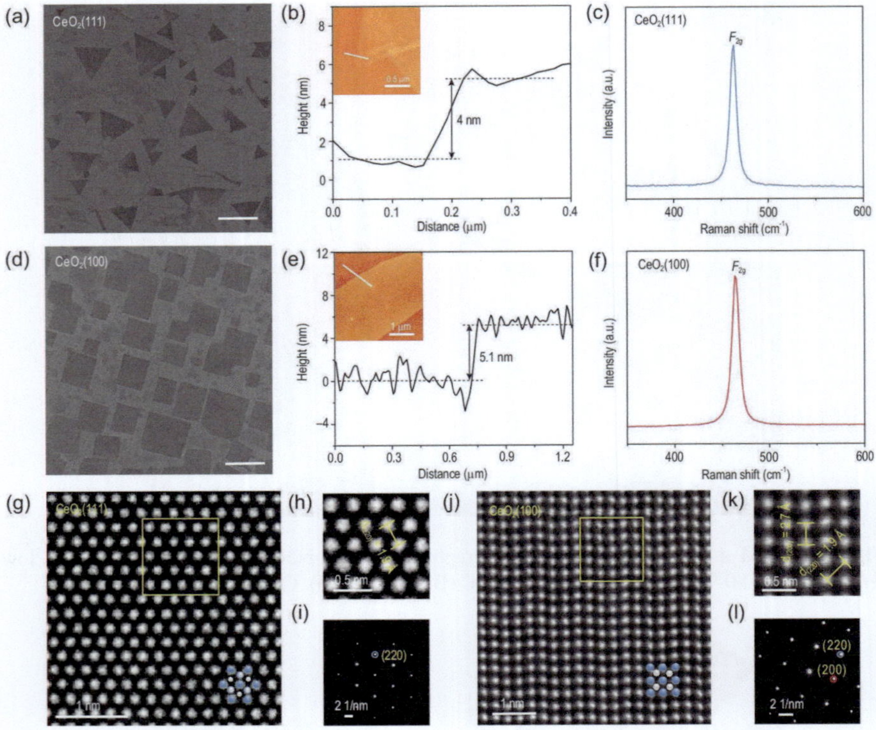

Fig. 4.14 Morphological and atomic morphology of the as-synthesized 2D CeO_2 single crystals: SEM images of trigonal and square 2D CeO_2 single crystals; scale bar: 5 μm. **b** and **e** AFM images of the trigonal CeO_2 single crystal with a thickness of 4 nm and the square CeO_2 single crystal with a thickness of 5.1 nm. **c** and **f** Typical Raman spectra of trigonal and square 2D CeO_2 single crystals. **g** and **j** HAADF-STEM images of 2D $CeO_2(111)$ and $CeO_2(100)$ single crystals, respectively (blue atoms: Ce; white atoms: O). (h and k) The magnified STEM images from the yellow box regions in **g** and **j**, respectively. **i** and **l** SAED patterns of 2D $CeO_2(111)$ and $CeO_2(100)$ single crystals, respectively. Reprinted with permission from ref. (Li et al. 2022). Copyright 2021, Oxford University Press

4.2.9.1 Viscosity-Mediated Assembly Process

In recent times Yu et al. (2022b) reported a general strategy for synthesising RE-based NPs with central divergent dendritic mesopores involving viscosity-mediated assembly process. The endowed morphological property enhanced the abundant optical, magnetic, and catalytic characteristics of RE metal oxide NPs. Its performance may be further improved by the controlled addition of mesoporous channels, which may expose additional RE-active sites and increase the loading capacity. More than 40 different types of dendritic mesoporous RE-based NPs of desired composition, for example, single or multiple RE elements, high-entropy compounds, etc. were created, each with the appropriate phase (amorphous hydroxides, crystalline

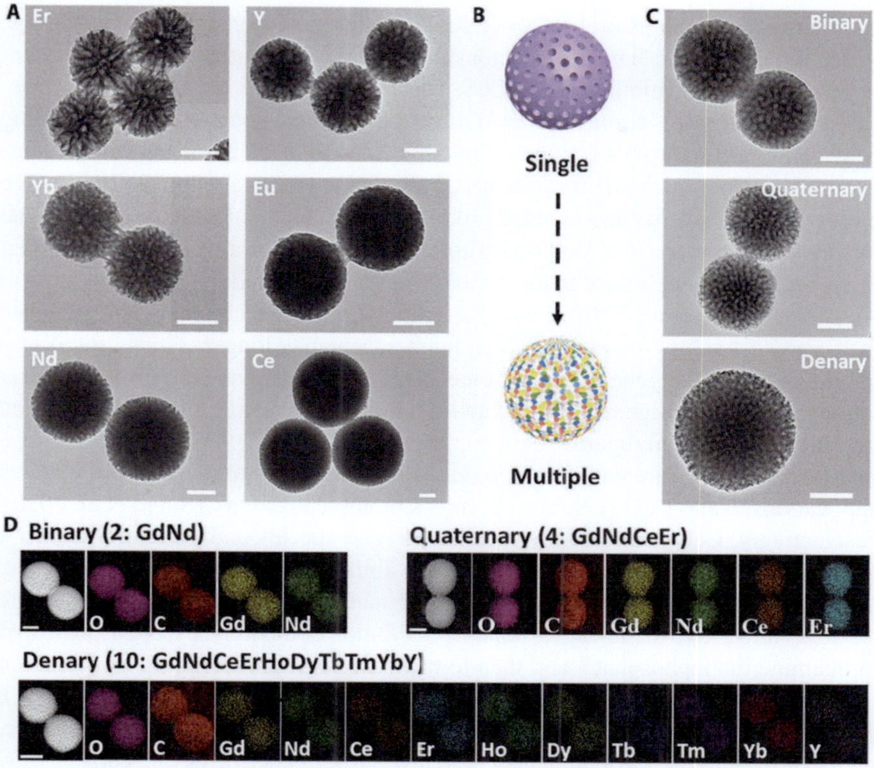

Fig. 4.15 General strategy for synthesis of dendritic mesoporous (DM)-RE$(OH)_x$ NPs with controllable composites. (A) TEM images of the obtained DM-Er(OH)x, DM-Y(OH)x, DM-Yb(OH)x, DM-Eu(OH)x, DM-Nd(OH)x, and DM-Ce(OH)x, respectively. Scheme illustrations (B), TEM images (C), and element mappings (D) of DM-RE(OH)x with multiple RE elements. Scale bars, 100 nm. Reproduced with permission from ref. (Yu et al. 2022b). Copyright 2022, AAAS

oxides, and fluorides), particle diameter (80 to 500 nm), pore size (3 to 20 nm), and architecture (Fig. 4.15).

4.3 Evolution of RE and Mixed Metal Oxides NPs in Versatile Fields

4.3.1 Biomedical Application

The medicinal use of metals is known from the historic age of Sushruta Samhita, the ancient book on Indian Ayurvedic treatment methodology written around seventh century BC and was prevalent until the discovery of antibiotics by Nobel laureate Sir Alexander Fleming in the 1920s. Ag_2O, ZnO, CuO, Fe_2O_3, TiO_2, etc. exhibit antimicrobial activity (Rao et al. 2017). Reducing the metal particle size to nanorange (<100 nm), however, has been demonstrated as an efficient and reliable tool for improving biocompatibility. AgS, CuS, and FeS nanoparticles, and Zn, Cu, Mn-based MOFs are very commonly used in various medical applications such as drug delivery and antimicrobial activities. Gold NPs (Au NPs) are attracting considerable interest as viable materials for cancer nanomedicine (for example, Table 4.1). The most well-known gold nano-enabled product now used in clinical settings for cancer treatment is Aurimune (CYT-6091)9. Tumour necrosis factor-alpha, a strong vascular disruptor, is safely delivered to the tumour site via PEGylated (polyethylene glycol) Au NPs (Koonce et al. 2015). Synthetic artificial organ development has benefited from nanotechnology by making them more acceptable, longer lasting, and robust in bio systems and regenerative medicine. Nanotechnology can be utilized for better immunosuppression, gene delivery, and contrast agent during manufacturing and transplanting artificial organs.

In recent years, rare earth metal oxide NPs have been explored for bio medical applications. Nano-CeO_2 exhibits enormous biological activity (Singh et al. 2020). Fascinatingly, CeO_2 can destroy reactive oxygen species and therefore acts as a free radical scavenger. Moreover, due to the ability of CeO_2 to adsorb phosphate, it can be used to design phosphatase-mimic nanozyme, conversion of phosphate prodrugs into active therapeutics, dephosphorylation of thiamine pyrophosphate first to thiamine monophosphate, and then to thiamine, thus producing the free vitamin form, chemotherapeutics for advanced cancer therapy, etc. (Rozhin et al. 2021). Therefore, cerium oxide has a wide range of potential therapeutic uses in the future, such as drug delivery, the treatment of oxidative stress-related diseases, redox therapy for cancer, adjuvant in antiviral therapy, prebiotic and immune-modulator, carrier and restriction enzyme mimetic in gene therapy, and modulator of signal transduction in neurology (Singh et al. 2020). Rare earth gadolinium-based nanoparticles have facilitated the development of advance bio imaging probes with excellent performance (Table 4.1). Iron oxide NPs have been applied as strong T2 MRI contrast agents (Kim et al. 2018; Jeon et al. 2021).

Table 4.1 Clinically approved Gd-based contrast agents (Bellin and Molen 2008)

Brand name	Chemical name	Approval year
Magnevist	Gadopentetate (Gd-DTPA)	FDA (1988)
Multi hance	Gadobenate (Gd-BOPTA)	FDA (2004)
Omniscan	Gadodiamide (Gd-DTPA-BMA)	FDA (1993)
Dotarem clariscan	Gadoterate (Gd-DOTA)	FDA (2013) FDA (2019)
Pro hance	Gadoteridol (Gd-HP-DO3A)	FDA (2003)
Gadavist	Gadobutrol (Gd-HP-DO3A)	FDA (2011)
Eovist primovist	Gadoxetate (Gd-EOB-DTPA)	FDA (2008)

4.3.2 Environmental Remediation

Nanotechnology has found extensive applications in environmental remediation. Heavy metals, petroleum hydrocarbons, pesticides, radioactive materials, as well as newly developing pollutants including pharmaceuticals and personal care items, can contaminate water supplies. Engineered nanomaterials are used in nanoremediation to purify contaminated air, water, and soil (Guerra et al. 2018) (Fig. 4.16). This procedure is more affordable and efficient than the majority of traditional ones like thermal treatment, pump-and-treat, chemical oxidation, etc. Due to the high surface-area to mass ratio nanomaterials surface present a random distribution of active sites which enable these materials for adsorption-based removal of pollutants. Also, NPs can diffuse in tiny spaces, and exhibit exceptional electronic and catalytic behaviour, which have improved their use in water and soil remediation. Nanomaterial-based membranes are used in the nanofiltration of water, because the membrane pores potentially hold large components in water effluents and interact selectively to separate the minor compounds. Metal oxides, carbon nanotubes, quantum dots, and biopolymers are just a few of the nanomaterials used for reclaiming contaminated water, soil, and air. Many metal oxide NPs including Fe_2O_3, ZnO, TiO_2, CeO_2, and carbon-based nanomaterials are utilized for water purification (Chatterjee et al. 2022). Due to their stability in suspension medium and high adsorption capacity, iron NPs have been used to clean colours in waste water from the textile, paint, and paper industries. Heavy metals like chromium (VI), which can be eliminated by metal/metal oxide-based nanomaterials, are another important class of water contaminants in addition to dyes. Titanium dioxide NPs are widely employed as photo catalyst for pharmaceutics and micro pollutants removal in water. CNTs can remove heavy metal ions such as Pb^{2+}, Cu^{2+}, Ni^{2+}, and Zn^{2+} from soil. However, pH, organic matter content, and the presence of silt and clay particles all have a role in the immobilization of heavy metals. Additionally, CNTs can improve soil quality by removing contaminants such as crude oil, hexachlorocyclohexane, Cr (VI), Cd, and total petroleum hydrocarbons, while boosting microbial activity and plant growth (Guerra et al. 2018). The inclusion of CNTs into membrane filtration, separation columns, and an aqueous dispersion is an example of application strategies for CNTs. Air pollution another significant problem the world is facing this century. The six most common and harmful outdoor air pollutants that affect the human health and climate change include particle matter (PM10 and PM2.5), nitrogen oxides (NO_x), sulphur dioxide (SO_2), carbon monoxide (CO), lead, and ground-level ozone. The most significant greenhouse gas produced by human activity is carbon dioxide (CO_2). A number of solutions have been explored to address this issue, including the use of graphene oxides, CNTs, etc. with highly reactive surface sites. Mesoporous silica materials are also found suitable for this application due to its ordered and tuneable porous structure, thermal stability, high surface area, and large pore volume.

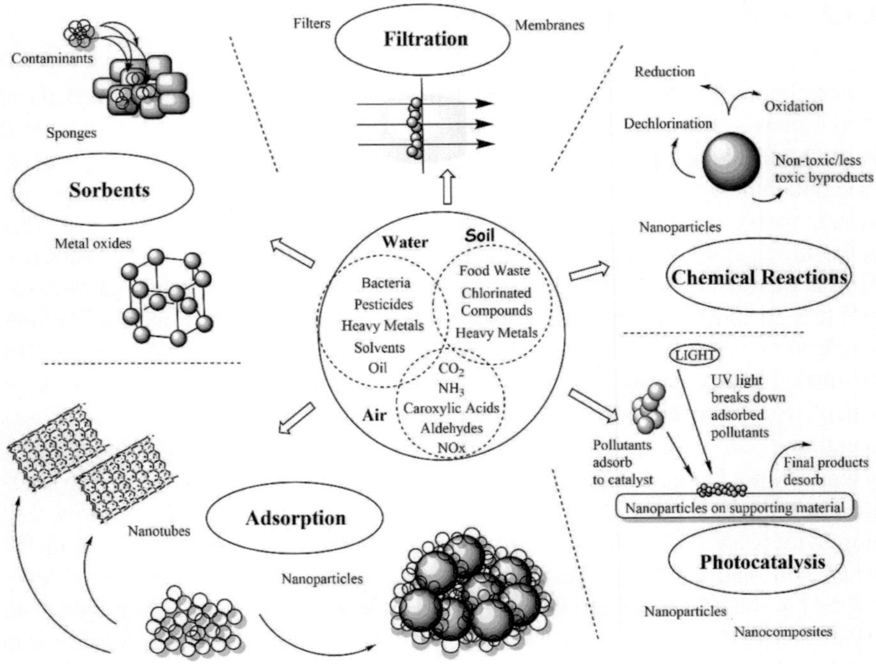

Fig. 4.16 Environmental remediation approach. Reproduced with from Ref. (Guerra et al. 2018). Licensee MDPI

4.3.3 *Nanocatalysis*

NPs as catalytically active materials have become indispensable in industrial sector. Due to their high surface-to-volume ratio, abundance of active surface atoms, and distinctive electronic structures as compared to their bulk counterparts, metal, metal oxides, and carbon-based NPs show considerable potential in heterogeneous catalysis. For example, Au, which was otherwise known to be a least reactive metal, showed high activity towards CO oxidation in nano-dimension. This breakthrough discovery by Professor Masatake Haruta in 1982 showed that when gold was deposited as NPs of smaller than 5 nm on 3d transition metal oxides became highly active for CO oxidation even at 200 °C (Valden et al. 1998). After this finding, researches from all over the world have explored the nanocatalysts in various aspects including material synthesis, characterization, and application (Bellin and Molen 2008). Thus, the field has experienced an explosive growth during the last few decades. Supported Au NPs also found several applications including direct oxidation of propylene-to-propylene oxide using O_2 and H_2, and subsequently O_2 alone in the presence of H_2O. Ag and Au can be used as catalysts for solar-driven photocatalytic water splitting, and CO_2 reduction. Pt-, Ir-, and Ru-based nanocatalysts comprising of CeO_2-based redox materials are indispensable in automobile industry for designing catalytic converters. Low over-potential of these noble metal nanocatalysts also make them suitable for half-reactions of importance for the conversion and storage of renewable energy, like electro catalytic oxygen reduction reaction, hydrogen evolution reaction, and oxygen evolution reaction, etc. Apart from expensive noble metals, Cu-based NPs are efficient catalysts for the production of methanol from the syngas ($CO + H_2$) and transition metal and metal oxide-based nanocatalysts found application in pharmaceutical and fine chemical industry (Astruc 2020). However, metal NPs are thermodynamically unstable due to their high surface energy and are susceptible to migration and coalescence during catalysis, particularly at high reaction temperatures. Typically, a sharp decline in catalytic activity and selectivity occurs concurrently with these structural alterations. Use of stable metal oxide supports and encapsulation of metal NPs in metal organic framework organic cages, nanoshells or nanopores is expected to improve the stability and lifetime of nanocatalysts.

4.3.4 Energy Storage for Electronics and Computation

The field of energy storage devices has drawn considerable attention in various nanomaterials, including carbon and layered materials like MoS_2, WS_2, clay, and others. Nanostructured materials offer several advantages: (1) they possess a large surface area, promoting electrochemical reactions or molecular adsorption at solid–liquid or solid–gas interfaces; (2) they develop high crystallinity and/or porous structures, enhancing electron or ion transport and electrolyte diffusion, ensuring efficient electrochemical processes; and (3) they produce optical effects that improve optical absorption in solar cells (Zhang et al. 2013). Rare earth elements such as Eu and Yb oxide-based materials are used in up-conversion materials, while Sm and Eu-based materials are utilized in low-voltage cathodoluminescent devices. Eu_2O_3 is employed in high-density optical storage devices, lasers, televisions, and fluorescent lamps as red and blue phosphors. Sm_2O_3 is used to prepare calcium chloride crystals for optical lasers and is a neutron absorber in nuclear reactors and infrared-absorbing glass. Additionally, Sm_2O_3 finds special applications in the glass and ceramics industries (Pitale et al. 2011). Carbon-based nanomaterials are applied in numerous fields with their diverse chemical compositions and forms. Advanced manufacturing techniques such as printing, spray coating, and roll-to-roll assembly of carbon-based nanomaterials have enabled the creation of wearable, flexible, and foldable energy storage devices (Jiang et al. 2022). Computation technology is also on the threshold of a pattern change by nanotechnology. Quantum computers use the quantum characteristics of materials: properties that only display at a scale of a few dozen nanometres. Thus, a quantum computer requires a diverse set of building blocks compared to a standard one. Researchers all over the world are involved in creating such building blocks. A group of Dutch investigators have produced a wire with a diameter of about 20 nm and then fitted it with tiny aluminium electrodes. At a temperature of 0.02 K above absolute zero, they confirmed superconductivity in this wire, and with the help of an external electric field, they created a quantum dot containing exactly one ‘electron–hole’ (Ridderbos et al. 2018).

4.4 Conclusions

The synthesis and application of RE and mixed metal oxide NPs have been reviewed in this chapter. The field of nanoscience and technology has shown an exponential growth in the past few decades due to their huge application in diverse fields. A variety of synthetic techniques have been reported so far and new synthetic methods are still being explored. Bottom-up synthetic methods are advantageous in terms of obtaining better control over the chemical homogeneity, shape, size, and size distribution of the NPs, which is crucial factor in controlling the properties of the nanomaterials. Metal and metal oxide NPs are essential in medical diagnosis, green energy production and storage, industrial catalysis, light emitting devices, etc. Downsizing of particle size towards metal atom cluster and single atom catalyst is the current trend for better metal atom utilization efficiency. However, the toxicological effects of metal and metal oxide NPs have to be kept in mind. Over usage of organic solvents and chemicals during nanoparticle synthesis and purification is another concern. In spite of these facts, metal and metal oxide NPs are indispensable in modern day science and technology.

Acknowledgements We wish to specially acknowledge all the researchers whose work is described in this chapter for their valuable contributions. D.M thanks University Grants Commission (UGC), New Delhi for the Dr. D. S. Kothari postdoctoral research fellowship and the Department of Chemistry, University of Burdwan for support.

Author Contributions The manuscript was written through contributions of all authors. All authors have given approval to the final version of the manuscript.

References

Alammar, T., Noei, H., Wang, Y., Grünert, W., Mudring, A.-V.: ACS Sustain. Chem. & Eng. **3**(1), 42–54 (2015)

Astruc, D.: Chem. Rev. **120**(2), 461–463 (2020)

Bai, J., Xu, Z., Zheng, Y., Yin, H.: Mater. Lett. **60**(9), 1287–1290 (2006)

Bellin, M.-F., Van Der Molen, A.J.: Eur. J. Radiol. **66**(2), 160–167 (2008)

Chatterjee, P., Mukherjee, D., Sarkar, A., Chakraborty, A.K.: Appl. Nanosci. **12**(10), 3031–3043 (2022)

Choudary, B.M., Mulukutla, R.S., Klabunde, K.J.: J. Am. Chem. Soc. **125**(8), 2020–2021 (2003)

de Jesús, I.-Sánchez. J., López-Luke, T., Ramírez-García, G., Sidhik, S., Córdova-Fraga, T., de Jesús, B.-A., et al.: J. Magn. Magn. Mater. **465**, 406–411 (2018)

de Jesus, R.A., de Assis, G.C., de Oliveira, R.J., Bilal, M., Bharagava, R.N., Iqbal, H.M.N., et al.: 21—Metal Oxide Nanoparticles for Environmental Remediation. In: Iqbal, H.M.N., Bilal, M., Nguyen, T.A., Yasin, G., editors. Biodegrad. Biodeterior. Nanoscale. Elsevier. p. 529–60 (2022)

Escudero, A., Calvo, M.E., Rivera-Fernández, S., de la Fuente, J.M., Ocaña, M.: Langmuir **29**(6), 1985–1994 (2013)

Feynman Richard, P.: There's plenty of room at the bottom: an invitation to enter a new field of physics. Handb. Nanosci., Eng., Technol.: CRC Press. (2012)

Guerra, F.D., Attia, M.F., Whitehead, D.C., Alexis, F.: Molecules **23**(7), 1760 (2018)

Herrera-Rodríguez, A., López-Luke, T., Romero Arellano, V.H., Doddoji, R., Medina, J.Z., Ramos, I.S., et al.: Opt. Mater. **133**, 112874 (2022)
Ishida, T., Murayama, T., Taketoshi, A., Haruta, M.: Chem. Rev. **120**(2), 464–525 (2020)
Jeon, M., Halbert, M.V., Stephen, Z.R., Zhang, M.: Adv. Mater. **33**(23), 1906539 (2021)
Jiang, X., Cao, C., Feng, W., Li, F.: J. Mater. Chem. B **4**(1), 87–95 (2016)
Jiang, Z., Guan, L., Xu, X., Wang, E., Wang, C.: ACS Appl. Electron. Mater. **4**(11), 5144 (2022)
Katta, L., Sudarsanam, P., Thrimurthulu, G., Reddy, B.M.: Appl. Catal. B **101**(1), 101–108 (2010)
Kim, T.K., Lee, K.J., Cheon, J.Y., Lee, J.H., Joo, S.H., Moon, H.R.: J. Am. Chem. Soc. **135**(24), 8940–8946 (2013)
Kim, D., Kim, J., Park, Y.I., Lee, N., Hyeon, T.: ACS Cent. Sci. **4**(3), 324–336 (2018)
Koonce, N.A., Quick, C.M., Hardee, M.E., Jamshidi-Parsian, A., Dent, J.A., Paciotti, G.F., et al.: Int. J. Radiat. Oncol. *Biology*Physics **93**(3):588–96 (2015)
Li, Z., Zhang, Y.: Nanotechnol. **19**(34), 345606 (2008)
Li, L., Lu, F., Xiong, W., Ding, Y., Lu, Y., Xiao, Y., et al.: Natl. Sci. Rev. **9**(5), 153 (2022)
López, Ó., Padrón, J.M.: Catalysts **12**(2) (2022)
López-Luke, T., De la Rosa, E., Campos Villalobos, I., Rodriguez, R.A., Ángles-Chávez, C., Salas, P., et al.: J. Lumin. **145**, 292–298 (2014)
Ma, Z., Wei, L., Zhou, W., Jia, L., Hou, B., Li, D., et al.: RSC Adv. **5**(107), 88287–88297 (2015)
Malek Khachatourian, A., Golestani-Fard, F., Sarpoolaky, H., Vogt, C., Toprak, M.S.: Ceramics International **41**(2, Part A):2006–14 (2015)
Melchionna, M., Fornasiero, P.: Mater. Today **17**(7), 349–357 (2014)
Mukherjee, D., Rao, B.G., Reddy, B.M.: Appl. Catal. B **197**, 105–115 (2016)
Mukherjee, D., Devaiah, D., Venkataswamy, P., Vinodkumar, T., Smirniotis, P.G., Reddy, B.M.: New J. Chem. **42**(17), 14149–14156 (2018)
Murray, C.B., Norris, D.J., Bawendi, M.G.: J. Am. Chem. Soc. **115**(19), 8706–8715 (1993)
Parashar, M., Shukla, V.K., Singh, R.: J. Mater. Sci. Mater. Electron. **31**(5), 3729–3749 (2020)
Peng, X., Wickham, J., Alivisatos, A.P.: J. Am. Chem. Soc. **120**(21), 5343–5344 (1998)
Peng, X., Manna, L., Yang, W., Wickham, J., Scher, E., Kadavanich, A., et al.: Nature **404**(6773), 59–61 (2000)
Pitale, S.S., Nagpure, I.M., Kumar, V., Ntwaeaborwa, O.M., Terblans, J.J., Swart, H.C.: Mater. Res. Bull. **46**(7), 987–994 (2011)
Priecel, P., Lopez-Sanchez, J.A.: ACS Sustain. Chem. & Eng. **7**(1), 3–21 (2019)
Rao, B.G., Mukherjee, D., Reddy, B.M.: Chapter 1—Novel approaches for preparation of nanoparticles. In: Ficai, D., Grumezescu, A.M., editors. Nanostructures Nov. Ther. Elsevier. pp. 1–36 (2017)
Reddy, B.M., Bharali, P., Saikia, P., Khan, A., Loridant, S., Muhler, M., et al.: J. Phys. Chem. C **111**(5), 1878–1881 (2007)
Reddy, B.M., Bharali, P., Thrimurthulu, G., Saikia, P., Katta, L., Park, S.-E.: Catal. Lett. **123**(3), 327–333 (2008)
Reddy, B.M., Thrimurthulu, G., Katta, L., Yamada, Y., Park, S.-E.: J. Phys. Chem. C **113**(36), 15882–15890 (2009)
Reddy, B.M., Vinodkumar, T., Naga Durgasri, D., Rangaswamy, A.: Proceedings of the national academy of sciences. India Sect. a: Phys. Sci. **87**(1), 155–161 (2017)
Ridderbos, J., Brauns, M., Shen, J., de Vries, F.K., Li, A., Bakkers, E.P.A.M., et al.: Adv. Mater. **30**(44), 1802257 (2018)
Rodriguez-Liviano, S., Becerro, A.I., Alcántara, D., Grazú, V., de la Fuente, J.M., Ocaña, M.: Inorg. Chem. **52**(2), 647–654 (2013)
Rozhin, P., Melchionna, M., Fornasiero, P., Marchesan, S.: Nanostructured ceria: biomolecular templates and (Bio)Applications. Nanomater. [Internet]. **11**(9):p. 2259 (2021)
Schubert, U.: Chemistry and Fundamentals of the Sol–Gel Process. Sol-Gel Handb. p. 1–28 (2015)
Senanayake, S.D., Stacchiola, D., Rodriguez, J.A.: Acc. Chem. Res. **46**(8), 1702–1711 (2013)
Singh, K.R.B., Nayak, V., Sarkar, T., Singh, R.P.: RSC Adv. **10**(45), 27194–27214 (2020)

Sudarsanam, P., Mallesham, B., Reddy, P.S., Großmann, D., Grünert, W., Reddy, B.M.: Appl. Catal. B **144**, 900–908 (2014)
Sui, R., Charpentier, P.: Chem. Rev. **112**(6), 3057–3082 (2012)
Tan, C., Cao, X., Wu, X.-J., He, Q., Yang, J., Zhang, X., et al.: Chem. Rev. **117**(9), 6225–6331 (2017)
Taniguchi, N.: Proceedings of the International Conference on Production Engineering, Tokyo: pp. 18–23 (1974)
Thompson, D.: Gold Bulletin **40**(4), 267–269 (2007)
Thrimurthulu, G., Rao, K.N., Devaiah, D., Reddy, B.M.: Res. Chem. Intermed. **38**(8), 1847–1855 (2012)
Tian, J., Tan, K.B., Liao, Y., Sun, D., Li, Q.: Chemosphere **292**, 133446 (2022)
Trovarelli, A.: Catal. Rev. **38**(4), 439–520 (1996)
Trovarelli, A., Llorca, J.: ACS Catal. **7**(7), 4716–4735 (2017)
Valden, M., Lai, X., Goodman, D.W.: Sci. **281**(5383), 1647–1650 (1998)
Venkataswamy, P., Rao, K.N., Jampaiah, D., Reddy, B.M.: Appl. Catal. B **162**, 122–132 (2015)
Vinodkumar, T., Naga Durgasri, D., Reddy, B.M.: Int. J. Adv. Eng. Sci. Appl. Math. **5**(4), 224–231 (2013)
Wei, W., Bai, F., Fan, H.: iSci. **11**:272–93 (2019)
Wu, G., Zhang, L., Cheng, B., Xie, T., Yuan, X.: J. Am. Chem. Soc. **126**(19), 5976–5977 (2004)
Xu, R., Wang, D., Zhang, J., Li, Y.: Chemistry—An Asian J. 1(6):888–93 (2006)
Yang, D., Wang, L., Sun, Y., Zhou, K.: J. Phys. Chem. C **114**(19), 8926–8932 (2010)
Yoshihara, J., Campbell, C.T.: J. Catal. **161**(2), 776–782 (1996)
Yu, H.-C., Zhuo, Q.-H., Shi, J.-T., Chu, K.-H.: Org. Electron. **110**, 106646 (2022a)
Yu, H., Wang, W., Liu, M., Zhao, T., Lin, R., Hou, M., et al.: Sci. Adv. **8**(30):eabq2356 (2022b)
Zhang, B., Sun, L.: Chem. Soc. Rev. **48**(7), 2216–2264 (2019)
Zhang, Q., Uchaker, E., Candelaria, S.L., Cao, G.: Chem. Soc. Rev. **42**(7), 3127–3171 (2013)
Zhang, Z., Liu, J., Gu, J., Su, L., Cheng, L.: Energy Environ. Sci. **7**(8), 2535–2558 (2014)
Zheng, Y.-R., Gao, M.-R., Gao, Q., Li, H.-H., Xu, J., Wu, Z.-Y., et al.: Small **11**(2), 182–188 (2015)
Zhou, K., Wang, X., Sun, X., Peng, Q., Li, Y.: J. Catal. **229**(1), 206–212 (2005)

Chapter 5
Metal–Organic Frameworks (MOFs) Materials: Synthesis and Application as Catalyst in Recent Organic Transformation

Ravindra Singh, Simant Kumar Srivastav, Roona Singh, Tapas Ghatak, and Prem Lama

5.1 Introduction

Metal–organic frameworks are crystalline porous substances containing metal nodes and organic linkers. The materials are prepared by the reaction of organic ligands with metal ions or metal clusters. The synthesis of 2D/3D structures of MOFs materials is constructed through acid–base interaction, hydrophobic interaction, electrostatic interaction, hydrogen bonding, and π-π stacking via reaction between metal nodes and organic linkers.

Last two decades, metal–organic frameworks(MOFs), have received great interest. They exhibited diverse applications in the field of, catalysis (Corma et al. 2010; Jiao et al. 2018; Li et al. 2018; Liu et al. 2014; Wu et al. 2016) removal of heavy

R. Singh (✉) · R. Singh
Department of Chemistry, Maharani Shri Jaya Government Post-Graduate College, Bharatpur, Rajasthan 321001, India
e-mail: iitk.ravindra@gmail.com

R. Singh
e-mail: singhroona1993@gmail.com

S. K. Srivastav
Department of Chemistry, University of Allahabad, Prayagraj, Uttar Pradesh, India
e-mail: simant.iitk@gmail.com

T. Ghatak
Department of Chemistry, School of Advanced Sciences, Vellore Institute of Technology, Vellore, India
e-mail: tapas.ghatak@vit.ac.in

P. Lama
Scientist and Assistant Professor-AcSIR, CSIR—Indian Institute of Petroleum, Dehradun, India
e-mail: prem.lama@iip.res.in

R. S. Ningthoujam and A. K. Tyagi (eds.), *Handbook of Materials Science, Volume 3*, Indian Institute of Metals Series, https://doi.org/10.1007/978-981-96-8229-4_5

metals from wastewater (Zhao et al. 2019), gas storage and separation (Wang & Zhao 2017), sensing (Hu et al. 2014), adsorption of various materials (Esrafili et al. 2018), biomedicine (Abánades Lázaro & Forgan 2019), energy storage applications electrocatalysis, and electrochemical energy conversion (Zhang et al. 2019), because of their unique features such as large surface areas, high porosity, a variable cavity for precise host–guest interactions, highly arranged, and variable metal nodes, and organic linkers that support modification of inner cavity size. Various MOFs show luminescent properties, which are used for detecting selective metal ions in trace amounts with the advanced spectrophotometric method by "turn on" and "turn off" mechanisms (Qu et al. 2013).

The luminescent property of MOFs is primarily governed by the charge transfer occurring within metal ions, often serving as luminescent centers. Specifically, the interaction between metal ions and organic ligands plays a crucial role in detecting various metal ions with exceptional selectivity through luminescent phenomena (Rath & Vittal 2020).

The properties, structure, and functions of the MOFs materials mainly depend on the metal ions and organic linker. The MOFs pores and shape can be changed with the different organic linkers and various metal ions in the synthesis of the materials. Moreover, MOFs can be modified through post-synthesis modification, to change the interaction between host and guest molecules. The properties, structure, and functionalities of MOF materials are primarily dictated by the specific combination of metal ions and organic linkers utilized. By employing different organic linkers and various metal ions during synthesis, the pores and shape of MOFs can be tailored accordingly. Furthermore, post-synthetic modifications offer a means to adjust the interaction between host and guest molecules to enhance the versatility of MOFs.

The advantage of MOFs, it has better porosity and higher surface area than carbon-based material zeolites which are used as adsorbent materials, in the heterogeneous catalysis process. MOFs materials as catalysts are very prominent materials to perform energy-efficient processes with stability and excellent selectivity under mild reaction conditions. Recently, MOF materials as catalysts have utilized and enhanced the activity in recent organic transformations due to their merits such as stability in the solution phase, unsaturated coordination sites for binding the substrate, size selectivity, high density of transition metals, and activation of oxidant by Lewis acid behavior, pore size, and surface area. The various catalytic applications are achieved by the modifications of the structural properties in MOF materials such as the dispersion of metal species on its periphery, porosity, and surface area. The flexibilities in the design of MOFs materials are the key factor in the field of heterogeneous catalysis.

Last few decades, the concentration of the greenhouse gas CO_2 has increased because of energy requirement dependence on fossil fuels. The CO_2 emissions are increasing continuously which is caused by anthropogenic and industrial activity. Climate change has been observed due to release the of CO_2 increasing exponentially. Thus, CO_2 emission is one of the major problems for the world. Last few years, metal–organic frameworks and MOF materials used as heterogeneous catalysts in the conversion of carbon dioxide to cyclic carbonates, heterocycles, oligomers, olefins, methanol, ethanol, and methane to reduce the greenhouse gas CO_2 concentration and overcome serious environmental problems. In the recent CO_2 conversion to above mentioned product various methods such as terminal alkynes carboxylation by carbon dioxide, photo-electrocatalysis, and the cycloaddition of epoxides, electro-catalysis, and photo-catalysis are used. The coordination geometries around the metal ions, oxidation states the flexibility of organic linkers, nature of counter ions and solvents mainly affected the catalytic efficiency of metal–organic frameworks (MOFs) materials. Generally, organic linkers which contain one or more N-donor or O-donor sites such as pyrazine and 4.4′-bipyridyl derivatives, polyamines, oxalic acid, sulfonates, phosphonates, cyano compounds, and crown ethers, are used to form a bridge with the metal ions in the synthesis of MOFs. The high stability in the solution phase of metal–organic frameworks (MOFs) materials supports for utility as catalysts. This important activity mainly arises due to inner MOF cavities instead of the outer surface of the MOF which provide a platform for inclusion of the guest molecule in the MOF. The modified MOF materials have been synthesized by the pre- and post-synthetic methods which have improved physical and chemical properties. Recently, various synthesis methods have been reported in the literature, including mechanochemical syntheses, electrochemical methods, and solvothermal methods with microwave-induced heating or conventional heating. These methods, synthesis conditions, and precursors are responsible for the modification of MOFs with high porosity that can facilitate mass transportation, morphology, and large surface area which generate the specific activity of MOFs. However, the stability of MOFs and MOF-modified materials as catalysts shows good efficiency in organic synthesis under harsh conditions. The high temperatures and acidic/basic environments are the problems in recyclability and high catalytic activity. To overcome the stability issue, many researchers have a great interest in the synthesis of MOF-based carbon-containing materials and metal/metal oxides which have good recyclability and higher tolerance capacity against difficult reaction conditions. Herein, we have presented and discussed the various MOFs and MOF-modified materials catalyst systems with their main merits in the recent organic transformation reactions.

5.2 Synthesis of MOFs-Based Nano-Material

5.2.1 Solvothermal Method

It is the most popular and valuable method for the synthesis of MOFs and MOF-based catalysts, so far. In this method, MOF materials and MOFs catalyst are formed in the solution of metal ions and appropriate organic linkers, via a self-assembly approach.

To date, the most widely adopted and esteemed method for synthesizing MOFs materials catalysts involves a solution-based approach. In this method, the catalyst is formed through the self-assembly of metal ions and suitable organic linkers. The reaction is performed in closed vessels at high pressures and temperatures more than the solvent's boiling point (Chen & Li 2016). Recently, Chen et al. synthesized a novel Zr-MOF, NNU-28 material which is showing photo-reduction activity of photo-reduction carbon dioxide under the visible light to generate formate (Chen et al. 2016). Initially, the carboxylate ligand was prepared through the Sonogashira coupling reaction. Later, orange color crystals of the catalyst NNU-28 were synthesized with the reaction between the carboxylate ligand and $ZrCl_4$ in the DMF solvent, and benzoic was used as a linker by the solvothermal method. Kitagawa et al. reported the preparation of MOFs material. They have used chemical modulators to form the desired shape and size of the MOFs material catalyst. Notably, the organic modulators are accountable for preparing MOFs nano-materials with unique shapes and sizes through very slow crystallization of the material (Fig. 5.1). The structural properties of MOFs are governed by the formation of coordination bonds between organic linkers (such as benzoic, lauric, and acetic acid) and metal ion precursors (Tsuruoka et al. 2009). Wang et al. synthesize Hf-based MOFs by adjusting the synthesis process of the metal ion precursor in an aqueous solvent. They employed acetic acid as an organic linker to regulate the coordination atmosphere surrounding the metal center (He et al. 2017).

Bharadwaj and co-workers have reported Zn-based MOF. They have synthesized a MOF with a reaction mixture of ligand (H_3L) and $Zn(NO_3)_2{\cdot}6H_2O$ in DMF, H_2O, and EtOH solvent mixture, The solution was heated at 90 °C under high pressure in a stainless-steel autoclave for 72 h. Finally, yellow block-shaped crystals were obtained by cooling of reaction mixture up to room temperature (Verma et al. 2017). (Scheme 5.1).

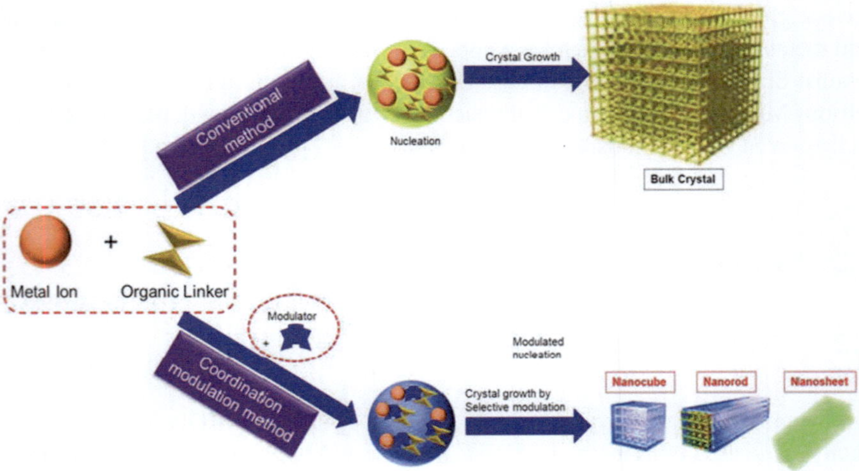

Fig. 5.1 Use of different chemical linkers for fabricating MOF material. Modified and adapted from Ref. (Tsuruoka et al. 2009)

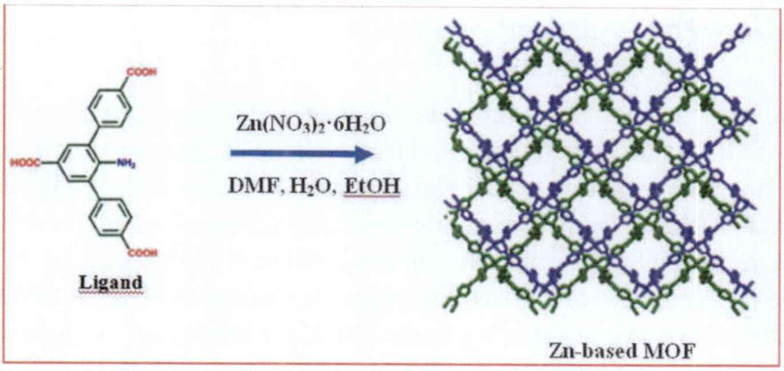

Scheme 5.1 Synthesis of Zn-based MOF using ligand (H_3L) and $Zn(NO_3)_2{\cdot}6H_2O$ in DMF, H_2O, and EtOH solvent mixture. Modified and adapted from Ref. (Verma et al. 2017)

Johnson and co-workers prepared polymers using linkers, methyl ether azide, and polyethylene glycol, in 2019. Notably, the linkers were first reacted with polyethylene glycol, and methyl ether azide. Interestingly, the length of polymer chains directly affected the morphology and size of the chemical linkers. It was observed that the polymer chain length is inversely proportional to a molecular weight of the chemical linker (Gu et al. 2019). PolyMOF-5 was formed, using PEG5k-L_4, zinc nitrate hexahydrate, and terephthalic acid in DMF solvent by solvothermal method. Moreover, the polyUiO-66 MOF was synthesized by using PEG10k-L_4.

Zhang et al. synthesized metal organic frameworks, by nucleation and growth method, in 2018. The MOF was prepared by gradually adding a solution of metal ions into the solution containing the ligand, as depicted in Fig. 5.2 (Wang et al. 2018). It was noticed that the preparation of MOFs is modified with the extent of super-saturation of the chemical reaction mixtures.

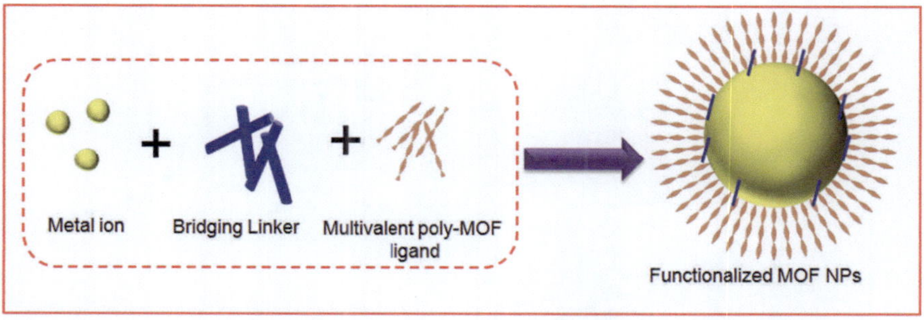

Fig. 5.2 Formation of functionalized MOF NPs with multivalent ligands as organic modulators

5.2.2 *Microwave-Assisted Synthesis*

Microwave-assisted method is much better than conventional hydro/solvothermal synthesis of MOFs due to their quick and facile synthesis. In this method, microwave radiations are for irradiation of the reaction mixture in an energy-efficient mode. This method is derived from the reaction between microwave radiation and the solution of the reactant. So, the selection of the solvent is a very crucial point in the microwave-assisted method due to the microwave absorption capacity of the specific solvent. The microwave-assisted synthesis is mainly controlled by the temperature inside the vessels, reaction time, irradiation power, and composition of the chemical reactive mixtures. Wang et al. synthesized Cu-based metal organic frameworks by microwave-assisted method. The Cu-MOF explores the catalytic activity for the reaction of terminal alkynes with carbon dioxide at mild conditions, carboxylation product was obtained. The reaction mixture was carried out with a sealed container in the microwave oven at 60 °C. Choi and coworkers found that the crystalline property and morphology of the MOF-5 were changing with variations in the solvent amount and substrate composition, power level, temperature, and irradiation time. They have investigated that the size and quality of the crystal varied after increasing the concentration of Zn^{2+} ion in the reaction mixture (Choi et al. 2008).

Tonigold et al. synthesized Co-MOF by microwave heating and solvothermal method in DMF solvent at 120°C (Fig. 5.3). Notably, the Co-MOF was prepared by microwave-assisted method. Due to its smaller size, informality, and crystalline properties, this method demonstrates enhanced efficiency in converting cyclohexene to tert-butyl-2-cyclohexenyl-1-peroxide in the presence of ^{t}BuOOH. This enhanced performance can be attributed to the lesser crystallite size of the Co-MOF (Tonigold et al. 2009).

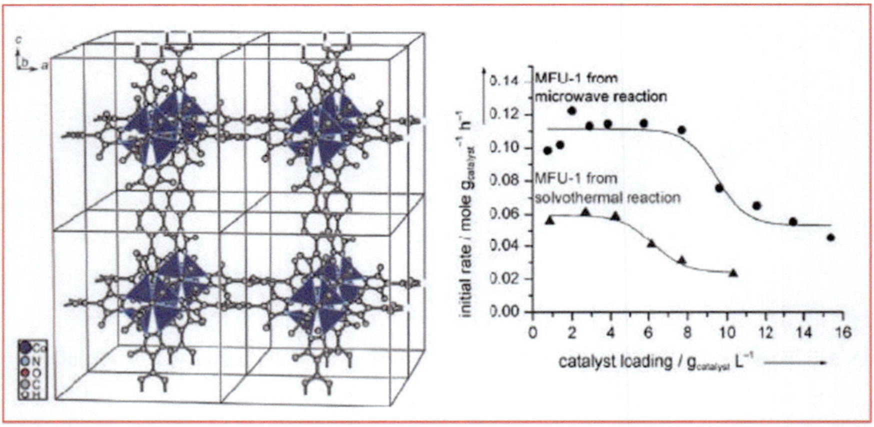

Fig. 5.3 (*left*) Crystal packing in [$Co_4O(bdpb)_3$] shows the large channels. (*right*). The plot exhibited that the microwave-assisted materials had superior oxidative activity in comparison to the solvothermal method. Modified and adapted from Ref. (Tonigold et al. 2009)

Recently, El-Shall et al. have prepared MIL-101 MOF with copper, palladium or Pd–Cu nanoparticles by microwave-assisted method to reduce the metal precursors, showing excellent oxidation ability toward carbon monoxide (El-Shall et al. 2009). Hong et al. synthesized iron oxide-based MOF with Microwave radiation, in 2006. The reaction was carried out using hydrochloride acid. for iron ion precursor $FeCl_3$. $6H_2O$ and $FeCl_2$ metal salt were used. Further, the reaction mixture was neutralization by sodium hydroxide (Hong et al. 2006).

5.2.3 *Mechanochemical Synthesis*

The mechanochemical synthesis method is environment-friendly. The reaction was performed in solvent-free conditions at room temperature (~25 °C), shown in Fig. 5.4. Mechanochemical synthesis is based on that the chemical bonds of the reactants are cleavaged by mechanical force which is responsible for the high yield of product in an organic transformation in shorter reaction times. Recently, the mechanochemical synthesis method of MOFs is mainly used compared to solvothermal methods in the solid state due to the advantage of solvent-free conditions. Last few decades, three different mechanochemical methods have been reported in the literature for the preparation of MOFs, **(i)** liquid-assisted grinding (LAG), this method for reducing the reaction time, a small portion of liquid phases is used to increase the mobility of the reactant. **(ii)** Neat grinding (NG), this method is based on a solvent-free approach. **(iii)** Ion and liquid-assisted grinding (ILAG), in this process salt additives are used along with a catalytic liquid to enhance the MOF construction.

Chen and coworkers have synthesized copper-based MOF-505 using a liquid-assisted mechanochemical method. It had been noticed that the porosity and crystallinity of copper-based MOF-505 were influenced by the amount of the assisted solvent used in the reaction, compared to the griding time which showed better CO_2 capture applications (Chen et al. 2018).

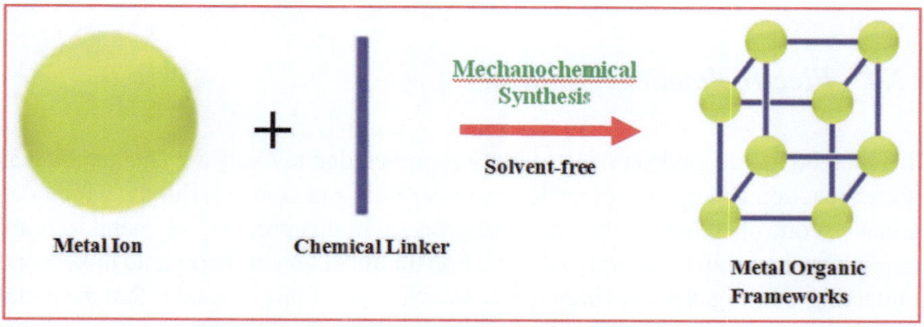

Fig. 5.4 Formation of MOF by mechanochemical method under solvent-free conditions (Singh et al. 2022)

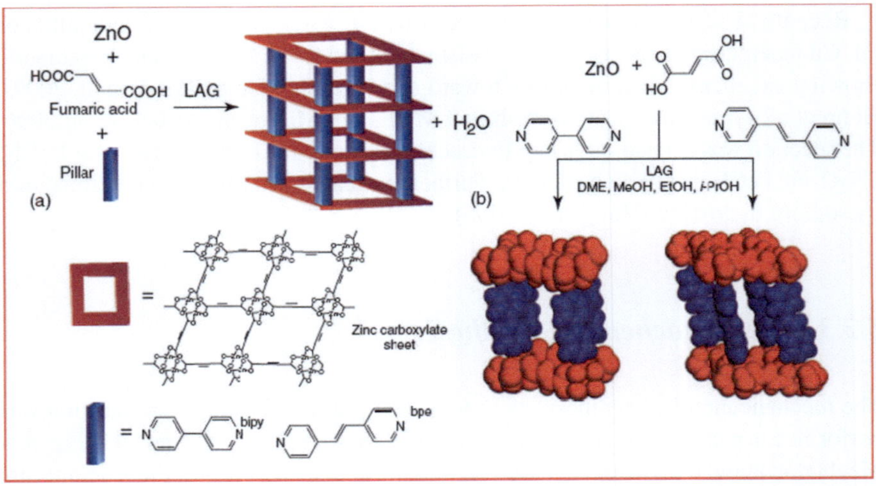

Fig. 5.5 Synthesis of MOFs by mechanochemical liquid-assisted grinding synthesis, metal oxide as starting material: **a** Scheme of the reaction and **b** zinc fumarate-based MOFs, prepared by liquid-assisted grinding which has low-porosity pillared. (Modified and adapted from Ref. (James et al. 2011))

Frišcic et al. have synthesized metal–organic frameworks by the liquid-assisted grinding (LAG) mechanochemical method through kneading of copper(II) chloride with ligand dace in DMSO or water which gives 1D coordination polymer (Friščič& Jones 2009). LAG can improve the mechanochemical reactivity and form molecular inclusion, which is used to prepare porous MOFs from the reaction mixture. The advantage of neat grinding is that the reactants do not react quickly. Recently, MOFs have been synthesized by the LAG mechanochemical method from the reaction mixture of ZnO, fumaric acid (H_2fum), and using 4,4'-dipyridyl or trans-1,2-(4,-pyridyl)ethylene as bridging ligands in the presence of DMF via one-pot reaction process (James et al. 2011) (Fig. 5.5).

5.2.4 *Electrochemical Synthesis*

This method has an environment-friendly approach due to many advantages such as shorter reaction times, ease of scale-up, under mild reaction conditions, which can be easily, controlled during the chemical process. In this process, the metal ions are supplied by the anodic solution to the reaction mixture containing organic linkers and conducting salt (Fig. 5.6). An important characteristic of this method is that the metal oxidation state can be adjusted either by applying a suitable current (potentiometry mode) or voltage (amperometry mode) to the electrode in the solution (Li et al. 2014).

Hartmann and co-workers had synthesized the HKUST-1 MOF from the solution of trimesic acid and methyl tributyl ammonium methylsulfate (MTBS) as an

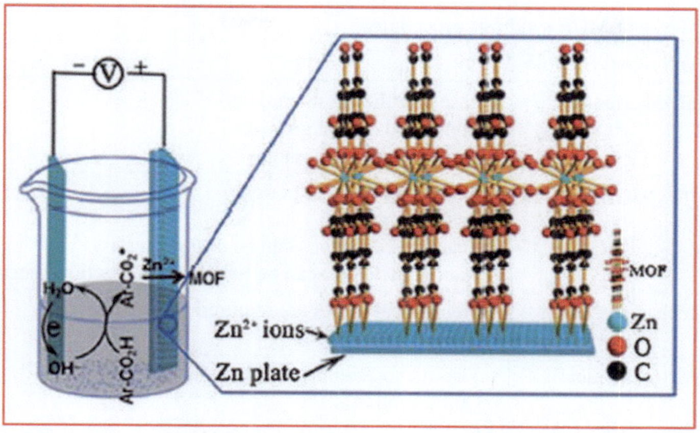

Fig. 5.6 Electrochemical synthesis of MOF (Reproduced and modified from Ref. (Li et al. 2014))

electrolyte in EtOH or EtOH/water (50:50 v/v) solvent mixtures, 15 V voltage, and 0.05 mA current was applied between two copper electrodes during 2 h in the reaction mixture (Li et al. 2014). It was noticed that HKUST-1 synthesized by using only ethanol solvent has a higher specific surface area than ethanol/water mixtures used in the reaction medium. Kulandainathan et al. investigated that the particle size of HKUST-1 MOF was highly influenced by the current density and the concentration of the supporting electrolyte concentration. The HKUST-1 MOF catalyzed the transformation of p-nitrophenol to p-aminophenol with $NaBH_4$ (Senthil Kumar et al. 2013). The electrosynthesis was performed in a methanol solution containing trimesic acid (organic linker) and tetrabutylammonium tetrafluoroborate as supporting electrolytes by applying 10–20 V voltage among two copper electrodes during 15 min in the reaction mixture. It was observed that crystallinity, surface area, and yields of MOFs depended on the reaction time, applied voltage, and supporting electrolyte concentration which were used during synthesis. Gascon et al. reported that the electrosynthesis method was affected by conditions like temperature, voltage–current density, solvent, and supporting electrolyte in the preparation of NH2-MIL-53(Al), HKUST-1, and ZIF-8 (Martinez Joaristi et al. 2012).

5.2.5 *Sonochemical Synthesis*

The sonochemical synthesis method is an environmentally friendly process that is carried out at room temperature. In this method, high-energy ultrasound waves are applied to a reaction mixture that contains metal ions and organic linkers. Jang et al. prepared an MOF by a sonochemical approach which has a smaller size than the MOF synthesized via the solvothermal approach (Jang et al. 2020). Kwangsun Yu and co-workers have synthesized MOF-525 and MOF-545 with the sonochemical route.

Table 5.1 Merits of MOF synthesis procedures

S. no	Synthesis procedure	Merit
1	Solvothermal	• Ease to utilize in the industry • Controlled crystal growth • Large temperature range
2	Microwave-assisted	• Less crystallization period • Energy capable • Reaction conditions can be changed easily • Nanoparticle size can be tuned with a reaction concentration
3	Mechanochemical	• Mechanical forces are required • Solvent-free method
4	Electrochemical	• The method can be used in industrial application • Quick synthesis
5	Sonochemical	• Homogenous morphology and crystal size are obtained • Pure phase utilizes

The MOF-525 was synthesized by the reaction of a solution of Zirconyl chloride octahydrate, and benzoic acid in the dimethylformamide. Then solution was treated with TCPP and sonication was applied for 5–60 min. The produced Zr-based MOFs have uniform size, and high purity, and are crystalline in nature. The synthesis time was reduced by using the sonochemical synthesis method. MOF-525 and MOF-545 demonstrated remarkable catalytic efficiency in both the adsorption of bisphenol-A and the hydrolysis of dimethyl-4-nitrophenyl phosphate. This efficacy can be attributed to their smaller particle size, higher concentration of defect sites, and/or larger Brunauer–Emmett–Teller surface areas (Yu et al. 2021). The merits of various synthetic procedures are shown in Table 5.1.

5.3 Application of MOF in Various Organic Transformation Reactions

5.3.1 Oxidation Reactions

The MOF-based materials have exhibited significant catalytic activity in oxidizing various substrates, including hydrocarbons, primary and secondary alcohols, and oxidative amine coupling reactions. The organic transformation is important because of resulting functional groups like carboxylic acids, epoxides or imines, ketones, and aldehydes, which can be used for further organic synthesis. The MOF-catalyzed oxidation reaction of primary alcohols to aldehydes showed high selectivity. MOF-derived materials like metal-based N-doped carbon were also synthesized. It has been exploited for the oxidation of hydrocarbons, amines, and epoxidation reactions.

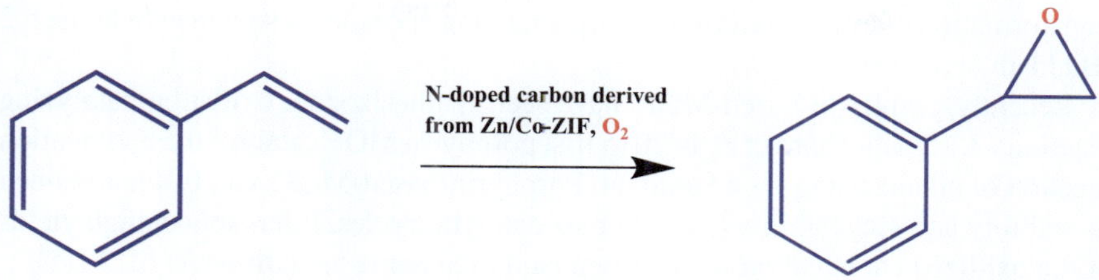

Scheme 5.2 Catalytic activity of Co-NP-based MOF for oxidation of alkenes to epoxide (Reproduced from Ref. (Yu et al., 2014))

(i) Epoxidation Reactions: Nitrogen-doped carbon MOF-derived material has been reported in the literature for epoxidation reactions of alkene. Yu et al. (2014), prepared ZIF-67-based Co-NPs, containing nitrogen-doped carbon which exhibited excellent catalytic activity designed for epoxidation of styrene with ~84% selectivity and ~ 91% yield, the reaction was performed under air in DMF solvent, as depicted in Scheme 5.2. In the synthesis of the ZIF, mesopores were produced during the annealing process of MOF. Furthermore, the high activity and selectivity of the material are closely associated with its nitrogen content and composition, which are influenced by temperature. The catalyst heated at 700ºC for 5 h in an argon atmosphere exhibited the highest activity, attributed to its elevated nitrogen content. This nitrogen content played a crucial role in the activation process of O_2 molecules (Yu et al. 2014).

Lin and coworkers have reported a ZnS catalyst which is supported by N-doped-C prepared from a Zn-based chiral MOF. It was formed by the heating of Zn-based chiral MOF at 700 ºC for 3 h under a nitrogen atmosphere. It showed catalytic activity for the organic transformation of cis-cyclooctene to epoxycyclo octane via a selective photocatalytic oxidation process. The product was obtained with 66% yield and 93% selectivity, which is achieved due to the synergistic effect of zinc sulfide with N-doped material, resulting from a zinc-based chiral MOF (Lin et al. 2018).

In recent years, Haddadi et al. reported organic–inorganic polyoxometalate-based MOFs for heterogeneous epoxidation of alkenes to react with H_2O_2 as an oxidant in CH_3CN solvent.

The PW-MOF, PMo-MOF, HSiW-MOF, and HPMo-MOF were synthesized using the hydrothermal method. The reaction was performed with an aqueous solution of polyoxometalates or polyoxometalate as precursors, copper chloride, and an organic moiety for linkage with various molar ratios. It was used for the conversion of olefins to the corresponding epoxide. The catalyst showed high recyclability and stability. The catalytic activity was maintained after reuse at least four times for epoxidation of cyclooctene, which showed high stability of the catalyst. The effect of temperature was also studied, and they demonstrated that temperature played a significant role in achieving the optimum products. The conversion of cyclooctene to the corresponding epoxide was improved with increasing the temperature, however, selectivity was decreased. It was found that 50 ºC temperature was most favorable for the transformation of cyclooctene to the corresponding epoxide, as shown in Scheme 5.3 (Haddadi et al. 2015).

Recently, porphyrin-based-MOFs have been synthesized for oxidation-catalyzing reactions. L.-S. Bouchard et al. in 2015 first porphyrin MOF catalyze the epoxidation reaction of alkene using O_2 an oxidant. Porphyrin-based-MOFs catalyst maintained its stability and recyclability even after six catalytic cycles. It has shown high yields of the oxidized chemical product, as depicted in Scheme 5.4 (Brown et al. 2015).

(ii) Oxidation of Amine: Wei et al. prepared a novel MOF, that contained anthracene-based organic linkers. The indium-containing metal—organic framework was synthesized using a solvothermal procedure. It showed catalytic activity for the photooxidation of benzylamines and sulfides, using molecular oxygen or hydrogen peroxide as an oxidant. Imines or sulfoxides were produced by extremely selective photocatalysis of the amine and sulfides. The photocatalytic reactions were carried out via the generation of superoxide radicals by the one-electron transfer from MOF to oxidants. This MOF has exhibited excellent potential for the synthesis of valuable organic chemicals. Imines are the main intermediate species for the preparation of pharmaceuticals, dyes, and chemicals that are used in the agriculture field. H. Wei et al. synthesized imines from amines via oxidative coupling reaction by the indium-based MOF, benzylamine as a model substrate, and molecular oxygen as an oxidant (Wei et al., 2019). The photocatalytic activity was performed toward benzylamine under visible light. The high photocatalytic activity was observed by using DMSO solvent. The various benzylamine derivatives were used to produce imines through a photocatalytic oxidative coupling reaction (Scheme 5.5).

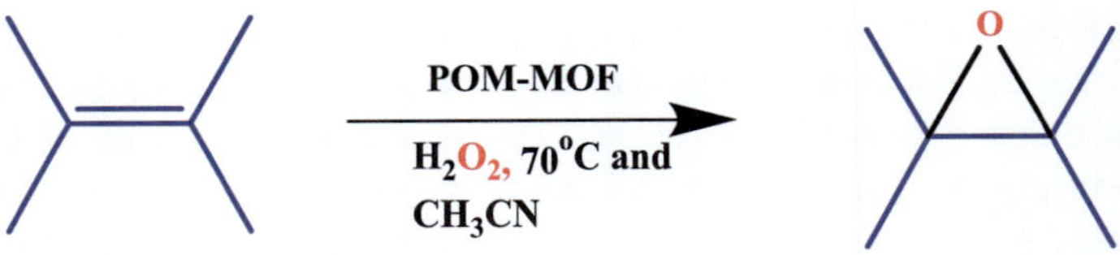

Scheme 5.3 Catalytic activity of POM-MOF for conversion of alkenes to epoxide (Reproduced from Ref. (Haddadi et al. 2015))

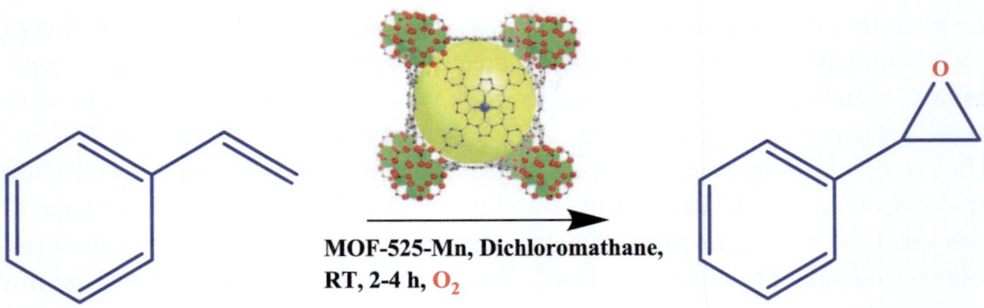

Scheme 5.4 Conversion of styrene to epoxide using manganese porphyrin-based MOF and O_2 as an oxidant. (Reproduced and modified from Ref. (Brown et al. 2015))

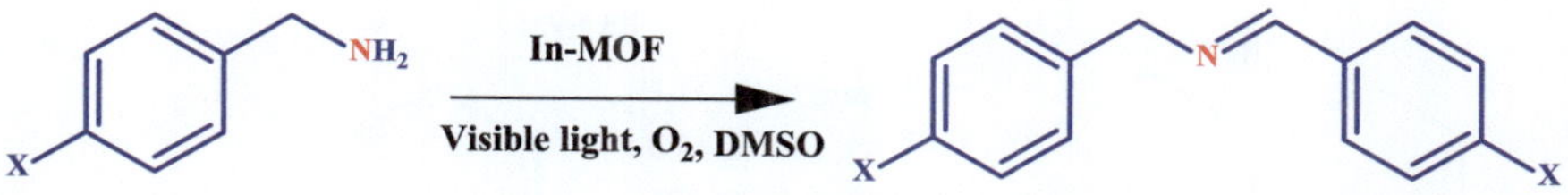

Scheme 5.5 Photocatalytic oxidative coupling reaction of various amines (X = -H, $-CH_3$, $-OCH_3$, -F, -Cl) (Reproduced and modified from Ref. (Wei et al. 2019))

It was noticed that the production of imines from benzylamines derivatives was affected by the electron-releasing property and position of the substituents. The conversion rate was higher in the case of electron-releasing groups ($-CH_3$/$-OCH_3$) than the electron-attracting substituents (-F/-Cl) present in the phenyl group of benzylamine.

(iii) Oxygenation of Sulfides: Wei et al. synthesized sulfoxides from sulfides by photocatalytic oxidation which are used in biological activities as a pesticide and pharmaceutical agent. In this oxidation, thioanisole was used as model a substrate and In-based MOF as a catalyst. Generally, H_2O_2 and O_2 were utilized as oxidants for photocatalytic oxidation of sulfides. More than 99% conversion of thioanisole into sulfoxides was obtained in 4 h when hydrogen peroxide was used as an oxidant under N_2 conditions, as shown in Scheme 5.6. The reaction was performed in $CHCl_3$/CH_3OH solvent mixture. The 55% conversion was achieved if the reaction was carried out under O_2 atmosphere. However, conversion was very low when the reaction was carried out under aerobic conditions. It was noticed that the electronic property of groups on sulfides played a crucial role in photocatalytic oxidation reactions. The higher conversion rate and selectivity were achieved in the case of electron-releasing groups compared to electron-donating groups (Wei et al. 2019).

(iv) Oxygenation of Alcohol: Lee and et al. synthesized carbon and copper-containing MOF by the pyrolysis of HKUST-1 at 800ºC in the argon atmosphere for 6 h, which has been shown excellent oxidation of benzyl alcohols and its derivatives to corresponding benzaldehyde. This catalyst showed good conversion of primary alcohol, however, less conversion of secondary alcohol (Kim et al. 2016). Recently, Yao prepared Fe_3O_4@C catalyst by heating MIL-88Bat 600 ºC under an inert condition. The catalyst was very efficient for the production of benzaldehyde with excellent conversion (~99%) and selectivity (~91%), shown in Scheme 5.7. The catalyst exhibited good catalytic activity and selectivity of electron-donating ($-CH_3O$, $-NH_2$,

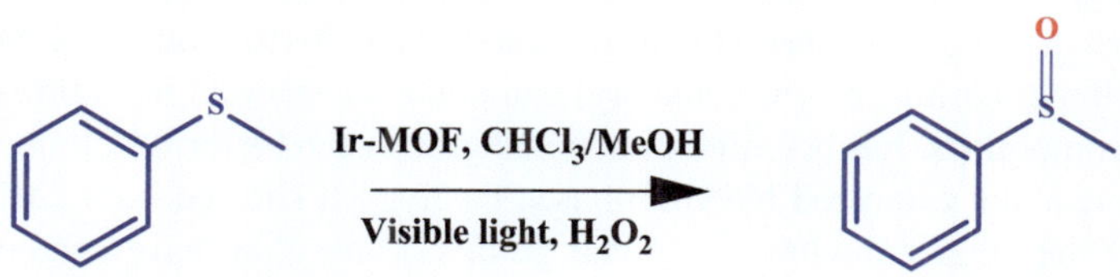

Scheme 6 Photocatalytic oxidation; conversion of sulfides to sulfoxides (Reproduced and modified from Ref. (Wei et al. 2019))

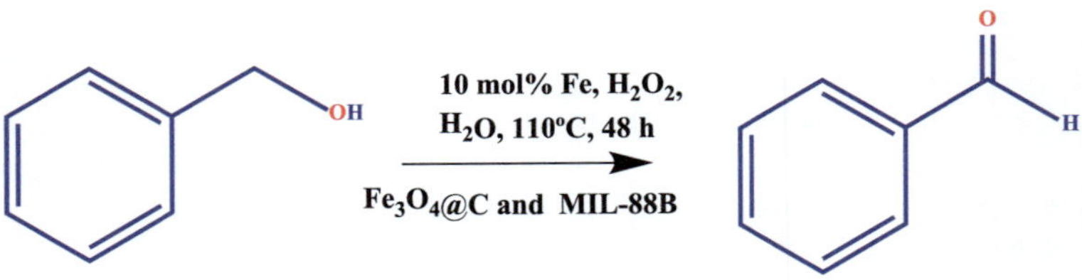

Scheme 5.7 Conversion of benzyl alcohol to benzaldehyde (Reproduced and modified from Ref. (Yao et al. 2016))

-CH_3) as well as electron-withdrawing (-N_2O, -F, -Cl, -Br and -$OCOCH_3$) substituent present at benzyl alcohols. The catalyst also exhibited the conversion of secondary alcohols and other derivatives to corresponding carbonyl compounds with very good yield and selectivity (~ 95–99%) (Yao et al. 2016).

Zhong et al. synthesized a Co@C-N catalyst by the heating of ZOF-67at 800 °C in the argon atmosphere for 8 h which has been shown oxidation of alcohols to ester. The catalyst was highly active for the production of ester with excellent conversion. The catalyst showed high catalytic activity toward oxidative esterification of aliphatic alcohol, benzyl alcohol, and heterocyclic alcohol derivatives. The high selectivity and good conversion were achieved in the presence of air under room temperature and pressure. The reaction time was varied with the derivatives of alcohol (Zhong et al. 2015). Jiang and co-workers have prepared a Co-CoO@N-doped carbon catalyst which was used for conversion of alcohol to produce esters. The catalyst showed oxidation of alcohol to ester in a step reaction. The catalyst showed excellent recyclability for esterification of alcohol. High selectivity and conversion were achieved when the reaction was performed under 0.1 kPa O_2 and potassium carbonate was used as a base, at 80°C for 24 h. The proposed mechanism of the esterification reaction of alcohol is that first benzyl alcohol is oxidized to corresponding aldehydes, and then a reaction occurs with aliphatic alcohols and the aldehydes (Zhou et al. 2015).

5.3.2 Reduction Reactions

Various MOF-based catalysts have been reported for the hydrogenation of numerous functional groups present attached to the substrate. Many intermediates or fine chemicals have been prepared by the reduction of ketones, aldehydes, and esters. Moreover, the reduction of olefins, alkynes, and aryl compounds converted into corresponding aliphatic compounds. Further, nitro-functional groups can be formed amines via the reduction reaction catalyzed by MOF-based catalyst. It showed high chemoselectivity over the other reducible functional groups. Long et al. have synthesized an N-doped Co@CN catalyst which was used to catalyze the hydrogenation reaction formed by primary amines and secondary imines from nitrile compounds. In this reaction, iso-propanol was used as the hydrogen source. The catalyst exhibited the

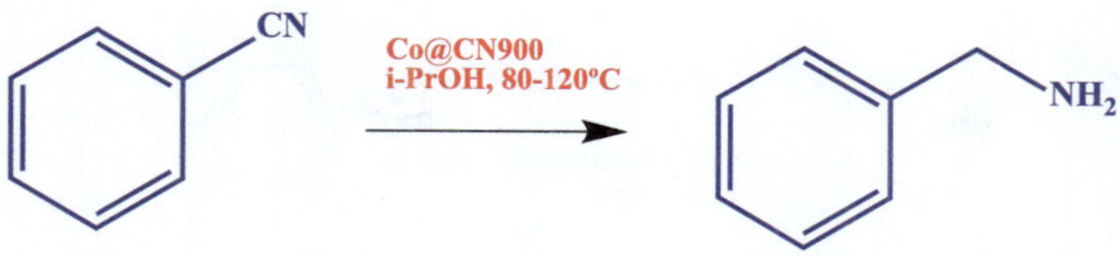

Scheme 5.8 Hydrogenation reaction of benzonitrile (Reproduced and modified from Ref. (Long et al. 2017))

high selectivity of the conversion reaction of benzonitrile to benzyl amines. It also shows good selectivity (~90%) of reduction of aliphatic nitrile to primary amine, as depicted in Scheme 5.8 (Long et al. 2017).

Recently, Li and co-workers have prepared a bimetallic catalyst. Initially, in a helium environment, the MOF precursor was heated at 200 °C for 2 h. Subsequently, after solvent removal, the reaction mixture underwent carbonization at 500 °C for 8 h under inert conditions. The resultant, N-doped carbon material was obtained which was charged with H_2 at 200 °C for 8 h before use as a catalyst. The bimetallic catalysts were catalyzed by hydrogen transfer reaction of conversion of benzonitrile to N-benzylidenebenzylamine. The reduction of various benzonitrile derivatives also catalyzes to convert corresponding N-benzylidenebenzylamines at ~ 100°C for 36–60 h (Long et al. 2016). Li and colleagues synthesized a Co–Ni@NC MOF catalyst for the reduction of phenol compounds. This catalyst facilitated the conversion of phenol to cyclohexanol within 12 h, achieving a remarkable selectivity of 99.9%. Under mild conditions, this catalyst achieved high selectivity. The reaction was completed in one step. The phenol was directly converted to cyclohexanol without the production of reactive species. Moreover, the catalyst also performed the hydrogenation reaction of various phenols containing alkyl groups and dihydroxybenzenes to cyclic alcohol derivatives (Li et al. 2017a, b).

Liao et al. developed a metal catalyst containing MOF which showed the catalytic activity for the synthesis of amines from the nitro compounds by reduction reaction. The reduction of 4-nitrophenol to 4-aminophenol was analyzed with UV–Vis spectroscopy because of the unique absorption peaks of 4-nitrophenol as well as 4-aminophenol. The reduction of 4-nitrophenol from industrial wastewater is very important due to its toxic behavior (Liao et al. 2016). Further, aniline and its derivatives are utilized for the synthesis of pigments and polyurethane polymers.

5.3.3 Metal-Catalyzed Cross-Coupling Reactions

The organic transformation was achieved through the metal-catalyzed cross-coupling method to the generation of carbon–nitrogen and carbon–carbon bonds. These reactions are very important due to the synthesis of versatile organic compounds that are utilized for the production of natural products and pharmaceuticals. The preparation

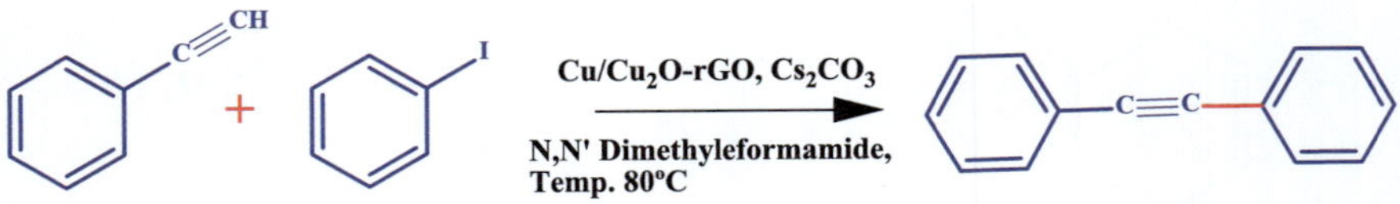

Scheme 5.9 Reaction catalyzed by Cu/Cu_2O-rGO MOF (Reproduced and modified from Ref. (Sun et al. 2018))

of Cu and Pd-based MOF-derived catalysts is very desirable to synthesize an efficient catalyst.

Recently, Wang and co-workers have been developed Pd-NPs MOF containing nitrogen-doped carbon material. It was employed for Suzuki–Miyaura coupling reactions, where the reaction between phenyl bromide and phenylboronic acid was conducted at room temperature in a water–ethanol solvent mixture. The reaction utilized 0.2 mol% Pd and potassium carbonate (K_2CO_3) as a base. The resultant, corresponding biphenyl was obtained after one hour. It has been observed that coupling reactions are also obtained between various substrates containing electronic-withdrawing and electronic-donating substituents with good yields (90–99%). However, the catalyzing the coupling reaction of an aryl chloride formed the biphenyl, obtaining a lower yield (~14%) (Zhang et al. 2015).

Dong et al. reported a MOF-5-NPC-Pd which was synthesized by the carbonization of MOFunder N_2atmosphere at 900 °C then Pd impregnated and reduced. MOF-5-NPC-Pd showed high catalytic activity in the Suzuki–Miyaura reaction between the coupling of phenyl bromide and phenylboronic acid forming biphenyl with a high yield (98.6%). The reaction was performed, using 0.2 mol% Pd, K_2CO_3 (base) at ~25 °C in the H_2O/EtOH mixture. The high catalytic activity may be exhibited to of high surface area (Dong et al. 2016). Yu et al. (2014) Zheng and co-workers synthesized Cu-based MOF which was fabricated with grapheme by solvothermal reduction method. In this process, MOFs-Cu_2O were decorated on the surface of graphene. The reaction between phenylacetylene and iodobenzene was catalyzed by MOFs-Cu_2O-rGO, resulting in 1,2-diphenylethyne with a high yield (~91%), as shown in Scheme 5.9. The excellent catalytic activity was achieved owing to the synergistic effect of copper and Cu_2O on the surface of graphene. In this process, graphene might accelerate the electron transfer between copper and copper(I)oxide for the reversible oxidation–reduction reaction (Sun et al. 2018).

5.3.4 Acid and Base-Catalyzed Reactions

MOF-derived materials catalyzed various acid–base reactions like the Strecker reaction, Knoevenagel condensation reaction, and acetalization reaction. Recently, MOFs and MOF-based materials have been used for acid-catalyzed biomass conversion reactions. As per C–C bond formation is concerned the Knoevenagel reaction is very important due to its utility in drug formation. In this reactionα-, β-unsaturated compounds were synthesized with the reaction between active methylene compounds and aldehydes or ketones.α-Aminonitriles compounds also have great importance due to their application toward the preparation of heterocycles, amino acids, and natural products. α-Aminonitriles compounds were synthesized by Strecker reaction (Enders & Shilvock 2000).

A tungstate zirconia catalyst, WOX/ZxO_2, was synthesized for the acid-catalyzed acetalization reaction, converting benzaldehyde to acetals. The tungstate-Zr-based MOF WOX@UiO-66 was prepared through a dual-solvent approach. The active form of the catalyst was produced by calcination at 650ºC for 6 h in air. The catalyst exhibited high activity attributed to its high surface area and the presence of the tetragonal phase of zirconium oxide. The tetragonal phase was produced in the process of derivatization of the MOF under optimized reaction conditions (Wang et al. 2016). Generally, nitrogen-containing material provides the Lewis basicity for base-catalyzed reactions. Li et al. prepared an N-doped material through the pyrolysis method of the Al-based MOF-253 at 800ºC for 4 h under argon conditions, with 2,2'-bipyridyl organic linkers. The material showed strong Lewis basicity properties due to the presence of pyrrolic and pyridine moiety. This material catalyzed the Knoevenagel reaction, involving the reaction of benzaldehyde and malononitrile in toluene solvent at 90ºC. Additionally, benzaldehydes containing various groups such as p-NO2, p-Cl, p-Br, p-OMe, o-OMe, and p-OH, underwent reaction with malononitrile, yielding the corresponding benzylidene malononitrile derivatives with very high yields (~99%) (Li et al. 2017a, b).

In 2018, Saha et al. have been synthesized CuO-NPs containing MOF by calcination method at 650ºC temperature. This MOF catalyzed the Strecker reaction under solvent-free conditions, yielding α-Aminonitriles with good yields. The reaction was carried out in two steps, in the first step the condensation reaction occurred between carbonyl and amine compounds via activation and the imine intermediate. In the next step, the nucleophilic addition reaction completed by the cyanide compound attacks on imine intermediate and produces α-Aminonitriles product (Scheme 5.10) (Singha et al. 2018).

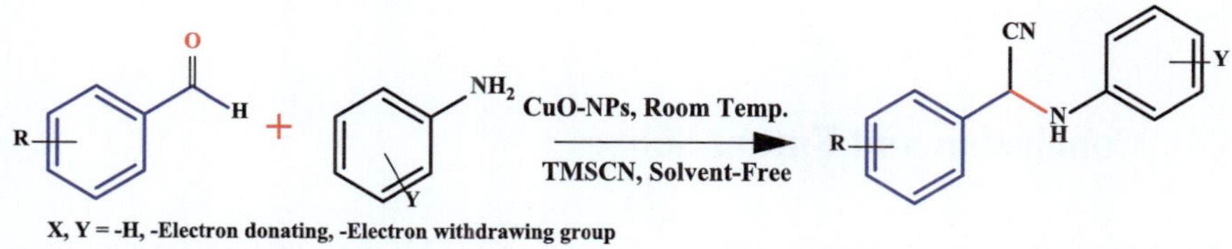

Scheme 5.10 Synthesis of α-aminonitriles (Strecker reaction) (Reproduced and modified from Ref. (Singha et al. 2018))

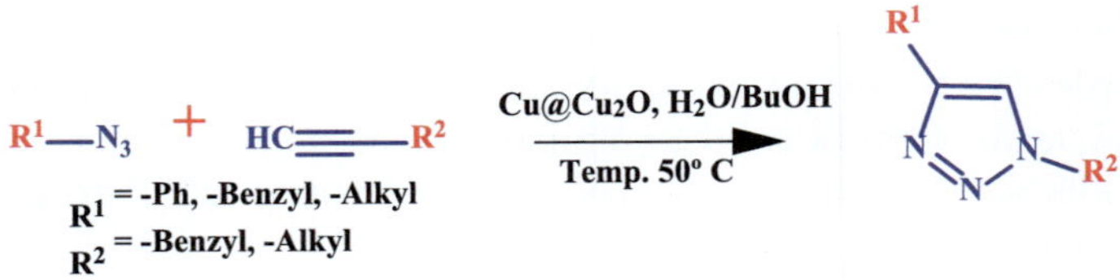

Scheme 5.11 Synthesis of 1,2,3-triazoles compound through Huisgen cycloaddition reaction catalyzed by Cu@Cu_2O-based MOF(Reproduced and modified from Ref. (Kim et al., 2018))

5.3.5 Miscellaneous Reactions

The organosilicon compounds are important precursors for the synthesis of various organic compounds which were synthesized by the hydrosilylation method. Recently, Bennedsen et al. synthesized a Co–Ni-containing MOF which catalyzed the hydrosilylation of ketones. The catalyst was prepared with ZIF-67 and Ni-precursor by carbonization method at 800°C for 3 h in the Ar- atmosphere. The conversion of cyclohexanone to hydrosilylation product with the reaction of Me_2PhSiH was catalyzed by Co–Ni containing MOF. The high yield of the product was observed. The catalytic activity was decreased after seven recycling experiments. However, less conversion of sterically hindered reactants like ^{t}Bu-substituted cyclohexanone hydrosilylation product was observed (Bennedsen et al. 2018).

Recently, Kim et al. synthesized Cu@Cu_2O containing MOF for the catalyst of the Huisgen cycloaddition reaction. The Cu@Cu_2O-based MOF catalyzed the cycloaddition reaction between azides and alkynes to synthesize 1,2,3-triazoles. The triazole compounds are precursors for the production of pharmaceutical compounds and agrochemicals. This catalyst also facilitated the reaction between benzyl azide and phenylacetylene under mild reaction conditions, the 1,4-disubstituted 1,2,3-triazole compound was formed with 100% yield. A notable feature of the catalyst is its ability to convert substrates containing both electron-donating and electron-withdrawing groups into corresponding triazole compounds, as depicted in Scheme 5.11 (Kim et al. 2018).

5.4 Conclusion and Future Scopes

The advancements in metal–organic frameworks (MOFs) nanomaterials have presented significant challenges as catalysts for various organic transformations over the past two decades. Great attention has been paid by the scientific community to developing MOF-efficient catalysts. The efficiency of the catalyst is mainly based on the surface functional groups, different size cavities with high surface area, and various functional groups that govern the good absorption efficiency. The structural properties and functions of the MOFs materials are mainly affected by metal ions and organic linkers. The MOF's pores and shape can vary with the use of different organic linkers and various metal ions. The various functional groups containing MOF provide the platform for host–guest interaction for organic transformation reactions. The structural modification of the MOFs material which acts as catalysts, can be tailored for the desired chemical transformation like acid- and base-catalysis, Huisgen cycloaddition reaction, coupling reactions, oxidations and reductions of various substrate, hydrosilylation reaction, Strecker reaction, etc.

This chapter, an overview of the synthesis and application of metal–organic frameworks and MOF-based noble material in the field of organic transformation, especially oxidation reaction, reduction, hydrogenation, transition metal-catalyzed cross-coupling reaction, Huisgen cycloaddition reaction and acid–base catalyzed reactions, hydrosilylation reaction, and Strecker reaction, is presented and discussed. Generally, most of the chemicals are performed in harmful solvent like CH_2Cl_3, $CHCl_3$, CCl_4, Benzene, and DMF. The scientific approach toward the use of environmentally harmless solvents has been made. An interest also has been made in applying the atom-economically approach to reduce the by-products. This chapter gives the futuristic approach to preparing efficient MOF-based catalysts for various organic transformations for the synthesis of valuable fine chemicals. These chemicals are used in various fields such as agrochemicals, pharmaceuticals, and textile industries.

References

Abánades Lázaro, I., Forgan, R.S.: Application of zirconium MOFs in drug delivery and biomedicine. Coord. Chem. Rev. **380**, 230–259 (2019). https://doi.org/10.1016/J.CCR.2018.09.009

Bennedsen, N.R., Kramer, S., Mielby, J.J., Kegnæs, S.: Cobalt–nickel alloy catalysts for hydrosilylation of ketones synthesized by utilizing metal–organic framework as template. Catal. Sci. Technol. **8**(9), 2434–2440 (2018). https://doi.org/10.1039/C8CY00150B

Brown, J.W., Nguyen, Q.T., Otto, T., Jarenwattananon, N.N., Glöggler, S., Bouchard, L.S.: Epoxidation of alkenes with molecular oxygen catalyzed by a manganese porphyrin-based metal–organic framework. Catal. Commun. **59**, 50–54 (2015). https://doi.org/10.1016/J.CATCOM.2014.09.040

Chen, D., Xing, H., Wang, C., Su, Z.: Highly efficient visible-light-driven CO2 reduction to formate by a new anthracene-based zirconium MOF via dual catalytic routes. J. Mater. Chem. A **4**(7), 2657–2662 (2016). https://doi.org/10.1039/C6TA00429F

Chen, J., Li, Y.: The road to MOF-Related functional materials and beyond: desire, design, decoration, and development. Chem. Rec. **16**(3), 1456–1476 (2016). https://doi.org/10.1002/TCR.201500304

Chen, Y., Wu, H., Liu, Z., Sun, X., Xia, Q., Li, Z.: Liquid-Assisted mechanochemical synthesis of copper based MOF-505 for the separation of CO2 over CH4 or N2. Ind. Eng. Chem. Res. **57**(2), 703–709 (2018). https://doi.org/10.1021/ACS.IECR.7B03712/SUPPL_FILE/IE7B03712_SI_001.PDF

Choi, J.S., Son, W.J., Kim, J., Ahn, W.S.: Metal–organic framework MOF-5 prepared by microwave heating: Factors to be considered. Microporous Mesoporous Mater. **116**(1–3), 727–731 (2008) https://doi.org/10.1016/J.MICROMESO.2008.04.033

Corma, A., García, H., Xamena, L.I., F. X.: Engineering metal organic frameworks for heterogeneous catalysis. Chem. Rev. **110**(8), 4606–4655 (2010). https://doi.org/10.1021/CR9003924/ASSET/CR9003924.FP.PNG_V03

Dong, W., Zhang, L., Wang, C., Feng, C., Shang, N., Gao, S., Wang, C.: Palladium nanoparticles embedded in metal–organic framework derived porous carbon: synthesis and application for efficient Suzuki-Miyaura coupling reactions. RSC Adv. **6**(43), 37118–37123 (2016). https://doi.org/10.1039/C6RA00378H

El-Shall, M.S., Abdelsayed, V., Khder, A.E.R.S., Hassan, H.M.A., El-Kaderi, H.M., Reich, T.E.: Metallic and bimetallic nanocatalysts incorporated into highly porous coordination polymer MIL-101. J. Mater. Chem. **19**(41), 7625–7631 (2009). https://doi.org/10.1039/B912012B

Enders, D., Shilvock, J.P.: Some recent applications of α-amino nitrile chemistry. Chem. Soc. Rev. **29**(5), 359–373 (2000). https://doi.org/10.1039/A908290E

Esrafili, L., Safarifard, V., Tahmasebi, E., Esrafili, M.D., Morsali, A.: Functional group effect of isoreticular metal–organic frameworks on heavy metal ion adsorption. New J. Chem. **42**(11), 8864–8873 (2018). https://doi.org/10.1039/C8NJ01150H

Friščič, T., Jones, W.: Recent advances in understanding the mechanism of cocrystal formation via grinding. Cryst. Growth des. **9**(3), 1621–1637 (2009). https://doi.org/10.1021/CG800764N/ASSET/IMAGES/MEDIUM/CG-2008-00764N_0001.GIF

Gu, Y., Huang, M., Zhang, W., Pearson, M.A., Johnson, J.A.: PolyMOF nanoparticles: dual roles of a multivalent polyMOF ligand in size control and surface functionalization. Angew. Chem. Int. Ed. **58**(46), 16676–16681 (2019). https://doi.org/10.1002/ANIE.201910542

Haddadi, H., Hafshejani, S.M., Farsani, M.R., Babahydari, A.K.: Heterogeneous epoxidation of alkenes with H2O2 catalyzed by a recyclable organic–inorganic polyoxometalate-based framework catalyst. New J. Chem. **39**(12), 9879–9885 (2015). https://doi.org/10.1039/C5NJ01661D

He, T., Xu, X., Ni, B., Wang, H., Long, Y., Hu, W., Wang, X.: Fast and scalable synthesis of uniform zirconium-, hafnium-based metal–organic framework nanocrystals. Nanoscale **9**(48), 19209–19215 (2017). https://doi.org/10.1039/C7NR06274E

Hong, R.Y., Pan, T.T., Li, H.Z.: Microwave synthesis of magnetic Fe3O4 nanoparticles used as a precursor of nanocomposites and ferrofluids. J. Magn. Magn. Mater. **303**(1), 60–68 (2006). https://doi.org/10.1016/J.JMMM.2005.10.230

Hu, Z., Deibert, B.J., Li, J.: Luminescent metal–organic frameworks for chemical sensing and explosive detection. Chem. Soc. Rev. **43**(16), 5815–5840 (2014). https://doi.org/10.1039/C4CS00010B

James, S.L., Adams, C.J., Bolm, C., Braga, D., Collier, P., Frišcic, T., Grepioni, F., Harris, K.D.M., Hyett, G., Jones, W., Krebs, A., Mack, J., Maini, L., Orpen, A.G., Parkin, I.P., Shearouse, W.C., Steed, J.W., Waddell, D.C.: Mechanochemistry: opportunities for new and cleaner synthesis. Chem. Soc. Rev. **41**(1), 413–447 (2011). https://doi.org/10.1039/C1CS15171A

Jang, M.S., Yu, K., Lee, J., Ahn, W.S.: Sonochemical synthesis of rho-ZMOF catalyst for an enhanced CO2 cycloaddition reaction. Mater. Lett. **277**, 128387 (2020). https://doi.org/10.1016/J.MATLET.2020.128387

Jiao, L., Wang, Y., Jiang, H.L., Xu, Q.: Metal-Organic frameworks as platforms for catalytic applications. Adv. Mater. **30**(37), 1703663 (2018). https://doi.org/10.1002/ADMA.201703663

Kim, A., Muthuchamy, N., Yoon, C., Joo, S. H., Park, K.H.: MOF-Derived Cu@Cu2O Nanocatalyst for Oxygen Reduction Reaction and Cycloaddition Reaction. Nanomaterials., **8**(3), 138. (2018). https://doi.org/10.3390/NANO8030138

Kim, B.R., Oh, J.S., Kim, J., Lee, C.Y.: Robust Aerobic alcohol oxidation catalyst derived from metal-organic frameworks. Catal. Lett. **146**(4), 734–743 (2016). https://doi.org/10.1007/S10562-016-1700-2/METRICS

Li, A., Shen, K., Chen, J., Li, Z., Li, Y.: Highly selective hydrogenation of phenol to cyclohexanol over MOF-derived non-noble Co-Ni@NC catalysts. Chem. Eng. Sci. **166**, 66–76 (2017a). https://doi.org/10.1016/J.CES.2017.03.027

Li, G., Zhao, S., Zhang, Y., Tang, Z.: Metal-Organic frameworks encapsulating active nanoparticles as emerging composites for catalysis: recent progress and perspectives. Adv. Mater. **30**(51), 1800702 (2018). https://doi.org/10.1002/ADMA.201800702

Li, W.J., Lü, J., Gao, S.Y., Li, Q.H., Cao, R.: Electrochemical preparation of metal–organic framework films for fast detection of nitro explosives. J. Mater. Chem. A **2**(45), 19473–19478 (2014). https://doi.org/10.1039/C4TA04203D

Li, X., Zhang, B., Fang, Y., Sun, W., Qi, Z., Pei, Y., Qi, S., Yuan, P., Luan, X., Goh, T. W., Huang, W.: Metal–Organic-Framework-Derived carbons: applications as solid-base catalyst and support for Pd nanoparticles in tandem catalysis. Chem. – Eur. J., **23**(18), 4266–4270 (2017). https://doi.org/10.1002/CHEM.201605852

Liao, Y.T., Chen, J.E., Isida, Y., Yonezawa, T., Chang, W.C., Alshehri, S.M., Yamauchi, Y., Wu, K.C.W.: De novo synthesis of gold-nanoparticle-embedded, nitrogen-doped nanoporous carbon nanoparticles (Au@NC) with enhanced reduction ability. Chem. Cat. Chem **8**(3), 502–509 (2016). https://doi.org/10.1002/CCTC.201501020

Lin, X., Liu, B., Huang, H., Shi, C., Liu, Y., Kang, Z.: One-step synthesis of ZnS-N/C nanocomposites derived from Zn-based chiral metal–organic frameworks with highly efficient photocatalytic activity for the selective oxidation of cis-cyclooctene. Inorg. Chem. Front. **5**(3), 723–731 (2018). https://doi.org/10.1039/C7QI00693D

Liu, J., Chen, L., Cui, H., Zhang, J., Zhang, L., Su, C.Y.: Applications of metal–organic frameworks in heterogeneous supramolecular catalysis. Chem. Soc. Rev. **43**(16), 6011–6061 (2014). https://doi.org/10.1039/C4CS00094C

Long, J., Shen, K., Chen, L., Li, Y.: Multimetal-MOF-derived transition metal alloy NPs embedded in an N-doped carbon matrix: highly active catalysts for hydrogenation reactions. J. Mater. Chem. A **4**(26), 10254–10262 (2016). https://doi.org/10.1039/C6TA00157B

Long, J., Shen, K., Li, Y.: Bifunctional N-Doped Co@C catalysts for base-free transfer hydrogenations of nitriles: controllable selectivity to primary amines versus imines. ACS Catal. **7**(1), 275–284 (2017). https://doi.org/10.1021/ACSCATAL.6B02327/SUPPL_FILE/CS6B02327_SI_001.PDF

Martinez Joaristi, A., Juan-Alcañiz, J., Serra-Crespo, P., Kapteijn, F., Gascon, J.: Electrochemical synthesis of some archetypical Zn 2+, Cu 2+, and Al 3+ metal organic frameworks. Cryst. Growth des. **12**(7), 3489–3498 (2012). https://doi.org/10.1021/CG300552W/SUPPL_FILE/CG300552W_SI_001.PDF

Qu, K., Wang, J., Ren, J., Qu, X.: Carbon dots prepared by hydrothermal treatment of dopamine as an effective fluorescent sensing platform for the label-free detection of Iron(III) Ions and dopamine. Chemistry—A Eur. J., **19**(22), 7243–7249 (2013). https://doi.org/10.1002/CHEM.201300042

Rath, B.B., Vittal, J.J.: Water stable Zn(II) Metal-Organic framework as a selective and sensitive luminescent probe for Fe(III) and chromate Ions. Inorg. Chem. **59**(13), 8818–8826 (2020). https://doi.org/10.1021/ACS.INORGCHEM.0C00545/SUPPL_FILE/IC0C00545_SI_001.PDF

Senthil Kumar, R., Senthil Kumar, S., Anbu Kulandainathan, M.: Efficient electrosynthesis of highly active Cu3(BTC)2-MOF and its catalytic application to chemical reduction. Microporous Mesoporous Mater. **168**, 57–64 (2013). https://doi.org/10.1016/J.MICROMESO.2012.09.028

Singh, R., Khangarot, R.K., Singh, A.K., Kumar, K.: Metal organic frameworks based nanomaterial: synthesis and applications; removal of heavy metal ions from waste water. Energy, Environ., Sustain., 377–392 (2022). https://doi.org/10.1007/978-981-16-8599-6_16

Singha, S., Saha, A., Goswami, S., Dey, S.K., Payra, S., Banerjee, S., Kumar, S., Saha, R.: A Metal-Organic framework to CuO nanospheres of uniform morphology for the synthesis of α-Aminonitriles under solvent-free condition along with crystal structure of the framework. Cryst. Growth des. **18**(1), 189–199 (2018). https://doi.org/10.1021/ACS.CGD.7B01085

Sun, W., Gao, L., Sun, X., Zheng, G.: A novel route with a Cu(II)-MOF-derived structure to synthesize Cu/Cu2O NPs@graphene: the electron transfer leads to the synergistic effect of the Cu(0)–Cu(I) phase for an effective catalysis of the Sonogashira cross-coupling reactions. Dalton Trans. **47**(16), 5538–5541 (2018). https://doi.org/10.1039/C8DT00465J

Tonigold, M., Tu, Y., Bredenkötter, B., Rieger, B., Bahnmüller, S., Hitzbleck, J., Langstein, G., Volkmer, D.: Heterogeneous catalytic oxidation by MFU-1: A Cobalt(II)-Containing Metal-Organic Framework. Angew. Chem. Int. Ed. **48**(41), 7546–7550 (2009). https://doi.org/10.1002/ANIE.200901241

Tsuruoka, T., Furukawa, S., Takashima, Y., Yoshida, K., Isoda, S., Kitagawa, S.: Nanoporous nanorods fabricated by coordination modulation and oriented attachment growth. Angew. Chem. **121**(26), 4833–4837 (2009). https://doi.org/10.1002/ANGE.200901177

Verma, A., De, D., Tomar, K., Bharadwaj, P.K.: An amine functionalized metal-organic framework as an effective catalyst for conversion of CO2 and biginelli reactions. Inorg. Chem. **56**(16), 9765–9771 (2017). https://doi.org/10.1021/ACS.INORGCHEM.7B01286/SUPPL_FILE/IC7B01286_SI_001.PDF

Wang, P., Feng, J., Zhao, Y., Wang, S., Liu, J.: MOF-Derived tungstated zirconia as strong solid acids toward high catalytic performance for acetalization. ACS Appl. Mater. Interfaces **8**(36), 23755–23762 (2016). https://doi.org/10.1021/ACSAMI.6B08057/SUPPL_FILE/AM6B08057_SI_001.PDF

Wang, X.G., Cheng, Q., Yu, Y., Zhang, X.Z.: Controlled nucleation and controlled growth for size predicable synthesis of nanoscale metal-organic frameworks (MOFs): a general and scalable approach. Angew. Chem. Int. Ed. **57**(26), 7836–7840 (2018). https://doi.org/10.1002/ANIE.201803766

Wang, Y., Zhao, D.: Beyond equilibrium: metal-organic frameworks for molecular sieving and kinetic gas separation. Cryst. Growth des. **17**(5), 2291–2308 (2017). https://doi.org/10.1021/ACS.CGD.7B00287/ASSET/IMAGES/MEDIUM/CG-2017-00287C_0015.GIF

Wei, H., Guo, Z., Liang, X., Chen, P., Liu, H., Xing, H.: Selective photooxidation of amines and sulfides triggered by a superoxide radical using a novel visible-light-responsive metal-organic framework. ACS Appl. Mater. Interfaces **11**(3), 3016–3023 (2019). https://doi.org/10.1021/ACSAMI.8B18206/SUPPL_FILE/AM8B18206_SI_002.CIF

Wu, Y.P., Wu, X.Q., Wang, J.F., Zhao, J., Dong, W.W., Li, D.S., Zhang, Q.C.: Assembly of Two Novel Cd3/(Cd3 + Cd5)-Cluster-Based metal-organic frameworks: structures, luminescence, and photocatalytic degradation of organic dyes. Cryst. Growth des. **16**(4), 2309–2316 (2016). https://doi.org/10.1021/ACS.CGD.6B00093/SUPPL_FILE/CG6B00093_SI_001.PDF

Yao, X., Bai, C., Chen, J., Li, Y.: Efficient and selective green oxidation of alcohols by MOF-derived magnetic nanoparticles as a recoverable catalyst. RSC Adv. **6**(32), 26921–26928 (2016). https://doi.org/10.1039/C6RA01617K

Yu, G., Sun, J., Muhammad, F., Wang, P., Zhu, G.: Cobalt-based metal organic framework as precursor to achieve superior catalytic activity for aerobic epoxidation of styrene. RSC Adv. **4**(73), 38804–38811 (2014). https://doi.org/10.1039/C4RA03746D

Yu, K., Lee, Y.R., Seo, J.Y., Baek, K.Y., Chung, Y.M., Ahn, W.S.: Sonochemical synthesis of Zr-based porphyrinic MOF-525 and MOF-545: Enhancement in catalytic and adsorption properties. Microporous Mesoporous Mater. **316**, 110985 (2021). https://doi.org/10.1016/J.MICROMESO.2021.110985

Zhang, L., Feng, C., Gao, S., Wang, Z., Wang, C.: Palladium nanoparticle supported on metal–organic framework derived N-decorated nanoporous carbon as an efficient catalyst for the Suzuki coupling reaction. Catal. Commun. **61**, 21–25 (2015). https://doi.org/10.1016/J.CATCOM.2014.12.004

Zhang, L., Liu, H., Shi, W., Cheng, P.: Synthesis strategies and potential applications of metal-organic frameworks for electrode materials for rechargeable lithium ion batteries. Coord. Chem. Rev. **388**, 293–309 (2019). https://doi.org/10.1016/J.CCR.2019.02.030

Zhao, J., Liu, X., Wu, Y., Li, D.S., Zhang, Q.: Surfactants as promising media in the field of metal-organic frameworks. Coord. Chem. Rev. **391**, 30–43 (2019). https://doi.org/10.1016/J.CCR.2019.04.002

Zhong, W., Liu, H., Bai, C., Liao, S., Li, Y.: Base-free oxidation of alcohols to esters at room temperature and atmospheric conditions using nanoscale Co-based catalysts. ACS Catal.

5(3), 1850–1856 (2015). https://doi.org/10.1021/CS502101C/SUPPL_FILE/CS502101C_SI_001.PDF

Zhou, Y.X., Chen, Y.Z., Cao, L., Lu, J., Jiang, H.L.: Conversion of a metal–organic framework to N-doped porous carbon incorporating Co and CoO nanoparticles: direct oxidation of alcohols to esters. Chem. Commun. **51**(39), 8292–8295 (2015). https://doi.org/10.1039/C5CC01588J

Chapter 6
Nanostructured Materials for Visible-Light-Driven Organic Synthesis: A New Catalytic Solution for a Sustainable World

Pratibha Saini, Krishan Kumar, Surendra Saini, Mukul Sethi, Priyanka Meena, Aditya Gurjar, Anshu Dandia, Wolfgang Weigand, and Vijay Parewa

6.1 Introduction

Advancements in organic synthetic chemistry have revolutionized human existence, fostering the fabrication of a plethora of fundamental products and chemicals, ranging from fuels to medicines (Clayden et al. 2012). Traditional approaches often entail intricate processes and harsh conditions, demanding high temperatures and pressures. These conditions not only demand specialized equipment and processes but also result in substantial energy utilization and the production of toxic by-products, posing environmental risks and escalating costs (Grzybowski et al. 2009). However, harnessing visible light for organic transformations offers a more benign method, capitalizing on light irradiation to circumvent the complexities associated with conventional synthesis routes. This embrace of photosynthesis represents a keystone in sustainable chemical reactivity, aligning with the objectives of Green Chemistry (Dandia et al. 2020a).

P. Saini · W. Weigand
Institute Inorganic and Analytical Chemistry, Friedrich Schiller University Jena, Jena, Germany

K. Kumar · S. Saini · M. Sethi · P. Meena · A. Gurjar · A. Dandia · V. Parewa (✉)
Centre of Advanced Studies, Department of Chemistry, University of Rajasthan, Jaipur, India
e-mail: parewavijay.parewa@gmail.com; parewavijay@uniraj.ac.in

P. Saini
Institute for Physical Chemistry (IPC), Friedrich Schiller University Jena, Jena, Germany

R. S. Ningthoujam and A. K. Tyagi (eds.), *Handbook of Materials Science, Volume 3*,
Indian Institute of Metals Series, https://doi.org/10.1007/978-981-96-8229-4_6

6.2 Photocatalysis: Complementary Tool in the Organic Synthesis

In response to the ever-growing need for eco-friendly and sustainable organic methods, visible-light-induced organic synthesis has appeared as a dominant avenue for creating several valuable organic compounds (Xuan and Xiao 2012). Solar energy stands out as the foremost and abundant renewable resource globally. Leveraging clean, non-toxic, affordable, sustainable, limitless, and readily accessible energy sources (such as sunlight or visible light) holds immense appeal for environmentally friendly organic transformations. This methodology addresses energy demands and associated issues efficiently. Hence, researchers worldwide are committed to utilizing renewable energy sources across various domains through the lens of "Green Chemistry", making it a vital objective.

Typically, the majority of reactant molecules cannot directly absorb visible light. To address this, photocatalytic systems serve as intermediary agents, facilitating energy transfer between the reactant molecules and visible light. These systems can be either homogeneous or heterogeneous. Numerous meticulously crafted photocatalytic systems, or photocatalysts, have been innovated by diverse research teams specifically for driving visible-light-induced organic reactions. So far, a diverse array of photocatalytic systems, totaling five different types, have been devised and utilized in groundbreaking catalytic conversions (Djurišić et al. 2020; Kudo 2006). These encompass molecular homogeneous photocatalytic systems, exemplified by ruthenium- and iridium-transition metal-based polypyridine complexes, as well as metal-free organic dyes like acridinium salts, Rhodamine B, Eosin Y, Rose Bengal, and Methylene blue, among others. Additionally, heterogeneous semiconductors such as ZnO, TiO_2, Fe_2O_3, WO_3, Cu_2O, $SrTiO_3$, Ag_3PO_4, $BiPO_4$, $NaTaO_3$, and CdS have been employed, along with plasmonic-metal NPs and other inventive photoelectric materials.

Each type of photocatalyst, whether homogeneous or heterogeneous, possesses its own set of advantages and disadvantages (Pelizzetti and Serpone 2012). For example, homogeneous photocatalysts offer a dense concentration of redox sites per unit volume in solution, enabling effective catalysis of photochemical reactions. In contrast, heterogeneous photocatalysts are characterized by their reusability and the ease with which their properties can be tailored through simple adjustments to their morphology. Although homogeneous photocatalysts, particularly solution-phase photocatalysts, have shown potential and garnered abundant favorable arrives, their practical application in organic transformations remains formidable. This is due to their costly nature, limited availability, and potential toxicity, risk of product adulteration, and their non-recyclability or non-reusability. Furthermore, their poor efficiency in harvesting visible light further obscures their widespread use. These downsides render their utility inadequate for viable applications in photocatalytic transformations. To address these challenges, the substitution of costly photocatalysts with a unique scale of affordable, heterogeneous alternatives is crucial for advancing environmentally friendly organic transformations driven by visible light.

Likewise, researchers are particularly focused on expanding the series of visible light-reactive photocatalysts with low charge carrier recombination rates to enhance photocatalytic efficiency.

The elementary challenge in current endeavors to enterprise highly efficient visible-light-responsive photocatalysts lies in advancing solar energy utilization to develop the parting and transportation of photo-induced e^-/h^+ pairs. Due to the numerous disadvantages associated with conventional photocatalysts, containing challenging separation procedures, low efficiency, elevated cost, toxicity, poor photostability, incomplete adsorption capacity, propensity for aggregation, complex synthesis approaches, low quantum yield, rapid charge carrier recombination, reduced curvature, diminished conductivity, limited charge transfer, and scarcity of active sites for photocatalytic reactions, it is imperative to advance efficient photocatalysts for catalytic transformations in organic synthesis to address these limitations (Pirsaheb et al. 2018).

6.3 Nanotechnology: The Competent Tool for Visible-Light-Induced Organic Synthesis

Ever since the initiation of Nanotechnology in 1959 by the American physicist and Nobel Prize laureate Richard Feynman, it has triggered interdisciplinary interest and acquired substantial research attention owing to its extraordinary properties. This interdisciplinary field is offered to the development and construction of materials at the nanometer scale (ranging from 1 to 100 nm) that demonstrate distinctive characteristics such as expansive surface areas and rapid charge transport capabilities. In recent times, an array of research papers has underscored the extensive potential of nanotechnologies in adapting and designing novel photocatalysts with enriched photocatalytic implementation under light exposure (Saini et al. 2022). Nanostructured photocatalysts offer not just a higher abundance of active sites and extensive surface area compared to bulk materials, but also inherently improved photocatalytic activity, enabling superior performance in green catalytic transformations. Subsequently, the incorporation of nanostructured material reimbursements, including durability and upfront separation from the reaction mixture via phase separation, with visible-light-absorbing photocatalysts presents an intriguing solution to tackle the challenges (Dandia et al. 2020b) (Fig. 6.1).

As advancements in nanoscience and nanotechnology continue, nanomaterials are increasingly employed as photocatalysts in various organic transformations. While significant strides have been made in recent decades, the primary challenge remains the quick recombination of photo-induced e^-/h^+ pairs, which directly impacts the efficiency, selectivity, and reactivity of the photocatalyst in catalyzing organic reactions. Furthermore, crucial characteristics of a photocatalytic system (photocatalyst) for efficiently utilizing visible light include possessing the anticipated band-gap, facilitating efficient e^-/h^+ transportation, exhibiting appropriate morphology, minimizing

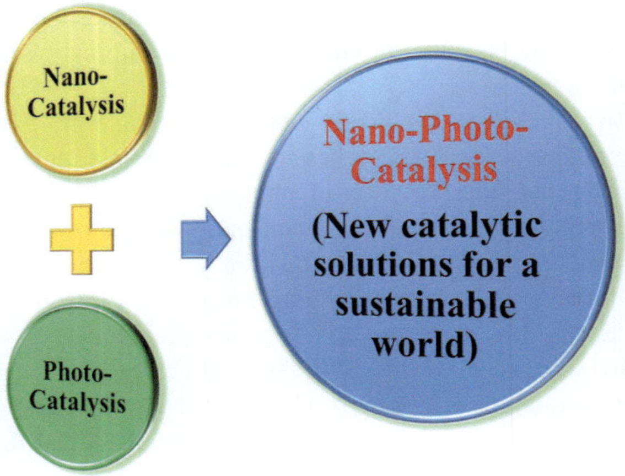

Fig. 6.1 Nano-photo-catalysis: new catalytic solution for organic synthesis

rates of photogenerated electron and hole recombination, offering a high surface area, achieving high quantum yields, and demonstrating good thermo-, photo-, and chemical stability, along with reusability (Dietzek-Ivanšić et al. 2022; König 2017) (Fig. 6.2).

Several heterogeneous semiconductors possess these favorable characteristics. In the course of photocatalytic processes, these photocatalysts absorb visible light or sunlight, activating a charge separation mechanism that generates photogenerated holes and electrons. These reactive species subsequently participate in catalytic transformations (Fox and Dulay 1993). Among the diverse applications for such materials, utilizing nanostructured materials as a nano photocatalyst for visible-light-induced organic reactions stands out as an exceptionally appealing field. It aligns well with the principles of "green chemistry," embodying the core mission of integrating sustainability criteria into chemical production (Radhika et al. 2019).

The nano photocatalyst is essentially comprised of two components: the light absorber and the active sites. Generally, the light absorbers encompass semiconductors, quantum dots (QDs) or carbon quantum dots (CQDs), and organic photosensitizers, while the dynamic locations are situated either on nano supports surface or within the porous structure (for example graphene, polymers, metal–organic frameworks, and covalent organic frameworks, silica-coated magnetic nano-spheres) of these light absorbers. The inclusion of these nano supports within the photocatalytic system offers a plethora of notable rewards, for example thermal, optical, and chemical stability, as well as the ability for functionalization, easy manufacturing, and an elevated surface area to volume ratio, and modest recovery. Current years have seen a surge in research focused on nanostructured photocatalytic systems. These materials, with their exceptional mechanical and optical assets compared to larger counterparts, have initiate wide use in numerous applications, including oxygen and hydrogen production, organic synthesis, photoredox reactions, solar fuel generation,

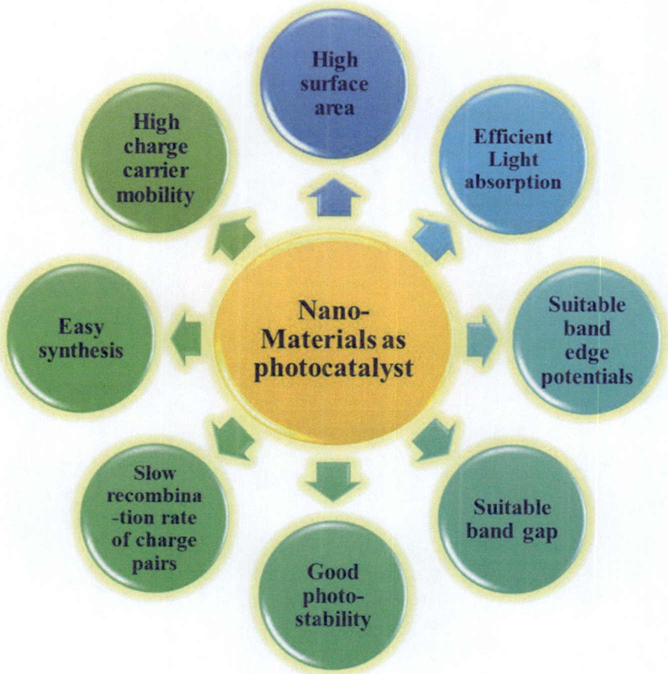

Fig. 6.2 Nanomaterials in photocatalysis: key points

pollutant degradation, photooxidation processes, and mitigation of toxic gases, etc. (Tong et al. 2012) (Fig. 6.3).

Nanostructured materials boast an array of remarkable traits, for example, extensive surface areas, incomparable chemical and thermal stability, straightforward synthesis, easy recovery, flexibility for functionalization, surface irregularities, and an affluence of low-coordinated sites vital for competent photocatalysis under visible light for organic conversions. They provide as a promising link between homogeneous and heterogeneous catalysis, revolutionizing the landscape of chemical methods with their innovative capabilities (Sharma et al. 2015).

6.3.1 The Fundamental Mechanisms of Nano-photocatalysis

Photocatalysis driven by visible light is a multifaceted procedure that follows multiple steps. Massive research efforts have focused on dissecting these stages and uncovering the underlying mechanisms governing heterogeneous photocatalysis (Zhang et al. 2018). Figure 6.4 presents a visual representation of how nano photocatalysts activate when exposed to light. As per accessible data, the procedure includes three main steps for inducing organic transformations under visible light: First, visible light

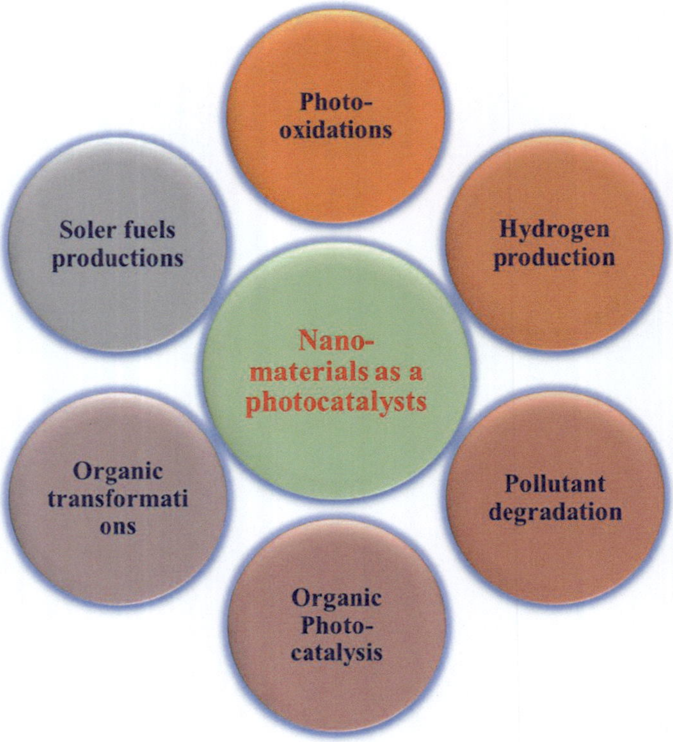

Fig. 6.3 Harnessing nanostructures for photocatalysis

triggers the creation of photo-excited charge carriers by absorbing energy exceeding the material's bandgap (Step 1). These carriers then relocate to the surface of the nanostructure-based photocatalyst or a related co-catalyst (Step 2), where they take part in organic transformations (Step 3). If not engaged in photocatalysis, the charge carriers recombine, releasing energy in the form of luminescence or heat.

6.3.2 Tailoring the Photocatalyst's Properties of Nanostructured Materials

Despite the substantial advancements in nano-photocatalysis research over the past decade, there remain critical concerns that must be focused before nano-photocatalysis can be widely adopted as a sustainable and effective organic transformation. Currently, a diversity of semiconducting nanostructured materials has been utilized as nano photocatalysts for catalytic transformations (Xu et al. 2019). An ideal photocatalyst should possess merits for instance stability, affordability, non-toxicity, and high photo-catalytic activity under visible light. The surface structure, particle

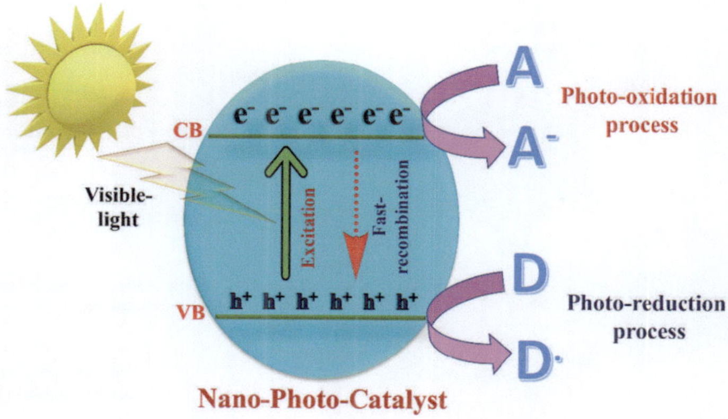

Fig. 6.4 An illustrative representation outlining the core mechanisms driving photocatalysis in nanostructured photocatalysts upon exposure to light

size, active surface area, and chemical composition of nanostructured materials are foremost factors in photocatalysis. Normally, the effectiveness of nano photocatalysts relies on two primary aspects: (i) efficient harvesting of light and (ii) effective separation and transmission of photogenerated electron–hole (e^-/h^+) pairs for catalytic transformations. Several findings have targeted to enhance light absorption, (e^-/h^+) pairs partition, and transmission effectiveness over the past years (Luciani et al. 2020).

Carbon-based materials, along with surface-functionalized types achieved through purposeful doping of heteroatoms, for example, B, P, S, and N, into the carbon matrix, emerge as promising contenders. These alterations can significantly modify the electronic/chemical attributes to fulfil specific purpose requirements (Ge et al. 2019).

Various methods have been carefully devised to improve the photocatalytic activity of nano photocatalysts by generating energetic locations on the surface of different light absorbers, aiming to produce highly efficient photocatalysts and suppress electron–hole recombination (Fig. 6.5). For instance, approaches such as creating surface vacancies, engineering crystal facets, adjusting bandgaps, incorporating or immobilizing co-catalysts and metal nanoparticles (including noble metals like Pt, Au, Pd, Ag, Ru, and non-precious metals such as Fe, Cu, Co, Ni, and Mo), employing semiconductor combinations, utilizing heterogeneous or homogeneous single-site catalysts, controlling morphology, introducing functional groups, incorporating organic linkers/metal nodes in metal–organic frameworks (MOFs), establishing multiphase or multicomponent heterojunctions, and developing composites with carbon materials, particularly graphene-based nano-assemblies, have been pursued to expand the reactivity of nano photocatalysts and alleviate electron–hole recombination (Cheng et al. 2021; Chen and Bai 2020).

Doped metals supported on various substrates serve as electron sinks or reservoirs, enhancing charge carrier mobility in semiconductors or the likelihood of flourishing

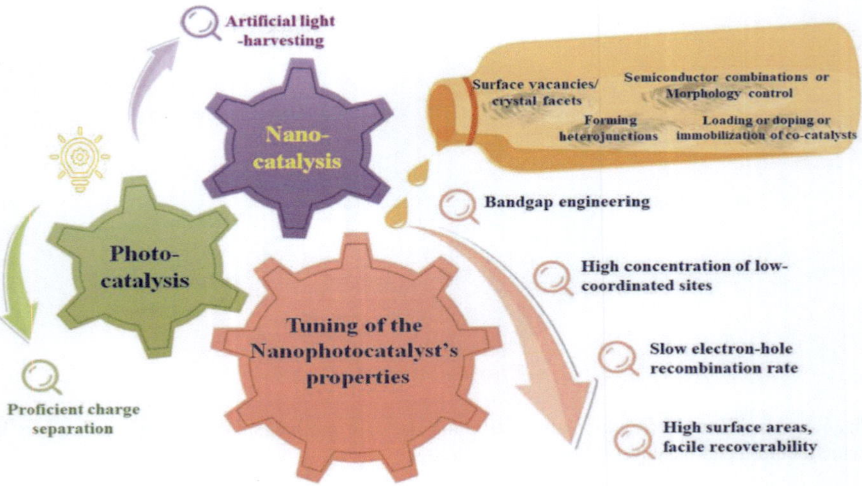

Fig. 6.5 Overview of the different methods for tuning the photocatalytic properties of nano-photocatalysts

mass diffusion of (e^-/h^+) pairs to the responsive photocatalytic interface. In short, the fundamental challenges for enlightening the performance of nanomaterials contain: (a) increasing light absorption, (b) maximizing surface area, (c) achieving high crystallinity, and (d) enhancing charge carrier separation. Additionally, optimal band positioning, strong chemical stability of the semiconductor material and affordability offer significant advantages.

6.4 The Focus of This Chapter

Even with the rising interest from research communities, instances of visible-light-induced organic transformations catalyzed by nanostructured materials under moderate conditions are still quite rare (Ghosh 2018). The emergence of these extraordinary nano-photocatalysts designed for employing visible light states opportunities for groundbreaking advancements in photocatalytic organic transformations. The flourishing body of research literature from diverse research groups underscores the central role nano-photocatalysts are floated to play in utilizing visible light for organic transformations. Subsequently, there is considerable eagerness for a comprehensive overview of this subject. We are assured that introducing such a chapter at this juncture will stimulate additional advancements in this area of research.

We have compiled a systematic examination of the literature regarding photocatalytic organic transformations utilizing nanostructured materials reactive to visible light. Rather than extensively ongoing numerous discoveries, our focus is on providing strategic insights into the fabrication of nanostructured photocatalysts and highlighting the importance of this burgeoning field. The combination of photocatalysis and nanocatalysis extends a fundamental framework for enabling visible-light-induced organic transformations, including methods such as C–H bond activation, coupling reactions, bond formation, hydroxylation, and others. This chapter will be organized into two main portions.

6.4.1 Metallic Nanomaterials as Photocatalysts for Organic Reactions

This section outlines the organic transformations facilitated by metallic nanomaterials acting as photocatalysts under visible-light exposure. Metallic nanomaterials have developed as a modern category of photocatalysts capable of catalyzing various chemical transformations through photocatalysis (Tahir et al. 2020). A notable benefit of metallic nanomaterials as nano-photocatalysts is their capacity to absorb light across a wide range of wavelengths, covering a substantial portion of the solar spectrum rather than being limited to specific wavelengths. Recent research from numerous groups advises that the light absorption properties of metallic nanomaterials as photocatalysts are depending upon factors for example nanoparticle shape, size, and particle–particle interactions, all of which directly affect their photocatalytic effectiveness.

Due to their appealing characteristics for example efficiency, practicality, cost-effectiveness, capability to produce charge carriers under exposure of light, and proper bandgap, CdS NPs emerge as an auspicious photo-responsive material for driving light-induces chemical reactions. Koohgard and colleagues (Firoozi et al. 2018) designed and characterized CdS NPs for photocatalytic activation of C–H bonds of t-amines at ambient environment. The C–H activation of t-amines into the analogous product (modern classes of quinolones) was achieved via a condensation reaction utilizing 10 mol% of CdS NPs as a remarkably efficient and reusable nano photocatalyst under visible light irradiation. Several substrates with different functional groups constructed the chosen compounds with high yields in a slight time frame. Significantly, there was no conversion without CdS NPs or light irradiation. According to the anticipated mechanism, photogenerated charge carriers are firstly generated from the CdS nano photocatalyst. The photogenerated electron combines with oxygen, while the photogenerated hole interacts with N,N-dimethylaniline (A) to build a sequence of different intermediate. The desired compound was achieved through the aromatization of (G) (Scheme 6.1). Moreover, the ability to reuse the photocatalyst for multiple runs is a notable benefit of this method.

Firoozi and Hosseini-Sarvari (2021) introduced a novel approach for the photo-induced activation of sp^3 and sp^2 C–H bonds in tertiary amines and aryl sulfonyl chlorides, employing CdS NPs as nanomaterials. They have successfully synthesized various compounds, including sulfonamides, aryl sulfonyl compounds, and

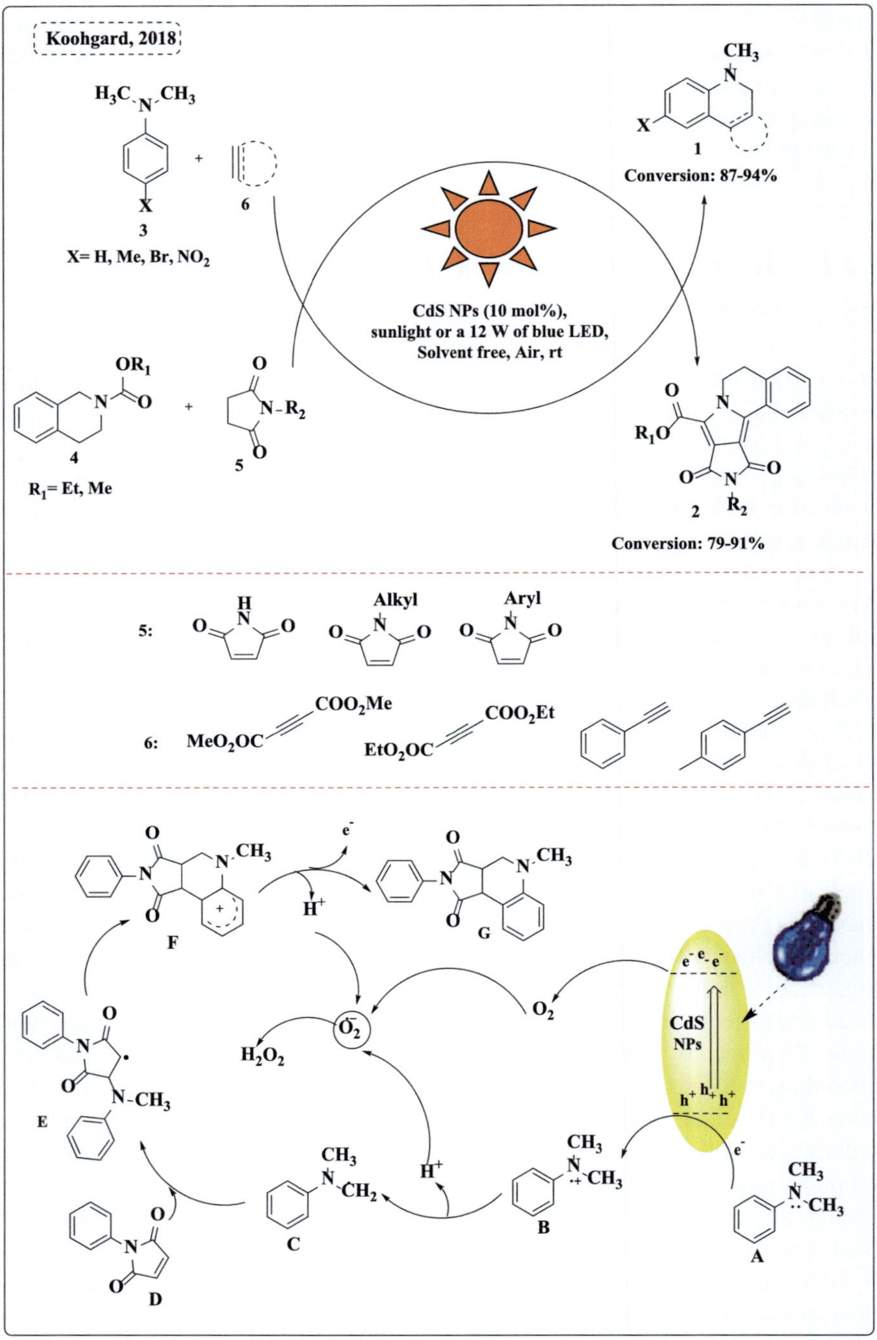

Scheme 6.1 A comprehensive depiction detailing the overall reaction scheme and the proposed mechanism for the photocatalytic C–H activation using CdS photocatalyst

vinyl sulfones, under optimized distinct circumstances. Under varying specifications utilizing CdS NPs, the authors done the exclusive formation of diverse products including sulfonamides, aryl sulfonyl compounds, and vinyl sulfones. Substrates featuring diverse degrees of substitution yielded these products in high yields. The conversion required the loading of 0.1 mmol of CdS nano photocatalyst. The remarkable photocatalytic efficiency of CdS nano photocatalyst is attributed to its superior surface area and suitable bandgap energy. The reaction was conducted in a solvent system comprising EtOH/H_2O (3:1 v/v, 2 mL) under an air atmosphere. No conversion occurred without visible light or the presence of CdS nano photocatalyst. Authors proposed various reaction pathways with slight modifications in reaction conditions, as illustrated in Scheme 6.2. Furthermore, the ability to reuse the photocatalyst is a significant benefit of this method.

Dandia et al. (2020c) conducted the synthesis and characterization of Cu-doped ZnS nanoparticles (Cu: ZnS NPs) in 2019. These nanoparticles were explored as a nano photocatalyst to selectively produce 2-substituted azoles from either 1,2-diaminobenzene or 2-mercaptoaniline and benzyl bromides when exposed to visible light. Through various analytical techniques, the prepared photocatalyst was confirmed to possess a consistent morphology, an initiate mesoporous structure, and copper doping integrated within the ZnS matrix. The inclusion of Cu doping is theorized to modify the forbidden band, establishing a distinctive trapping reservoir (referred to as the mid-gap band) for visible-light-induced e^-/h^+ pairs. This mid-gap band serves as a site for trapping visible-light-induced e^-/h^+ pairs, consequently diminishing the rate of e^-/h^+ pair recombination.

The outstanding photocatalytic efficacy of used nanomaterial can be accredited to various factors. These include exceptional stability, suitable bandgap, recyclability, the creation of highly responsive sites, robust adsorption capabilities, and the existence of the mid-gap band, which prolongs the relaxation period of e^-/h^+ pairs. A significant advantage of this method, in comparison to others, is the creation of an amide intermediate and quinazoline-4(3H)-ones. Additionally, this method demonstrates compatibility with a broad scale of different reactants. A mechanism is introduced by the N-arylation of o-phenylenediamine with benzyl bromide, ensuing in the construction of the intermediate 3a′. Upon irradiation, the nano photocatalyst makes photo-induced e^-/h^+ pairs. Upon photoexcitation, electrons migrate into 3a′, initiating the formation of the radical intermediate (Y). Concurrently, $O_2^{\bullet-}$ is generated by capturing electrons from O_2. $O_2^{\bullet-}$ subsequently reacts with the activated reactant, facilitating the formation of the amide intermediate via Csp^3–H oxidation. Ultimately, the desired product is synthesized through the intramolecular cyclization of the final intermediate, with excess water being eliminated during the process. Furthermore, the ability to reuse the photocatalyst is a significant advantage of this method (Scheme 6.3).

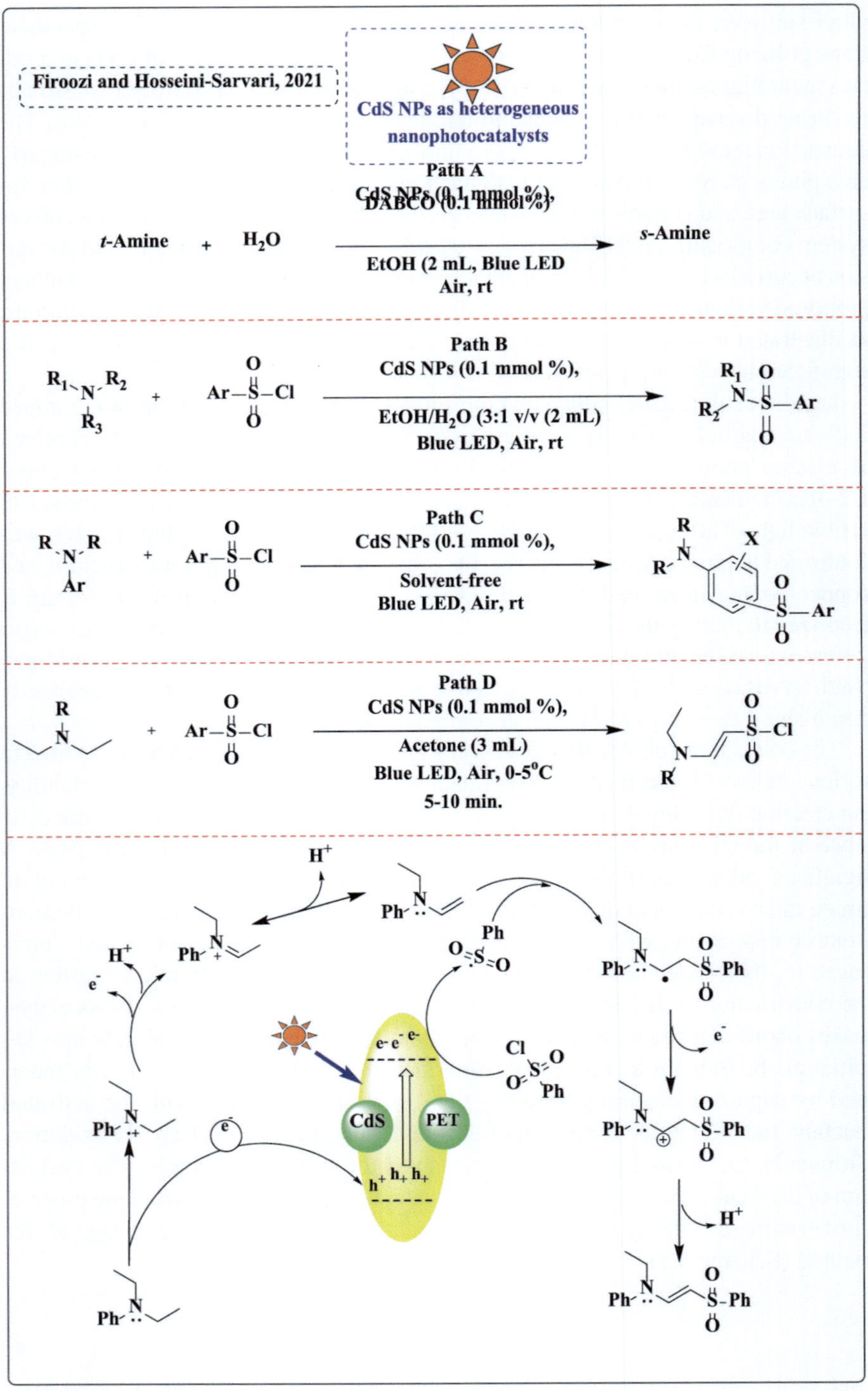

Scheme 6.2 An outlined reaction scheme featuring diverse pathways and the proposed route for sp^3 and sp^2 C–H bonds using CdS and visible-light irradiation

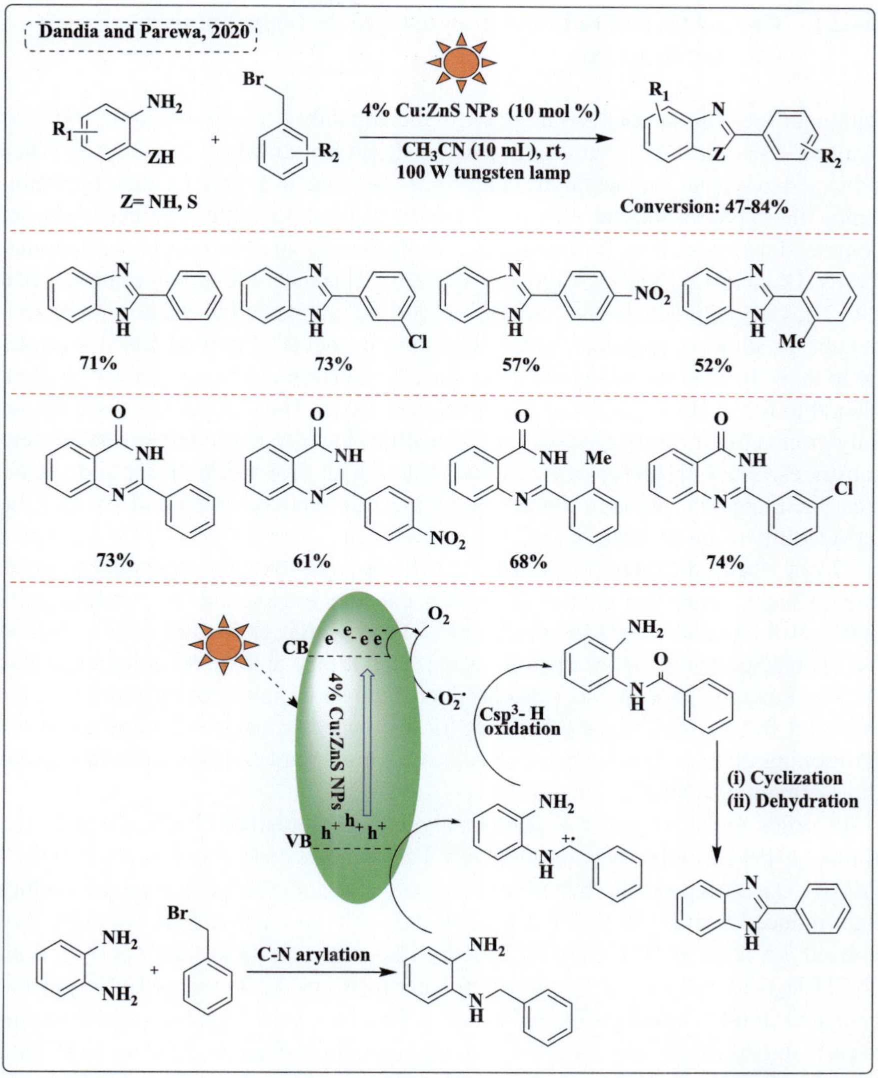

Scheme 6.3 A visual representation of the overall reaction scheme and the envisioned route for the selective production of 2-substituted azoles employing 4% Cu: ZnS NPs

6.4.2 *Carbon-Based Nanomaterials as Nanophotocatalysts for Organic Reactions*

Within this section, we delve into instances of photo-induced organic reactions accelerated by carbon-based nanomaterials. Here, the art of fine-tuning the nano-photocatalyst unveils pathways to sustainable reaction conditions.

6.4.2.1 Carbon Quantum Dots Catalyzed Visible-Light-Induced Organic Reactions

Although metallic nanomaterials possess the capability to drive organic transformations in photocatalytic setups, they face a significant drawback: instability. When subjected to typical photocatalytic conditions, they tend to aggregate, thereby diminishing their photocatalytic efficacy. Therefore, most metallic nanomaterials are combined with a solid support system, typically consisting of carbonaceous nanomaterials (Dandia et al. 2020d; Ajiboye et al. 2020). This includes various materials like g-C_3N_4, carbon quantum dots, carbon nanotubes, graphene, nano onions, nanodiamonds, in addition to metal oxides. The improvements of carbon-based supports lie in their flexible porous nature and adjustable surface chemistry, rendering them favorable for visible-light-induced organic reactions. These support structures not only grant a platform for chemical functionalities but also exceed their role as mere substrates. Their extensive surface area, replete with diverse chemistry and vacancies, facilitates the uniform dispersion of metallic nanomaterials and enables the recyclability of photocatalysts.

Given their distinctive attributes such as non-toxicity, abundance, cost-effectiveness, competent photon absorption, capacity to generate electron hole pairs under visible light, size-dependent optical assets, high extinction coefficients in the visible spectrum, and extremely responsive surface area, carbon quantum dots (CQDs) emerge as talented supports for photocatalysis. Their stability under photoirradiation further underscores their superiority over homogeneous photocatalysts, positioning them as key elements in achieving remarkable performance in organic transformations (Dandia et al. 2021).

In order to investigate the photocatalytic capabilities of CQDs, Duarah and Karak (2019) developed a novel approach. They synthesized a starch-modified HPU/Pd@RCD nanocomposite (hyperbranched polyurethane) for facilitating the visible-light-induced Suzuki–Miyaura of aryl boronic acid and aryl iodide. Notably, they utilized 7.5 wt% of HPU/Pd@RCD-nanocomposite as a nano photocatalyst, with ^{i}PrOH-H_2O (1:1, 4 mL) acting as the solvent medium, 1.5 mmol of K_2CO_3 as the base, and sunlight as the visible light source. This innovative approach enabled them to carry out the reaction under ambient conditions and without the need for additional ligands.

Under these environments, the reaction is facilitating the formation of biphenyls via the Suzuki–Miyaura coupling route. The CQDs were synthesized using an in situ polymerization method. Pointedly, without visible light or the HPU/Pd@RCD-nanocomposite, no conversion was observed, highlighting the essential contribution of both visible light and the nano photocatalyst in facilitating this reaction. Characterization of the synthesized HPU/Pd@RCD-nanocomposite was conducted using various analytical methods. Leveraging these optimized conditions, they explored the substrate scope for synthesizing biphenyls and their derivatives. The authors argue that the collaborative effect of the utilized Pd@RCD and the starch-based hyperbranched polyurethane matrix is pivotal in facilitating this bond-forming reaction. Furthermore, the HPU/Pd@RCD-nanocomposite demonstrated the ability to be reused across multiple cycles without any reduction in its photocatalytic efficiency (Scheme 6.4).

Expanding their research into photo-induced organic transformations, the Pillai group (Chakraborty et al. 2019) investigated a visible-light-driven C–C coupling between PhPyr and $PhSO_2$. They utilized InP/ZnS QDs as the single photocatalyst, conducting the reaction at room temperature. The InP/ZnS QD nano photocatalyst, characterized by its high surface area, presents several rewards containing extraordinary photocatalytic activity, minimal toxicity, reusability, strong visible-light absorption capabilities, and an appropriate bandgap. In this experiment, they utilized 5 μM of InP/ZnS QDs, alongside 3 equivalents of cesium acetate in benzene, and a 1 W blue LED for visible light. It is valuable to emphasize that this transformation fails to occur under light exposure without the inclusion of a photocatalyst. This method stands out because it relies solely on InP/ZnS QD as a nano photocatalyst, eliminating the need for any extra cocatalysts, sacrificial oxidants, or reductants in C–C bond-forming reactions (Scheme 6.5).

The Maiti team unveiled an inventive dual-phase technique (Chandrashekar et al. 2019) for crafting aryl boronate structures, employing a Sandmeyer-type borylation of amines with MPA-capped quantum dots (QDs) which serve as a heterogeneous photocatalyst in a mildly acidic medium. The newly prepared water-soluble MPA-capped QDs photocatalyst underwent rigorous scrutiny through diverse

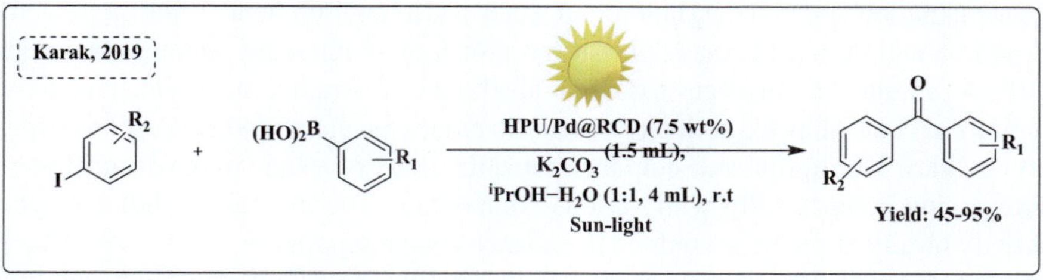

Scheme 6.4 A novel starch-modified HPU/Pd@RCD nanocomposite (hyperbranched polyurethane) designed for facilitating visible-light-induced C–C coupling reactions under ambient conditions

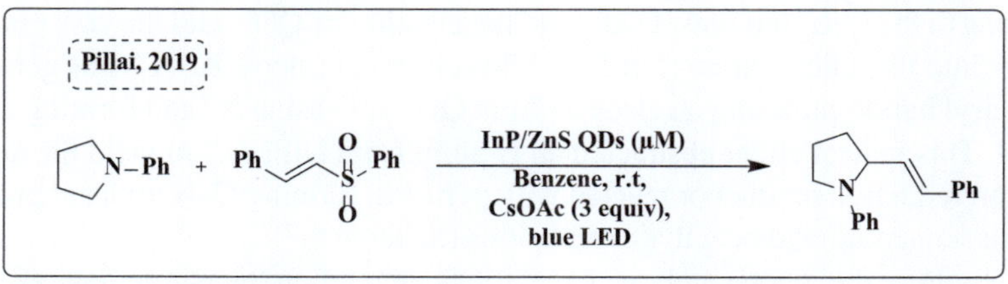

Scheme 6.5 Visible-light-driven C–C coupling using InP/ZnS QD as a nanophotocatalyst

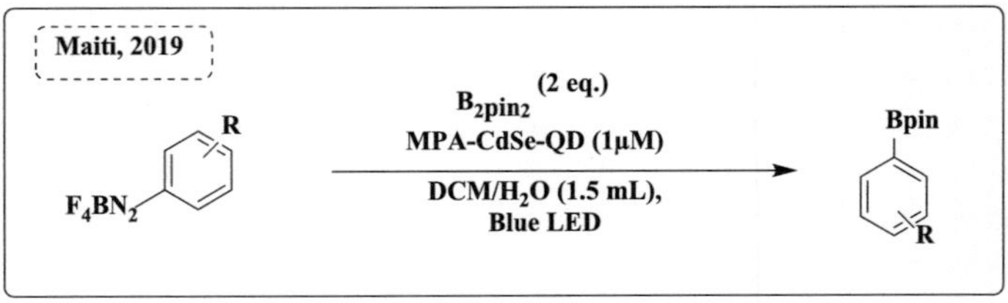

Scheme 6.6 Mildly acidic conditions facilitate the Sandmeyer-type borylation in the presence of MPA-CdSe-QD nanomaterial

analytical means. Following meticulous experimentation, they pinpointed the fundamental conditions: a dual-phase photocatalytic setup with MPA and CdSe QDs in a mild acidic condition. After the optimized parameters, they deftly transmuted a variety of substrates—both electron-withdrawing and electron-donating groups—into their desired counterparts with yields ranging from moderate to commendable. Moreover, there is no product without visible light irradiation and photocatalyst. Crucially, this photocatalytic metamorphosis boasts a trifecta of virtues: simplicity, eco-friendliness, and a strikingly high turnover number (TON). Notably, the pathway to this chemical alchemy traverses through an electron transfer journey in the realm of excited states (Scheme 6.6).

In Pal et al. (2017) uncovered an interesting discovery involving the catalytic triggering of C–H bonds using ZnSe/CdS core/shell QDs nanomaterial. This groundbreaking approach facilitated the stimulation of carbon–halogen bonds to carry out dehalogenation and C–H arylation in a water-based environment. After meticulous experimentation, it was finalized that 0.6 mol% of synthesized nanomaterial with DIPEA (8 Equiv.) as an electron donor under blue LED irradiation provided the most optimized conditions for both photocatalytic reduction and arylation reactions using aryl halides. This optimized setup facilitated the study of a wide range of substrates, showcasing compatibility with various compounds. The incredible photocatalytic activity observed can be accredited to enhance charge separation, significant surface area, uniform dispersion of carbon dots in solution, and improved photostability. Markedly, the desired product was only gotten under the presence of QDs and blue-light irradiation, highlighting their indispensable role in the process. The reaction pathway proceeded as follows: Initially, under visible light irradiation, photogenerated charge carriers were generated from quantum dots. Next, DIPEA supplied electrons to the quantum dots, leading to the creation of $QD^{\bullet -}$ and the conversion of DIPEA into its radical cation ($DIPEA^{\bullet +}$). In subsequent steps, $Ar\text{-}X^{\bullet -}$ was generated by the aryl halide capturing an electron from $QD^{\bullet -}$, releasing X^- and forming an aryl radical. This radical either abstracted an H-atom from $DIPEA^{\bullet +}$ to yield the dehalogenation reduction product or reacted with pyrroles, forming C–H arylated products through sequential photocatalytic oxidations (Scheme 6.7).

To explore the photocatalytic capabilities of carbon quantum dots (CQDs), Boukherroub and collaborators (Maaoui et al. 2016) produced a new nanocomposite of Prussian blue/carbon dots (PB/CD) for the activation of amine bonds. In their

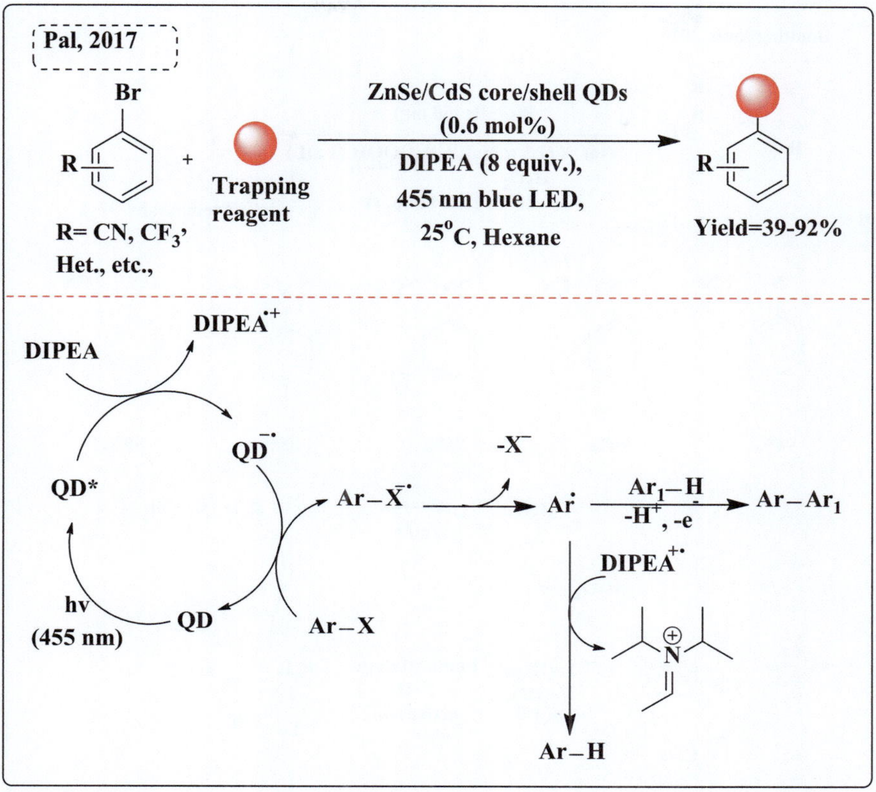

Scheme 6.7 Visible-light-induced C–H arylation using ZnSe/CdS core/shell QDs nanomaterials

research, they employed 50 mg of PB/CD nanocomposite as a nano photocatalyst, NaCN (1.5 mmol)/acetic acid (1 mL) as a cyanide source, MeOH (10 mL) as a solvent, H_2O_2 as an oxidizing agent, and a 20 W LED lamp for visible light irradiation. In this optimized environment, oxidative cyanation of tertiary amines to yield α-aminonitriles occurred via visible-light induced of C–H functionalization. The integration of CQDs was achieved by alkaline hydrolysis of D-fructose with NaOH. Prominently, it was performed that the desired outcome did not form in the absence of synthesized nanomaterial and light irradiation, underscoring the indispensable roles of visible light and the nano photocatalyst in this reaction. Moreover, the reusability of the PB/CD nanocomposite was a substantial benefit of this method (Scheme 6.8).

6.4.2.2 Perovskite Catalyzed Visible-Light-Induced Organic Reactions

In contrast to other semiconductor photocatalysts like quantum dots (QDs) and TiO_2, perovskites deliver benefits such as straightforward synthesis and compatibility with

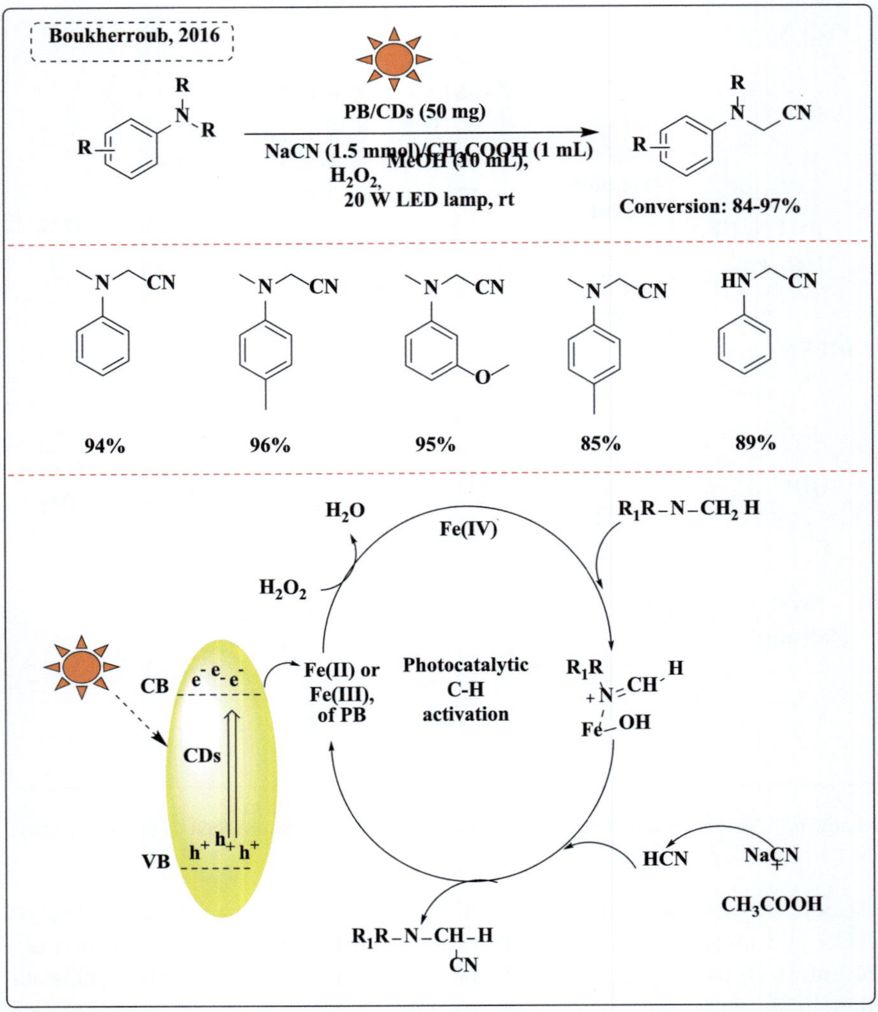

Scheme 6.8 Visible-light-driven activation of amine bonds facilitated by the PB/CD nanocomposite

low-temperature solution-based procedures. Moreover, they allow for easy manipulation of their bandgap by adjusting their halide configurations (Dandia et al. 2020e). The versatility of perovskite heterogeneous photocatalysts enables a wide array of visible-light-driven chemical reactions. Their essential characteristics, including high activity, efficient extinction coefficients, conductivity, stability, reusability, adjustable redox potentials, extended electron–hole pair lifetimes, bandgap engineering, mobility, and light absorption, offer vast potential for future applications in eco-friendly catalytic processes.

Dai et al. (2020) published a remarkable study on lead-free perovskite-catalyzed organic reactions using visible light. They developed $Cs_3Bi_2Br_9$ halide perovskites incorporated into mesoporous SBA-15 silica ($Cs_3Bi_2Br_9$/SBA-15) as a nano photocatalyst (Scheme 6.9). This catalyst enabled the stimulation of hydrocarbons to create useful oxygenates in the presence of atmospheric oxygen. Optimization experiments revealed that perovskite, when loaded at 10 wt%, achieved a maximum transformation rate of 12,600 μmol $gcat^{-1}$ h^{-1} with superior selectivity. Especially, aromatic hydrocarbons exhibited considerably higher conversion rates compared to aliphatic ones under optimal circumstances. This method showed compatibility with various substrates, allowing for the synthesis of significant oxygenates like alcohols and aldehydes/ketones. Crucially, no transformations occurred without light or $Cs_3Bi_2Br_9$ halide perovskites present. The incredible photocatalytic efficiency of $Cs_3Bi_2Br_9$/SBA-15 (2–5 nm) can be assigned to several factors. These include their heightened capacity for absorbing visible light, huge surface area, greater charge separation, increased availability of active sites, and appropriate band gap. A multitude of analytical techniques, including isotopic labeling and density functional theory, were commissioned to investigate the structure of these catalysts.

Wu and his team (2018) introduced an innovative approach to cross-dehydrogenative coupling reactions. They used 1 mol% nanocrystalline lead-halide

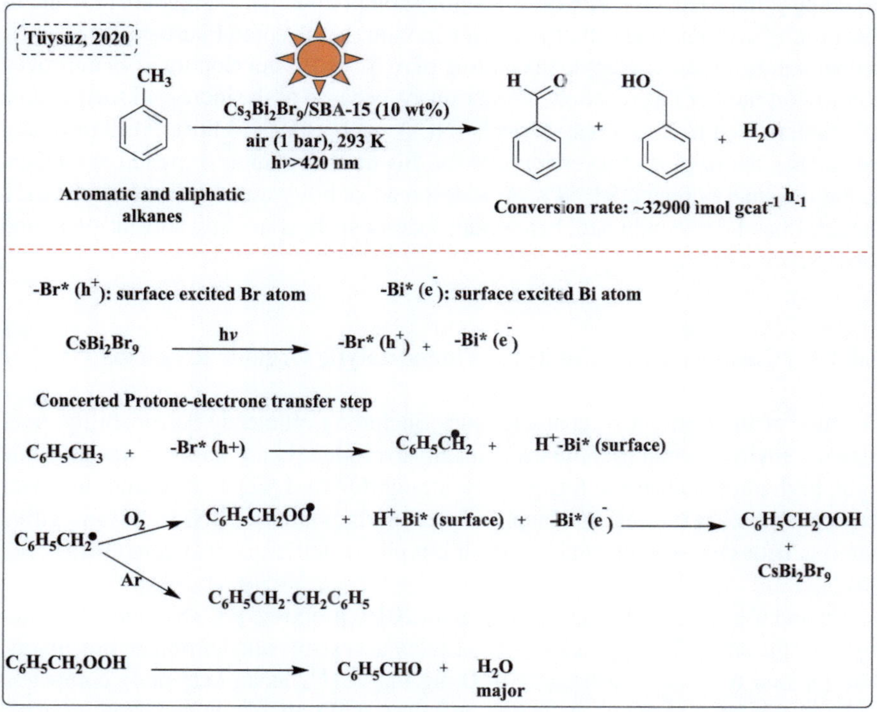

Scheme 6.9 Visible-light-driven C–H oxidation reaction using $Cs_3Bi_2Br_9$ nanomaterials

perovskite photocatalysts ($CsPbBr_3$) and exposed them to visible light. After thorough experimentation, it was observed that the most effective improved environment for the cross-dehydrogenative coupling between tertiary amines and phosphite esters (yielding 50–96% isolated yields) was achieved using $CsPbBr_3$ perovskite photocatalyst in CH_2Cl_2, with or without an oxygen environment. It is worth noting that the bromide-rich $CsPbBr_3$ perovskite photocatalyst yielded the utmost yields (98%) under optimized environment. Importantly, the desired product did not form when $CsPbBr_3$ perovskite or light irradiation was absent. The ability to reuse the synthesized nanomaterial is a notable advantage of this method (Scheme 6.10).

Zhu et al. (2019) suggested an alternative methodology to photocatalyst design, aiming to create a substantially reactive system for C–C bond development reactions via C–H initiation. In their findings, they utilized colloidal perovskite nanocrystals, specifically $CsPbBr_3$ (P1 with sizes ranging from 2 to 100 nm), as a competent photocatalyst. In order to achieve optimal results in their reaction, they employed 1.0 mg of synthesized photocatalyst, 1.0 mmol of 2,6-lutidine, 0.1 mmol of bis(2-chloroethyl)amine hydrochloride in DCM (Scheme 6.11). The employed photocatalysts underwent characterization consuming numerous analytical tools. Particularly, the required compound failed to develop without synthesized photocatalyst or light exposure. With these improved conditions, they magnificently converted a diversity of substrates, encompassing both electron-withdrawing groups (EWGs) and electron-donating groups (EDGs), into the analogous compounds. They argue that the manifestation of extraordinary charge carrier motion, a substantial surface area, and a proper band gap can accelerate the parting of e^-/h^+ pairs, developing in an improved conversion rate for this reaction. Mysteriously, because of the increased surface area of smaller-sized $CsPbBr_3$ nanocrystals (NCs), such as P2 (14 nm), P3 (9 nm), and P4 (6 nm), the original conversion rate for this transformation is greater associated to the originally created P1 NCs. A widespread stability evaluation of synthesized material has been conducted, examining factors such as acidity, solvent type, and ions (Scheme 6.12).

6.4.2.3 Carbon Nitride Catalyzed Photocatalytic Organic Reactions

Because of their attractive characteristics, including efficiency, affordability, non-toxicity, environmental friendliness, capacity to produce charge carriers under visible light, high photon absorption capability, straightforward preparation from inexpensive materials, appropriate bandgap, and great surface area, $g\text{-}C_3N_4$ stands as a gifted heterogeneous nano photocatalyst for various photocatalytic conversions (Wang et al. 2012).

Hu and Cai's research teams (Geng et al. 2019) proposed $g\text{-}C_3N_4$ nano photocatalyst for the sp^3 C–H oxygenation of excitedly accessible alkyl aromatic precursors. The method elaborates utilizing 10–20 mg of $g\text{-}C_3N_4$ as an extremely competent and reusable heterogeneous nano photocatalyst, willingly obtainable alkyl aromatic precursors for example ethers, amines, and alkylarenes, molecular O_2 as an oxidant. This approach assisted the production of useful isoindolinones, isoquinolinones,

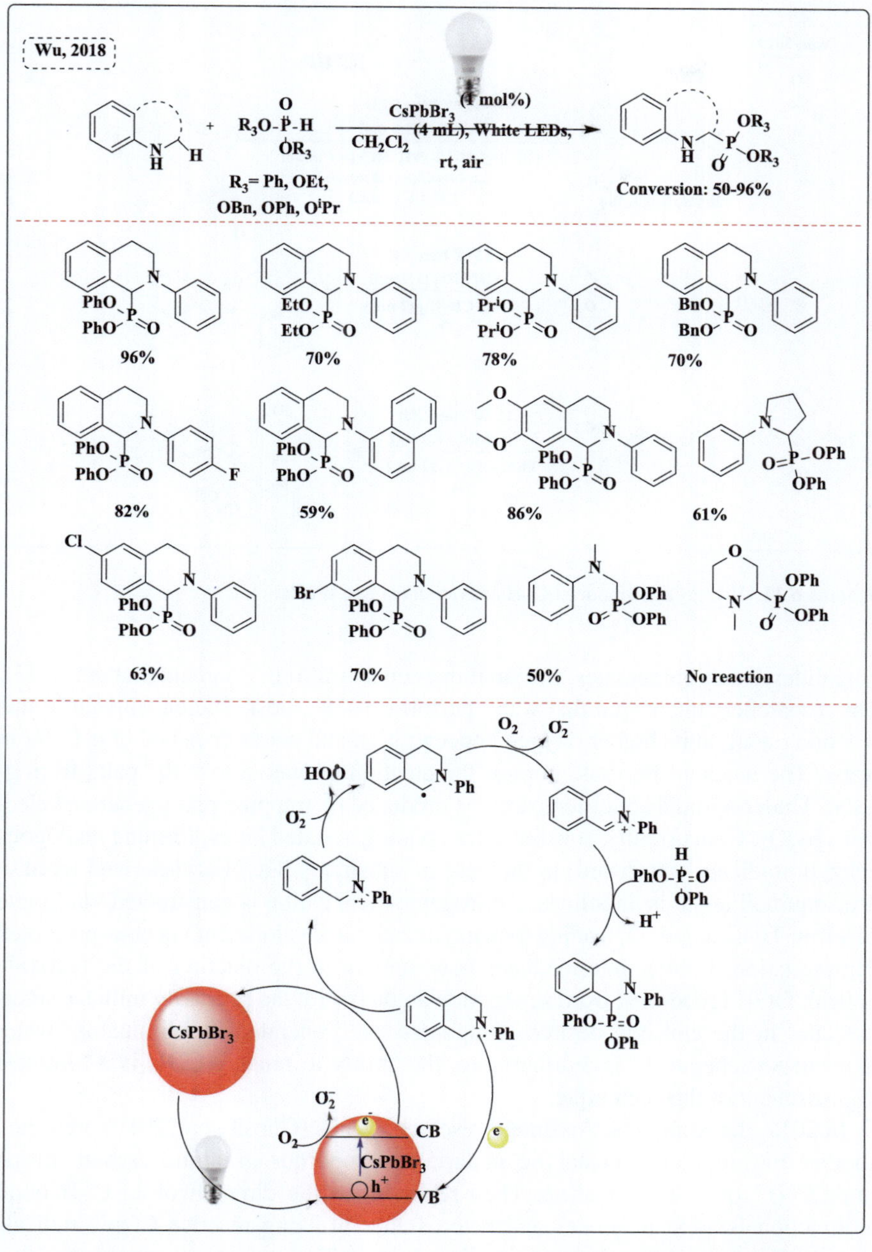

Scheme 6.10 Photocatalytic of C–H functionalization using $CsPbBr_3$ perovskite photocatalyst

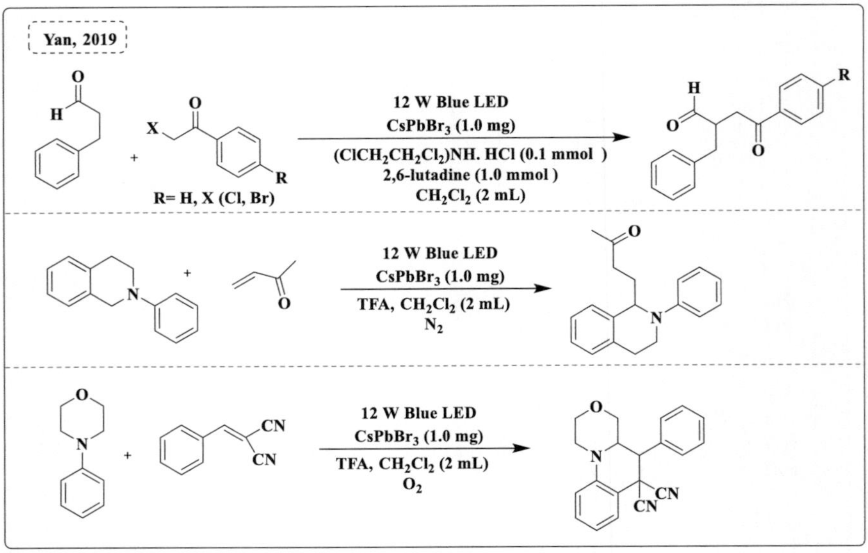

Scheme 6.11 Visible-light-induced C–H activation using $CsPbBr_3$

phthalides, isochromannones, and xanthones under mild environment (Scheme 6.13). The researchers uncovered that water performs as the most efficient solvent in this reaction owing to its higher oxygen concentration and better dispersal of g-C_3N_4 in water. The mechanistic route implies the creation of energetic e^-/h^+ pairs from g-C_3N_4. Under optimal conditions, $O_2^{\cdot-}$ is produced by trapping photogenerated electrons by O_2. Concurrently, substrates trap photogenerated holes, forming analogous benzylic radicals. Afterward, in the case of substrates like N-aryl-derived tetrahydroisoquinolines or isoindolines, a nitrogen radical cation is constructed via Single Electron Transfer (SET), leading to benzylic radical development via base-promoted deprotonation. Further intermediates may start from the reaction of the peroxide radical $^{\bullet}OOH$ (generated from superoxide anions) with the primarily initiated intermediate. In the end, the desired compounds are generated by eliminating water molecules (Scheme 6.14). Furthermore, the ability to reuse g-C_3N_4 is a foremost improvement of this technique.

In 2019, the König and Antonietti research groups (Ghosh et al. 2019) presented another intriguing photocatalytic material, mesoporous graphitic carbon nitride (mpg-CN), for C–H activation. They reported on the direct twofold C–H bond biofunctionalization of arenes and heteroarenes utilizing reactive $C(sp^3)$ radicals and strong nucleophiles (Schemes 6.15 and 6.16). The optimal environments intricate consuming 12 mg of mpg-CN under a N_2 atmosphere and visible light irradiation. No transformation ensued in the absence of light or mpg-CN. With the utilization of these optimized conditions and environments, a diverse range of substrates, encompassing both electron-withdrawing groups (EWGs) and electron-donating groups

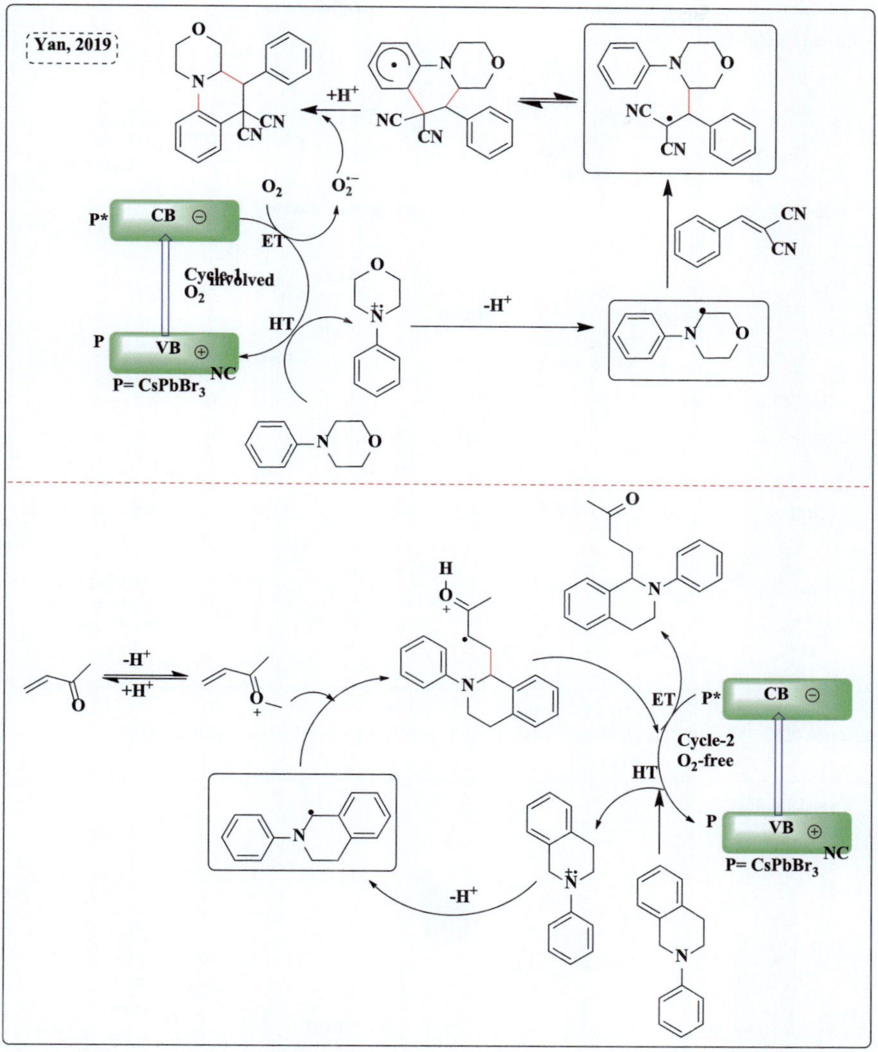

Scheme 6.12 The suggested mechanism for $CsPbBr_3$ catalyzed C–H activation under visible-light irradiation

(EDGs), underwent successful transformation, resulting in the formation of analogous compounds. This broad substrate scope highlights the versatility and efficacy of the approach across various chemical functionalities. Moreover, pharmaceutically relevant compounds were biofunctionalized to assess the viability of this approach. The envisioned pioneering reactivity of synthesized catalyst was substantiated through thorough mechanistic investigations.

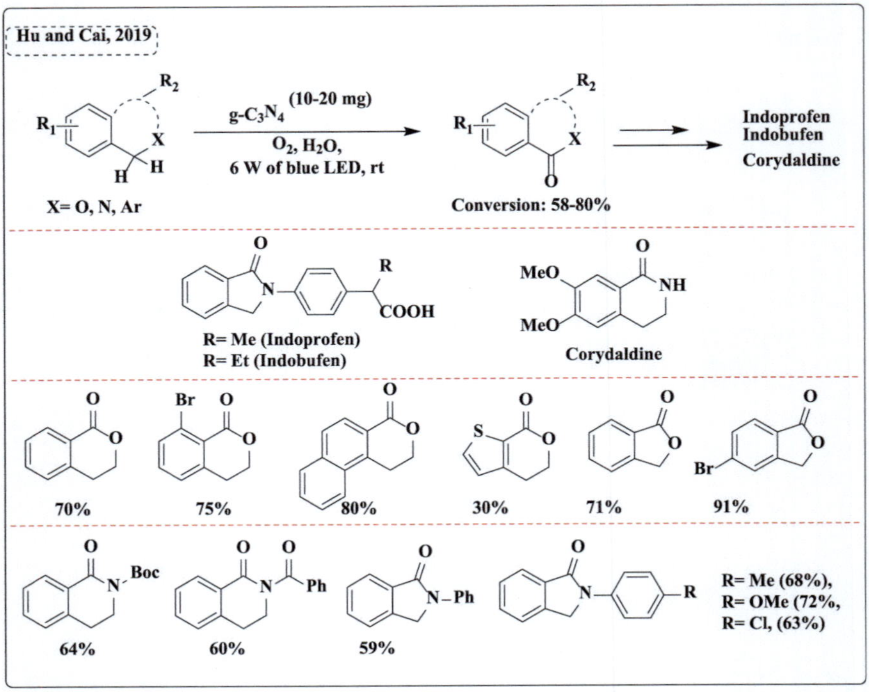

Scheme 6.13 Visible-light-driven sp^3 C–H oxygenation over $g\text{-}C_3N_4$ nanomaterials

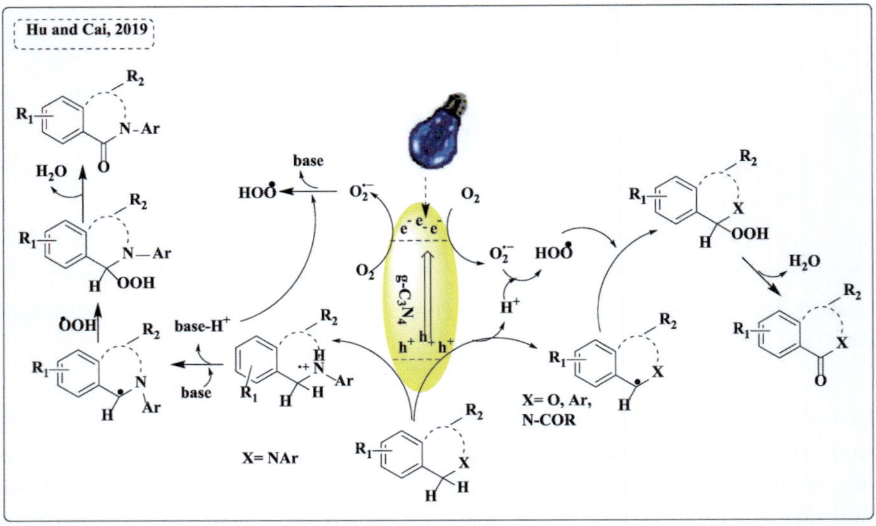

Scheme 6.14 Probable route for Visible-light-driven sp^3 C–H oxygenation employing $g\text{-}C_3N_4$

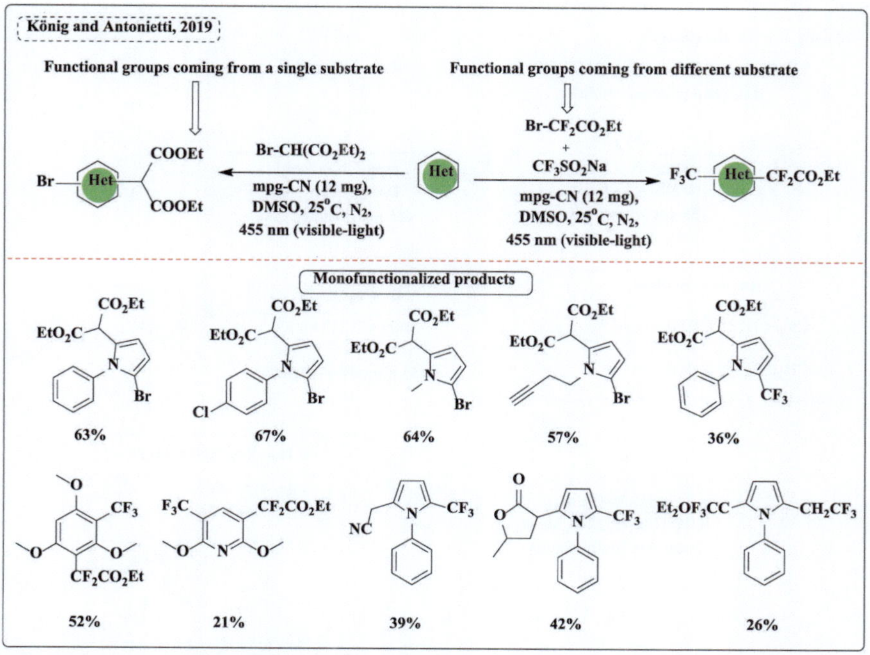

Scheme 6.15 Visible-light-driven C–H biofunctionalization in the presence of mpg-CN

In their endeavor to advance metal-free semiconductor photocatalysis (SPC) for sp^2 C–H functionalization, the teams under the leadership of Zheng et al. (2019) developed and analyzed hexagonal boron carbon nitride (h-BCN) as a nanophotocatalyst for the C–H amination of arenes under green environment. Employing 12 mg of h-BCN as nano photocatalyst, along with molecular oxygen as the oxidant and electron intermediary, this approach obviates the need for additional metals and oxidants. The notable reactivity of synthesized photocatalyst in activating "N-nucleophiles" to yield N-arylated products is endorsed to its appropriate bandgap, advanced surface area, and stability. Analytical studies underscore the pivotal function of $O_2^{\bullet-}$ and H_2O_2 for this C–H activation. The mechanistic process starts with h-BCN absorbing blue light, which generates charge carriers. This leads to the formation of oxygen radicals by using molecular oxygen. In the meantime, the excited hole induces an arene to form an arene carbon cation radical, which serves as an electrophile for nitrogen nucleophiles. Intermediates are then produced through a series of reactions between the nucleophile and the initial intermediate via direct C–H aryl amination, followed by deprotonation steps. Finally, the desired product is obtained through the oxidative aromatization. As well, the recyclability of h-BCN emerges as a substantial improvement of this approach (Scheme 6.17).

Broadening their exploration of visible-light-driven organic photocatalysis, the Zhang et al. (2021) has introduced a semiheterogeneous metal-free catalytic system. This system features PCN/Cl_4NHPI (polymeric carbon nitrides/N-hydroxy

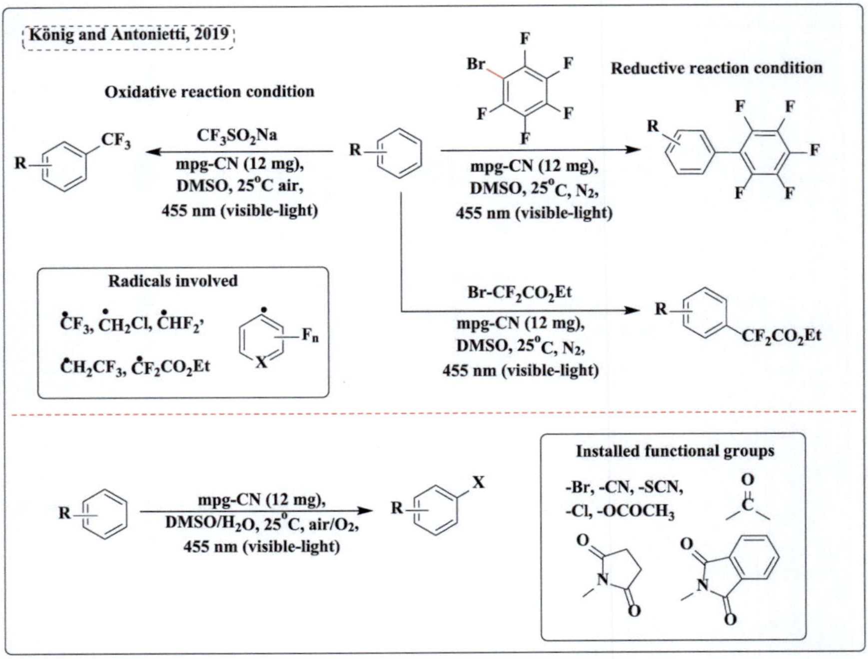

Scheme 6.16 Application of mpg-CN-catalyzed C–H biofunctionalization

tetrachloro-phthalimide) as a nano photocatalyst. It enables the tertiary benzylic C(sp^3)-H hydroxylation, facilitating the synthesis of tertiary benzylic alcohols. The characterized PCN were analyzed using diverse analytical methods. Remarkably, the target product was not obtained without PCN/Cl_4NHPI catalytic system or exposure to light (Scheme 6.18). The method utilized 10 mg of PCN as a photocatalyst, 0.3 equivalents of Cl_4NHPI as a hydrogen atom transfer (HAT) reagent, and O_2 as an oxidant in acetonitrile solvent. This setup enabled the direct production of tertiary benzylic alcohols driven by solar energy. Mechanistic studies (Scheme 6.19) uncovered that the PCN triggered O_2 to react with N-hydroxy tetrachloro-phthalimide to producing an N-hydroxy tetrachlorophthalimide radical, which served as a HAT reagent. Moreover, the ability to reuse PCN constitutes a notable benefit of this method.

In 2018, Savateev and Antonietti co-groups embarked on an intriguing exploration into the utilization of precisely engineered g-C_3N_4-based materials to expedite the stimulation of C–H bonds under visible-light irradiation (Savateev et al. 2018). Their research unveiled the remarkable effectiveness of potassium poly(heptazine imide), dubbed K-PHI, in C–H functionalization with elemental sulfur. Particularly, this procedure unfolds by forming polysulfides as intermediate species under visible-light irradiation. After thorough experimentation, the ideal conditions were identified as employing 10 mg of K-PHI in conjunction with 0.96 mg of elemental sulfur in acetonitrile (Scheme 6.20).

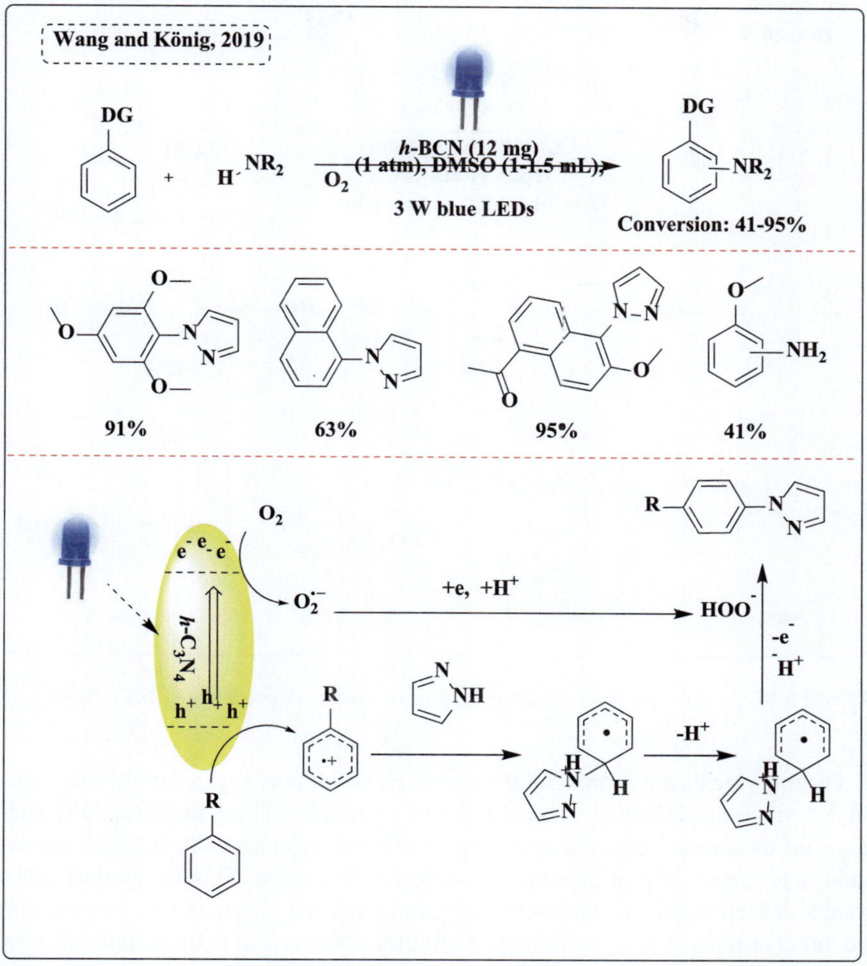

Scheme 6.17 Visible-light-induced sp^2 C–H activation using h-BCN

The produced nanomaterial underwent thorough characterization employing various analytical techniques. It is worth mentioning that the desired product failed to materialize in the absence of K-PHI or light. The outstanding activity of synthesized catalyst can be ascribed to its exceptional stability, the existence of a suitable band of delocalized states for the extra electron, and its elevated definite surface area with diverse chemical assets. Understanding the mechanism unveiled that the activation of the bond ensued from the oxidation of methylarene by the hole, accelerated by light in K-PHI*. This procedure ultimately led to the creation of a K-PHI$^{\bullet -}$ under optimized conditions (Scheme 6.21).

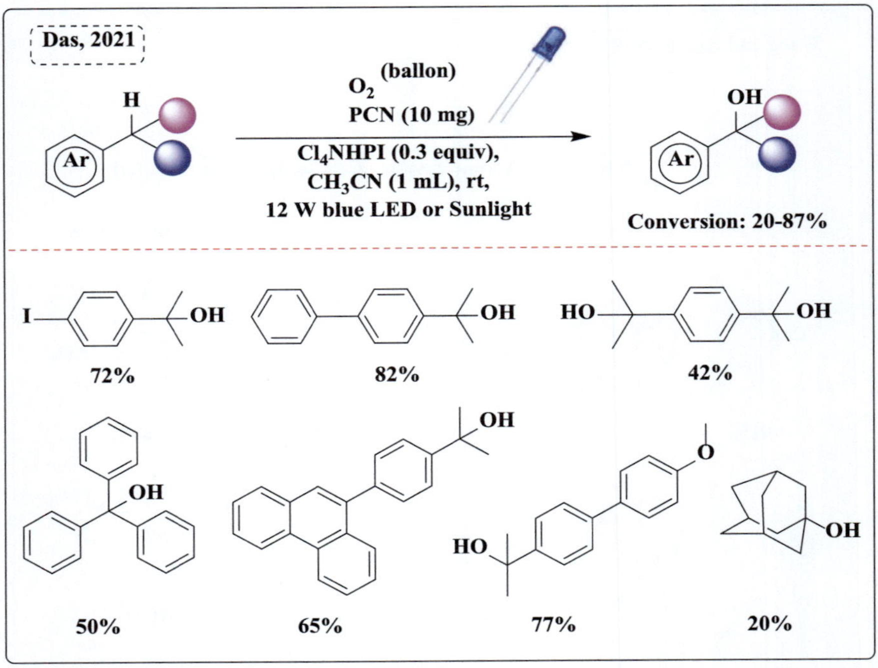

Scheme 6.18 A schematic depiction for photocatalytic hydroxylation using PCN/Cl_4NHPI

Diversifying their research endeavors into photo-induced organic transformations, the König group (Das et al. 2021) presented a groundbreaking photocatalytic technique for the direct (het)arylation of C(sp^3)–H bonds adjacent to the nitrogen atom of amides, leveraging mpg-CN in tandem with nickel catalysis. This technique displays notable selectivity for C–H functionalization targeting amide bonds, empowering the direct utilization of a plethora of voluntarily obtainable amides, containing pharmaceuticals and agrochemicals, for late-stage functionalization with incomparable chemo- and regioselectivities (Scheme 6.22). Through meticulous kinetic and mechanistic experiments, the authors propose that this process followed a unique energy-transmission route (EnT) under visible light irradiation. The experimental setup entailed 10 mg of mpg-CN as a vastly competent reusable nano photocatalyst, in conjunction with 5 mol% of $NiCl_2$•glyme, N,N-dimethyl acetamide (DMA) serving as dual roles as a C–H precursor and system medium, 2,2′-bipyridine as the optimal ligand system, and 2,6-lutidine as a base. Remarkably, the desired transformation ceases to progress in the absence of the Ni-cocatalyst with mpg-CN, even under light exposure. Two potential routes have been proposed to underlie this reaction, as outlined in Scheme 6.23. The paramount advantage of this strategy, in contrast to other approaches, stems from the utilization of mpg-CN as a heterogeneous catalyst under green conditions.

Expanding their research scope to include photocatalytic organic conversions, the Varma research group (Verma et al. 2017) engineered a novel visible-light responsive

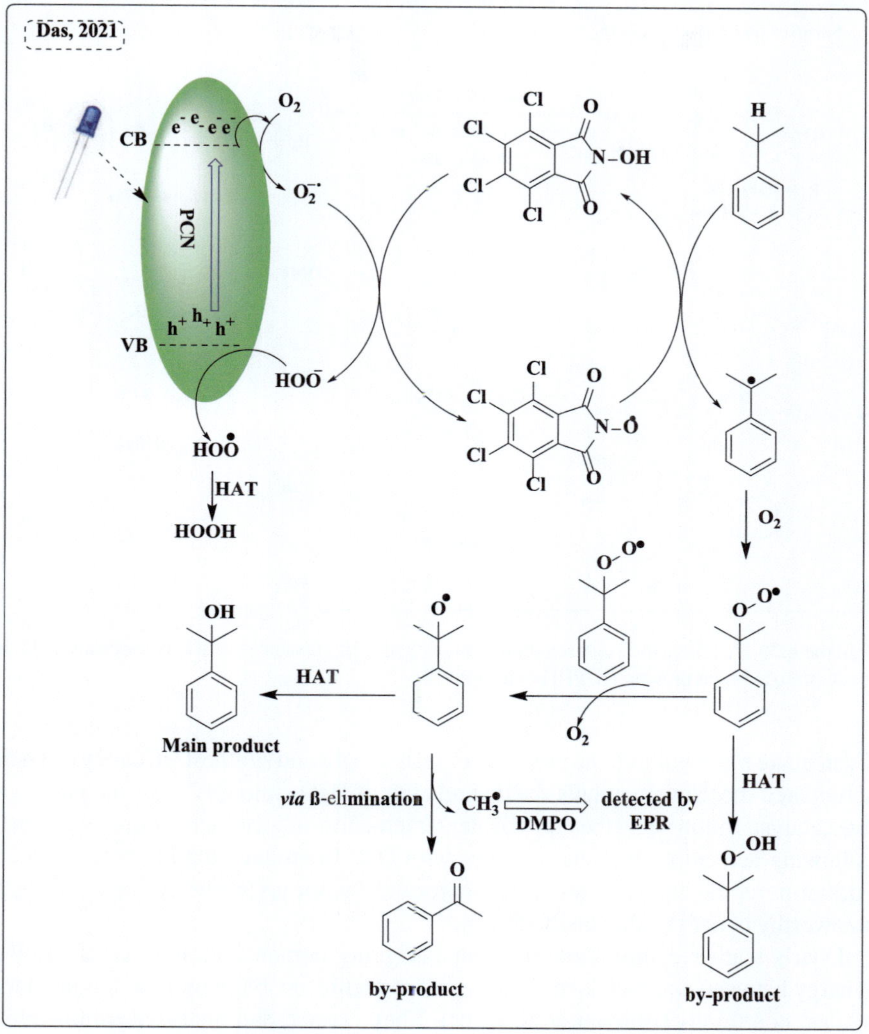

Scheme 6.19 Anticipated route for visible-light-induced C(sp^3)–H hydroxylation using PCN/Cl_4NHPI

bimetallic copper-silver catalyst supported on g-C_3N_4 (known as CuAg@g-C_3N_4). This nano photocatalyst was purposefully designed to trigger the C–H hydroxylation of benzene under mild environment. The characterized CuAg@g-C_3N_4 underwent scrutiny through numerous analytical methods. In their experimentation, they employed 100 mg of the CuAg@g-C_3N_4 as a nano photocatalyst with H_2O_2 as an oxidant in MeCN solvent (Scheme 6.24). The significant performance of this nano photocatalyst is endorsed to the synergistic interplay between Cu and Ag, accelerating reactant activation and $OH^{\bullet}$ creation. Remarkably, the activation of photocatalysis

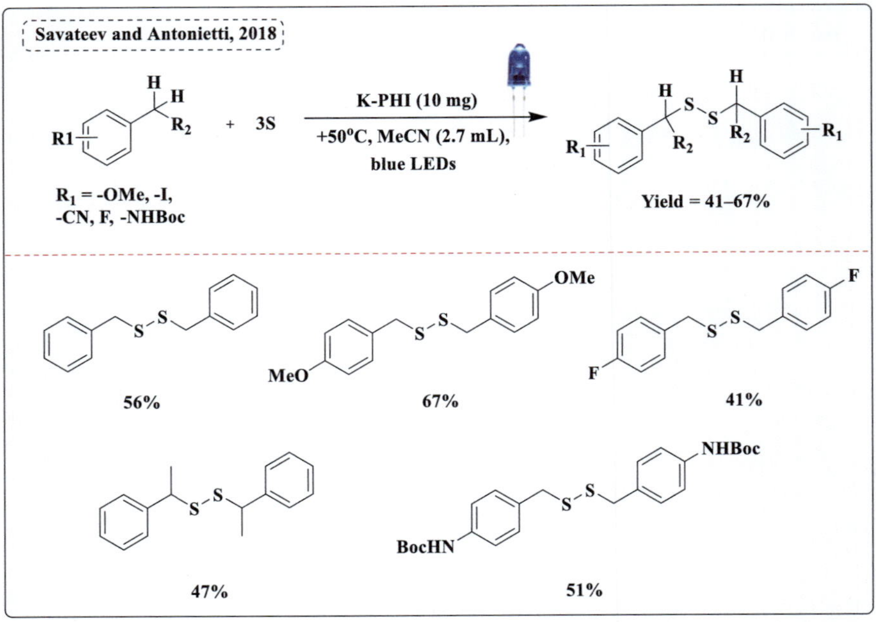

Scheme 6.20 A schematic representation featuring specific instances of the photocatalytic C–H functionalization employing (K-PHI) in acetonitrile

did not take place without the presence of visible light and synthesized catalyst. Their anticipated mechanism entails the degradation of H_2O_2 into $OH^{\bullet}$ over the photoactive catalyst, followed by the non-covalent interaction-induced activation of benzene. Following activation, benzene engages with $OH^{\bullet}$ to produce the ultimate product. Furthermore, the ability to reuse the photocatalysts for up to five cycles stands as a noteworthy benefit of this methodology.

Dhar's team recently showcased an intriguing demonstration of collaborative synergy between two metals in a photocatalytic setup, enabling hydroxylation under modest conditions (Bhuyan et al. 2018). They created and analyzed a bimetallic CuAu@g-C_3N_4 catalyst system utilizing a range of analytical techniques. Employing 30 mg of CuAu@g-C_3N_4 as a nano photocatalyst, along 5.0 mL of methyl cyanamide as a solvent, 01 mmol of H_2O_2 as an oxidant, they achieved an outstanding conversion rate of up to 99% under visible-light irradiation (Scheme 6.25). This enriched catalytic activity can be accredited to the cooperative effect between both metals, which advances the hydroxylation of benzene by generation of $OH^{\bullet}$. Moreover, UV-DRS analysis displayed the superior performance of the catalytic system, endorsed to the existence of very small NPs of Cu and Au (5–10 nm). Substantially, without visible light or the photocatalyst, there was no transformation of benzene to phenol.

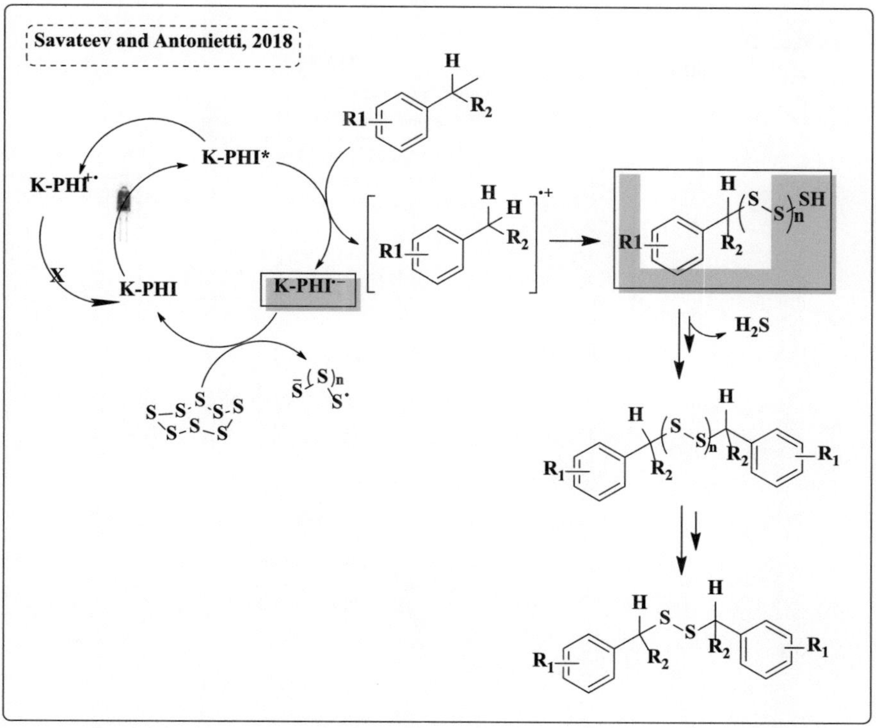

Scheme 6.21 Probable mechanism elucidating C–H activation facilitated by K-PHI

6.4.2.4 Carbon Nano-onions (CNOs) Catalyzed Visible-Light-Induced Organic Reactions

The limitations inherent in homogeneous photocatalysts, notably the challenges of separating them from reaction mixtures, highlight the imperative for advancing heterogeneous, reusable photocatalytic systems. Carbon nano-onions (CNOs) stand out among various supporting nanomaterials, pointing considerable promise for visible-light-driven nano photocatalysis. Their special attributes, including extraordinary active surface area, pronounced curvature, well-defined graphitic structure, exceptional conductivity, lack of internal porosity, tunable bandgap, and elevated stability, render them decidedly appropriate for a diversity of photocatalytic reactions. Acknowledging the effectiveness of heterogeneous catalytic systems in driving organic conversions, researchers have focused their devotion toward investigating the reactivity of CNOs-based recyclable nanomaterials, such as Ni_4Cu@CNOs, in photocatalysis.

In 2021, the Khorsandi et al. (2021) showcased a novel application wherein NiCu nanoparticles anchored on carbon nano onions (CNOs) demonstrated remarkable efficacy. They were able to convert aldehydes into ketones through a mechanism driven by visible-light-driven C–H bond activation. The obtained nanomaterials

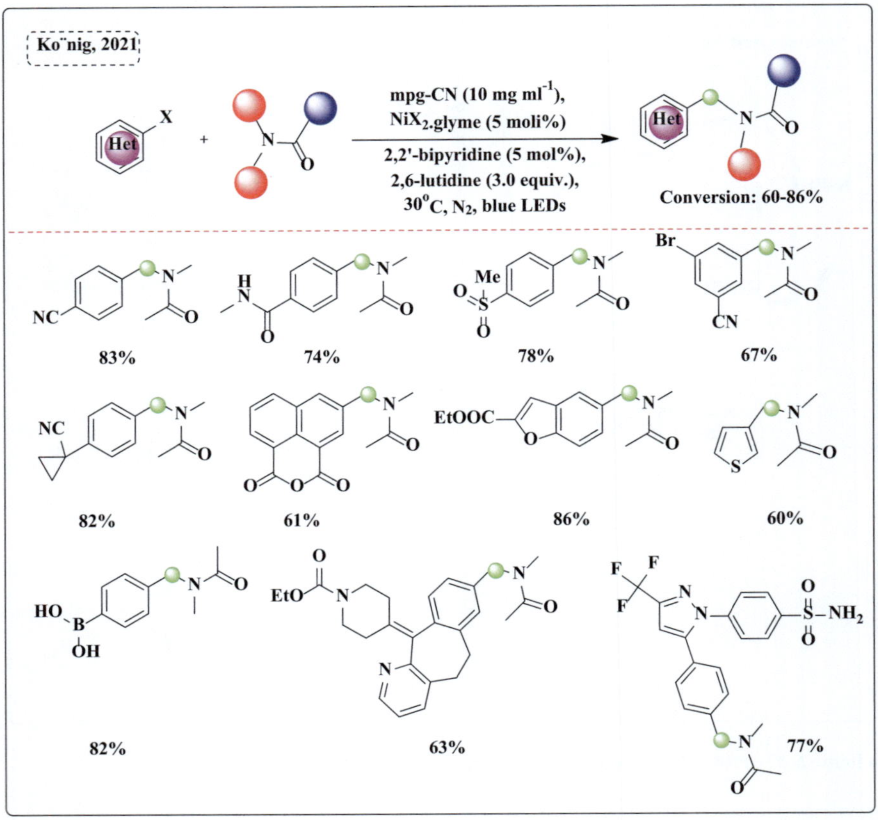

Scheme 6.22 Photocatalytic activation of C–H bonds using mpg-CN

were considerably analyzed with diverse analytical techniques. After careful fine-tuning, the optimal environments were identified: a catalytic blend consisting of 20 mg of a Ni–Cu bimetallic synergy, with 2 mmol of potassium carbonate which serving as a base acetone. These environments demonstrated highly effective in converting aryl aldehydes into diaryl ketones. Further investigation into the range of substrates demonstrated that benzaldehyde derivatives with acceptor substitutions displayed superior reaction yields under optimized conditions, surpassing those with donor substitutions. Additionally, the synthesis of medically significant compounds possessing anti-Alzheimer and anti-HIV properties underscored the versatility of this reaction. They credit the achievement of the reported reaction to the collaborative effect of the bimetallic system within the photocatalytic setup. Crucially, activation occurred solely in the presence of synthesized nanomaterials and visible light. This groundbreaking reactivity of synthesized nanomaterials was interpreted through in-depth mechanistic tests. The ability to reuse the $Ni_4Cu@CNOs$ photocatalyst for up to five cycles underscores a notable reward of this process (Scheme 6.26).

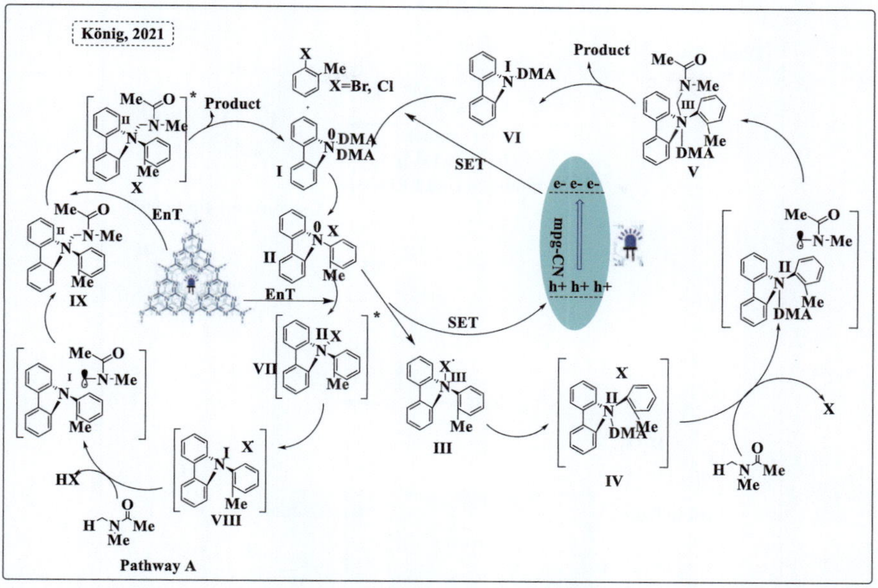

Scheme 6.23 The proposed mechanism for C–H functionalization reaction using mpg-CN

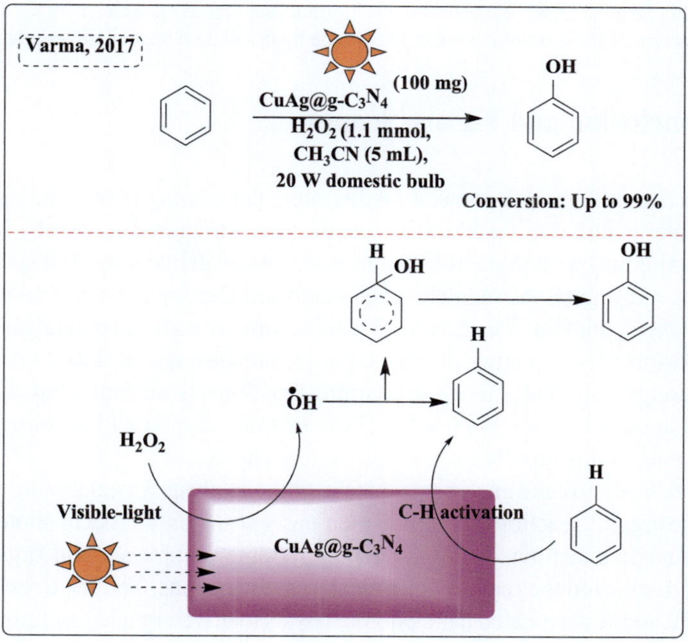

Scheme 6.24 Visible-light-driven hydroxylation facilitated by the CuAg@g-C_3N_4

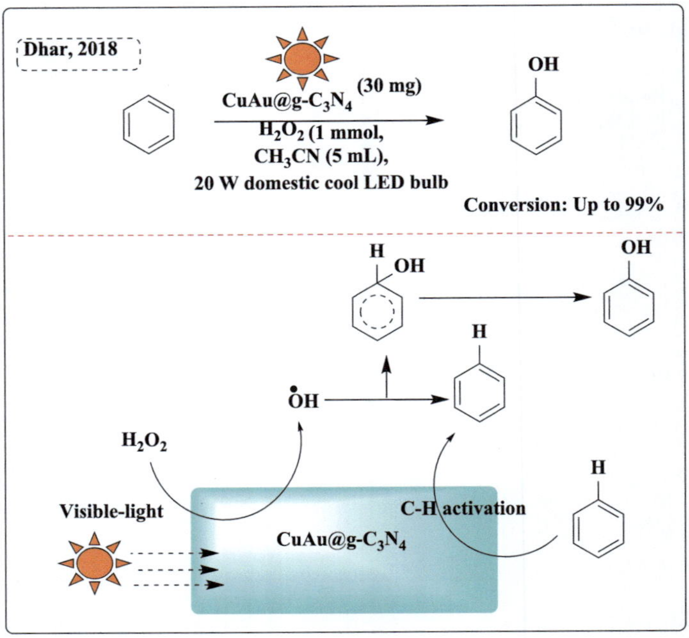

Scheme 6.25 Scheme depicting the overall reaction and anticipated mechanism for the cooperative catalytic reaction of the bimetallic system in benzene hydroxylation under visible-light exposure

6.5 Conclusion and Future Scopes

We offer a comprehensive review of the literature concerning photocatalytic organic transformations utilizing visible light-reactive nanostructured materials. Instead of merely cataloguing a plethora of findings, we focus on delineating strategic methodologies for crafting nanostructured photocatalysts, thereby elevating awareness of this burgeoning domain. The fusion of photocatalysis with nano-catalysis offers a perfect platform for directing diverse organic transformations under visible light, encompassing C–H bond activation, coupling reactions, bond formation, hydroxylation reactions, and beyond. We firmly believe that this chapter will encourage further advancements in this rapidly evolving field of research.

The transformative advancements in nano photocatalysts for catalyzing green and sustainable organic reactions are still in their nascent stages. Classical photocatalysts like Ru- and Ir-transition metal-based polypyridine complexes and organic dyes have long dominated the realm of chemical transformations. Yet, the quest for more economical and sophisticated nano photocatalysts to drive organic reactions has only just begun. Likewise, the critical hurdles in photo-driven organic transformations, involving the selectivity of ideal photocatalysts with appropriate band gaps and site

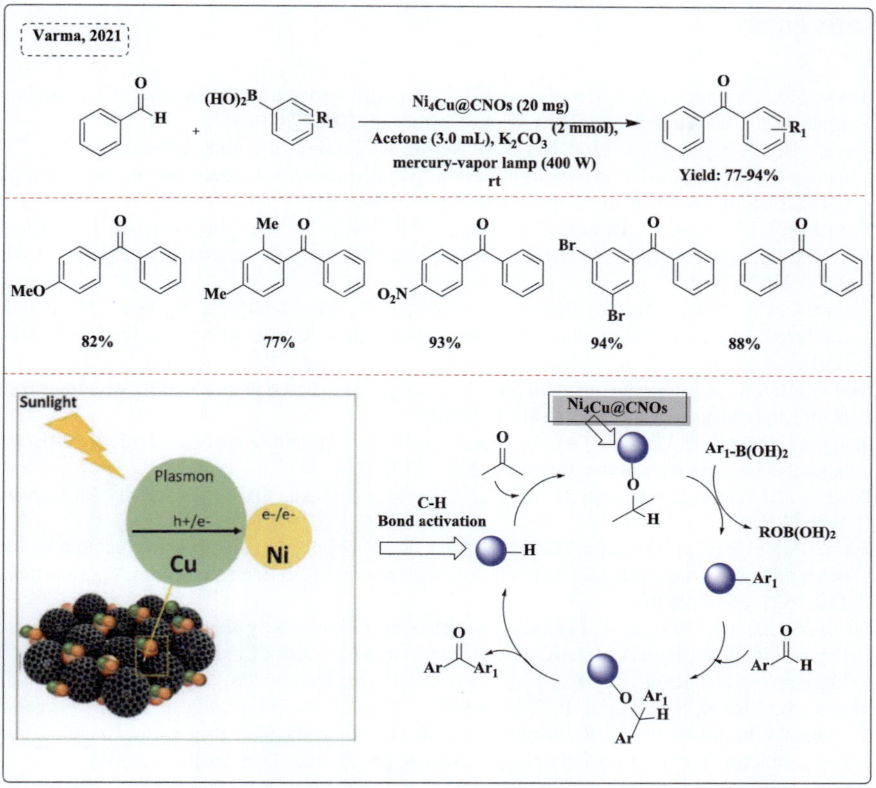

Scheme 6.26 Visible-light-driven C–H functionalization bond over Ni_4Cu@CNOs nanomaterials

specificity, have yet to be fully delivered. Given the promising potential of photo-driven organic transformations through nano-photocatalysis, continued works are needed to drive advancements in this promptly evolving research domain.

Acknowledgements P.S. extends her appreciation to the funding programs, Thuringian Scheme for Young Female Researchers and Young Female Artists, ProChance[career] Programme for the Promotion of Equal Opportunities in Academia as well as the Walter Benjamin Programme (DFG, German Research Foundation, project number: 081410-54) and the Deutsche Forschungsgemeinschaft via the TRR 234 CataLight (DFG, project number: 364549901—TRR 234 [B6] for providing the financial support. V.P. express their gratitude to the Royal Society of Chemistry for their support through the RSC Research Fund grant (R24-4460070889).

Consent for Publication Not Applicable.

Conflicts of Interest The author affirms that there are no conflicts of interest, whether financial or otherwise, to declare.

References

Ajiboye, T.O., Kuvarega, A.T., Onwudiwe, D.C.: Graphitic carbon nitride-based catalysts and their applications: a review. Nano-Struct. Nano-Objects **24**, 100577 (2020)

Bhuyan, B., Devi, M., Bora, D., Dhar, S.S., Newar, R.: Design of a photoactive bimetallic Cu-Au@g-C_3N_4 catalyst for visible light-driven hydroxylation of the benzene reaction through C-H activation. Eur. J. Inorg. Chem. **2018**, 3849–3858 (2018)

Chakraborty, I.N., Roy, S., Devatha, G., Rao, A., Pillai, P.P.: InP/ZnS quantum dots as efficient visible-light photocatalysts for redox and carbon-carbon coupling reactions. Chem. Mater. **31**, 2258–2262 (2019)

Chandrashekar, H.B., Maji, A., Halder, G., Banerjee, S., Bhattacharyya, S., Maiti, D.: Photo-catalyzed borylation using water-soluble quantum dots. Chem. Commun. **55**, 6201–6204 (2019)

Chen, Y., Bai, X.: A review on quantum dots modified g-C_3N_4-based photocatalysts with improved photocatalytic activity. Catalysts **10**, 142 (2020)

Cheng, L., Zhang, H., Li, X., Fan, J., Xiang, Q.: Carbon-graphitic carbon nitride hybrids for heterogeneous photocatalysis. Small **17**, 2005231 (2021)

Clayden, J., Greeves, N., Warren, S.: Organic Chemistry, 2nd edn. Oxford University Press, New York (2012)

Dai, Y., Poidevin, C., Ochoa-Hernández, C., Auer, A.A., Tüysüz, H.: A supported bismuth halide perovskite photocatalyst for selective aliphatic and aromatic C-H bond activation. Angew. Chem. **132**, 5837–5845 (2020)

Dandia, A., Saini, P., Sharma, R., Parewa, V.: Green organic synthesis by photochemical protocol. In: Inamuddin, B.R., Asiri, A.M. (eds.) Green Sustainable Process for Chemical and Environmental Engineering and Science, Chap. 7, pp 155–198. Elsevier Inc. (2020a)

Dandia, A., Parihar, S., Sharma, R., Rathore, K.S., Parewa, V.: Nanocatalysis in green organic synthesis. In: Inamuddin, B.R., Asiri, A.M., (eds.) Green Sustainable Process for Chemical and Environmental Engineering and Science, Chap. 4, pp. 71–103. Elsevier Inc. (2020b)

Dandia, A., Bansal, S., Sharma, R., Mahawar, D.K., Rathore, K.S., Meena, M.L., Parewa, V.: Nanoporous Cu doped ZnS nanoparticles are an efficient photocatalyst for the chemoselective synthesis of 2-substituted azoles *via* CN arylation/CSp3-H oxidation/cyclization/dehydration sequence in visible light. J. Photochem. Photobiol. **389**, 112242 (2020c)

Dandia, A., Gupta, S.L., Saini, P., Sharma, R., Meena, S., Parewa, V.: Structure couture and appraisal of catalytic activity of carbon nitride (g-C_3N_4) based materials towards sustainability. Curr. Res. Green Sustain. Chem. **3**, 100039 (2020d)

Dandia, A., Saini, P., Sharma, R., Parewa, V.: Visible light-driven perovskite-based photocatalysts: a new candidate for green organic synthesis by photochemical protocol. Curr. Res. Green Sustain. Chem. **3**, 100031 (2020e)

Dandia, A., Saini, P., Sethi, M., Kumar, K., Saini, S., Meena, S., Parewa, V.: Nanocarbons in quantum regime: an emerging sustainable catalytic platform for organic synthesis. Catal. Rev. 1–55 (2021)

Das, S., Murugesan, K., Villegas Rodríguez, G.J., Kaur, J., Barham, J.P., Savateev, A., König, B.: Photocatalytic (het) arylation of C (sp3)-H bonds with carbon nitride. ACS Catal. **11**, 1593–1603 (2021)

Dietzek-Ivanšić, B., Griesbeck, A., Strehmel, B., Wächtler, M.: Photochemistry in Germany. ChemPhotoChem **6**, e202200103 (2022)

Djurišić, A.B., He, Y., Ng, A.M.: Visible-light photocatalysts: prospects and challenges. APL Mater. **8**, 030903 (2020)

Duarah, R., Karak, N.: Hyperbranched polyurethane/palladium-reduced carbon dot nanocomposite: an efficient and reusable mesoporous catalyst for visible-light-driven C-C coupling reactions. Ind. Eng. Chem. Res. **58**, 16307–16319 (2019)

Firoozi, S., Hosseini-Sarvari, M.: Nanosized CdS as a reusable photocatalyst: the study of different reaction pathways between tertiary amines and aryl sulfonyl chlorides through visible-light-induced N-dealkylation and C-H activation processes. J. Org. Chem. **86**, 2117–2134 (2021)

Firoozi, S., Hosseini-Sarvari, M., Koohgard, M.: Solvent-free and room temperature visible light-induced C-H activation: CdS as a highly efficient photo-induced reusable nano-catalyst for the C-H functionalization cyclization of t-amines and C-C double and triple bonds. Green Chem. **20**, 5540–5549 (2018)

Fox, M.A., Dulay, M.T.: Heterogeneous photocatalysis. Chem. Rev. **93**, 341–357 (1993)

Ge, J., Zhang, Y., Park, S.J.: Recent advances in carbonaceous photocatalysts with enhanced photocatalytic performances: a mini-review. Materials **12**, 1916 (2019)

Geng, P., Tang, Y., Pan, G., Wang, W., Hu, J., Cai, Y.: A gC3N4-based heterogeneous photocatalyst for visible light-mediated aerobic benzylic C-H oxygenations. Green Chem. **21**, 6116–6122 (2019)

Ghosh, S.: Visible-Light-Active Photocatalysis: Nanostructured Catalyst Design, Mechanisms, and Applications, 1st edn. Wiley (2018)

Ghosh, I., Khamrai, J., Savateev, A., Shlapakov, N., Antonietti, M., König, B.: Organic semiconductor photocatalyst can bifunctionalize arenes and heteroarenes. Science **365**, 360–366 (2019)

Grzybowski, B.A., Bishop, K.J., Kowalczyk, B., Wilmer, C.E.: The 'wired' universe of organic chemistry. Nat. Chem. **1**, 31–36 (2009)

Khorsandi, Z., Metkazini, S.F.M., Heydari, A., Varma, R.S.: Visible light-driven direct synthesis of ketones from aldehydes via CH bond activation using NiCu nanoparticles adorned on carbon nano onions. Mol. Catal. **516**, 111987 (2021)

König, B.: Photocatalysis in organic synthesis-past, present, and future. Eur. J. Org. Chem. **2017**, 1979–1981 (2017)

Kudo, A.: Development of photocatalyst materials for water splitting. Int. J. Hydrog. Energy **31**, 197–202 (2006)

Luciani, G., Imparato, C., Vitiello, G.: Photosensitive hybrid nanostructured materials: the big challenges for sunlight capture. Catalysts **10**, 103 (2020)

Maaoui, H., Kumar, P., Kumar, A., Pan, G.H., Chtourou, R., Szunerits, S., Jain, S.L.: A Prussian blue/carbon dot nanocomposite as an efficient visible light active photocatalyst for CH activation of amines. Photochem. Photobiol. Sci. **15**, 1282–1288 (2016)

Pal, A., Ghosh, I., Sapra, S., König, B.: Quantum dots in visible-light photoredox catalysis: reductive dehalogenations and C-H arylation reactions using aryl bromides. Chem. Mater. **29**, 5225–5231 (2017)

Pelizzetti, E., Serpone, N. (eds.): Homogeneous and Heterogeneous Photocatalysis. Springer Science & Business Media (2012)

Pirsaheb, M., Asadi, A., Sillanpää, M., Farhadian, N.: Application of carbon quantum dots to increase the activity of conventional photocatalysts: a systematic review. J. Mol. Liq. **271**, 857–871 (2018)

Radhika, N.P., Selvin, R., Kakkar, R., Umar, A.: Recent advances in nano-photocatalysts for organic synthesis. Arab. J. Chem. **12**, 4550–4578 (2019)

Saini, P., Meena, S., Mahawar, D.K., Dandia, A., Parewa, V.: Principles and concepts of nanocatalysis. In: Rai, V.K., Singh, M., Rai, A. (eds.) Advanced Nanocatalysis for Organic Synthesis and Electroanalysis, Chap. 1, pp. 1–23. Bentham Science Publishers (2022)

Savateev, A., Kurpil, B., Mishchenko, A., Zhang, G., Antonietti, M.: A "waiting" carbon nitride radical anion: a charge storage material and a key intermediate in direct C-H thiolation of methylarenes using elemental sulfur as the "S"-source. Chem. Sci. **9**, 3584–3591 (2018)

Sharma, N., Ojha, H., Bharadwaj, A., Pathak, D.P., Sharma, R.K.: Preparation and catalytic applications of nanomaterials: a review. RSC Adv. **5**, 53381–53403 (2015)

Tahir, M.B., Iqbal, T., Rafique, M., Rafique, M.S., Nawaz, T., Sagir, M.: Nanomaterials for photocatalysis. In: Tahir, M.B., Rafique, M., Rafique, M.S. (eds.) Nanotechnology and Photocatalysis for Environmental Applications, Chap. 5, pp. 65–76. Elsevier Inc. (2020)

Tong, H., Ouyang, S., Bi, Y., Umezawa, N., Oshikiri, M., Ye, J.: Nano-photocatalytic materials: possibilities and challenges. Adv. Mater. **24**, 229–251 (2012)

Verma, S., Nasir Baig, R.B., Nadagouda, M.N., Varma, R.S.: Hydroxylation of benzene via C-H activation using bimetallic CuAg@g-C_3N_4. ACS Sustain. Chem. Eng. **5**, 3637–3640 (2017)

Wang, X., Blechert, S., Antonietti, M.: Polymeric graphitic carbon nitride for heterogeneous photocatalysis. ACS Catal. **2**, 1596–1606 (2012)

Wu, W.B., Wong, Y.C., Tan, Z.K., Wu, J.: Photo-induced thiol coupling and C-H activation using nanocrystalline lead-halide perovskite catalysts. Catal. Sci. Technol. **8**, 4257–4263 (2018)

Xu, C., Anusuyadevi, P.R., Aymonier, C., Luque, R., Marre, S.: Nanostructured materials for photocatalysis. Chem. Soc. Rev. **48**, 3868–3902 (2019)

Xuan, J., Xiao, W.J.: Visible-light photoredox catalysis. Angew. Chem. Int. Ed. **51**, 6828–6838 (2012)

Zhang, J., Tian, B., Wang, L., Xing, M., Lei, J. (eds.): Mechanism of photocatalysis. In: Photocatalysis: Fundamentals, Materials and Applications, Chap. 1, pp. 1–15. Springer, Singapore (2018)

Zhang, Y., Qin, S., Claes, N., Schilling, W., Sahoo, P.K., Ching, H.V., Das, S.: Direct solar energy-mediated synthesis of tertiary benzylic alcohols using a metal-free heterogeneous photocatalyst. ACS Sustain. Chem. Eng. **10**, 530–540 (2021)

Zheng, M., Ghosh, I., König, B., Wang, X.: Metal-free semiconductor photocatalysis for sp^2 C-H functionalization with molecular oxygen. ChemCatChem **11**, 703–706 (2019)

Zhu, X., Lin, Y., San Martin, J., Sun, Y., Zhu, D., Yan, Y.: Lead halide perovskites for photocatalytic organic synthesis. Nat. Commun. **10**, 1–10 (2019)

Chapter 7
Enzyme Catalysis: A Green Chemistry Perspective

Sucheta Chatterjee

List of Abbreviations

E_a	Activation energy
API	Active pharmaceutical ingredient
Ala	Alanine
ADH	Alcohol dehydrogenase
GABA	γ-aminobutyric acid
Ab	Antibody
Abzymes	Antibody-enzymes
Bmim [BF_4]	1-butyl-3-methylimidazolium tetrafluoroborate
Bmim [PF_6]	1-butyl-3-methylimidazolium hexafluorophosphate
CAL-B	*Candida antarctica* B
((*S*)-CCMA)	(*S*)-2-Carboxyethyl-3-cyano-5-methylhexanoic acid
CNDE	2-Carboxyethyl-3-cyano-5-methylhexanoic acid ethyl ester
DAI	Desalanine-(B30)-insulin
DPP-4	Dipeptidyl peptidase-4
ES	Enzyme–substrate
Ep-PCR	Error-prone polymerase chain reaction
GI	Glucose isomerase
GDH	Glucose dehydrogenase
HHDH	Halohydrin dehalogenase
HFCS	High-fructose corn syrup
HMG-CoA	Hydroxymethylglutaryl-coenzyme A
KR	Kinetic resolution
Lys	Lysine

S. Chatterjee (✉)
Bio-Organic Division, Bhabha Atomic Research Centre, Trombay, Mumbai, India
e-mail: sucheta@barc.gov.in

Homi Bhabha National Institute, Anushaktinagar, Mumbai, India

R. S. Ningthoujam and A. K. Tyagi (eds.), *Handbook of Materials Science, Volume 3*,
Indian Institute of Metals Series, https://doi.org/10.1007/978-981-96-8229-4_7

PIKAT	Parallel interconnected kinetic asymmetric transformations
PPL	Porcine pancreatic lipase
PGCCA	Presidential Green Chemistry Challenge Awards
RTIL	Room temperature ionic liquid
SDM	Site-directed mutagenesis
SCF	Supercritical fluid
$SCCO_2$	Super critical carbon dioxide
ToGI	Thermus oshimai glucose isomerase
Thr	Threonine
TS	Transition state
VA	Vinyl acetate

7.1 Introduction

The twelve precious principles of green chemistry proposed by Paul Anastas and Warner in the year 1998 (Fig. 7.1) proved to be a landmark event in the history of chemical research and industry, as it brought about a radical change in our way of thinking (Anastas and Warner 1998).

These principles which preach of *waste prevention*, *risk minimization*, *reduction/elimination of derivatization steps* and *solvent usage*, *maximization of atom economy*

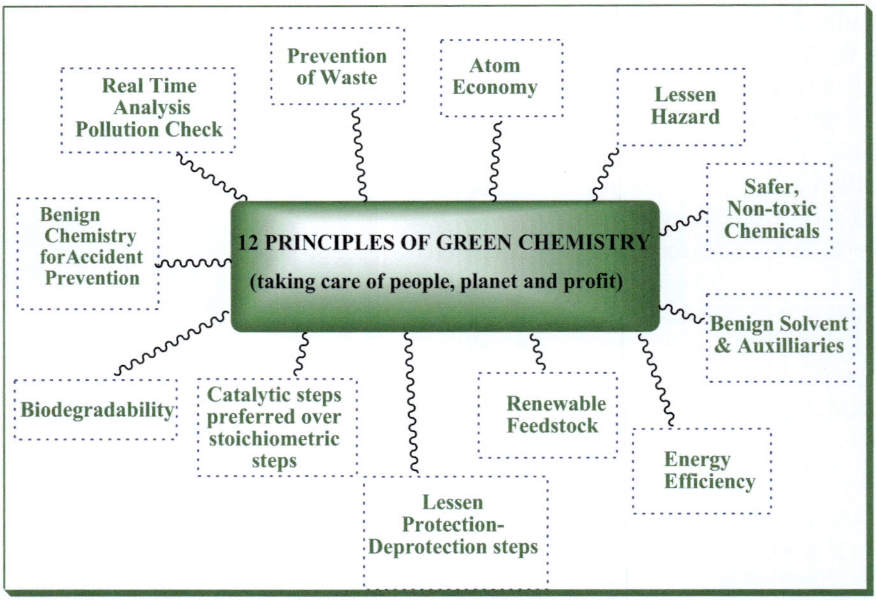

Fig. 7.1 Principles of green chemistry (adapted from reference (Anastas and Warner 1998))

and *energy efficiency*, served as guidelines for designing cleaner and sustainable processes with an eye toward economic, environmental and safety aspects.

7.2 Biocatalysts as Potential Flag Bearers of Green Chemistry

The true potential of enzymes as exponents of green chemistry were realized once it was established that these macromolecules complied with all the principles required for a green process (vide infra). The immense popularity of biocatalysts stems from the economic, environmental and operational advantages these offer (Faber 2017). Reusable, recyclable and environment-friendly biocatalysts can help process chemists/ biotechnologists to design operationally simple, atom-economic and energy efficient organic reaction protocols, obtaining the desired products selectively in higher yields with minimum waste generation, as described below.

Enzymes are environmentally benign—The plant, animal and microbial kingdom offers us enzymes in bounty. Derived from renewable sources, it is not difficult to accept why these biocompatible catalysts are far superior options than conventional heavy, toxic and precious metal catalysts which have been used since ages for the syntheses and industrial productions of fine chemicals (Sheldon and Pereira 2017; Wu et al. 2021). Unlike chemical catalysts which pile on hazardous wastes, enzymes are completely biodegradable and produce less side products.

Enzymes have higher catalytic efficiency and are required in low amounts—Chemical catalysts are required in 5–50 mol% concentrations and enhance reaction rates by a factor of $10–10^3$. The rate enhancement in an enzyme-mediated reaction is about $10^8–10^{17}$ times that of the corresponding uncatalyzed reaction. The consolidated representative reaction progress profiles of uncatalyzed and catalyzed (chemical/ enzyme) reactions are shown in (Fig. 7.2).

Biocatalysts, with superior catalytic abilities, are thereby required in miniscule (0.005–0.05 mol%) concentrations.

A wide variety of organic reactions can be catalyzed by enzymes—Barring a few exceptions, like the pericyclic reactions (Cope rearrangement, Diels–Alder reactions, etc.) (Sanz-Cervera et al. 1993; Oikawa et al. 1995), enzymes can accomplish almost all types of reactions possible in nature and even carry out selective reactions of non-activated centers in organic molecules (which is otherwise impossible by any other methodology). In-fact, growing a suitable antibody (Ab) with the reaction *transition state mimic* as the substrate, Antibody-enzymes (Abzymes) are being generated to carry out even these pericyclic classes of organic reactions (Faber 2017). Miraculously, enzyme engineering has also made reactions possible which are otherwise unknown to nature (Wu et al. 2021), and some of them have potential for large scale applications as well. For example, a rare carbene-transfer reaction (Coelho et al.

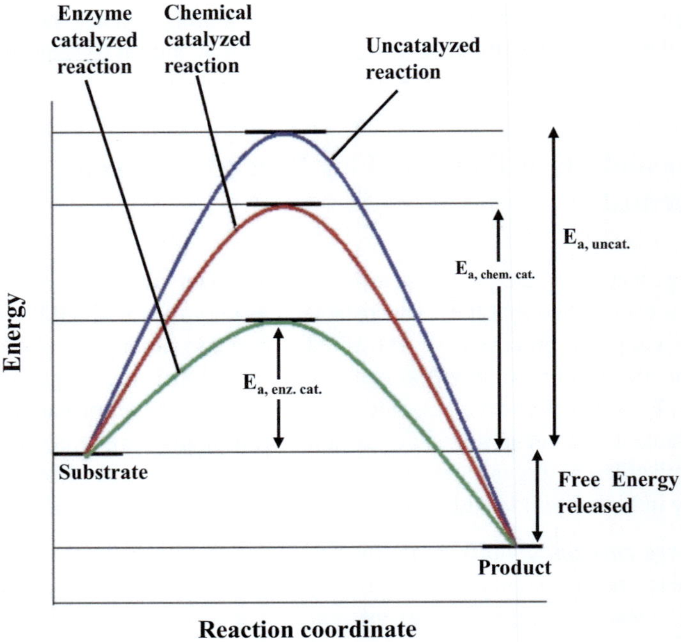

Fig. 7.2 A typical energy diagram of enzyme/chemical-catalyzed and uncatalyzed reactions

2013), where styrene is converted to a cyclopropane derivative with ethyl diazoacetate using an engineered P450 monooxygenase (derived from the P450 enzyme BM3 of *Bacillus megaterium*) (Scheme 7.1).

Enzymes are highly selective—The extremely high selectivity enzymes display is undoubtedly the most desirable trait of these super catalysts. Enzymes are (a) *Chemo-selective*—under a given condition, a specific enzyme transforms a particular functional group in a substrate leaving all other sensitive functional groups intact (unlike chemical catalysis); (b) *Enantio- and diastereoselective*—*L*-amino acids are the building bricks of these chiral catalysts, and hence any form of chirality present in the substrate is recognized by them, leading to a very high stereoselectivity. An enzyme will have a preference for only one enantiomer/diastereomer, which fits best onto it. For the ill-fitted isomer, there will be no or slower reaction; (c) *Regioselective*—enzymes can easily differentiate between two same functionalities located in

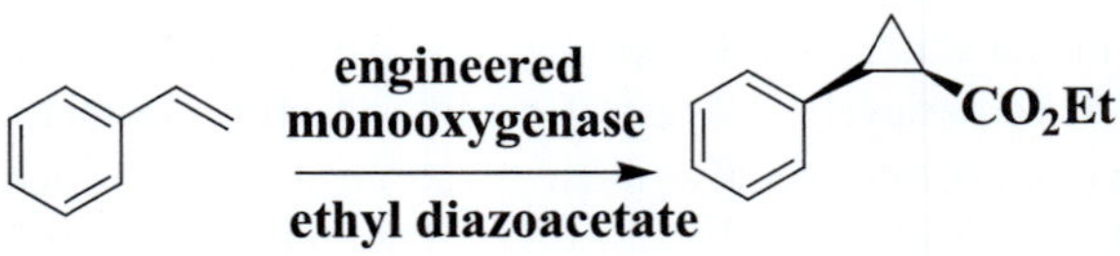

Scheme 7.1 Cyclopropanation of styrene using an engineered enzyme

two separate zones in the substrate and selectively functionalize one. The origin of enzyme stereoselectivity is depicted in Fig. 7.3. *The excellent stereo/regio/chemo-selectivity afforded by biocatalysts makes them indispensable tools for asymmetric/enantiomeric synthesis of key chiral intermediates for food & drug industries.*

Enzymes operate under mild condition—Some of the common operational challenges faced by chemical catalysis like the possibility of waste- or by-product generation, product/substrate decomposition, racemization, etc. due to harsh reaction conditions (elevated temperatures, pressures, extreme pH) can be averted by using biocatalysts.

Enzymes have wide substrate tolerance**—**One of the most attractive features of enzymes is their ability to accept non-natural man-made substrates, leading to their diverse applications (Wu et al. 2021; Coelho et al. 2013; Hughes and Lewis 2018). Had this not been the case—enzymes would have been of little practical use to us. The foregoing adequately establishes biocatalysts as powerful, efficient, energy-saving, environment-friendly and safe tools for future green chemical processes. Most of these attributes have their origins to the unique enzyme structure as explained below.

It is possible to run multi-enzymatic cascade reactions**—**Majority of the enzymes are compatible among themselves under ambient reaction conditions (as found with cytosolic reactions in living systems). Hence, cascade reactions involving multi-enzymes can be run smoothly in one-pot without hampering each other's activities. In these reactions, the end product of preceding enzymatic reaction serves as the

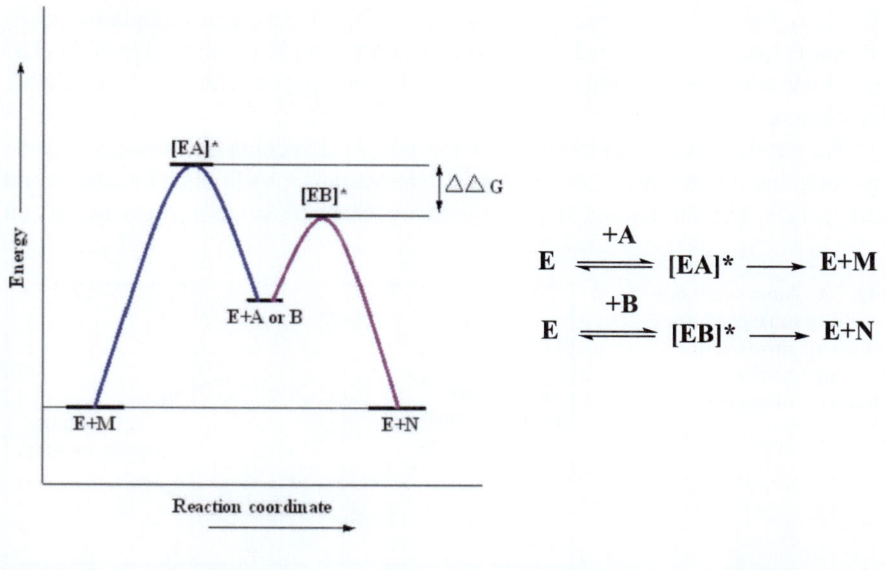

Fig. 7.3 Energy diagram for enzyme-mediated enantioselective reaction, where E: Enzyme; A, B: enantiomeric substrates; M, N: enantiomeric products; * denotes transition state; [EA], [EB]: diastereomeric enzyme–substrate complexes; ΔΔG: difference in free energy of the reactions

starting material of the succeeding reaction in linear, orthogonal or parallel routes. Linear cascades involve multistep and one-pot conversions. This reduces process time and waste generation, eliminates downstream processing steps and accumulation of unstable or toxic intermediates, leading to safer processes and better yields (since reversible equilibria are driven in the forward directions). In orthogonal cascades, the side-products are automatically by-passed. In parallel cascades, cofactors integrate two distinct biocatalytic reactions having two different substrates, furnishing two separate, isolable useful products. For example, use of alcohol dehydrogenases (ADH) and Baeyer–Villiger monooxygenases in tandem led to the parallel generation of non-racemic alcohols and sulfoxides via parallel interconnected kinetic asymmetric transformations (PIKAT) (Ricca et al. 2011).

7.3 What Makes Biocatalysts so Special?

Biocatalysts (average molecular weight of 10,000–2,00,00,000 Dalton) are basically protein molecules possessing three-dimensional architectures. The primary structures consist of a sequence of amino acids, covalently linked to each other by amide linkages forming the polypeptide structure (bond energy: ~ 100 kcal/mol). The secondary structure of enzymes, consisting of alpha (α) helices and beta (β) pleated sheets are formed due to extensive hydrogen-bonding (H-bond energy: 10–15 kcal/mol) between various closely associated, but *non-covalently bonded* amino acid residues in the polypeptide chain. For example, amino acid side chains like -NH_2 in lysine, -SH in cysteine, -OH in serine or -CO_2H in aspartic or glutamic acids, are not covalently linked and hence free to entangle in H-bonding (Fig. 7.4). The non-polar amino acid residues such as Ph, alkyl groups. contribute to hydrophobic interactions.

The various forces contributing to the tertiary structure of enzymes include hydrophobic interactions, van der Waals' interactions, ion–dipole interactions and salt bridges. The various stackings and interactions result in clefts, crevices, etc. on

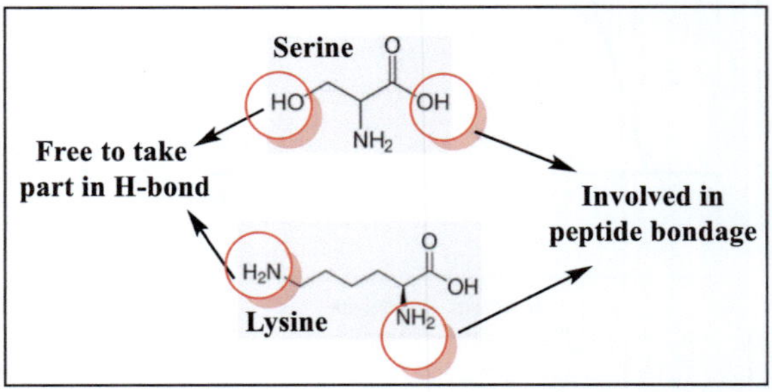

Fig. 7.4 Amino acid side chains in some enzymes which are involved in building secondary structures of enzymes

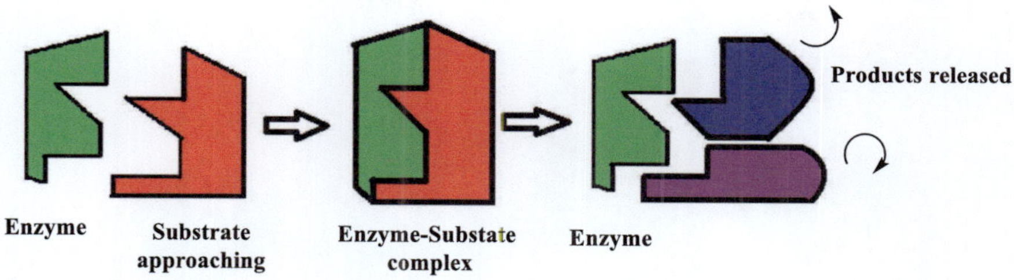

Fig. 7.5 Pictorial representation of the 'Lock and Key' mechanism

the enzyme surface. The secondary and tertiary structures of an enzyme is responsible for overall folding of the entire polypeptide chain giving it a compact, globular 3D form, making the enzyme more robust and stable, and thus, playing a significant role in enzyme catalysis.

Being made of L-amino acids, enzymes are chiral in nature and possess an active (catalytic + binding) site comprising mainly of hydrophobic pockets (with or without few polar sites), allowing them to bind onto the substrates and eventually catalyze a reaction (Kirby 1996). Biocatalytic reactions follow Michaelis–Menten kinetics involving the reversible formation of an enzyme–substrate (ES) complex. The complimentary groups of the substrate fit into these pockets. Proper matching of size and orientation of substrate groups with complimentary pockets of the enzyme results in faster reactions, while mismatch leads to a slower or no reaction. Emil Fischer's 'lock and key mechanism' (1890), can broadly explain both the substrate-specificity of enzymes and the high enantioselectivity of enzyme-mediated reactions (Fig. 7.5).

However, this theory considered enzymes as rigid molecules and hence could not explain the wide substrate (including substrates which are structurally dissimilar to the natural one) tolerance of enzymes. Rather, the 'induced-fit mechanism' (Koshland, 1958) which considered enzymes as non-rigid, soft, delicate spherical ball-like structures with the ability to modify its conformation during the approach of a substrate and wrap itself around the guest molecule efficiently (Fig. 7.6), could explain the 'wider substrate acceptance' as well as 'high reaction selectivity'.

In this model, the substrate is positioned firmly in space within the active site of the enzyme, thereby providing the enzyme's most active conformation required for an efficient transformation. The mono- or bi-layer of an essential water (discussed later) of the enzyme acts as a lubricant as well as an increase in its mobility and stability. This selective yet versatile property of enzymes has made them an integral and indispensable part of toolkit of synthetic organic chemists and biotechnologists (Wong and Whitesides 1994; Santiniello et al. 1992). *The unique structure and power-packed combination of selectivity and versatility make enzymes super catalysts.*

The superior catalytic powers of enzymes can be attributed to various factors as described below.

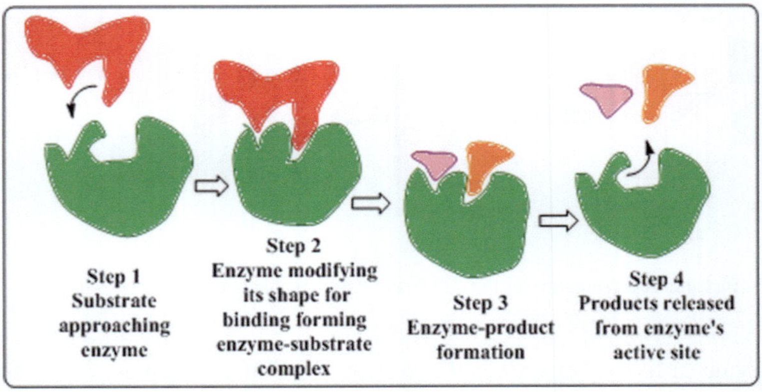

Fig. 7.6 Pictorial depiction of induced-fit mechanism

Proximity effect: The binding of substrates to the enzyme brings the reacting partners closer, thereby hastening the approach of equilibrium which leads to faster reaction.

Kinetic effect—bond weakening and stretching: Binding with the enzymes brings about weakening of existing covalent bonds in the substrates, thereby reducing activation energy (E_a) by stabilizing the reaction transition state. Reduction in E_a will effectively lead to faster reaction rates according to the Arrhenius equation ($k = A\,e^{-Ea/RT}$, where k = rate constant of a reaction, A = pre-exponential factor, R = universal gas constant, T = temperature in absolute scale, E_a = activation energy). A more product-like transition state (TS) leads to faster product formation.

Thermodynamic impetus: Exclusion of trapped water molecules brought about by the substrate binding maximizes the increase in entropy leading to a more spontaneous process.

7.4 Sources of Biocatalysts, Their Classification and Catalytic Properties

Since times immemorial, enzymes are being used in food and beverage sectors (cheese and bread making, wine and beer-brewing, etc.), as well as animal feed, leather, pulp and detergent industries (Polaina MacCabe 2007). Commercial production of enzymes for various applications dates back to late 19th and early twentieth centuries. Crude preparations or pure isolated forms of enzymes can be accessed from various natural sources like plants, animals and microorganisms. The fungal enzymes, used by the food industries, are easily available and can be used for biocatalytic chemical reactions as well.

Based on the type of reactions carried out by them, enzymes are classified into six broad categories, viz. (i) oxidoreductases: catalysts for oxidation and reduction reactions; (ii) hydrolases: catalysts for hydrolysis and formation of ester and amide bonds; (iii) transferases: catalysts for transfer of functional groups, viz. alkyl, formyl, glycosyl, hydroxymethyl, methyl, phosphate and sulfate groups; (iv) lyases: catalyzing addition and elimination reactions; (v) isomerases: used for intramolecular rearrangement; and (vi) ligases: responsible for the most important cellular processes of creating chemical bonds with nucleotide triphosphates.

Enzyme catalysis includes processes catalyzed by whole cells or isolated enzymes (Jones and Beck 1976). Whole-cell systems are usually used to carry out selective reactions with cofactor-requiring enzymes such as oxidoreductases, lyases and makes use of low-cost, readily available precursors/ enzyme sources. Major drawbacks include the requirement of expensive equipment, low output (undesirable side reactions due to uncontrolled metabolic processes), tedious and complex product isolation from substantial bi-products, and rest of the cell culture. Whole-cell enzymes are generally preferred for processes which need multiple cofactor-containing enzymes or where enzyme isolation is not feasible/practical. On the other hand, isolated enzymes offer several benefits, including use of simple reactors, higher productivity and easy product purification. The isolated enzymes are often very expensive, but several commercially available semi-purified enzyme preparations are affordable and are used to design new biocatalytic reactions.

Theoretically, an enzyme-catalyzed reaction can move both in forward and backward directions. The oxidoreductases, present in many microorganisms including common baker's yeast, are often used for the asymmetric reduction of ketones, but can catalyze enantioselective Baeyer–Villiger oxidation of cyclic ketones, oxidation as well as reduction of alkanes, alkenes, aromatics, etc. The oligosaccharides and polysaccharides are industrially synthesized using transferases. Isomerases are rarely used in industrial applications, an exceptional and remarkable application being the production of high-fructose corn syrup using glucose isomerase (GI). Among the hydrolases, lipases hydrolyze triglycerides into fatty acids and glycerol; proteases viz. α-chymotrypsin, papain and subtilisin hydrolyze/form peptide bonds. The lipases are widely used for kinetic resolution via esterification/aminolysis of α- or β-substituted acids as well as esterification/*trans*- esterification of secondary alcohols/amines and provide good to excellent enantioselectivity. In some studies, lipases have also been found to be promiscuous to carry out reactions akin to those expected. Some examples in this category include: (i) use of *Candida antarctica* lipase for Michael addition reactions (Svedendahl et al. 2005) and Cannizzaro-type reaction of substituted benzaldehydes (Arora et al. 2014) and (ii) Henry reaction of aromatic aldehydes and nitroalkanes using Lipase A from *Aspergillus niger* (Le et al. 2013).

7.5 Emerging Areas—Special Techniques in Biocatalysis and Some Applications

Despite these special attributes, regular utilization of biocatalysts for organic synthesis or industries has been plagued, partly because of conceptual unawareness and partly due to practical or operational obstacles. Conventionally enzymatic reactions were carried out in water, which is neither a good solvent for majority of the organic substrates/products, nor a preferred medium for organic transformations. Moreover, product isolation and enzyme recovery from water (containing various other products including proteins) would be extremely tedious. Consequently, barring a few streamlined applications like synthesis of β-lactam antibiotics (Buchholz 2016), enzyme-mediated reactions were mainly utilized for resolution of optically active intermediates in the process industries (Breuer et al. 2004).

The development of special techniques like solvent engineering, enzyme immobilization and use of modified and artificial enzymes has increased enzyme robustness and stability many-folds. Such special techniques, which facilitate enzyme's recovery-recycling-reuse and product isolation, have elevated biocatalysts to new heights of versatility, making them priceless assets for synthesis of precious materials. Some of the special techniques are briefly discussed below.

7.5.1 Non-aqueous Enzymology—Solvent Engineering

Water is vital to maintain enzymes in their catalytically active conformation. Water modulates the various weak interactions that support the secondary/tertiary structure of an enzyme. During the process of substrate binding, when H-bonds get disrupted, water immediately replenishes new H-bonds to the enzyme, thereby serving as a lubricant and giving enzymes the much-needed mobility to accommodate and adjust to non-natural and structurally diverse guest substrates. However, merely a mono-, bi- or a few layers of water around the enzyme surface (tightly bound to the enzyme, known as essential/bound/structural water), not the bulk water, is essential for the enzyme to be catalytically active. The essential water can be visualized as a part of the enzyme structure rather than the medium and is distinctly different from the surrounding bulk water in physico-chemical properties. Ironically, the bulk water can lead to enzyme denaturation under non-ambient temperature and pH, by rapturing hydrogen bonds and leading to a collapse of the enzyme's secondary structure. Hence, the bulk water may be replaced by a suitable organic medium to improve the enzyme's performance. The physico-chemical properties of water are tunable marginally within a narrow range, restricting modulation of catalytic efficiency and selectivity of the enzyme-catalyzed reactions. In contrast, when organic solvents or alternate media (ionic liquids, supercritical fluids) are used for biocatalytic reactions, there is a wide pool of solvents (with diverse physico-chemical parameters, e.g., dipole moments, solubility, boiling points) to be chosen. Such solvents can alter the non-covalent forces like H-bonding, hydrophobic interactions, etc. leading to subtle conformational changes in the enzyme structure. The enzymes adopt a rigid conformation in organic media, thereby providing better control over enzyme selectivity and specificity, widen substrate acceptability and increase operational range.

Regarding choice of organic solvents, the Log P parameter, based on solvent hydrophobicity serves as a reliable guide for choosing solvents which are compatible with enzymatic activities. Log P = $[S]_{oct}/[S]_{water}$ (where P = partition co-efficient of a given solvent in octanol and water).

It can be safely inferred that the water-immiscible hydrophobic solvents (toluene, hexane, di-isopropyl ether, etc.) are preferred for higher enzymatic activities, whereas water-miscible/polar/hydrophilic solvents (DMF, DMSO, acetone, MeOH) often denature an enzyme by stripping off its essential water. Nevertheless, such solvents may be required as co-solvents to enhance selectivity of hydrolytic enzymes, but their proportions have to be strictly controlled.

Due to better solubility of organic compounds in organic solvents, the reactions are faster, giving better yields, and product isolation is simple. Moreover, the enzymes can be recovered easily since they are insoluble in organic solvents. Due to differential solubility of enzymes and organic substrates in volatile organic solvents, enzyme inhibition phenomenon is dramatically reduced in such solvents. Using water-immiscible solvents, the stereochemical course of an enzyme-mediated reaction can be directed (modified/reversed) in the desired pathway. A major advantage of non-aqueous enzymology is the possibility of carrying out the reverse processes that are otherwise unfavorable in water. For example, in water, hydrolases like lipases and proteases can carry out hydrolysis of esters and proteins. But in an organic medium, synthesis (esterification and aminolysis) is favored over hydrolysis, and water-assisted undesired reactions like hydrolysis, polymerization, racemization, etc. can be avoided. With proper choice of organic solvents, it is possible to design new reactions and even alter enantioselectivity of a reaction (Editor-in-Chief: Murray Moo-Young 2019; Holtmann and Hollmann 2022). One such example is shown in Scheme 7.2 (Ortega-Rojas et al. 2017) where a lipase catalyzed double-bond addition reaction occurs (product A) when benzyl amine is added to an α, β- unsaturated ester in non-polar or hydrophobic solvent. However, changing the solvent to a polar or hydrophilic one results in aminolysis, giving an α, β- unsaturated amide (product B) instead.

The changes of reaction course and selectivity with solvent polarity are more pronounced with hydrolytic enzymes and more marked for the better-fit substrates

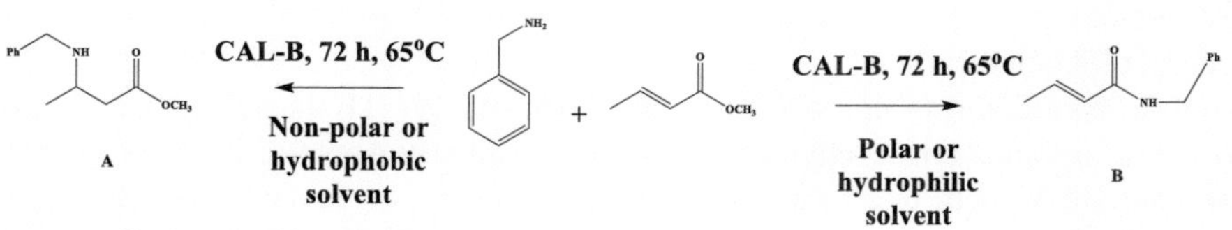

CAL-B: ***Candida antarctica*** **lipase B**

Scheme 7.2 Change of reaction course via solvent engineering

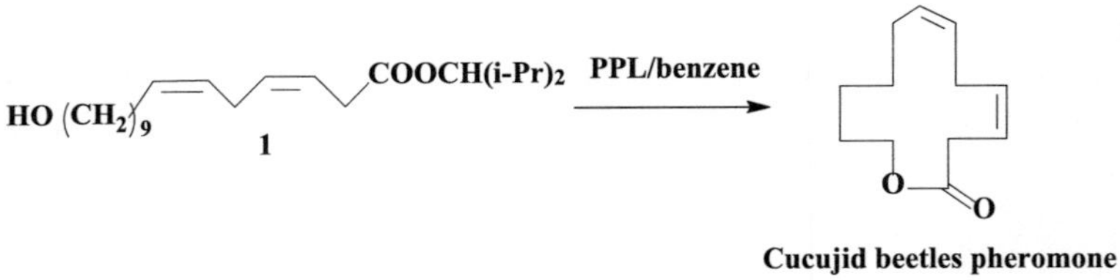

Scheme 7.3 Biocatalytic step of in the synthesis of Cucujid beetle pheromone

as compared to the ill-fitted ones. Even the exclusive specificity of protease for (L)-amino acid esters in water gets ruined when aqueous medium is replaced by organic solvents—a fact used for synthesis of non-natural peptides (made of D-amino acids).

Chattopadhyay et al. (Pawar et al. 1993; Chatterjee et al. 2014a) have worked extensively on various aspects of solvent engineering to develop new selective protocols for the syntheses of chiral as well as achiral bioactive molecules of contemporary interests. A few examples are schematically presented below, highlighting only the biocatalytic steps.

Synthesis of aggregation pheromone component of Cucujid beetle was challenging considering its macrolidic moiety, and the existence of a β,γ-unsaturated carbonyl functionality is susceptible to isomerization or decomposition under harsh chemical reaction conditions. The key macrocycle formation step was accomplished using porcine pancreatic lipase (PPL) in benzene at room temperature to obtain the pheromone in 78% yield, without any diolide formation. Importantly, no activated ester and/ or high dilution conditions were required in the macrolactonization step (Scheme 7.3) (Pawar et al. 1993).

Different stereomers of hept-6-ene-2,5-diol derivatives are the common structural motifs present in a host of biologically active oxygenated heterocyclic compounds, viz. tetrahydropyrans, diolides and macrolides. Chatterjee et al. developed an efficient, enantiomeric synthesis of hept-6-ene-2,5-diol derivative stereomers, and formulated a divergent strategy for the synthesis of (i) decarestrictine L, a cholesterol biosynthesis inhibitor, (ii) pyrenophorol, an antimicrobial and a potent nootropic and antidepressant agent, and (iii) stagonolide E, a fungal pathogen (Chatterjee et al. 2014a). The two key biocatalytic steps in the syntheses were the highly enantioselective *trans*-acetylation of (i) *rac*-sulcatol (2) in hexane solvent and (ii) diastereomers of compound 4 in solvent-free conditions; using vinyl acetate (VA) as the acyl donor in the presence of Novozym 435® [*Candida antarctica* lipase B (CAL-B) immobilized on acrylic resin] as the inexpensive, commercially available lipase catalyst (Scheme 7.4).

The *Penicillium turbatum*-derived macrolide, (-)-A26771B showed moderate anti-bacterial and anti-fungal activities, but it is perceived to be a potential broad spectrum antimicrobial agent, due to its additional keto moiety. In view of this, Chatterjee et al. developed its first chemoenzymatic synthesis (Chatterjee et al. 2014b) and instilled its two required stereogenic centers by kinetic resolution (KR), using the Novozym 435® and VA in di-isopropyl ether for the key acylation steps. In addition,

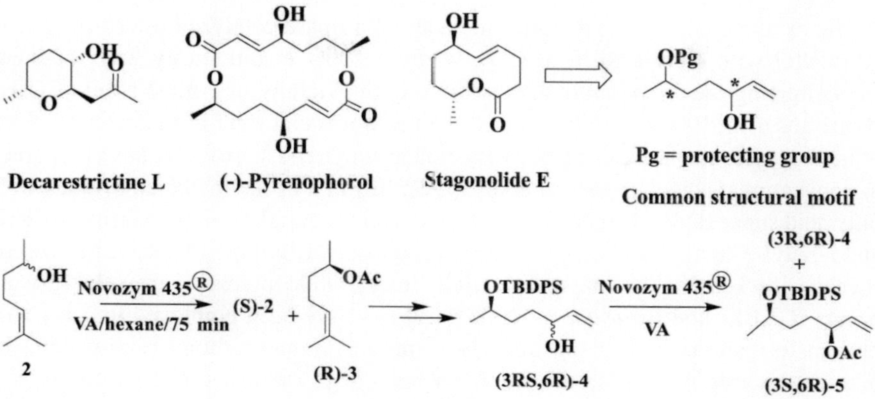

Scheme 7.4 Biocatalytic steps for the kinetic resolution of the common structural motif

the lipase, Novozym 435® was also instrumental in developing a new and relatively green acrylation protocol of a carbinol moiety with ethyl acrylate (under solvent-free conditions), thereby steering clear of acryloyl chloride, the conventional acrylating agent for a chemical process which is both hazardous and hygroscopic. The enzymatic strategy also revealed that the chosen lipase could carry out the acrylation step regioselectively at the MeCH(OH)-site in the presence of an unprotected allylic alcohol moiety. The adopted KR strategy ensured access to all possible stereoisomers of the antibiotic macrolide (Scheme 7.5).

Besides organic solvents, green media such as supercritical fluids (SCFs) and room temperature ionic liquids (RTILs) have also been exploited for biocatalytic reactions.

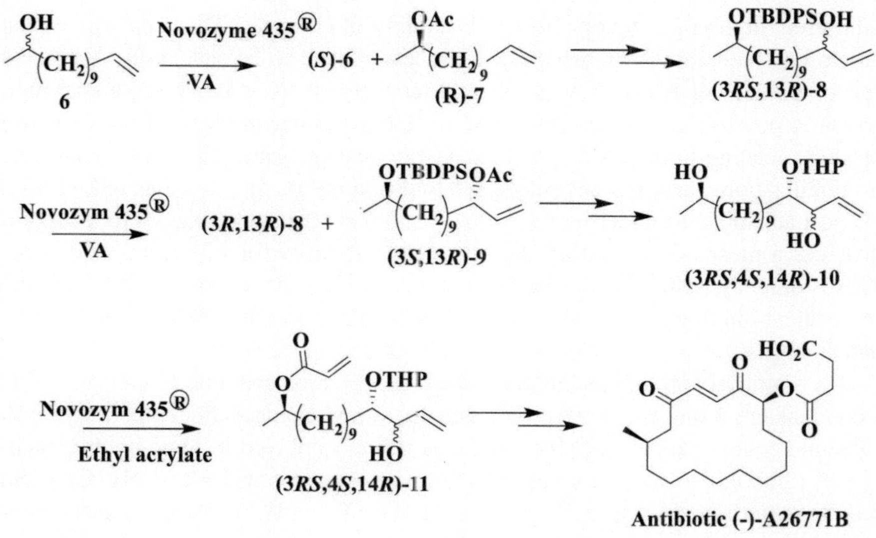

Scheme 7.5 Biocatalytic steps for the asymmetric synthesis of (-)-A26771B

A few examples in these categories include: (i) a lipase-catalyzed esterification of citronellol with oleic acid in $SCCO_2$, wherein 100% enantiopurity was obtained by optimizing the temperature and pressure in the vicinity of critical point of CO_2 (Ikushima et al. 1996); (ii) higher reaction rates were achieved for transesterification of methyl methacrylates in supercritical ethane, ethylene, fluoroform, etc. compared to conventional solvents (Kamat et al. 1992); (iii) in a lipase-mediated synthesis of fatty acid sugar esters, 1-butyl-3-methylimidazolium tetrafluoroborate [bmim][BF_4] and 1-butyl-3-methylimidazolium hexafluorophosphate [bmim][PF_6] have been used successfully keeping the enzymatic activity intact, while increasing the solubility of sugars (Ganske and Bornscheuer 2005); (iv) ILs have been employed to carry out biotransformations of polar substrates like amino-acids and carbohydrates which are otherwise sparingly soluble in organic solvents, *e.g.,* synthesis of Z-aspartame by a thermolysin-mediated reaction of carbobenzoxy-L-aspartate and L-phenylalanine methyl ester hydrochloride in [bmim][PF_6] (Erbeldinger et al. 2000).

In summary: *'Modification of enzyme properties by altering the medium (bulk solvent) or environment falls under the domain of medium engineering (alteration of solvent, co-solvent, temperature, pH* etc.*) which is an important tool for chemists to design new enzymatic reactions with better control on reaction outcomes with respect to selectivity and efficiency. Solvent engineering can be a means of improving biocatalytic performance in synthetic organic chemistry'.*

7.5.2 Enzyme Immobilization

Enzyme fixation or immobilization onto a carrier or support unit can efficiently counter some age-old industrial/laboratory problems associated with biocatalysis like inhibition, high price and operational instability of enzymes. The economic aspect has to be taken care of if biocatalytic processes have to be scaled-up. A process can be made cost-effective by reuse and recovery of the water-soluble enzymes, which is possible if they are immobilized. Higher concentrations of substrates or products leading to inhibition phenomena can be significantly reduced by enzyme immobilization. Enzymes are vulnerable to denaturation, due to undesired reaction or mechanical stress under harsh process conditions. Enzyme fixation to a support provides a means of increasing the operational stability for a longer time period. This is done by fixing an enzyme to a carrier (an inert matrix) by adsorption, ionic or covalent binding, cross-linking or gel-entrapment, each having their own merits and demerits.

The simple adsorption technique, though retains good enzymatic activity, suffers from leakage from the carrier unit due to minor changes in temperature, pH, substrate concentrations, etc. Ion-exchange resins employed in ionic binding techniques although stronger are susceptible to ion concentrations and pH. Covalent bonding to carrier molecules can prevent leakage efficiently, but use of harsher conditions for covalent bond formation may lead to some loss of enzyme activity. Cross-linking of enzymes to protein molecules (which may be its own self or others like α,ω-glutaraldehyde or isothiocyanate) makes the enzyme robust to sustain extreme conditions, but problems may arise due to diffusion and inaccessibility of substrates to the buried enzymes. Gel-entrapment method is a mild immobilization method, whereby biocatalysts can be physically trapped in a matrix providing easy entry and exit of substrates and products in order to maintain enzyme activity (Grunwald 2018).

A few examples of successful use of immobilized enzymes are manufacture of (i) allulose, a 'zero-calorie' sweetener from fructose, using a D-psicose epimerase; (ii) a chiral herbicide, dimethenamide-P using CAL-B lipase in an organic solvent with 99% enantiomeric excess; (iii) odanacatib, a potent cathepsin K inhibitor using immobilized CAL-B—where lactone to ester conversion was accomplished in organic solvents, and the carrier of the immobilized enzyme played an important role in cost reduction; and (iv) a cocoa butter substitute, using a *Rhizopus niveus* lipase. Some other examples include the use of immobilized enoate reductase and alcohol dehydrogenase for reduction of $C=C$ and $C=O$ bonds, respectively (Basso and Serban 2019).

7.5.3 Protein Engineering

Wild-type enzymes (natural unmutated genes) perform superbly in vivo on their natural substrates, but often fail in vitro in terms of productivity, lower selectivity and efficiency, as they have to work with non-natural substrates at higher concentrations, harsher conditions and non-natural media. The arrival of *biocatalysis engineering* which handles optimization of all relevant variables by engineering of the enzymes, proteins, substrate or reaction medium promises to provide economically viable and environmentally superior green processes. The new tools deliver the tailored enzymes of choice, which will cater to 'the need for speed', fulfilling the voluminous demand of billions across the globe within a short time span. Rational design and directed (molecular) evolution are the main techniques used for protein engineering. The former is based on use of X-ray-based protein structure, mechanistic information along with molecular modelling to predict the required structural changes to modify or include the desired properties. Directed evolution aims to generate enzymes that can adapt to reaction conditions and molecular structures of substrates. Various strategies, viz. recombinant DNA techniques, site-directed mutagenesis (SDM), DNA scuffling, error-prone polymerase chain reaction (ep-PCR), are used for this purpose. These strategies help in library generation of enzyme variants with differential substrate specificities in a short span of time for better process efficiencies with improved yields, lower enzyme loadings and easy purifications. In an alternative strategy, metabolic engineering is now being used to alter established industrial processes. The hydantoinase process for the synthesis of optically pure amino acids that is based on the separate use of immobilized hydantoinase, carbamoylase and racemase is now directly amenable using engineered *E. coli* cells (Behrens et al. 2011; Behrens et al. 2011).

7.6 Representative Examples of Industrial Applications of Biocatalytic Processes

Asymmetric Synthesis of Pregabalin

The active pharmaceutical ingredient (API), pregabalin ((*S*)-3-aminomethyl-5-methylhexanoic acid) is a lipophilic γ-aminobutyric acid (GABA) analog used for treating various CNS disorders like epilepsy-related seizures, fibromyalgia, neuropathic pain, anxiety disorders and sociophobia in adults. Due to its huge demand, several chemical and enzymatic routes have been developed for their industrial scale production. (*S*)-2-Carboxyethyl-3-cyano-5-methylhexanoic acid ((*S*)-CCMA) that can be synthesized from racemic 2-carboxyethyl-3-cyano-5-methylhexanoic acid ethyl ester (CNDE) is the key chiral precursor of pregabalin as well as gabapentin, another clinically used CNS drug (Fig. 7.7). The chemical synthetic routes for (*S*)-CCMA via resolution (Hoekstra et al. 1997), asymmetric hydrogenation (Burk et al. 2003) suffer from high catalyst cost, non-recyclability of undesired isomer, huge amount of wastages, catalyst poisoning and operational difficulties (viz. requirement of high-pressure reactors).

Instead, the commercially available Lipolase®-catalyzed kinetic resolution protocol of (*rac*)-CNDE produced (*S*)-CCMA (>99.5% enantiomeric excess), while the undesired (*R*)-CNDE was recycled, racemized and subjected to repeated batches of resolution (Martinez et al. 2008). The process worked at high substrate concentrations (up to 255 g/L) without any enzyme denaturation and inhibitions by using additives like calcium acetate (Scheme 7.6) (Yazbeck et al. 2004).

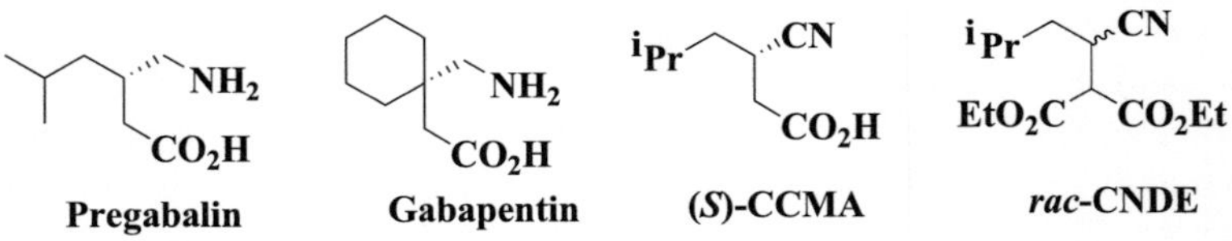

Fig. 7.7 Chemical structures of pregabalin, gabapentin, (*S*)-CCMA and CNDE

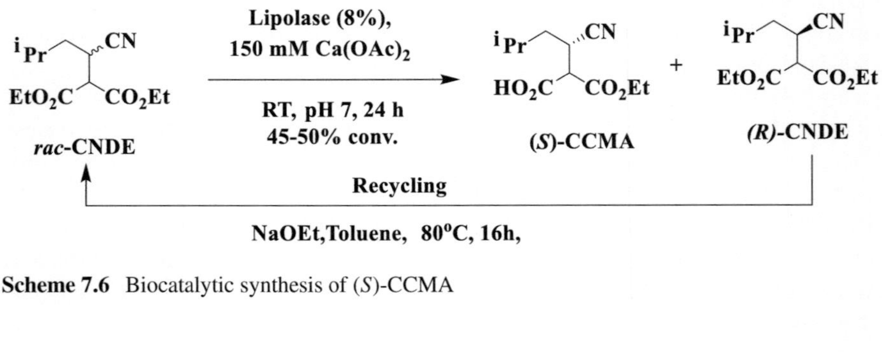

Scheme 7.6 Biocatalytic synthesis of (*S*)-CCMA

Asymmetric Synthesis of Statin Aliphatic Side Chain

Statins, an important class of hydroxymethylglutaryl-coenzyme A (HMG-CoA) reductase inhibitors, used for reducing the levels of LDL and triglycerides in blood, are prescribed to prevent cardiovascular diseases. The APIs, viz. atorvastatin, rosuvastatin, etc., bear a homochiral 3,5-dihydroxycarboxylic acid side chain connected to a (hetero)aromatic or carbocyclic residue (Fig. 7.8). Various chemical approaches involving (i) asymmetric catalytic (metal-catalyst) reduction of suitable β-keto esters followed by a Claisen condensation under a strongly basic/cryogenic conditions and finally a directed hydride reduction, (ii) chiral chromatography of the racemic key intermediate, (iii) a diastereoselective aldol reaction and (iv) a chiral pool approach using isoascorbic acid have been adopted for the synthesis of the statin side chain. Use of circuitous pathways, expensive chiral auxiliaries, heavy/precious metal catalyst, complex and harsh operational conditions as well as low selectivities are the major impediments of these processes.

Instead, biocatalytic synthesis of the key chiral intermediate, 6-cyano-3*R*,5*R*-dihydroxyhexanoate or its simple precursors I-III (Fig. 7.8) followed by chemoenzymatic approaches provided greener and efficient routes to the chiral side chain of statins. The key steps of the syntheses are the introduction of the two chiral centers of the statin side chain. Some of these are described below.

In one method, 3-hydroxyglutaronitrile (13), obtained from epichlorohydrin (12), was desymmetrized using a nitrilase to obtain (*R*)-3-hydroxy-4-cyanobutyrate (I) with impressive yield and optical purity (Scheme 7.7) (DeSantis et al. 2003).

In an outstanding example of a biocatalytic protocol showing excellent regio- and enantio-selectivities in carbonyl reduction, a sterically hindered diketo ester 14 was selectively reduced at its C-5 position by an alcohol dehydrogenase (ADH) with excellent enantio- and regioselectivity to give the intermediate II (Scheme 7.8) (Wolberg et al. 2000).

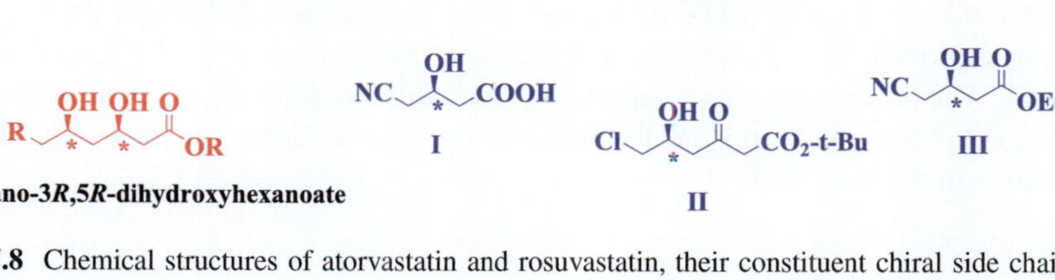

Fig. 7.8 Chemical structures of atorvastatin and rosuvastatin, their constituent chiral side chain and key synthons

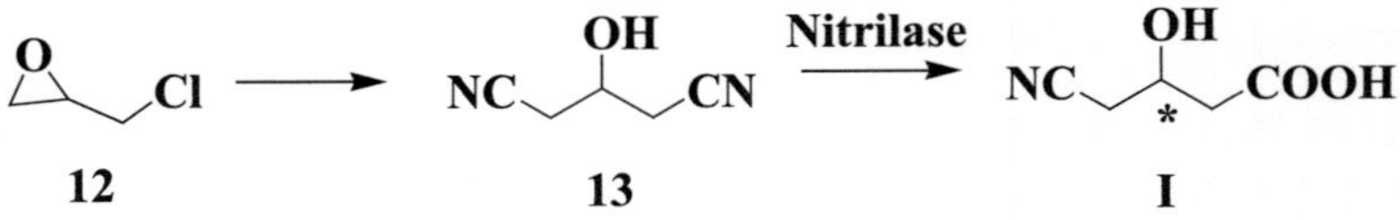

Scheme 7.7 Synthesis of the precursor I

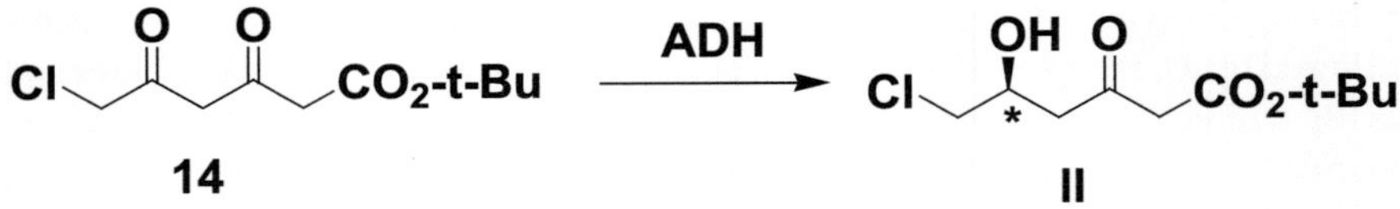

Scheme 7.8 Synthesis of the precursor II

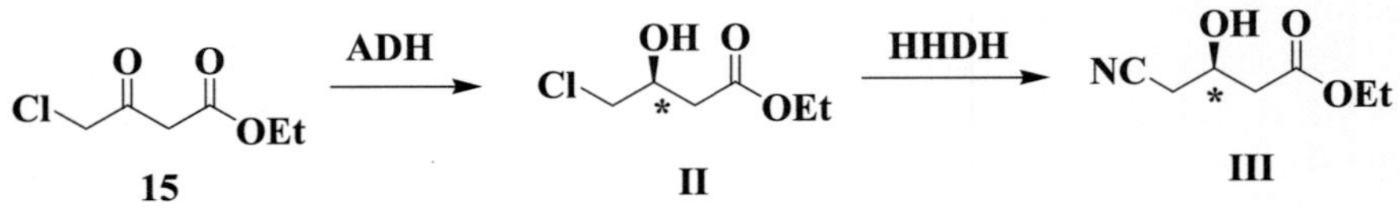

Scheme 7.9 Synthesis of the precursor III

On the other hand, the key intermediate II was obtained by enantioselective reduction of the chloroacetoacetate 15 using a recombinant ADH (from *Escherichia coli* strain) along with glucose dehydrogenase (GDH) (from *Bacillus megaterium)* for NADPH recycling (Kizaki et al. 2001; Ernst et al. 2005). Even the conversion of the chloro intermediate II-analog to the nitrilo intermediate III analog was successfully achieved using halohydrin dehalogenase (HHDH) (Scheme 7.9) (Fox et al. 2007).

Biocatalytic Synthesis of Sitagliptin

Sitagliptin, a dipeptidyl peptidase-4 (DPP-4) inhibitor, is an oral anti-hyperglycemic drug to treat type II diabetes mellitus in adults. The conventional chemical method involves hydrogenation under high pressures (250 psi) of an enamine derived from the ketone (16) over expensive Rh metal catalyst. However, the stringent requirement of removing Rh metal impurity, maintenance of high H_2 pressure and a crystallization step offset efficiency of the process (Hansen et al. 2009). Instead, an enzymatic transamination of the ketone (16) using a genetically engineered transaminase (with increased activity than wild type variant) directly gave the target drug with excellent (and improved) stereoselectivity and chemical purity (Scheme 7.10) (Savile et al. 2010). The process was much superior to the traditional chemical method and earned the joint proposers, Merck and Codexis, the prestigious Green Chemistry Challenge Awards in the year 2010.

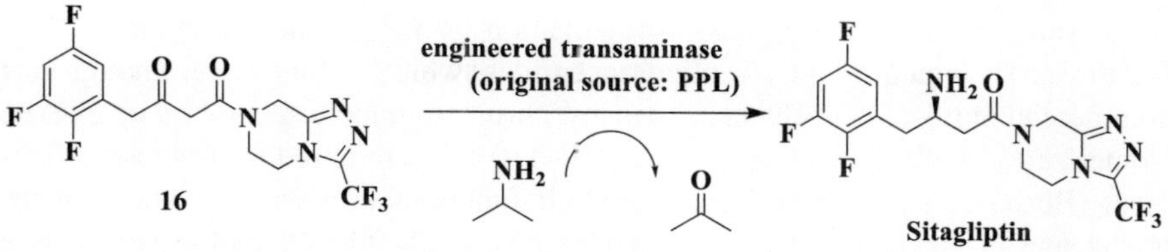

Scheme 7.10 Biocatalytic transamination of the precursor (Arora et al. 2014)

Enzymatic Glycosylation in Process Industries

Polysaccharides, an important class of natural compounds, have a crucial role to play in various cellular processes like cell–cell interactions, oncogenesis, secretion of glycoproteins, interactions of cell and surface with pathogens, etc. (Faber 2017). Carbohydrate chemistry has gained a lot of momentum in recent times owing to the novel coronavirus epidemic and the subsequent quest for suitable covid vaccines, since host–virus interactions are controlled by glycans. Extracting these natural compounds from natural sources is not practical due to high cost, complexity of the process and low concentration of desired substances. Carbohydrate synthesis is an extremely challenging task for organic chemists due to their constituent marginally differentiated multihydroxy functionalities, requiring extensive protection group strategies, and hence, leading to low overall yields. The inherent chemo-, regio-, stereo- selectivities of biocatalysts makes them superior tools for carbohydrate synthesis. Trans-glycosidase (class: hydrolytic enzymes), glycosyltransferase (class: transferase) and phosphorylases (intermediate between the two above classes) are most instrumental in forming glycosidic bonds. For instance, sucrases (a class of trans-glycosidases) are used in polymerizing glucose or fructose units forming dextrans (1,6-linkages) or mutans (1,3-glycosidic linkages) that are useful in thermoplastics and medicinal sectors. In another very prominent example, multiple sugar-converting enzymes are utilized for industrial production of high-fructose corn syrup (HFCS).

Green Synthesis of High-Fructose Corn Syrup

HFCS is a fructose-rich (55% by weight) liquid carbohydrate sweetener, used mostly in colas and processed foods. It is produced by a multistep bioprocess of (i) corn starch hydrolysis to polysaccharides using α-amylase (from various species of genus *Bacillus*), (ii) conversion to glucose by glucoamylase (saccharification) and (iii) the most important step of glucose to fructose conversion by the enzyme, glucose isomerase (GI) (Fig. 7.9).

Fig. 7.9 Enzymatic process for production of HFCS

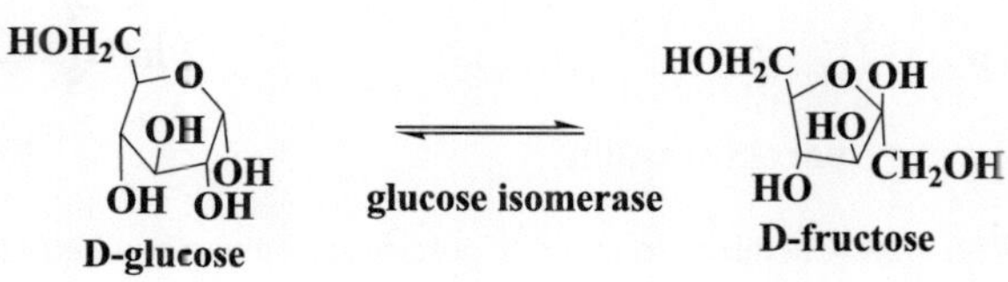

In water medium, the glucose isomerization by GI produces a syrup of 42% fructose, 51% glucose and 7% oligosaccharides, which is 'not sweet enough' for commercial applications. Hence, expensive chromatographic separation or glucose isomerization at elevated temperatures (100–110 °C) is required for fructose enrichment (Bhosale et al. 1996). Fortunately, GI functions efficiently in non-aqueous media, and an enzymatic isomerization of glucose in 85–90% ethanol–water mixture at ambient temperatures gave HFCS with 55% fructose content (Kalevi and Klibanov 1987).

Alternatively, an immobilized enzyme ToGI (Thermus oshimai glucose isomerase) on Celite 545 was used for large scale cost-effective production of HFCS with higher fructose enrichment. The enzyme preparation worked efficiently at 85 °C and physiological pH with good catalytic activity, superior operational stability and retainment of activity even after several batches of bioconversions (Kognou et al. 2022). Large-scale production and *biocatalyst engineering* have paved the way for more efficient variants of GI that show high thermal and acid stability, resistance to substrate/product inhibition, acceptance of other hexose substrates and offer increased fructose yields.

Enzymatic Peptide Synthesis

The following example of enzymatic peptide synthesis in the conversion of porcine insulin to human insulin is one of the most iconic examples ever in the history of 'enzyme achievements'. Insulin is the mainstay drug for a large human population suffering from type II diabetes and is vital for type I diabetes. Structurally, the inexpensive porcine insulin with an alanine (Ala) residue at lysine (Lys)-29 differs from that in human insulin, where (Lys)-29 is connected to threonine (Thr) (Fig. 7.10). Hence, conversion of porcine insulin to human insulin provides a plausible and unique method of developing a biocatalytic industrial process for peptide synthesis.

In one such successful attempts, a serine protease from *Achromobacter lyticus* M497-1 (highly specific for peptide bond at lysine residues on the carboxyl site) was used to hydrolyze the terminal Ala^{30} residue of porcine insulin to obtain desalanine-(B30)-insulin (DAI), which on coupling with threonine *tert*-butyl (OBu^t) ester using the same enzyme, followed by deprotection of the OBu^t group furnished human insulin (Morihara et al. 1980). Hydrolysis of single-chain Des-(B-3O)-insulin precursor with lysyl endopeptidase has also been utilized for enzymatic semi-synthesis of human insulin (Morihara and Ueno 1991). In one of the most

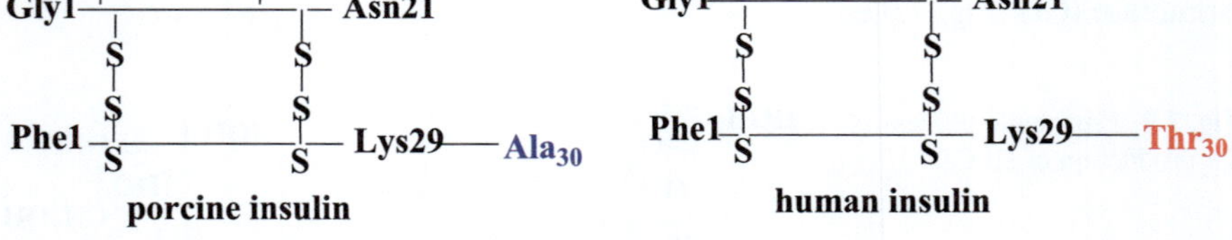

Fig. 7.10 Chemical structures of porcine and human insulins (adapted from reference (Faber 2017))

popular current methods, recombinant DNA technology can be used for high-capacity production of human insulin using *Escherichia coli*, yeast and especially transgenic plants-based expression systems (Baeshen et al. 2014).

Biocatalysts are undeniably the frontrunner of green chemistry as is evident from the number of US Presidential Green Chemistry Challenge Awards (PGCCA) won by the biocatalytic processes over the years (Table 7.1). The chosen examples clearly manifest the wide spectrum of biocatalytic applications.

A few illustrative examples of industrial uses of biocatalysts in chemical processes are provided in Table 7.2.

7.7 Conclusion

Given the ever-increasing human requirements of various fine and bulk organic chemicals (pharmaceuticals, agrochemicals, food additives, polymers, diesels, paints, etc.), coupled with depletion of natural resources and energy, it is imperative to produce these materials by suitable *clean and green* processes that minimize waste production and keep the environment safe by reducing heat generation and carbon emission. Biocatalysts offer potential solutions to this grave problem, due to their several unique attributes as discussed in the foregoing. The ultimate goal of biocatalytic technologies is to use these whole cell-based/ purified enzyme systems to produce valuable substances on an industrial scale in a cost-effective manner, largely protecting the objectives of 'green chemistry'. Figure 7.11 showcases how enzymes, 'the green chemistry tools', have knitted nature with the world of science, industry and technology.

Impressive progress notwithstanding, the regular use of enzymatic reactions in organic reactions or industries has been plagued by the common misconceptions about the enzymes as discussed previously. Besides factors like high costs and commercial unavailability of suitable enzymes, ironically '*selectivity*—the key attribute' of the enzymes, have restricted their extensive usage for long. The field of biocatalysis has evolved over the ages. The initial attempts of using large-scale production of different microbial systems, solvent engineering and use of immobilized enzymes/whole cell systems provided only partial success. During the past few decades, biocatalysts have made a welcome transition from *'a topic of academic interest with few technological applications'* to becoming an *'indispensable tool for innovative organic process chemistry/biotechnology'*. The advent of a host of genetic engineering techniques made directed evolution of enzymes simpler, and these are now being applied to modify the biocatalysts to suit the process requirements. Protein engineering has become faster and cheaper because of the availability of (i) X-ray crystal structure database of a large number of proteins and (ii) bioinformatics tools for DNA and genome sequencing. Together, these help in identifying the protein encoding gene and its recombinant expression bypassing microorganism growth, protein purification, sequencing and genomic DNA isolation. Indeed, the directed

Table 7.1 Some biocatalyst-mediated processes that have won the US PGCCA (*Source* https://www.epa.gov/greenchemistry/green-chemistry-challenge-winners (accessed on 19.01.2023)

Year	Awardee	Achievements
2001	Novozymes North America, Inc	Development of a technology (BioPreparation™) to separate contaminants from cotton before fabric-weaving using enzymes (pectate lyase mainly) instead of corrosive chemicals and saving billion gallons of water
2003	Richard a. gross, polytechnic university, USA	Development of highly regio-selective, energy efficient one-pot, lipase catalyzed condensation polymerization routes to synthesize polyesters in mild conditions
2004	Bristol-Myers squibb company, USA	Green synthesis of paclitaxel, API for the anticancer drug, Taxol® via plant cell fermentation, which eliminates use of hazardous materials, several organic solvents and drying steps
2004	Buckman laboratories international, Inc	Development of a new enzyme technology (Optimyze®) for removing sticky contaminants (polyvinyl acetates) via an esterase-catalyzed hydrolysis before recycling paper
2005	Metabolix, Inc., USA	Production of biodegradable natural plastics polyhydroxyalkanoates (PHAs) using genetically engineered microorganisms, with reduced solid waste and green-house gas emissions
2005	Archer Daniels Midland Company & Novozymes	Production of low trans fats and oils (NovaLipidTM) by enzymatic interesterification of vegetable oils using Lipozyme®, thereby using less resources—vegetables, chemicals and water
2005	Codexis, Inc	Production of the key chiral intermediate (ethyl (*R*)-4-cyano-3-hydroxybutyrate) for atorvastatin, using directed evolution of three biocatalysts. The process increases yield, improves worker safety, reduces wastes and uses less solvent
2009	Eastman chemical company	Development of a solvent-free biocatalytic synthesis of esters required for manufacturing of skin-safe cosmetics and personal care ingredients using immobilized lipases under mild conditions
2010	Merck & Co and Codexis, Inc	Development of an improved synthesis of sitagliptin, using an evolved transaminase capable of directly converting a precursor ketone to desired chiral amine. The overall process gave 10–13% increased yield and 19% reduction in waste generation
2010	LS9, Inc	Microbial production of biodiesel (Renewable Petroleum™ Fuels and Chemicals like UltraClean™ diesel) by engineering new metabolic pathways in *Escherichia coli* bacteria. The direct conversion of biomass to fuel eliminates benzene, sulfur and heavy metals normally found in petroleum-based diesel, resulting in an 85% decrease in greenhouse gas emissions

(continued)

Table 7.1 (continued)

Year	Awardee	Achievements
2012	Buckman International, Inc	Use of Maximyse® enzymes to increase wood fibrils in cellulose, for making high-quality and strong paper and paperboard, thereby avoiding costly wood pulp
2019	Kalion, Inc	Commercializing the first microbial fermentation process (involving a novel biosynthetic pathway expressed in *E. coli* and using 3 enzymes) to produce a biodegradable, non-toxic, high purity, low-cost sugar-derived product, glucaric acid, which finds application in water treatment, additives for polymer formulations, excipients for APIs, chelants for the detergent industry, concrete admixtures and corrosion inhibitors

Table 7.2 Some enzyme-mediated industrial processes with key biocatalytic steps and application

Sl no	Product [ref.]	Key biocatalytic step	Application
1	1,3-propanediol (Nakamura and Whited 2023) (Dupont and Genencor)	Use of cheaper carbon source, glucose (instead of natural source glycerol) for conversion to 1,3-propanediol by integrating three different pathways from 3 different organisms in *Escherichia coli* strain	The process for 1,3-propanediol, used for polyester manufacture is energy-efficient and results in less greenhouse gas emission
2	Higher alcohols (6–8 carbon) (Atsumi et al. 2008)	Use of 2-ketoacid decarboxylase for conversion of the intermediate, 2-ketoacid to aldehydes and subsequent reduction to higher alcohols	The higher alcohols are potential second-generation biofuels
3	Shikimic acid from glucose (Ran and Frost 2007)	Directed evolution used to modify substrate-specificity of an aldolase to obtain shikimic acid in higher yields from abundant pyruvate *vis-à-vis* phosphoenol pyruvate	Key intermediate to produce oseltamivir phosphate, API of Tamiflu used for preventing influenza
4	Montelukast (Tao Kazlauskas 2011)	Using gene scuffling, an improved variant of ketoreductase was developed for reducing ketones to chiral alcohols	LTD_4 receptor, used for treatment of asthma

evolution approach has become an integral part of industrial syntheses of fine chemicals and APIs. The activity, selectivity and stability of these '*near-perfect-catalysts*' could now be manipulated in a predictive manner to suit our needs.

Despite its phenomenal successes, the enormous potential of biocatalysts can only be realized after complete understanding of the key processes in protein biochemistry especially the hydrophobic and polar effects of the active sites of enzymes, protein folding and the effects of organic solvents on these factors. These would

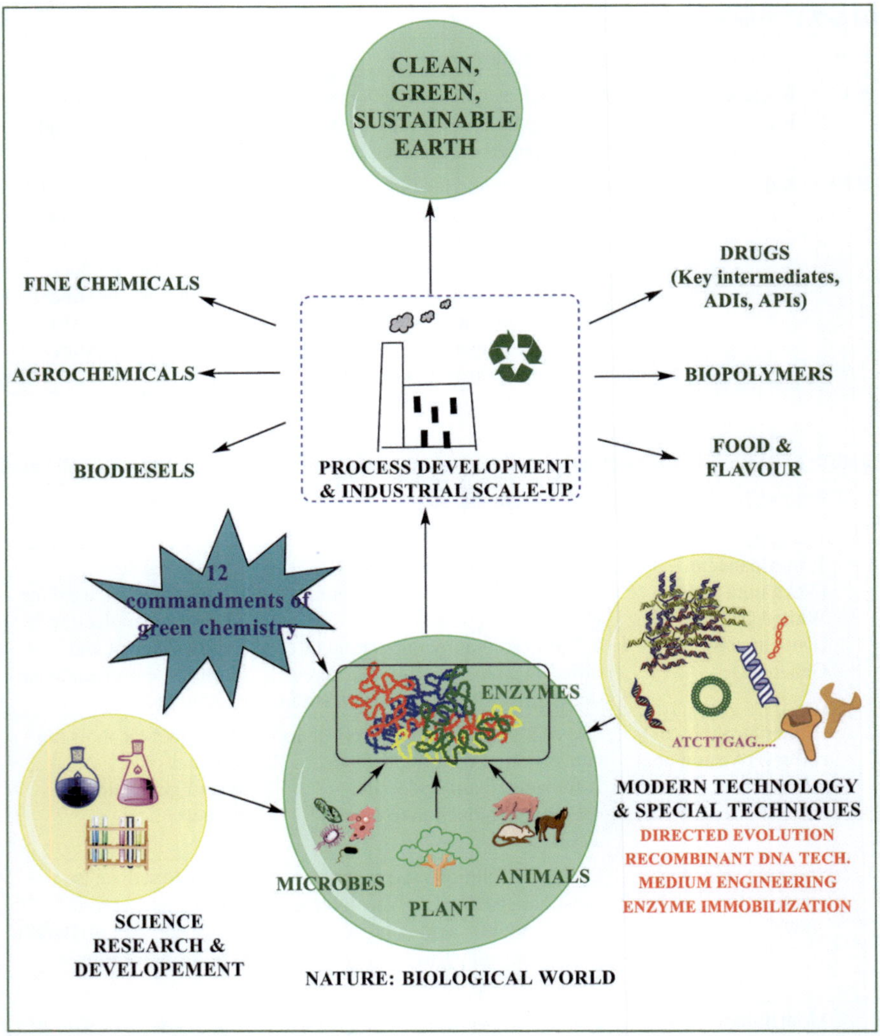

Fig. 7.11 Utilities of biocatalysts in modern sustainable process chemistry

enable us better prediction of the protein structures under dynamic industrial reaction conditions (co-solvents, temperature, pH) and allow to develop tailored enzymes redesigned for novel functions or expanded operating ranges.

Enantiopure syntheses by biocatalysts is likely to play a major role in pharmaceutical industries, as two different isomers—eutomer (isomer with desired or beneficial properties) and distomer (inactive, toxic, different or undesirable properties)—are to be treated as two separate entities. In recent years, the most important developments have been made in the formation of optically active alcohols, amines and amino acids by means of biocatalytic cascades, as well as in the integration of C–C bond-forming enzymes into cascade systems. New engineered enzymes including artificial metalloenzymes have to be designed to deliver complimentary stereoselectivity, address substrate/product/co-factor inhibition issues and to carry out unusual reactions like enantioselective silylations, borylation, etc. chemoenzymatic catalysis, that involves one-pot sequential reactions using both chemocatalysis and biocatalysis in water needs to be developed for industrial benefits. Overcoming societal and economic issues such as public acceptance of the use of engineered organisms in technology and the high costs associated with enzyme engineering provides other challenges. Addressing these issues is expected to erase the borders between classical and engineered biocatalysts.

Acknowledgements *I would like to acknowledge Dr. A.K. Tyagi and Dr. R. S. Ningthoujam for giving me the opportunity to share my understanding on the subject with the readers. I would like to express my sincere gratitude to Dr. B. S. Patro and Dr. D. Goswami for their kind support.*

I will forever be grateful to Dr. S. chattopadhyay for introducing me to the world of chemoenzymatic syntheses, imparting knowledge and giving valuable inputs whenever required.

Thank you Dr. subrata Biswas, Mrs. Kanchan Biswas, Sudipto Biswas, Snehit and Dr. Manisha Banerjee for your unconditional love, support and inspiration.

References

Anastas, P.T., Warner, J.C.: Green Chemistry: Theory and practice, p. 30. Oxford University Press, New York (1998)

Arora, B., Pandey, P.S., Gupta, M.N.: Lipase catalyzed Cannizzaro-type reaction with substituted benzaldehydes in water. Tetrahedron Lett. **55**, 3920–3922 (2014)

Atsumi, S., Hanai, T., Liao, J.C.: Non-fermentative pathways for synthesis of branched-chain higher alcohols as biofuels. Nature **451**, 86–89 (2008)

Baeshen, N.A., Baeshen, M.N., Sheikh, A., Bora, R.S., Ahmed, M.M.M., Ramadan, H.A.I., Saini, K.S., Redwan, E.M.: Cell factories for insulin production. Microb Cell Fact. 13:141, Published online 2014 Oct 2. (2014). https://doi.org/10.1186/s12934-014-0141-0

Basso, A., Serban, S.: Industrial applications of immobilized enzymes—A review Mol Catalysis 479:110607 (2019). https://doi.org/10.1016/j.mcat.2019.110607

Behrens, G.A., Hummel, A., Padhi, S.K., Schätzle, S., Bornscheuera, U.T.: Discovery and protein engineering of biocatalysts for organic synthesis. Adv. Synth. Catal. **353**, 2191–2215 (2011)

Bhosale, S.H., Rao, M.B., Deshpande, V.V.: Molecular and industrial aspects of glucose isomerase. Microbiol Mol Biol Rev **60**, 280–300 (1996)

Breuer, M., Ditrich, K., Habicher, T., Hauer, B., Keßeler, M., Stgrmer, R., Zelinski, T.: Industrial methods for the production of optically active intermediates. Angew. Chem. Int. Ed. **43**, 788–824 (2004)

Buchholz, K.: A breakthrough in enzyme technology to fight penicillin resistance-industrial application of penicillin amidase. Appl. Microbiol. Biotechnol. **100**, 3825–3839 (2016)

Burk, M.J., de Koning, P.D., Grote, T.M., Hoekstra, M.S., Hoge, G., Jennings, R.A., Kissel, W.S., Le, T.V., Lennon, I.C., Mulhern, T.A., Ramsden, J.A., Wade, R.A.: An enantioselective synthesis of (*S*)-(+)-3-aminomethyl-5-methylhexanoic acid via asymmetric hydrogenation. J. Org. Chem. **68**, 5731–5734 (2003)

Chatterjee, S., Ghadigaonkar, S., Sur, P., Sharma, A., Chattopadhyay, S.: A chemoenzymatic synthesis of hept-6-ene-2,5-diol stereomers: Application to asymmetric synthesis of decarestrictine L, pyrenophorol and stagonolide E. J. Org. Chem. **79**, 8067–8076 (2014a)

Chatterjee, S., Sharma, A., Chattopadhyay, S.: Chemoenzymatic synthesis of the macrolide antibiotic (-)-A26771B. RSC Adv. **4**, 42697–42705 (2014b)

Coelho, P.S., Brustad, E.M., Kannan, A., Arnold, F.H.: Olefin Cyclopropanation via carbene transfer catalyzed by engineered cytochrome p450 enzymes. Sci. **339**, 307–310 (2013)

DeSantis, G., Wong, K., Farwell, B., Chatman, K., Zhu, Z., Tomlinson, G., Huang, H., Tan, X., Bibbs, L., Chen, P., Kretz, K., Burk, M.J.: Creation of a productive, highly enantioselective nitrilase through gene site saturation mutagenesis (GSSM). J. Am. Chem. Soc. **125**, 11476–11477 (2003)

Editor-in-Chief: Murray Moo-Young (2019) Comprehensive biotechnology, 3rd Edition, Elsevier

Erbeldinger, M., Mesiano, A.J., Russell, A.J.: Enzymatic catalysis of formation of Z-aspartame in ionic liquid - An alternative to enzymatic catalysis in organic solvents. Biotechnol. Prog. **16**, 1129–1131 (2000)

Ernst, M., Kaup, B., Müller, M., Bringer-Meyer, S., Sahm, H.: Enantioselective reduction of carbonyl compounds by whole-cell biotransformation, combining a formate dehydrogenase and a (*R*)-specific alcohol dehydrogenase. Appl. Microbiol. Biotechnol. **66**, 629–634 (2005)

Faber, K.: Biotransformations in Organic Chemistry: A Textbook. Springer Int. Publ., Germany (2017)

Fox, R.J., Davis, S.C., Mundorff, E.C., Newman, L.M., Gavrilovic, V., Ma, S.K., Chung, L.M., Ching, C., Tam, S., Muley, S., Grate, J., Gruber, J., Whitman, J.C., Sheldon, R.A., Huisman, G.W.: Improving catalytic function by ProSAR-driven enzyme evolution. Nat. Biotechnol. **25**, 338–344 (2007)

Ganske, F., Bornscheuer, U.T.: Lipase-catalyzed glucose fatty acid ester synthesis in ionic liquids. Org. Lett. **7**, 3097–3098 (2005)

Grunwald, P.: Immobilized biocatalysts. Catalysts **8**, 386 (2018). https://doi.org/10.3390/catal8090386

Hansen, K.B., Hsiao, Y., Xu, F., Rivera, N., Clausen, A., Kubryk, M., Krska, S., Rosner, T., Simmons, B., Balsells, J., Ikemoto, N., Sun, Y., Spindler, F., Malan, C., Grabowski, E.J.J., Armstrong, J.D., III.: Highly efficient asymmetric synthesis of sitagliptin. J. Am. Chem. Soc. **131**, 8798–8804 (2009)

Hoekstra, M.S., Sobieray, D.M., Schwindt, M.A., Mulhern, T.A., Gote, T.M., Huckabee, B.K., Hendrickson, V.S., Franklyn, L.C., Granger, E.J., Karrick, G.L.: Chemical development of CI-1008, an enantiomerically pure anticonvulsant. Org. Process Res. Dev. **1**, 26–38 (1997)

Holtmann, D., Hollmann, F.: Is water the best solvent for biocatalysis? Mol Catal **517**, 112035 (2022). https://doi.org/10.1016/j.mcat.2021.112035

Hughes, G., Lewis, J.C.: Introduction: Biocatalysis in industry. Chem. Rev. **118**, 1–3 (2018)

Ikushima, Y., Saito, N., Arai, M.: Promotion of lipase-catalyzed esterification of *N*-valeric acid and citronellol in supercritical carbon dioxide in the near-critical region. J. Chem. Eng. Japan **29**, 551–553 (1996)

Jones, J.B., Beck, J.F.: In: (Eds. Jones, J.B., Sih, C.J., Perlman, D.) Application of biochemical systems in organic chemistry, Part 1, Wiley: New York, **X**, pp 236–401 (1976)

Kalevi, V., Klibanov, A.M.: Enzymatic production of high fructose corn syrup (HFCS) containing 55% fructose in aqueous ethanol. Biotechnol. Bioeng. **30**, 917–920 (1987)

Kamat, S., Barrera, J., Beckman, J., Russell, A.J.: Biocatalytic synthesis of acrylates in organic solvents and supercritical fluids: I. Optim. Enzym. Environment. Biotechnol. Bioeng **40**, 158–166 (1992)

Kirby, A.J.: Enzyme mechanisms, models, and mimics. Angew. Chem. Int. Ed. **35**, 707–725 (1996)

Kizaki, N., Yasohara, Y., Hasegawa, J., Wada, M., Kataoka, M., Shimizu, S.: Synthesis of optically pure ethyl (*S*)-4-chloro-3-hydroxybutanoate by *Escherichia coli* transformant cells coexpressing the carbonyl reductase and glucose dehydrogenase genes. Appl. Microbiol. Biotechnol. **55**, 590–595 (2001)

Kognou, A.L.M., Shrestha, S., Jiang, Z., Xu, C., Sun, F., Qin, W.: High-fructose corn syrup production and its new applications for 5-hydroxymethylfurfural and value-added furan derivatives: Promises and challenges. J. Bioresourc. Bioprod. **7**, 148–160 (2022)

Le, Z.G., Guo, L.T., Jiang, G.F., Yang, X.B., Liu, H.Q.: Henry reaction catalyzed by Lipase A from *Aspergillus niger*. Green Chem. Lett. Rev. **6**, 277–281 (2013)

Martinez, C.A., Hu, S., Dumond, Y., Tao, J., Kelleher, P., Tully, L.: Development of a chemoenzymatic manufacturing process for pregabalin. Org. Process Res. Develop. **12**, 392–398 (2008)

Morihara, K., Ueno, Y.: A new procedure for enzymatic semis ynthesis of human insulin by hydrolysis of single-chain Des-(B-3O)-Insulin precursor with lysyl endopeptidase. Biotechnol. Bioeng. **37**, 693–695 (1991)

Morihara, K., Oka, T., Tsuzuki, H., Tochino, Y., Takashi, K.: Achromobacter protease I-catalyzed conversion of porcine insulin into human insulin. Biochem. Biophys. Research Commun. **92**, 396–402 (1980)

Nakamura, C.E., Whited, G.M.: Metabolic Engineering for the microbial production of 1,3-propanediol. Curr Opin Biotechnol 14:454–459 (2023)

Oikawa, H., Katayama, K., Suzuki, Y., Ichihara, A.: Enzymatic activity catalysing exo-selective Diels-Alder reaction in solanapyrone biosynthesis. J. Chem. Soc. Chem. Commun. 1321–1322 (1995)

Ortega-Rojas, M.A., Rivera-Ramírez, J.D., Ávila-Ortiz, C.G., Juaristi, E., González-Muñoz, F., Castillo, E., Escalante, J.: One-pot lipase-catalyzed enantioselective synthesis of (*R*)-(-)-N-benzyl-3-(benzylamino)butanamide: The effect of solvent polarity on enantioselectivity. Molecules **22**, 2189 (2017). https://doi.org/10.3390/molecules22122189

Pawar, A.S., Chattopadhyay, S., Chattopadhyay, A., Mamdapur, V.R.: Novel synthesis of 1,4-dienic macrolide pheromone of *Cucujid* grain beetles. J. Org. Chem. **58**, 7535–7536 (1993)

Polaina, J., MacCabe, A.P.: (Ed.) Industrial enzymes. Structure, function and applications. Springer Dordrecht, (2007). eBook ISBN 978–1–4020–5377–1, https://doi.org/10.1007/1-4020-5377-0

Ran, N., Frost, J.W.: Directed evolution of 2-keto-3-deoxy-6-phosphogalactonate aldolase to replace 3-deoxy-D-arabino-heptulosonic acid 7-phosphate synthase. J. Am. Soc. **129**, 6130–6139 (2007)

Ricca, E., Brucher, B., Schrittwieser, J.H.: Multi-enzymatic cascade reactions: overview and perspectives. Adv. Synth. Catal. **353**, 2239–2262 (2011)

Santiniello, E., Ferraboschi, P., Grisenti, P., Manzocchi, A.: The biocatalytic approach to the preparation of enantiomerically pure chiral building blocks. Chem. Rev. **92**, 1071–1140 (1992)

Sanz-Cervera, J.F., Glinka, T., Williams, R.M.: Biosynthesis of the brevianamides: Quest for a biosynthetic Diels-Alder cyclization. J. Am. Chem. Soc. **115**, 347–348 (1993)

Savile, C.K., Janey, J.M., Mundorff, E.C., Moore, J.C., Tam, S., Jarvis, W.R., Colbeck, J.C., Krebber, A., Fleitz, F.J., Brands, J., Devine, P.N., Huisman, G.W., Hughes, G.J.: Biocatalytic asymmetric synthesis of chiral amines from ketones applied to sitagliptin manufacture. Sci. **329**(5989), 305–309 (2010)

Sheldon, R.A., Pereira, P.C.: Biocatalysis engineering: The big picture. Chem. Soc. Rev. **46**, 2678–2691 (2017)

Svedendahl, M., Hult, K., Berglund, P.: Fast carbon−carbon bond formation by a promiscuous lipase. J. Am. Chem. Soc. **127**, 17988–17989 (2005)

Tao, J., Kazlauskas, R. (Ed.): Biocatalysis for green chemistry and chemical process development. John Wiley and Sons (2011)

Wolberg, M., Hummel, W., Wandrey, C., Müller, M.: Highly regio- and enantioselective reduction of 3,5-dioxocarboxylates. Angew. Chem. Int. Ed. **39**, 4306–4308 (2000)

Wong, C.H., Whitesides, G.M.: Enzymes in synthetic organic chemistry. Elsevier Science (1994)

Wu, S., Snajdrova, R., Moore, J.C., Baldenius, K., Bornscheuer, U.T.: Biocatalysis: Enzymatic synthesis for industrial applications. Angew. Chem. Int. Ed. **60**, 88–119 (2021)

Yazbeck, D.R., Martinez, C.A., Hu, S., Tao, J.: Challenges in the development of an efficient enzymatic process in the pharmaceutical industry. Tetrahedron. Asymm. **15**, 2757–2763 (2004)

Yi, D., Bayer, T., Badenhorst, C.P.S., Wu, S., Doerr, M., Höhne, M., Bornscheuer, U.T.: Recent trends in biocatalysis. Chem. Soc. Rev. **50**, 8003–8049 (2021)

Chapter 8
Nanostructures of Transition Metal Oxides/Hydroxides: Their Synthesis and Use in Catalysis

Sakshi Kansal, Satvik Anshu, Puja De, and Amreesh Chandra

8.1 Introduction

The process of catalysis is now being utilised in a large number of areas ranging from petrochemical industries to cosmetics, pharmaceutical products, agro-industries to food industries, clean energy to hydrogen economics, cloth to defence industry, water and solid to air treatment (Lee and Yoo 2014). Our human body performs several bio-catalysis processes, using natural enzyme as a catalyst. It includes breaking down complex carbohydrates to simple sucrose, hydrolysis of ATP, conversion of NAD to NADH, and many more, which are vital for sustainable biological processes (Maenaka et al. 2012). Now, the development of several catalysts for the production of hydrogen gas, synthetic fuel and fuel cells is gaining speed (Maduraiveeran et al. 2019). Catalysis is generally described as a chemical kinetic process, where a substance known as a catalyst, either in the same chemical phase or different from that of the reactants, helps accelerate the chemical reaction by altering the energy required during the process. On the basis of the reactants–catalysts phase, catalysis can be of two broad types: (i) homogeneous catalysis and (ii) heterogeneous catalysis. There are several types of catalysts available based on noble metals (Ag, Au, etc.), transition metals (Fe, Ni, Cu, Co, etc.), precious metals (Pt, Pd, Rh, etc.) and non-metal (alpha-halo carbonyl compounds, ketones, imines, nitroxyl radicals, etc.). A high-performing catalyst must have efficient electron transfer capacity, high surface area with accessible active sites, structural stability, cost-effectiveness and non-toxic nature. If one focuses on areas where catalysis is extensively used, i.e. energy, dye degradation and water treatment mostly metal-based catalysts are being used.

S. Kansal · S. Anshu · A. Chandra (✉)
School of Energy Science and Engineering, Indian Institute of Technology Kharagpur, Kharagpur, India
e-mail: achandra@phy.iitkgp.ac.in

P. De · A. Chandra
Department of Physics, Indian Institute of Technology Kharagpur, Kharagpur, India

R. S. Ningthoujam and A. K. Tyagi (eds.), *Handbook of Materials Science, Volume 3*, Indian Institute of Metals Series, https://doi.org/10.1007/978-981-96-8229-4_8

Precious metal-based catalysts are extensively used in several processes ranging from refining crude oil into fuels, synthesis of high-grade pharmaceuticals and agrochemicals products to energy conversion reactions (Bullock 2017). To further advance their applications factors that need to be addressed are: cost, low abundance and toxicity. Transition metals are extensively utilised for catalytic applications as their incompletely filled d-orbitals, allow them to show variable oxidation states. This provides them with the capacity to exhibit multiple oxidation states and undergo fast redox reactions with electron transfer. This makes them quite useful for applications such as hydrogenation, metathesis, electrocatalysis and photocatalysis. The intrinsic catalytic properties of such transition-metal oxide/hydroxide nanomaterials depend on a number of factors, which include morphology, dimensionality, atomically dispersed metal sites, composition, metal oxidation state and functionality. Single transition metal-based oxides/ hydroxides allow maximum metal atom utilization, peculiar electronic structure and exposed catalytic active sites (Maduraiveeran et al. 2019). These oxides can be readily fabricated into quantum sizes which further enhance the intrinsic catalytic activity. The reaction kinetics can be further tuned by creating oxygen vacancies, using composites and doping with multiple cations and/ or anions.

8.2 Classification and Types of Catalysis

Mostly, the catalysis processes are broadly characterized under two broad headings viz, homogeneous and heterogeneous.

8.2.1 Homogeneous Catalysis

Homogeneous catalysis, by definition, refers to a catalytic process where the reactants and the catalysts are in the same chemical phase. The homogeneous catalysis can be sub-characterized into single-species and complex catalysis. When a single molecule or an ion catalyses the whole reaction, it is termed as single species catalysis. If several species are available to work as catalysts in conjugation with each other and interconvert into one another during reaction, it is defined as complex catalysis.

A common example of homogenous catalysis is enzyme-based reactions. Enzymes are natural homogeneous catalysts, containing several active sites on the surface where the reactants bind, before converting into products. Experimentally, the enzyme kinetics is governed by Michaelis–Menten mechanism, where an enzyme (E) reacts with the substrate (S) to form an enzyme–substrate complex (ES). Subsequently, this intermediate substance converts to form a product (P). The two-step process is represented below:

$$E + S \rightarrow ES \tag{8.1}$$

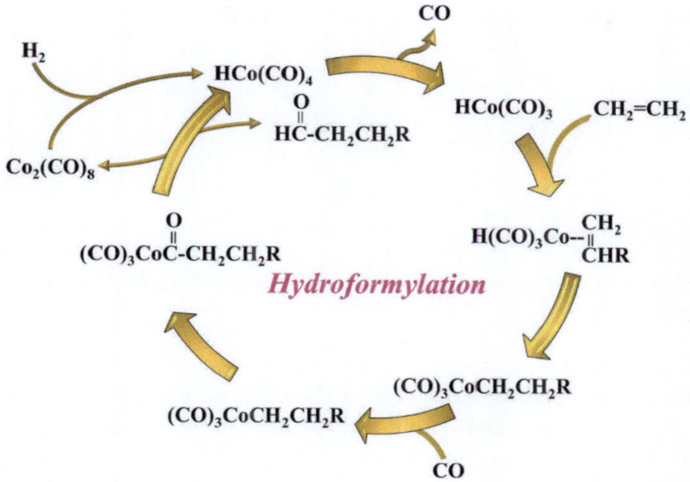

Fig. 8.1 Mechanism of hydroformylation reaction using transition metal-based catalyst

$$ES \rightarrow P \tag{8.2}$$

The rate of product formation is determined by Michaelis–Menten equation (Chapin et al. 1954). Apart from this, homogeneous catalysis is utilised in many pharmaceutical and polymer industries. Homogeneous catalysis process offers better selectivity and reactivity, economic efficiency and manageable reaction conditions compared to other catalysis processes. On an industrial scale, homogeneous catalysis is initiated using transition metal-based catalysts.

Transition metals have the unique ability to bind with a variety of ligands and form stable complexes. These stable complexes can undergo facile change from one oxidation state to another and transfer electrons or protons for catalysis. One of the widespread applications of homogeneous catalysis in industries is hydroformylation reaction where an olefin reacts with syngas to generate alcohols and carbonyl compounds. A mechanism of hydroformylation reaction using transition metal-based catalyst is highlighted in Fig. 8.1.

8.2.2 *Heterogeneous Catalysis*

Heterogeneous catalysis can be driven by many redox active solids during the production of variety of commercially important chemicals, pharmaceuticals and fuels (Hutchings 2009). Additionally, energy sector largely depends on heterogeneous catalysis. This ranges from petroleum refining to preparation of synthetic fuel. Also, the niche area of hydrogen energy needs economical and facile catalysts for the production of hydrogen gas during water-splitting.

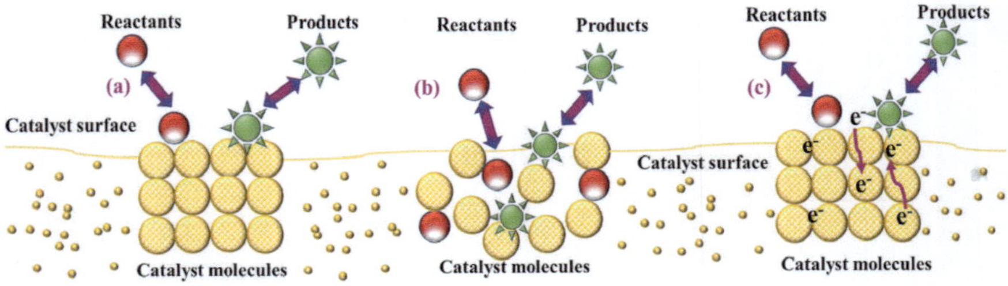

Fig. 8.2 Different types of heterogeneous catalysis **a** surface catalysis, **b** bulk-type-I catalysis and **c** bulk-type-I I catalysis

In general terms, heterogeneous catalysis involves catalyst, which is in a different chemical phase of the reactants and it does not solvate in the reaction mixture. For example, in *Haber's process,* the transition metals in solid state catalyse the reaction of ammonia gas formation. Another example is the *Contact process*, where sulphuric acid is formed on an industrial scale and vanadium oxide is used to accelerate the process. Similarly, nitric acid is formed using Pt-based catalyst through *Ostwald process*. Generally, heterogeneous catalysis is classified into three sub categories:

(a) *Surface catalysis:* It is the most common type of heterogeneous catalysis, where the whole process occurs at the surface of catalyst only. There is no diffusion of redox carriers inside the catalyst, as highlighted in Fig. 8.2a. The rate of catalysis is directly influenced by the catalyst surface area. Initially, the reactants are adsorbed on the surface of the catalyst. Subsequently, cleavage of the bonds leads to formation of reactive intermediate species. The reaction occurs between these intermediatory reactive species, leading to product. Finally, the product is desorbed from the surface of catalyst, leaving behind the catalyst in its original state.
(b) *Bulk-type-I catalysis:* In this type of heterogeneous catalysis, reactant molecules adsorb in the bulk of catalyst material, where they undergo redox reactions to transform into products, as highlighted in Fig. 8.2b.
(c) *Bulk-type-II catalysis:* In such cases, reactants molecules are adsorbed on the surface of catalyst. During reduction, the electrons, which are released at the surface, travel through the whole catalyst resulting in a simultaneous reduction of entire catalyst. The mechanism is schematically explained in Fig. 8.2c.

Several other factors govern the efficiency of a catalyst. These include chemical nature, structural stability, active surface area, redox activity, etc. This chapter focuses on heterogeneous catalysis as it is more economical and rapid than homogeneous catalysis. Almost all the important industrial and energy generation processes utilise heterogeneous catalysts.

8.2.3 Electrocatalysis

Electrocatalysis has strategic importance in future sustainable energy technologies. Electrocatalysis is the primary process in fuel cells, metal-air batteries and large-scale wastewater treatment (Zhang et al. 2021). It is a process, which involves the introduction of a catalyst at the electrode–electrolyte interface to alter the rate of an electrochemical reaction. The catalysts that are involved in such a catalysis process are termed as "electrocatalysts." Further, electrocatalysis can be both homogeneous and heterogeneous. Consider an electrochemical reaction that is occurring at an electrode–electrolyte interface:

$$O + ne^{-} \rightarrow R \tag{8.3}$$

The thermodynamic potential needed to drive this reaction can be calculated from the Nernst's equation:

$$E_{eq} = E_o + \frac{RT}{nF} log \frac{C_O}{C_R} \tag{8.4}$$

where E_{eq} is the cell potential , E_o standard equilibrium potential of redox couple, C_O and C_R are surface concentrations of oxidant and reductant, considering the activity coefficient to be 1. The general pathway of an electrochemical reaction is shown in Fig. 8.3.

Here, the potential is largely altered by the chemical and physical properties of the electrode materials and inherent to a particular electrochemical system. However, in addition to the theoretically calculated potential, an additional potential is required to overcome the kinetic barrier of an electrochemical reaction. This additional potential is termed as overpotential and is a key parameter that regulates the electrochemical reaction kinetics. The current received for an electrochemical reaction after applying the cell potential has a direct correlation with the overpotential. This is generally

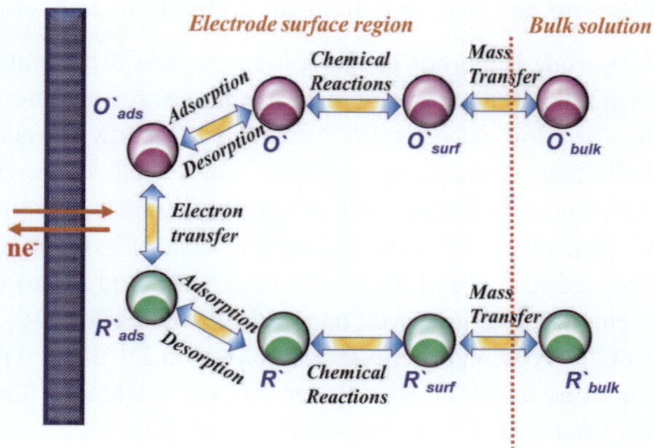

Fig. 8.3 General pathway of an electrochemical reaction

explained through Butler–Volmer model (Bard and Faulkner 2000). In simple words, overpotential is similar to activation energy i.e. minimum energy required to drive a chemical reaction. Similar to chemical kinetics, in an electrochemical reaction, materials that reduce the overpotential while enhancing the rate of electrochemical reactions are electrocatalysts.

For use in energy devices, an electrocatalyst should have high current densities with low overpotential requirements, capabilities to undergo facile electron transfer kinetics, long-term stability, ease of fabrication, economically viable and environmentally friendly (Sun et al. 2021).

8.3 Synthesis of Nanostructures of Transition Metal Oxides/Hydroxides-Based Catalysts

The synthesis methods used to obtain various nanostructures of catalysts can be categorized into two main categories: template-assisted or template-free methods. For template-assisted methods, to shape the initial structure, sacrificial templates (e.g. AAO (anode aluminium oxides) templates) are generally used. The templates are removed using either an acidic or a basic solution. In the following section, let us discuss some of synthesis strategies for obtaining hierarchical structures of TMOs and hydroxides, by dividing the hierarchical structure into three main categories viz., (a) 0-D, (b) 1-D and (c) 2-Dl.

8.3.1 Synthesis of 1-D Hierarchical Structures

One dimensional (1-D) nanostructures, including nanorods, nanowires and nanotubes, are the kind of materials that have two of their dimensions in the range of 1–100 nm. Such nanostructures can offer effective transport channels for both electrons and ions, making them useful in the field of catalysis.

Nanorod: Nanorods or nanopillars possess aspect ratios (length/diameter) of more than 10. Although the increase in surface area of nanorods and nanopillars is limited by their low aspect ratio compared to nanowires and nanofibers, the stability of individual nanorods/nanopillars allows them to provide a channel for electrolyte diffusion, which efficiently enhances the ionic conductivity. Let us take an example of Co_3O_4, which is extensively used because of its tunable surface area, electrochemical properties, size and shape. Chandra et al. used a This can be a simple hydrothermal method to prepare Co_3O_4 nanorods (Akhtar et al. 2019). In the beginning of synthesis, $CoCl_2.6H_2O$ and urea were dissolved in DI water with a molar ratio of 2:1 under vigorous magnetic stirring for 30 min. Thereafter, the obtained solution has to be kept static for 12 h and then transferred to a sealed borosilicate glass bottle, at 120 °C for 8 h. The obtained precipitate was washed and then dried overnight at 80 °C before calcining it at 500 °C in the air for 3 h (Akhtar et al. 2019).

Another widely explored metal oxide is Fe_2O_3, owing to its low cost, availability and eco-friendliness, and it is widely used for various energy conversion and storage applications. The most stable phase of iron oxides is hematite (α-Fe_2O_3), which shows various physical and chemical properties due to the size and shape effect. But, the major drawback of the α-Fe_2O_3 is its poor electrical conductivity (10–14 S cm^{-1}). To overcome this limitation, strategy of morphology tuning is being extensively explored. The presence of higher porosity and higher aspect ratio can significantly enhance the catalytic performance of the material. Chandra et al. recently reported rod-like morphology of α-Fe_2O_3 nanorods by calcining it at 400 °C. They used $FeSO_4.7H_2O$ as an iron source and oxalic acid as a precipitating agent to make the FeOOH precursor utilizing a straightforward and inexpensive wet chemical method. The corresponding SEM micrograph is highlighted in Fig. 8.4a (Halder et al. 2023).

Another rod-like morphology has been synthesized by Chandra et al. in which instead of using monometallic oxides they have explored binary metal oxides. Binary metal oxides, such as $MnCo_2O_4$ have a lower activation energy level for transporting electrons than monometallic oxides (such as Co_3O_4, MnO_2), leading to a higher electrochemical activity. Chandra et al. have established a cost-effective synthetic route for mass production of microrods of $MnCo_2O_4$ (Akhtar et al. 2016). For the synthesis, cobalt chloride and manganese chloride were mixed in the molar ratio of 2:1 and added dropwise to a pre-heated 0.4 M oxalic acid to form the solid solution of $Mn_{1/3}Co_{2/3}C_2O_4{\cdot}2H_2O$. For the removal of Cl^- and excess oxalic acid, the resultant precipitate was filtered and washed several times with ethanol and de-ionized water. The filtrate was dried in an oven at 90 °C for 8 h. Finally, $MnCo_2O_4$ microrods were procured after the powder was calcined at 400 °C or 800 °C in air for 4 h.

Nanowire: Nanowires have been extensively explored for catalytic applications on account of their high aspect ratio which can offer a lot of surface area for charging storage and effective charging transfer routes. For instance, Kana et al. synthesized the vertical CuO nanowire for catalytic applications by hydrothermal technique. Such CuO nanowires deliver the excellent property to absorb dyes from a polluted environment (Kana et al. 2019). For the synthesis of CuO nanowire, cetyltrimethylammonium bromide (CTAB) was used as a soft template (Kana et al. 2019). Wei et al. developed the 3D gully-network structure of Co_3O_4 nanowire arrays, grown on Ni foam through a facile and scalable hydrothermal procedure followed by annealing treatment. The direct growth of a 3D gully-network structure on Ni foam could ensure that every nanowire possesses good mechanical adhesion (Guo et al. 2018).

Binary metal oxide is also becoming popular because of its low cost, environmental safety and low toxicity. For example, $NiCo_2O_4$ shows high electrochemical activity and at least two orders of magnitude higher electrical conductivity than pure Co_3O_4 and NiO. Recently, Hao et al. developed novel spinel $NiCo_2O_4$ nanowires using a cost-effective template assisted hydrothermal technique followed by calcination. Nickel nitrate hexahydrate and cobalt nitrate hexahydrate were used as a source of nickel and cobalt, respectively and were dissolved in 40 mL deionized

Fig. 8.4 **a-f** Hierarchical morphologies of various transition metal oxides. (Halder et al. 2023; Anshu et al. 2023; Sharma et al. 2019, 2018a; Kansal et al. 2023a) (Adapted with permission, ACS, IOP, Elsevier Publishers)

water and ethylene glycol under continuous magnetic stirring. The templating agent, viz. 0.8106 g of urea was added into the prepared solution of nickel and cobalt salt. Subsequently, the solution was transferred into a 50-mL Teflon-lined autoclave maintained at 160 °C for 12 h. Thus, the obtained precursor was calcined at 350 °C for 2 h in the air to obtain the nanowire morphology of $NiCo_2O_4$ (Hao et al. 2018).

Nanotube: Tubular nanostructures provide higher gravimetric capacity than solid nanostructures (e.g. nanowires and nanorods), due to their higher surface area and lower density. Nanotubes are generally fabricated using template-assisted procedures, CVD, hydrothermal methods, etc. Nanotubes can be easily prepared using template-assisted method, by partially entrapping the precursors into a nanoporous template. The reactants hold on to the walls of nanochannels at their early infiltrating stage. For example, Yao et al. proposed the hard template hydrothermal method for the synthesis of Co_3O_4 nanotubes.(Yao et al. 2015) During the synthesis, guanidine hydrochloride was used as a precipitant and activated carbon was used as a hard template to control the structural growth and pore size distribution. The solution of cobalt sulphate heptahydrate, guanidine hydrochloride and activated carbon was placed in Teflon lined stainless steel autoclave at 160 °C for 16 h. After the filtration of the material, the dried product then calcinated at 500 °C for 4 h, with a slowly rising rate of 0.5 °C min^{-1}, to obtain Co_3O_4 nanotubes.(Yao et al. 2015) Liu et al. demonstrated a facile and reliable template free solvothermal route for the synthesis of NiO nanotubes. In a typical synthesis, a solution of nickel acetate and ethylene glycol was placed in an autoclave at 180 °C for 10 h. The washed and dried product was calcined at 350 °C for 2 h to obtain nanotubes of NiO (Liu et al. 2016).

8.3.2 Synthesis of 2-D/Pseudo 2-D Hierarchical Structures

The interest in 2-D nanostructures was initiated by the fascinating world of graphene. Two-dimensional (2D) nanostructured materials having fascinating properties, versatility and a wide range of applications have drawn great interest from researchers. The most interesting property of 2D nanostructured materials is the high active specific surface area without any bulk volume, which makes them great interest for catalysis applications. In addition, 2D materials exhibit high mechanical stability, strength, flexibility, transparency and chemical stability. Let us discuss 2-D nanostructured materials under two main sub-categories, i.e. nanosheets and nano-discs.

Nanosheets/Nanoflakes: These have length in micrometres and thickness in the scale nanometre range. Ultrathin Au-doped SnO_2 nanoflakes arrays were developed by Chandra et al. (Anshu et al. 2023) via one step hydrothermal method using hexadecyltrimethylammonium bromide (CTAB) as morphology modifier as highlighted in Fig. 8.4b (Anshu et al. 2023) Li et al. (Li et al. 2019) have established a synthesis protocol to obtain $MnCo_2O_4$ nanosheets using a two-step template assisted hydrothermal method with post-annealing treatment. In a typical synthesis method, manganese and cobalt acetate tetrahydrate with polyvinyl pyrrolidone was dissolved using distilled water and ethylene glycol, under constant stirring for 2 h. The molecules of polyvinyl pyrrolidone might bind to metal ions and distribute them uniformly in a cross-linked chelate. Subsequently, the mixture solution is transferred into an autoclave and heated at 180 °C for 12 h. Finally, the product was annealed at 600 °C for 4 h in air to obtain $MnCo_2O_4$ nanosheets. Co_3O_4 nanosheets demonstrated in Fig. 8.4 c fabricated by Chandra et al. using a simple one-step co-precipitation process (Sharma et al. 2019).

Nanopetals/Nanodiscs: Due to their superior chemical and thermal stability, natural abundance, environmental friendliness and low cost, MnO_2-based materials have been extensively used in electrocatalysis. Furthermore, to increase the electrochemical activity of manganese, novel morphologies with high specific surface area and large pore volume have been explored by various groups of researchers. For instance, a very fascinating nanodisc of Co_3O_4 was reported by Chandra et al. (Sharma et al. 2019) via co-precipitation method. During the synthesis, 0.1 M NaOH solution was added to the solution of 0.05 M cobalt nitrate dropwise, under constant stirring by maintaining the temperature at 70 °C. After the formation of a light brown colour precipitate, the solution was filtered and washed with ethanol several times. Finally, to get the nanodisc morphology of Co_3O_4, the dried powder was calcined at 600 °C for 2 h in air.

8.3.3 Synthesis of 3-D Hierarchical Structures

3D porous hierarchical structures provide large specific surface areas with stable and high mechanical stability. Typically, 3D electrodes are created by either organizing active materials into 3D nanostructures or employing metal foam as templates.

The 3-D porous spherical morphology of Cu_2O was synthesized by Chandra et al. (Sharma et al. 2018a) via a template-assisted precipitation approach. Sodium dodecyl sulphate was used as a soft template and hydrazine hydrate as a reducing agent. At first, sodium dodecyl sulphate was dissolved into deionized water under constant stirring to form the various stable micellar solution. Subsequently, an aqueous solution of copper sulphate, ammonia and NaOH was added to the as prepared solution at intervals of 20 min. Once the colourless solution turned to a light turbid blue, N_2H_4 was added dropwise. Following the addition of N_2H_4, the solution transformed to a light brown colour. The precipitate was collected by centrifugation and washed several times with excess deionized water and ethanol. Finally, the porous sphere of Cu_2O as highlighted in Fig. 8.4 (d) was obtained by drying the obtained product overnight at 70 °C in a vacuum oven. Chen et al. (Ci et al. 2015) proposed a controlled synthesis mechanism to fabricate the 3-D spherical microflower of NiO. The precursor of nickel chloride, urea and sodium dodecyl sulphate was prepared. Next, the mixed solution was transferred into an autoclave for hydrothermal treatment. The combined solution was then put into an autoclave for hydrothermal processing. Urea was hydrolysed to produce OH^- and CO_3^{2-}, which then reacted with Ni^{2+} resulting in the formation of the green precipitate of $NiCO_3 \bullet Ni(OH)_2 \bullet xH_2O$. During the annealing process, the resultant precipitate broke down into a black NiO powder with a 3-D microflower morphology.

Nanocubes: Cube-shaped nanoparticles can become quite useful in catalysis because the corners can act as highly reactive sites due to unsaturation. Additionally, the cube-shaped nanoparticles are formed by exposed facets i.e. crystal planes growing preferentially. This preferential growth can promote certain reactivity that cannot be observed in other shapes. Xu et al. (Chen et al. 2020) synthesized the novel $ZnCo_2O_4$ quasi-cubes through a simple hydrothermal process. In a typical synthesis procedure, zinc acetate, cobalt acetate and urea were first dissolved in distilled water with intense stirring, to form a clear solution. Following the addition of glycerine, the mixture was loaded into a Teflon-lined stainless-steel autoclave and maintained at 180 °C for 10 h. Once the autoclave was naturally cooled down to room temperature, the precipitate was harvested and washed with water and ethanol several times. The dried precipitate was calcined at 350 °C in air for 4 h. The porous quasi cubes of $ZnCo_2O_4$ were obtained when the increasing temperature rate of the calcination was maintained at 2 °C per minute. The SEM micrographs of various miscellaneous hierarchical nanostructures (3D) of different metal oxides are shown in Fig. 8.4e and f (Kansal et al. 2023a).

8.4 Applications of Transition Metal Oxides in Catalysis

8.4.1 Catalysis Process Important for Fuel Production and Energy Storage

Hydrogen Evolution Reaction

Growing worldwide energy demand as well as global warming increases the need to discover alternate energy sources that are renewable, clean and sustainable. Recently, molecular hydrogen gas has attracted a lot of attention as a prospective green energy fuel owing to its high calorific value, which is even greater than conventional fossil fuels. Additionally, the complete combustion of hydrogen gas releases only water as the by product. One of the cost-effective and sustainable processes to generate molecular hydrogen gas is electrochemical water splitting. The electrolysis of water was first reported in 1789, where in the process hydrogen gas was generated at the cathode by the electrochemical reduction of water molecules. However, the chemical reaction of hydrogen gas evolution is kinetically sluggish and requires additional potential in addition to the thermodynamic potential. This is commonly defined as overpotential (η). Hence, to minimize the overpotential suitable catalysts are required. Certain parameters are used to decide the efficiency of catalysts towards the hydrogen evolution reaction:

(a) Overpotential

The theoretical thermodynamic potential for hydrogen evolution reaction concerning reversible hydrogen electrode potential (RHE) is zero under standard pressure and temperature conditions. However, experimentally, an additional potential is required to complete the chemical reaction, which is calculated as:

$$\eta_{HER} = E_{app} - E_{HER} \tag{8.5}$$

where η_{HER} is the overpotential, E_{app} is the experimental potential applied and E_{HER} is zero with respect to RHE.

(b) Tafel Slope

Tafel slope is another parameter that influences the rate kinetics of the evolution reaction. It is an intrinsic property of a catalyst and is calculated using the following equation:

$$\eta_{HER} = b\log(\frac{j}{j_o}) \tag{8.6}$$

where b is the Tafel slope, j and j_o are the current density and exchange current density, respectively. Herein, the exchange current density (j_o) is basically the amount of current required to drive the evolution reaction when there is no additional potential needed, i.e. η_{HER} is zero. Further, it also depicts the catalytic efficiency of a catalyst, which means if a catalyst delivers a lower value of Tafel slope that means less overpotential is required to deliver the current density similar to the value of exchange current density. Hence, the kinetics of the overall reaction is enhanced.

(c) Faradic Efficiency

Faradic efficiency is another important parameter that defines the electrocatalytic activity of a catalyst toward HER, which can be measured in terms of how efficiently electrons, provided by the catalysts, participate in the chemical reduction of water molecules into hydrogen gas. Basically, for HER, it is calculated by measuring the ratio of the amount of experimentally generated hydrogen gas to theoretically generated hydrogen gas and can be represented by:

$$Faradic\ efficiency = \frac{n_{H_2}}{n_{H_2}\prime} = \frac{2Fn_{H_2}}{It} \tag{8.7}$$

where n_{H_2} is the mole of experimentally generated hydrogen gas, $n_{H_2}\prime$ is the mole of hydrogen gas produced theoretically, F is the Faraday constant, I is the constant reduction current and t is the time at the constant reduction current.

(d) Turnover Frequency (TOF)

Turnover frequency evaluates the catalytic rate of each site on the catalyst surface that can participate in the reaction in other words it is defined as the total number of molecules that converted into the product per catalytic site per unit time and given by the following equation for HER process:

$$TOF = \frac{jA}{4nF} \tag{8.8}$$

where j is the current density required to drive the reduction of water molecules to H_2, A is the area of working electrode and n is the number of moles of active catalyst.

All these parameters are the major evaluating factors to measure the activity of the catalyst materials towards the HER. Also, the same factors define the rate kinetics of other chemical processes such as oxygen evolution reaction (OER), oxygen reduction reaction (ORR), etc. Generally, over the last few decades, noble-metal based catalysts have been extensively used due to their high hydrogen binding energy, lower Tafel slope and their capability to deliver high current density close to that of thermodynamic potential. However, their high cost and limited resources restrain them from becoming cost-effective and sustainable options for catalysing the water splitting reaction on a large scale. In recent times, transition metal oxides-hydroxides on the account of their high redox activity, large specific surface area, tunable morphologies and appropriate natural abundance are widely preferred as electrocatalysts for HER process. The overall mechanism of hydrogen evolution reaction (HER) in an alkaline medium, which is driven by an electrocatalyst can be explained through Volmer-Heyrovsky pathway in the following two steps: (i) where the electrocatalyst activates and alters the O–H bond of the water molecules to generate H_{ads} on its surface and (ii) H_{ads} desorb and react with water to generate H_2 gas and shown in Fig. 8.5 (Kansal et al. 2023b).

$$E_m + H_2O + e^- \rightarrow E_m - H_{ads} + OH^- \text{(Volmer step)} \tag{8.9}$$

$$H_2O + E_m - H_{ads} + e^- \rightarrow H_2 + E_m + OH^- \text{(Heyrovsky step)} \tag{8.10}$$

where E_m is the transition metal-based oxides-hydroxides as a metal electrocatalyst and $E_m - H_{ads}$ is the adsorbed hydrogen on surface of an electrocatalyst.

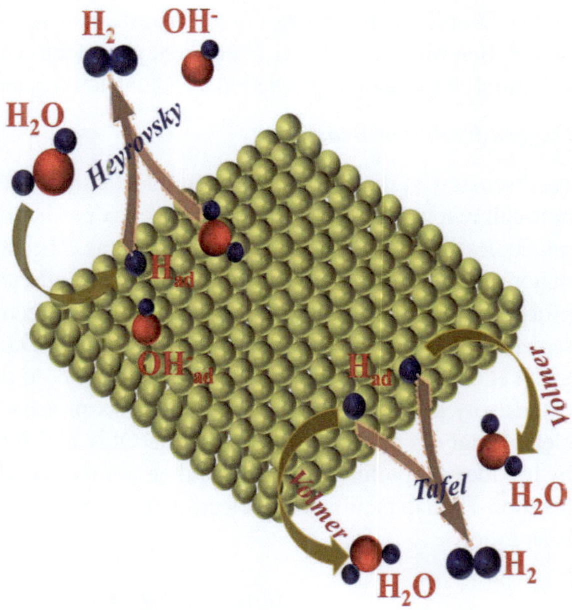

Fig. 8.5 Overall mechanism of HER

Several articles have been published where single, binary and ternary metal-based transition metal oxides/hydroxides are used as an efficient electrocatalysts for hydrogen evolution reaction. For instance, Mondal et al. prepared NiO-based electrocatalyst for HER and achieved an overpotential of 424 mV at 10 mA cm^{-2} current density in 1 M KOH (Mondal et al. 2018). Similarly, in Cu_2O based electrocatalyst reported by Paul et al., 148.7 mV was required to raise the current density up to 12.8 mA cm^{-2} (Paul et al. 2020). Researchers are exploring new ideas to enhance the catalytic activity of transition-metal-based electrocatalysts by tuning their morphologies, inducing defects, fabrication on carbon based substrate and doping with heteroatoms. Further, the performance of an electrocatalyst is greatly influenced by the surface area of the materials. Hence various morphologies such as nanorods, nanosheets, nanowires, nanocubes etc. are being explored. Recently, Tahira et al. fabricated 1-D nanowire morphology exhibited Co_3O_4–CuO composite as an efficient electrocatalyst for hydrogen evolution reaction, which show a Tafel slope of 65 mV dec^{-1} and at overpotential of 288 mV to achieve a current density of 10 mA cm^{-2} (Tahira et al. 2019). Chandra et al. synthesized a 2D-nanosheets morphology of Ni-Co Layered double hydroxide (LDH) using facile co-precipitation method. The 2D nanosheets exhibited a surface area of 123 m^2 g^{-1}. Ni-Co LDH shows high catalytic activity towards HER, with an overpotential requirement of 247 mV at the current density of 10 mA cm^{-2} (Kansal et al. 2023b). Wang et al. developed a microsphere nanostructural based $NiCo_2O_4$ and Co_3O_4 for hydrogen evolution reaction catalysis with high BET specific surface area of 168 m^2 g^{-1} and 172 m^2 g^{-1}, respectively. $NiCo_2O_4$ and Co_3O_4 microspheres display a lower overpotential of 141.1 mV and 155.6 mV to reach 10 mA cm^{-2}, respectively (Paul et al. 2020). Despite such advances in electrocatalysts field for HER, still there are major factors that hinders the overall efficiency of electrocatalysts are current collectors, structural deformation, prediction of actual active sites and electrolytes limitation.

Oxygen Evolution Reaction

As discussed in the previous section, electrochemical water splitting involves two half-cell reactions. At the cathode, hydrogen gas is liberated and at the anode water molecules are oxidised to generate oxygen gas. The oxygen evolution reaction is the major rate determining process that governs the efficiency of the overall electrolysis process of water. OER involves transfer of four electrons in order to oxidize water molecules to oxygen gas that make this process more kinetically sluggish as compared to HER. Along with this, fuel cells, which are common energy storage devices, extensively depend on the water electrolysis process. Similar to the HER process, electrocatalysts are used to catalyse the OER and for that transition metal oxides-hydroxides are preferred because of their ability to undergo fast and reversible redox reactions, which facilitates the electron transfer process. The overall reaction of OER is as follows:

$$4OH^- \rightarrow 2H_2O + O_2 + 4e^- \tag{8.11}$$

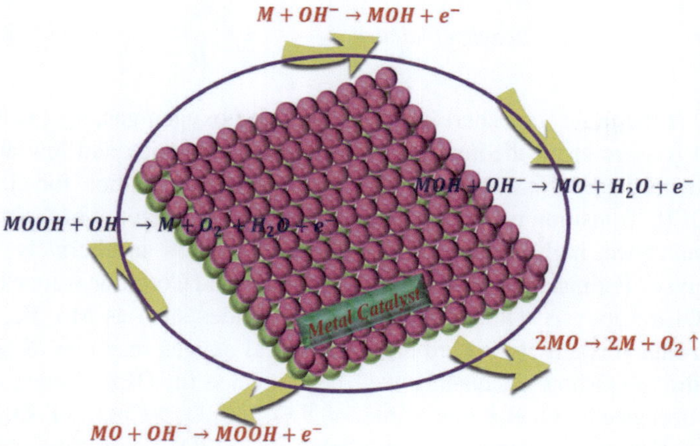

Fig. 8.6 Overall mechanism of OER

In the ideal case, to proceed the reaction, a thermodynamic potential of 1.23 V is required. However, the actual applied potential needed is much higher. Similar to the HER process, electrocatalysts are used to catalyze the OER and for that transition metal oxides-hydroxides are preferred because of their ability to undergo fast and reversible redox reactions which facilitates the electron transfer process. More understanding related to the OER process can be obtained from the following mechanism for alkaline medium and shown in Fig. 8.6 (Plevova et al. 2021). The major reactions are:

$$M + OH^{-} + \rightarrow MOH + e^{-} \tag{8.12}$$

$$MOH + OH^{-} \rightarrow MO + H_2O + e^{-} \tag{8.13}$$

$$2MO \rightarrow 2M + O_2 \tag{8.14}$$

$$MO + OH^{-} \rightarrow MOOH + e^{-} \tag{8.15}$$

$$MOOH + OH^{-} \rightarrow M + O_2 + H_2O + e^{-} \tag{8.16}$$

where M is the catalytic active site of transition metal oxide-hydroxide based electrocatalyst, M–O, M-OOH, M-OH are the metal–oxygen linkages. The parameters that are discussed above in the HER section are equally applicable for measuring the electrocatalytic activity of the catalyst for OER also. There is only a slight difference in calculating the faradic efficiency for OER as the oxidation process here involves the transfer of four electrons. It follows the equation:

$$Faradic\ efficiency = \frac{n_{O_2}}{n'_{O_2}} = \frac{4Fn_{O_2}}{It} \tag{8.17}$$

where n_{O_2} is the moles of experimentally generated oxygen gas, $n_{O_2}\prime$ is the moles of oxygen gas produced theoretically. The unavoidable price and low abundance of noble-metal based catalysts restrain their efficient application for electrocatalysts for OER. Transition metal oxides/hydroxides on the account of their low cost, natural abundance, high redox activity, are coming up as an alternative to noble metal catalysts. For instance, Ni(oxides), Cu(oxides) and Co(oxides) are extensively used for OER catalysis. Other transition metal oxides such as Mn, Fe, etc. also show high catalytic activity toward electrochemical oxygen reactions. Single metal oxides/hydroxides show adequately well performance for OER. Generally, metal oxide performance for OER is in the order of $NiO_x > CoO_x > CuO_x > FeO_x > MnO_x$, for instance Pawar et al. reported a two-dimensional (2D) electrocatalyst based on CuO for OER. The OER occurred at 1.48 V (vs RHE) with an overpotential of 250 mV. Tafel slope of 59 mV dec$^{-1\ \mathrm{was}}$ observed in a 1 M KOH solution for a current density of 10 mA cm^{-2} (Pawar et al. 2017). Similarly, Chandra et al. demonstrated the fabrication of a Cu-doped NiCo-LDH based electrocatalyst which shows an overpotential of 275 mV overpotential at 10 mA cm^{-2}.[31]Meng et al. presented OER activity of various structures of manganese oxides and highlighted how the electrocatalytic performance of manganese oxide varies on changing the crystallographic structures. The OER activity turns to follow an order of α-MnO_2 > β-MnO_2 > δ-MnO_2 with α-MnO_2 requires an overpotential of 490 mV in reaching 10 mA cm^{-2} (Meng et al. 2014). Generally, single transition metals containing oxides/hydroxides possess overpotential in the range of 0.9–1.3 V, which is still much higher for an efficient OER catalysis reaction (Wu et al. 2020). So, multi-transition metal based electrocatalysts are found to be more effective as they exhibit multiple reactive sites which improve the four-electron transfer process for OER. As earlier mentioned, Ni, Co, Cu, Mn, etc., have high catalytic activity towards HER and OER, composites of these metals are widely explored for catalysis. For instance, Lambert et al. demonstrated $Ni_xCo_{3-x}O_4$ nanostructured films, which show an overpotential of 0.96 V which is 76 mV lesser than that of pristine cobalt oxide (Lambert et al. 2015). More recently, Kazakova et al. fabricated a trimetallic (Fe, Co, Ni) oxide NPs with various combinations of $Fe_xNi_xCo_{1-2x}$ where $0.25 \leq x \leq 0.4$ and evaluated their performance for OER. Moreover, it was reported by the group that the trimetallic catalysts show higher catalytic activity and overpotential of nearly 0.8 V, which is lower compared to the bimetallic Fe–Ni, Fe-Co and Co–Ni catalysts (Kazakova et al. 2020). However, this is also not enough as a discrepancy of 0.8–0.9 V from the thermodynamic potential of 1.23 V is quite large. So, to further improve the performance of OER catalysts, morphology tuning and heteroatom doping is another way out, which not only increases the active surface area but also allows the facile electrolyte diffusion to the inner surfaces and high electron conductivity. Recently, Zhang et al. reported a dendritic core–shell nickel–iron-copper metal/metal oxide as the electrocatalyst for water oxidation, that shows remarkable activity with an overpotential of only 180 mV at a current density of 10 mA cm^{-2} (Zhang et al. 2018). Despite such advances, further development is needed in the field to reduce the electron and ion transfer resistance, enhancing surface area, making the catalysts cost-effective to procure sustainable energy production.

Fischer–Tropsch (F-T) Process

Fischer–Tropsch (F-T) process is used for obtaining synthetic fuel from conventional fossil fuels. This process has remarkable economic importance in converting scarce natural gas and coal to eco-friendly fuels and valuable chemicals. It is the key process in converting the syngas ($CO + H_2$) into a variety of aliphatic hydrocarbons and their derivatives such as olefins, paraffins, alcohols, etc. that can be used as liquid fuel. This process is catalyzed by a transition metal-based catalyst. There are several possible reactions involved in the overall conversion process of coal or natural gas to synthetic liquid fuel. The major reactions are:

i. Production of syngas through partial oxidation,

$$CH_4 + \frac{1}{2}O_2 \leftrightarrow CO + 2H_2 \tag{8.18}$$

or, steam reforming of natural gas

$$CH_4 + H_2O \leftrightarrow CO + 3H_2 \tag{8.19}$$

or, carbon dioxide reforming,

$$CH_4 + CO_2 \leftrightarrow 2CO + 2H_2 \tag{8.20}$$

ii. F-T process involves:
(a) F-T reaction

The formation of aliphatic hydrocarbons and their derivatives from syngas are represented by following reactions:

$$2nH_2 + nCO \leftrightarrow C_nH_{2n} + nH_2O(\text{Olefins}) \tag{8.21}$$

$$(3n+1)H_2 + nCO_2 \leftrightarrow C_nH_{2n+2} + (2n)H_2O(\text{Paraffins}) \tag{8.22}$$

$$(2n)H_2 + nCO \leftrightarrow C_nH_{2n+1}OH + (n-1)H_2O(\text{Alcohols}) \tag{8.23}$$

(b) Water gas shift reaction (WGS)

The by-product of F-T reaction, i.e. water combines with carbon monoxide of syngas and represented by following reaction:

$$CO + H_2O \leftrightarrow CO_2 + H_2 \tag{8.24}$$

Usually, syngas are produced following Eq. (8.18) and have a fixed stoichiometric ratio of hydrogen and carbon monoxide. If that is not the case sometimes, then the Eq. (8.24) is used to alter the H_2:CO ratio to 2:1, here the metal-oxide catalysts catalyse the water–gas shift reaction and F-T process. Water is a primary product of the F-T reaction and CO_2 can be produced by the WGS reaction. The basic chemistry involved in F-T synthesis process mainly converges into 5 steps: (i) Facile adsorption of CO over the surface of the catalysts, (ii) Chain initiation reaction involving dissociation of CO followed by its hydrogenation to produce an intermediate, i.e. active methyl species, (CH_2*), (iii) Chain growth step where more number of CO groups are adsorbed and simultaneously get hydrogenated and more CH_2 groups subsequently added and alkyl chain starts to develop (CH_2*), (iii) then the process of chain termination where final olefins, paraffin, alcohols are produced, (v) lastly, desorption of final products from the catalyst surface (Mirzaei et al. 2014).

$$CO + 2H_2 \leftrightarrow CH_2^* + nH_2O \tag{8.25}$$

For catalyzing the above F-T process, a catalyst should have a highly effective surface area for facile adsorption of CO and H_2, catalytic activity towards dissociation and hydrogenation of CO Fig. 8.7. represents the different types of F-T process. Also, a catalyst should not make a stable bond with carbon monoxide, oxygen or hydrogen, for instance in the case of metals as catalysts the metal oxides, which are formed as an intermediate should be easily reducible. Various transition metals show activity towards the F-T process. However, only certain metals can be effectively used. For instance, transition metals compounds of III, IV, V and VI groups (Ti, Ta, Mo, etc.) show good activity towards the dissociative adsorption of CO but, at the same time, they form irreversible and stable metal oxides, which decrease the rate of F-T process. In comparison, the metals belonging to the noble-metal category (Pt, Pd, Ir, etc.) do not allow dissociative adsorption of CO on their surface, hence are inactive towards F-T reaction. Group VIII, IX, X, metals (Fe, Ni, Co, Ru, etc.) are considered suitable for the overall catalyzing the F-T process. Amongst these, Ru shows exceptionally high activity towards F-T process. However, its high cost and low-abundance limit its large-scale industrial application as a catalyst for F-T process. Similarly, Ni is too active toward hydrogenation of CO process. It shows very high selectivity towards methane formation and chain propagation step gets hindered. Recently, transition metal oxides are preferred for catalysis, as they undergo fast reversible redox reactions with hydrogen and with carbon monoxide and revert to the original form without much structural changes. Also, they offer a higher surface area for efficient adsorption of CO and H_2. Hence, these metal oxides are much more suitable for catalyzing the conversion process. Most of the researches were carried out using iron oxide and cobalt oxide-based catalysts, which are frequently utilised for the CO-FT reaction. These catalysts simultaneously facilitate both CO hydrogenation activation and suppress methanation reaction. Along with this, these are highly active and selective to activate the C–O bond for CO_2 dissociation and also possess mesoporous and high surface area to accommodate a larger distribution of metal active sites for catalysis. Atharibonnroujeny et al. showed that the Co-oxides based catalysts have higher selectivity towards olefins (Eq. (8.21)) as Co oxides form a kinetically favoured Co_2-C phase, which promotes the C = C coupling (Athariboroujeny et al. 2019). Also, bifunctional metal oxide catalysts can be useful for enhancing the chemical and thermal stability, selectivity towards length of carbon chain and particular hydrocarbons. For instance, Cu-based oxides act as promoters when they are added with Fe-based catalysts for F-T process. Firstly, $Cu^{II}O$ gets reduced by the H_2 and Cu^0 produced, which act as positive feedback for more adsorption of H_2 on the surface of catalyst via acting as a centre for H_2 dissociation in F-T process (Eqs. 8.21, 8.22, 8.23). Also, Cu^0 nanoparticles promote the CO/CO_2 dissociative adsorption in F-T process (Zhu et al. 2020). Recently, Choi et al. showed that the various crystal structures of Cu-precursor used along with Fe/Co -based catalyst, also influence the rate and selectivity of F-T process. For instance, when delafossite ($CuFeO_2$) in which Cu exists in + 1 oxidation state, is used as the promoter in F-T process, there is a high selectivity for olefins (Eq. 8.20) especially with 5 carbon chain (C-5), whereas when spinel structure of Cu, i.e. $CuFe_2O_4$ where Cu exists in Cu^{II} is used as a promoter, major selectivity for paraffin is observed (Eq. 8.22) (Choi et al. 2017). Moreover, apart from Cu-based, Zhang et al. reported that when Zn is added to Fe/Co based catalysts, the formation of spinel $ZnFe_2O_4$ takes place which enhances the thermal stability of the catalysts. Also, Zn-based Fe/Co catalysts have higher selectivity towards C-2 or C-4 olefins (Eq. 8.21) (Zhang et al. 2015). Li et al. reported the effect of Ni addition in Fe-based oxides. The group presented the fabrication of $NiFe_2O_4$ and compared its catalytic activity and selectivity with the pristine Fe_2O_3-based catalyst. It was observed that the presence of Ni improves the rate of dissociation of CO and promotes its hydrogenation and has a much higher selectivity for smaller hydrocarbons ($<$C-5) than pristine Fe_2O_3-based catalyst (Li et al. 2014). Recently, Kim et al. reported the composite of Co and Fe oxides for catalysing F-T process. It was observed that $CoFe_2O_4$ shows 34.4% higher conversion efficiency for CO/CO_2 conversion and 38.8% more selectivity towards olefins than the pristine Fe_3O_4. This is due to the formation of a bimetallic carbide intermediate $(Fe_{1-x}Co_x)_5C_2$ on the surface of Fe and Co mixed oxide catalyst as soon as CO adsorbs on it. This intermediate inhibits the binding of Fe-CO, Co–CO and promotes CO hydrogenation activity (Kim et al. 2020). Hence, the conversion of natural gas and coal to synthetic fuel is a great opportunity to reduce the dependency on fossil fuels and limit global warming. Still, significant experimental and theoretical modulation is needed to select the suitable and stable catalysts, efficient conversion steps with less amount of intermediates formation, optimization of the precise selectivity to increase the yield of desired products.

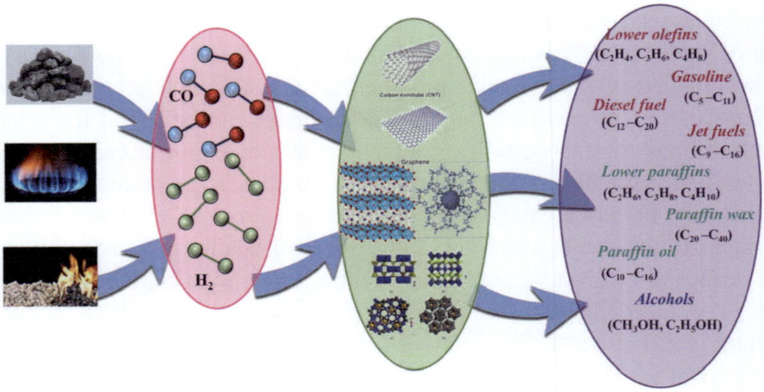

Fig. 8.7 Different types of Fischer–Tropsch (F-T) processes

8.4.2 *Catalysis Process Important for Environmental Remediation*

The majority of environmental science involves the use of catalysts such as in wastewater treatment (removal of dyes, organic pollutants, etc.) and green air remediation (the degradation of harmful air pollutants including CO, NO_x, hydrocarbons (HC), volatile organic compounds (VOC) and particulate matter).

Wastewater Treatment

Wastewater treatment has a tremendous importance for environment remediation, owing to the current increment in water demand, the rapid industrialization process and the wide use of pesticides, herbicides and drugs, which are sources of major pollutants such as phenols, organic dyes and heavy metals. Phenolic compounds are present in the effluents of various industries such as oil refining, petrochemicals, pharmaceuticals, coking operations, resin manufacturing, plastics, paint, pulp, paper and wood products. Discharge of these compounds without treatment may lead to serious health risks to humans, animals and aquatic systems. Out of all the phenols, nitrophenols are the most persistent organic contaminants commonly found in rivers or seawater. Pesticides, plasticizers, dyes and pharmaceutical companies all emit nitrophenols as industrial waste. Agencies working round the clock to regulate and protect the environment, such as the United State Environment Protection Agency [USEPA] have listed nitrophenols as "Priority pollutants". The regulated levels of nitrophenols in natural water should be $< 10\ \text{ng L}^{-1}$ (Hasan et al. 2016). Amongst the three isomeric nitrophenols, research showed that p-nitrophenol (p-NP) is more harmful in causing various disorders. It is toxic and causes adverse health effects both for animals and humans. The pollutant is highly soluble in water bodies and in industrial wastewater. Due to this fact, regular water filtration methods to remove pollutants from water bodies are not effective. Therefore, it is very important to look for an effective reduction of the p-NP. The reduction of pollutants can be broadly classified based on two factors: (a) aiming for the complete removal of the pollutant or (b) conversion of the pollutant into other usable forms. It is more convenient to convert the pollutant into a usable form. The catalytic reduction of p-NP to p-aminophenol (p-AP), which is non-toxic, has a commercial value in pharmaceutical and cosmetic industries. Based on the current research demand namely a greener approach, low toxicity and safe operation, direct catalytic environment-friendly conversion routes have been studied for the hydrogenation of p-NP. One method for generating p-aminophenol involves utilizing sodium borohydride, a mild agent, to directly hydrogenate p-nitrophenol, which can be done in an aqueous solution.

$$NaBH_{4(s)} + 2H_2O \rightarrow 4H_{2(g)} + NaBO_{2(aq)} \tag{8.26}$$

However, this reaction is kinetically sluggish and researchers found that the presence of a suitable catalyst could accelerate the hydrolysis process. In recent years, transition metal-based nanoparticles have been investigated as a suitable catalyst. Metal oxides/ hydroxides like CuO, LDH, Co_3O_4, MnO_2, etc., have been gaining attention recently owing to their highly porous structure with tunable size, morphology and composition. The investigation of the conversion of p-NP to p-AP using metal oxides as catalysts shows limited reports. A recent article on layered double hydroxides (LDHs) from Chandra et al. showed the use of Ni-Co based LDH, with nanosheets morphology for hydrogenation of p-NP to p-AP. The Langmuir–Hinshelwood model is used to construct an overall catalytic mechanism for the p-nitrophenol conversion reaction based on the kinetic and thermodynamic analyses of the reaction by the authors as highlighted in Fig. 8.8. They claim that Ni-Co LDH shows excellent catalytic properties with an apparent rate constant (k_{app}) = 0.31 min^{-1} at 25 °C to k_{app} = 0.65 min^{-1} at 55 °C and upto 97% retention of k_{app} after 5 successive cycles at room temperature. The excellent catalytic activity was a direct result of the lamellar morphology, which provides a higher surface area of 123 m^2 g^{-1} (Kansal et al. 2023b).

Another work by Chandra et al. on p-NP reduction to p-AP using copper (II) oxide emphasized the cost-effectiveness and industrial feasibility of the synthesis protocol adopted to stabilize CuO particles with flake-like morphology. The CuO nanoflakes have a catalytic rate constant of 0.0487 s^{-1}. The high-rate constant of the metal oxide was supported by the lowest zeta potential of -40.37 mV for the CuO nanoflakes (Sharma et al. 2018b). An interesting work on $MnCo_2O_4$ for catalysis application has also been reported by Chandra et al. (Akhtar et al. 2016). In the work, the catalytic response is tuned by the modulation of mesoporous structures of these materials. The idea of varying formation temperature is used. A shift in the particle growth mechanism was observed from Oswald ripening to digestive ripening with a change in the temperature. The calcination temperatures of 400 and 800 ºC were used in the work. The sample calcined at 800 °C reduced p-nitrophenol in 8 min at 1 mg mL^{-1} catalyst concentration, while the sample obtained at 400 °C could complete the task only after 16 min. The superior catalytic response has been explained by analyzing the change in zeta potential, which develops on the surface of different porous structures. The eradication of phenol from wastewater is

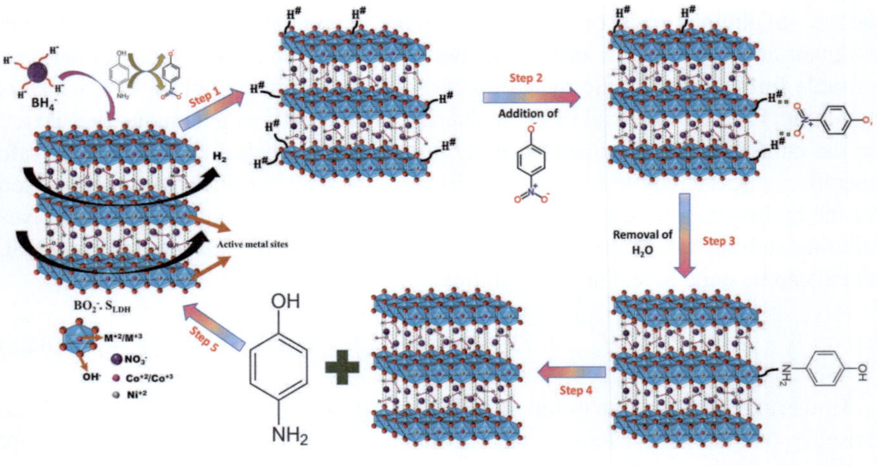

Fig. 8.8 Mechanism of p-nitrophenol reduction (Kansal et al. 2023b) (Adapted with permission, Elsevier Publishers)

important in conjunction with the development of numerous cutting-edge methods for its reduction. An allowable level of phenol discharge has become stricter globally as phenol has become the main environmental problem. This analysis has demonstrated that the list of current technologies has indicated a potential partial resolution to this issue.

Reduction of Harmful Air Pollutants

The harmful or toxic gases released by the industries are also a topic of great concern. Researchers around the globe are trying to deal with the degrading air quality day by day. Pollutants such as nitrogen oxides (NO_x), carbon monoxide and various volatile organic compounds (VOCs) are primarily emitted from industrial practices and automobiles. The prime solution to the problem is to address it with efficient and cost-effective technology, i.e. catalysis. However, the major challenge faced by researchers today is the lack of efficient catalysts. Nanomaterials have greater surface areas or more pores than bulk materials, which is advantageous for the adsorption, diffusion and increased activity of the reactants. As a result, materials that are porous or are in the nanorange show higher catalysis than compared to bulk materials. The synthesis of porous or nano-ranged catalysts and their usage in the environment is being experimentally investigated.

Acid rain, photochemical smog and haze are the direct consequences of these NO_x emissions. Coal power plants, automobiles and chemical industries are common sources of NO_x emissions. Out of various NO_x removal techniques (e.g. three-way catalysis, NO_x storage reduction (NSR), continuously regenerating trap (CRT) and selective catalytic reduction (SCR)) SCR is a major technique for the elimination of NO_x. The techniques involve the use of reducing agents such as urea, hydrocarbons, ammonia, light hydrocarbons, etc. Currently, the use of NH_3-SCR is the most suitable technique. In this process, NO_x is brought down to N_2 and H_2O with the help of NH_3

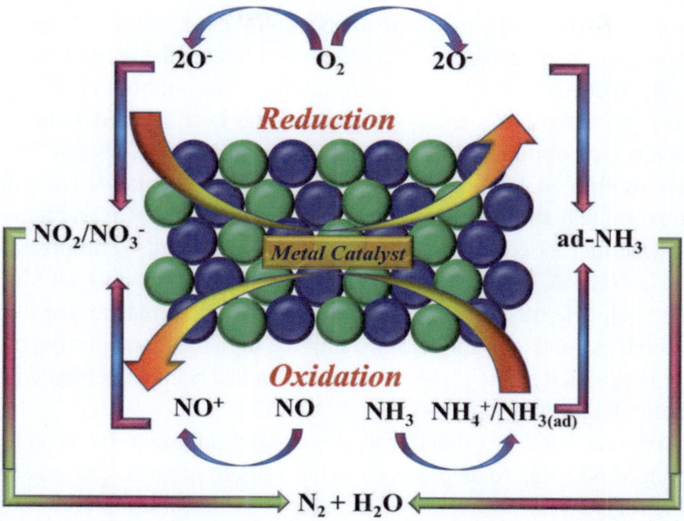

Fig. 8.9 Mechanism of NO_x reduction

addition. The reaction process is given as follows and highlighted in Fig. 8.9 (Wang et al. 2021).

$$4NH_3 + 4NO + O_2 \rightarrow 4N_2 + 6H_2O \tag{8.27}$$

$$4NH_3 + 2NO_2 + O_2 \rightarrow 3N_2 + 6H_2O \tag{8.28}$$

$$4NH_3 + 6NO \rightarrow 5N_2 + 6H_2O \tag{8.29}$$

$$8NH_3 + 6NO_2 \rightarrow 7N_2 + 12H_2O \tag{8.30}$$

Due to the complicated chemical makeup of fuel gas and the various particle size distributions, the NO_x oxidation rate under normal circumstances is often quite low. As a result, considerable work has gone into creating suitable catalysts for NO_x oxidation. Many researchers have investigated the SCR of NO_x with hydrogen or ammonia with the help of transition metal oxides. Despite the fact that noble metals like platinum have a significant NO_x oxidation activity, their expensive price severely limits their use. Due to their inexpensive price, high level of stability and strong catalytic activity, metal oxide catalysts have become more popular. Also, metal oxides have good thermal stability and high surface area, which is suitable for the adsorption of nitrogen. This makes metal oxides a good candidate for NO_x catalytic oxidation. Nikolov et al. have investigated carbon-modified alumina-supported copper oxide catalysts have been investigated. They report the catalytic activity enhancement in the presence of γ-Al_2O_3 in CuO catalyst (Nickolov et al. 2003). Researchers from Chen's group examined how the Ce/Zr molar ratio affected the SCR of NO_x with NH_3 over a Fe_2O_3-WO_3/$Ce_xZr_{1-x}O_2$ monolith catalyst. The authors claim that Fe_2O_3-WO_3/$Ce_xZr_{1-x}O_2$ monolithic catalyst delivered the best NH_3-SCR activity. More than 90% NO_x was eliminated in the temperature range of 247–454 °C. The high performance of NH_3-SCR of the Fe_2O_3-WO_3/$Ce_xZr_{1-x}O_2$ catalyst was attributed to the good textural and structural qualities, greater surface Fe, Ce and active oxygen and all of these factors combined (Xu et al. 2014). Recently, composite materials are trending in NH_3-SCR of NO_x. Yao et al. recently reported work on MnO_2/CeO_2-based composite. The material provides exceptional performance due to its unique properties, such as good low-temperature activity, high mechanical strength and high oxygen storage capacity. The possible mechanism has been presented with the help of a schematic diagram in Fig. 8.9 (Yao et al. 2017).

Carbon monoxide (CO) is also a critical industrial pollutant. It can be regarded as the redundant of NO, due to its presence in the vehicle exhaust. CO was not considered suitable for SCR because it is consumed in presence of O_2. Nobel metals, metal oxides, perovskite-type oxides have been investigated for catalytic oxidation of CO. Amongst the various metal oxides cobalt oxides are among high-performance catalyst for CO oxidation (Xie et al. 2009). In general, properties like redox ability, surface area and oxygen defects are known to dominate catalytic activity of the material. Teng et al. demonstrated that by utilizing the nanocasting method and solid porous materials (such as KIT-6) as hard templates, well-ordered mesoporous Co_3O_4 could be produced. The resulting material demonstrated appreciable catalytic activity in CO oxidation (Teng et al. 2011). Researchers have made remarkable progress in various catalytic applications of metal oxides/hydroxides for the eradication of atmospheric pollutants such as NO_x and CO. The development of new, highly effective catalytic materials is still a crucial step in the process of purifying the atmosphere. Each technique has its own distinct advantages, including adsorption, biodegradation, heterogeneous catalysis, photocatalysis and plasma treatment. The use of coupling technologies that integrate the benefits of the aforementioned techniques would ensure high efficacy in the removal of air contaminants. Table 8.1 highlights the comparison of metal oxides/hydroxides with several metal phosphides/nitrides.

8.5 Conclusions and Perspective

In summary, catalysis is the most prominent process in industries ranging from food, cosmetics to pharmaceuticals, procuring and storing energy and environmental remediation. For several decades, various catalysts based on noble metals, non-metals, transition metals have been used to catalyse chemical reactions of economic importance. Transition metal oxides (TMOs)-based catalysts have now been recognized as the most promising catalysts for large-scale catalysis due to their low cost and excellent properties. TMO-catalysed HER and OER mechanisms in an alkaline environment are discussed. TMOs are believed to follow an associative pathway through

Table 8.1 Comparison of metal oxides/hydroxides with several metal phosphides/nitrides

Properties	Metal compounds		
	TMOs	Metal nitrides	Metal phosphides
Composition	MO, M_xO_y, $M(OH)_2$	MN, M_2N, M_4N	MS_2, M_xS_y
Crystal structure	Complex structures such as perovskites, spinels, or layered structures	Simple structures such as rock salt, zincblende, or wurtzite	Crystalline structure either tetragonal or orthorhombic
Electronic properties	Insulators/ Semi-conductors	Metallic/Semi-metallic	Metallic due to presence of free electrons
Chemical stability	Stable in acid/ alkaline medium	Unstable in acid medium	Stable in acid/alkaline medium
Adsorption strength	Higher adsorption strength towards acidic molecules such as CO, SO_2, NO_x	Higher adsorption strengths for basic molecules such as NH_3	Higher adsorption strengths for non-polar molecules
Surface properties	Complex surface chemistry due to presence of surface defects such as oxygen vacancies	Simpler surface chemistry, with fewer surface defects and active sites, high surface area compared to TMOs	Highly reduced surface, fewer defects with similar surface area compared to TMOs
Common catalytic applications	Reduction of NO_x emissions, water–gas shift reaction	Decomposition of ammonia, hydrogenation of organic compounds	Hydro-deoxygenation of bio-oils, oxygen reduction reaction(ORR)

$2e^-$ and $4e^-$ processes, respectively. TMOs provide large surface area and facile electron transfer making them useful for electrocatalysis processes. The transition metal oxide catalysts are also an important Fischer–Tropsch (F-T) process, where especially Fe and Co based oxides are primarily used for catalysis. TMOs-based catalysts have higher catalytic activity for the activation of C-O bond for CO/CO_2 dissociation and also possess high surface area to accommodate larger distribution of metal active sites for catalysis. Further, Cu-based catalysts have specific selectivity for olefins when Cu(I) catalysts are present, when Cu (II) is used as catalysts paraffins have higher selectivity. Further, introducing Zn and Ni with Fe/Co based catalysts increases the thermal stability and shows high selectivity towards lower olefins. The use of TMOs for wastewater treatment is described in detail. Higher surface area and redox activity allow efficient adsorption of harmful contaminants on the surface of catalyst. The subsequent reduction or neutralizing agents are able to convert them into less harmful substitutes.

Author Contributions: "The manuscript was written through contributions of all authors. All authors have given approval to the final version of the manuscript."

References

Akhtar, M.A., Sharma, V., Biswas, S., Chandra, A.: RSC Adv. **6**, 96296 (2016)

Akhtar, M.A., Chowdhury, A., Chandra, A.: J. Phys. d. **52**, 155501 (2019)

Anshu, S., Priya, S., Mandal, D., Rahul, R., Priya, S., De, P., Sharma, V., Singh, T., Chandra, A.: J. Phys. D. Appl. Phys. **56**:205501 (2023)

Athariboroujeny, M., Raub, A., Lablokov, V., Chenakin, S., Kovarik, L., Kruse, N.: ACS Catal. **9**, 5603 (2019)

Bard, A.J., Faulkner, L.R.: Electrochemical Methods: Fundamentals and Applications, 2nd edn. John. Wiley and Sons, UK (2000)

Bullock, R.M.: Chem **2**, 444 (2017)

Chapin, D.M., Fuller, C.S., Pearson, G.L.: J. Appl. Phys. **25**, 676 (1954)

Chen, H., Du, X., Sun, J., Mao, H., Wu, R., Xu, C.: Appl. Surf. Sci. **515**, 146008 (2020)

Choi, Y.H., Jang, Y., Park, H., Kim, W.Y., Lee, Y.H., Choi, S., Lee, J.S.: Appl. Catal. B **202**, 605 (2017)

Ci, S., Wen, Z., Qian, Y., Mao, S., Cui, S., Chen, J.: Sci. Rep. **5**, 1919 (2015)

Guo, C., Yin, M., Wu, C., Li, J., Sun, C., Jia, C., Li, T., Hou, L., Wei, Y.: Front. Chem. **6**, 636 (2018)

Halder, J., Biswas, S., Chowdhury, A., Mandal, D., Kansal, S., Priya, S., De, P., Srivastava, A.K., Chandra, A.: J. Phys. Chem. C **30**, 14623 (2023)

Hao, C., Zhou, S., Wang, J., Wang, X., Gao, H., Ge, C.: Ind. Eng. Chem. Res. **57**, 2517 (2018)

Hasan, Z., Cho, D.W., Chon, C.M., Yoon, K., Song, H.: Chem. Eng. J. **298**, 183 (2016)

Hutchings, G.J.: J. Mater. Chem. **19**, 1222 (2009)

Kana, N., Kaviyarasu, K., Khamliche, T., Magdalane, C.M., Maaza, M.: J. Environ. Chem. Eng. **7**, 103255 (2019)

Kansal, S., Singh, P., Biswas, S., Chowdhury, A., Mandal, D., Priya, S., Singh, T., Chandra, A.: Int. J. Hydrog. Energy **48**, 21372 (2023b)

Kansal, S., RR, Anshu, S., Priya, S., Mandal, D., Srivastava, A.K., Chandra, A.: ACS Sustainable Chem. Eng. **12**:13511 (2023)

Kansal, S., Halder, J., Mandal, D., Srivastava, R.R, A,K., Chandra, A.: Mater. Sci. Eng. B. **293**: 116438 (2023)

Kazakova, M.A., Morales, D.M., Andronescu, C., Elumeeva, K., Selyutin, A.G.: Catal **357**, 259 (2020)

Kim, K.Y., Lee, H., Noh, W.Y., Shin, J., Han, S.J., Kim, S.K., An, K., Lee, J.S.: ACS Catal. **10**, 8660 (2020)

Lambert, T.N., Vigil, J.A., White, S.E., Davis, D.J., Limmer, S.J., Burton, P.D., Coker, E.N., Beechem, T.E., Brumbach, M.T.: Chem. Commun. (Camb.) **51**, 9511 (2015)

Lee, D.W., Yoo, B.R.: J. Ind. Eng. Chem. **20**, 3947 (2014)

Li, T., Wang, H., Yang, Y., Xiang, H., Li, Y.: Fuel Process. Technol. **118**, 117 (2014)

Li, J., Xiong, D., Wang, L., Hirbod, M.K.S., Li, X.: J. Energy Chem. **10**, 4913 (2019)

Liu, A., Che, H., Mao, Y., Wang, Y., Mu, J., Wu, C., Bai, Y., Zhang, X., Wang, G.: Ceram. Int. **42**, 11435 (2016)

Maduraiveeran, G., Sasidharan, M., Jin, W.: Prog. Mater. Sci. **106**, 100574 (2019)

Maenaka, Y., Suenobu, T., Fukuzumi, S.: J. Am. Chem. Soc. **134**, 367 (2012)

Meng, Y., Song, W., Huang, H., Ren, Z., Chen, S.Y., Suib, S.L.: J. Am. Chem. Soc. **136**, 11452 (2014)

Mirzaei, A.A., Arsalanfar, M., Bozorgzadeh, H.R.: Samimi. A **2**, 179 (2014)

Mondal, A., Paul, A., Srivastava, D.N., Panda, A.B.: Int. J. Hydrog. Energy **43**, 21665 (2018)

Nickolov, R., Stankova, N., Khristova, M., Mehandjiev, D.: J. Colloid Interface Sci. **265**, 121 (2003)

Paul, A.M., Sajeev, A., Nivetha, R., Gothandapani, K., Bhardwaj, P., Govardhan, K., Raghavan, V., Jacob, G., Sellapan, R., Jeong, S.K., Grace, A.N.: Diam. Relat. Mater. **107**, 107899 (2020)

Pawar, S.M., Pawar, B.S., Hou, B., Kim, J., Ahmed, A.T.A., Chavan, H.S., Jo, Y., Cho, S., Inamdar, A.I., Gunjakar, J.L., Kim, H., Cha, S.N., Im, H.S.: J. Mater. Chem. **5**, 12747 (2017)

Plevova, M., Hnat, J., Bouzek, K.: J. Power. Sources **507**, 230072 (2021)

Sharma, V., Singh, I., Chandra, A.: Sci. Rep. **8**, 1307 (2018a)
Sharma, V., Singh, I., Chandra, A.: Mater. Lett. **211**, 285 (2018b)
Sharma, V., Biswas, S., Sundaram, B., Halder, P., Dubey, B., Chandra, A.: ACS Sustainable Chem. Eng. **7**, 5385 (2019)
Sun, Y., Chen, G., Xu, Z.J.: ACS Catal. **11**, 13947 (2021)
Tahira, A., Ibupoto, Z.H., Willander, M.: Int. J. Hydrog. Energy **44**, 26148 (2019)
Teng, Y., Kusano, Y., Azuma, M., Haruta, M., Shimakawa, Y.: Catal. Sci. Technol. **1**, 920 (2011)
Wang, B., Song, Z., Sun, L.: Chem. Eng. J. **409**, 128136 (2021)
Wu, X., Tang, C., Cheng, Y., Min, X., Jiang, S.P., Wang, S.: Chem **26**, 3906 (2020)
Xie, X., Li, Y., Liu, Z.Q., Haruta, M., Shen, W.: Nature **458**, 746 (2009)
Xu, H., Cao, Y., Wang, Y., Fang, Z., Lin, T., Gong, M., Chen, Y.: Sci Bull **59**, 3956 (2014)
Yao, M., Hu, Z., Xu, Z., Liu, Y.: J. Alloys Compd. **644**, 721 (2015)
Yao, X., Ma, K., Zou, W., He, S., An, J., Yang, F., Dong, L.: Chinese J. Catal. **38**, 146 (2017)
Zhang, P., Li, L., Nordlund, D., Chen, H., Fan, L., Zhang, B., Sheng, X., Daniel, Q., Sun, L.: Nat. Commun. **9**, 381 (2018)
Zhang, S., Chen, M., Zhao, X., Cai, J., Yan, W., Yen, J.C., Chen, S., Yu, Y., Zhang, J.: Electrochem Energ. Rev. **4**, 336 (2021)
Zhang, J., Lu, .S, Su, X., Fan, S., Ma, Q., Zhao, T.: J. CO_2 Util. **12**:95 (2015)
Zhu, M., Tian, P., Ford, M.E., Chen, J., Xu, J., Han, Y.F., Wachs, I.E.: ACS Catal. **10**, 7857 (2020)

Chapter 9
Semiconductor and Plasmonic Nanomaterials Immobilized Over 2D GO/r-GO/g-C_3N_4 Photocatalyst for Enhanced Photocatalytic Degradation of Organic Pollutants

Kijay Bahadur Singh, Neelam Gautam, Deen Dayal Upadhyay, Anil K. Yadav, and Gajanan Pandey

9.1 Introduction

Many material chemists have been working on the development of quite-efficient, long-lasting solar radiation and encouraged semiconductor nano-photocatalysts to directly utilize solar energy and convert it into chemical energy for improved photocatalytic applications over the past few decades (Ding et al. 2015). A range of different photocatalytic composite materials have already been created for solar energy intensive degradation procedures due to their excellent properties, such as an appropriate band gap to exhibit strong visible range reactions with excellent photocatalytic performance, non-toxicity, and good chemical and thermal stability (Fu et al. 2013; Han et al. 2012). The plasmonic materials such as materials consisting of Ag, Au, etc., as well as semiconductor-based photocatalysts should also be composed of materials that are abundant on Earth, manufactured quickly with a specific porosity and thereby following a greener process (Darabdhara et al. 2016).

Due to their ease of use, affordability, high stability, benign make-up, and suitability as a model catalyst for photocatalysis, semiconductors are typically used in band positions (Fujishima and Honda 1972; Marschall 2014). However, because of

K. B. Singh · N. Gautam · D. D. Upadhyay · G. Pandey (✉)
Department of Chemistry, Babasaheb Bhimrao Ambedkar University, Lucknow, UP, India
e-mail: pandeygajanan@bbau.ac.in

A. K. Yadav
Department of Physics, Babasaheb Bhimrao Ambedkar University, Lucknow, UP, India

R. S. Ningthoujam and A. K. Tyagi (eds.), *Handbook of Materials Science, Volume 3*,
Indian Institute of Metals Series, https://doi.org/10.1007/978-981-96-8229-4_9

the fast annihilation of photogenerated e^--h^+ pairs, that causes lower quantum efficiency leading to the depressed photocatalytic performance, semiconductors' photocatalytic activity is still much below the threshold required for meaningful applications (Shams-Ghahfarokhi and Nezamzadeh-Ejhieh 2015). To overcome this limitation material chemists have implemented various strategies including mesoporosity, heterocomposites, structure-changing, doping of noble metal, mixed co-catalysts, and combinations with other semiconductor or support materials (Huang et al. 2019). In addition to that, GO is a high-demand photocatalyst support or surface provider that not only increases the efficacy of photocatalyst dispersion but also quickens the rate transmittance of photoexcitedcharged species due to its outstanding physical characteristics, including its promising electronic along with great mechanical strength. Photocatalysts have received mechanical support from GO nanosheets, which have a single atomic layer, and an electric charge carrier transport (Xiao and Wang 2021; Hu et al. 2017).

Because they have tuneable properties as a result of the hybridization of the specific features of the parent samples, hybrid photocatalysts can be fabricated by intermixing or lading graphene oxide nanosheets with semiconductor materials such as TiO_2, Bi_3O_3, SnO_2, MnO_2, ZnO, α-Fe_2O_3, WO_3, MoS_2, CdSe and g-C_3N_4 have attracted a lot of attention recently (Rathnasamy et al. 2018; Li et al. 2010). Recent years have seen significant advancements attained in the fabrication of GO-enabled semiconductor photocatalysts. To hasten further advancements in this exciting study field, we think that a thorough examination of the topic is important. The recent reports on the fabrication of semiconductor and plasmonic material doped GO, r-GO, and g-C_3N_4 hybrids are briefly discussed in this chapter along with various methods of modification for their novel and workable applications, such as the degradation of dyes, nitroaromatics, and other organic pollutants (Pant et al. 2013; Zhou et al. 2020).

Over the past few decades, numerous environmental problems have arisen on a world-wide scale. These include air pollution brought on by agricultural waste burning, climate change brought on by the greenhouse effect, and water pollution brought on by improperly handled industrial effluent. Clean water standards in daily life are a key concern addressing especially improved treatment technologies. For a variety of purposes, including irrigation, infrastructure, transportation, and power generation, good water quality is required (Demirbas 2009). The industrial and agricultural wastes, as well as religious rituals, are the main contributors to inorganic and organic contaminants in the water bodies.

Water supplies can have long-term negative impacts from carcinogenic substances such as heavy metals, colorants or dyes, non-biodegradable organic toxicants, and antibiotic drugs (Nazri and Sapawe 2020). In addition, germs and microorganisms will become more resistant to antibiotics as a result of pharmaceutical waste, such as left-over antibiotics (Bel et al. 2009). These pharmaceutical wastes are often produced by inadequate wastewater treatment systems, hospital and industrial wastewater, and live creatures' excreta (Bel et al. 2009; Wang et al. 2019a). In addition, organic pollutants are an issue in the modern world because they are refractory chemicals that may persist for a long time in the lower atmospheric region and cover large enough distances before being settled or redeposited. Organic toxicants were found by researchers to exist far from their field origins. Such sources of these compounds include combustion, transportation, and agriculture. To assure enhanced water quality for human consumption, a lot of effort has been undertaken to remove these impurities.

The traditional methods used in sewage treatment facilities to treat wastewater include coagulation, sedimentation, and adsorption. The combination of microscopic flocs, that tend to absorb the dispersed organic materials to create large clumps, is known as coagulation. Contrarily, adsorption is a process when a combination of substances acts as both an adsorbate and an adsorbent near the boundary between different phases of materials. These methods are useful and often employed in the remediation of toxic substances present in wastewater. However, these methods were ineffective in treating pharmaceutical discharge, textile effluents, and heavy metals (Zhou et al. 2020; Deng and Zhao 2015). For instance, the adsorption procedure may result in subsidiary waste. Only the pollutant gets adsorbed over its surface, rather than mineralizing it into smaller non-toxic species like water and carbon dioxide. The used adsorbent has to be cleaned and regenerate before being reused. Because it may mineralize the pollutants through redox reactions into safer molecules, the advanced oxidation process (AOP) is a good alternative to mineralize these organic toxicants present in wastewater.

9.1.1 Advanced Oxidation Process

The advanced oxidation pathway breaks down organic and inorganic materials, which are refractory contaminants in wastewater, using oxidants including the superoxide radical ($^{-}O_2^{\bullet}$), hydroxyl radical ($OH^{\bullet}$), and also sulphate radical ($SO_4^{\bullet}$) (Deng and Zhao 2015). For the first time, AOP was used as a method of the treatment of drinking water in the twentieth century (Glaze 1987). After then, this technology was utilized to treat wastewater for home use. The physicochemical characteristics of the pollutants encountered, the ambient factors and the type of AOP employed all affect the treatment's efficacy. The creation of these radicals must be sufficient for the purification process to proceed. However, because the produced radicals have a limited duration and a low concentration, they are less effective for other uses, such as disinfecting pathogens (Deng and Zhao 2015). This method has the benefits of being adaptable on a small scale, high and non-selective degradation, and quick reaction speeds. By producing less harmful compounds, this method also aids in mineralizing the contaminants (Huang et al. 1993). The most often employed advanced oxidation techniques in research publications are the photolysis methodology (Wammer et al. 2013), photo-Fenton (Michael et al. 2010), sonolysis (Bel et al. 2009), and photocatalysis (Hu et al. 2015).

Commonly employed AOP that produces radicals utilizing semiconductors, plasmonic materials, and light energy is photocatalysis. When the intensity of incident light radiation is greater than the semiconductor's band gap energy, generation of charged species such as electron–hole pairs and radicals can occur. However, these need to be tuned to slow down the fast annihilation of photo-generated electron/hole pairs, which causes lower quantum yield and depressed photocatalytic performance, and also to improve and preserve their photocatalytic activity against toxic organic chemicals by hybridizing with someimportant materials such as GO, polymer, g-C_3N_4 or plasmonic metals.

9.1.2 Basic Principle of Photocatalysis

The advanced oxidation processes (AOPs) are now the most traditional and efficient technologies for wastewater treatment. More research has been done on AOPs-based approaches globally. Among all of them, the photocatalysis process is the most practical since it doesn't require the use of additional chemicals or gases and is economical and environmentally beneficial. The hazardous dye is simply broken down using this process into non-hazardous form and then into metabolites. Fujishima and Honda discovered the photocatalytic water splitting event in a titanium dioxide electrode (Fujishima and Honda 1972). In the photocatalytic process, the solar radiation and the catalysts work together to support and speed up the reaction. Chemical reactions that are triggered by solar radiation are also referred to as photo-induced reactions.

Basic concepts of the photocatalysis process are briefly explained. When a photocatalyst is exposed to solar radiation or photon source with energy equal to or higher than the photocatalyst's band gap, holes (h^+) and electrons (e^-) are created in the valence band and conduction band, respectively. This entails oxidizing the water molecule with h^+in the valence band while reducing the absorbed oxygen with e^- in the conduction band to create a superoxide radical. These reactive oxygen species (ROS) quickly break down the dangerous substance into smaller, non-toxic metabolites (Rathnasamy et al. 2018; Sonkusare et al. 2020). The process that takes place in the photocatalytic reaction pathway for the breakdown of organic pollutants may be explained by the following Eqs. (9.1–9.4)

$$\text{Photocatalyst} + hv(\geq \text{band gap energy}) \rightarrow h^+_{(VB)} + e^-_{(CB)}(\text{Charge carriers}) \quad (9.1)$$

$$H_2O + h^+_{(VB)} \rightarrow {}^{\bullet}OH + H^+(\text{Hydrogen ions}) \quad (9.2)$$

$$O_2 + e^-_{(CB)} \rightarrow O_2^{\bullet -}(\text{Superoxide}) \quad (9.3)$$

$$\text{Dyes} + {}^{\bullet}OH/O_2^{\bullet -} \rightarrow \left(CO_2 + H_2O + NH_4^+ + NO_3^- + SO_4^{2-} + Cl^-\right) \text{(Byproduct)} \quad (9.4)$$

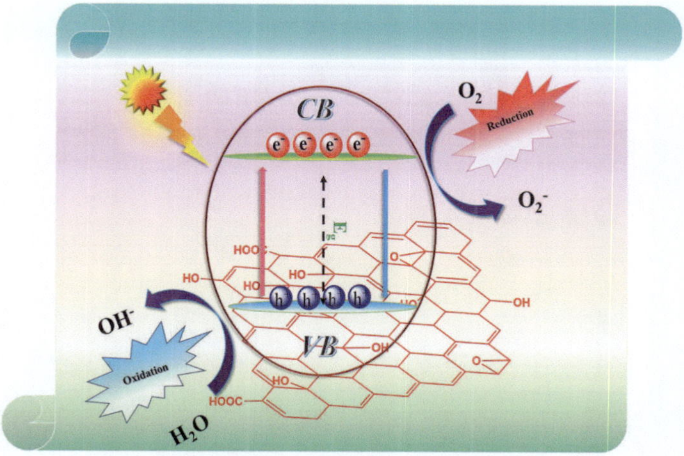

Fig. 9.1 Schematic illustrative representation of the mechanism of semiconductor/plasmonic material photocatalysis

In a photocatalytic process, the catalyst and photon energy from the light's wavelength plays two key roles. Semiconductor materials are frequently utilized as catalysts that operate as a sensitizing agent for the solar radiation-driven redox process because of their electronic structure which is defined by a top of the valence band and an empty conduction band (Khan et al. 2015). The mechanism of semiconductor photocatalysis is shown schematically in Fig. 9.1.

The most prominentsteps involved in semiconductor photocatalysis are listed below:

I. When solar radiation or light energy (hv) falls on the surface of photocatalyst material and if the incoming ray's energy is equal to or greater than the semiconductor's band gap energy, the valence band (VB) electrons in the semiconductor are stirred up and go to the conduction band (CB).
II. Holes (h^+) would remain in the valence band of the semiconductor. By interacting with water molecules, these valence band holes can lead to the oxidation of donor molecules and the generation of hydroxyl radicals ($OH^•$), which havea strong oxidizing capability and thus, it plays important role in the photolytic deterioration of dyes.
III. When reactive species in the solvent interact with the electrons in the conduction band, superoxide ions (O_2^-) are produced. These electrons power the redox reactions.

9.1.2.1 Elucidation of Oxidation Mechanism

On the photocatalyst surface, there is water. This water is oxidized, resulting in the production of hydroxyl ($OH^•$) radicals, as a result of the electrons migrating towards the conduction band of the material encouraged due to the incident radiation (agents

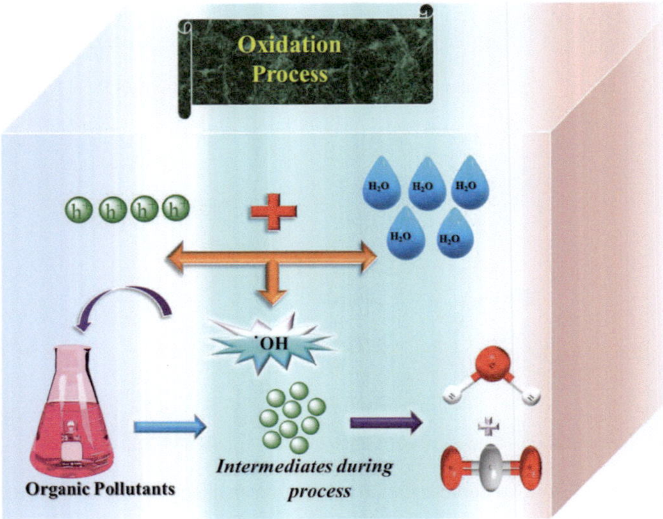

Fig. 9.2 Oxidation mechanism illustrated schematically

with significant oxidative breaking power). The hydroxyl radicals subsequently react with the chemical component contained in the organic pollutant. If there is oxygen accessible during the course of this process, the intermediate radicals in the toxic organic pollutants and the oxygen molecules may engage in radical-based chain reactions and utilize oxygen. As a result, the dye gradually breaks down and takes on the forms of carbon dioxide and water (Fujishima and Honda 1972; Saravanan et al. 2017). Organic materials may directly interact with the positive holes under these circumstances, leading to oxidative destruction. In Fig. 9.2, the entire oxidation process was displayed.

9.1.2.2 Elucidation of Reduction Mechanism

Since there isn't any hydrogen available, oxygen reduction takes place instead of the fact that oxygen may be swiftly depleted. The generation of superoxide radicals takes place when dissolved oxygen species interact with the electrons of the conduction band. These superoxide radical anions interact with the intermediates formed in the oxidative process, resulting in the production of peroxide or the conversion of hydrogen peroxide into water molecules. The reduction step is presumably more likely to happen in toxic organic materials than in water. Therefore, a large quantity of organic dye tends to produce more positive holes. This lessens carrier recombination while increasing photocatalytic activity. In Fig. 9.3, the entire reduction procedure was displayed.

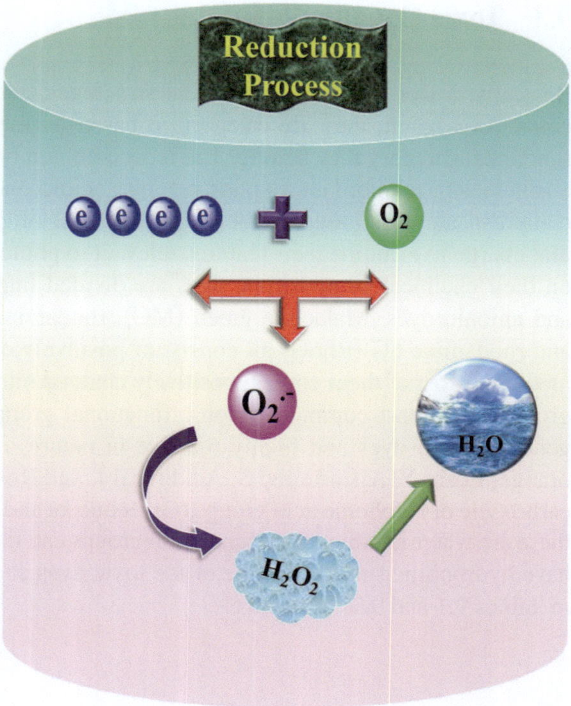

Fig. 9.3 Reduction mechanism illustrated schematically

9.2 Limitations of Pure Plasmonic and Semiconductor-Based Photocatalyst

- Previous reports indicated a lower photo-assisted degradation of organic pollutants present in wastewater using the un-doped nanomaterials.
- The use of UV radiation-responsive photocatalytic materials necessitates an expensive experimental procedure and an accumulation of photocatalyst materials that reduce the surface area of nanomaterial and subsequently diminish the effectiveness of the photo-assisted degradation process.
- Another limitation of the current photocatalysts is that they are not universally applicable.
- The perovskite material, i.e. $BiFeO_3$, really possesses a small bandgap in the presence of a rapid recombination rate barrier, which reduces its photodegradation efficiency.

With the development of nanotechnology and sophisticated materials, a crucial method to overcome these obstacles has been developed.

9.3 Toxic Organic Pollutants

Basically, a dye molecule's chemical structure determines its characteristics and color. As a result, they are categorized based on their color, application, and the functional category they belong. The most common types of dyes used in the textile business are azo, mordant, sulphur, reactive, acidic, and basic dyes. Following dissociation in aqueous-based applications, dyes are often classed using their molecular charge to examine their features relevant to photodegradation processes. Based on their chemical composition, dyes are divided into two categories viz. cationic and anionic dyes. Malachite green (MG), rhodamine B (RhB), safranin O (SO), and rhodamine 6G (Rh6G) all consist of positively charged species inthe aqueous medium. Most of them contain positively charged nitrogen (N^+) because the onium group is the most common cationic functional group. Whereas, anionic dyes are generally acid dyes and highly reactive in nature, including Phenol red, Methyl orange, Eosin Y, Alizarin red S, Acid red 14, and Rose Bengal (PR) and consist of carboxylic or sulphonic acid groups (Grzechulska and Morawski 2002). That is why these are water-soluble; these functional groups can interact with photocatalysts that have hydrophilic surfaces. Some of the toxic dyes and organic pollutants are listed in Tables 9.1 and 9.2, respectively.

9.4 Impact of 2D Material Support

9.4.1 Role of GO Support

Owing to extremely high electronic conductivity, graphene serves as a fantastic electron sink that catches and transports the photo-excited electrons in photocatalytic processes. Thus, one of the crucial processes in photocatalytic reactions, the recombination of electrons-holes in semiconductors, is substantially delayed by graphene. Additionally, graphene has a smaller negative Fermi level than the majority of semiconductor-based nanomaterials, which contributes to the advantageous alignment of the band gap of the graphene and the semiconductors. Since electrons transfer towards lower potential and the holes towards the higher potential in photocatalysts, the creation of active radical species and the movement of electrons from the conduction band (CB) of the semiconductors to graphene cause the breakdown of organic contaminants (Marschall 2014). Structural change observed during conversion of graphene to graphene oxide is illustrated in Fig. 9.4.

Table 9.1 Some dyes with chemical structures

Dyes	Structure	$\square_{max}$
Methylene Blue (MB)	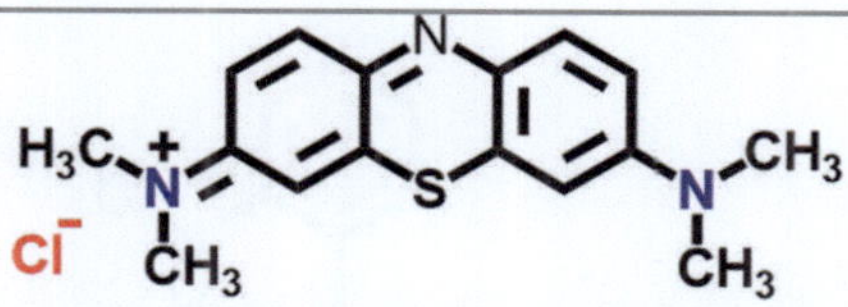	664
Crystal Violet (CV)	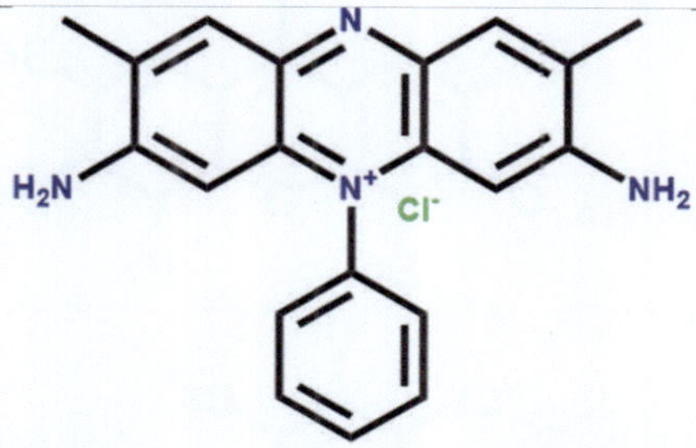	573
Rhodamine B (RhB)	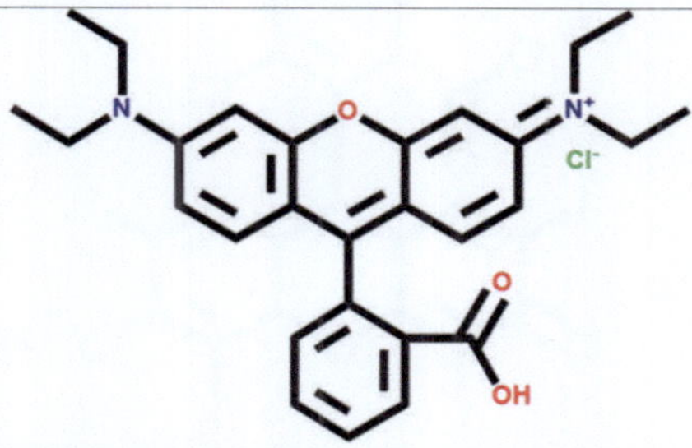	553
Safranin O (SO)	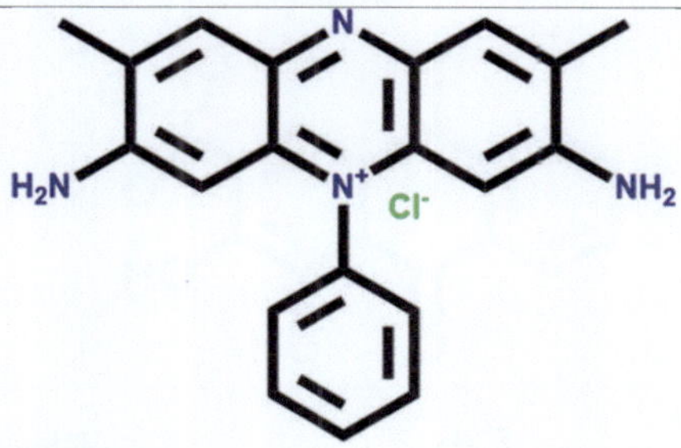	520

(continued)

Table 9.1 (continued)

Dyes	Structure	□$_{max}$
Eosin Y (EY)		518
Rose Bengal (RB)	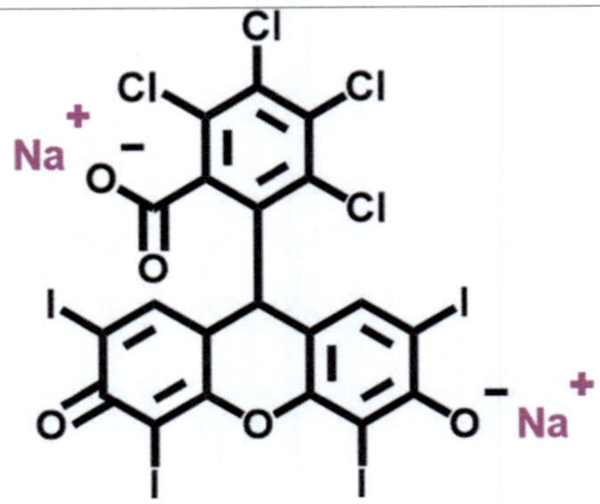	550
Alizarin Red S (ARS)	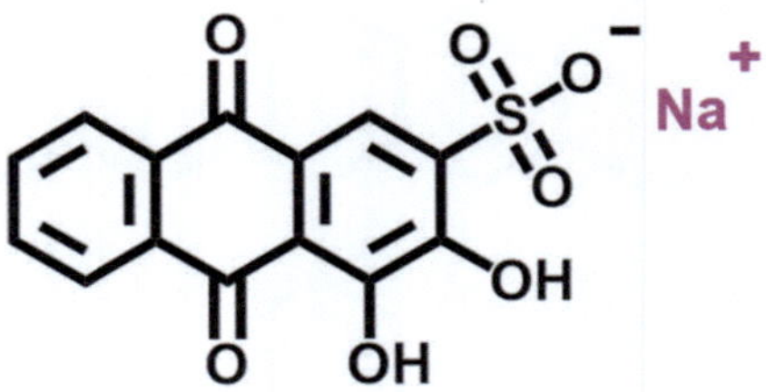	426
Malachite Green (MG)	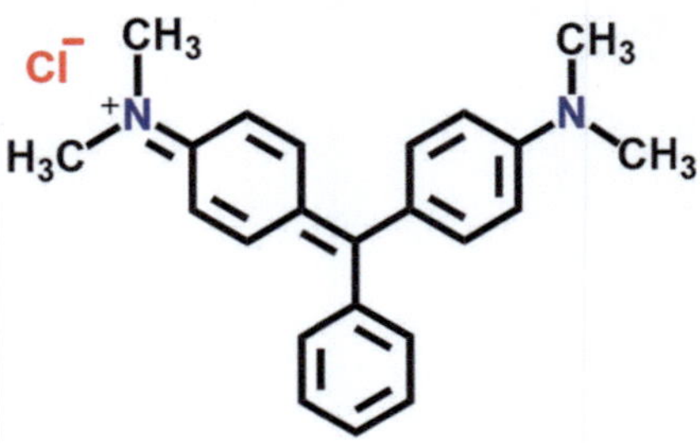	614

Table 9.2 Some toxicphenolic compounds

Nitroarene compounds	Structure	$\square_{max}$
p-Nitrophenol	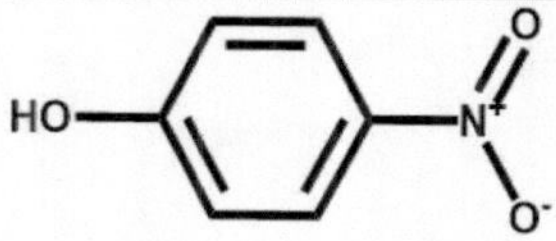	317
Picric acid 2, 4, 6 trinitrophenol	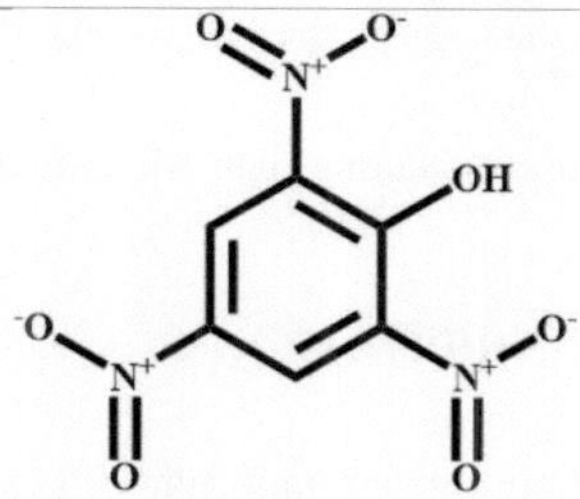	415
2-Nitrobenzoic acid (2 NBA)	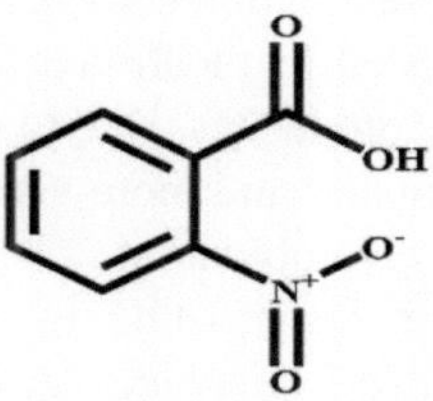	412
4-Nitrotoluene	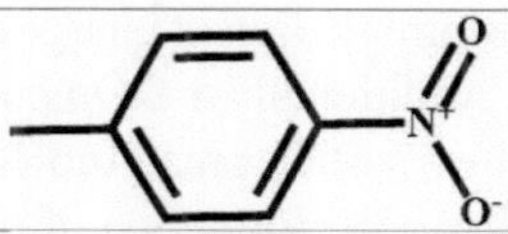	385
4-Chloronitrobenzene	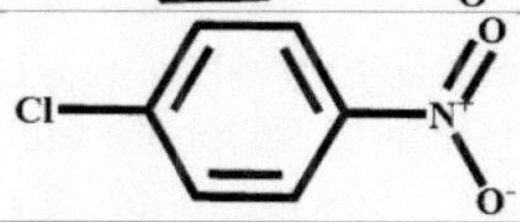	269
2-Choloro-4-nitrophenol	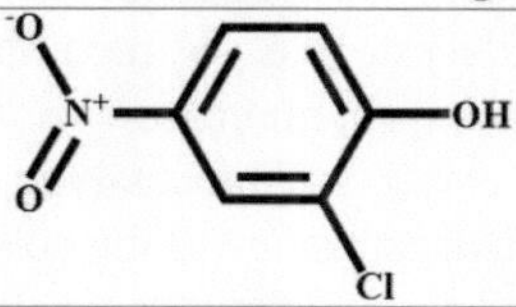	210
2, 4-Dinitrophenylhydrazine	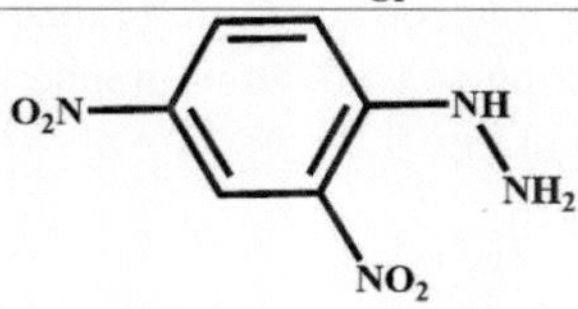	520

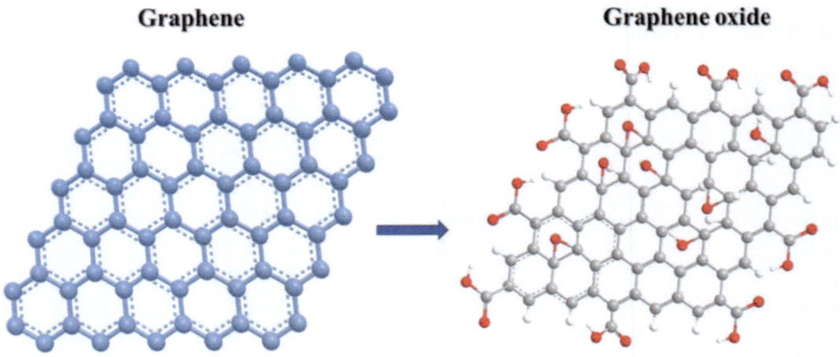

Fig. 9.4 Structural change observed during conversion of graphene to graphene oxide

9.4.2 *Role of r-GO Support*

The hybridization of semiconductor nanomaterials with graphene oxide derivatives, notably r-GO is a hot topic of research in recent years. The enhanced photocatalytic capacity and effective photo-assisted degradation of a variety of organic contaminants are provided by r-GO based on composites. Due to these adjustments, the band gap of pure photocatalysts is now smaller and more visible light active. In addition, the enhancement lowers the rate of recombination.

Due to the outstanding properties of r-GO, as previously indicated, the r-GO-based composites will serve as an efficient electron acceptor, boosting photoinduced charge transfer, preventing recombination of charge carriers, and thereby, improving photocatalytic performance. Additionally, r-GOincorporation later decreases aggregation and raises the composites' surface area. Additionally, nucleation sites are regulated by r-GO for the well-dispersed formation of photocatalysts, later on boosting photocatalytic degradation efficacy by exposing a large number of active sites. The addition of r-GO to the semiconductor photocatalysts also enhances the specific surface area of the composites, allowing mobility of electron mobility on the catalysts' surface and reducing the rate of e^{-}-h^{+} recombination. Moreover, r-GO shields semiconductor nanostructures from photo-corrosion and alters their lattice arrangement, which subsequently impacts photocatalytic activity. The interaction of the r-GO functional groups with the structure of the pollutants will result in a synergistic impact of photocatalysts with good adsorption capabilities (Wang et al. 2019a). Structural change observed during conversion of graphene to reduced graphene oxide via graphene oxide is shown in Fig. 9.5.

9.4.3 *Role of g-C_3N_4*

A non-metallic conjugated semiconductor with triazine or heptazine as its main structural unit and layer structure resembling graphite is known as graphitic carbon

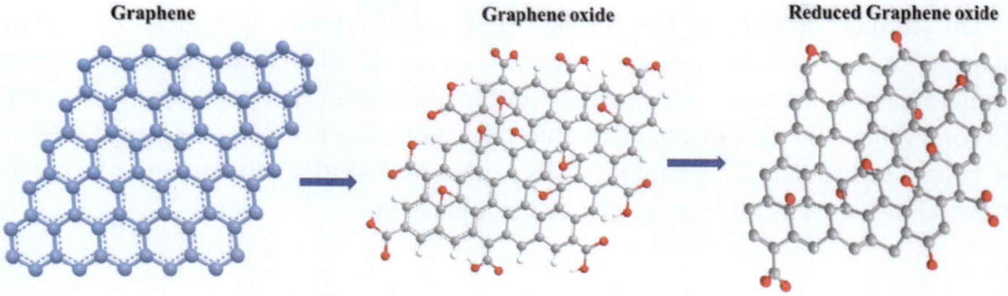

Fig. 9.5 Structural change observed in graphene, GO, and r-GO

nitride (g-C_3N_4).Owing to its beneficial characteristics, including good physico-chemical stability, nontoxicity, applicable band gap energy of about 2.7 eV with visible light absorption, and composition with elements C and N that are abundant on Earth, g-C_3N_4 is currently regarded as an emerging material in the field of photocatalysis. Therefore, g-C_3N_4-based materials provide a potentially attractive choice for photocatalytic applications that are concerned with the environment and energy. However, in addition to being used for industrial purposes, g-C_3N_4 unavoidably has a number of flaws that limit its viability as an effective photocatalyst. The principal drawbacks of g-C_3N_4 are (i) photoexcited recombination; (ii) poor absorption of visible light; (iii) vulnerability to H_2O_2 poisoning; (iv) a very small specific surface area; and (v) low intrinsic quantum efficiency. Numerous studies focus on creating heterostructures or creating composite materials to address the concerns with the use of g-C_3N_4 in photocatalysis (Wang et al. 2019b). Fabrication of g-C_3N_4 from various precursors are schematically represented in Fig. 9.6.

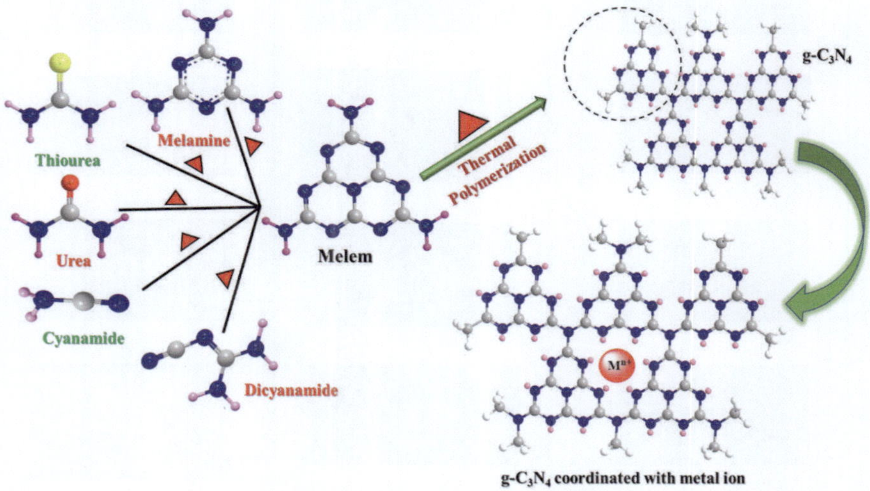

Fig. 9.6 Schematic representation of fabrication of g-C_3N_4 from various precursors

Due to their capacity to use a broad spectrum of visible light, hence using the majority of solar radiation, these materials, such as GO, r-GO, and g-C_3N_4, supported semiconductor and plasmonic materials and are the active photocatalysts in use today. To comprehend the true mechanism of its operation, let's examine the many varieties of binary and ternary GO, r-GO, g-C_3N_4 supported semiconductor, and plasmonic materials, as well as their technique of manufacturing.

9.5 Classification of Photocatalytic Materials

GO, r-GO, and g-C_3N_4 hybridized semiconductor and plasmonic nanomaterials are the most employed photocatalyst these days in the degradation of toxic organic pollutants in water. Some of the most important hybrid materials are listed in Fig. 9.7.

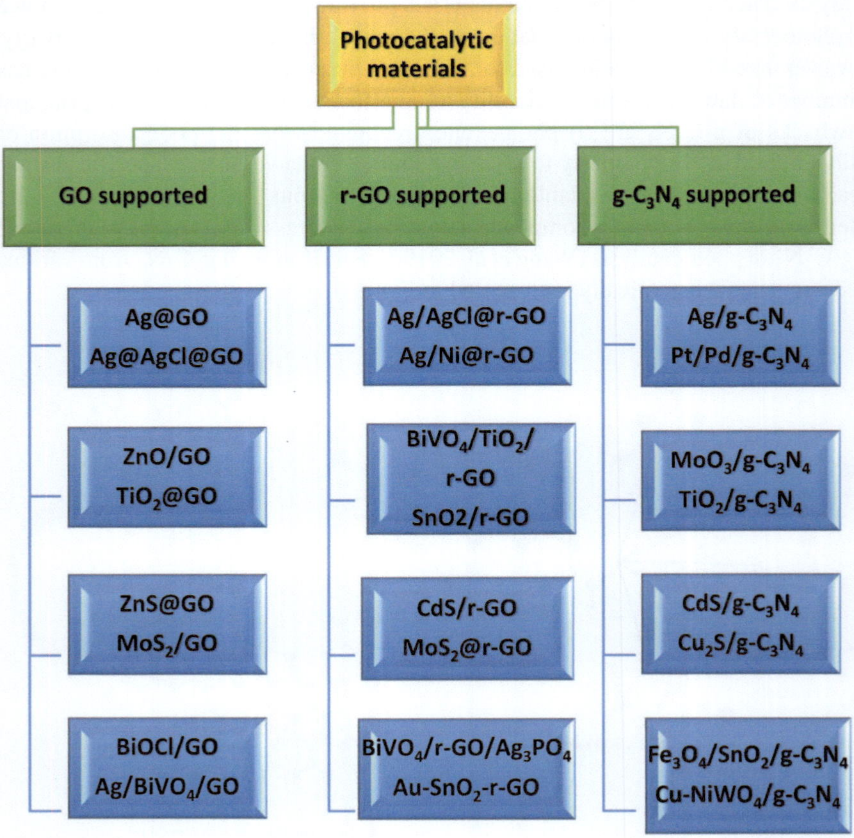

Fig. 9.7 Various examples of hybrid semiconductor materials

9.5.1 GO-Supported Semiconductor/Plasmonic Materials

Plasmonic and semiconductor nanomaterials show dramatic improvement in photocatalytic performance when immobilized or hybridized with GO sheets. To illustrate this, some of the recent studies have been elaborated here.

9.5.1.1 Metal-Doped GO Supported

Cuiling et al. (2019) reported that this study describes the successful fabrication of visible light range hybrid material Ag@AgCl-GO employing photoplasmic reduction approach anddeposition precipitation method at ambient temperature and observed that on increasing UV exposure time during photoplasmic reduction, Ag^0 particle crystallinity gradually rose and stabilized after 90 min. Ag/AgCl-GO photocatalyst demonstrated strong visible-light response capabilities of Ag (0) nanoparticles due to itstendency of surface plasmon resonance (SPR) absorption. However, graphene oxide sheets that carry Ag/AgCl, had a significant impact on the crystal structure of Ag/AgCl, which reduced the crystallite size and improved nanocomposite dispersity.

Impressive photocatalytic performance was demonstrated by Ag/AgCl-GO hybrid in the visible light-assisted deterioration of RhB, which was significantly higher than that of AgCl-GO and Ag/AgCl. This is because of the surface plasmon resonance (SPR) effect of Ag (0) nanoparticles along with the synergistic effect coming from AgCl and GO. The RhB degradation efficiency was 98.43% following a 100-min exposure to visible light.

Thi et al. (2021) also successfully created an Ag/GO nanohybrid and examined its photocatalytic capabilities in a 4-nitrophenol hydrogenation reaction. Silver nitrate was reduced using sodium borohydride as a reducing agent on graphene oxide (GO) to create the Ag/GO nanocomposite. According to the SEM micrographs, the AgNPs on the surface of GO have a somewhat homogeneous distribution of sizes and had an average grain size of roughly 30 nm. Based on the 4-nitrophenol hydrogenation reaction, the catalytic capability was examined. The outcomes demonstrated the outstanding catalytic performance of the Ag/GO nanocomposite, with a reaction constant of $18.19 \times 10^3\ s^{-1}$ and a reduction reaction yield of 97%. In addition, this Ag/GO nanocomposite photocatalyst is readily recoverable and may be recycled three times without suffering any appreciable reduction in efficiency. Ag/GO nanocomposite is additionally secure, less poisonous, and environmentally beneficial. However, the catalytic efficiency of Ag/GO nanohybrid significantly decreased after each catalytic cycle because AgNPs clump together on the GO surface during the synthesis of Ag/GO nanocomposite. Ag/GO nanocomposite fabrication and recycling must be optimized in the future to improve the catalytic activity of recycled forms.

9.5.1.2 Metal Oxide Doped GO

Semiconductor oxides such as ZnO andTiO_2 are extensively employed in photo-assisted organic pollutants elimination because of adequate band gap energy, natural abundance, and non-toxic nature.

The liquid phase deposition process was employed by Jiang et al. (2011) for fabricating graphene oxide hybridized TiO_2 (GO-TiO_2).Here, TiO_2 nanoparticles were grown over 2D graphene oxide sheets. As-fabricated GO-TiO_2 nanohybrid was examined for photocatalytic performance in photo-assisted deterioration of the toxic methyl orange dye as well as reduction of noxious inorganic contaminant Cr(VI) into Cr(III).The fabricated GO-TiO_2 nanohybrid displayed significantly superior photoactivity than that of pure TiO_2 catalysts due to the increased nanocomposite surface area (80 m^2g^{-1}), stronger capacity to adsorb pollutants, and rapid charge transfer (commercial TiO_2). Additionally, because of TiO_2 nanoparticle's high work function and rapid transfer of photoinduced e^- from the conduction band to GO, their recombination is prevented. As a result, the transferred photo-induced electrons help to increase the photocatalytic activity of the nanocomposite by reducing Cr(VI) to Cr(III) and oxidizing organic dyes. Other azo dyes like methyl red, acid orange 7, andorange G, were degraded using the synthesized graphene oxide-TiO_2 nanocomposite, and a considerably improved performance was seen. It has been described in the literature thatsemiconductor and plasmonic nanoparticles and GO hybrid materials have also shown promising results in the removal of toxic aromatic pollutants from the water as well as the mineralization of volatile organic pollutants in the air. Zhang et al. (2010) provided details on the hydrothermal reduction of GO to create a graphene-TiO_2 nanocomposite in this context. They altered the amount of graphene in the nanocomposite in order to remove MB from water and benzene from the air, and they examined the comparative photocatalytic activity of various compositions.

Similarly, liquid-phase deterioration of organic pollutants employing nanocomposites with higher graphene content exhibit reduced activity to remove contaminants. Graphene is able to absorb light across the entire solar spectrum. As a result, the enhanced photocatalytic activity of GO-TiO_2 nanohybrid has been connected to the higher solar radiation utilization percentage and rapid charge transfer that leads to the production of reactive $^{\bullet}OH$ radical that causesthe mineralization of toxic organic pollutants into CO_2 and H_2O.

Several other semiconductors have also been hybridized with GO and examined for photocatalytic performance in environmental clean-up. Li et al. (2010) demonstrated in 2009 that ZnO is a superior photocatalytic material due to its higher photoinduced charge generation efficiency and longer charge carrier lifespan than TiO_2. This study has had a significant impact on the creation of nanocomposite materials based on ZnO and GO as effective nanohybrids for the photo-assisted degradation of organic toxicants from water. Bai et al. (2017) described the hydrothermal approach for the synthesis of ZnO-GO nanocomposites and observed their effectiveness in the detoxification of mycotoxin by dissolving deoxynivalenol (DON) from water while exposed to UV radiation. In the field or while being stored, several Furasium species generally infect wheat, maize, barley, rice, oats and other food grains. DON, one of these mycotoxins, has the potential to be quite dangerous for the aquatic environment. ZnO, which is photoexcited by UV light, has a considerably higher photocatalytic activity when it is hybridised with graphene. Graphene accelerates the photocatalytic process by inhibiting contaminant adsorption and photogenerated electron–hole recombination, the synergistic effects of ZnO and graphene work together to remove the adsorbed pollutant by transferring photoinduced e^- from the conduction band of zinc oxide to graphene. The trap experiment results indicated that the superoxide radicals ($^{\bullet}O_2{}^-$) and holes (h^+) were the active species formed during photodegradation process. The shading effect, which prevents ZnO from absorbing light energy, and as a result a littleproduction of charge carriers takes place which leads to the lowers photocatalytic activity. As the proportion of graphene in the nanohybrid rises, the adsorption phenomenon becomes more prominent, though at the same time, increased graphene results in less photocatalytic activity.

9.5.1.3 Metal Sulphide Doped GO Supported

ZnS-CdS/GO hybrid composites were created by Wang et al. (2015) utilising a light-aided method for the synthesis. The ZnS and CdS nanostructures impregnated into the surface of GO nanosheets have demonstrated a photocatalytic hydrogen conversion efficiency that is almost two-fold that of the bare ZnS-Cd Sheterostructure. When 2(wt%) Pt nanoparticles are usedas co-catalysts, the hydrogen production capacity of the obtained heterostructure is drastically boosted to 1.68 and 0.78 mmol h^{-1} with UV and visible light irradiation, respectively.

Hu et al. (2017) proposed a simple one-step solvothermal process used to create the flower-like CuS/GO composites with increased photocatalytic activity. CuS microflowers were shown to have been uniformly deposited on the surface of GO sheets. Additionally, the GO could increase the capacity of composites for MB adsorption and increase the separation efficiency of photogenerated electron–hole pairs. A photocatalytic test revealed that CuS/GO had a four times greater photocatalytic efficiency for MB under a Xenon light than pure CuS. The results of five separate trials suggested that the CuS/GO photocatalysts showed good stability and anti-photo-corrosion properties. This improved photocatalytic activity was attributed to the photocatalysts' ability to absorb dye molecules and the efficient prevention of photogenerated electron–hole pair recombination in CuS with the addition of GO. In light of this, CuS/GO composites may have a use for the photodegradation of organic contaminants.

9.5.1.4 Mixed/Metal Oxide/Sulphide Doped GO

By using electrospinning and in-situ sol-gel techniques, new PNGB (GO/BiOCl/PAN) fibres were created by Zhang et al. (2016) on the basis of heterojunction band alignment of BiOCl and GO. It was shown that the as synthesised heterojunction nanofibers had nanoscale interfacial contacts. The heterojunction nanofibers made of

GO and BiOCl had the betterefficiency for the photo-assisted degradation of organic pollutants among the manufactured fibres. The increased photocatalytic performance of PNGB could be due to the nanostructure interface interaction of these heterojunction nanofibers. The PNGB composite nanofibers were also easily reusable due to their nanostructure and flexibility. Moreover, experiments on recycling showed that the nanofibers were stable. This result demonstrated a way for making photocatalysts linked to nanofibers that had good photocatalytic performance. We anticipate that this study will increase interest in nanosized heterojunctions for use in photocatalytic water.

Further, Ouyang et al. (2021) also fabricated three component Ag/$BiVO_4$/GO nanohybrid via hydrothermal technique was used to successfully fabricated. The ternary Ag/$BiVO_4$/GO nanohybrid demonstrated increased photodegradation efficiency (90.43%), binary $BiVO_4$/GO (82.65%) and OTC (34.88%) than pure $BiVO_4$ under visible irradiation of light. The investigations on trapping experiment showed that the two major speciesthat were active during the photocatalytic during the photocatalytic degradation process were $\bullet O_2^-$ and h^+. Ag/$BiVO_4$/GO nanocomposite has a high photocatalytic efficiency and was attributed to the quickly moving photoinducede^- from the $BiVO_4$ to the Ag and GO interfaces, which aided in the separation of photogenerated charges and stopped electron–hole pair recombination. This novel method for fabricating efficient ternary Ag/$BiVO_4$/GO for photocatalysts and make wastewater treatment applications more feasible.

9.5.2 r-GO-Supported Semiconductor/Plasmonic Materials

9.5.2.1 Metal-Doped rGO Ssupported Semiconductor/Plasmonic Materials

A core–shell nanocomposite of magnetically recoverable Au@Ni/rGO nanostructures with a size of less than 8 nm was synthesized by Darabdhara et al. (2019) with magnetically recoverable property and described the colorimetric detection and photocatalytic degradation of phenolic compound for the first time. Additionally, the Au@Ni/rGO nanohybrid demonstrated enzymatic response toward the oxidation of TMB and colorimetric detection of phenolic compounds with a minuscule detection limit of 1.68 M with exceptional sensitivity. The Au@Ni/rGO nanocomposite decomposed 2-nitrophenol, 2-chlorophenol, and phenol by more than 87% when exposed to solar radiation, demonstrating outstanding photo-responsive activity. The magnetic property of this nanocomposite, which aids in its recovery after each cycle, permits it to be reused for up to six cycles without losing any activity, makes it special. After being used, the nanocomposite's characteristics show no significant change, showing its outstanding stability and sustainability. Luo et al. (2013) successfully created ternary Ag/AgCl/rGO nanohybrid photocatalyst via hydrothermal immobilization of $Ag[(NH_3)_2]^+$ into GO dispersion and subsequent in-situ oxidation of Ag/GO at ambient temperature by freshly prepared $FeCl_3$ solution. In the hydrothermal process, metallic Ag is shown to develop in situ on the rGO surface along with the efficient reduction of GO, whilst the production of a highly linked Ag/AgCl/rGO nanohybrid photocatalyst occurred due to oxidation of Ag(0) by ferric chloride solution. The results of the photocatalytic tests revealed that the formed Ag/AgCl/rGO hybrid photocatalyst exhibited many-fold higher photodegradation performance compared to that of physically mixed Ag/AgCl/rGO and bare Ag/AgCl. The Ag/AgCl/rGO exhibits the greatest photocatalytic activity (3.2 wt% rGO). The engineered Ag/AgCl/rGO heterostructures have increased photocatalytic activity, which may be ascribed to the interaction between the effectively separated photoinduced electron–hole pairs through closely connected rGO sheets separation of and the concentration of organic pollutants deposited on the surface of rGO nanosheets. The extremely efficient Ag/AgCl/rGO nanocomposites might potentially find extensive usage in a range of sectors, such as wastewater treatment and air purification, given the straightforward and ecologically benign manufacturing process.

9.5.2.2 Metal Oxide Doped rGO Supported Semiconductor/Plasmonic Materials

Xiao et al. (2021) presented a hierarchical TiO_2 nanostructure with the Miller plane (001) as the principal face by a straightforward hydrothermal process using ethanol, isopropanol, and HF solution. The precipitation of Ag_2O and graphene coated TiO_2nanoflowers were fabricated using the wet chemical co-precipitation approach and the hydrothermal method, respectively. The ternary composite material rAT (r-GO/Ag_2O/TiO_2) has broken down RhB solution under solar radiation and analysis revealed that the 92.3% degradation was achieved in only 70 min. Wang and his co-worker concluded that the rAT photocatalyst exhibits a synergistic impact on aqueous RhB and has a greater sun light sensitivity. The composite material improves charge separation and broadens the light response range. It also demonstrates how the three aspects listed below may be used to characterise the good photocatalytic performance: unusual hierarchical flower-type pattern, Z-scheme heterojunction of Ag_2O, TiO_2, and rGO, and huge specific surface area in two-dimensional graphene. These exceptional properties of formed hierarchical flower-type rAT nanohybrid have considerable real-life application in the realm of photocatalysis.

With the ideal r-GO addition doses of 0.5 percent, reaction pH 3, and hydrothermal temperatures of 200 °C, $BiVO_4$/TiO_2/r-GO fabrication was described by Wanget. al. (2019a) After 120 min, four derivatives of tetracycline antibiotics were effectively broken down using $BiVO_4$/TiO_2/rGO, removing 96.2%, 97.5%, 98.7%, and 99.6% of the tetracycline, chlortetracycline, oxytetracycline, and doxycycline, respectively. The degradation of toxic pollutants follows pseudo-first order kinetics. In contrast to pure $BiVO_4$ and $BiVO_4$/TiO_2, the $BiVO_4$/TiO_2/rGO hybrid photocatalyst showed improved breakdown activity for all the four-tetracycline based antibiotics when exposed to visible light. $BiVO_4$nanoparticles's response to visible light was substantially improved by the addition of r-GO, which also served as a superb centre for electron transport and storage, enhancing photocatalytic performance even further.

The breakdown procedure proved that the tetracycline antibiotics' breakdown process produced unstable intermediates. This study established novel strategy for effective remediation of particular toxicants in the aqueous medium and demonstrated the good photochemical catalysis effect of $BiVO_4/TiO_2$/r-GO.

9.5.2.3 Metal Sulphide Doped rGO Supported Semiconductor/ Plasmonic Materials

Due to their fascinating electrical properties of MX_2 type semiconductor dichalcogenides of transition (where M represents transiton metal like Mo and W while X is a chalcogen (oxygen family member), are used in the removal of contaminants. Additionally, the narrow band gap of semiconductors like MoS_2 (~1.86 eV) enables the visible light capturing tendency. However, the meterials like MoS_2 possessing tiny band gap limits their use in potential utility because of the fast annihilation of photogenerated electrons and holes that it causes. As a result, we can say single material would not contain all the requirements for photocatalytic activity. The development of MoS_2 nanocomposites using minerals like GO is one of the exciting techniques that could provide nanocomposites with a variety of potential benefits. Ding et al. (2015) fabricated a comparable MoS_2-reduced graphene oxide (r-GO) nanocomposite was using a ordinary hydrothermal manufacturing approach. The effective charge carrier separation in the synthesized nanohybrid was confirmed from photoluminescence analysis, which revealed low recombination of photoinduced charge carriers. The degradation of the RhB dye, due to rapid formation of superoxide radicals ($^{\bullet}O_2^-$), which are active oxidising agents, takes place when visible light is irradiated and also electrons from MoS_2's CB are transferred to r-GO. Therefore, the improved photoactivity of fabricated nanohybrid for the removal of pollutants under visible radiation might be due to the quick generation of charge carriers as well as the suppression of their recombination caused by the growth of the heterointerface between MoS_2 and r-GO. The capability of macrostructured, self-standing paper for environmental cleaning has been established. For the production of such a hybrid photocatalyst, a basic exfoliation approach involving evaporation under room temperature has been employed. These 2D-2D type hybrid materials are created by combining layers together with a lot of face-to-face interaction. These materials allow for more powerful electrical interactions among both the 2D material layers. The 2D-2D nanocomposites perform better due to improved charge separation and higher pollutant adsorption in interfacial regions. Additionally, this nanocomposite's efficiency in photocatalyzing the conversion of organic nitro compounds into corresponding amino compounds has been proven. Due to its extensive utility in the removal of contaminants, metallic sulphides, such as cadmium suphide (CdS) and zinc sulphide (ZnS), are of great scientific interest in contrast to dichalcogenides. A intriguing semiconductor material, CdS is effective under visible light and has a narrow band gap of about 2.4 eV. The application of undoped CdS is severely constrained by its flaws, including rapid charge recombination, insufficient pollutants adsorption, and also low photostability. The development of nanocomposite CdS with substances like r-GO can efficiently separate and transfer photoinduced electrons and holes due to the lower negative Fermi energy level. Chen et al. (2013) produced a nanocomposite comprised of descrete CdS nanospheres covered over sheets of r-GO utilising a simple hydrothermal manufacturing process (r-GO). A channel for quick transmission of photoinduced charges is established by the interfacial engagement of CdS nanospheres and the sheets of r-GO. The CdS-r-GO nanocomposite with homogenous CdS nanospheres equally spaced across RGO nanosheets. With different r-GO loadings, these nanocomposites were found to be extremely effective at selectively reducing nitro-aromatic molecules to corresponding amino derivatives in aqueous medium. The photocatalytic performance of the nanocomposite containing 5% r-GO is the highest for the reduction of 4-nitroaniline. r-GO's superior pollutant adsorption efficiency and improved photogenerated charge separation give transmitted electrons a chance to reduce the adsorbed species.

9.5.2.4 Metal/Metal Oxide/Sulphide Doped rGO Supported Semiconductor/Plasmonic Material

Patra et al. (2019) and his co-worker converted MoS_2/GO composite into a MoS_2/MoO_{3-x}/r-GO hybrid by employing hydrothermal method of fabrication, thereafter calcination in a nitrogen (inert) atmosphere. By diffusing oxygen atoms from the exfoliated GO layers, which significantly reduced the 2D MoS_2 layer and produced a heterojunction of MoO_{3-x}/MoS_2, it was possible to use visible and near IR photons by the plasomon resonance of 2D MoO_{3-x}. Compared to the unaltered MoS_2/GO composite, the calcined MMR composites exhibit increased photocatalytic activity. These findings show that atomic oxygen diffusion produces an interfacial and plasmonic alteration in the nanocomposite. The MoS_2/GO composite's GO content can be optimised, and the extent of oxidation in MoO_{3-x} can be managed to further enhance photocatalytic activity. These findings can serve as a basis for developing such photochemical devices that can create H_2 by acting as both a cocatalyst and a photocatalyst. Highly abundant photocatalyst possessing plasmonic behaviour and a cocatalytic property is highly needed for commercially successful H_2 generation.

Mohanta et al. (2020) successfully fabricated an Au-SnO_2-rGO ternary nano-heterojunction using a straightforward hydrothermal method. Under UV-LED illumination, the fabricated nano-photocatalyst showed an improved photo-assisted mineralization of clothianidin. With a pseudo-first order kinetic reaction rate of 0.0309 min^{-1}, over 97% of degradation was attained after 120 min of UV-LED exposure. Based on LC/HRMS data, two potential degradation mechanisms were suggested. Superoxide anion radical ($^{\bullet}O_2^{-}$) was identified as the main reactive oxygen species in charge of the photodegradation of organic pollutants by photogenerated reactive species trapping tests and the ESR spectroscopy. In comparison to their pristine counterparts, the Au-SnO_2-rGO ternary nano-heterojunction demonstrated improved charge separation, proving the heterojunction effect as a key mechanism for increased photocatalytic efficiency. The expanded light-harvesting range caused by Au's SPR effect and the decreased charge recombination caused by rGO's electron-sink function work together to improve photocatalytic performance. Their work suggested a fresh method for fabricating ternary nano-heterojunctions in a range of configurations and a deeper understanding of hetero junction photocatalysts.

9.5.3 g-C_3N_4 Supported Semiconductor/Plasmonic Materials

9.5.3.1 Metal Doped g-C_3N_4 Supported Semiconductor/Plasmonic Materials

g-C_3N_4 and Ag frequently form heterojunctions as cocatalysts for semiconductor-mediated photocatalysis, which is remarkable. In a work by Yang et al. (2013) it was claimed that the thermal polymerization of urea and concurrent photodeposition were straightforward methods for producing Ag doped g-C_3N_4 (Ag@g-C_3N_4). The Ag@g-C_3N_4 nanocomposite displayed an effective utilisation of visible radiaion due to the excitation in SPR of fabricated Ag nanostructures as well as the rapid transition and induction of charge carriers that are theoretically comparable with the gold co-catalysts. Local electromagnetic fields improved due to metallic silvers' SPR effect, which accelerated the production of electrons and holes in g-C_3N_4. The inclusion of metallic Ag into g-C_3N_4 sheets enables Ag nanoparticles to quickly scavenge photogenerated electrons on g-$C_3N_4$and provide a Schottky barrier that inhibits charge recombination.

DFT calculations were used by Gao et al. (2016) to examine single metals, such as Pt and Pd, immobilized on g-C_3N_4 and used as photocatalytic materials for the reduction of CO_2 via hydrogen. The noble atom serves as the reaction's active site, and the hydrogen supply, which comes from the hydrogen generation reaction, is provided by g-C_3N_4. A single-atom catalyst has good CO_2 reduction activity in addition to being able to maximise the efficiency of noble-metal atoms. The Pt/g-C_3N_4 photocatalyst is capable of effectively reducing CO_2 to CH_4, with a reaction barrier of 1.16 eV, whereas the Pd/g-C_3N_4 photocatalyst prefers to convert CO_2 to HCOOH, with a reaction barrier of 0.66 eV. The mono metal incorporated in g-C_3N_4 effectively boost the absorption of visible radiation, made these hybrid materials important photocatalysts for the CO_2 reduction assisted by visible light. These findings offer a fresh pointer for experimental verification of the process of harvesting solar energy to turn CO_2 into fuel.

9.5.3.2 Metal Oxide Doped g-C_3N_4supported Semiconductor/ Plasmonic Materials

Numerous research teams have created various g-C_3N_4-based semiconductor heterojunction types up to this point for use in solar energyharvesting and environmental cleanup. TiO_2 has been identified as a viable option among the various metal oxides for creating heterojunctions with g-C_3N_4 and the effective synthesis of TiO_2/g-C_3N_4nanocomposites has been demonstrated in numerous research to enhance the photoactivity. In the initial investigation, Zhou et al. (2012) revealed the simple electrochemical method for creating TiO_2 nanotube array/g-C_3N_4 heterostructured photocatalyst. According to Zhou et al. invesitigation a straightforward ball milling process was successful in producing TiO_2/g-C_3N_4 nanohybrids. Rather than direct physical combining the two distinct materials, TiO_2 and the g-C_3N_4, it was discovered that the TiO_2 nanostructures with average crystallite size of 20–40 nm were incorporated in the the g-C_3N_4 monolayer with 0.468 nm thickness to giving rise to a core–shell–nanostructure.

Huang et al. (2019) generated the MoO_{3-x}/g-C_3N_4 nanocomposite employing a moderate solvothermal method. The finest MoO_{3-x}/g-C_3N_4 hybrid (MCN-20) demonstrated greater photocatalytic perperty in LED light source than pure MoO_{3-x}, g-C_3N_4, which can eliminate 92% methyl orange in 100 min and neutralise 100% E. coli within 15 min, correspondingly. The promoted photocatalytic efficiency of MoO_{3-x}/g-C_3N_4 composite is associated with a number of aspects, including: (i) dark-blue MoO_{3-x} in the composite expanded the range of light that it could absorb; (ii) oxygen vacancies in the Z-type MCN hybrid photocatalyst served as recombination sites, improving the removal percentage of photoinduced charge carriers; (iii) active species ($^{\bullet}O_2^-$, h^+) were found to significantly conduct photocatalytic degradation; and (iv) 1D MoO_{3-x} furnished efficacious transport pathway for e^- to enhance the lifespan of charge carriers.

This study established a straightforward photochemical degradation strategy to incorporate oxygen deficiency into the tranasition metal oxide/g-C_3N_4 combination, which is advantageous for enhancing photocatalytic performance.

9.5.3.3 Metal Sulphide Doped g-C_3N_4 Supported Semiconductor/ Plasmonic Materials

The construction of g-C_3N_4 in conjunction with a narrow band gap metallic sulphide semiconductor based photocatalyst, like CdS, has become a significant field of analysis nowadays. For instance, Fu and co-workers developed a inorganic–organic composite (CdS/g-C_3N_4) comprising the both visible-light sensitive semioconductors utilising in-situ precipitation-deposition method (Fu et al. 2013). The CdS/g-C_3N_4 absorption edges may be altered by varying the weight percentage of CdS and g-C_3N_4. As a result, the CdS/g-C_3N_4 nanohybrid' optical absorption moved to the longer wavelength range. It is widely known that bare CdS is especially vulnerable to photocorrosion due to susceptibility of sulphide ion (S^{2-}) to easily to get self-oxidise by the light-generated holes during photocatalytic processes. Since CdS is prone to corrosion, it is predicted that the trasnmission of holes from CdS to connected another semiconductor (g-C_3N_4) at the interfaces of heterojunction will be the most suitable solution to address this issue and boost CdS stability. Even more fascinating than their findings is the fact that the CdS/g-C_3N_4 hybrid photocatalyst shown exceptional stabilities over 5 subsequent degradation cycles in comparison to naked CdS. The successful charge segregation at the Type II heterojunction produced by the interwooven band orientations of the CdS nanostructures and g-C_3N_4 sheets in the CdS/g-C_3N_4 system was credited with the greatly higher activity and stability. In fact, it is believed that the hybrid system's use of CdS and g-C_3N_4 helps to concurrently address the drawbacks of each component.

Zhou et al. (2020) reported the hydrothermal growth approach in depositing the 0D plasmonic $Cu_{2-x}S$ nanodots on 2D g-C_3N_4 nanosheets. The developed CSCNNs catalysts displayed favourable chemical stability as well as outstanding electrical conductivity. The powerful light absorption tendency and effective charge segregation are more noticeable. As a result, the refractory antibiotic LVX may reach 100% degradation rate in 20 min and exhibit excellent degradation in the UV-vis-NIR broad range. The LSPR effect of the right amount of $Cu_{2-x}S$ loading dramatically increased the catalyst's absorption edge, while simultaneously promoting charge carrier migration and the separation of electrons and holes. These tactics will result in greater changes to TMS and 2D materials, and the investigation of mechanisms will aid in understanding how these two materials interact. What's more, this kind of inexpensive LSPR photocatalyst has a lot of promise for widespread use in environmental protection.

9.5.3.4 Mixed Metal/Metal Oxide/Sulphide Doped g-C_3N_4

Prabavathi et al. (2021) have successfully created a Sm_6WO_{12}/g-C_3N_4 heterojunction photocatalyst using a straightforward method. The Sm_6WO_{12}/g-C_3N_4heterojunction significantly increased the photocatalytic degradation efficiency, which is approximately 20.52 and 2.93 times more than that of single Sm_6WO_{12} and g-C_3N_4, respectively. $O_2^{\bullet-}$ and $^{\bullet}OH$ were identified as the primary reactive species for the photo-assisted deterioration of levofloxacin over Sm_6WO_{12}/g-C_3N_4 composite based on the analysis of the radical trapping studies. Based on the GC–MS analysis, the primary degradation products or intermediates and potential levofloxacin photodegradation mechanism were proposed. The quick recombination that was quenched, the reducing bandgap, and the enormous catalytic active sites may be the reason behind the impressive photocatalytic efficiency.

To effectively declining the band gap and lengthen the lifetime of photogenerated e^- and h^+ of the fabricated photocatalyst, Cu was doped into the $NiWO_4$ crystal. We successfully combined g-C_3N_4 and Cu-$NiWO_4$ to create the Cu-$NiWO_4$/g-C_3N_4 direct Z-type heterostructure (Truc et al. 2019). Photoinduced electrons in Cu-$NiWO_4$ conduction band and photoinduced holes in g-C_3N_4 valence band tended to recombine in the system to maintain electrons in g-C_3N_4 conduction band and holes in Cu-$NiWO_4$ valence band. The developed direct Z-scheme, generated a large quantity of accessible electrons and holes, which enabled the photocatalytic breakdown of hexane. Cu-$NiWO_4$/g-C_3N_4 had a maximum hexane degradation of 96.8%. The established Cu-$NiWO_4$/g-C_3N_4 photocatalyst has now shown good stability during extended n-hexane elimination, giving it the best chance of success in real-world VOC removal systems.

9.6 Factors Influencing Photocatalytic Degradation

Photocatalytic degradation of toxic pollutants and various factors influencing the photo assisted degradation are illustrated in Fig. 9.8.

9.6.1 Effect of pH

Since pH can affect how rapidly dyes react, the parameter is essential for the photocatalytic process. This parameter may have an effect on the adsorption of dye onto a semiconductor surface since the surface of the catalyst will rely on the pH of the supplied solution. According to Zhu et al. (2012) the pH solution has a considerable influence on the surface charge of the photocatalyst. Since pH has an impact on both the photocatalytic process and the overall rate of degradation, it is an important consideration in photocatalysis. Adsorbent surfaces have a positive charge that may either attract anions or repel cations.

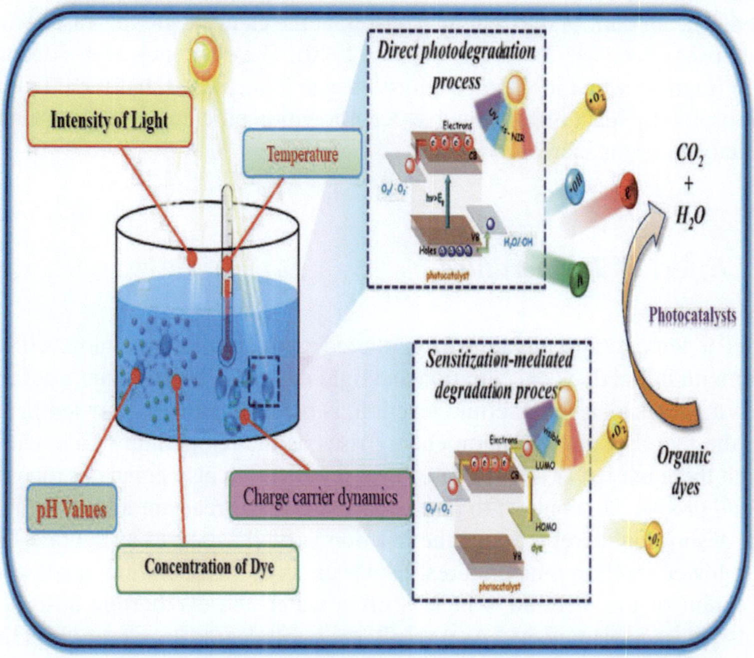

Fig. 9.8 Graphical representation of factors influencing the photo-assisted degradation of organic pollutants

9.6.2 Effect of Initial Concentration of Dye

The rate constant for decolorization will decline as dye concentration increases, as was previously claimed and reported. Under two circumstances, (1) initial concentration can affect dye photodegradation, (2) the photocatalysts may get surrounded by active dye ions if the dye degradation process is carried out at a high dye concentration. In conclusion, a decline in photocatalytic effectiveness will occur when dye concentration is increased. The initial dye concentration will increase as the distance to the UV source (photons) gets closer (Hanafi and Sapawe 2020).

9.6.3 Effect of Catalyst Load

The correct loading of the catalyst is required to prevent the overuse of dye degradation catalyst in catalysis. The size or design of the reactor, as well as the photodegradable process, determine the quantity of catalyst needed for a certain process, such as the Haber process, the initialization phase of the hydrocracking reaction, etc. The light intensity and dye selection also have a big impact on how well dyes are removed. The amount of dye that may adsorb at the catalyst site increases when the catalyst dosage is raised (Nazri and Sapawe 2020). The efficiency of dye decolorization will improve when the photocatalyst dosage is increased, but the efficiency of the photocatalytic reaction and the dye decolorization efficiency will decrease if the photocatalyst dose is increased further.

9.6.4 Effect of Temperature

There is a temperature range where photodegradation occurs most efficiently, together with initial dye concentration and light intensity. The reactant's adsorption, which is a spontaneous exothermic reaction, is better favored by low temperatures, despite the fact that the activation energy rises as the temperature falls near 0 °C. Although there are fewer active sites, lower temperature also enhances the removal of the end product. In contrast to photodegradation and reactant adsorption, slower product desorption thereby delays the reaction as well as serves as the rate-limiting phase at lower reaction temperatures. However, as the temperature approaches the boiling point of the medium, which is often water, the exothermic adsorption of reactants is less advantageous. As a result, the photodegradation process is limited. Charge–carrier recombination is also strongly enhanced when the reaction temperature goes over 80 °C. Due to their increased kinetic energy, dye molecules may escape from the photocatalyst surface at higher temperatures, which would decrease the effectiveness of photodegradation (Barakat et al. 2013). As a result, the limiting step at high temperatures is the adsorption of dye molecules. As a result, it is thought that a reaction temperature range of 20–80 °C is optimal for the effective photodegradation of dye molecules (Shams-Ghahfarokhi and Nezamzadeh-Ejhieh 2015).

9.6.5 Effect of Light Intensity

It has been demonstrated that at low-light conditions, the photodegradation rate rises linearly with intensity of light. The 2,4-dichlorophenoxyacetic acid (2,4-D) is photodegraded more effectively between 100 and 600 lx of light (Kundu et al. 2005). Though the efficiency rose nonlinearly, the photo-assisted deterioration of benzene over nitrogen-doped TiO_2 was improved. The rate of photo-assisted degradation does not alter with change in intensity of light at high light intensities. The segregation of photo-excited electrons and holes is assumed to compete with their recombination at low light intensities, preventing the production of reactive radicals. As light intensity rises, the electron–hole production pathway takes over, leading to a higher photodegradation rate. Due to the constant total number of photodegradation active sites despite the rising light intensity, the reaction rate reaches its maximum value (Sun et al. 2018).

9.7 Future Aspects

Graphene-based nanocomposites have made considerable strides in the removal of pollutants over the years, but there are still a number of challenges associated with their manufacturing and application in catalysis. GO/r-GO/g-C_3N_4-based nanocomposites have been manufactured using a variety of processes, but there is still a need for concentrated efforts to achieve large-scale practical applications due to the growing need for high efficiency and optimised parameters. The industrialization of graphene is a task that is eagerly anticipated in the coming years since its applications may be severely restricted by its low bulk density. Therefore, rather than using a random combination, deliberate steps should be taken to build specific hybrid nanocomposite structures with the presumed qualities. Additionally, creating thin, uniform, high-quality, cost-effective material is challenging such as GO/r-GO/g-C_3N_4nanocomposites with distinct electrical, optical, and thermal properties and changeable layer thickness. It is crucial to note that the hybridization of GO/r-GO/g-C_3N_4 with various semiconductor and plasmonic nanomaterials has greatly enhanced the thermal, electrical, and mechanical qualities of hybrid materials. The remarkable photocatalytic capacity of these nanocomposites to remove organic contaminants was observed.

In the majority of the studies, the visible active photocatalyst was made using a simple or modified sol–gel method. Some of the main issues that the researchers tried to address included the low sensitivity towards longer wavelengths, excessive electron–hole recombination, and inadequate number of active sites or low surface area.

The aforementioned issues have been the subject of several investigations employing a variety of techniques, including doping cations or anions into the pure semiconductor, composing multiple semiconductors, and sensitising using plasmonic nanostructures. Intraband states are injected into the band gap when cations are doped ideally, which lowers the resultant band gap and boosts absorption of longer wavelength light radiation. However, excessive cationic doping boosts recombination sites and decreases the stability of the lattices.

The creation of unique dopants and inexpensive preparation methods is one of the best chances for addressing the photocatalyst's reusability and repeatability problems. Additionally, the fabrication of plasmonic metals possessing resonance frequency in the range of shorter wavelength would result in greater photocatalytic performance through higher production and incorporation of the hot electrons due to the stronger localised electric field. The synergistic effects of composite creation and plasmonic nanostructure on the photocatalytic performance and charge transfer system also need to be thoroughly studied.

References

Ai, C., Yang, S., Zhang, F., Shao, X., Xu, J.: Ag/AgCl-GO: A composite for degradation of Rhodamine B in dye wastewater. Adv. Powder Technol. **30**(12), 3193–3202 (2019)

Bai, X., Sun, C., Liu, D., Luo, X., Li, D., Wang, J., Wang, N., Chang, X., Zong, R., Zhu, Y.: Photocatalytic degradation of deoxynivalenol using graphene/ZnO hybrids in aqueous suspension. Appl. Catal. B **204**, 11–20 (2017)

Barakat, N.A., Kanjwal, M.A., Chronakis, I.S., Kim, H.Y.: Influence of temperature on the photodegradation process using Ag-doped TiO2 nanostructures: negative impact with the nanofibers. J. Mol. Catal. A: Chem. **366**, 333–340 (2013)

Chen, Z., Liu, S., Yang, M.-Q., Xu, Y.-J.: Synthesis of uniform CdS nanospheres/graphene hybrid nanocomposites and their application as visible light photocatalyst for selective reduction of nitro organics in water. ACS Appl. Mater. Interfaces. **5**(10), 4309–4319 (2013)

Chiu, Y.-H., Chang, T.-F.M., Chen, C.-Y., Sone, M., Hsu, Y.-J.: Mechanistic insights into photodegradation of organic dyes using heterostructure photocatalysts. Catalysts **9**(5), 430 (2019)

Darabdhara, G., Boruah, P.K., Borthakur, P., Hussain, N., Das, M.R., Ahamad, T., Alshehri, S.M., Malgras, V., Wu, K.C.-W., Yamauchi, Y.: Reduced graphene oxide nanosheets decorated with Au–Pd bimetallic alloy nanoparticles towards efficient photocatalytic degradation of phenolic compounds in water. Nanoscale **8**(15), 8276–8287 (2016)

Darabdhara, G., Das, M.R.: Dual responsive magnetic Au@ Ni nanostructures loaded reduced graphene oxide sheets for colorimetric detection and photocatalytic degradation of toxic phenolic compounds. J. Hazard. Mater. **368**, 365–377 (2019)

De Bel, E., Dewulf, J., De Witte, B., Van Langenhove, H., Janssen, C.: Influence of pH on the sonolysis of ciprofloxacin: biodegradability, ecotoxicity and antibiotic activity of its degradation products. Chemosphere **77**(2), 291–295 (2009)

Demirbas, A.: Agricultural based activated carbons for the removal of dyes from aqueous solutions: a review. J. Hazard. Mater. **167**(1–3), 1–9 (2009)

Deng, Y., Zhao, R.: Advanced oxidation processes (AOPs) in wastewater treatment. Current Pollution Reports **1**(3), 167–176 (2015)

Ding, Y., Zhou, Y., Nie, W., Chen, P.: MoS2–GO nanocomposites synthesized via a hydrothermal hydrogel method for solar light photocatalytic degradation of methylene blue. Appl. Surf. Sci. **357**, 1606–1612 (2015)

Fu, J., Chang, B., Tian, Y., Xi, F., Dong, X.: Novel C 3 N 4–CdS composite photocatalysts with organic–inorganic heterojunctions: in situ synthesis, exceptional activity, high stability and photocatalytic mechanism. J. Mater. Chem. A **1**(9), 3083–3090 (2013)

Fujishima, A., Honda, K.: Electrochemical photolysis of water at a semiconductor electrode. Nature. **238** (5358), 37–38 (1972)

Gao, G., Jiao, Y., Waclawik, E.R., Du, A.: Single atom (Pd/Pt) supported on graphitic carbon nitride as an efficient photocatalyst for visible-light reduction of carbon dioxide. J. Am. Chem. Soc. **138**(19), 6292–6297 (2016)

Glaze, W.H.: Drinking-water treatment with ozone. Environ. Sci. Technol. **21**(3), 224–230 (1987)

Grzechulska, J., Morawski, A.W.: Photocatalytic decomposition of azo-dye acid black 1 in water over modified titanium dioxide. Appl. Catal. B **36**(1), 45–51 (2002)

Han, L., Wang, P., Dong, S.: Progress in graphene-based photoactive nanocomposites as a promising class of photocatalyst. Nanoscale **4**(19), 5814–5825 (2012)

Hanafi, M.F., Sapawe, N.: Performance of nickel catalyst toward photocatalytic degradation of methyl orange. Materials Today: Proceedings **31**, 257–259 (2020)

Hu, X.-S., Shen, Y., Zhang, Y.-T., Nie, J.-J.: Preparation of flower-like CuS/reduced graphene oxide (RGO) photocatalysts for enhanced photocatalytic activity. J. Phys. Chem. Solids **103**, 201–208 (2017)

Hu, Z.-T., Liu, J., Yan, X., Oh, W.-D., Lim, T.-T.: Low-temperature synthesis of graphene/Bi2Fe4O9 composite for synergistic adsorption-photocatalytic degradation of hydrophobic pollutant under solar irradiation. Chem. Eng. J. **262**, 1022–1032 (2015)

Huang, C., Dong, C., Tang, Z.: Advanced chemical oxidation: its present role and potential future in hazardous waste treatment. Waste Manage. **13**(5–7), 361–377 (1993)

Huang, L., Zhang, F., Li, Y., Wang, H., Wang, Q., Wang, C., Xu, H., Li, H.: Chemical reduction implanted oxygen vacancy on the surface of 1D MoO3− x/g-C3N4 composite for boosted LED light-driven photoactivity. J. Mater. Sci. **54**(7), 5343–5358 (2019)

Jiang, G., Lin, Z., Chen, C., Zhu, L., Chang, Q., Wang, N., Wei, W., Tang, H.: TiO2 nanoparticles assembled on graphene oxide nanosheets with high photocatalytic activity for removal of pollutants. Carbon **49**(8), 2693–2701 (2011)

Khan, M.M., Adil, S.F., Al-Mayouf, A.: Metal Oxides as Photocatalysts. Elsevier **19**, 462–464 (2015)

Kundu, S., Pal, A., Dikshit, A.K.: UV induced degradation of herbicide 2, 4-D: kinetics, mechanism and effect of various conditions on the degradation. Sep. Purif. Technol. **44**(2), 121–129 (2005)

Li, Y., Xie, W., Hu, X., Shen, G., Zhou, X., Xiang, Y., Zhao, X., Fang, P.: Comparison of dye photodegradation and its coupling with light-to-electricity conversion over TiO2 and ZnO. Langmuir **26**(1), 591–597 (2010)

Luo, G., Jiang, X., Li, M., Shen, Q., Zhang, L., Yu, H.: Facile fabrication and enhanced photocatalytic performance of Ag/AgCl/rGO heterostructure photocatalyst. ACS Appl. Mater. Interfaces. **5**(6), 2161–2168 (2013)

Marschall, R.: Semiconductor composites: strategies for enhancing charge carrier separation to improve photocatalytic activity. Adv. Func. Mater. **24**(17), 2421–2440 (2014)

Michael, I., Hapeshi, E., Michael, C., Fatta-Kassinos, D.: Solar Fenton and solar TiO2 catalytic treatment of ofloxacin in secondary treated effluents: evaluation of operational and kinetic parameters. Water Res. **44**(18), 5450–5462 (2010)

Mohanta, D., Ahmaruzzaman, M.: A novel Au-SnO2-rGO ternary nanoheterojunction catalyst for UV-LED induced photocatalytic degradation of clothianidin: Identification of reactive intermediates, degradation pathway and in-depth mechanistic insight. J. Hazard. Mater. **397**, 122685 (2020)

Nazri, M.K.H.M., Sapawe, N.: A short review on photocatalytic toward dye degradation. Materials Today: Proceedings **31**, A42–A47 (2020)

Ouyang, K., Yang, C., Xu, B., Wang, H., Xie, S.: Synthesis of novel ternary Ag/BiVO4/GO photocatalyst for degradation of oxytetracycline hydrochloride under visible light. Colloids Surf., A **625**, 126978 (2021)

Pant, H.R., Pant, B., Kim, H.J., Amarjargal, A., Park, C.H., Tijing, L.D., Kim, E.K., Kim, C.S.: A green and facile one-pot synthesis of Ag–ZnO/RGO nanocomposite with effective photocatalytic activity for removal of organic pollutants. Ceram. Int. **39**(5), 5083–5091 (2013)

Patra, K.K., Ghosalya, M.K., Bajpai, H., Raj, S., Gopinath, C.S.: Oxidative disproportionation of MoS2/GO to MoS2/MoO3–x/RGO: integrated and plasmonic 2D-multifunctional nanocomposites for solar hydrogen generation from near-infrared and visible regions. J. Phys. Chem. C **123**(35), 21685–21693 (2019)

Prabavathi, S.L., Saravanakumar, K., Park, C.M., Muthuraj, V.: Photocatalytic degradation of levofloxacin by a novel Sm6WO12/g-C3N4 heterojunction: Performance, mechanism and degradation pathways. Sep. Purif. Technol. **257**, 117985 (2021)

Rathnasamy, R., Santhanam, M., Alagan, V.: Anchoring of ZnO nanoparticles on exfoliated MoS2 nanosheets for enhanced photocatalytic decolorization of methyl red dye. Mater. Sci. Semicond. Process. **85**, 59–67 (2018)

Saravanan, R., Gracia, F., Stephen, A.: Basic principles, mechanism, and challenges of photocatalysis, pp. 19–40. Springer, In Nanocomposites for visible light-induced photocatalysis (2017)

Shams-Ghahfarokhi, Z., Nezamzadeh-Ejhieh, A.: As-synthesized ZSM-5 zeolite as a suitable support for increasing the photoactivity of semiconductors in a typical photodegradation process. Mater. Sci. Semicond. Process. **39**, 265–275 (2015)

Sonkusare, V.N., Chaudhary, R.G., Bhusari, G.S., Mondal, A., Potbhare, A.K., Mishra, R.K., Juneja, H.D., Abdala, A.A.: Mesoporous octahedron-shaped tricobalt tetroxide nanoparticles for photocatalytic degradation of toxic dyes. ACS Omega **5**(14), 7823–7835 (2020)

Sun, P., Zhang, J., Liu, W., Wang, Q., Cao, W.: Modification to LH kinetics model and its application in the investigation on photodegradation of gaseous benzene by nitrogen-doped TiO_2. Catalysts **8**(8), 326 (2018)

Thi, P.T., Van Khai, T., Diem, T.X., Nghia, C.N.T., Ngan, T.T.T., Vinh, N.N., Hien, N.M.: Synthesis of Ag/GO nanocomposite with promising photocatalytic ability for reduction reaction of p-nitrophenol. Mater. Res. Express **8**(10), 105009 (2021)

Truc, N.T.T., Pham, T.-D., Van Thuan, D., Tran, D.T., Nguyen, M.V., Dang, N.M., Trang, H.T.: Superior activity of Cu-NiWO4/g-C3N4 Z direct system for photocatalytic decomposition of VOCs in aerosol under visible light. J. Alloy. Compd. **798**, 12–18 (2019)

Wammer, K.H., Korte, A.R., Lundeen, R.A., Sundberg, J.E., McNeill, K., Arnold, W.A.: Direct photochemistry of three fluoroquinolone antibacterials: norfloxacin, ofloxacin, and enrofloxacin. Water Res. **47**(1), 439–448 (2013)

Wang, C.-C., Yi, X.-H., Wang, P.: Powerful combination of MOFs and C3N4 for enhanced photocatalytic performance. Appl. Catal. B **247**, 24–48 (2019b)

Wang, W., Han, Q., Zhu, Z., Zhang, L., Zhong, S., Liu, B.: Enhanced photocatalytic degradation performance of organic contaminants by heterojunction photocatalyst BiVO4/TiO2/RGO and its compatibility on four different tetracycline antibiotics. Adv. Powder Technol. **30**(9), 1882–1896 (2019a)

Wang, X., Yuan, B., Xie, Z., Wang, D., Zhang, R.: ZnS–CdS/Graphene oxide heterostructures prepared by a light irradiation-assisted method for effective photocatalytic hydrogen generation. J. Colloid Interface Sci. **446**, 150–154 (2015)

Xiao, H., Wang, T.: Graphene oxide (rGO)-metal oxide (TiO2/Ag2O) based nanocomposites for the removal of rhodamine B at UV–visible light. J. Phys. Chem. Solids **154**, 110100 (2021)

Yang, Y., Guo, Y., Liu, F., Yuan, X., Guo, Y., Zhang, S., Guo, W., Huo, M.: Preparation and enhanced visible-light photocatalytic activity of silver deposited graphitic carbon nitride plasmonic photocatalyst. Appl. Catal. B **142**, 828–837 (2013)

Zhang, Y., Park, M., Kim, H.-Y., Park, S.-J.: In-situ synthesis of graphene oxide/BiOCl heterostructured nanofibers for visible-light photocatalytic investigation. J. Alloy. Compd. **686**, 106–114 (2016)

Zhang, Y., Tang, Z.-R., Fu, X., Xu, Y.-J.: TiO2− graphene nanocomposites for gas-phase photocatalytic degradation of volatile aromatic pollutant: is TiO2−graphene truly different from other TiO2− carbon composite materials? ACS Nano **4**(12), 7303–7314 (2010)

Zhou, L., Liu, Z., Guan, Z., Tian, B., Wang, L., Zhou, Y., Zhou, Y., Lei, J., Zhang, J., Liu, Y.: 0D/2D plasmonic Cu2-xS/g-C3N4 nanosheets harnessing UV-vis-NIR broad spectrum for photocatalytic degradation of antibiotic pollutant. Appl. Catal. B **263**, 118326 (2020)

Zhou, X., Jin, B., Li, L., Peng, F., Wang, H., Yu, H., Fang, Y.: A carbon nitride/TiO 2 nanotube array heterojunction visible-light photocatalyst: synthesis, characterization, and photoelectrochemical properties. J. Mater. Chem. **22**(34), 17900–17905 (2012)

Zhu, H., Jiang, R., Fu, Y., Guan, Y., Yao, J., Xiao, L., Zeng, G.: Effective photocatalytic decolorization of methyl orange utilizing TiO2/ZnO/chitosan nanocomposite films under simulated solar irradiation. Desalination **286**, 41–48 (2012)

Chapter 10
Ionic Liquid as an Efficient Catalytic Material for the Organic Synthesis

Kirtikumar C. Badgujar, Vivek C. Badgujar, and Bhalchandra M. Bhanage

10.1 Introduction

Industries are always looking for the new significantly improved, greener and cleaner process for the production of organic compounds and moieties (Qureshi et al. 2014). Till date, the main agenda of the industries is to use the conventional volatile organic solvent for the bulk production of organic compounds (Potdar et al. 2015). However, volatile organic solvents are not acceptable as greener and cleaner solvents which possess the several operational and environmental issues (Potdar et al. 2015). These volatile organic solvents possess several non-green features such as lesser reaction yield, lesser selectivity, high-volatility, toxicity, flammability, recyclability and distillation issues in certain organic reactions and processes (Badgujar and Bhanage 2023). Considering the safety and environmental legislations, it is well understood that the conventional solvents are no longer acceptable for the organic process at industrial scale (Badgujar and Bhanage 2023; Schindl et al. 2019). In order to reduce the negative impact of conventional solvents on the environment, various kinds of the solvent system are used which involves the supercritical CO_2, solvent-free system, fluorous solvents and ionic liquids (Sekharan et al. 2021).

Ionic liquids are well known as the "Newton's Liquid" consisting of an organic cations and organic/inorganic anions (Badgujar et al. 2022a). The cation is of generally larger size while anion is smaller or larger size. Ionic liquids are also recognized as "Room-Temperature Molten-Salts" that exits in liquid phase below the temperature 100 °C (Badgujar et al. 2019). The ionic liquids can perform multiple

K. C. Badgujar · B. M. Bhanage (✉)
Department of Chemistry, Institute of Chemical Technology, Matunga, Mumbai, India
e-mail: bm.bhanage@gmail.com; bm.bhanage@ictmumbai.edu.in

K. C. Badgujar
Department of Chemistry, SIES College of Arts, Science and Commerce, Mumbai, India

V. C. Badgujar
Department of Chemistry, Pratap College of Arts, Science and Commerce, Amalner, India

R. S. Ningthoujam and A. K. Tyagi (eds.), *Handbook of Materials Science, Volume 3*,
Indian Institute of Metals Series, https://doi.org/10.1007/978-981-96-8229-4_10

functions such as solvent and catalysis also and hence recognized as multitasking media (Badgujar and Bhanage 2015a). This astonishing feature reported the use of ionic liquid in various sectors such as organo catalytic synthesis, catalysis, biocatalysis, fluid mechanics, biorefineries, electro-chemistry, extraction process, fuel cell, polymer and material science (Olivier-Bourbigou and Morvan 2010). These ionic liquids have become absolutely popular in the last three decades because of their exclusive feature of functioning as multitasking agents as contrast to other catalytic systems like hetero-polyacids, zeolites, simple metal salt, functionalized nanomaterials and solid acid catalysts (Badgujar et al. 2023). Thus, these ionic liquids become an exciting potential candidate to serve as the media or catalyst for the organic synthesis (Badgujar et al. 2022a).

Figure 10.1 depicted the related Scopus literature survey for the organic catalysis using ionic liquid which indicates that the ionic liquid is dynamic catalytic material for organic synthesis (https://www.scopus.com/search/form.uri?display:basic#basic). An ionic liquid catalyst can play diverse kinds of roles in organic synthesis. Ionic liquids can act as acids that can promote the proton transfer phenomenon like traditional catalyst (Badgujar et al. 2023). In certain microwave-assisted organic synthesis, ionic liquid may promote the absorption of energy and corresponding transfer of heat (Martínez-Palou 2010). Further supported ionic liquids are used for organic synthesis in biphasic media (Kusumawati and Shirai 2022). Thus, ionic liquids are reported in literature for various strategic organic synthesis.

In view of this, the present book-chapter will provide a survey of the usefulness of ionic liquid catalysis for efficient organic synthesis. Moreover, this book-chapter will also provide the future outlook for the successful use of the ionic liquid for catalysis at scale-up level.

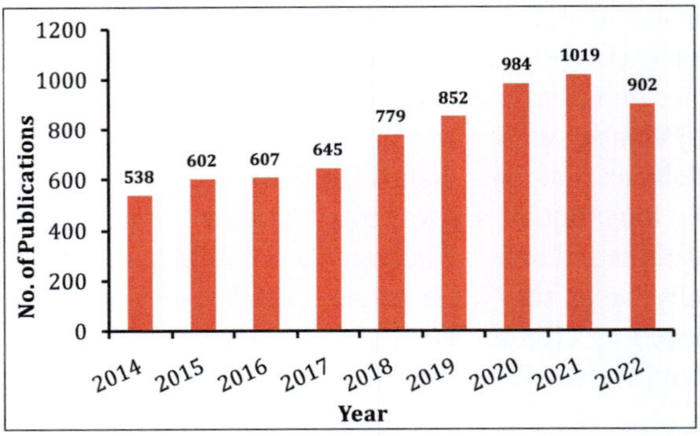

Fig. 10.1 Scopus literature survey for the Scopus search key: (organic AND catalysis) AND (ionic AND liquid)

10.2 Ionic Liquid's Green Properties

"Transforming World: The Year 2030 Agenda for the Sustainable Development" revealed by United Nations which directs the essentially three factors such as ecological, economic and social for sustainable development (https://sustainabledevelopment.un.org/post2015/transformingourworld/publication). The phrase "Green and Sustainable" can be somewhat applicable for the ionic liquids due to their excellent green physico-chemical characteristics (Hagen et al. 2022). These ionic liquids offer the polar-organic reaction media wherein most of the reaction moieties can be well solubilized for the effective reaction processing (Hagen et al. 2022; Poole and Atapattu 2021; Badgujar et al. 2022b). Moreover, as a catalyst, use of these ionic liquids showed improved reaction rate, selectivity and yield of the corresponding products (Badgujar et al. 2022b). Ionic liquids are not at all volatile and hence less toxic compared to the volatile organic solvents (Barrueto et al. 2022; Kulshrestha et al. 2021). It is also notable to mention that the use of the ionic liquid reduces the energy requirement by dissolving the substrates at mild reaction conditions (Klähn et al. 2011). Further, long chain arms of the ionic liquids or active polar centers of the ionic liquids can play essential roles in stabilizing the various reactive intermediates (Hajipour and Rafiee 2015). The recent third generation of ionic liquids is almost pleasing the objectives of the biodegradability; further these generation ionic liquids are simple and synthesized at moderately low-cost or can be obtained from the biomass-based bioproducts (Jopp 2020; Carrozza et al. 2019). Aromatic-organic acids as the lignin derived products, amino acids and choline can be used as the greener counterpart for the production of the industrial scale-up synthesis of ionic liquid (Badgujar et al. 2022a, b). Moreover, these ionic liquids offer high ionic-conductivity that can be helpful for the proton transfer mechanism to carry out reaction (Badgujar et al. 2022a, 2019). The better temperature-stability and liquid-range can offer the wide range of the working reaction system (Barrueto et al. 2022; Kulshrestha et al. 2021; Jopp 2020). Further, availability of the several possible designing-combinations and dual nature of catalysis-solvation fulfills the demands of use at a variety of industrial sectors (Badgujar and Bhanage 2023). Thus, the immense tunable solvent-catalytic features obtained via different generations of the ionic liquid are representing the positive movement of the use of ionic liquids for attaining the goal of sustainability.

10.3 Ionic Liquid for Catalysis

10.3.1 Acylation and Alkylation

Friedel–Crafts acylations and alkylation are the most significant organic reactions which possess the extensive applications in commercial organic synthesis (Cai 2014; Lin et al. 2020; Zhikai et al. 2014; Shaohua et al. 2011; Chen et al. 2009). The alkylation reactions have great importance due to the production of a diversity of long-chain fine chemicals, whereas acylation reactions are broadly applied to manufacture the pharmaceutical drug intermediate such as ibuprofen and naproxen (Cai 2014; Lin et al. 2020; Zhikai et al. 2014; Shaohua et al. 2011; Chen et al. 2009). The major substrate reagent utilized for the acylation and alkylation reaction is the acid halide and alkyl halide in the volatile organic solvents respectively (Cai 2014; Lin et al. 2020; Zhikai et al. 2014; Shaohua et al. 2011; Chen et al. 2009). Traditionally, these reactions are carried out by catalysis such as the Lewis acid aluminum trichloride. Other catalysts used for these reactions are ion-exchange resins, Brønsted acids, zeolites, mesoporous materials or any kind of the acid catalysts may be in solid or liquid forms are used. However, the use of this acid catalysis suffered from various drawbacks such as severe environmental issues, disposal of catalyst, reaction equipment corrosion, deactivation and reactivation of the catalyst (Cai 2014; Lin et al. 2020; Zhikai et al. 2014; Shaohua et al. 2011; Chen et al. 2009). In search to defeat above said drawbacks, ionic liquids can be the better choice and has received much attention.

Cai (2014) carried out the acylation of anisole with acid chloride which offers moderate yield 65.3% of acylated product in the ionic liquid $[BMIM]BF_4$ while the conventional catalyst $AlCl_3$ offered only 51.8% yield of the corresponding product (Table 10.1, entry 1). The higher yield is attributed to more acidity and appropriate solvation phenomenon. Lin et al. (2020) performed Friedel–Crafts acylation (benzoylation) using a substituted-aryl imidazolium-based different ionic liquids possessing various electron donating and withdrawing substituent (Table 10.1, entry 2). Further they claimed that use of the substituted-aryl imidazolium-based ionic liquids offers excellent catalytic activity along with efficient recycling of the ionic liquid. Among the tested ionic liquids, the nitro substituted-aryl imidazolium-based ionic liquid $[NO_2\text{-Ph-IM}]FeCl_4$ displayed excellent catalytic activity than other electron donating substituted-aryl imidazolium-based ionic liquids $–CH_3$-Ph-IM and $–OCH_3$-Ph-IM. The substituted electron-withdrawing group ($-NO_2$) of the aryl ring permitted an elevated Brønsted acidity that improve the yield of the reaction. Zhikai et al. (2014) mentioned N-acylation of amines to the corresponding acetamides using an heteropolyanion-based ionic liquid $[MIMPS]_3PW_{12}O_{40}$ that offers excellent Bronsted acidity and superior yield of product with reusability (Table 10.1, entry 3). Further, the developed protocol is said to be efficient and greener one.

Interestingly Shaohua et al. (2011) carried out the alkylation of p-cresol with the tert-butanol to synthesize 2-tert-butyl-p-cresol with the multiple-SO_3H functioned ionic liquid that showed excellent yield and selectivity with the operational simplicity,

Table 10.1 Ionic liquid catalysis for acylation and alkylation reaction at the prescribed reaction condition

No	Substrate	Product	Catalyst: Yield or conversion (%)	Refs.
1	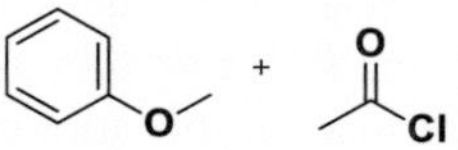	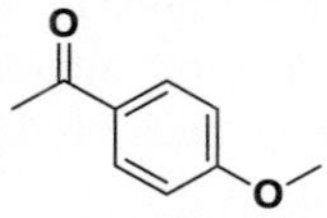	[BMIM]BF_4: 65.3 $AlCl_3$: 51.8	Cai (2014)
2	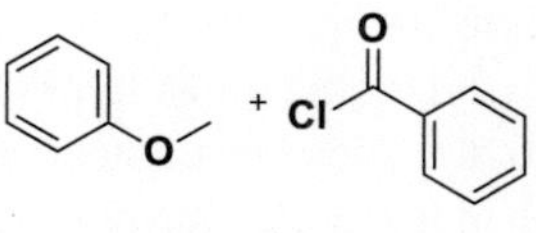	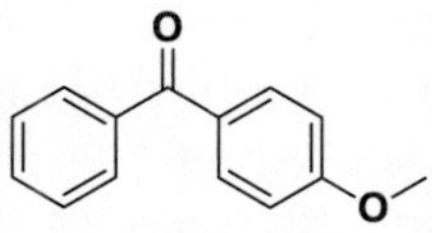	[X-Ph-IM]$FeCl_4$: X: NO_2: 78 X: OCH_3: 72 X:CH_3: 32 X: H: 57	Lin et al. (2020)
3	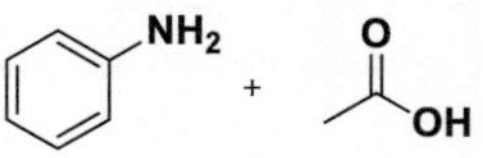	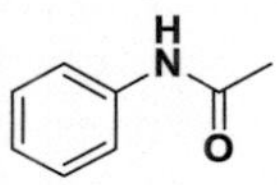	$[MIMPS]_3PW_{12}O_{40}$:95	Zhikai et al. (2014)
4	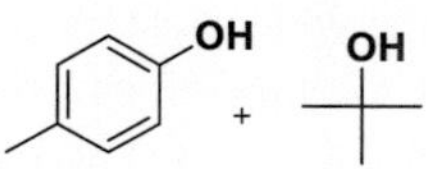	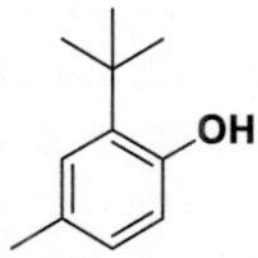	Multiple-SO_3H functioned IL:85.3	Shaohua et al. (2011)
5	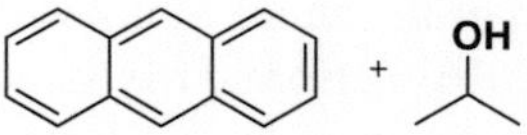	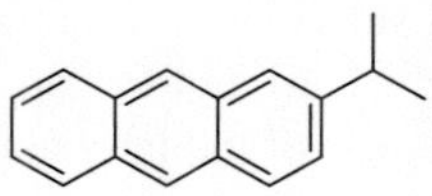	[EMIM]C1-$AlCl_3$:74.5 $AlCl_3$:52.3 $FeCl_3$:1.3	Chen et al. (2009)

high catalytic activity and stability as compared to respective traditional ionic liquids and Lewis acid catalyst (Table 10.1, entry 4). Similarly, Chen et al. (2009) reported Friedel–Crafts alkylation of anthracene to 2-isopropylanthracene with an excellent yield of 74.5% by using an ionic liquid [EMIM]C1-$AlCl_3$ (Table 10.1, entry 5). Whereas the analogous Lewis acids such as $AlCl_3$ and $FeCl_3$ displayed sluggish yield of the product. This might be attributed to the higher molar fraction of $AlCl_3$ in ionic liquid than that of sole use of the $AlCl_3$ which enhances the acidity and corresponding reaction progress. Moreover, this method also offered easy separation of product and recyclability of ionic liquid catalyst for clean and environmentally friendly strategy.

10.3.2 Amidation, Aminolysis and Transamidation

The amide bond is the most utilized functionality in the various kinds of pharmaceuticals and drug intermediates, polymers, biomedicine, herbal and neutracitical products (Fu et al. 2014, 2015a; Patil and Pratap 2016; Malhotra et al. 2008; Nakhate and

Yadav 2018; Muskawar et al. 2015). It has been known that, nearly one-fourth of the synthetic drugs like phenacetin, levitiracetam, armodafinil, acetazolamide, valsartan, diltiazem and captopril is having at least one amide bond functionality (Fu et al. 2014, 2015a; Patil and Pratap 2016; Malhotra et al. 2008; Nakhate and Yadav 2018; Muskawar et al. 2015). Recently, it is mentioned that about two-thirds of the primary reactions occupy amide formation in a commercial medicinal synthetic laboratory. Conventionally, amide bond formation is achieved by the reaction of aminolysis, amidation and transamidation reaction.

Amidation: Traditionally amidation reaction is performed with the activated carboxylic acid derivatives (esters, anhydrides and acyl halides) with primary/secondary amines (Fu et al. 2015a; Patil and Pratap 2016). These reactions possess the major disadvantages like lesser yield/selectivity, use of harsh reaction conditions, poor atom-economy, difficulty in down-streaming process, etc. (Fu et al. 2015a; Patil and Pratap 2016). In context to this, ionic liquid can be the better solution as the catalyst for amidation reaction. Fu et al. (2015a) proposed microwave-promoted phenyl amide synthesis catalyzed by heteropoly anion based ionic liquids $[PyPS]_3PW_{12}O_{40}$ (Table 10.2, entry 1). Practically, protocol is applicable to the wide substrate arrays with the moderate to excellent yields along with the reusable ionic liquid for synthesis. The reaction mechanism indicated that catalytic reaction initiated by the activation of $C = O$ of esters by use of sulfonic acid group possessing ionic liquid. Then both reactants tend to form the dipolar adducts that undergo the proton-exchange, dehydration or de-alcoholization and subsequent formation of the amide along with the regenerated catalyst. Interestingly Patil and Pratap (2016) performed synthesis of the monoethanol amide by amidation of the fatty acid ester using the biodegradable [CH][Cl] as an efficient ionic liquid catalyst that offers excellent yield and 100% selectivity (Table 10.2, entry 2). The choline-based ionic liquid can be well recognized as green and biobased ionic liquid. More green approach in this ionic liquid can be incorporated by the change of ionic liquid anion from halide to amino-acid.

Aminolysis: The aminolysis of esters or carboxylic acids functionality is very common fundamental organic reactions. Aminolysis reactions are carried out conventionally by homogeneous catalysis by use of metal halides, sulfonic acid and metal triflates (Malhotra et al. 2018; Nakhate and Yadav 2008). While some of the heterogeneous catalysts that are active at the higher reaction temperature involve the resins like amberlyst-15, modified silica/alumina, zirconia-silicate, graphene oxide and zeolites which may be utilized for the aminolysis reaction (Malhotra et al. 2008; Nakhate and Yadav 2018). However, this catalyst possesses drawbacks such as higher energy activation requirement, long working time of reaction, lesser conversion and moisture sensitivity. Malhotra et al. (2008) carried out an aminolysis of the epoxides using the ionic liquid 1-ethyl-pyridinium trifluoroacetate [EtPy] TFA under mild operating reaction conditions to offer higher yield of β-aminoalcohols (Table 10.2, entry 3). Moreover, the Nakhate and Yadav (2018) utilized graphene-oxide supported SO_3H-modified imidazolium-base ionic liquid as a proficient and reusable catalyst for regio-selective ring opening of styene oxide with aniline (Table 10.2, entry 4). In this case, the nucleophile attacks on epoxide carbon from the more substituted

Table 10.2 Ionic liquid catalysis for amidation, aminolysis, transamidation and halogenations reaction at the prescribed reaction condition

No	Substrate	Product	Catalyst: Yield or conversion (%)	Refs.
1	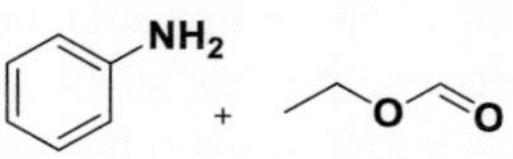	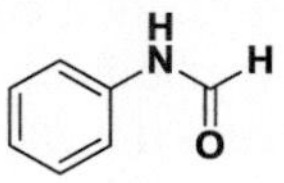	$[PyPS]_3PW_{12}O_{40}$:96 Without catalyst:33	Fu et al. (2015a)
2	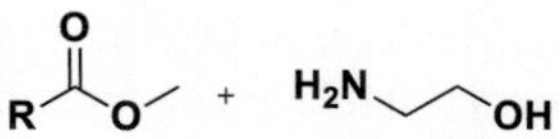	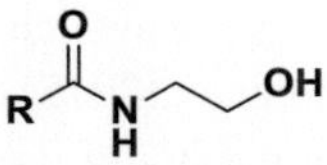	[CH]Cl: 98	Patil and Pratap (2016)
3	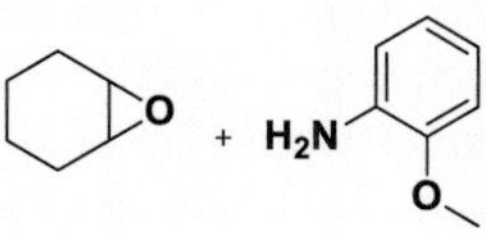	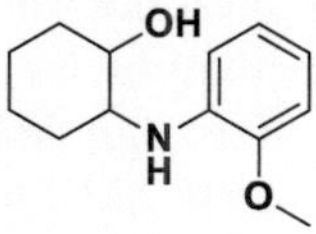	[EtPy]TFA: 90	Malhotra et al. (2008)
4	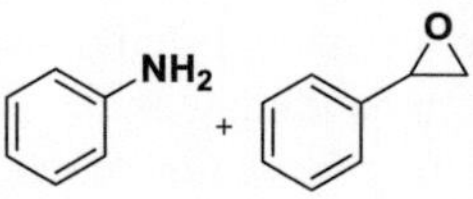	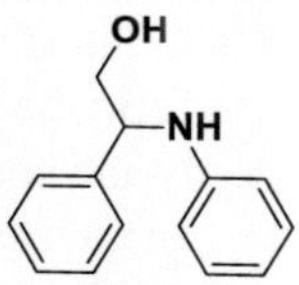	PTS-IM-3@GO:94 GO:29	Nakhate and Yadav (2018)
5	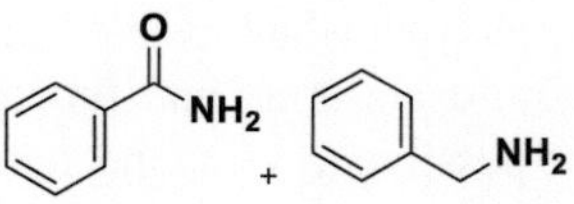	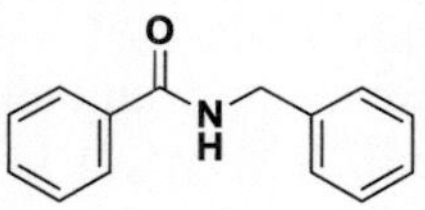	$[DAFBIM]ClO_4$: 93 $[DAFBiIM]BF_4$: 45 $LiClO_4$:24 Without catalyst:5	Muskawar et al. (2015)
6	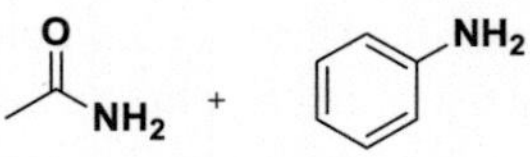	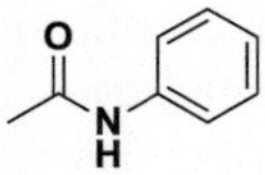	$[PyPS]_3PW_{12}O_{40}$:83 Without catalyst:0	Fu et al. (2014)
7	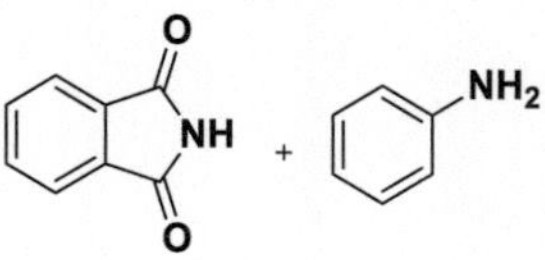	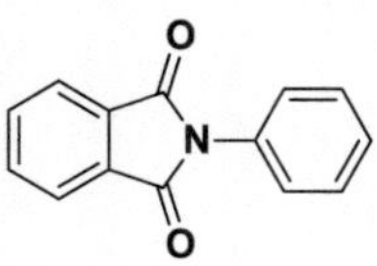	$[PyPS]_3PW_{12}O_{40}$:82	Fu et al. (2014)
8	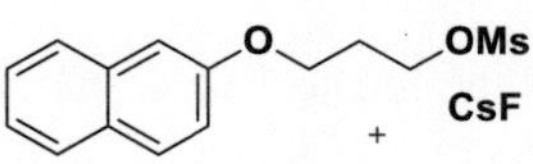	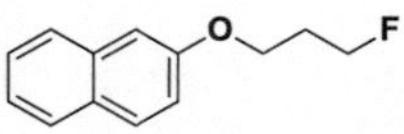	[MIM-t-BuOH]OMs: 97 [BMIM]OMs: 37 t-BuOH:22	Shinde and Lee (2008)

side which indicated formation of an electron deficient species during of ionic liquid catalysis.

Transamidation: Presently, the transamidation reaction is receiving enormous attention due to its attractive route for the synthesis of amide compounds as compared to that of the conventional routes applied for the amidation (Muskawar et al. 2015; Fu et al. 2014). These transamidation reactions demand harsh reaction operating conditions by using various metal-based homogeneous and heterogeneous catalysts such as transition metal triflates, metal acetates, metal oxides, sulfated tungstate and other organo-metallic compounds (Muskawar et al. 2015; Fu et al. 2014). These all are metal-consisting catalysts which possess disposing issues and severe environmental concerns. Such typical reaction conditions and highly unfavorable catalyst limits the reaction scope in synthesis of organic compounds at the larger scale. However, in search of the greener synthetic methodologies, the ionic liquids may find the great scope in the transamidation reaction (Muskawar et al. 2015; Fu et al. 2014). Very few research articles are available for the transamidation reaction catalyzed by ionic liquids. Muskawar et al. (2015) designed stable amide functionalized benzimidazolium-perchlorate ionic liquid [DAFBIM]ClO_4 for the transamidation of primary carboxamides, wherein various substrates acetamide, benzamide and p-nitrobenzamide showed excellent activity under mild operating reaction conditions (Table 10.2, entry 5). The catalyst offers significant thermo-stability up to 203 °C and reusability. Other catalysts such as [DAFBiIM]BF_4 and $LiClO_4$ offer sluggish yield may be attributed to lower acidity and lesser hydrogen bond formation ability than that of the ionic liquid [DAFBIM]ClO_4. Since reaction mechanism demonstrated that [DAFBIM]ClO_4 ionic liquid tends to form the hydrogen bonding with the amide functionality to initiate the catalytic activity and corresponding formation of reaction intermediate. Fu et al. (2014) mentioned an environmentally friendly procedure for the transamidation of carboxamides with various primary amines using the heteropolyanion-based ionic liquids $[PyPS]_3PMo_{12}O_{40}$ as catalysts which offered operational simplicity, potential recyclability with broad substrate array and functional group tolerance features (Table 10.2, entry 6,7). In the possible reaction mechanism, the ionic liquid catalyzed cycle for transamidation reaction started with activation of $C = O$ group of amide moiety through coordination with the acidic ionic liquid. In the later stage, this activated amide is found to much reactive to nucleophilic attack of amine that involves the possible proton-exchange and succeeding removal of leaving group amine to produce new amide.

10.3.3 Halogenation

The fluorinated moieties played a very crucial role in the medical field, oncology, molecular imaging area, radio-pharmacy and determination of the reaction mechanism in the organic chemistry (Shinde and Lee 2008). Nucleophilic halogenation says fluorination by use of the alkali metal fluorides (using the conventional fluoride sources) is widely used in organic synthetic transformation. Conversely, the partial solubility and lower nucleophilicity restrict the use of alkali metal fluorides in the organic synthesis. Apart from this, these reactions demand the harsh operating condition, expensive, skillful developed phase transfer catalyst, quaternary ammonium fluorides, crown ethers and/or polymeric supported immobilized active catalyst (Shinde and Lee 2008). These catalysts offered lower solubility, lower yield-selectivity, critical solvent selectivity and catalyst disposable issues. Interestingly, imidazolium ionic liquids are reported to offer excellent synthetic catalytic activity as well as optional reaction media to improve the dynamic activity of various alkali metal fluorides for the fluorination reaction. Shinde et al. (2008) carried out nucleophilic fluorination catalyzed by functionalized hybrid ionic liquid [BMIM-t-BuOH]OMs that enhance the nucleophilic reactivity of fluoride anion with higher selectivity to yield 97% of desired product (Table 10.2, entry 8). The synergistic effect of the functionalized ionic liquid not only offered high chemo-selectivity in the fluorination reaction but also boosted nucleophilic reactivity of anion fluoride.

10.3.4 Addition Reaction

The Diels–Alder reaction in the organic synthesis has received enormous attention due to its wide substrate applicability (Meracz and Oh 2003; Zheng et al. 2010). It is used to synthesize six-member carbocyclic organic moieties in a single concerted step for the synthesis of naturally occurring and biologically active moieties (Meracz and Oh 2003; Zheng et al. 2010). Although these reactions are well carried out by the heat or light in non-polar media, still these reactions find great scope in the organic chemistry due to yield and selectivity or product stereochemistry issues (Meracz and Oh 2003; Zheng et al. 2010). Hence, various kinds of reaction medium also tested like water, organic polar solvents, surfactants and lithium amides. However, in recent years, ionic liquids have not only been used as catalysts but also acted as a solvent for the Diels–Alder reactions that can offer better yield and selectivity than conventional methodology. Meracz and Oh (2003) carried out asymmetric Diels–Alder synthesis reaction in the ionic liquid [HBUIM]BF_4 (Table 10.3, entry 1). The improved reaction rate, excellent stereoselectivity and enantioselectivity of the Diels–Alder reactions are attributed to imidazolium based ionic liquid at room temperature. As compared to this, the conventional catalyst/solvent $ZnCl_2$/ether system offered very sluggish yield of the product. Similarly, Zheng et al. (2010) proposed Diels–Alder reaction of asymmetric aza compounds catalyzed by ionic liquid [EMIM]Pro to offer yield up to 93% of desired product along with six reuses and without loss of significant catalyst activity of ionic liquid catalyst (Table 10.3, entry 2). The anionic as well as cationic portion of ionic liquid greatly offers feasible catalytic molecular environments for the Diels–Alder reaction. Since, ionic liquid as catalyst can play important role in maintaining an effective dipole moments, electrostatic force and Lewis acidity that favors the Diels–Alder reaction (Meracz and Oh 2003; Zheng et al. 2010). The Diels–Alder reactions broadly examined in different kind of solvents which indicating that, solvent having hydrogen-bond donating capacity and cohesive solvent feature greatly affect kinetic as well as stereo-chemical fate of the Diels–Alder reactions & product. Thus, ionic liquids can greatly improve the selectivity of Diels–Alder product (Meracz and Oh 2003; Zheng et al. 2010).

Table 10.3 Ionic liquid catalysis for addition, elimination and etherification reaction at the prescribed reaction condition

No	Substrate	Product	Catalyst: Yield or conversion (%)	Refs.
1	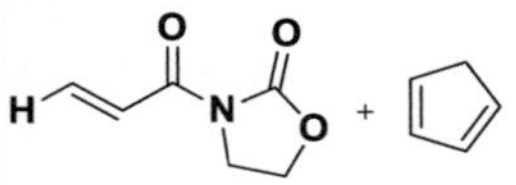	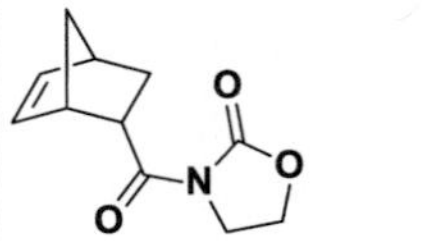	[HBUIM] BF_4: 65 $ZnCl_2$/ether:13	Meracz and Oh (2003)
2	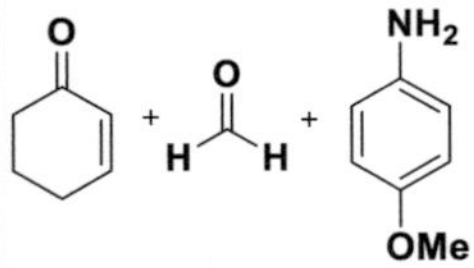	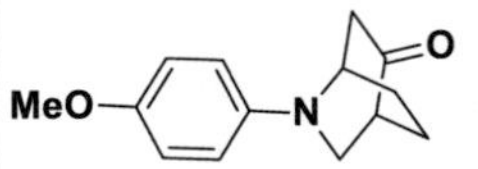	[EMIM]Pro: 93	Zheng et al. (2010)
3	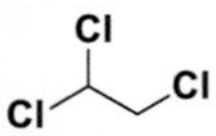	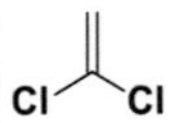	[Ph_4P]Cl: 99.5	Zhang et al. (2019)
4	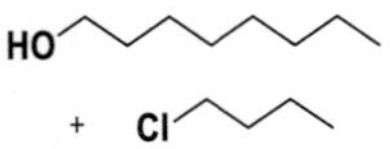		[Bu_4N]Br: 96 Blank:4	Bender et al. (2010)
5	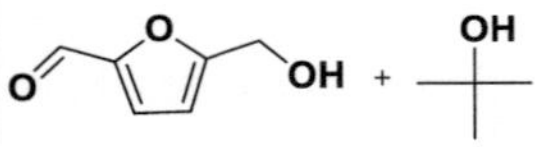	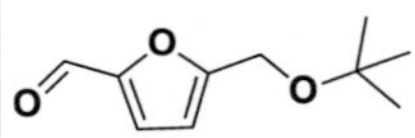	[C_4C_1IM]$CrCl_4$-H β- zeolite:46	Salminen et al. (2013)
6	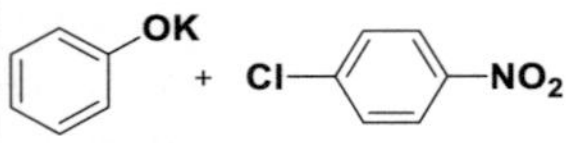	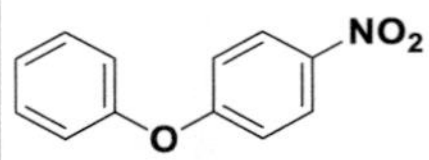	[TDHP]Br: 95	Yadav et al. (2008)

10.3.5 Elimination Reaction

The most commonly referred organic reaction is the elimination reaction wherein the polar/non-polar solvent and strong acid/base catalyst plays a vital role in the reaction and governs the fate of the product, yield and selectivity (Schindl et al. 2019; Sekharan et al. 2021). Conventionally, these reactions required higher reaction temperature and acidic/basic reagents to carry out the reaction. Most of these reactions are carried out in the homogenous reaction medium with acid or base wherein product yield/selectivity ratio greatly affected (Zhang et al. 2019). Moreover, use of the large quantity of the acid and bases for the elimination reaction finds the down streaming, recyclability and other important ecological issues like disposal (Badgujar et al. 2022b). Further, use of the inappropriate or weak base may tend to offer mixed elimination and substitution product. Ionic liquids can be the better solution for all above concerns to carry out the elimination reaction. Zhang et al. (2019) studied catalytic performance of the supported ionic liquid catalysts Tetraphenylphosphonium chloride [Ph_4P]Cl for the dehydrochlorination reaction of 1, 1, 2-trichloroethane to generate 1, 1-dichloroethylene with 99.5% conversion (Table 10.3, entry 3). Further, they claimed that catalyst remains stable for 1100 h in the chlorinated reaction medium. In this case, the reaction follows elimination unimolecular conjugate base reaction mechanism catalyzed by the ionic liquid $[Ph_4P]Cl$. The function of the anion chloride is to accept the proton while cation accepts the departing leavening group. The anion Cl abstract the highly acidic α-hydrogen atom of 1, 1, 2-trichloroethane and then causes breaking of the C–H bond, later on ionic liquid cation co-ordinate with the β-carbon attached chlorine atom and causes removal of it with formation of double bond (Zhang et al. 2019).

10.3.6 Etherification Reaction

The ether functionality has an enormous broad range of applications in synthetic organic chemistry (Bender et al. 2010; Salminen et al. 2013; Yadav et al. 2008). Ethers are widely used in various industrial sectors such as agrochemical sector for the synthesis of herbi- and fungicides, pharma-sector, food-industries, synthetic chemical industries, personal care products such as perfumery, fragrances, domestic appliances products like disinfectants/cleaning agents, precursors for polymers/plastic industries, chemical engineering and mechanical engineering sector (Bender et al. 2010; Salminen et al. 2013; Yadav et al. 2008). These ethers are synthesized by dehydration reaction using homogeneous Bronsted acids and Lewis acid catalysis wherein various water molecules are liberated (Bender et al. 2010; Salminen et al. 2013; Yadav et al. 2008). These methods possess several drawbacks such as deactivation of the catalyst, leaching, difficult down-streaming process and no recyclability. These difficulties inverse the process costing and demands for the suitable media that can be used as solvent as well as catalyst (Bender et al. 2010; Salminen et al. 2013; Yadav et al. 2008). In view of this, the ionic liquid can be a suitable option that can be used as catalyst as well as solvent for the etherification reaction.

Bender et al. (2010) tested various conventional phase-transfer catalysts, for *O*-alkylation reaction. Among all tested ionic liquids, quaternary tetra-butyl ammonium catalysts $[Bu_4N]Br$ offered better yield of the product *O*-alkylation than immidazolium [EMIM]Br, pyridine [4-MBP]Br and pyrrolidinium base ionic liquids (Table 10.3, entry 4). This may be attributed to the reason that, catalyst like $[Bu_4N]Br$ with fixed positive charge displayed high conversion rate for etherification reactions, while catalysts involving with aromatic π- system were not so active due to resonance. However, without catalyst reaction, it offered sluggish yield.

Salminen et al. (2013) used ionic liquid functionalized with H-Beta zeolites $[C_4C_1IM]CrCl_4$-H beta- zeolite) for the etherification of 5-hydroxy-methyl-furfural into the tert-butoxy-methylfurfural that offered the conversion of 46% (Table 10.3, entry 5). The ionic liquid demonstrated excellent selectivity and improved catalytic surface area due to hybrid modification of H-Beta zeolites with ionic liquid. Yadav and Motirale (2008) studied various catalytic synthesis of the p-nitro-diphenyl-ether from p-nitro-chloro-benzene using potassium phenoxide (Table 10.3, entry 6). The ionic liquid Tetradecyl-(trihexyl)-phosphonium-bromide [TDHP]Br was reported to be the most suitable catalyst that offers almost 95% conversion rate as compared to polyethylene glycol, ethyl (trihexyl) phosphonium bromide, tetradecyl (trihexyl) phosphonium-chloride and tetradecyl (trihexyl) phosphonium decanoate. Also, this ionic liquid offered exceptional selectivity, stability and recyclability in reaction. However, the blank reaction without catalyst just offered traces of the product. These etherification reactions catalyzed by ionic liquid have been widely known as phase transfer catalysis that help to migrate anionic reactant from one liquid phase to another liquid phase (Yadav et al. 2008).

10.3.7 Condensation Reaction

Condensation reactions are found to have very wide applications that are useful in organic combinatorial synthetic routes. Nevertheless, the effectiveness of the condensation reactions is governed by specific electronic as well as steric crowding surrounded in the reactants (Khaligh 2012; Dong et al. 2008; Zhao et al. 2004, 2013; Wang et al. 2014a). Various kinds of condensation reactions are reported in literature to synthesize different kinds of organic specialty moieties. Most condensation reactions are performed by the Lewis or Bronsted acids or by using mineral acids or bases which possess several issues related to recyclability and product separation (Khaligh 2012; Dong et al. 2008; Zhao et al. 2013, 2004; Wang et al. 2014a). Ionic liquid can be the better solution to overcome the above-mentioned issues for condensation reaction.

Pechmann reaction: Coumarin derivatives have attracted significant attention because of their wide synthetic application in various industrial sectors such as food, agro-chemicals, cosmetics, insecticides, optical brightening and dispersed fluorescent agents (Khaligh 2012). Moreover, these coumarin derivatives find scope of specialty organic chemical synthesis such as fluorocoumarins, coumarones and 2-acyl-resorcinols. In medicinal chemistry also, coumarin derivatives display a significant role as anti-thrombotic, anti-cancer, anti-coagulant, anti-HIV agents, estrogen-like effective drugs. Coumarins can be produced by different synthetic routes which involve the Perkin reaction, Knoevenagel reaction, Reformatsky reaction and flash vacuum-pyrolysis reaction. Among these reactions, Pechmann reaction is extensively used as it utilizes the simplest and cheaper starting material such as phenols and β-ketoesters. Khaligh (2012) utilized 3-Methyl-1-sulfonic acid imidazolium hydrogen sulfate [MSIM]HSO_4 for the synthesis of coumarin products via Pechmann condensation reaction between the resorcinol and ethyl acetoacetate that offers the yield of 94% (Table 10.4, entry 1). The absence of the ionic liquid did not lead to the reaction. Moreover, the ionic liquid was simply well reused without loss of significant catalyst activity. The ionic liquid in this reaction acts as the promoter and makes the $C = O$ group more prone toward the nucleophilic attack by resorcinol and for further dehydration reaction.

Table 10.4 Ionic liquid catalysis for condensation reactions at the prescribed reaction condition

No	Substrate	Product	Catalyst: Yield or conversion (%)	Refs.
1	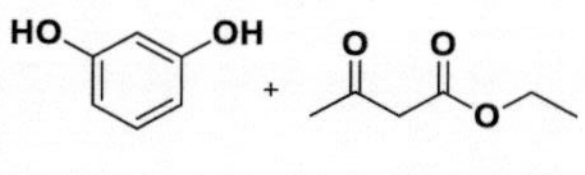	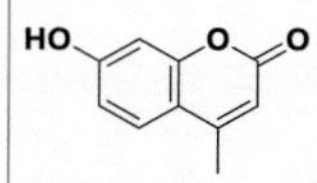	[MSIM]HSO_4: 94 Without catalyst:0	Khaligh (2012)
2	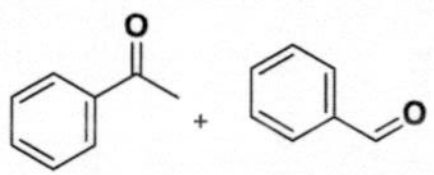	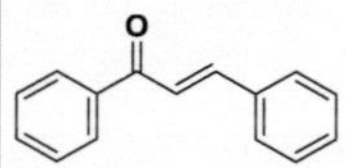	[TMPSA]HSO_4: 95 Without catalyst:5	Dong et al. (2008)
3	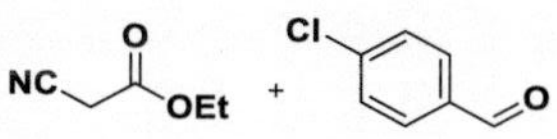	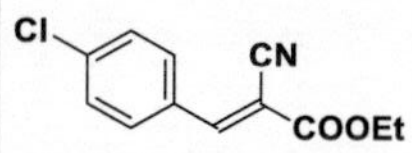	HMTA–AcOH–H_2O:97	Zhao et al. (2013)
4	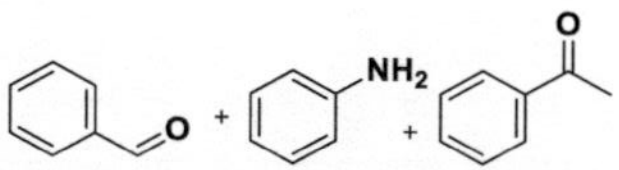	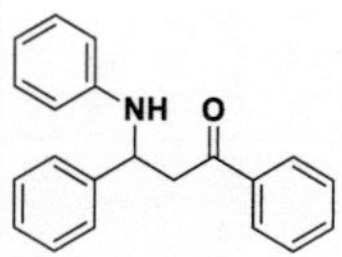	[HMIM]TFA:83 [BMIM][HSO_4]:0	Zhao et al. (2004)
5	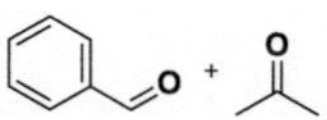	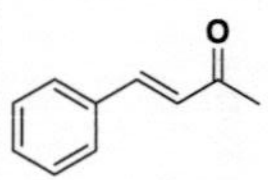	[N_{2222}]$EtNHC_3SO_3$: 98 [P_{4444}]IM:88 [P_{4444}]ARG:87	Wang et al. (2014a)

Claisen–Schmidt condensation reaction: Chalcones and its various derivatives have received significant importance due to their pharmacological uses. These chalcones are acting as the essential precursor for the natural product synthesis such as flavonoids which displayed great biomedicinal activities as anti-inflammatory, anti-malignant and anti-hyperglycemic action (Dong et al. 2008). Conventionally, these chalcones synthesized by the Claisen–Schmidt condensation using basic/acidic reaction media in the homogeneous reaction conditions. Dong et al. (2008) used reusable SO_3H-functionalized ionic liquids [TMPSA]HSO_4 as catalysts for the production of the chalcones through the Claisen–Schmidt condensation reaction that offers 95% yield, whereas without catalyst trace yield was reported (Table 10.4, entry 2). Moreover, the chalcones can be simply separated through decanting from catalysts that can be easily allowed catalyst reusability numerous times without affecting catalytic activity. Thus, the immiscibility of chalcone compounds makes it easy for separation during work-up procedure.

Knoevenagel condensation reaction: The Knoevenagel condensation is a useful transformation for the production of electrophilic olefins using an active methylene and carbonyl moieties (Zhao et al. 2013). It is useful for the production of various organic moieties such as alkenes, coumarin that is widely used in the personal care products, cosmetics, perfumeries and drug chemical synthesis. Conventionally, Knoevenagel reaction performed in harmful volatile organic media with sophisticated catalytic system. However, use of the ionic liquid may be useful to perform Knoevenagel condensation reaction with role of catalyst in addition to solvent.

Zhao et al. (2013) reported application of the protic-ionic-liquid solvent–catalytic (HMTA–AcOH–H_2O) system for Knoevenagel condensation reaction involving reaction of aromatic aldehydes with ethyl 2-cyanoacetate that offered 97% yield whereas, the isolated systems (AcOH–H_2O) or (HMTA–H_2O) offered sluggish yield (Table 10.4, entry 3). Furthermore, these HMTA–AcOH–H_2O solvent–catalyst systems worked smoothly, cleanly and can be easily recyclable with significant retention of the catalytic activity and hence recognized as eco-friendly in ongoing revolution of chemical processes. According to the reaction mechanism, the protic-ionic-liquid promotes the protonation, removal of proton of active methylene group, attack on carbonyl carbon and subsequent elimination reaction.

Mannich reactions: Mannich reactions possess great importance in the production of the medicinally active nitrogenous compounds. Moreover, most of the natural products can be well synthesized artificially by starting with the Mannich reaction that offered the β-amino carbonyl compounds (Zhao et al. 2004). Zhao et al. (2004) tested various Brønsted acidic ionic liquids for the multicomponent Mannich reaction of an aldehydes, amines and ketones in which [HMIM]TFA offered maximum yield of the Mannich product with considerable reusability and without substantial loss of activity (Table 10.4, entry 4). The author mentioned several advantages of the use of ionic liquid protocol such as improved yield, elevated reaction rate, easy preparation of ionic liquid, inexpensive ionic liquid, easy recyclability and greener approach than conventional methodologies.

Aldol condensation reaction: Aldol condensation reaction is the most effective reaction for the formation of the C–C bond which offered α, β-unsaturated carbonyl compound (Wang et al. 2014a). These compounds are most commonly observed and extracted as polyol architecture from natural products (Wang et al. 2014a). Conventionally, these compounds can be well synthesized by the use of strong acidic or basic catalysts in the conventional organic solvents which make the process non-green by means of environmental concern of disposing of catalyst and organic solvent (Wang et al. 2014a). Moreover, applying the strong acid/base causes the reverse and side reactions that decrease the effective yield of desired product. In view of this, the ionic liquid may be the most helpful catalyst as well as solvent that can be utilized for the aldol condensation reaction.

Wang et al. (2014a) tested various quaternary ammonium and tetrabutylphosphonium ionic liquids for aldol condensation reactions between the aromatic aldehydes with acetone under microwave irradiation (Table 10.4, entry 5). The aldol condensation reactions carried out with quaternary ammonium ionic liquid catalyst $[N_{2222}]EtNHC_3SO_3$ displayed excellent yield (98%) than tetrabutylphosphonium ionic liquids catalyst with excellent recovery and reusability of the ionic liquid catalyst. Further, use of the ionic liquid provides the ionic environment to carry out the reaction in the microwave irradiation that decreases the long-hour working reaction time. Moreover, the author claimed to extend the same catalyst applications for the Henry reaction also.

10.3.8 Esterification, Transesterification and Ester Production

In the organic synthesis, an esterification reaction has received very wide importance due to its applications in the various major industrial sectors such as home-care products, cosmetic, toiletries, pharma and flavor-fragrance-food-beverages (Zhang et al. 2018; Yang et al. 2010; Fan et al. 2013). The extraction to get ester compounds from the herbal smelling resources possesses severe drawbacks like sluggish yield with high processing cost (Badgujar and Bhanage 2014a, 2015b, c). Traditionally, these essential esters are manufactured by reaction of esterification/ transesterification reaction wherein strong mineral acids/bases at elevated temperature range from 150 to 200 °C are utilized (Badgujar and Bhanage 2014b; Badgujar et al. 2017). These traditional methodologies having numerous shortcomings like use of hazardous chemical catalyst, sluggish yield-selectivity, lesser reaction rate, corrosion to reactors, heavy down streaming process, side-product impurity and environmental issues (Badgujar and Bhanage 2014a, b, 2015b, c; Badgujar et al. 2017). While looking to these disadvantages, ionic liquid may be a possible sustainable solution for the production of esters. Zhang et al. (2018) performed esterification reaction of acetic acid and benzyl alcohol catalyzed by the acidic ionic liquid functionalized with ethyl bridged organo-silica hollow nanospheres $[C_3IM][SO_3CF_3]$-Si(Et)Si HNSs

(Table 10.5, entry 1). This catalyst has nanometer sites, hollow opening with large BET surface area that can be permeable for the reactants such as acetic acid/ benzyl alcohol. Moreover, the catalyst also offered excellent catalytic activity-stability, efficient mass transfer, reusability with minimum leaching and negligible modification of nanohybrids (Zhang et al. 2018). The report mentioned that, Brønsted acidic site, density of ionic liquid and textural properties of ionic liquid-nanohybrids influence the esterification reaction activity in forward direction.

Dimethyl carbonates receive significant attention due to its spectacular features of safeness, non-corrosiveness and as eco-friendly building block chemical for production of the polymeric materials like polycarbonate (Yang et al. 2010). It has also been applied as an electrolyte for the lithium-ion batteries because of its high dielectric constant. Furthermore, it is also used as a fuel blending agent to boost the octane number that can reduce particulate emission, other harmful oxides hence

Table 10.5 Ionic liquid catalysis for ester production and multicomponent reactions at the prescribed reaction condition

No	Substrate	Product	Catalyst: Yield or conversion (%)	Refs.
1	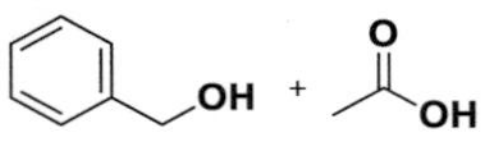	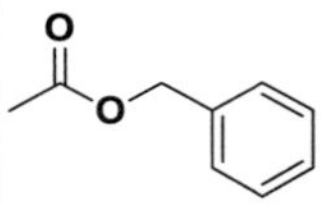	$[C_3IM][SO_3CF_3]$-Si(Et)Si HNSs:85 Amberlyst-15:70 Blank:30 Zeolite:25	Zhang et al. (2018)
2		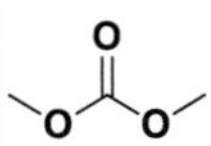	$[C_4DABCO]OH$: 81 Without catalyst:6	Yang et al. (2010)
3	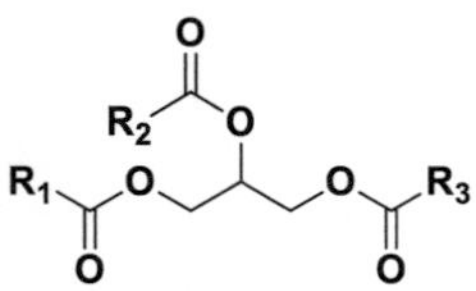	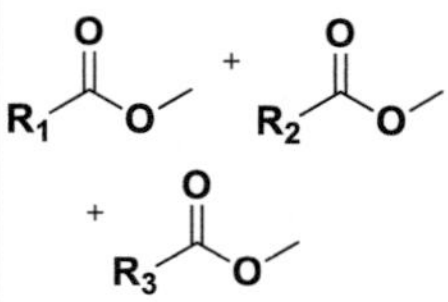	[Ch][OH]:95	Fan et al. (2013)
4	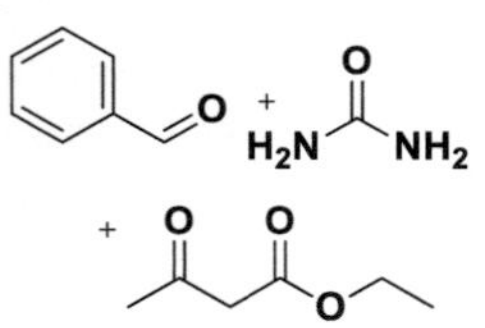	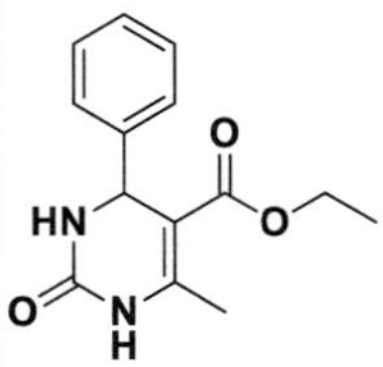	$[PyPS]_3PW_{12}O_{40}$:95 Without catalyst:8	Fu et al. (2015b)
5	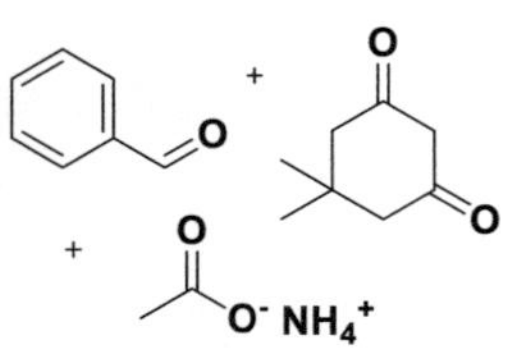	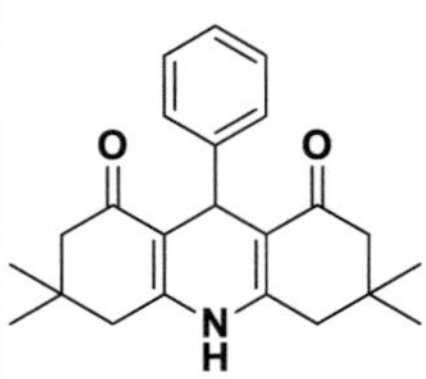	[HBET]LAC:90 Betain:50 [Choline][LAC]:25 Without catalyst:0	Zhu et al. (2017)

received significant attention in oil refinery and chemical industries (Yang et al. 2010). Conventional synthesis of it involves the use of hazardous and explosive gas process like methanolysis of the phosgene and carbonylation of methanol that restrict the potential synthesis of the dimethyl carbonate. More recently, some ionic liquids have been reported to offer competitive dimethyl carbonate yield with safer process and greener protocols. Yang et al. (2010) synthesized dimethyl carbonate using DABCO-derived basic ionic liquids [C_4DABCO]OH with the excellent catalytic activity (81% yield) under gentle reaction conditions (Table 10.5, entry 2). However, without catalyst trace yield was reported. Fan et al. (2013) revealed use of the [Ch]OH as a suitable reaction catalyst for the biodiesel production with several reuse and excellent catalytic activity (Table 10.5, entry 3). The ionic liquid [Ch]OH offered better reactivity in the methanol medium for the biodiesel production. The use of the ionic liquid has several benefits for biodiesel production such as it initiates the reaction, offered sufficient sites for the catalysis, recyclability and better product separation from the reaction media.

10.3.9 Multicomponent Reaction

Multicomponent reactions gained a wide attention in the synthetic organic chemistry due to the combinatorial synthesis and molecular diversity (Fu et al. 2015b; Zhu et al. 2017). These reactions endorse the formation of several new bonds in one step that can offered several advantages like simple operating procedure, multibond formation ability in one-step, high efficiency, time as well as and energy saving that can minimize waste generation (Fu et al. 2015b; Zhu et al. 2017). Considering the industrial applications, these reactions play an effective role for the construction of drug molecule libraries with structurally diverse moieties that can have applications in designing of new synthetic drugs (Fu et al. 2015b). Various catalytic systems involving Brønsted acids/ bases, metal complexes, Lewis acids, organo-catalysts and heterogeneous catalysts or synthetic chemical regents are widely applied for the multicomponent reactions. These may involve the expensive/harmful catalysts/ reagents with stoichiometric amount that provided sluggish yields, purification issues and use of the huge amount of volatile organic solvents and harsh reaction conditions. The development of green, clean and eco-friendly synthetic protocol still finds great scope in the synthesis of multicomponent reactions (Zhu et al. 2017). The unexpected properties of ionic liquid make it facile to use as reaction medium/catalyst for multicomponent reactions.

Fu et al. (2015b) proposed an efficient, robust, recyclable path for the production of dihydropyrimidinone compound via Biginelli reaction that is catalyzed by heteropolyanion-base ionic liquids $[PyPS]_3PW_{12}O_{40}$ which was found to be applicable for the diverse kind of the substrates (Table 10.5, entry 4). The reaction involves imine bond formation initiated by ionic liquid from an aldehyde and urea, that following the addition of nucleophile carbanion obtained from 1,3-dicarbonyl compounds by cyclodehydration to give desired product. Zhu et al. (2017) studied various betainium-based ionic liquids for the catalytic synthesis of the Hantzsch reactions for the production of acridinediones (Table 10.5, entry 5). The H-betainium lactate [HBET]LAC ionic liquid offered maximum catalytic activity in gentle reaction conditions as compared to that of Betain and [Choline]LAC ionic liquid. In this case, the hydrogen bond donor ability of ionic liquid tends to surge the enol content and initiate the reaction. These ionic liquids are non-expensive, simple to fabricate and displayed excellent industrial potential utility considering the biocompatibility and recyclability.

10.4 Ionic Liquid Catalysis for Biomass Valorization

Ionic liquids can be able to dissolve the lignocellulosic biomass that offered several novel prospects for the generation of value added as well as energy chemicals with potential of renewability (Badgujar et al. 2019; Badgujar and Bhanage 2015a; Olivier-Bourbigou and Morvan 2010). The lignocellulose biomass feedstock is an optional and sustainable energy resource for the future energy security of the nation. These lignocelluloses can be well utilized for the biotransformation of fermenting sugars into the potential sustainable platform chemicals using catalytic transformations (Fig. 10.2) (Wang et al. 2014b).

Ionic liquid can be the most suitable solvent as well as catalyst that can be used for the transformation of the cellulose, lignin, hemicelluloses to platform chemicals. Use of the ionic liquid for the biomass transformation can be offered several benefits such as:

- Highly crystalline nature of the cellulose makes it more unwilling for dissolution and subsequent depolymerization for chemical transformation. The solubilization of biomass is a great challenge in the conversion of the biomass that can be well solved by the use of the ionic liquid. Ionic liquid disrupts the H-bonding

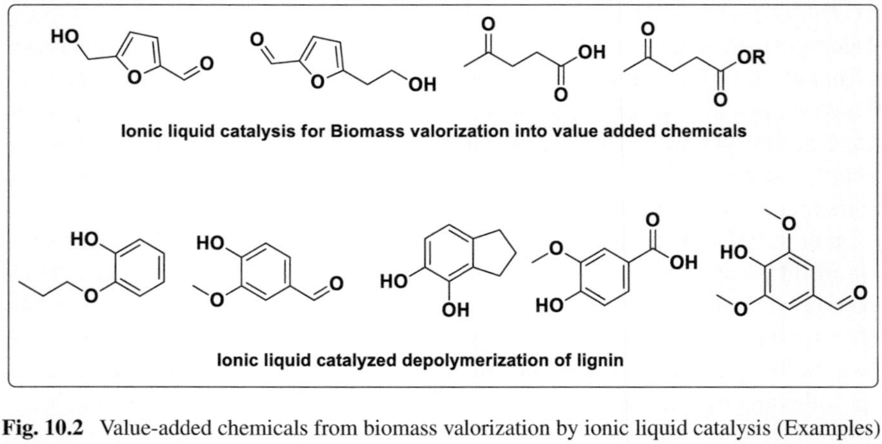

Fig. 10.2 Value-added chemicals from biomass valorization by ionic liquid catalysis (Examples)

of cellulosic network, hydrogen bonding of lignin monomer, related β-glycosidic bonds, β-O-4 ether linkage and π-π stacking interaction (Badgujar and Bhanage 2015a).

- Moreover, the ionic liquids can be worked as the suitable reaction medium for the biomass transformation. These ionic liquids can be well utilized as catalyst also and performed dual role in the biotransformation (Badgujar et al. 2019; Badgujar and Bhanage 2015a; Olivier-Bourbigou and Morvan 2010).
- Besides this, the lignocellulosic conversion in ILs can be performed at milder reaction conditions that can save energy requirement and help to make more sustainable modern society (Badgujar et al. 2023).

Thus, the ionic liquid can have a remarkable role in the treatment of the lignocellulosic biomass and further transformation. Conversely, there are numerous challenges associated with use of the ionic liquid for the biomass processing that requisites hard core research to address it. Kumar et al. (2021) performed catalytic transformation of glucose into the levulinic acid by using the 2-phenyl-2-imidazoline-based ionic liquid $[C_4SO_3HPHIM]H_2PO_4$ that offered 98% glucose conversion to levulinic acid along with four recycles. Guo et al. (2020) mentioned that glucose can be efficiently transformed into the 5-HMF and 5-EMF by using Brønsted–Lewis acidic di-cationic ionic liquids with an excellent yield. Mehata et al., (2021) reported room temperature lignin depolymerization by use of protic and metal based ionic liquid $[PRO]_2MnCl_4$ to obtain value product such as vanillin and other possible proposed products such as syringaldehyde, propanolcatechol, vanillic acid and ethylguaiacol.

10.5 Ionic Liquid's Physico-Chemical Properties for Catalysis

Over the last two decades, the publication/articles of the ionic liquid catalysis have increased dramatically which is indicating the potential use of the ionic liquid for the catalysis. Basically, in catalysis, three things are important involving the dissolution of starting compound, end product separation and precise catalyst–substrate interaction. Apart from this, several physico-chemical properties are responsible for the efficient role of ionic liquid in catalysis which is listed below (Wilkes 2004; Greaves and Drummond 2008; Wilkes 2002; Zhen et al. 2012; Hallet et al. 2011; Badgujar et al. 2021; Wagh et al. 2016; Chiappe and Pieraccini 2005).

Melting point and vapor pressure: The lower end of liquid range is mentioned by the melting point of the ionic liquid while these ionic liquid possesses the wide liquid range that offers suitable working conditions. The low vapor pressure of the ionic liquid offers suitability to use at higher temperature condition that causes the ionic liquid to be very striking than conventional solvents (Schindl et al. 2019; Poole and Atapattu 2021).

Density: The density of an ionic liquid is governed by the type of cation/anion used. It reduced when bulkiness of the organic cation increases. Further, minor structural change tends to offer ionic liquid with desirable density. Density can play an important role in the extraction process wherein ionic liquid is used as one extracting solvent. Since ionic liquid presents the miscibility gap with water. Further, density can also tends to play an essential role in biphasic catalysis (Badgujar and Bhanage 2015a; Olivier-Bourbigou and Morvan 2010; Barrueto et al. 2022).

Viscosity: Most of the ionic liquids are viscous in nature and tend to cause the problem in substrate dissolution as well as effective mass transfer. The viscosity of an ionic liquid is majorly attributed to hydrogen bond formation and van der Waals forces. Viscosity can be reduced appropriately by the rise of reaction temperatures (Olivier-Bourbigou and Morvan 2010; Greaves and Drummond 2008).

Polarity and solvation strength: The polar solvent has better solvation ability. The solvation ability depends on the charge separation of the polar species. The ionic liquid is associated with cation and anion that offers solvation of active charged species or reactive intermediate of reaction. Dielectric constant, dipole moment and polarizability are correlated with polarity and solvation ability (Schindl et al. 2019; Wilkes 2002).

Thermal decomposition: Thermal decomposition can be mentioned as the possible upper limit in liquid range stability. Generally, ionic liquid possesses the sufficient thermal stability measured by thermal gravimetric analysis (TGA) for the reaction chemistry. Ionic liquid with higher TGA onset temperatures is said to be more stable compared to lower onsets. Further higher TGA onset temperature offers long-term stability at elevated temperature (Wilkes 2004; Greaves and Drummond 2008).

Empirical solvent parameters: Three different models are utilized to determine the solvation ability of the ionic liquid, namely the Hansen solubility model, Kamlet–Taft polarity and COSMO-RS86 model. The Kamlet–Taft model is broadly used to explain the possible solvation phenomenon that consists of the term α (H-bond acidity/Hydrogen bond donor), β (H-bond basicity/Hydrogen bond acceptor) and π*(dipolarity/polarizability). These α, β, π* well determined by the UV–VIS spectroscopy. These parameters are responsible for the solute–solvent interaction involving the solubility and solvation phenomenon. These terms revel polarizability of ionic liquids, hydrogen bond acidity/basicity, interface of pi-bonding and non-bonding electrons, cohesive and bond dispersion exchanges. Ionic liquids possessing the anion [OAc], [OFo], [Cl] having the β value more than 0.82 showed better polarity, H-bonding ability and dissolution ability while anions like [BF_4], [PF_6] are displaying the poorer β value that may affect solvation phenomenon due to lesser H-bonding ability (Badgujar et al. 2022a).

10.6 Future Outlook and Conclusion

The revolution of the use of the ionic liquid for organic catalytic synthesis is started from the recognition that ionic liquid can be used as reaction media or solvent. However, later ionic liquids are recognized as effective catalysts for the organic synthesis. Various physico-chemical features such as wide liquid range, designing of ionic liquid, solvation ability, dissolution ability, supporting tendency with better surface area of the ionic liquid are accountable for the catalytic applications. The present book-chapter summarized various types of examples for organic synthesis that can be performed by the use of ionic liquid catalyzed reactions along with plausible reaction mechanism and justification. However, for the large-scale practical applications from the laboratory to industrial scale-up synthesis, there is a need to address several challenging issues such as synthesizing of ionic liquid from the renewable resources that can offer the more sustainable way. Designing of the biodegradable and non-toxic ionic liquid along with the environmental impact assessment is still needed for the industrial regulatory affairs to clear granting of ionic liquid use at large scale. Moreover, the infrastructural updating with corrosion-free large-scale equipment/reactors at the industrial large scale will be challenging.

Acknowledgements This work is not supported by any kind of funding.

Declaration of Interest None

References

Badgujar, K.C., Bhanage, B.M.: Synthesis of geranyl acetate in non-aqueous media using immobilized Pseudomonas cepacia lipase on biodegradable polymer film: Kinetic modelling and chain length effect study. Process. Biochem. **49**(8), 1304–1313 (2014)

Badgujar, K.C., Bhanage, B.M.: Application of lipase immobilized on the biocompatible ternary blend polymer matrix for synthesis of citronellyl acetate in non-aqueous media: kinetic modelling study. Enzym. Microb. Technol. **10**(57), 16–25 (2014)

Badgujar, K.C., Bhanage, B.M.: Immobilization of lipase on biocompatible co-polymer of polyvinyl alcohol and chitosan for synthesis of laurate compounds in supercritical carbon dioxide using response surface methodology. Process. Biochem. **50**(8), 1224–1236 (2015)

Badgujar, K.C., Bhanage, B.M.: Solvent stability study with thermodynamic analysis and superior biocatalytic activity of Burkholderia cepacia lipase immobilized on biocompatible hybrid matrix of poly (vinyl alcohol) and hypromellose. J. Phys. Chem. B. **118**(51), 14808–14819 (2015)

Badgujar, K.C., Badgujar, V.C., Bhanage, B.M.: Application of ionic liquids for value-addition of lignin. In: Biomass, biofuels, biochemical, pp. 221–241. Elsevier (2021)

Badgujar, K.C., Badgujar, V.C., Bhanage, B.M.: Ionic liquids for bioenergy production. In: Ionic liquid-based technologies for environmental sustainability, pp. 235–256. Elsevier, (2022)

Badgujar, K.C., Badgujar, V.C., Bhanage, B.M.: Recent update on use of ionic liquids for enzyme immobilization, activation, and catalysis: A partnership for sustainability. Curr. Opin. Green Sustain. Chem., **36**, 100621 (2022)

Badgujar, K.C., Badgujar, V.C., Bhanage, B.M.: Synthesis of alkyl levulinate as fuel blending agent by catalytic valorization of carbohydrates via alcoholysis: recent advances and challenges. Catalysis Today **408**, 9–21 (2023)

Badgujar, K.C., Bhanage, B.M.: Factors governing dissolution process of lignocellulosic biomass in ionic liquid: Current status, overview and challenges. Biores. Technol. **178**, 2–18 (2015a)

Badgujar, K.C., Bhanage, B.M.: Editorial overview: Ionic liquids as a potential media for sustainable applications. Curr. Opin. Green Sustain. Chem. **40**, 100739 (2023)

Badgujar, K.C., Wilson, L.D., Bhanage, B.M.: Recent advances for sustainable production of levulinic acid in ionic liquids from biomass: Current scenario, opportunities and challenges. Renew. Sustain. Energy Rev. **102**, 266–284 (2019)

Badgujar, V.C., Badgujar, K.C., Yeole, P.M., Bhanage, B.M.: Immobilization of Rhizomucor miehei lipase on a polymeric film for synthesis of important fatty acid esters: kinetics and application studies. Bioprocess Biosyst. Eng. **40**, 1463–1478 (2017)

Barrueto, A., Hernández, P., Jiménez, Y.P., Morales, J.: Properties and application of ionic liquids in leaching base/precious metals from e-waste. A review. Hydrometall, **212**, 105895 (2022)

Bender, J., Jepkens, D., Husken, H.: Ionic liquids as phase-transfer catalysts: Etherification reaction of 1-Octanol with 1-Chlorobutane. Org. Process. Res. & Dev. **14**, 716–721 (2010)

Cai, M.: Wang X activity of imidazolium-based ionic liquids as catalysts Forfriedel-crafts acylation of aromatic compounds. Asian J. Chem. **26**, 5981–5984 (2014)

Carrozza, C.F., Papa, G., Citterio, A., Sebastiano, R., Simmons, B.A., Singh, S.: One-pot bio-derived ionic liquid conversion followed by hydrogenolysis reaction for biomass valorization: A promising approach affecting the morphology and quality of lignin of switchgrass and poplar. Bioresour. Technol. **294**, 122214 (2019)

Chen, M., Luo, Y., Li, G., He, M., Xie, J., Li, H., Yuan, X.: Alkylation of anthracene to 2-isopropylanthracene catalyzed by Lewis acid ionic liquids. Korean J. Chem. Eng., **26**, 1563–1567 (2009)

Chiappe, C., Pieraccini, D.: Ionic liquids: solvent properties and organic reactivity. J. Phys. Org. Chem. **18**(4), 275–297 (2005)

Dong, F., Jian, C., Zhengh, F., Kai, G., Zuliang, L.: Synthesis of chalcones via Claisen-Schmidt condensation reaction catalyzed by acyclic acidic ionic liquids. Catal. Commun. **9**, 1924–1927 (2008)

Fan, M., Huang, J., Yang, J., Zhang, P.: Biodiesel production by transesterification catalyzed by an efficient choline ionic liquid catalyst. Appl. Energy **108**, 333–339 (2013)

Fu, R., Yang, Y., Lai, W., Ma, Y., Chen, Z., Zhou, J., Chai, W., Wang, Q., Yuan, R.: Efficient and green microwave-assisted multicomponent Biginelli reaction for the synthesis of Dihydropyrimidinones catalyzed by Heteropolyanion-based ionic liquids under solvent-free conditions. Synth. Commun. **45**(4), 467–477 (2015)

Fu, R., Yang, Y., Ma, Y., Yang, F., Li, J., Chai, W., Wang, Q., Yuan, R.: Microwave-promoted direct amidation of unactivated esters catalyzed by heteropolyanion-based ionic liquids under solvent-free conditions. Tetrahedron Letts. **56**, 4527–4531 (2015)

Fu, R., Yang, Y., Chen, Z., Lai, W., Ma, Y., Wang, Q., Yuan, R.: Microwave-assisted heteropolyanion-based ionic liquids catalyzed transamidation of non-activated carboxamides with amines under solvent-free conditions. Tetrahedron **70**, 9492–9499 (2014)

Greaves, T.L., Drummond, C.J.: Protic ionic liquids: properties and applications. Chem. Rev. **108**, 206–237 (2008)

Guo, W., Zuo, M., Zhao, J., et al.: Novel Brønsted-Lewis acidic di-cationic ionic liquid for efficient conversion carbohydrate to platform compound. Cellulose **27**, 6897–6908 (2020)

Hagen, M.L., Harper, J.B., Croft, A.K.: Recent advances in the use of ionic liquids as solvents for protein-based materials and chemistry. Curr. Opin. Green Sustain. Chem. **36**, 100637 (2022)

Hajipour, A.R., Rafiee, F.: Recent progress in ionic liquids and their applications in organic synthesisOrg. Prep. Proced. International. **47**(4), 249–308 (2015)

Hallet, J.P., Welton, T.: Room-temperature ionic liquids: Solvents for synthesis and catalysis. Chem. Rev. **111**, 3508–3576 (2011)

https://sustainabledevelopment.un.org/post2015/transformingourworld/publication
https://www.scopus.com/search/form.uri?display:basic#basic
Jopp, S.: Carbohydrate Based Ionic Liquids (CHILs): synthesis and applications. Eur. J. Org. Chem. **2020**, 6418–6428 (2020). https://doi.org/10.1002/ejoc.202000714
Khaligh, N.G.: Synthesis of coumarins via Pechmann reaction catalyzed by 3-methyl-1-sulfonic acid imidazolium hydrogen sulfate as an efficient, halogen-free and reusable acidic ionic liquid. Catal. Sci. Technol. **2**, 1633–1636 (2012)
Klähn, M., Lim, G.S., Seduraman, A., Wu, P.: On the different roles of anions and cations in the solvation of enzymes in ionic liquids. Phys. Chem. Chem. Physics. **13**(4), 1649–1662 (2011)
Kulshrestha, A., Gehlot, P.S., Kumar, A.: Paramagnetic surface active ionic liquids: synthesis, properties, and applications. Mater. Today Chem., **21**, 100522 (2021)
Kumar, K., Kumar, M., Upadhyayula, S.: Catalytic conversion of glucose into levulinic acid using 2-phenyl-2-imidazoline based ionic liquid catalyst. Molecules. **26**(2), 348 (2021)
Kusumawati, E.N., Shirai, M.: Recent progress in biphasic catalytic systems using ionic liquids. Curr. Opin. Green Sustain. Chemistry. **1**(35), 100613 (2022)
Lin, Y.J., Wu, Y.P., Thul, M., Hung, M.W., Chou, S.H., Chen, W.T., Lin, W., Lin, M., Reddy, D.M., Wu, H.R., Ho, W.Y., Luo, S.Y.: Tunable Aryl Imidazolium recyclable ionic liquid with dual Brønsted-lewis acid as green catalyst for Friedel-crafts acylation and Thioesterification. Molecules. **25**(2), 352(2020)
Malhotra, S.V., Andal, R.P., Kumar, V.: .Aminolysis of epoxides in ionic liquid 1-Ethylpyridinium Trifluoroacetate as green and efficient reaction medium, synthetic communications. An Int. J. Rapid Commun. Synth. Org. Chem., **38**, 4160–4169 (2008)
Martínez-Palou, R.: Microwave-assisted synthesis using ionic liquids. Mol. Divers. **14**, 3–25 (2010)
Mehta, M.J., Kulshrestha, A., Sharma, S.: Kumar ARoom temperature depolymerization of lignin using a protic and metal based ionic liquid system: an efficient method of catalytic conversion and value addition. Green Chem. **23**, 1240–1248 (2021)
Meracz, I., Oh, T.: Asymmetric Diels-Alder reactions in ionic liquids. Tetrahedron Lett. **44**, 6465–6468 (2003)
Muskawar, P., Thenmozhi, K., Bhagat, P.R.: Designing of thermally stable amide functionalized benzimidazoliumperchlorate ionic liquid for transamidation of primary carboxamides. Appl. Catal. A: Gen. **493**, 158–167 (2015)
Nakhate, A.V., Yadav, G.D.: Graphene-oxide-supported SO_3H-functionalized imidazolium-based ionic liquid: Efficient and recyclable heterogeneous catalyst for Alcoholysis and Aminolysis reactions. Chemistry Select **3**, 4547–4556 (2018)
Olivier-Bourbigou, H., Morvan, L.M.D.: Ionic liquids and catalysis: Recent progress from knowledge to applications. Appl. Catal. A **373**, 1–56 (2010)
Patil, P., Pratap, A.: Choline Chloride catalyzed amidation of fatty acid ester to monoethanolamide: A green approach. J. Oleo Sci. **65**(1), 75–79 (2016)
Poole, C.F., Atapattu, S.N.: Determination of physicochemical properties of ionic liquids by gas chromatography. J. Chromatogr. A, **1644**, 461964 (2021)
Potdar, M.K., Kelso, G.F., Schwarz, L., Zhang, C., Hearn, M.T.: Recent developments in chemical synthesis with biocatalysts in ionic liquids. Molecules. **20**, 16788–16816 (2015)
Qureshi, Z.S., Deshmukh, K.M., Bhanage, B.M.: Applications of ionic liquids in organic synthesis and catalysis. Clean Technol. Environ. Policy **16**, 1487–1513 (2014)
Salminen, E., Kumar, N., Virtanen, P., Tenho, M., Arvela, P., Mikkola, I.: Etherification of 5-hydroxymethylfurfural to a biodiesel component over ionic liquid modified zeolites. Top. Catal. **56**, 765–769 (2013)
Schindl, A., Hagen, M.L., Muzammal, S., Gunasekera, H.A.D., Croft, A.K.: Proteins in Ionic Liquids: Reactions, applications, and futures. Front Chem. **7**, 347 (2019)
Sekharan, T.R., Katari, O., Rahman, S.N., Pawde, D.M., Goswami, A., Chandira, R.M., Shunmugaperumal, T.: Neoteric solvents for the pharmaceutical industry: An update. Drug Discovery Today **26**(7), 1702–1711 (2021)

Shaohua, B., Nannan, Q., Jing, Z., Jianguo, Y.: Alkylation of p-Cresol with tert-Butanol catalyzed by novel multiple-SO3H functioned ionic liquid. Chin. J. Chem. Eng. **19**, 64–69 (2011)

Shinde, S.S., Lee, B.S.: Chi DC synergistic effect of two solvents, Tert-alcohol and ionic liquid, in one molecule in nucleophilic fluorination. Organic Lett. **10**, 733–735 (2008)

Wagh, K.V., Badgujar, K.C., Patil, N.M., Bhanage, B.M.: Recent trends of ionic liquids for the synthesis of 5-hydroxymethylfurfural. Curr. Org. Chem. **1**, 20(7), 736–751 (2016)

Wang, C., Liu, J., Leng, W., Gao, Y.: Rapid and efficient functionalized ionic liquid-catalyzed aldol condensation reactions associated with microwave irradiation. Int. J. Mol. Sci., **15**, 1284–1299 (2014)

Wang, H., Block, L.E., Rogers, R.D.: Chapter 1: Catalytic conversion of biomass in ionic liquids, in catalysis in ionic liquids: From catalyst synthesis to application, pp. 1–19 (2014)

Wilkes J.S.: A short history of ionic liquids—from molten salts to neoteric solvents. Green Chem., **4**, 73–80 (2002)

Wilkes, J.S.: Properties of ionic liquid solvents for catalysis. J. Mol. Catal. a: Chem. **214**, 11–17 (2004)

Yadav, G.D., Motirale, B.G.: Ionic liquid as catalyst for solid-liquid phase transfer catalyzed synthesis of p-Nitrodiphenyl ether. Ind. Eng. Chem. Res. **47**, 9081–9089 (2008)

Yang, Z., He, L., Dou, X., Chanfreau, S.: Dimethyl carbonate synthesis catalyzed by DABCO-derived basic ionic liquids via transesterification of ethylene carbonate with methanol. Tetrahedron Lett. **51**, 2931–2934 (2010)

Zhang, P., Jiang, Z., Cui, Y., Xie, G., Jina, Y., Guo, L., Xu, X., Zhang, Q., Li, X.: Catalytic performance of ionic liquid for dehydrochlorination reaction: Excellent activity and unparalled stability. Appl. Catal. B: Environ. **255**, 117757 (2019)

Zhang, P., Sun, Y., Zhang, Q., Guo, Y., Song, D.: Upgrading of pyrolysis biofuel via esterification of acetic acid with benzyl alcohol catalyzed by Brønsted acidic ionic liquid functionalized ethylbridged organosilica hollow nanospheres. Fuel, **228**, 175–186 (2018)

Zhao, G., Jiang, T., Gao, H., Han, B., Huang, J., Sun, D.: Mannich reaction using acidic ionic liquids as catalysts and solvents. Green Chem. **6**, 75–77 (2004)

Zhao, S., Wang, X., Zhang, L.: Rapid and efficient Knoevenagel condensation catalyzed by a novel protic ionic liquid under ultrasonic irradiation. RSC Adv. **3**, 11691–11697 (2013)

Zhen, L., Yingwei, Z., Feng, H.: Catalysis and applications of task-specific ionic liquids. Sci. Sin. Chim. **4**, 502–524 (2012)

Zheng, X., Qian, Y., Wang, Y.: Direct asymmetric aza Diels-Alder reaction catalyzed by chiral 2-pyrrolidinecarboxylic acid ionic liquid. Catal. Commun. **11**, 567–570 (2010)

Zhikai, C., Fu, R., Chai, W., Zheng, H., Sun, L., Lu, Q., Yuan, R.: An eco-benign and highly efficient procedure for N-acylation catalyzed by heteropolyanion-based ionic liquids using carboxylic acid under solvent-free conditions. Tetrahedron **70**, 2237–2245 (2014)

Zhu, A., Liu, R., Dua, C., Lia, L.: Betainium-based ionic liquids catalyzed multicomponent Hantzsch reactions for the efficient synthesis of acridinediones. RSC Adv. **7**, 6679–6685 (2017)

Chapter 11
Metal–Organic Framework (MOF) as Catalysts for Nitroarenes

Debashis Acharya, Taponitya Samantaray, Niharika Das, Manoj Kumar Sahoo, Jared Kinyon, Nityananda Agasti, Anirudha Jena, Jatin K. Sinha, Rojalin Sahu, and Raghabendra Samantaray

11.1 Introduction

Arenes containing nitro groups represent a significant group of chemically analogous compounds widely utilized across industries like pesticides, explosives, pharmaceuticals, and chemical production. However, these nitro-based substances pose a substantial risk to both the environment and public health due to their carcinogenic, toxicological, and mutagenic properties. As a result, they are classified as major pollutants in water systems (Sadeghzadeh et al. 2018). The strong electron-withdrawing nature of nitro groups creates a highly reactive environment, leading to numerous undesired chemical reactions, especially in aquatic ecosystems. Notable nitroarenes (NAs) that directly threaten the environment include nitrobenzene (NB), 2,4,6-trinitrotoluene (TNT), and 2,4,6-trinitrophenol (TNP) (Remya and Kurian 2019).

To address these challenges, several conventional methods have been employed, such as photo-catalytic and electrochemical reductions, as well as catalytic processes aimed at converting nitro groups into amines. This conversion significantly reduces the toxicity of NAs. Catalytic reduction of nitro aromatic compounds is the most

D. Acharya
School of Applied Sciences, Kalinga Institute of Industrial Technology (KIIT), Bhubaneswar, India

M. K. Sahoo
Defence Research Development Establishment (DRDE), Defence Research and Development Organisation–DRDO, Ministry of Defence, Government of India, Gwailor, India

N. Agasti
Department of Chemistry, Deen Dayal Upadhyaya College (University of Delhi), Dwarka, New Delhi, India

T. Samantaray · N. Das · J. Kinyon · A. Jena · J. K. Sinha · R. Sahu · R. Samantaray (✉)
School of Physics, University of Hyderabad, Hyderabad, India
e-mail: raghabendra.samantarayfch@kiit.ac.in

R. S. Ningthoujam and A. K. Tyagi (eds.), *Handbook of Materials Science, Volume 3*,
Indian Institute of Metals Series, https://doi.org/10.1007/978-981-96-8229-4_11

commonly practiced method and is considered efficient compared to alternative techniques. This approach is advantageous as it consumes less energy and eliminates the necessity for harmful organic solvents, making it a superior choice in various aspects (Yusuf et al. 2022).

Thus far, a broad range of reducing agents has been utilized in the reduction of nitroarenes, including hydrazines, hydrides, molecular hydrogen (H_2), and various reductants based on metals (Mouchaham et al. 2018). Recent studies have explored the use of copper (Cu), silver (Ag), gold (Au), and palladium (Pd) nanoparticles as catalysts to convert aromatic nitro compounds into their respective amino derivatives, aiming to mitigate the toxicity associated with nitro groups (Har-Kedar and Bleehen 1976). Heterostructure nanomaterials are gaining increased attention due to their multifunctionality, selectivity improved catalytic performance, and stability compared to monometallic nanomaterials (Tseng et al. 2023). However, these metal nanoparticles (NPs), specifically transition metal NPs, are less favored due to their toxicity, higher oxidation potential, costliness, and instability. Another drawback is their tendency to agglomerate, which diminishes their catalytic efficiency by reducing the surface-to-volume ratio and consequently weakening their ability to reduce nitro groups (Orlandi et al. 2016).

In response to these challenges, scientists have introduced a novel material known as metal–organic frameworks (MOFs), composed of metal linkers and organic ligands. MOFs possess a porous metal–inorganic complex structure resembling cage. As heterogeneous catalysts, MOF materials exhibit distinct surface properties, well-dispersed active sites, and customizable pore structures, and these properties render them an optimal choice for catalytic applications (Mouchaham et al. 2018). Yang et al. synthesized a simple, environmentally friendly, and cost-effective MOF composite Cu-BTC@Fe_3O_4 for the reduction of nitroarenes in an aqueous environment. This catalyst demonstrated the ability to be employed repeatedly for six successive cycles without any discernible reduction in its efficacy (Whelan et al. 2021). Kaur et al. successfully fabricated metal oxide incorporated MOF, by incorporating inexpensive ZnO nanoparticles which significantly enhanced the catalytic performance of the MOF, thereby replacing the traditionally utilized corrosive and reactive catalysts, for the first time (Tsuzuki et al. 2021).

Taking these factors into account, within this chapter, we have provided an overview of MOFs, highlighting particular framework designs for catalytic activity. We have delved into the analysis of kinetic parameters and engaged in a comprehensive discussion on the catalytic mechanism for the reduction of nitro-based compounds. Moreover, we have expounded upon the utilization of MOF-composites as a hetero junction platform aimed at bolstering catalytic performance.

11.2 MOFs as Catalyst

To date, a myriad of MOF structures has been reported. However, only a limited number have exhibited promising properties under non-reactive conditions, where alterations can be implemented on the MOFs without changing their porous structure.

MOFs, which are made up of metal cations (Mn^{2+}) and organic linkers (mostly carboxylate-based), are generally delicate substances. For instance, MOF-5, a Zn^{2+} terephthalate, quickly breaks down in water, while HKUST-1, a Cu^{2+} trimesate, slowly deteriorates in water at room temperature (An et al. 2022). This instability in water greatly restricts the utility of MOFs, particularly for real-world applications that necessitate direct interaction with water such as the separation processes from flue gas, which might contain significant amounts of water (Xiao et al. 2022), or catalysis for water splitting. Much of the catalysis are conducted in an aqueous medium. Therefore, when creating a MOF that can withstand degradation in an environment where water or humidity could be a major concern, chemical stability becomes a key and fundamental factor to consider.

When designing robust MOF for catalytic applications, it is essential to consider the coordinate bonds (Xiao et al. 2022). These bonds form the core structure of a MOF, and the most effective strategy to prevent the disruption of cation–ligand bonds is to enhance the interaction between these two elements. This can be accomplished through several intuitive steps: Firstly, use organic linkers that have highly complex functional groups with high pKa values (greater than that of typical carboxylic acids (~4.5)). Finally, use metal cations that are acidic to highly acidic and have a higher oxidation state, which should also exhibit a strong polarizing power.

Alternative approach focuses on protecting the existing or enhanced coordinate bonds. In this method, the interaction between cation and ligand is shielded from potential attacks by competing species. This involves a range of factors such as dimensionality, nuclearity, network topology, and the existence of defects (Yang et al. 2018). A particular molecular configuration or topology is vital for potential applications, and these metal–ligand bonds can be safeguarded in several ways. This includes introducing bulky or hydrophobic groups to serve as a defense against competing species, adorning the interior of the MOF, and applying or embedding a hydrophobic layer on the exterior of the MOF (An et al. 2022; Li et al. 2019).

Temperature is a critical factor in catalytic applications, particularly when dealing with thermophilic organisms and products. Indeed, the heat resistance of the MOFs must also be considered. Additionally, since most catalysis occurs in an aqueous medium, the hydrothermal stability of the MOF must also be considered alongside its thermal stability (Kottappara et al. 2020; Huang et al. 2023).

When subjected to heat, the structure of MOFs gradually deteriorates, resulting in effects such as amorphization, melting, dehydration of metal-oxo-clusters, or dehydrogenation or graphitization of linkers. The strategies to preserve thermal stability (Huang et al. 2023) are somewhat restricted. In the context of catalysts, measures like preferring aromatic moieties over aliphatic linkers or opting for phosphonate-

and sulfate-based MOFs instead of carboxylate or phenolate linkers can be implemented. Other elements that impact thermal and hydrothermal stability encompass framework density, network topology, interpenetration, and the existence of defects.

Another crucial factor to consider when evaluating the robustness of MOFs is their mechanical stability. The primary source of mechanical instability in MOFs is destruction driven by capillary forces. A practical solution to this problem is the application of supercritical CO_2 activation. In this procedure, the guest solvent molecules, which are strongly bound within the pores, can be substituted with liquid CO_2. This results in a solvent surface tension of zero, thereby nullifying the capillary force (Butova et al. 2016).

11.3 Design and Synthesis of MOF for Catalysis

The process of designing and synthesizing MOFs for catalysis is a multifaceted process that requires a deep understanding of various factors. It involves the careful selection of metal nodes and organic ligands, which are key to achieving the desired catalytic properties. The metal center chosen can greatly influence the catalytic activity and selectivity, while the design of the ligand can determine the stability and accessibility of active sites as shown in Fig. 11.1. A variety of synthesis techniques are employed as illustrated in Fig. 11.2 in the fabrication of MOFs, including solvothermal, hydrothermal, and microwave-assisted methods. These techniques allow for controlled structures and properties in the resulting MOFs. To analyze these MOFs, characterization techniques, such as X-ray diffraction, nitrogen adsorption–desorption isotherms, and various spectroscopic methods, are utilized.

Recent studies have highlighted the importance of design engineering and synthesis protocols in the development of MOF-derived electro catalysts. Theoretical calculations and energy applications of these MOFs have been systematically analyzed. Moreover, factors such as morphology, structure, metal ions, ligands, and pore characteristics can significantly affect their performance as catalysts. These

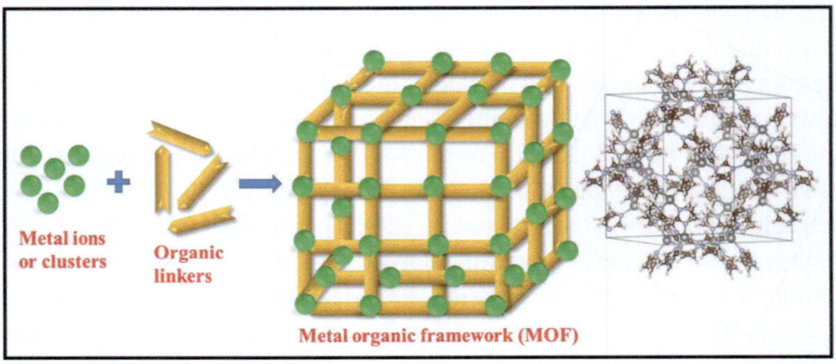

Fig. 11.1 Structure of metal–organic framework (MOF)

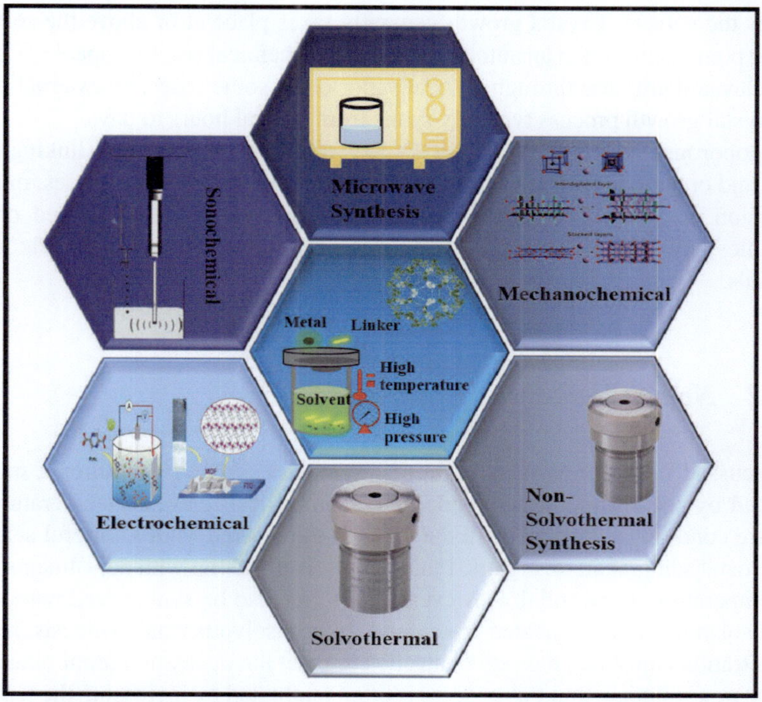

Fig. 11.2 Synthesis of metal–organic framework (MOF)

characteristics can be controlled by adjusting parameters such as pH, reaction time, temperature, solvent, and metal ion concentration. MOFs have found substantial applications in various fields such as energy storage, sensing, and gas storage, adsorption due to their high surface area, controllable pore size and desirable chemical functionalities (Feng and Li 2017). Future research should focus on the synthesis of MOF derivatives according to the confinement–pyrolysis–oxidation strategy to synthesize MOF derivatives with high activity even at high GHSV.

11.3.1 Solvothermal Synthesis

MOFs are typically synthesized using solvents under high-temperature and high-pressure conditions (Angel et al. 2021; Bosch et al. 2014). As the temperature increases, the pressure within the reaction vessel rises, facilitating the formation of a honeycomb-like structure where crystal nucleation occurs. Under such conditions, MOFs crystal growth is achieved, typically resulting in the formation crystals ranging from the nanometer to micrometer scale.

This synthetic technique is commonly referred to as solvothermal synthesis, a broader term than hydrothermal synthesis, which specifically applies when water is used as the solvent. Crystal growth generally takes place at or above the solvent's boiling point, achieved in an autoclave—a sealed chemical reactor capable of generating elevated pressure through solvent vapor or, in some cases, an external pump. The crystal growth process typically spans from several hours to days.

Nanoporous and microporous MOFs are formed by successfully linking metal nodes and organic ligands, resulting in a reticular structure. In some cases, multiple nucleation events occur, leading to the aggregation of surface-adsorbed organic molecules onto metal clusters, ultimately forming these inorganic–organic hybrid materials.

11.3.2 Non-Solvothermal Synthesis

As discussed in the previous section, the synthesis of MOFs requires a medium provided by solvent molecules, under either ambient or elevated temperature and pressure conditions. To maximize the yield of the desired MOFs, careful selection of the metal salt precursor, organic linker, and solvent, along with adjustments to pH and temperature, is essential. However, MOFs can also be synthesized without the use of solvents or sophisticated equipment. In non-solvothermal synthesis, reagent concentrations must be precisely controlled to meet the nucleation requirements for precipitate formation; and this process is often facilitated by increasing the temperature. Some of these alternative methods are described in the following sections (Kaur et al. 2018).

11.3.3 Mechanochemical Synthesis

In this method, little to no solvent is used for MOFs synthesis and instead, mechanical energy is used to drive the reaction. This approach has been adopted to enable MOF production without the need for specialized equipment like autoclaves, gaining popularity for its ability to facilitate rapid, high-yield reactions even in the absence of solvents. The reactions between solid precursors, such as metal clusters and organic linkers, are typically initiated solely through mechanical means—for instance, using the kitchen grinder method. In this method, the organic ligands and metal salts are placed in a stainless steel kitchen grinder jar with minimal or no solvent. Notably, the resulting MOFs can be ground further at different stages, and the products are analyzed using various characterization techniques to understand the reaction mechanism and evaluate the yield. This method is recognized as a simple, convenient, and cost-effective approach to MOF synthesis, eliminating the need for solvents or advanced equipment (Pournoori et al. 2023; Li et al. 2022).

11.3.4 Microwave Synthesis

The microwaves approach of MOFs synthesis is considered unique as it utilizes microwave radiation or energy in frequencies ranging from 300 to 300,000 MHz which facilitates the reaction. Microwave radiations impact on reactant precursors leading to heating via—(a) action on polar molecules, (b) action on free ions. In both cases, polar molecules and ions align themselves with the alternating field. The resistance causes an electric current to flow through an electrolyte solution, resulting in generation of heating in samples. In a typical microwave heating process, water is commonly used as resistance in water is caused by intrinsic hydrogen bonding, which predominantly prevent the dipoles from easily aligning with the fluctuating field. However, the efficiency of this process or heating depends on a number of factors such as (a) the degree of rotational freedom, (b) the dipole moment of a polar molecule, etc. Furthermore, the interaction of microwaves with water/solvent is contingent upon its physical state. The most pronounced effect occurs when the solvent is in its liquid state. In contrast, in the solid and gaseous phase, the effect is negligible due to the absence of hydrogen bonding and the reduced density of solvent molecules. In the solid state, such as ice, the influence of microwaves is minimal, as the molecules are securely fixed within a crystalline lattice structure, inhibiting their realignment or movement.

Thus, synthesizing MOFs by microwave treatment presents a distinct advantage by considerably minimizing the reaction time. Although the activation energy required for a reaction remains constant, microwaves supply the necessary energy to surmount this barrier more efficiently, thereby expediting the process in comparison to traditional heating techniques (Kong and Li 2021; Konnerth et al. 2020).

11.3.5 Sonochemical Synthesis

The synthesis of MOFs primarily relies on the nucleation and subsequent growth of MOF crystals, and thus, it demands a substantial input of energy, typically in the form of heat. In this approach, ultrasound is employed to generate the necessary heat within the precursor materials. The effect of ultrasound is based on the phenomenon of cavitation—the formation and collapse of cavities (or bubbles) within a liquid or colloid medium. Sound waves induce pressure fluctuations that create vapor bubbles in the liquid phase, decreasing local pressure and causing localized boiling and the release of dissolved gases. This results in the formation of numerous microscopic cavities within the liquid medium.

These newly formed bubbles possess internal pressures exceeding the vapor pressure of the surrounding liquid, enabling them to expand and then rapidly collapse. This collapse occurs within milliseconds and typically happens in microscopic crevices of the liquid–air interface (Tang et al. 2023). The violent implosion of these bubbles produces extremely high local temperatures and pressures, supplying the energy required for MOF synthesis within a remarkably short time frame—often under one hour.

This technique offers several significant advantages for MOF synthesis. Most notably, the creation of resonant bubbles acts as microscopic agitators, enhancing the contact between reactants and thereby increasing the interfacial area for the reaction. Additionally, the intense heat and pressure generated through cavitation help break apart particle agglomerates, improving surface accessibility for reactants (Wu and Zheng 2013).

Beyond this, sonication can induce larger structural changes, such as removing adsorbed or solvated layers from material surfaces, breaking chemical bonds in liquid polymer chains, and generating nucleation sites characterized by physical or chemical imbalances. This method was first applied to MOF synthesis in 2008, notably for the production of $Zn_3(BTC)_2$ (Dhakshinamoorthy and Garcia 2012).

11.3.6 Electrochemical Synthesis

In the electrochemical synthesis of MOFs, a direct metal source, typically in the form of a metal plate electrode, is used alongside a solution containing an electrolyte and an organic linker dissolved within it. A power supply, such as a potentiostat, is connected to the metal electrode (serving as the anode), which facilitates the oxidation of the metal into metal cations. These cations then coordinate with the organic ligands present in the solution. This method allows for continuous MOF synthesis and eliminates the need to introduce additional anions into the system—a significant advantage for large-scale MOF production, as it reduces both time and complexity. Protic solvents are often included to prevent metal deposition on the cathode surface (Shen et al. 2023; Toyao et al. 2013).

11.4 Reduction of Nitroarenes by MOFs

These reactions entail the transformation of the nitro group ($-NO_2$) into an amino group ($-NH_2$) in the presence of a reducing agent. Such reactions play a crucial role in the production of aniline derivatives, which are widely used in pharmaceuticals, dyes, and various other sectors. Conventional reduction techniques, such as hydrogenation with metal catalysts or stoichiometric reductions, often face challenges like low selectivity, severe reaction conditions, and the production of undesired byproducts. However, employing MOFs as catalysts for nitroarene reduction could lead to better selectivity, gentler reaction conditions, and increased catalytic efficiency, which is illustrated in Fig. 11.3.

The selection of an appropriate MOF as a catalyst for nitroarene reduction is influenced by several factors. These include the specific nitroarene being used, the

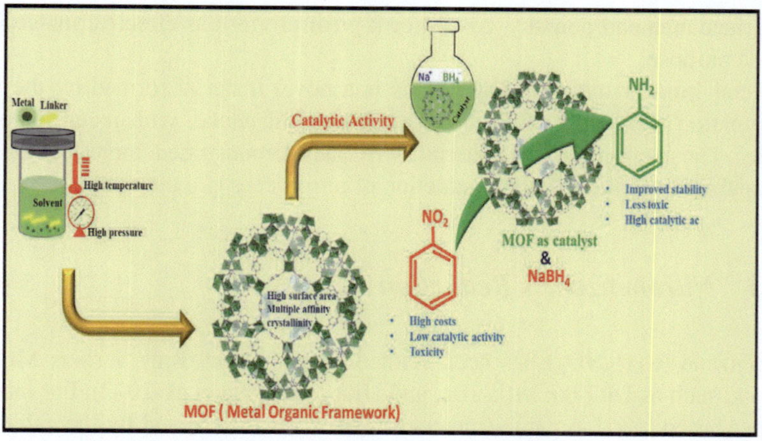

Fig. 11.3 Efficient reduction of nitroarene to nitroamine

preferred reaction conditions, and the characteristics of the MOF. Here are some MOFs that have been studied for their potential in nitroarene reduction:

i. UiO-66: A type of MOF that has been effectively utilized as a catalyst in the reduction processes of nitrobenzene and 4-nitroaniline. Furthermore, UiO-66-NH_2 has proven to be an outstanding support for the creation of highly dispersed nanoparticles. The magnetic nanoparticles synthesized from UiO-66-NH_2, TTACP, Ni, and Pd (UiO-66-NH_2/TTACP/Ni@Pd) have demonstrated remarkable catalytic efficiency in the reduction of 2-nitroaniline and 4-nitrophenol (Poshti et al. 2023).
ii. MIL-101: Another MOF has been employed in the reduction of 2-nitrotoluene, 4-nitroaniline and nitrobenzene. The adsorption capabilities of 2-amino-4-nitrotoluene and 2-nitroaniline on MIL-101 (Cr) were recorded to be 25.0 and 54.3mg/g, respectively. This higher adsorption capacity can be ascribed to the more optimal pore size of MIL-101 (Cr) compared to that of MIL-100 (Fe, Cr) (Kadam and Tilve 2015).
iii. HKUST-1: A copper-based MOF has been utilized in the reduction of nitrobenzene. This catalyst has demonstrated exceptional performance in the reduction process, with sodium borohydride ($NaBH_4$) serving as the reducing agent (Mo et al. 2020).
iv. ZIF-8: This MOF has been employed in the reduction of 4-nitroaniline. The nano-ZIF-8@chitosan microspheres (ZIF-8@CSM) were synthesized using a combination of the inverse emulsion method and in-situ growth method. The efficacy of ZIF-8@CSM in adsorbing p-hydroxybenzoic acid from an aqueous solution was also examined (Deenadayalan et al. 2016).
v. Hierarchical mesoporous MOF derived from Co-MOF with di-ligands Ni-MOF: This included the Co-MOF-derived hierarchical mesoporous MOF and di-ligands Ni-MOF among others. These MOFs, characterized by their high

surface area and porosity, have shown great promise as effective materials for this purpose.

vi. Metal–biochar hybrid (MBC): This is a novel material formed by the amalgamation of biochar (BC), a carbonaceous substance, with metal nanoparticles. The resulting hybrid material, MBC, has demonstrated impressive catalytic capabilities in the selective reduction of nitroarenes (Lu et al. 2023).

11.4.1 Nitrobenzene's Reduction

Nitrobenzene ($C_6H_5NO_2$) has been studied for reduction using various MOFs as catalysts, such as UiO-66, MIL-101, and HKUST-1 (Wu et al. 2022). For instance, a copper-based metal–organic framework (MOF) was designed by the calcination of Cu–MOF at 700 °C, denoted as Cu@C (Mo et al. 2020). This catalyst showed superior catalytic performance toward the reduction of nitrobenzene, using sodium borohydride ($NaBH_4$) as the reducing agent.

i. **4-Nitroaniline's reduction**

($C_6H_6N_2O_2$): This compound has been investigated for reduction using MOFs like UiO-66 (Pournoori et al. 2023), MIL-101 (Li et al. 2022), and ZIF-8 (Dhakshinamoorthy and Garcia 2012; Wu and Zheng 2013; Tang et al. 2023; Shen et al. 2023) as catalysts. For instance, a study reported the use of a palladium-based MOF for the reduction of 4-nitroaniline to p-phenylenediamine. The MOF demonstrated high catalytic activity and selectivity, and the reaction could be carried out under ambient conditions (Li et al. 2021).

ii. **2-Nitrotoluene's reduction**

($C_7H_7NO_2$): Some studies have explored the reduction of 2-nitrotoluene using MOFs such as UiO-66 and MIL-101 as catalysts [36/38]. The synthesized MOFs were used as catalysts in numerous reactions catalyzed by Lewis acid catalysts which are very significant in organic synthesis but have not been studied using MOF catalysts.

iii. **3-Nitroaniline's reduction**

($C_6H_6N_2O_2$): This compound has been studied for reduction using MOFs like UiO-66 and MIL-101 as catalysts. The resulting Ni@C nanocomposites served as highly efficient and magnetically recyclable catalysts for the hydrogenation of diverse functionalized nitro compounds to the corresponding anilines under relatively milder conditions (Li et al. 2023).

iv. **Functionalized nitroarenes reduction**

This process involves the reduction and reductive addition (also known as formal hydroamination) of functionalized nitroarenes. This is achieved using a straightforward and bench-stable iron (III) catalyst and silane. Additionally, nanocomposites derived from MOFs, specifically those based on nickel, have proven to be highly effective catalysts. They facilitate the hydrogenation of a wide range of functionalized nitro compounds, converting them into their corresponding anilines. The reaction can be represented as follows:

$$\mathbf{RNO_2 + 3H_2 \rightarrow RNH_2 + 2H_2O}$$

Here, R represents the rest of the molecule that is attached to the nitro group. The iron (III) catalyst plays a crucial role by facilitating the transfer of hydrogen from silane to the nitro group, culminating in the creation of an amine group (Yang et al. 2021).

v. **Nitroarenes reduction for wastewater remediation**

This process involves the reduction of nitroarenes in a water-based medium, facilitated by heterogeneous catalysts, which is a more environmentally friendly approach. A new hybrid material, known as the metal–biochar hybrid (MBC), is formed when biochar (BC), a carbonaceous material, is combined with metal nanoparticles. This hybrid material has demonstrated significant catalytic activity in the selective reduction of nitroarenes. Additionally, silver nanoparticles that have been immobilized on an amine-functionalized MOF, specifically Ag/UiO-66-NH_2, have proven to be a robust catalyst for the reduction of 4-nitrophenol (Dhakshinamoorthy et al. 2017). The reaction can be represented as follows:

$$\mathbf{RNO_2 + 6H_2 \rightarrow RNH_2 + 2H_2O}$$

Here, R represents the rest of the molecule that is attached to the nitro group. The MBC serves as a catalyst, enabling the transfer of hydrogen from water to the nitro group, which leads to the creation of an amine group.

11.4.2 Catalytic Mechanism of Reduction of Nitro-Based Compounds

The catalytic mechanisms of MOF-catalyzed nitroarene reduction reactions indeed involve the interaction of the nitroarene substrate with the active sites within the MOF structure. These active sites can be metal centers, open metal sites, or functional groups on the ligands (Luz et al. 2015).

The reduction process may proceed via various pathways. One such pathway is direct hydrogenation, where hydrogen is added to a compound from a source other than molecular H_2. This process is often used in laboratory and industrial organic synthesis to reduce organic compounds, e.g., ketones to alcohols, and imines to amines.

Another pathway is transferring hydrogenation, which involves the transfer of hydrogen from a donor compound to nitroarene. This process can be facilitated by various catalysts, including those based on ruthenium and rhodium complexes.

In the context of MOFs, a study demonstrated that a copper-based MOF showed superior catalytic performance toward the reduction of nitrobenzene, using sodium borohydride ($NaBH_4$) as the reducing agent (Jicsinszky et al. 2023).

Electron transfer mechanisms can also play a role in these reactions. For instance, confinement effects in MOFs have been reported to promote the stability of CO_2 hydrogenation catalysts, such as molecular complexes of immobilization effect, active sites in size effect, stabilization in the encapsulation effect, and electron transfer and interfacial catalysis in the synergistic effect (Tang et al. 2017). The structure of the MOF, including pore size, surface area, and accessibility of active sites, plays a crucial role in determining the reaction mechanism and catalytic performance. For example, two-dimensional (2D) MOF-based heterostructure increases the contact surface and charge transfer, which can greatly enhance catalytic performance. Moreover, MOFs with high surface area and porosity are promising materials for this purpose (Kovacic and Somanathan 2014).

11.4.3 Evaluation of Kinetic Parameters

Evaluating the kinetic parameters of MOFs as catalysts for nitroarenes involves studying the rate of the reaction and how it changes with various factors. These factors can include the concentration of the reactants, temperature, and the specific MOF used as a catalyst. Here, some key points have been studied:

i. **Acidity of MOFs**: The performance of MOFs as catalysts can be affected by their acidity. For example, the catalytic abilities of MOFs are connected to their physical and chemical characteristics, including aspects like particle size, defect chemistry, and porosity. These properties can potentially be manipulated through coordination modulation (Kefayati et al. 2023).
ii. **Metal nodes**: The metallic components in MOFs can serve as catalytic active sites. When considering MOFs in the role of catalysts, there are primarily two types of active sites. The most prevalent type originates from the metal nodes that form the framework of the MOFs, which can be synthesized using the hydrothermal technique (Desai et al. 2023).
iii. **Surface area and active sites dispersion**: The catalytic performance of MOFs can be influenced by their surface area and the distribution of active sites. For instance, effective catalysts are vital in a range of organic reactions and polymerization processes. MOFs with their extremely high porosity, expansive surface area, scattered polymetallic sites, and adjustable linkers, present themselves as potential catalysts (Raptopoulou 2021).
iv. **Micro- and nanoscale pores**: MOFs can employ pores at the micro and nano scale as sites for catalytic reactions and can leverage the interaction between host and guest to control the reaction's activity and selectivity. For example, 2D MOF nanosheets are characterized by a multitude of accessible active sites,

meso/macroporosity, and a quicker electron transfer, all of which are critical factors in the creation of high efficiency electrocatalysts (Hu et al. 2019).

These factors are vital in comprehending the kinetics of nitroarene reduction when MOFs are used as catalysts. Nonetheless, additional research is required to thoroughly grasp these factors and enhance the efficiency of MOF catalysts.

11.5 Nano-MOFs for NA Reduction

(a) **Palladated nanohydroxyapatite MOF**: This MOF-derived catalyst was used for the proficient catalytic reduction of nitroarenes. The catalyst was produced using a simple method, leading to the creation of nanorods with a distinct morphology. These nanorods acted as an outstanding and diverse base for the construction of a metal–organic framework composite. The final catalyst exhibited superior catalytic performance in the reduction of nitro compounds, using $NaBH_4$ as the reducing agent, and achieved excellent yields (Das et al. 2022).

(b) **Co-MOF-derived hierarchical mesoporous MOF**: This MOF has been used for the reduction of nitroarenes. The hybrid derived from MOFs, characterized by its unique hollow and porous structures, could provide a large surface area and abundant isolated active sites for the adsorption/activation of CO_2 and redox reactions (Isaeva et al. 2018).

(c) **Di-ligands Ni-MOF**: This MOF has also been used for the reduction of nitroarenes. The catalyst consists of ultra-small nickel nanoclusters that are generated in situ through the reduction of $Ni(OAc)_2$ by $NaBH_4$. These highly reactive Ni nanoclusters are stabilized by nitroarenes coordinated on the surface, which prevents further growth and aggregation of the Ni nanoclusters (Dinadayalane et al. 2022).

(d) **Palladium-anchored MOFs**: These MOFs were utilized as catalysts for the hydrogenation of aromatic aldehydes in the presence of visible light. The catalyst consists of ultra-small palladium nanoclusters that are generated in situ through the reduction of $Pd(OAc)_2$ by $NaBH_4$. Further growth and clustering of these Pd nanoclusters are prevented by nitroarene coordination to their surface (Wibowo et al. 2021).

(e) **MOF-supported Copper nanoparticles**: These were used as steady catalysts for the selective semi hydrogenation of acetylene. When biochar (BC), a carbonaceous material, is merged with metal nanoparticles, it results in a novel hybrid material, namely, a metal-BC hybrid (MBC) (Singh et al. 2023). This hybrid demonstrates effective catalytic performance in the selective reduction of nitroarenes.

(f) **Molybdenum-infused MOF-derived catalysts**: These catalysts exhibited considerable promise in the reduction of nitrophenols in the solution phase. The infusion of molybdenum not only amplified the catalytic attributes of the MOFs but also boosted their resilience and longevity, making them apt for multiple uses without substantial activity degradation. Moreover, the distinct porous configuration of the MOFs ensures superior accessibility to the substrate, thereby enhancing the total efficiency of the reaction.

11.5.1 MOF Composites as Heterojunction Platform

Heterojunctions create electronic mismatches, which play a crucial role in catalytic processes. The concept involves integrating heterojunctions into MOF structures by coupling MOFs with other MOFs or COFs. Hybrid MOF-on-MOF architectures, especially MOF-COF hybrids, enhance photocatalytic performance by leveraging the combined porosity and crystallinity of these two porous materials (Kaushal et al. 2021). The structural diversity of such heterojunctions can be further expanded by incorporating alternative MOFs—for instance, ZIF-90@ZIF-8 systems, where ZIF-90 serves as the host and ZIF-8 as the guest (Kassem et al. 2021).

Additionally, MOFs offer tunable geometries and properties, enabling precise control over pore structures through the choice of metal nodes and organic linkers. This makes them excellent hosts for incorporating functional guest materials. As research advances, MOF-based composites are emerging as highly effective platforms for heterojunctions. The following section presents examples that highlight the advantages of these heterojunction systems in enhancing catalytic efficiency.

i. **Hybrid heterostructures of MOF–COF**

The effective integration of MOFs and COFs, each with distinct porosity and crystal structures, generates notable electronic mismatches that facilitate the formation of active sites for catalysis. These integrated hybrid systems are formed through modifications in covalent bonding, showing great promise for the development of efficient photocatalysts. Moreover, this combination enhances light absorption and promotes effective charge separation, ultimately leading to improved photocatalytic performance (Healy et al. 2020).

ii. **Complexes of ZIF-90@ZIF-8**

To construct a MOF-on-MOF heterojunction, ZIF-90 is combined with ZIF-8, where ZIF-90 serves as the host and ZIF-8 as the guest. The resulting binary composite, ZIF-90@ZIF-8, successfully integrates the beneficial properties of both ZIF structures, presenting promising potential for enhanced catalytic performance (Kaushal et al. 2021).

iii. **MOF-based composites with various components**

MOFs, with their versatile architectures and tunable pore structures, also serve as excellent hosts for functional materials like nanoparticles, quantum dots, polyoxometalates, enzymes, and polymers. When integrated into hybrid systems, these combinations often exhibit enhanced or entirely novel properties that surpass those of the individual components, unlocking broader applications across various catalytic processes (Toyao et al. 2013).

iv. **N-doped C @ CoO/MoC heterojunction catalyst**

Moreover, a heterojunction catalyst featuring uniformly distributed CoO/MoC embedded within MOF-derived nitrogen-doped carbon has been developed to enable efficient dual functionality (Wang et al. 2023). This well-integrated structure, with atomic-level variations, significantly enhances both the stability and activity of the catalyst, making it highly suitable for a wide range of catalytic applications.

11.5.2 Recent Advances in MOF-Catalyzed Nitroarene Reduction

In recent years, research on the reduction of nitroarenes using MOFs as catalysts has expanded considerably. MOFs such as UiO-66, MIL-101, and ZIF-8 have exhibited outstanding catalytic activity and selectivity in the reduction of nitro-based compounds. These MOFs possess unique active sites and structural features that facilitate the adsorption and activation of nitro substrates on their surfaces. Compared to conventional catalysts used for nitro reduction, MOFs demonstrate significant advantages in terms of selectivity, stability, and reusability (Mohammadi et al. 2023).

A recent review underscores the effectiveness of MOF-derived catalysts in the selective hydrogenation of nitroarenes, even in their pyrolyzed forms (Shin et al. 2023). It clearly demonstrates the versatility of pyrolyzed MOF-based systems, featuring catalysts such as Cu, Co, and Ni, along with nitrogen-doped carbon materials, transition-metal phosphides and sulfides, and metal nanoparticles embedded within porous frameworks, all of which contribute to synergistic catalytic effects (Melillo et al. 2023).

In another study, the catalytic performance of MOF-derived materials was investigated, particularly focusing on a nickel-based system in which Ni nanoparticles were uniformly embedded within nitrogen-doped carbon shells. The role of amino groups in these materials was thoroughly analyzed, revealing that they functioned both as a nitrogen source and as anchoring sites for the Ni nanoparticles (Chen et al. 2019). Additionally, the capacity of MOFs to preserve active sites during pyrolysis was highlighted in a publication on SpringerLink, which described the application of biochar-supported metal catalysts for the chemoselective reduction of nitroarenes. These metal–biochar hybrids (MBCs) demonstrated outstanding catalytic activity, significantly accelerating the reduction process (Singh et al. 2023).

The magnetic nature of a copper-based MOF, Cu-BTC@Fe_3O_4 (where BTC = 1,3,5-benzenetricarboxylate), has been studied to evaluate the catalytic performance for the reduction of nitroarenes under moderate aqueous conditions. The reducing agent used was sodium borohydride. This reaction produced a substantial quantity of the matching anilines. Furthermore, the catalyst's magnetic nature allows for facile separation with an external magnet and reuse up to six times (Whelan et al. 2021).

The magnetic properties of a copper-based MOF, Cu-BTC@Fe_3O_4 (where BTC refers to 1,3,5-benzenetricarboxylate), have been explored to assess its catalytic efficiency in the reduction of nitroarenes under mild aqueous conditions. Sodium borohydride was employed as the reducing agent, resulting in a significant yield of the corresponding anilines. Moreover, the inherent magnetism of the catalyst enables easy separation using an external magnet and allows for reuse across six consecutive cycles (Whelan et al. 2021).

Investigations into MOF-based composites incorporating metal oxides or metal nanoparticles (M/MO@MOF) have demonstrated significant advancements in heterogeneous catalysis. For instance, integrating ZnO nanoclusters into a cobalt-based MOF, $[Co(OBA)(H_2O)_2]$, results in a stable ZnO-MOF composite that serves as an efficient catalyst for the conversion of p-nitrophenol to p-aminophenol (Tsuzuki et al. 2021). The synergistic interaction between the ZnO nanoclusters and the MOF framework is key to the enhanced catalytic performance, achieving an impressive conversion efficiency of up to 97.78%. The catalytic role of the MOF is further emphasized by the π-π interactions between p-nitrophenol and the MOF's organic linkers, which facilitate effective substrate adsorption and boost catalytic activity.

Similarly, cobalt nanoparticles embedded within a carbon framework derived from ZIF-67, known as Co@ZDC@mC-700, have exhibited outstanding catalytic activity in nitroarene reduction using hydrazine hydrate as the reducing agent. In this system, the mesoporous carbon (mC) produced from ZIF-67 confines the Co nanoparticles, preventing agglomeration and metal leaching, thereby enhancing the catalyst's stability and reusability. Moreover, the spacing between the Co nanoparticles and the mesoporous carbon structure allows reactants to accumulate efficiently, improving contact with active sites and promoting effective catalytic transformations (Yin et al. 2023).

In another study, palladium nanoparticles incorporated within UiO-67 (Pd/UiO-67) demonstrated excellent catalytic performance for nitrobenzene reduction at room temperature under hydrogen pressure. This high activity and durability are attributed to the effects of nanoconfinement and electron donation from the MOF framework (Yin et al. 2023). A similar strategy has been employed with gold nanoparticles, forming Au@MOF systems that effectively catalyze the reduction of 4-nitrophenol and 2,4-dinitrophenol in the presence of sodium borohydride and water (Lv et al. 2019).

MOFs additionally act as proficient photosensitizers in the photocatalytic conversion of nitroarenes to valuable chemicals. For example, a zinc-based MOF, $[Zn_3(OH)_2(ADBEB)_2]\cdot 3DEF$, synthesized from 4,4′-(anthracene-9,10-diylbis(ethyne-2,1-diyl))dibenzoic acid and $Zn(NO_3)_2\cdot 6H_2O$, demonstrated significant photocatalytic activity in transforming nitroarenes into anilines using hydrazine hydrate as a reducing agent (Ebadi et al. 2023). Photogenerated electrons inside the MOF framework are responsible for the reduction of nitroarenes. Cu-β-CD-MOF, a photoactive MOF made of copper and beta-cyclodextrin (β-CD), has a rod-like shape and high thermal stability. This material has shown effective photocatalytic performance in the reduction of nitroarenes, which is primarily due to its narrow band gap. Adding metals to MOFs has been shown to improve their efficiency in visible light-driven photocatalytic applications (Li et al. 2022). The Ni(II)-porphyrin-based metal–organic framework, $[Ni_3(Ni\text{-}HTCPP)_2(\mu_2\text{-}H_2O)_2(H_2O)_4(DMF)_2]\cdot 2DMF$, shows strong catalytic efficiency in the visible-light-driven reduction of various nitroarenes, including 4-nitrotoluene, 1-chloro-4-nitrobenzene, 1-bromo-4-nitrobenzene, 1,3-dinitrobenzene, and 2,6-dinitrotoluene. All of these reductions are performed at room temperature with sodium borohydride as the reducing agent. The photocatalytic activity of this Ni(II)-MOF is mostly due to the transport of photoexcited electrons from the porphyrin units to the Ni-oxo nodes in the framework (Begum et al. 2016).

Similarly, a bimetallic porphyrin MOF composite combining indium and palladium (In/PdPMOF) demonstrated selective and effective catalytic reduction of nitroaromatic compounds with deionized water as the hydrogen source. The improved catalytic performance of this material is attributed to the synergistic interaction of the two metal centers, which facilitates efficient separation and migration of photogenerated charge carriers (Yan et al. 2021). Platinum-decorated, amino-functionalized titanium-based MOF (Pt/Ti-MOF-NH_2) shows efficient photocatalytic reduction of nitrobenzene in visible light. The inclusion of platinum considerably improves catalytic efficiency, owing to electron transport from the MOF's organic linker to the Pt nanoparticles, which operate as a cocatalyst (Guo et al. 2022).

Additionally, a polyoxometalate-based MOF (POMOF), namely $\{[Co(H_2O)]_2[Co_2(H_2O)_6(TPT)][Co(TPT)PW_{11}O_{39}]\}\cdot 3H_2O$, has shown potential photocatalytic activity for nitrobenzene reduction under moderate circumstances. The efficiency of this system is attributed to the coordination between Co(II)-substituted Keggin-type anions ($[PCoW_{11}O_{39}]^{5-}$) and the ligand TPT (2,4,6-tri(4-pyridyl)-1,3,5-triazine), along with π-π interactions among TPT units—both of which play a crucial role in facilitating the photocatalytic process. Similarly, a polyoxometalate-based metal–organic framework (ZnW-TPT) has demonstrated excellent photocatalytic reduction of nitroarenes into their corresponding anilines under moderate circumstances, with hydrazine hydrate serving as the reducing agent. The material is created by combining a zinc-substituted Keggin-type polyanion, $[SiZnW_{11}O_{39}]^{6-}$, with a light-responsive organic linker, 2,4,6-tri(4-pyridyl)-1,3,5-triazine (TPT). As in comparable systems, direct coordination between the polyanion and the TPT ligand is critical for improving photocatalytic activity.

11.6 Summary

11.6.1 Challenges and Future Perspectives

Despite the significant progress in MOF-catalyzed nitroarene reduction, several challenges remain. The optimization of MOF structures and active sites to enhance catalytic efficiency and selectivity is an ongoing research focus. The development of scalable synthesis methods and the exploration of new MOF materials with improved properties are also important for practical applications. Furthermore, understanding the reaction mechanisms and the role of reaction conditions in MOF catalysis will contribute to the rational design of catalysts with enhanced performance. Future research efforts should also focus on the integration of MOFs with other catalytic systems and the exploration of new reaction pathways for nitroarene reduction.

11.6.2 Outlook

In conclusion, the use of MOFs as catalysts for nitroarene reduction offers a promising approach to overcome the limitations of traditional catalysts. The unique properties of MOFs, including high surface area, tunability, and diverse active sites, make them attractive candidates for efficient and selective catalysis. Recent advancements in the design, synthesis, and characterization of MOF catalysts have demonstrated their potential in nitroarene reduction reactions. However, further research is needed to address the remaining challenges and unlock the full potential of MOFs in this field. The continued exploration of MOF-catalyzed nitroarene reduction will contribute to the development of sustainable and efficient synthetic methodologies to produce valuable compounds.

References

Kovacic, P., Somanathan, R.: Nitroaromatic compounds: Environmental toxicity, carcinogenicity, mutagenicity, therapy and mechanism. J. Appl. Toxicology. **34**(8), 810–24 (2014)

Orlandi, M., Brenna, D., Harms, R., Jost, S., Benaglia, M.: Recent developments in the reduction of aromatic and aliphatic nitro compounds to amines. Org. Process. Res. & Development. **22**(4), 430–45 (2016)

Kottappara, R., Pillai, S.C., Vijayan, B.K.: Copper-based nanocatalysts for nitroarene reduction-A review of recent advances. Inorg. Chem. Commun. **1**(121), 108181 (2020)

Dhakshinamoorthy, A., Garcia, H.: Catalysis by metal nanoparticles embedded on metal–organic frameworks. Chem. Soc. Reviews. **41**(15), 5262–84 (2012)

Kadam, H.K., Tilve, S.G.: Advancement in methodologies for reduction of nitroarenes. RSC Advan. **5**(101), 83391–407 (2015)

Wu, B., Zheng, N.: Surface and interface control of noble metal nanocrystals for catalytic and electrocatalytic applications. Nano Today. **8**(2), 168–97 (2013)

Begum, R., Rehan, R., Farooqi, Z.H., Butt, Z., Ashraf, S.: Physical chemistry of catalytic reduction of nitroarenes using various nanocatalytic systems: past, present, and future. J. Nanoparticle Research. **18**, 1–24 (2016)

Yang, S., Zhang, Z.H., Chen, Q., He, M.Y., Wang, L.: Magnetically recyclable metal–organic framework@ Fe3O4 composite-catalyzed facile reduction of nitroarene compounds in aqueous medium. Appl. Organomet. Chemistry. **32**(3), e4132 (2018)

Kaur, R., Chhibber, M., Mahata, P., Mittal, S.K.: Induction of catalytic activity in ZnO loaded cobalt based MOF for the reduction of nitroarenes. Chemistry Select. **3**(12), 3417–25 (2018)

Mouchaham, G., Wang, S., Serre, C.: The stability of metal–organic frameworks. Met. Org. Fram.: Appl. Sep. Catal. **23**, 1–28 (2018)

Lv, X.L., Yuan, S., Xie, L.H., Darke, H.F., Chen, Y., He, T., Dong, C., Wang, B., Zhang, Y.Z., Li, J.R., Zhou, H.C.: Ligand rigidification for enhancing the stability of metal–organic frameworks. J. Am. Chem. Society. **141**(26), 10283–93 (2019)

Del Angel, R., Mouchaham, G., Nouar, F., Tissot, A., Serre, C.: Robust and environmentally friendly MOFs. Met. - Org. Fram. Biomed. Environ. Field, 1–31 (2021)

Wibowo, A., Marsudi, M.A., Pramono, E., Belva, J., Parmita, A.W., Patah, A., Eddy, D.R., Aimon, A.H., Ramelan, A.: Recent improvement strategies on metal-organic frameworks as adsorbent, catalyst, and membrane for wastewater treatment. Molecules. **26**(17), 5261 (2021)

Bosch, M., Zhang, M., Zhou, H.C.: Increasing the stability of metal-organic frameworks. Adv. Chem. **2014**(182327.10), 1155 (2014)

Healy, C., Patil, K.M., Wilson, B.H., Hermanspahn, L., Harvey-Reid, N.C., Howard, B.I., Kleinjan, C., Kolien, J., Payet, F., Telfer, S.G., Kruger, P.E.: The thermal stability of metal-organic frameworks. Coord. Chem. Reviews. **15**(419), 213388 (2020)

Kong, X.J., Li, J.R.: An overview of metal–organic frameworks for green chemical engineering. Engineering. **7**(8), 1115–39 (2021)

Huang, Q., Yang, Y., Qian, J.: Structure-directed growth and morphology of multifunctional metal-organic frameworks. Coord. Chem. Reviews. **1**(484), 215101 (2023)

Dinadayalane, T., Lazare, J., Alzaaqi, N.F., Herath, D., Hill, B., Campbell, A.E.: Structures, properties, and applications of nitrogen-doped graphene. In: Theoretical and computational chemistry, vol. 21, pp. 211–248. Elsevier, (2022)

Feng, S.H., Li, G.H.: Hydrothermal and solvothermal syntheses. In: Modern inorganic synthetic chemistry, pp. 73–104. Elsevier, (2017)

Desai, A.V., Lizundia, E., Laybourn, A., Rainer, D.N., Armstrong, A.R., Morris, R.E., Wuttke, S., Ettlinger, R.: Green synthesis of reticular materials. Adv. Funct. Materials. **1**, 2304660 (2023)

Pournoori, N., Delavari, H., Madah, M.: Radio wave/microwave-involved methods for cancer diagnosis. In: Electromagnetic waves-based cancer diagnosis and therapy, pp. 1–64. Academic Press, (2023)

Har-Kedar, I., Bleehen, N.M.: Experimental and clinical aspects of hyperthermia applied to the treatment of cancer with special reference to the role of ultrasonic and microwave heating. In: Advances in radiation biology, vol. 6, pp. 229–266. Elsevier, (1976)

Jicsinszky, L., Rossi, F., Solarino, R., Cravotto, G.: Comparison of the conventional and mechanochemical syntheses of cyclodextrin derivatives. Molecules. **28**(2), 467 (2023)

(a) Tsuzuki, T.: Mechanochemical synthesis of metal oxide nanoparticles. Commun. Chem. **4**(1), 143 (2021); (b) Samal, M., Panda, J., Biswal, B.P., Sahu, R.: Kitchen grinder: a tool for the synthesis of metal–organic frameworks towards size selective dye adsorption. Cryst. Eng. Comm. **20**(18), 2486-2490 (2018)

Li, Z., Dong, J., Zhang, H., Zhang, Y., Wang, H., Cui, X., Wang, Z.: Sonochemical catalysis as a unique strategy for the fabrication of nano-/micro-structured inorganics. Nanoscale Adv. **3**(1), 41–72 (2021)

Das, R.S., Tiwari, B.K., Chemat, F., Garcia-Vaquero, M.: Impact of ultrasound processing in alternative protein systems: Protein extraction, nutritional effects and associated challenges. Ultrason. Sonochemistry. **21**, 106234 (2022)

Butova, V.V., Soldatov, M.A., Guda, A.A., Lomachenko, K.A., Lamberti, C.: Metal-organic frameworks: structure, properties, methods of synthesis and characterization. Russ. Chem. Reviews. **85**(3), 280 (2016)

Raptopoulou, C.P.: Metal-organic frameworks: Synthetic methods and potential applications. Materials. **14**(2), 310 (2021)

Yusuf, V.F., Malek, N.I., Kailasa, S.K.: Review on metal-organic framework classification, synthetic approaches, and influencing factors: Applications in energy, drug delivery, and wastewater treatment. ACS Omega. **7**(49), 44507–31 (2022)

Sadeghzadeh, S.M., Zhiani, R., Emrani, S.: The reduction of 4-nitrophenol and 2-nitroaniline by the incorporation of Ni@ Pd MNPs into modified UiO-66-NH 2 metal–organic frameworks (MOFs) with tetrathia-azacyclopentadecane. New J. Chem. **42**(2), 988–94 (2018)

Chen, M.L., Zhou, S.Y., Xu, Z., Ding, L., Cheng, Y.H.: Metal-organic frameworks of MIL-100 (Fe, Cr) and MIL-101 (Cr) for aromatic amines adsorption from aqueous solutions. Molecules. **24**(20), 3718 (2019)

Tang, J., Zhang, S., Chen, X., Zhang, L., Du, L., Zhao, Q.: Highly efficient catalytic reduction of nitrobenzene using Cu@ C based on a novel Cu–MOF precursor. Catalysts. **13**(6), 956 (2023)

Yang, S., Wang, Y., Li, H., Zhan, Y., Ding, X., Wang, M., Wang, X., Xiao, L.: Synthesis of nano-ZIF-8@ chitosan microspheres and its rapid removal of p-hydroxybenzoic acid from the agro-industry and preservatives. J. Porous Mater. **28**, 29–38 (2021)

Mohammadi, P., Heravi, M.M., Mohammadi, L., Saljooqi, A.: Preparation of magnetic biochar functionalized by polyvinyl imidazole and palladium nanoparticles for the catalysis of nitroarenes hydrogenation and Sonogashira reaction. Scientific Reports. **13**(1), 17375 (2023)

Melillo, A., Franconetti, A., Alvaro, M., Ferrer, B., Garcia, H.: Metal nodes of metal-organic frameworks can activate molecular hydrogen. Chemistry–A Eur. Journal. **29**(1), e202202625 (2023)

Kaushal, S., Kaur, G., Kaur, J., Singh, P.P.: First transition series metal–organic frameworks: synthesis, properties and applications. Materials Advan, **2**(22), 7308–7335 (2021)

Tang, B., Song, W.C., Yang, E.C., Zhao, X.J.: MOF-derived Ni-based nanocomposites as robust catalysts for chemoselective hydrogenation of functionalized nitro compounds. RSC advances. **7**(3), 1531–9 (2017)

Li, X., She, W., Wang, J., Li, W., Li, G.: A highly efficient LaOCl supported Fe–Fe3 C-based catalyst for hydrogenation of nitroarenes fabricated by coordination-assisted pyrolysis. Catal. Sci. & Technol. **11**(13), 4627–35 (2021)

Xiao, W.Z., Xiao, L.P., Yang, Y.Q., Xu, Q., He, W.Q., Zhang, J., Wang, R.Y., Zhao, X., Zhai, S.R., Sun, R.C.: Fully exposed silver nanoparticles stabilized on pH-responsive lignin-reactors for enhanced 4-nitrophenol reduction. J. Environ. Chem. Engineering. **10**(3), 107945 (2022)

Wu, Z., Li, Y., Zhang, C., Huang, X., Peng, B., Wang, G.: Recent advances in metal-organic-framework-based catalysts for thermocatalytic selective oxidation of organic substances. Chem Catalysis, (2022)

Kassem, A.A., Abdelhamid, H.N., Fouad, D.M., Ibrahim, S.A.: Catalytic reduction of 4-nitrophenol using copper terephthalate frameworks and CuO@ C composite. J. Environ. Chem. Engineering. **9**(1), 104401 (2021)

Lu, X., Song, C., Qi, X., Li, D., Lin, L.: Confinement effects in well-defined metal-organic frameworks (MOFs) for selective CO2 hydrogenation: A review. Int. J. Mol. Sciences. **24**(4), 4228 (2023)

An, H., Hu, Y., Song, N., Mu, T., Bai, S., Peng, Y., Liu, L., Tang, Y.: Two-dimensional heterostructures built from ultrathin CeO_2 nanosheet surface-coordinated and confined metal–organic frameworks with enhanced stability and catalytic performance. Chemical Science. **13**(10), 3035–44 (2022)

Remya, V.R., Kurian, M.: Synthesis and catalytic applications of metal–organic frameworks: a review on recent literature. Int. Nano Letters. **9**, 17–29 (2019)

Isaeva, V.I., Nefedov, O.M., Kustov, L.M.: Metal–organic frameworks-based catalysts for biomass processing. Catalysts. **8**(9), 368 (2018)

Konnerth, H., Matsagar, B.M., Chen, S.S., Prechtl, M.H., Shieh, F.K., Wu, K.C.: Metal-organic framework (MOF)-derived catalysts for fine chemical production. Coord. Chem. Rev. **1**(416), 213319 (2020)

Shen, L., Qian, Y., Lyu, Z., Kim, D.H., Kang, D.J.: Two-dimensional metal-organic frameworks and their derivative electrocatalysts for water splitting. Applied Sciences. **13**(16), 9343 (2023)

Kefayati, Z., Malmir, M., Heravi, M.M.: A proficient catalytic reduction of nitroarenes and organic dyes using MOF-derived palladated nano-hydroxyapatite. Res. Chem. Intermed. **13**, 1–20 (2023)

Poshti, E.S., Malmir, M., Heravi, M.M.: Employing a magnetic metal organic framework-derived graphene oxide enhanced with incorporated copper: An exceptional catalyst for the selective reduction of organic dyes and nitro compounds. Diam. Relat. Mater. **22**, 110539 (2023)

Hu, A., Lu, X., Cai, D., Pan, H., Jing, R., Xia, Q., Zhou, D., Xia, Y.: Selective hydrogenation of nitroarenes over MOF-derived Co@ CN catalysts at mild conditions. Mol. Catal. **1**(472), 27–36 (2019)

Ebadi, M., Asikin-Mijan, N., Md. Jamil, M.S., Iqbal, A., Yousif, E., Md Zain, A.R., Aziz, T.H., Rahimi Yusop, M.: Palladium nanoparticles on chitosan-coated superparamagnetic manganese ferrite: A Biocompatible heterogeneous catalyst for nitroarene reduction and allyl carbamate deprotection. Polymers. **15**(1), 232 (2023)

Singh, P., Mukherjee, A., Mahato, A., Pramanik, A., Dhak, D.: A review on chemoselective reduction of nitroarenes for wastewater remediation using biochar supported metal catalysts: Kinetic and mechanistic studies. Chemistry Africa. **6**(2), 561–78 (2023)

Li, S., Zhao, L., Yao, Y., Gu, Z., Liu, C., Hu, W., Zhang, Y., Zhao, Q., Yu, C.: MOF-on-MOF heterostructures with core–shell and core–satellite structures via controllable nucleation of guest MOFs. Cryst Eng Comm. **25**(2), 284–9 (2023)

Mo, D., Wang, Z., Sun, K., Xie, X., Zhang, J., Cai, K.: Core–shell metal–organic frameworks and hierarchical host–guest structures toward water-stable luminescence of lanthanide complexes in encoding beads. J. Mater. Chem. C. **8**(32), 11110–8 (2020)

Guo, C., Duan, F., Zhang, S., He, L., Wang, M., Chen, J., Zhang, J., Jia, Q., Zhang, Z., Du, M.: Heterostructured hybrids of metal–organic frameworks (MOFs) and covalent–organic frameworks (COFs). J. Mater. Chem. A. **10**(2), 475–507 (2022)

Yin, R., Ma, S., Ying, J., Lu, Z., Niu, X., Feng, J., Xu, F., Zheng, Y., Liu, W., Cao, X.: MOF–derived N-doped C@ CoO/MoC heterojunction composite for efficient oxygen reduction reaction and long-life Zn–air battery. Batteries. **9**(6), 306 (2023)

Li, W., Guo, Z., Yang, J., Li, Y., Sun, X., He, H., Li, S., Zhang, J.: Advanced strategies for stabilizing single-atom catalysts for energy storage and conversion. Electrochem. Energy Rev. **5**(3), 9 (2022)

Yan, X., Chen, L., Song, H., Gao, Z., Wei, H., Ren, W., Wang, W.: Metal–organic framework (MOF)-derived catalysts for chemoselective hydrogenation of nitroarenes. New J. Chem. **45**(39), 18268–76 (2021)

Li, J., Xu, Y., Liang, L., Ge, R., Yang, J., Liu, B., Feng, J., Li, Y., Zhang, J., Zhu, M., Li, S.: Metal-organic frameworks-derived nitrogen-doped carbon with anchored dual-phased phosphides as efficient electrocatalyst for overall water splitting. Sustain. Mater. Technol. **1**(32), e00421 (2022)

Li, J., Wang, B., Qin, Y., Tao, Q., Chen, L.: MOF-derived Ni@ NC catalyst: synthesis, characterization, and application in one-pot hydrogenation and reductive amination. Catal. Sci. & Technol. **9**(14), 3726–34 (2019)

Tseng, P.S., Chang, L.X., Ou, Y.S., Chou, C.M., Tsao, C.S., Wu, Y., Chou, J.P., Chen, P.J., Wang, C.Y.: ZIF-67 derived Co nanoparticles on ZIF-Derived carbon for hydrogen spillover and storage. Appl. Surf. Science. **30**(638), 158097 (2023)

Luz, I., Rösler, C., Epp, K., Llabres i Xamena, F.X., Fischer, R.A.: Pd@ UiO-66-type MOFs prepared by chemical vapor infiltration as shape-selective hydrogenation catalysts. Eur. J. Inorg. Chemistry. **2015**(23), 3904–3912 (2015)

Dhakshinamoorthy, A., Asiri, A.M., Garcia, H.: Metal organic frameworks as versatile hosts of Au nanoparticles in heterogeneous catalysis. ACS Catalysis. **7**(4), 2896–919 (2017)

Shin, C., Kim, J., Huh, S.: Fluorescent and catalytic properties of a 2d lamellar Zn metal-organic framework with sql network structure. Molecules. **28**(17), 6357 (2023)

Whelan, É., Steuber, F.W., Gunnlaugsson, T., Schmitt, W.: Tuning photoactive metal–organic frameworks for luminescence and photocatalytic applications. Coord. Chem. Reviews. **15**(437), 213757 (2021)

Deenadayalan, M.S., Sharma, N., Verma, P.K., Nagaraja, C.M.: Visible-light-assisted photocatalytic reduction of nitroaromatics by recyclable Ni (II)-porphyrin metal–organic framework (MOF) at RT. Inorganic Chem. **55**(11), 5320–7 (2016)

Wang, S., Li, S., Feng, H., Yang, W., Feng, Y.S.: Visible-light-driven porphyrin-based bimetallic metal-organic frameworks for selective photoreduction of nitro compounds under mild conditions. ACS Appl. Mater. & Interfaces. **15**(3), 4845–56 (2023)

Toyao, T., Saito, M., Horiuchi, Y., Mochizuki, K., Iwata, M., Higashimura, H., Matsuoka, M.: Efficient hydrogen production and photocatalytic reduction of nitrobenzene over a visible-light-responsive metal–organic framework photocatalyst. Catal. Sci. & Technol. **3**(8), 2092–7 (2013)

Chapter 12
Heterogeneous Electrophilic Catalysis

Anand S. Burange

12.1 Introduction

Catalysis has proven to be indispensable in both fine industries and automotive applications (Roessler et al. 1996). Very few chemical processes are carried out using non-catalytic pathways (Budzinski 1997). Particularly within the realm of heterogeneous catalysis, supported metal catalysis holds significant appeal, as it allows for the precise dispersion of metal onto support through suitable techniques. For instance, typically in the case of platinum supported on carbon (Pt/C), the fraction of catalytically active atoms (Pt) exposed, close to 100% can be achieved. Among various metal groups, precious group metals also known as platinum group metals (PGMs) have numerous applications in chemical sciences. PGM includes Pd, Pt, Rh, Ir, Ru, and Os metals (Table 12.1).

PGMs are further classified as light and heavy triads. The light triad includes Pd, Rh, and Ru, whereas Os, Ir, and Pt are the members of the heavy triad class. They are known for their low crustal abundance typically in the range of 0.1–10 ng g^{-1} (Rao et al. 2000). Industrially, carbon is one of the most used supports due to its stability in acidic and basic mediums. In the context of the economic-viability of industrial processes, recovery, refining, and recycle are the major points to be considered. When compared to oxide supports, carbon can be readily burnt off to yield ashes, facilitating the economic recovery of precious metals. Hence, it is preferred over other inorganic supports in industrial applications (Degussa n.d.).

Heterogeneous catalysts have several plus points over homogeneous like ease of separation, reusability, and thermal/chemical stability. However, homogeneous catalysis is purely molecular catalysis; and their mechanisms are easy to understand and investigate and show higher catalytic activity/selectivity compared to heterogeneous

A. S. Burange (✉)
Department of Chemistry, John Wilson Education Society's, Wilson College (Autonomous), Mumbai, Maharashtra, India
e-mail: anand.burange@wilsoncollege.edu

R. S. Ningthoujam and A. K. Tyagi (eds.), *Handbook of Materials Science, Volume 3*, Indian Institute of Metals Series, https://doi.org/10.1007/978-981-96-8229-4_12

Table 12.1 Platinum group metals with their crustal abundance in ngg^{-1}

Ru (1.0)	Rh (< 0.5)	Pd (0.1–3)
Os (<0.1)	Ir (0.3–5)	Pt (1–10)
Platinum group metals		

(Luque and Burange 2022). There are various attempts to link the gap between homo- and heterogeneous catalysis. By bridging the mentioned gap, one can club all the positive features of both hetero- and homogeneous catalysis. Besides this, one can synthesize reusable, stable heterogeneous catalysts as effectively as homogeneous ones. Some of the approaches for the same include the use of Zr/Hf butterfly peroxides onto polyoxometalate surfaces (POM) (Carraro et al. 2011) and metal nanoparticles (NPs) (Astruc et al. 2005). Various metal (MOs) and mixed-metal oxide (MMOs) catalysts displayed considerable activities for the oxidation of a variety of substrates like alkylaromatics, alcohols, and alkanes (Burange et al. 2012).

The function of MOs and MMOs in heterogeneous catalysis varies based on the particular reaction, reaction conditions, and the chemical composition of the catalyst material (Burange and Gawande 2016). Zirconia-based heterogeneous mixed metal oxides received appreciable attention in the last few decades especially in oxidation catalysis (Maksimchuk et al. 2005). Moreover, there are references claiming the presence of surface-anchored peroxo sites as potential oxidants (Dengel and Griffith 1989) where limited information is available concerning their structural and electronic properties. Earlier interestingly, Stanciu et al. observed the activation of oxygen by Zr(IV)bis(diamido) species (Stanciu et al. 2007). A reaction of Zr(IV)bis(diamido) complex with dry dioxygen produced stable orange-brown crystals of Zr(IV) bisperoxo compound (Fig. 12.1). In later studies, peroxo-bridged Hf^{IV} complexes were reported by Falber et al. where the crystal structure of unusual peroxo complex was confirmed (Falber et al. 2008).

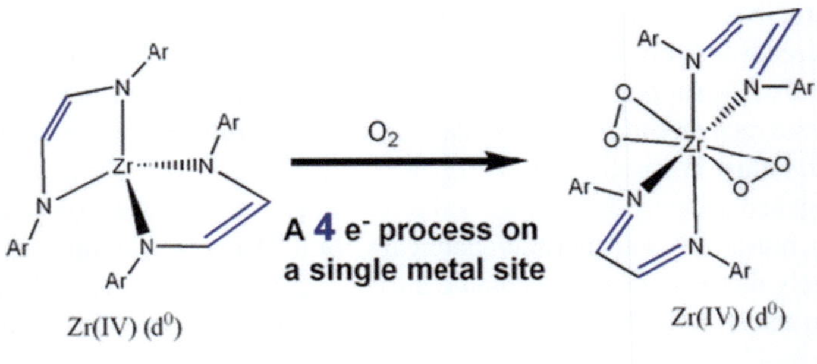

Fig. 12.1 Synthesis of Zr(IV) bisperoxo complex [(Stanciu et al. 2007), Copyright ACS Publications, reproduced with permission]

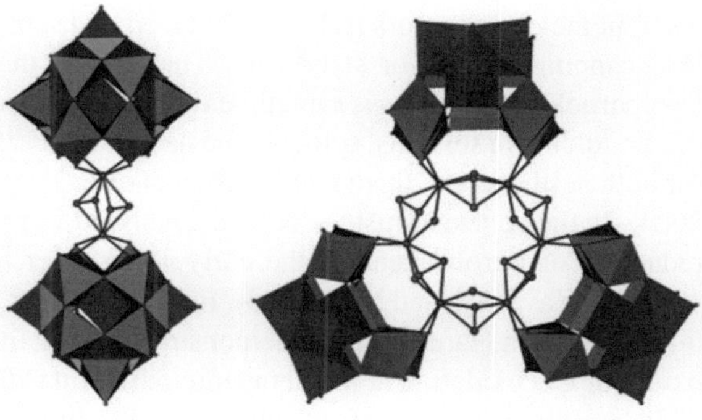

1-3 **4,5**

$K_{12}[Zr_2(O_2)_2(\alpha\text{-}SiW_{11}O_{39})_2]\cdot 25H_2O$	$\mathbf{Zr_2Si}$	**(1)**
$K_{12}[Zr_2(O_2)_2(\alpha\text{-}GeW_{11}O_{39})_2]\cdot 25H_2O$	$\mathbf{Z_2Ge}$	**(2)**
$K_{10.5}Rb_{1.5}[Hf_2(O_2)_2(\alpha\text{-}SiW_{11}O_{39})_2]\cdot 21H_2O$	$\mathbf{Hf_2Si}$	**(3)**
$K_{18}[Zr_6(O_2)_6(OH)_6(\gamma\text{-}SiW_{10}O_{36})_3]\cdot 59H_2O$	$\mathbf{Zr_6Si}$	**(4)**
$K_{18}[Hf_6(O_2)_6(OH)_6(\gamma\text{-}SiW_{10}O_{36})_3]\cdot 59H_2O$	$\mathbf{Hf_6Si}$	**(5)**

Fig. 12.2 Combined polyhedral/ball and stick representation of $[\{M(O_2)(\alpha\text{-}XW_{11}O_{39})\}_2]^{12-}$ [M = Zr, X = Si (1), Ge (2); M = Hf, X = Si(3)], and of $[M_6(O_2)_6(OH)_6(\gamma\text{-}SiW_{10}O_{36})_3]^{18-}$ [M = Zr (4) or Hf (5)] [(Budzinski 1997), Copyright Wiley Publications, adapted with permission]

12.2 Heterogenized Peroxo Species

In the past, polyoxometalates (POMs) consisting mainly of Mo^{VI} and W^{VI} metals have been extensively studied. Recently, the incorporation of tetravalent Zr and Hf ions into the framework of POM has been investigated. This incorporation of larger ions allows for achieving coordination numbers of up to seven, including terminal water molecules. The formed junctions of M-(μ-O)-M (M = $Zr^{4+}$and Hf^{+4}) were also found to promote the formation of oligomerized structures (Gaunt et al. 2003; Kikukawa et al. 2008; Falber et al. 2009). Hetero-polytungstates containing incorporated bimetallic peroxides (Zr^{IV} and Hf^{IV}) (Mal et al. 2010; Bassil et al. 2008), featuring mono-/bis-μ-η^2:η^2 structural rearrangements of peroxo ligand for zirconium or hafnium center within the POM framework are commonly known as butterfly-type cores (Fig. 12.2).

These butterfly peroxides onto POM surfaces displayed excellent catalytic activities for the oxidation of L-methionine and alcohols (Carraro et al. 2011).

12.3 Stabilization of Metal NPs

Catalytic activities of metal NPs (sometimes referred to as nanoclusters/giant clusters/colloids) were recorded quite high even at mild reaction conditions due to their high surface area with surface defects (El-Sayed 2001). In metal nanoparticle synthesis, the bottom-up approach is favored over the top-down approach.

wherein metal precursors (salts) are reduced and subsequently stabilized using suitable capping ligands or stabilizers. The area at the forefront of utilizing metal nanoparticles as catalysts is sometimes termed as semi-heterogeneous. Nanoparticles can be soluble in solvents; soluble gold nanoparticles, for instance, were investigated for both aesthetic and therapeutic purposes over 2500 years ago (Daniel and Astruc 2004). In the course of history, Rapino and Nord were the first to report the catalytic reduction of nitrobenzene in the early 1940s over metal NPs (Rapino and Nord 1941a, b; Kavanagh and Nord 1943). In the late 1980s, the most significant contribution came from Haruta, and he demonstrated the activity of gold NPs supported on oxide for CO oxidation at low temperature (Haruta et al. 1987, 1989; Haruta 2002).

In liquid-phase catalysis, the selectivity for supported metal nanoparticles is generally lower compared to homogeneous protocols. Homogeneous catalysis involves soluble nanoparticles acting as catalysts. A significant challenge posed by metal nanoparticles is their tendency to agglomerate, which is addressed through the use of stabilizers. Stabilization of metal NPs during the synthesis occurred by electrostatic, electrosteric, and steric effects (or combined effect) well depicted in Fig. 12.3.

Polymers as well as the dendrimers both are well-documented for metal NP stabilization (Astruc et al. 2005). Polymers stabilize the metal NPs either by steric interaction and/or by weak binding with the heteroatoms of the polymeric framework where heteroatoms act as ligands. Two major polymer families used for NP stabilization are poly(2,5-dimethylphenylene oxide) and poly(vinylpyrrolidone). Other polymers employed include polyacrylonitrile, polyacrylic acid (Demir et al. 2004), polyurea (Ley et al. 2003), polysiloxane (Chauhan et al. 2004), polysilane (Kidambi et al. 2004), PEG (Pillai and Sahle-Demessie 2004), polyamides (Tabuani et al. 2003), etc. One of the examples where palladium NPs adsorbed on block copolymer (polyacrylic acid) (Demir et al. 2004; Gröschel et al. 2005) is shown in Fig. 12.4. It was reported as an excellent catalytic material for the hydrogenation reaction. On

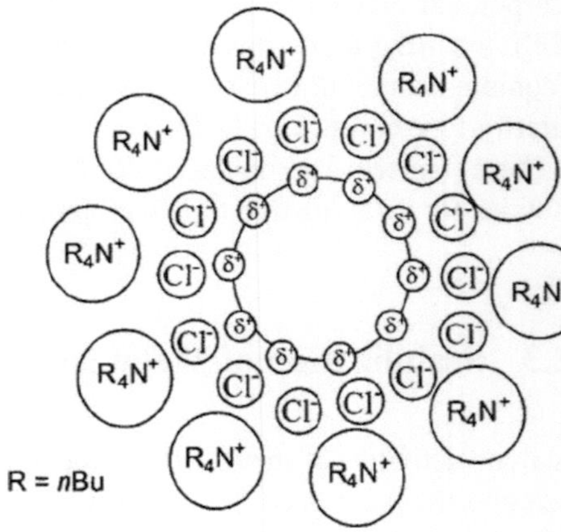

Fig. 12.3 Stabilization of metal NPs obtained by reduction of metal chloride in the presence of tetra-N-alkylammonium cations [(Astruc et al. 2005), Copyright Wiley, reprinted with permission]

comparison between various other polymers used as stabilizers, water-soluble polymers exhibited high selectivity for the olefin's hydrogenation (cyclic versus non-cyclic) (Bradley 1994). The widely used surfactant, sodium dodecyl sulfate (SDS), also serves as a nanoparticle stabilizer in catalysis (Yang et al. 2004).

In homogeneous catalysis, filtration of the catalyst presents a significant challenge. The issue is addressed using modern nanofiltration techniques (Dijkstra et al. 2002), well documented in the literature along with the non-conventional separation methods by using supercritical fluids, fluorous media, and ionic liquids (ILs) (Tzschucke et al. 2002). In these types of separation strategies, dendrimers play a crucial role (Frechet and Tomalia 2001). Dendrimers are macromolecules, their structures are well-defined on a molecular level and have a 1.0 value for polydispersity. Dendrimers resembling the shapes of molecular trees or cauliflowers showed similar behaviors as molecular boxes (Jansen 1994). Such dendrimers showed efficacy in the stabilization of metal NPs by entrapping them with the help of heteroatoms in the dendrimer framework (Astruc et al. 2005). Representation of possible sites/locations for the metal complexes in dendrimer or dendron functionalized with a ligand of choice is given below (Fig. 12.5) (Jesús and Flores 2008).

The solubility of the dendrimer in various media like fluorous, aqueous, and organic can be altered by choosing the right dendrimer terminal groups. For instance, polar terminal groups like amino, hydroxyl, etc. enhance the solubility in aqueous

Fig. 12.4 Pd NP adsorbed on polyacrylic acid block copolymer as hydrogenation catalyst: the Pd NP stabilized through the block copolymer [(Gröschel et al. 2005), Copyright Wiley, reprinted with permission]

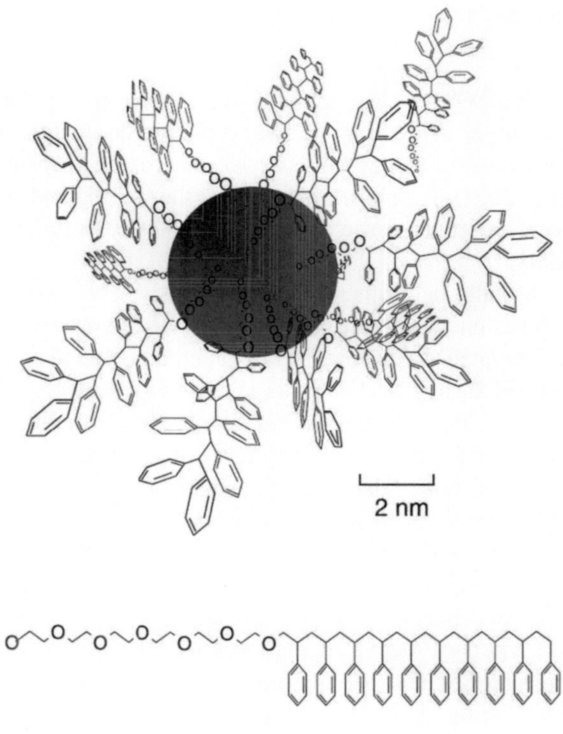

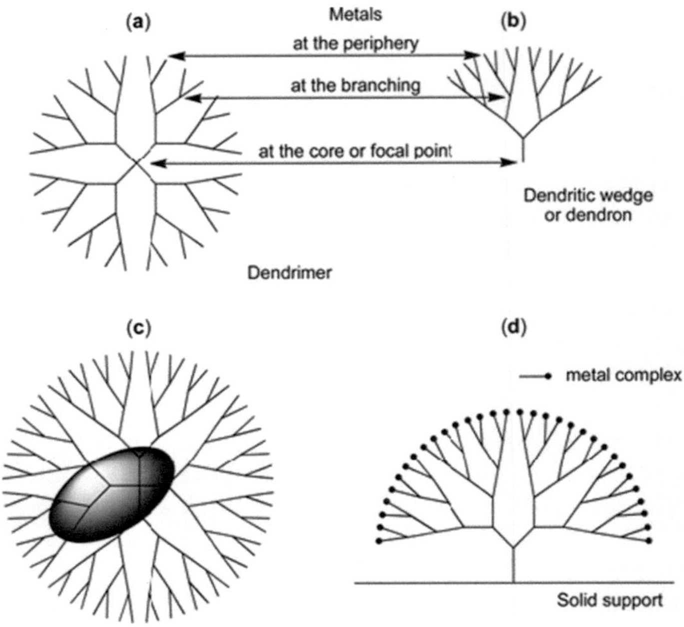

Fig. 12.5 Dendrimer: **a** Possible locations for the metal sites; **b** dendron; **c** a dendrimer encapsulated metallic NP; **d** a dendron-solid support composite [(Jesús and Flores 2008), Copyright ACS Publications, Reprinted with permission]

media. Whereas the reaction of a dendrimer with an amino-terminal group with carboxylic acid (decanoic acid) forms amide and thereby solubility in organic solvent (toluene) increases [well established in case of poly(amidoamine) dendrimer] (Astruc et al. 2005). The pioneering work, in the area of dendrimer-stabilized NPs, as catalysts were reported by the research groups namely Esumi, Crooks, and Tomalia (Zhao et al. 1998).

Metallo-dendrimers used in catalysis have well-defined structures with tailored dimensions which are correlated with the number of layers or dendrimer generation abbreviated as Gn. A variety of generations can be synthesized by employing precise control during the dendrimer synthesis process.

Two major families of commercial dendrimers used in catalysis are Poly(amidoamine) and Poly(propylene imine) abbreviated as PAMAM and PPI, respectively. PPI dendrimers are also referred to as Astromol and DAB-Am-x where DAB stands for diaminobutane core and x represents the primary amino end group number. Selected generations of PPI and PAMAM dendrimers along with their number of end primary amine/terminal nitrogen atoms are enlisted in Table 12.2. PPI dendrimers are smaller in size as well as more stable compared to PAMAM. G3 and G4 PAMAM dendrimers were established as good stabilizers over G2 by preventing metal NP agglomeration (Astruc et al. 2005). PAMAM dendrimer has been explored as a complexing agent for a variety of bivalent (Pt^{2+}, Pd^{2+}, and Cu^{2+})

Table 12.2 Dendrimer generations and their respective end groups

Dendrimer	Generation (Gn)	No. of primary amine end groups
PPI	G1	4
PPI	G2	8
PPI	G3	16
		No. of terminal nitrogen atoms
PAMAM	G1	8
PAMAM	G2	16
PAMAM	G3	32
PAMAM	G4	64

and trivalent metal ions (Au^{3+}, Fe^{3+}, and Ru^{3+}) (Astruc et al. 2005). Crooks pioneered a strategy of developing dendrimer encapsulated catalysts. Dendritic supports have many advantages, for instance, their degree of homogeneous character due to the solvation of dendrimer in a chosen solvent. Another advantage is the ease the recyclability while using such supports in catalysis. In nanoscience and nanotechnology, more specifically in nanocatalysis arena, achieving uniform nanostructured material is a major challenge. The activity of certain materials as catalysts is governed by their structures and particle size range known as structure-sensitive reactions. However, when the catalytic activities are not dependent on structures of catalyst, material is named structure-insensitive reactions by Boudart. However, structure-sensitive reactions are the function of particle size. For particle size ranges 1–5 nm, dramatic structural changes are observed and thus alter and give rise to structural sensitive reactions (Boudart 1985).

12.4 Electrophilic Monodisperse Metal NPs as Catalysts

Comprehending the impact of the size of metal NPs on catalytic activities is relatively straight forward compared to understanding the effect of nanoparticle shape. While controlling size is generally achievable, controlling shape is more challenging. The shape of nanoparticles in the catalysis field is also sometimes referred to as morphology. In the case of Pd, Pt, and Ag, morphology could be altered by the addition of foreign ions (Wiley et al. 2005). During the metal ion reduction process, the chloride ions and oxygen selectively formed single crystalline products with truncated tetrahedra and cuboctahedra morphology. Reduction of Pt (II) species tends to form Pt NPs, whereas the presence of iron chloride induces anisotropic growth of Pt by slowing down the reduction process forming Pt nanowires (Chen et al. 2004). The role of various morphologies/shapes and other aspects related to bridging the gap is well discussed in the recent review (Burange et al. 2021). The first report on the catalytic application of purely heterogeneous NPs exhibiting activity like homogeneous catalysis was reported by Somorjai and Toste's group (Witham et al. 2010). Cyclization reactions like hydroalkoxylation and synthesis of nitrogen or oxygen-containing heterocycles via π-bond activation were carried out using Pt halides which follow a homogeneous pathway. Somorjai et al. carried out the above-mentioned reactions using heterogeneous electrophilic catalysts. Typically, in catalyst preparation, capping was assisted using 4th generation PAMAM dendrimers, i.e., G4OH to yield Pt NPs (Pt_{40}) of size 1.0 nm. However, with the use of PVP, sizes of 1.5, 2.9, and 5.0 nm were achieved. In the next step of the catalyst preparation method, synthesized nanoparticles were supported onto mesoporous silica supports (SBA-15 and MCF-17). The latter was found useful for the support of larger NPs (5.0 nm). On comparison, in terms of catalytic activity (discussed below), it was evident that PAMAM dendrimers capped NPs supported on SBA-15 found better activity for different catalytic reactions along with recyclability. SBA-15 support not only made the catalyst recyclable but also helped in preventing aggregation. Based on the used precursors, catalysts were designated as metal/capping agent/support. For instance, Pt NPs with G4OH dendrimer on SBA-15 were labeled as Pt/G4OH/SBA-15. To introduce an electrophilic nature, the materials were treated with a mild oxidant, $PhICl_2$ (Witham et al. 2010). Reagents such as hypervalent iodine species, N-halosuccinimides (NXS), and iodobenzene dichloride ($PhICl_2$) were reported as electrophilic halogenating reagents (Whitfield and Sanford 2007). The treatment of catalyst materials with $PhICl_2$ and N-chlorosuccinimide (NCS) reagents oxidized metal NPs by forming metal-halogen bonds which led to an increase in the oxidation state of the metal and thus making catalyst material electrophilic in nature (Scheme 12.1).

Pt surfaces with Cl_2 etching and Pd catalysts with $PhICl_2$ showed the formation of surface Pt-Cl and Pd-Cl bonds, respectively (Whitfield and Sanford 2007; Dona et al. 2009). Similar observations were recorded when the above-mentioned virgin catalysts, i.e., metal NPs capped with dendrimer supported onto SBA-15 and MCF-17 upon treatment with $PhICl_2$. To comprehend the effect of these surface modifications (M-Cl bonds), virgin and oxidant-treated (electrophilic) catalysts were evaluated for the catalytic activity for hydroalkoxylation of reactant **1** to benzfuran (**2**) (Scheme 12.2). Hydroalkoxylation of **1** to benzfuran (**2**) via a homogeneous pathway using $PtCl_2$ showed excellent yield (97%) at 100 °C in 15 h. Under identical reaction conditions, when the reactions were carried out over heterogeneous catalysts Pt/PVP/

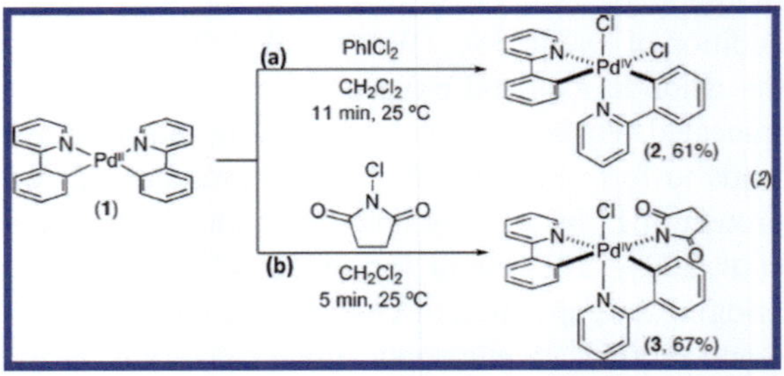

Scheme 12.1 Oxidation of Pd^{II} $(PhPy)_2$ **a** with $PhICl_2$ and **b** with NCS [(Whitfield and Sanford 2007), Copyright *ACS* Publications, Reprinted with permission]

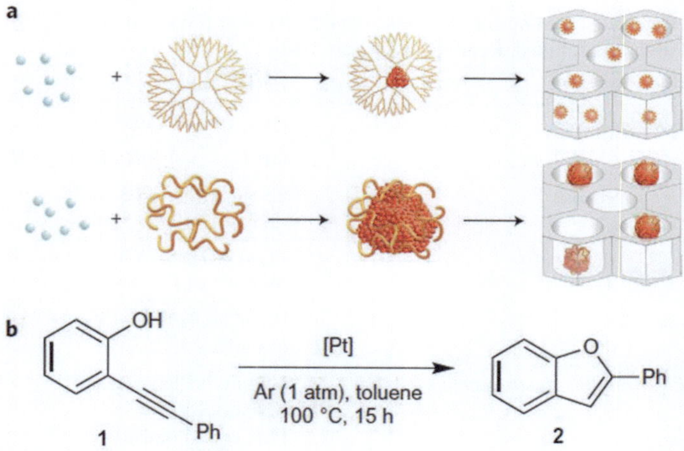

Scheme 12.2 Syntheses and reactivity of NPs: **a** Top scheme: Pt ions loaded onto a PAMAM dendrimer and reduced to form a dendrimer-encapsulated NP. Sonication was employed for the deposition of NPs onto the SBA-15 (mesoporous silica). Bottom scheme: polyvinylpyrrolidone was used to encapsulate the NPs before the deposition on SBA-15. In both cases, NPs were synthesized before being loaded onto SBA-15. **b** Hydroalkoxylation of (1) over Pt NPs. To induce electrophilicity, PtNPs were treated with the mild oxidant $PhICl_2$. The sample, Pt_{40}/G4OH/SBA-15 NPs, required further reduction under an H_2 atmosphere at 100 °C for 24 h before the reaction. The treatment with oxidant produced catalytically active NPs capable of activating the π-bond in (1), leading to the hydroalkoxylation to form benzofuran 2 [(Witham et al. 2010), Copyright Macmillan Publishers Limited, Reprinted with permission]

SBA-15(2.9 nm) and Pt_{40}/G4OH/SBA-15, only 9 and 10% yields of benzfuran were achieved respectively (Table 12.3) (Witham et al. 2010). The reason for the least catalytic activities of these catalysts is the lesser availability of active metal sites owing to the heterogeneity of the protocol. But when these materials were treated with $PhICl_2$ oxidant, the activity was surprisingly found equivalent to the homogeneous protocol ($PtCl_2$). Among all the catalysts, Pt_{40}/G4OH/SBA-15 treated with $PhICl_2$ exhibited excellent catalytic activity (Table 12.3, entry 7).

12.4.1 Effect of Oxidation State and Electrophilicity on Catalytic Activity

In the homogeneous protocol, the highest excellent catalytic activity of catalysts was recorded where Pt exists in + 2 and + 4 oxidation states. In the case of Pt_{40}/G4OH/SBA-15 and Pt/PVP/SBA-15, no appreciable yield of said product was obtained (≤10%). So, it was hypothesized that the reason behind the least activity of heterogeneous systems is the presence of an overabundance of zero-valent Pt species.

Table 12.3 Catalyst screening for hydroalkoxylation reaction [(Witham et al. 2010), Copyright Macmillan Publishers Limited, Reprinted with permission]

Entry	Protocol	Entry	Catalyst/time	Yield (%)
1	*Homogeneous*	1	$PtCl_2$ (5 mol%)/2 h	97
2		2	$PdCl_2$ (4 mol%)/15 h	50
3	*Heterogeneous*	3	2.9 nm Pt/PVP/SBA-15 (2.5 mol%)/15 h	9
4		4	Pt_{40}/G4OH/SBA-15 (4 mol%)/15 h	10
5		5	Pd_{40}/G4OH/SBA-15 (4 mol%)/ 15 h	10
6	*Electrophilic-Heterogeneous*	6	2.9 nm Pt/PVP/SBA-15 (2.5 mol%)/15 h $PhICl_2$ (7.5 mol%)	> 99
7		7	Pt_{40}/G4OH/SBA-15 (4 mol%)/15 h $PhICl_2$ (12 mol%)	98
8		8	Pd_{40}/G4OH/SBA-15 (4 mol%)/15 h $PhICl_2$ (12 mol%)	48

The overabundance of zero-valent Pt was verified by the X-ray photoelectron spectroscopy (XPS) technique showing >90% of Pt species in a metallic (0) oxidation state. To support this hypothesis, the above-mentioned catalysts were treated with $PhICl_2$ oxidant to increase the oxidation state of Pt. After treatment with $PhICl_2$, > 25% of Pt NPs were oxidized in the case of 1.5 nm Pt/PVP proved by XPS analysis. After the treatment with $PhICl_2$, catalytic activity of Pt_{40}/G4OH/SBA-15 was significantly improved to 98% yield (Table 12.3, entry 7). The treatment with oxidant showed significant improvement in catalytic activity compared to unoxidized samples (Table 12.3, entries 6–7) than homogeneous $PtCl_2$ catalyst (Table 12.3, entry 1) for the above-mentioned reaction (Scheme 12.2).

To check the recyclability, catalysts were filtered from the reaction mass by using a simple glass microfiber filter, reduced, and retreated with $PhICl_2$. Pt_{40}/G4OH/SBA-15 was recycled for four consecutive cycles, and the activity was found consistent with > 90% yield. The turnover number (TON) for the said catalyst per metal basis was observed to be 400. Interestingly, the single batch of the mentioned catalyst displayed sustainable activity for one month. Similar results were also observed for Pd catalysts. The catalytic activities of virgin Pd_{40}/G4OH/SBA-15 and $PhICl_2$-treated Pd_{40}/G4OH/SBA-15 were compared with $PdCl_2$ for hydroalkoxylation reaction (Table 12.3, entries 2, 5, and 8). $PhICl_2$-treated Pd_{40}/G4OH/SBA-15 (electrophilic heterogeneous catalysts) displayed comparable activity to homogeneous $PdCl_2$. In both Pt and Pd catalysts, it was required to reduce metal NPs using H_2 (1 atm.) at 100 °C before treatment with oxidant $PhICl_2$ to generate electrophilic activity in situ.

Observations revealed that smaller dendrimer-capped metal nanoparticles exhibit greater stability in higher oxidation states compared to their larger counterparts. The enhanced stability is ascribed to the documented stabilizing effect of PAMAM dendrimers in the literature. In the scenario of dendrimer-capped nanoparticles, one cannot dismiss the possibility of dendrimer decomposition. Indeed, under certain reaction conditions, dendrimers may undergo decomposition. Dendrimer decomposition did not alter the catalytic activity which was evidenced by sustained catalytic activity of Pt_{40}/G4OH/SBA-15 after the recycling. Intriguingly, the cyclization of a phenyl urea derivative (Scheme 12.3) recorded better results over electrophilic heterogeneous catalysts compared to homogeneous routes (Pt catalysts). For the reaction displayed below (Scheme 12.3), only 8% yield of product (methylene-bridged bisindole) was achieved when 5 mol% of $PtCl_2$ and $PtCl_4$ were used as homogeneous catalysts. However, $PhICl_2$-treated Pt_{40}/G4OH/SBA-15(4.0 mol%) and Pt/PVP/SBA-15(2.9 nm) (2.5 mol%) electrophilic catalysts showed 43 and 48% of product respectively (Scheme 12.3).

In metal cluster-based materials, metal atoms are not always strongly bonded to the support which makes leaching a significant concern. Recyclability serves as a strong indicator of heterogeneity, as confirmed by leaching studies. A consistent recyclability was observed in the case of $PhICl_2$ treated Pt_{40}/G4OH/SBA-15 catalyst. A thorough three-phase test was carried out using a Wang-resin bound substrate to ascertain whether nanoparticles from the mentioned catalyst were leaching and responsible for the pronounced activity, akin to homogeneous protocols. Experiments with Wang-resin bound substrate showed 26% conversion in the presence of $PtCl_2$ salt (homogeneous), whereas less than 2% conversion was achieved in the presence of Pt/PVP/SBA-15 (2.9 nm) and Pt_{40}/G4OH/SBA-15 catalysts. Results indicated no role of leached species in catalysis and were further supported by Transmission electron microscopy (TEM) analysis and filtrate transfer test. TEM images of Pt_{40}/G4OH/SBA-15 after and before the reaction did not validate any appreciable agglomeration as well as leaching (Fig. 12.6).

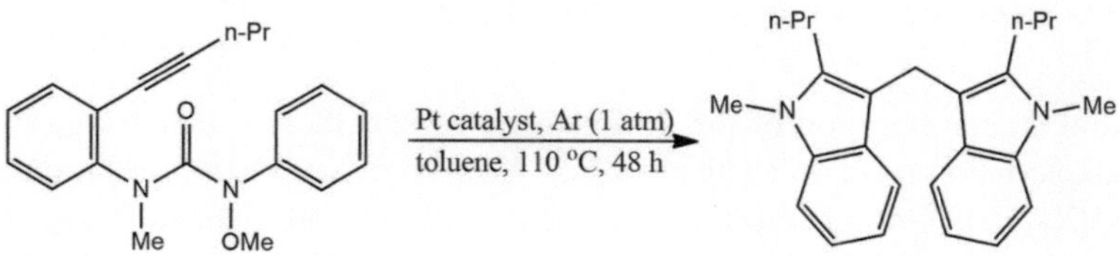

Scheme 12.3 Cyclization of substituted phenyl urea over Pt catalysts [(Witham et al. 2010), Copyright Macmillan Publishers Limited, Adapted with permission]

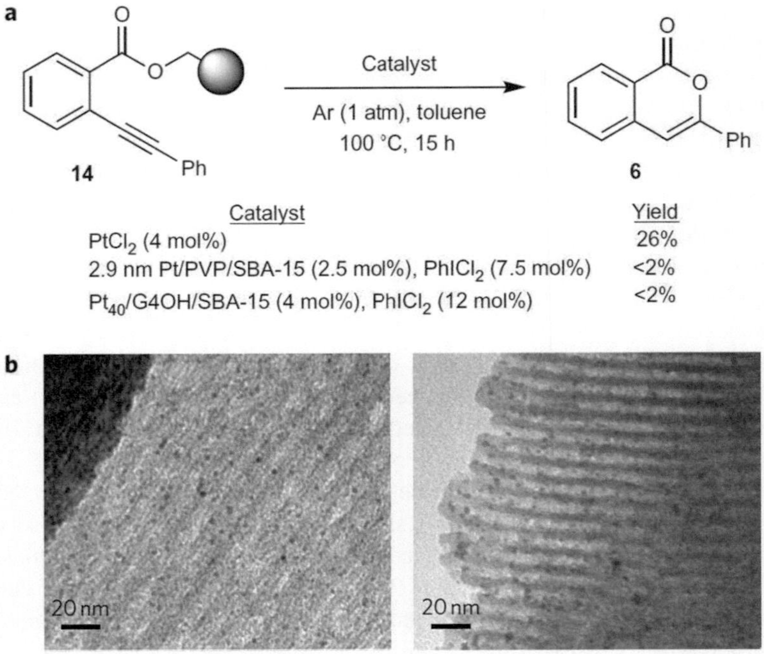

Fig. 12.6 Results from catalyst heterogeneity tests: **a** A three-phase test to determine the presence of a homogeneous catalyst by leaching or release and capture. The reaction rate between the heterogeneous catalyst and polymer-bound substrate should be negligible. Yields were determined by NMR versus internal standard; **b** transmission electron microscope (TEM) images of Pt_{40}/G4OH/SBA-15 before reaction (left), and after treatment with $PhICl_2$ and reaction with 1 (right). No leaching or aggregation was observed [(Witham et al. 2010), Copyright Macmillan Publishers Limited, Adapted with permission]

12.4.2 Mechanistic Investigation of Hydroalkoxylation Over Electrophilic Pt_{40}/G4OH/SBA-15

Structural characterization of the said catalyst reveals the mechanistic part of the hydroalkoxylation reaction (Scheme 12.2). Near-edge X-ray absorption fine structure (NEXAFS), X-ray absorption spectroscopy (XAS), and extended X-ray absorption fine structure (EXAFS) techniques were used to get information regarding the structure and composition of material.

Before the treatment of an oxidant, i.e., $PhICl_2$, the catalyst sample was pre-reduced in a hydrogen atmosphere. Sample Pt_{40}/G4OH/SBA-15 after activation with $PhICl_2$, verified the presence of Pt^0 species along with $Pt^{(II)}$ and $Pt^{(IV)}$ chlorides by XAS studies. Kinetic studies revealed that the presence of $Pt^{(IV)}$ species increases the conversion rate in the hydroalkoxylation reaction (Li et al. 2011). Pt species evolution under given reaction conditions and the yield of the desired hydroalkoxylation reaction as a function of time were analyzed by NEXAFS depicted in Fig. 12.7.

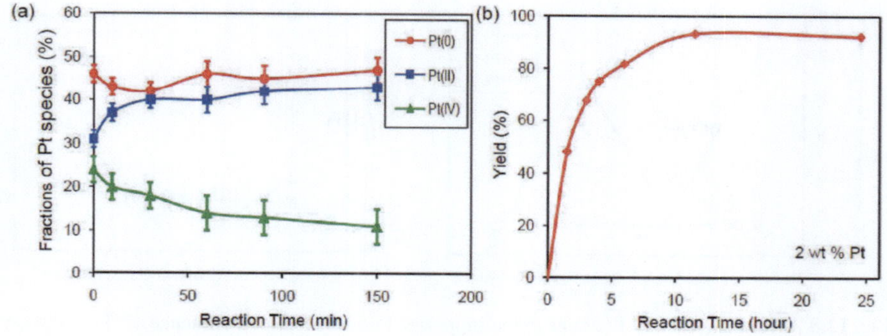

Fig. 12.7 **a** Pt species evolution in the supported catalyst using NEXAFS measurements under reaction conditions. The catalyst sample, 2 wt. % Pt catalyst was pre-oxidized by $PhICl_2$. **b** The product yield as a function of time obtained in a batch reaction under identical conditions [(Li et al. 2011), Copyright ACS Publications, Reprinted with permission]

Pt_{40}/G4OH/SBA-15 treated with molecular hydrogen (24 h, 100 °C) and three equivalents of $PhICl_2$ (45 min, 100 °C) were designated as reduction and oxidation respectively (Scheme 12.4).

These treatments altered the Pt-Cl and Pt-Pt coordination numbers (C.N.) in the material. Typically, in EXAFS measurements at the PtL_3 edge (Fig. 12.8a), after reduction 1, the Pt-Pt C.N. was determined to be 5, indicating a Pt cluster size of ca. 1.0 nm. However, following the treatment with the $PhICl_2$ resulted in a decrease in the Pt-Pt C.N. < 1, accompanied by an increase in the Pt-Cl C.N. The effect of these treatments on C.N. and fractions of Pt species is represented (Fig. 12.8).

Relative catalytic activities of each catalyst after treatments were compared by quantifying their rate constants, where the rate constants were calculated by using a first-order chemical kinetic model for a unimolecular reaction. Oxidation 2 showed significantly higher catalytic activity over oxidation 1, attributed to higher fractions of $Pt^{(IV)}$ species. Especially for π-bond activation reactions, $PtCl_4$ displayed greater activity in the homogeneous protocol. A similar observation was recorded for $Pt^{(IV)}$ species due to its more electrophilic nature than $Pt^{(II)}$ formed after oxidation 2 (Fig. 12.8b). Because of the small-sized (~1.0 nm) dendrimer encapsulated Pt clusters, C.N. and the oxidation state of Pt could be easily altered by oxidation–reduction treatment confirmed by XAS studies. $Pt^{(IV)}$ species were found more catalytically active compared to $Pt^{(II)}$ species for the chosen reaction.

Scheme 12.4 Sample treatments and codes/designations

Sample treatments

$$\text{Reduction 1} \xrightarrow{PhCl_2} \text{Oxidation 1}$$

$$\text{Oxidation 1} \xrightarrow{H_2} \text{Reduction 2}$$

$$\text{Reduction 2} \xrightarrow{PhICl_2} \text{Oxidation 2}$$

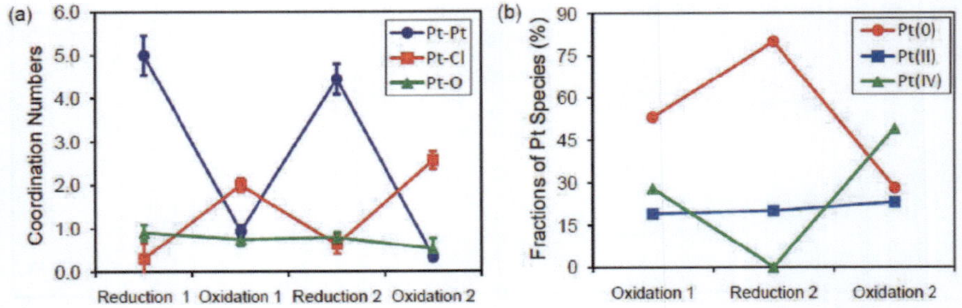

Fig. 12.8 **a** Average C.N. of Pt atoms in the supported Pt catalyst after a sequence of H_2 reduction and $PhICl_2$ oxidation treatments in the toluene derived from EXAFS analysis. **b** The fractions of $Pt^{(0)}$, $Pt^{(II)}$, and $Pt^{(IV)}$ chloride species of the Pt catalyst were derived from NEXAFS analysis [(Li et al. 2011), Copyright ACS Publications, Adapted with permission]

12.4.3 Stability Against Leaching Study and Controlled Experiment Studies

Other than the heterogeneity test discussed above, a few controlled experiments were set up to prove the heterogeneity of the catalyst. The possibilities of leaching involve either the diffusion of Pt ions from inside the dendrimers into reaction media or by desorption from the used silica support. Disparities in the polarities of the solvent used for the reaction and catalyst material were used for the successful leaching inhibition. The solvent used for the reaction is toluene, non-polar in nature while support SBA-15 is polar. Polar materials have generally less solubility in the non-polar medium. Pt ions should theoretically prefer a polar environment over a non-polar one. Validation of this hypothesis was done by comparing product yield (Scheme 12.2) as a function of time in non-polar solvents like toluene, dioxane, and dioxane-water systems (Fig. 12.9). Along with it, the loss of Pt content was also measured by using the ICP-MS technique (Table 12.4).

No leaching was observed in non-polar solvent toluene. However, dioxane is comparatively polar where a subtle Pt loss was noticed after the reaction (Table 12.4, entry 1). To confirm the hypothesis of Pt leaching into a polar medium, highly polar

Table 12.4 Measurement of Pt content using the ICP-MS technique [(Li et al. 2011), Copyright ACS publications, Reprinted with permission]

Entry	Reaction solvent	Pt content loss in the supported catalyst (wt.%)	Solution Pt concentration (ppm)
1	Dioxane	0.04	41.5
2	Dioxane + 2.5% H_2O	0.10	50.2
3	Dioxane + 5.0% H_2O	0.12	81.0

The Pt content in the unused catalyst is 0.632 wt.%; data obtained from ICP-MS measurements

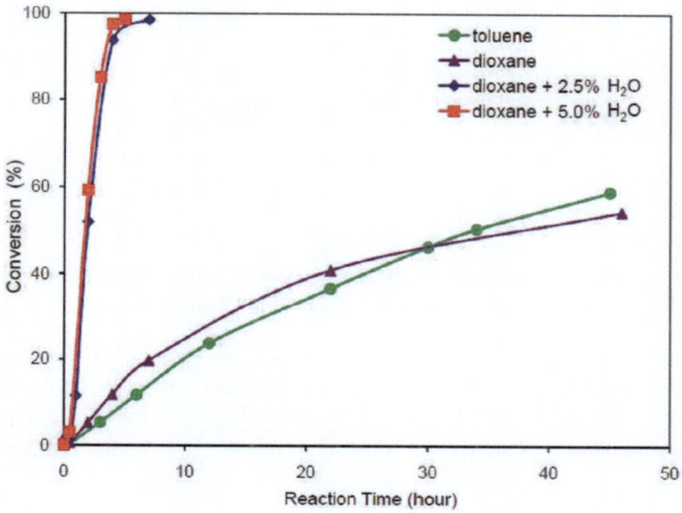

Fig. 12.9 Product yield as a function of reaction time for the reactions. The Pt loading in the catalyst is 1 wt% [(Li et al. 2011), Copyright ACS Publications, Reprinted with permission]

water was added in trace amounts along with dioxane. It resulted in increased leaching of Pt species into the reaction media, as evidenced by the enhanced conversion rate (Fig. 12.10) (Li et al. 2011).

Kim et al. explored Pd NPs supported on SiO_2 and Pd^{IV} oxo-clusters and their electrophilic counterparts (Kim et al. 2014). One of their studies includes the synthesis of a highly Lewis-acidic Pd@SiO_2 catalyst (Kim et al. 2014), wherein TEM analysis revealed the formation of spherical particles with the average diameter of 5.0 ± 0.3 nm (Fig. 12.10a). Formation of Pd@SiO_2 yolk-shell structure was achieved by using the water-in-oil microemulsion method (hydrothermal condition) followed by the treatment with molecular hydrogen. Yolk-shell Pd@$SiO_2$of 5.1 ± 0.4 nm size of Pd cores with large vacancies in the silica shell and 30 ± 2.6 nm average diameter of yolk-shell particle (Fig. 12.10b). Unsupported Pd NPs and yolk-shell Pd@SiO_2 were further treated with oxidant in situ, and their catalytic activity was compared for hydroalkoxylation reaction (Scheme 12.5).

Initially, the hydroalkoxylation was carried out at an elevated temperature (140 °C), but the activity of Pd@SiO_2 activated using $PhICl_2$ was surprisingly high. The conversion data at these temperatures, with and without oxidant, are well tabulated in Table 12.5.

$Ph(OAc)_2$ was incapable of activating Pd@SiO_2, thus no formation of product was observed (Table 12.5, entry 7). However, N-chlorosuccinimide (NCS) was found as effective as $PhICl_2$ with the formation of a lesser amount of electrophilic chlorination product (Table 12.5, entries 8–9). Under given reaction conditions from Table 12.5, the least conversion was observed in the presence of Pd salts (Tables 12.5, 12.1–12.2, 10–11), while no conversion was achieved in the presence of Pd NPs without

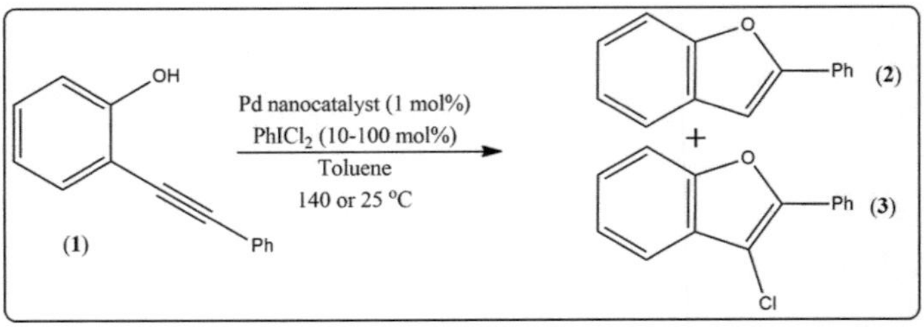

Scheme 12.5 Hydroalkoxylation of 2-phenylethynylphenol (1) to (2) and electrophilic chlorination of (2) to (3) [(Kim et al. 2014), Copyright RSC, adapted with permission]

Table 12.5 Hydroalkoxylation of (1) using homogeneous and heterogeneous catalysts (Scheme 12.5)[a]. [(Kim et al. 2014), Copyright RSC, adapted with permission]

Entry	Temperature (ºC)	Catalyst (1 mol%)	Oxidant (X mol%)	Cyclization (%)[b] (2 + 3)	Conversion (%)[b] (2:3)
1	140	$Pd(OAc)_2$	–	10	10:0
2	140	$PdCl_2$	–	7	7:0
3	140	Pd NPs	–	0	–
4	140	Pd NPs	$PhICl_2$ (25)	100	80:20
5	140	$Pd@SiO_2$	–	0	–
6	140	$Pd@SiO_2$	$PhICl_2$ (25)	100	76:24
7	140	$Pd@SiO_2$	$PhI(OAc)_2$ (100)	0	–
8	140	Pd NPs	NCS (25)	100	90:10
9	140	$Pd@SiO_2$	NCS (25)	100	94:6
10	25	$Pd(OAc)_2$	–	9	9:0
11	25	$PdCl_2$	–	< 1	–
12	25	Pd NPs	$PhICl_2$ (25)	0	–
13	25	$Pd@SiO_2$	$PhICl_2$ (25)	100	78:22
14	25	$Pd@SiO_2$	$PhICl_2$ (10)	100	95:5
15[c]	25	$Pd@SiO_2$	$PhICl_2$ (10)	0	0

[a] All the reactions were carried out in sealed vial in toluene
[b] determined by ^{1}H NMR
[c] Excess surfactants were added

oxidant activation (Table 12.5, entry 3). Electrophilic activation of $Pd@SiO_2$ using 10 mol% of $PhICl_2$ was found effective over 25 mol%, with better selectivity of (2) [Table 12.5, entry 14] at 25 ºC. When reaction (Scheme 12.5) was carried out over $Pd@SiO_2$ in the presence of 10 mol% of $PhICl_2$ at 25 ºC with the addition of excess surfactants, no conversion was observed (Table 12.5, entry 15) (Kim et al. 2014). XPS helps to identify changes on material surfaces in terms of oxidation states and chemical environment. When $Pd@SiO_2$ was subjected to XPS analysis without any oxidant treatment, the palladium 3d core level displayed two major peaks for Pd $3d_{5/2}$ at 335.8 eV for $Pd^{(0)}$ species. After the treatment with an oxidant, the peak at binding energy (BE) 335.8 eV shifted to higher BE at 338.2 eV attributed to $Pd^{(IV)}$ species (Fig. 12.11). It confirmed the generation of electrophilic $Pd^{(IV)}$ species. Thus, a highly Lewis acidic surface could be generated over $Pd@SiO_2$ material with the oxidant treatment. During the reaction in non-polar solvent toluene at 140 °C, after stirring for 1 h, $Pd^{(IV)}$ fractions increased rapidly with an increase in the amount of $PhICl_2$ (Fig. 12.11).

Thus, over non-reducible support like silica, the formation of $Pd^{(IV)}$ surface species enhanced the activity for the said hydroalkoxylation reaction. And NCS was found more efficient compared to $PhICl_2$ for the activity. In further studies, Kim et al. reported Pd NPs for C-H halogenation reactions wherein a controlled oxidation state of Pd was found effective for the said reaction (Kim et al. 2014). As discussed above, Pd NPs were prepared by the decomposition of complex $Pd(acac)_2$ in the presence of oleylamine and trioctylphosphine designated as Pristine, which was further oxidized and reduced (Fig. 12.12a). The materials with and without treatments were well characterized using XPS analysis where changes in an oxidation state and Pd fractions were studied (Fig. 12.12).

XPS data revealed the presence of $Pd^{(IV)}$ species after treatment with oxidant ($PhICl_2$) (Fig. 12.12b) which on further reduction with H_2 at 60 °C regenerates Pd(0) species (Fig. 12.12c). Characterization in terms of change in coordination number, Pd fractions, and other details was obtained using XANES and EXAFS. The Pd–Pd co-ordination number of Pd NPs before treatment, i.e., Pristine was found to be 4.7 ± 0.4 which decreased to 1.0 ± 0.1 after treatment with oxidant. The surface potential of the samples was measured using Zeta potential. The surface potential of the Pristine material was found to be –20.3 mV which changed to –87.3 mV which on reduction further dropped to –4.1 mV after oxidation with $PhICl_2$.

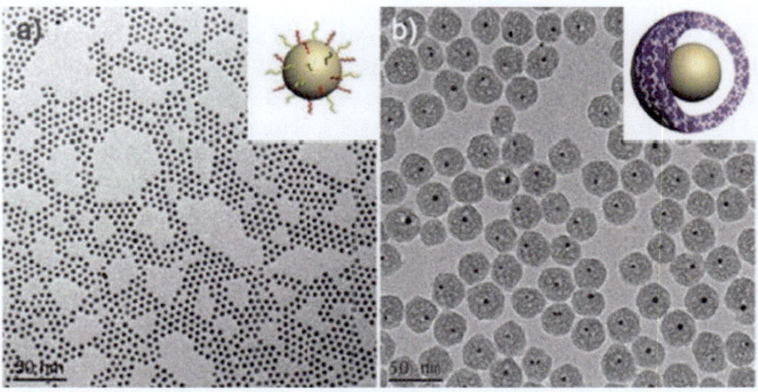

Fig. 12.10 TEM images: **a** Pd NPs; **b** $Pd@SiO_2$ yolk-shell nanocatalyst [(Kim et al. 2014), Copyright RSC, adapted with permission]

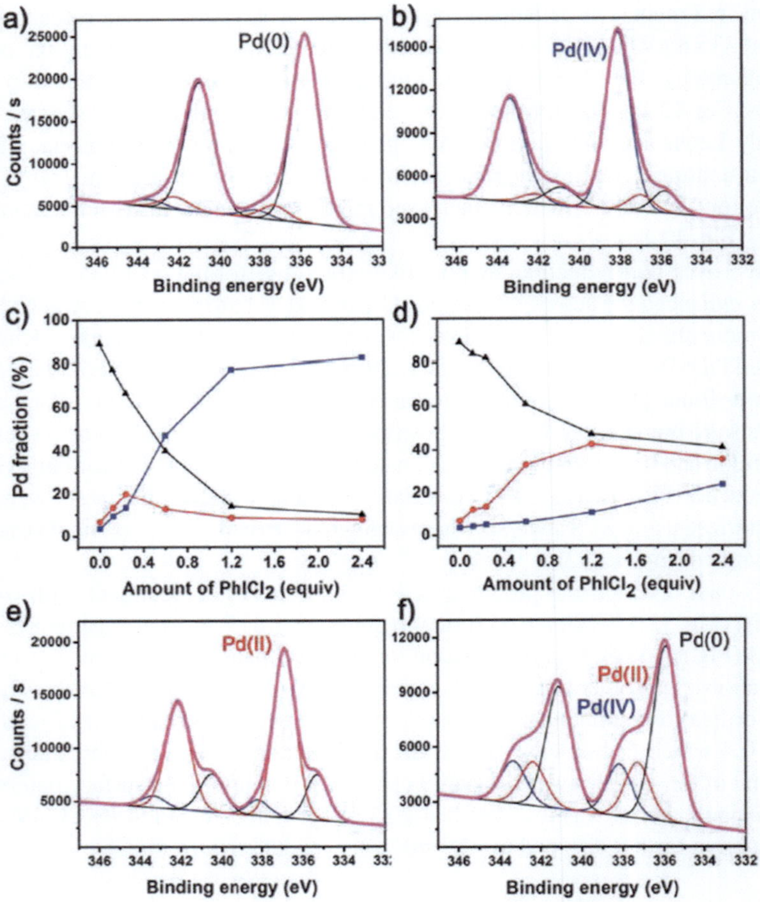

Fig. 12.11 XPS spectra of the palladium 3d level (pink) and deconvolution peaks of the Pd species [$Pd^{(0)}$ (black), $Pd^{(II)}$ (red), and $Pd^{(IV)}$ (blue); **a** Pristine Pd NPs; **b** Pd NPs after the treatment with oxidant $PhICl_2$ (2.4 equiv.) at 140 °C/1 h; fraction of the Pd species after treatment with variable amounts of $PhICl_2$ **c** for 1.0 h and **d** for 12 h; XPS spectra of the Pd 3d level for Pd NPs after treatment **e** with $PhI(OAc)_2$ (2.4 eq.) and **f** with NCS (2.4 eq.) [(Kim et al. 2014), Copyright Royal Society of Chemistry, Reprinted with permission]

12.5 Conclusion and Future Prospects

Among various approaches in bridging the mentioned gap between homo- and heterogeneous catalysis, supported electrophilic metal NPs found excellent for cyclization reactions. Electrophilicity on metal NPs was induced using chlorinated oxidants forming surface metal chlorine bonds. The catalytic activities of a few such electrophilic heterogeneous catalysts were found even better compared to the reported homogeneous protocols. So far, very little literature is available in this

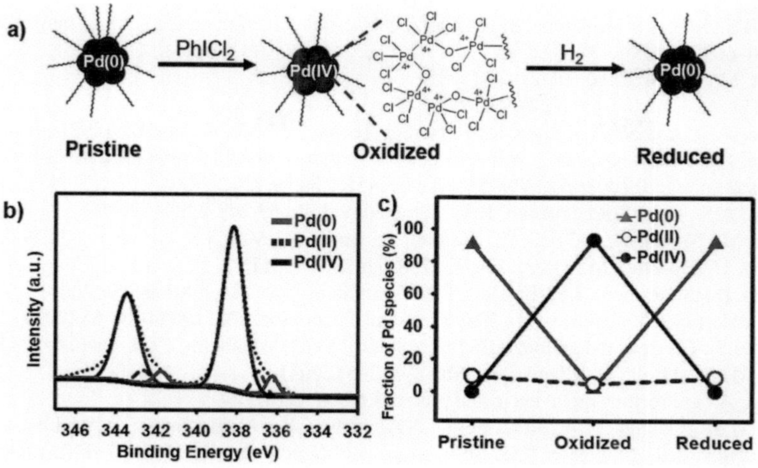

Fig. 12.12 **a** Oxidation and reduction of palladium NPs. **b** XPS spectrum of the Pd 3d level and deconvolution peaks of the Pd species oxidized by $PhICl_2$. **c** The fraction of the Pd species derived from XPS analysis of each sample. [(Kim et al. 2014), Copyright Wiley, Reprinted with permission]

arena. To develop this research further, there is still scope to develop simpler ways to synthesize and stabilize NPs of the required size for sustainability along with scale-up. The magnitude of electrophilicity incorporated into catalyst along with metal support interaction is still not thoroughly studied. The application of electrophilicity and modification for industrial catalysts is the future; however, it needs more investigations.

Acknowledgements and dedication The author ASB is thankful to Wilson College (Autonomous) for the support. This chapter is dedicated to the memory of Professor Gábor A. Somorjai (1935–2025), for his contribution to the modern surface science and catalysis.

References

(a) Dijkstra, H.P., van Klink, G.P.M., van Koten, G.: Acc. Chem. Res. **35**, 798–810 (2002). (b) van Klink, G.P.M., Dijkstra, H.P., van Koten, G.: C. R. Chim. **6**, 1079–1085 (2003). (c) Müller, C., Nijkamp, M.G., Vogt, D.: Eur. J. Inorg. Chem. 4011–4021 (2005). (d) Berger, A., Gebbink, R.J.M.K., van Koten, G.: Top. Organomet. Chem. **20**, 1–38 (2006)

(a) El-Sayed, M.A.: Acc. Chem. Res. **34**, 257–264 (2001); (b) Crooks, R.M., Zhao, M., Sun, L., Chechik, V., Yeung, L.K.: Acc. Chem. Res. **34**, 181–190 (2001); (c) Roucoux, A., Schulz, J., Patin, H.: Chem. Rev. **102**, 3757–3778 (2002); (d) Moiseev II., Vargaftik, M.N.: Russ. J. Gen. Chem. **72**, 512–522 (2002); (e) Bell, A.T.: Science **299**, 1688–1691 (2003); (f) Moreno-Manas, M., Pleixats, R.: Acc. Chem. Res. **36**, 638–643 (2003)

(a) Astruc, D., Lu, F., Aranzaes, J.R.: Angew. Chem. Int. Ed. **44**, 7852–7872 (2005); (b) Gröschel, L., Haidar, R., Beyer, A., Reichert, K., Schomäcker, R.: Catalysis Letts. **95**, 67–75 (2004)

Bassil, B.S., Mal, S.S., Dickman, M.H., Kortz, U., Oelrich, H., Walder, L.: J. Am. Chem. Soc. **130**, 6696–6697 (2008)
Boudart, M.: J. Mol. Catal. **30**, 27–38 (1985)
Budzinski, A.: Europa Chemie **24**, 4 (1997)
Burange, A.S., Ahmad, A., Luque, R.: Catal. Sci. Technol. **11**(13), 4315–4326 (2021)
Burange, A.S., Gawande, M.B.: Encyclopedia of Inorganic and Bioinorganic Chemistry (2016). https://doi.org/10.1002/9781119951438.eibc2458(Chapter)
Burange, A.S., Kale, S.R., Jayaram, R.V.: Tetrahedron Lett. **53**(24), 2989–2992 (2012)
Carraro, M., Nsouli, N., Oelrich, H., Sartorel, A., Sorar, A., Mal, S.S., Scorrano, G., Walder, L., Kortz, U., Bonchio, M.: Chem. Eur. J. **17**, 8371–8378 (2011)
Chauhan, B.P.S., Rathore, J.S., Bandoo, T.: J. Am. Chem. Soc. **126**, 8493–8500 (2004)
Chen, J., Herricks, T., Geissler, M., Xia, Y.: J. Am. Chem. Soc. **126**, 10854–10855 (2004)
Bradley, J.S.: Clusters and colloids (Ed.: G. Schmid), VCH, Weinheim, 1994, chap. 6, p. 459
Daniel, M.-C., Astruc, D.: Chem. Rev. **104**, 293–346 (2004)
Degussa, A.G.: Precious metal cycle and precious metal recovery
Demir, M.M., Gulgun, M.A., Menceloglu, Y.Z., Erman, B., Abramchuk, S.S., Makhaeva, E.E., Khokhlov, A.R., Matveeva, V.G., Sullman, M.G.: Macromolecules **37**, 1787–1792 (2004)
(a) Newkome, G.R., Moorefield, C.N., Vögtle, F., (eds.).: Dendrimers and dendrons: Concepts, syntheses, applications. Wiley-VCH:Weinheim, Germany (2001). (b) Frechet, J.M.J., Tomalia, D.A., (eds.): Dendrimers and other dendritic polymers. Wiley-VCH, New York (2002)
Dengel, A.C., Griffith, W.P.: Polyhedron **8**, 1371–1377 (1989)
Doná, E., Cordin, M., Deisl, C., Bertel, E., Franchini, C., Zucca, R., Redinger, J.: J. Am. Chem. Soc. **131**, 2827–2829 (2009)
Falber, A., Burton-Pye, B.P., Radivojevic, I., Todaro, L., Saleh, R., Francesconi, L.C., Michael Drain, C.: Eur. J. Inorg. Chem. 2459–2466 (2009)
Falber, A., Todaro, L., Goldberg, I., Favilla, M.V., Drain, C.M.: Inorg. Chem. **47**, 454–467 (2008)
Gaunt, A.J., May, I., Collison, D., Holman, K.T., Pope, M.T.: J. Mol. Struct. **656**, 101–106 (2003)
Haruta, M.: CATTECH **2002**(6), 102–115 (2002)
Haruta, M., Kobayashi, T., Sano, H., Yamada, N.: Chem. Lett. **16**, 405–408 (1987)
Haruta, M., Yamada, N., Kobayashi, T., Iijima, S.: J. Catal. **115**, 301–309 (1989)
Jansen, J.F.G.A., de Brabander-van den Berg, E.M.M., Meijer, E.W.: Science **266**, 1226–1229 (1994)
Jesús, E., Flores, J.C.: Ind. Eng. Chem. Res. **47**, 7968–7981 (2008)
Kavanagh, K.E., Nord, F.F.: J. Am. Chem. Soc. **65**, 2121–2125 (1943)
Kidambi, S., Dai, J.-H., Lin, J., Bruening, M.L.: J. Am. Chem. Soc. **126**, 2658–2659 (2004)
Kikukawa, Y., Yamaguchi, S., Tsuchida, K., Nakagawa, Y., Uehara, K., Yamaguchi, K., Mizuno, N.: J. Am. Chem. Soc. **130**, 5472–5478 (2008)
(a) Kim, M., Lee, S., Kim, K., Shin, D., Kim, H., Song, H.: Chem. Commun. **50**, 14938–14941 (2014); (b) Kim, K., Jung, Y., Lee, S., Kim, M., Shin, D., Byun, H., Cho, S., Song, H., Kim, H.: Angew. Chem. **129**, 1–6 (2017)
Knecht, M.R., Weir, M.G., Myers, V.S., Pyrz, W.D., Ye, H., Petkov, V., Buttrey, D.J., Frenkel, A.I., Crooks, R.M.: Chem. Mater. **20**, 5218–5228 (2008); b) Ye, H., Scott, R.W.J., Crooks, R.M.: Langmuir. **20**, 2915–2920 (2004); c) Jones, C.W.: Top. Catal. **53**, 942–952 (2010)
Ley, S.V., Mitchell, C., Pears, D., Ramarao, C., Yu, J.-Q., Zhou, W.-Z.: Org. Lett. **5**, 4665–4668 (2003)
Li, Y., Liu, J., Witham, C.A., Huang, W., Marcus, M.A., Fakra, S.C., Alayoglu, P., Zhu, Z., Thompson, C.M., Arjun, A., Lee, K., Gross, E., Toste, F.D., Somorjai, G.A.: J. Am. Chem. Soc. **133**, 13527–13533 (2011)
Luque, R., Burange, A.S.: Heterogeneous catalysis. (2022). eISBN:9780841299450ACS in Focus https://doi.org/10.1021/acsinfocus.7e5032
Maksimchuk, N.V., Melgunov, M.S., Mrowiec-Białon, J., Jarzëbski, B., Kholdeeva, O.A.: J. Catal. **235**, 175–183 (2005)

Mal, S.S., Nsouli, N.H., Carraro, M., Sartorel, A., Scorrano, G., Oelrich, H., Walder, L., Bonchio, M., Kortz, U.: Inorg. Chem. **49**, 7–9 (2010)

Pillai, U.R., Sahle-Demessie, E.: J. Mol. Catal. A **222**, 153–158 (2004)

(a) Rao, C.R.M., Reddi, G.S.: Trends Anal. Chem. **19**(9), 565–586 (2000); (b) Saternus, M., Fornalczyk, A.: METABK **52**(2), 267–270 (2013)

Rapino, L.D., Nord, F.F.: J. Am. Chem. Soc. **63**, 2745–2749 (1941a)

Rapino, L.D., Nord, F.F.: J. Am. Chem. Soc. **63**, 3268 (1941b)

(a) Roessler, F.: Chimia **50**, 106 (1996); (b) Auer, E., Freund, A., Pietsch, J., Tacke, T.: App. Catalysis A: Gen. **173**, 259–271 (1998)

Stanciu, C., Jones, M.E., Fanwick, P.E., Abu-Omar, M.M.: J. Am. Chem. Soc. **129**, 12400–12401 (2007)

Tabuani, D., Monticelli, O., Chincarini, A., Bianchini, C., Vizza, F., Moneti, S., Russo, S.: Macromolecules **36**, 4294–4301 (2003)

(a) Tzschucke, C.C., Markert, C., Bannwarth, W., Roller, S., Hebel, A., Haag, R.: Angew. Chem., Int. Ed. **41**, 3964–4000 (2002). (b) Alonso, F., Beletskaya, I.P., Yus M.: Tetrahedron, **61**, 11771–11835 (2005)

Whitfield, S.R., Sanford, M.S.: J. Am. Chem. Soc. **129**, 15142–15143 (2007)

Wiley, B., Sun, Y., Mayers, B., Xia, Y.: Chem. Eur. J. **11**, 454–463 (2005)

Witham, C.A., Huang, W., Tsung, C., Kuhn, J.N., Somorjai, G.A., Toste, F.D.: Nat. Chem. **2**, 36–41 (2010)

Yang, C.C., Wan, C.C., Wang, Y.Y.: J. Colloid Interface Sci. **279**, 433–439 (2004)

(a) Zhao, M., Sun, L., Crooks, R.M.: J. Am. Chem. Soc. **120**, 4877–4878 (1998); (b) Zhao, M., Crooks, R.M.: Angew. Chem. **111**, 375–377 (1999); (c) Balogh, L., Tomalia, D.A.: J. Am. Chem. Soc. **120**, 7355–7356 (1998); (d) Balogh, L., Swanson, D.R., Tomalia, D.A., Hagnauer, G.L., McManus, A.T.: Nano Lett. **1**, 18–21 (2001); (e) Esumi, K., Susuki, A., Aihara, N., Usui, K., Torigoe, K.: Langmuir **14**, 3157–3159 (1998)

Chapter 13
A Review of Dimethyl Ether Synthesis Routes and Recent Technological Advancements

Samruddhi Mane, Akash Bhatkar, Siva Prasad Mekala, Archana Ramakrishnan, Prabu Marimuthu, Marimuthu Manikandan, Nazreen Parayil Valappil, and Raja Thirumalaiswamy

Abbreviations

CAPEX	Capital Expenditure
CD	Catalytic Distillation
RD	Reactive Distillation
SAPO	Sulphonated alumina
CZA	CuO–ZnO–Al_2O_3
CZZ	CuO–ZnO–ZrO
DME	Dimethyl Ether
DWC	Dividing-Wall Column
GHSV	Gas Hourly Space Velocity
LHSV	Liquid Hourly Space Velocity
LPG	Liquefied Petroleum Gas
OPEX	Operating expense
WGS	Water Gas Shift
MTG	Methanol to gasoline

S. Mane · A. Bhatkar · S. P. Mekala · A. Ramakrishnan · P. Marimuthu · M. Manikandan · N. P. Valappil · R. Thirumalaiswamy
Catalysis and Inorganic Chemistry Division, CSIR-National Chemical Laboratory, Pune 411008, Maharashtra, India

S. Mane · A. Bhatkar · S. P. Mekala · R. Thirumalaiswamy (✉)
Academy of Scientific and Innovative Research (AcSIR), Ghaziabad 201002, India
e-mail: t.raja@ncl.res.in

R. S. Ningthoujam and A. K. Tyagi (eds.), *Handbook of Materials Science, Volume 3*, Indian Institute of Metals Series, https://doi.org/10.1007/978-981-96-8229-4_13

13.1 Introduction

Since the human civilization, fuels played a crucial role in our growth and revolutions. Wood is supposed to be the first known source of solid fuel we know. Later, according to developments and various revolutions like First Industrial Revolution which changed the whole scenario by adding more and more sources of energy. Coal obtained from fossil fuels is considered as one of the best sources of energy in terms of heat and gases. Our overall daily activities are revolving around energy and various energy sources. One of the most significant C1 building components in synthesis chemistry is thought to be carbon monoxide. In the past several years, various advanced conversion techniques have been discovered and published, and most of them have already reached the industrial scale. When carbon monoxide and hydrogen are mixed in various ratios, it is termed as synthesis gas or syngas as one can synthesize a variety of chemicals from this class of mixtures of gases depending on operating conditions and various methods deployed in the process.

Carbon monoxide (CO) is a diatomic gas which is colourless, odourless and tasteless in nature. In the year 300 BC, Aristotle observed that burning coal produces toxic fumes, which he observed as colourless and odourless gas (Peng et al. 2019). De Lassone, a French chemist in 1776, heated zinc oxide with coke and noticed how some of the gases evolved and assumed, incorrectly, that the gaseous output was hydrogen. This was mainly concluded because of its complete combustion and blue-coloured flame. It took until 1800 to discover the composition of CO. W. C. Cruikshank, a Scottish chemist, revealed the constituents of carbon monoxide, where he claimed that the compound is comprised primarily of carbon and oxygen molecules. Italian physicist, F. Fontana in the year 1780 discovered the water gas shift reaction, and the world realized its wide applications by the twentieth century as hydrogen production and metallurgical reactions. The first application of syngas was studied in the year 1897, by L. Gattermanna German chemist and Julius A. Koch who developed and demonstrated the method for conversion of benzene into toluene with the help of the formylation reaction. In this reaction, CO and Hydrochloric acid act as an acylating agent which was later recognized as the well-known Gattermann–Koch synthesis reaction. At the start of the nineteenth century, a remarkable discovery by Fischer and Tropsch suggested a new way to prepare hydrocarbons by the reaction of CO with hydrogen on the Fe/ZnO catalyst and Co/Cr_2O_3 catalysts and here the era starts where industries started synthesizing various hydrocarbons of their desire. The Fischer–Tropsch method is well adopted by the chemical industry in the fields of gas-to-liquid (hydrocarbon) technology and coal gasification followed by liquefaction (Mahmoudi et al. 2017).

There are several sources to obtain syngas or synthetic gas. Mostly, it is obtained from the gasification of natural gases; along with this, solid biomass, solid waste gasification unit and industrial gases are the reported sources for syngas (El-Nagar and Ghanem 2019; MillánOrdóñez et al. 2021; Uriarte et al. 2016). Various chemicals can be synthesized by using synthetic gas as shown in Fig. 13.1.

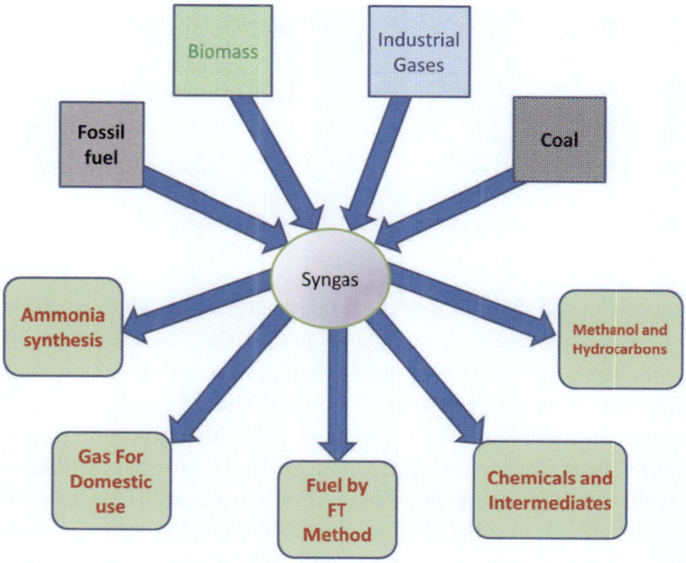

Fig. 13.1 Sources and utilization of syngas

Methanol, formaldehyde, mixed alcohols and oxygenated products, hydrogen gas and olefins are the most widely synthesized and practically commercialized chemicals reported so far. Synthesis gas is composed of CO, hydrogen, some amount of methane and traces of moisture, and nitrogen along with carbon dioxide gas. The actual composition of the gases is variable, and it is mostly dependent on the source from where syngas is produced.

Chemists consider methanol as one of the first and most important building blocks in synthesis chemistry, which has the capacity to synthesize a wide variety of synthetic materials that we require in our daily life at various stages recently; also it is used and reported as one of the forms of alternative fuel and a fuel additive as well. Formaldehyde, acetic acid and DME are the major chemicals that are being synthesized with the help of methanol. Therefore, methanol is considered as a building block of synthesis chemistry. Also, it is reported as an alkylating agent in various processes. Methanol has also recently been reported for use in fuel cells; and methanol reforming is an emerging branch for hydrogen synthesis and therefore considering all the aspects of this chemical it is well understood that it will play a key role in the energy sector in near future. Therefore, researchers are keenly interested in methanol synthesis and optimization of this technology. As syngas is a promising source of single carbon starting material in the form of CO and CO_2, efforts should be made to develop a promising, sustainable and highly efficient technology to convert syngas into value-added chemicals like methanol. It can even help to reduce the carbon footprints and will eventually reduce the impacts of global warming. This can be done by switching the alternative source of fuels and utilization of greenhouse gases to fine chemicals. The synthesis of methanol from synthesis gas which is broadly a

$$CO_2 + 3H_2 \rightleftharpoons CH_3OH + H_2O \quad \triangle H_{298}=-41.17kJmol^{-1} \ldots\ldots (1)$$

$$CO_2 + H_2 \rightleftharpoons CO + H_2O \quad \triangle H_{298}=+49.47kJmol^{-1} \ldots\ldots (2)$$

$$CO + 2H_2 \rightleftharpoons CH_3OH \quad \triangle H_{298}=-90.64kJmol^{-1} \ldots\ldots (3)$$

Schematic 13.1 Overall reactions in syngas hydrogenation to methanol

combination of $CO/CO_2/H_2$ is the second major industrial achievement in terms of catalysis whereas ammonia synthesis stands first (MillánOrdóñez et al. 2021).

The uses and production of methanol are reported and listed for millennia, by ancient Egyptians. They were practising for an embalming process, where the combination of materials created during the destructive distillation (pyrolysis) of wood is used. Up until Robert Boyle developed pure methanol by further distillation in 1661, it was not published. Reports say that Eugene Peligot, a French researcher and analytical chemistry professor in collaboration with Jean-Baptiste Dumas, carried out some experiments with wood spirit (methanol) (MillánOrdóñez et al. 2021), during which they discovered methyl radical and they both together coined the terminology "methyl alcohol"; in this research they achieved a successful synthesis of gaseous DME and many other ethers.

In the early years, researchers keenly focused on copper as the material for the production of methanol; and during that time, copper was a primary contender for the catalytic processes which continues for years and made researchers focus mainly on copper-based catalyst. The first industrial methanol production via catalytic technology was started almost 99 years back by BASF. Later BASF commercialized the technology for syngas hydrogenation to methanol in the year 1923. The expected reactions in the composition of syngas containing carbon monoxide, carbon dioxide and hydrogen are as follows.

13.2 Classification

Syngas hydrogenation to methanol via intermediates can be carried out by various catalytic methods, such as

1. Electrocatalytic method
2. Thermal homogeneous catalytic method
3. Thermal heterogeneous catalytic method
4. Photocatalytic method.

Considering all current available facts and parameters, quoting the statement, "heterogeneous hydrogenation catalysis and electro-catalysis are the only greatest and promising fields for large-scale CO_2 reduction to CH_3OH (methanol)" will not be wrong. Electricity availability, its source and price will determine future of the sustainable electrochemical catalysis for efficient methanol synthesis. Therefore, in this review we focused mainly on recent leading catalysts and catalytic conditions for various processes in which they operate considering the advantages and drawbacks in sustainability, emerging areas for syngas hydrogenation, future scopes of process and various conditions of operation, selectivity, stability and productivity in the thermal heterogeneous catalytic method. In the upcoming years according to surveys, about a 10% increment is expected only for methanol production. It is because of its chemical importance and recent advances towards alternative fuel approaches going all over the globe. Recently, battery applications of formaldehyde are also emerging, which is expected to add more advances in syngas hydrogenation by contributing at least about 5% extra demand solely for formaldehyde. Considering this entire scenario together, it will increase global demand for methanol as a chemical commodity at a rate of around 6.5% in the next decade. Therefore, the methanol to gasoline (MTG) by catalytic technology is a forefront route which is still under exploration. This article summarizes recent notable advances in heterogeneous thermal catalytic hydrogenation reactions of syngas conversion to methanol, numerous barriers of this process that must be considered, and the various constraints which cause lower rate of conversion and selectivity of methanol which may be useful for researchers in upcoming years to flourish this technology in a well-settled and planned way to be the competitive process in the near future.

From Eqs. (1) and (3), one can easily notice that the synthesis of methanol by employing the hydrogenation process of CO_2 and CO is mildly exothermic, whereas Eq. (2) is a reverse water gas shift (RWGS) reaction, which reported to be slightly endothermic since ΔH is positive for the reaction. The catalyst deployed for this process at the industrial level is a combination of Cu/ZnO/Alumina which is considered the standard catalyst. This catalyst formulation was developed by a British chemical company, named Imperial Chemical Industries (ICI) in the 1960s, which took methanol synthesis in an advanced way and opened doors of possibility for researchers, for which they got the patent with patent number: EP0329292A2 (European Patent) and GB-1159035-A (US patent). Methanol synthesis at the commercial level from syngas started back in the 1940s. Various attempts were already made to scale up this process. Later in the 1960s, this process became widespread. This catalyst formulation by ICI exhibited higher activity than the ZnO/Cr_2O_3 predecessor combination of catalysts proposed by BASF also. ICI catalyst exhibits the easiest operating conditions at minimal reaction conditions in terms of pressure and temperature (<100 bar, < 573 K); therefore it helped in increasing profits and efficiency for industries. Its feasibility and minimal operating conditions greatly helped in reducing operating costs (OPEX).

The original technique was based on a $ZnO–Cr_2O_3$ catalyst that could hydrogenate CO to methanol at high pressures of 240–300 bar and temperatures of 350–400 °C, and was commercialized by BASF in 1923. The main drawback of this catalyst was low selectivity along with undesired side products like methane and a few lower hydrocarbons with almost 5% selectivity. This attracted researchers to work further with the scope to reduce side products and increase the selectivity; meanwhile, researchers carried out various studies to demonstrate the actual contribution of each parameter like the feed ratio of H:C, investigating the role of CO versus CO_2 in reaction kinetics and various aspects of reaction kinetics. A major breakthrough was cracked by ICI, in the 1960s simply by producing the more active Cu/ZnO catalyst for sulphur-free synthesis. This catalyst exhibited a promising performance, in terms of minimal operating conditions like lower pressure and temperature, such as 60–80 bar and 250–280 °C respectively. Such improved performance by catalysts in a significant way helped in the reduction of OPEX for industries. The lower reaction temperature here helped to improve the overall selectivity by the reduction in numerous side reactions where light hydrocarbons were one of the products. This was a drawback for the BASF technology.

R. Guil-López et al. reported that it is difficult to activate the CO_2 molecule due to its higher thermodynamic stability ($\Delta_f G^0 = -394.38$ kJ mol^{-1}) and along with the kinetic inertness of the same molecule, which act as an add-on for the inertness. CO_2 is a triatomic linear molecule with non-polar characteristics. However, it is having two reactive sites namely electron-rich and electron-deficient sites in terms of oxygen and carbon centres, respectively. The electron-deficient centre of CO_2 represents its tendency to react with electron-rich centre or towards nucleophiles and while the oxygen atom reveals exactly opposite behaviour. Liu and co-workers, Chinchen and co-workers carried out certain experiments and greatly altered the aspects to look at the reaction pathway for syngas conversion via hydrogenation pathway to methanol synthesis over $Cu/ZnO/Al_2O_3$ catalysts. They carried out an isotope labelling experiment to trace the source of carbon in methanol, and they found that the source of carbon in methanol under $CO/CO_2/H_2$ composition was in fact CO_2, not CO, even though CO is more electronically prone to hydrogenation reaction. Along with these studies, they found a linear relationship between copper metal content and surface area in catalysts with the partial pressure of CO_2 under differential conditions for methanol synthesis.

Out of the above two reported catalysts, the first generation methanol synthesis was carried out at a relatively high temperature and pressure with the catalyst composed of ZnO/Cr_2O_3 material, which is used to be operated at 350 °C and 250–350 bar, having 20–75% of Zn contents. It shows high activity and selectivity; it is well resistant to sulphur toxins coming from the syngas of coal gasification, whereas coal is considered a source of syngas. For the next generation catalyst, the sulphur toxic content was removed separately, so that the Cu catalyst can be used easily without getting poisoned in order to achieve the proper selectivity and conversion.

Since DME is considered as the alternative fuel of the future, various attempts are being carried out to check its compatibility with the existing sources of energy. Moreover, DME is famous as a substitute option for domestic fuel LPG or for the automobile sector in terms of diesel fuel. This is due to the close resemblance in their properties. Some of the properties are mentioned below in Graph 13.1 with a comparison of six fuels. Important properties like cetane number, flash point, lower heating value (LHV) and higher heating value (HHV) are mentioned in Graph 13.1 (Seddon 2011).

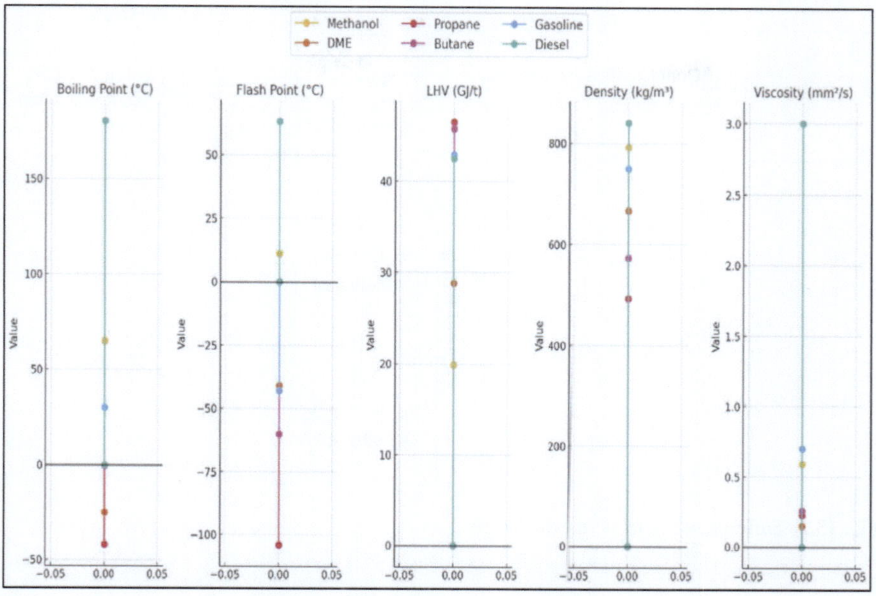

Graph 13.1 Comparative properties for DME and other fuels (Seddon 2011)

Due to close similarities in terms of cetane number, DME can be a great replacement for diesel with just some little modification in the existing equipment. Due to the big margin in boiling point between the DME and diesel, we require bit modifications in the existing facility, whereas DME and LPG (containing propane majorly) do not require such modifications as their properties are nearly close to each other. It is a great relief as it may decrease the direct dependency on natural resources like crude oil and coal for domestic gas production and utilization.

DME is a class of organic compounds which is projected to be an option of one of the alternative fuels. There are various sources of production of DME, broadly classified into direct and indirect methods depending on the starting material used. Both these methods are very well known, and they are carried out in the reactor system with the catalyst. DME and the best catalyst for the process is a fascinating part of catalysis chemistry for researchers (Fig. 13.2).

DME synthesis is reported to be carried out in two main ways

(a) *Direct route of DME synthesis*
(b) *Indirect route of DME synthesis.*

The precursor of DME, syngas, is typically created by gasifying a variety of materials, including biomass, waste, coal and natural gas that has been partially oxidized or reformed (El-Nagar and Ghanem 2019). The synthesis gas (holds the composition of CO_2, CO, H_2) can be converted into dimethyl ether by 2 well-known methods: Indirect Process (two-step process) and Direct Process (one-step process). In the traditional indirect route of synthesis, first the methanol is generated using a

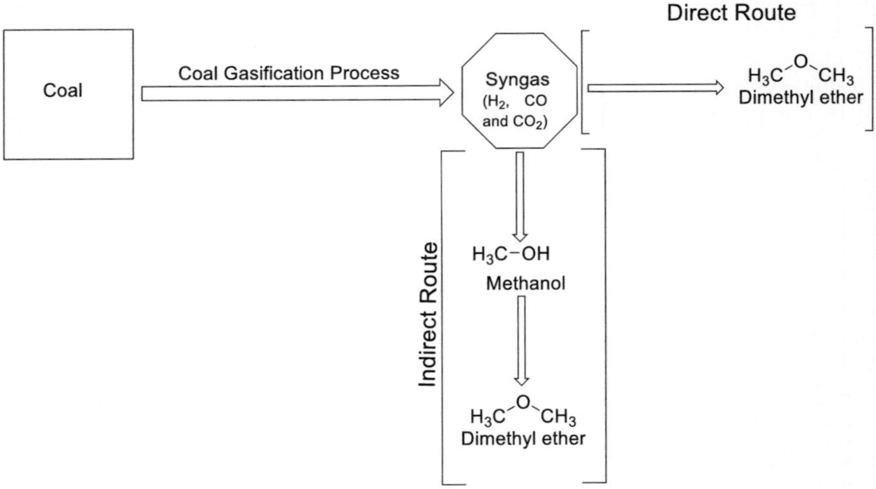

Fig. 13.2 Different ways to synthesize DME

copper-based catalyst (often $Cu/ZnO/Al_2O_3$), and DME is then produced using a solid acid catalyst (most frequently $Cu/ZnO/Al_2O_3$) from the generated methanol. The literature survey conferred that γ-Al_2O_3, zeolite and SAPO are separately reported as the support. In the direct synthesis approach, a heterogeneous catalyst is supposed to perform the synthesis of methanol followed by dehydration operations in the same reactor (Peng et al. 2019; Mahmoudi et al. 2017; El-Nagar and Ghanem 2019; Uriarte et al. 2016). The conversion of syngas to methanol is not possible during indirect DME synthesis due to thermodynamic constraints. High syngas conversion requires high pressure and a big recycling flow (Park et al. 2021). The constraint of direct DME synthesis can be overcome by continuously converting methanol to DME, which enables higher CO conversion and improved selectivity towards DME. Therefore, studies that explained the approach towards single-step DME synthesis are thermodynamically preferable than the indirect synthesis method (Catizzone et al. 2017; Merkouri et al. 2022).

Because it involves the extremely exothermic reactions for methanol production, along with its dehydration and the gas–water transfer reaction, the one-step synthesis of DME is a relatively exothermic reaction. The process of synthesizing DME from CO_2 which is having content of syngas includes equilibrium reactions. The direct DME synthesis technique simplifies reactor modeling, enabling higher carbon monoxide conversion and reducing the manufacturing costs of DME compared to the indirect synthesis route.

13.3 Synthesis of Materials

13.3.1 Copper Catalysts for Methanol Production

Many catalysts with various catalytic systems are explored and well-recognized for methanol synthesis since the development of the process, but copper-based catalysts hold an important place. To increase the stability of the catalyst, researchers focused strongly on the distribution of active metal which is copper in order to increase its life. The commonly used metals in methanol synthesis catalysts consist of Cu, Al, Cr, Zn and Zr combinations. On the other hand, researchers focused more recently on metals like Mn, Ce, Ti, Ga, Co, Mg and some other transition metals. Generally, the catalysts for the synthesis of the methanol are synthesized by using the co-precipitation method solution of the above-mentioned metal salts.

CuO–ZnO–Alumina (CZA) a well-known and well-investigated example of a catalyst for the production of methanol normally contains 50–70% of copper, 20–50 mol% of Zinc and 05–20% Alumina. In the case of the CZA catalyst, researchers scrutinized and confirmed their claim about the catalyst and proved their fine outcomes for this process. Active Cu sites present in the catalysts play a vital role in the boom of methanol synthesis and yield for dimethyl ether. CZA catalysts are typically synthesized through conventional co-precipitation technique by using respective metal salts for the formation of structures like hydrotalcite (Flores et al. 2011). In Cu–ZnO–Al_2O_3, the metal Cu clusters serve as the Cu's active sites for WSG reactions and the formation of methanol. Metals like Cu, their distribution and the type of support used have an effect on the selectivity and the overall conversion (Merkouri et al. 2022; Flores et al. 2011). ZnO here plays a very important role in providing a precise geometric region among Cu nanoparticles, which results in a better distribution of metallic copper particles and ceases the sintering of particles of copper, and also helps to reduce active site poisoning. Al_2O_3 eventually maintains the shape of the catalyst in addition to this; it even provides increased surface area. On the other hand, the utilization of Al_2O_3 in the CZA catalyst system can give rise to certain disadvantages because of the water, which is a side product formed during methanol production by CO_2 hydrogenation reaction due to the hydrophilicity of the Al_2O_3. Researchers used promoters such as ZrO_2 to get around this issue which can decrease the overall hydrophilic character of the catalytic system (Merkouri et al. 2022), so ZrO_2 helps to maintain the overall hydrothermal balance of the catalyst, which actively keeps CuO–ZnO throughout the reaction medium containing water. Since water adsorption can make this reaction a bit difficult to move forward, Al_2O_3 shape catalysts are appropriate for assuaging Cu metal sintering, as it acts as the best active site.

The specific role of Al_2O_3 in increasing the activation and adsorption of CO molecules is well understood and is being researched; along with this, its role of inhibiting the sintering of active metal is well known among researchers; however, additionally, it can increase the steadiness of the catalyst (Flores et al. 2011). The intention of including M^{3+} (e.g., Al^{3+}) ions into CuO–ZnO is primarily to facilitate the overall distribution of copper material and enhance the overall surface area. This is done by adding extra thermal and chemical stability to enhance the overall life of the catalyst (Mondal and Yadav 2019; Azizi et al. 2014). Yang et al. discovered that Cu/ZnO catalyst with doping of Zr has greater conversion in the hydrogenation process of CO at lower temperatures (523 K), despite the fact that Cu/ZnO exhibits higher conversion. The activity of Zr–Cu/ZnO is relatively more towards hydrogenation of CO_2, for which the selectivity of methanol is reported around 60.45% at 523 K. Additionally, the hydrogenation for syngas feed comprising the mixture of CO and CO_2 demonstrated the activity and selectivity better in the case of Zr–Cu/ZnO catalyst, where values of conversion at moderate temperature of 250 °C show 28% and 55% selectivity and conversion respectively (Yang et al. 2006).

Recently, researchers are primarily interested in the synthesis of MeOH and DME with the aid of using the CO_2 hydrogenation process (Etim et al. 2020). The inferior performance of CZA has limited interest of researchers in the hydrogenation of CO_2 to methanol. Although Al_2O_3, which has a higher hydrophilicity than $Cu/ZnO/ZrO_2$ was used in this experiment, it prevented CO_2 hydrogenations from occurring (Yang et al. 2006; Din et al. 2019).

For the DME synthesis by the CO_2 hydrogenation process, Cu/ZrO_2 has referred to a well-explored catalyst in the synthesis of methanol than the combination of catalysts with copper and zinc oxide (Mahmoudi et al. 2017; Din et al. 2019; Catizone et al. 2018). In order to synthesize CuO–ZnO–ZrO_2 (CZZM)-based catalyst with MxOy, here, M is metal with higher oxidation states like W, Cr and molybdenum (Mo) that are used for the CO_2 hydrogenation to MeOH, Wang et al. used co-precipitation. The results showed that the inclusion of additional WO_3 and MoO_3 increased the overall selectivity and yield towards methanol with CZZ catalyst, but marginally get diminished due to doping of Cr_2O_3. The actual variance for the yield of methanol claimed by the authors could be directly linked with CO_2 adsorption ability and the BET surface area of catalysts. Furthermore, to this, the surface contents of Cu and Zn also explained the trend of methanol selectivity by variation of active metal contents (Wang et al. 2019). Cu/ZnO methanol synthesis catalyst was modified by Guil-López et al. by employing the co-modification of Aluminium and Gallium, combined together around 3%. Methanol is created in a flow reactor at around 250 °C with 3 MPa pressure. The use of syngas combination was reported to have (CO_2/CO/H_2:4.5/22/58.8 along with 14.7% of the nitrogen in volume percent) (Guil-López et al. 2020). The copper–zinc-based catalyst modified with aluminum and gallium exhibited overall stable performance and demonstrated significantly higher syngas conversion compared to the unmodified parent catalyst. The mono-substituted systems with Al or Ga catalysts have shown equivalent methanol synthesis in this regard; however, Al and Ga co-substituted catalysts have shown a higher rate of synthesis (Guil-López et al. 2020). Meshkini et al. synthesized the class of newly modified CZA catalysts for the production of methanol with the help of the co-precipitation technique. Here, they introduced the incorporation of s-block elements like Mg and Ba along with the inclusion of a few transition metals oxides like Mn, Zr, Ce, W and Cr-based oxides to the parent catalyst system. The hydrogenation of the CO/CO_2 to methanol is the catalytic process which was conducted at around 513 K with a pressure of 5 MPa with a syngas feed ratio containing CO, CO_2 and hydrogen in the ratio of 1/1/14.5. A new catalyst with the content of Mn and Zr for methanol has is been recently discovered to showcase the maximum interest. Studies showed that Mn and Zr help greatly to increase the dispersion of active copper on support, which ultimately leads to an increase in the stability of the CZA catalyst during the reaction (Wang et al. 2024).

Peter et al. demonstrated the Cu- and γ-alumina-based catalytic system by using the impregnation of metals like ZnO, ZrO_2 and MgO. The outcomes demonstrated that the catalyst's performance was much improved by the tiny Cu^0 particle's excellent dispersion and the active involvement of the promoters that led to this activity enhancement. It has been confirmed that a better response can be achieved in case of cease of methanol synthesis reaction with respect to temperature on Cu–ZnO–ZrO_2–MgO/Al_2O_3; however it booms opposite reactions called WGS reaction along with reactions of methanation (Ren et al. 2015; Warmuth et al. 2024).

13.3.2 Solid Acid-Based Catalytic Systems for Dehydration of Methanol

Most reported catalysts for dehydration reactions are usually zeolites, γ-Al_2O_3 and SiO_2-based catalysts which are used for methanol dehydration processes. For methanol dehydration catalyst design, solid acid catalysts with well-defined acidity are typically preferred, where porosity plays a critical role. Mesoporous materials (2–50 nm) are commonly employed, while microporous materials (2 nm) have also attracted significant research interest. The acidity of such solid acid catalysts plays a vital role in DME synthesis as it is considered to be an important parameter of the catalyst. It is also found that microporous material containing a higher surface density of acidic sites gets easily poisoned due to the formation of side products during the course of the reaction along with coke formation (Rutkowska et al. 2015).

Coke deposition is more common in the case of zeolites having microporous channels than that mesoporous zeolite material. It may be the reason behind the early deactivation of such materials along with this, and for mesoporous materials, less diffusion restrictions are expected. The higher the Brønsted acidity of the catalyst, more are the chances for coke formation which results in instant deactivation of the catalyst; therefore, quoting the sentence that excessive Brønsted acidity of the catalyst can deactivate the dehydration catalyst will not be wrong. To overcome problems like this, some of the researchers also claimed that reducing such strong acidic sites can diminish the coke deposition and it will enhance the life span of the catalyst (Tokay et al. 2012).

For the synthesis of DME from methanol dehydration, usually the catalysts based of zeolite-Y, ferrierite, mordenite and zeolites of protonic and non-protonic forms (HZSM-5 and ZSM-5) more commonly employed zeolite catalyst known for its acidic function (Ramos et al. 2005). An entirely different type of material that hierarchically structured micro-mesoporous ZSM-5 zeolite was tested for dehydration of the methanol by Rutkowska et al. Their research conclusions suggested that conventional ZSM-5 catalyst gets easily deactivated and also, it is not so durable as that of micro-mesoporous zeolites which show promising results (Hassanpour et al. 2010).

Alumina content plays an important role in the acidity of zeolite-based materials; therefore, by altering the alumina content in Si/Al ratio, one can control the acidity of zeolites (Tokay et al. 2012; Hassanpour et al. 2010). Hassanpour et al. attempted to synthesize ZSM-5 with different ratios to test it for methanol conversion. The synthesized (Silica/Alumina = 25–250) ratio catalyst was employed for the reaction of 573 K and 1.6 Mpa. They found that with Si/Al ratio acidity increased up to 125, later it decreased drastically. As a result, they discovered that the protonated ZSM-5 catalyst having the ratio of 125 for (Si/Al) was the best catalyst for the synthesis of methanol. It also had better DME selectivity and decreased side products (such as ethylene and propylene) as the Si/Al ratio increased (Hassanpour et al. 2010).

For (Si/Al = 10, 25 and 45) zeolites and γ-Al_2O_3, the effect of water is reported by Catizzone et al. They studied the catalyst for MeOH to DME reaction. Results claimed the water resistance and acidity of zeolites make them a unique and efficient catalyst for methanol dehydration. They observed a trend which claimed Si/Al ratio $10 > 25 > \gamma\text{-}Al_2O_3 > 45$ for higher temperatures and $10 > 25 > 45 > \gamma\text{-}Al_2O_3$ for lower temperatures which are around 200 °C (Catizzone et al. 2019).

Another attempt to develop a mesoporous catalyst based on STA-SiO_2 for the study of W/Si atomic ratio variation loading was done by Ciftci et al. According to the authors, the best ratio for methanol conversion is 0.33 for which they claimed above 60% of methanol conversion along with DME selectivity of 99% having space–time of 0.27 s.g.cm^{-3} at 250 °C (Ciftci et al. 2010). Keshavarz et al. reported using a sol–gel produced nano-crystalline-Al_2O_3 (AI-N) catalyst with 3.9 nm pore size for the methanol dehydration process. In the case of Al-N catalyst system, with addition of surfactant containing cationic properties the crystalline nature of catalyst reduced but it increased the overall surface area and medium acidic sites which made it highly active for methanol dehydration reactions. They found conversion more than 80% than catalyst without cationic surfactant (Keshavarz et al. 2010).

Another attempt to synthesize DME from methanol dehydration was done by Hosseini et al. who carried out a reaction on γ-Al_2O_3 at 250 to 400 °C and 2.0 bar pressure with alumina catalyst having pore size 2.9 nm. During this study, they observed very less conversion for the temperature below 270 °C but around 300 °C it shoots to 82.0% which remained almost unchanged up to 350 °C. At 300 °C, selectivity and yield reached 99.99% and 81.9% (Hosseini and Khosravi-Nikou 2016).

Mesoporous alumina silicate catalysts were prepared by Varisli et al. with an atomic ratio of Al/Si varying in the range of 0.03–0.18 with the hydrothermal method. This catalyst when employed for methanol dehydration reaction was found to be optimum when Al/Si ratio was around 0.09. For the silica source, Silicotungstic acid (STA) was impregnated over an alumina silicate network. The catalyst showed better activity with 80% of methanol conversion and higher DME selectivity at around 250 °C (Varisli et al. 2009).

Tang et al. stated that materials including composite materials, like ZSM-5/MCM-41, which is highly porous in nature, are considered to be highly stable and active than ZSM-5, which is solely microporous in nature, more active and stable. ZSM-5/MCM-41 showed better performance by methanol conversion at around 80% and DME selectivity at around 100%. In addition to this, overall lifespan also increased even after use in the range of temperature of 190–300 °C (Tang et al. 2012).

Rutkowska et al. explained the lower coke formation rate in the case of hierarchical ZSM-5 (interconnected micro/mesopores) as compared to ZSM-5 containing microporous structures. The hierarchical structure provides safer access for reactants from the acidic sites of the catalyst in the form of array (Hassanpour et al. 2010).

In a variety of composite zeolite materials having BFZ, Y zeolite, cores of Beta zeolite, polycrystalline shells and Y zeolite polycrystalline shells, Wang et al. examined the methanol to DME processes. The authors performed direct DME synthesis over CZA and H-zeolite catalyst systems which are indicated by HBFZ or HY. Therefore it is found that CZA/HBFZ exhibits more stability and activity than that CZA/HY. CO was converted up to 94.2% for which selectivity was achieved nearly to sixty eight percent over CZA/HBFZ catalyst at 1500 h^{-1}, 5.0 MPa, 250 °C reaction conditions (Wang et al. 2013).

13.3.3 Direct DME Synthesis with Bi-functional Catalyst/Hybrid Catalysts

Multiple routes for bi-functional and hybrid catalyst syntheses are reported in the literature. Methods like sol–gel, co-precipitation, impregnation, physical mixing, precipitating sedimentation, solid fusion method and wet chemical methods are well-known and explored methods. Among all these, co-precipitation methods are widely employed for catalyst synthesis (El-Nagar and Ghanem 2019; Catizzone et al. 2017). For direct DME synthesis, a bi-functional catalyst can be employed in a single reactor. This catalyst possesses some well-known metal oxides for the reaction like CuO, ZrO_2, ZnO and Al_2O_3 which can impart acidic function to it along with some alumina and silica-based supports such as γ-Alumina, Alumina-silica, mesoporous silica, zeolite-based and SAPOs. This catalyst can be used either as a powder or in pellet form depending upon the reaction condition and reactor design that is involved in the process (Catizzone et al. 2017; Hassanpour et al. 2010; Tang et al. 2012; Wang et al. 2013).

Due to variable syngas composition and catalyst variations, in the case of single-step DME synthesis from syngas, the operating temperature range may vary from 200 °C to 400 °C with pressure ranging 20–100 bar (Mahmoudi et al. 2017; Varisli et al. 2009; Tang et al. 2012; Wang et al. 2013). Since this is a reversible exothermic reaction, the reaction temperature governs the key role here. By thermodynamic limitation of an exothermic reaction, with rising temperature for the reactor system, conversion of carbon monoxide gets reduced. Also, due to the exothermicity of the reaction parameter, Cu sintering can occur, which acts as an actual loss of catalyst activity during reaction (Mahmoudi et al. 2017; Park et al. 2021; Wang et al. 2013). Ereña et al. reported that temperatures between 250 and 300 °C are optimal for obtaining the greatest oxygenated product concentration (Ereña et al. 2005). If the reaction temperature increased above 300 °C, it caused a decrease in the selectivity of DME by adding some undesired compound with hydro cracking reactions which run parallel over the catalytic surface and deactivate the catalyst early during higher temperature (Wang et al. 2013; Ereña et al. 2005). Moreover, it is observed that considering the high activity of the catalyst and stability of this reaction, there should be a combination of active sites where methanol is synthesized and some other sites where dehydration reaction occurs (Ereña et al. 2005).

Bayat et al. demonstrated the synthesis of DME by direct conversion of syngas containing $CO/CO_2/H_2$ in various combinations by using a bi-functional catalyst. In this work silicotungstic acid was impregnated over mesoporous alumina (STA-MA) which was then employed for dehydration purposes along with Cu-based commercial catalyst as a methanol synthesis catalyst. They were successful in achieving DME selectivity 0.9 with 50 bar pressure and 275 °C reaction condition, for which feed contained was CO/CO_2: 4/1. The highest yield of 0.55 was achieved for the feed with hydrogen, carbon monoxide and carbon dioxide in of 50:10:40 compositions. Also, they found that increments in the CO_2/CO ratio of feed can cause a dropdown in the selectivity of DME because it triggers the RWGS reaction (Eq. 2) due to which more water is produced and which can decrease the catalytic performance (Bayat and Dogu 2016).

Celik et al. prepared a bi-functional catalyst by mixing silicotungstic acid (STA) which they added in mesoporous material [TRC-75(L)] along with commercially available methanol synthesis Cu–Zn catalysts. They carried out reactions with various compositions of syngas containing CO, CO_2, and H_2 with above-mentioned bi-functional catalyst. The authors claimed around 85% selectivity for DME with 10% of CO_2 operated within the range of 250–275 °C with 5 MPa pressure. Since STA used is a kind of heteropoly acid (HPA), it has good strengths of Brønsted acidic sites, which even strengthened the acidic sites of [TRC-75(L). HPA is considered to be a solid acid catalyst which is highly prone to alcohol dehydration reactions. The main reason behind their incorporation at mesopores of dehydration catalyst is due to their lowest surface area (Celik et al. 2013).

A deactivation model was also developed by Peláez et al. for the process of direct synthesis of the DME. Here, the authors claimed the use of well-explored CZA and γ-Alumina catalysts with ratio variations. The results explained that with increasing temperature, there is a sudden decrease in activity, and this phenomenon is reversible which is considered to be due to water formed in the reaction that causes adsorption on the catalyst surfaces which ultimately leads to blocking the active sites. The drop in conversion observed at 250 °C can be attributed to water adsorption on the catalyst surface. As the temperature increases, water gets desorbed which subsequently leads to increase in conversion. This correlation is enough to prove the above statement. The optimum reaction temperature is 270 °C with a higher temperature; catalyst starts instant deactivation due to the sintering of active metal sites (copper sites) and coke formation at higher temperatures (Wang et al. 2019).

The Cu–ZnO–Al_2O_3–ZrO_2 (CZAZ) catalyst, with a variable Zr ratio as a methanol synthesis catalyst, was prepared with a co-precipitation technique by Song et al., and this catalyst was then mixed physically with γ-Al_2O_3 which is considered as dehydration catalyst. The role of Zr as a promoter in methanol synthesis catalyst is well known. DME selectivity achieved was 79.3 mol% with space–time yield of 85.8 C-mol% at 5 MPa pressure, 270 °C temperature for syngas composition of hydrogen, carbon monoxide and carbon dioxide along with methane in the ratio of 52:24:23:1. With increment of Zr as promoter from 0 to 10%, it is observed that there is a trend of first increased and then decreased CO conversion. For 3 mol% of Zr loading catalyst, the highest surface area was observed. The highest conversion of CO was achieved at 79.3% for CZAZ (3)/A catalyst, whereas the lowest conversion of CO was achieved at 67.8% for CZAZ (10)/A catalyst (Ren et al. 2015; Celik et al. 2013; Wang et al. 2014).

Hua et al. proposed the use of copper zinc-based metal oxide where metals like Ti, Al, Zr and Cr, as methanol synthesis catalysts, are prepared with the co-precipitation process followed by mixing with HZSM-5 as dehydration catalyst for syngas to dimethyl ether conversion. The authors observed the yield of DME as 33.9%, selectivity as 61.4% and CDO conversion as 55.2% at 3 MPa pressure and 250 °C temperature for their best catalyst CZZ/HZSM-5. It was found that CZZ/HZSM-5 > CZA/HZSM-5 > CZC/HZSM-5 > CZT/HZSM-5; this trend was followed for catalytic activity. The data was studied for the composition of syngas containing 10 molar % of nitrogen as co-feed along with hydrogen, carbon monoxide and carbon dioxide in the molar range of 36:36:18:10 (molar %). Water Gas Shift (WGS) was found actively for all the catalysts; therefore, a considerable amount of CO_2 and H_2 was observed during the course of the reaction (Hua et al. 2017).

The beautiful piece of work is demonstrated by Cai et al.; the researchers tried to co-relate the effect of a hybrid catalyst made up of various particle sizes on zeolite. In this study, the reaction parameters were studied at 260 °C temperature and 2 MPa pressure for H_2/CO: 2/1 feed ratio. The authors claimed the observation of the trend of CZA-ZSM-5@95 > CZA-ZSM-5@110 > CZA-ZSM-5@80 > CZA-ZSM-5@65 > CZA-ZSM-5@500 > CZA-ZSM-5@800 for decreasing CO conversion. The lowest CO conversion was found for the ZSM-5@500 and ZSM-5@800 zeolite-based catalyst. The catalyst with smaller zeolite crystal size shows higher CO conversion along with high DME selectivity (Cai et al. 2016).

Evaluation of catalytic behaviour of CZZ FER hybrid catalyst was studied by Bonura et al. where they employed samples of ferrite with variations in acidity and difference in grain size for hydrogenation reaction of CO_2 to DME. These reactions were carried out at 220–260 °C temperature with 3.0 MPa pressure of CO_2. During the reaction, water is generated as one of the products which along with exothermicity causes the sintering of active metal (Cu) by absorbing water. The ferrite particle size is actively involved in the activation of CO_2 followed by the dehydration of methanol to dimethyl ether (Bonura et al. 2018).

Xu et al. studied the effect of the combination of HZSM-5 and the (CZA) commercial JC207 ($CuOZnOAl_2O_3$) and as bi-functional catalyst employed for the synthesis and dehydration of methanol, respectively. Here, the author used HZSM-5 as a catalyst for dehydration reaction with variation in the Si/Al ratios where they employed four different combinations. Out of all, the JC207/HZSM-5 performed better than all other synthesized catalysts with the combination of [Si/Al = 38]. The results were recorded for the 250–270 °C temperature range and 3 barG pressure (Xu et al. 2011).

Some experiments were carried out for syngas hydrogenation by S. Asthana et al. with Cu-based catalysts having different compositions of ZnO, and MgO along with γ-Al_2O_3 as a methanol dehydration catalyst. Out of the series of catalysts CZM (20)-γ-Al_2O_3 hybrid catalyst showed the best result as 83% selectivity towards DME along with 37% of CO conversion (Mao et al. 2005). Mao et al. suggested that MgO helped bringing down stronger acidic sites into weaker sites. They also observed that stronger acidic sites can cause the degradation of DME to lower hydrocarbons (Phienluphon et al. 2015).

Core–shell-like catalyst is a new form of material that emerged recently as a catalyst of interest for many researchers and is used in direct CO_2 hydrogenation to DME. Core–shell catalyst has two parts basically; the core part is responsible for the methanol synthesis reaction, whereas the methanol dehydration catalyst is at the shell side. Here, the concept of methanol formed in cores will diffuse outwards into the shell side, where it will dehydrate to form DME. This catalyst system has been developed as a bi-functional catalyst and the main aim behind it is to reduce thermodynamic limitations by providing different reaction zones for different reactions in the same catalyst. The function of both reactions in a single catalyst particle explains the link of the reaction which is governed by their individual functions (Etim et al. 2020). Catalysts such as Cr/ZnOSAPO-46 (Bonura et al. 2018), CuO-ZnO@HZSM-5 (Bonura et al. 2018), Cu/ZnO/Al_2O_3/SAPO-11 (Phienluphon et al. 2015), CuO-ZnO-Al_2O_3@SiO_2-Al_2O_3 (Wang et al. 2014) and Cr-ZnO-S-H-ZSM-5 (Mao et al. 2005) are reported as core–shell catalysts.

Wang et al. studied the effect of CZA catalyst with SiO_2-Al_2O_3 catalyst as a hybrid catalyst of core–shell type prepared by physical mixing for conversion of syngas and for the DME selectivity. The catalyst as core–shell catalyst performed exceptionally well in FBR with nominal operating reaction conditions of 5 MPa and 260 °C temperature, with 1500 mL/(h.gcat). For these conditions, 28.3% of CO selectivity along with 51.6% of CO conversions by core–shell type of catalyst was obtained, whereas selectivity of 71.1% and 61.9% conversion were observed for these physically mixed catalysts. This exhibits the best suitable conditions of physical mixing catalyst (Wang et al. 2013).

T. Lan et al. prepared a core–shell kind of catalyst by a physical coating method. They prepared Cu/ZnO/Al_2O_3/SAPO-11 catalyst by the above-mentioned technique later; this catalyst was deployed for comparison of catalytic activity of the catalyst at 260° C temperature with 5 MPa pressure in a fixed bed reactor. Here, for reaction feed, H_2/CO/CO_2:58.10/33.80/5.10 along with argon as co-feed having a 3.09 mol% ratio was taken into consideration. The results indicated that the arrangement of capsule comprised core–shell capsule arrangement for the catalyst CZA and SAPO-11 was found to be highly prone with 92% CO selectivity and 90.3% of C-mol conversion over 64.9% of selectivity and 46.6% C-mol conversion than that of physical mixing arrangement of catalyst. Also, DME yield is found to be more in the core–shell capsule structure of the catalyst over a physically mixed catalyst (83.1% vs. 30.2%) (Poto et al. 2021).

Liu et al. used a packed CuO–ZnO–Al_2O_3 catalyst prepared with HZSM-5 membrane packed system, in direct DME production with hydrogenation of CO_2 process. The authors claimed selectivity towards 48.5% DME selectivity and 48.3% conversion of CO_2. Overall, 23.4% of DME yield was obtained at 270 °C at 3 MPa and a feed ratio of 1:3 for CO_2/H_2 (Liu et al. 2015).

Researcher Yang et al. demonstrated the synthesis of a double-shell capsule type of catalyst composed of Cr/ZnO and S–Z catalyst, reported especially in the synthesis of dimethyl ether from methanol. Here, syngas was employed with co-feed "Argon" gas in the composition of $CO/CO_2/H_2$:32.6/56/59.22 mol% at around 5 MPa pressure with a temperature range of 300–350 °C. Similarly, a core–shell type of catalyst was also prepared for Cr/ZnO–S–Z with a dual-layer method. Here, Cr/ZnO catalyst with an intermediate layer was synthesized with Silicate-1 zeolite on Cr/ZnO and then on surface protonated zeolite with Si:Al ratio of 50 formed. For this, gradual selectivity improvement was observed from 45.42% to 50.84% with an increment of temperature 300–350 °C. However the overall selectivity of the methanol gets reduced (Yang et al. 2011).

13.4 Materials Characterization

The above-mentioned catalyst is characterized by means of various techniques which are usually involved in the characterization of a heterogeneous catalyst. The ratio of acidic and basic sites in the solid acid catalyst was measured by TPD techniques, whereas all the catalysts were studied by XRD, XPS and Raman spectroscopy for their characteristic identification. SEM and TEM were employed to generate the data on particle size and morphology of the material.

Many authors studied surface area and porosity studies by BET surface area technique. For thermal stability and weight loss characterization, a few researchers additionally involved thermo gravimetric analysis (TGA) and differential thermal analysis/differential scanning calorimetry (DTA/DSC) techniques.

13.5 Types of Reactors Used in DME Synthesis

Due to the limitation of the thermodynamic parameter to the reaction, researchers are still trying to develop and study the reactors and their effects on this process. The fixed bed reactor (FBR) is extensively used and deployed in this process. Along with this, various other reactors were also employed and used to achieve the desired selectivity and yields. During scale-up of this process from lab scale to industrial scale, various challenges are there. Challenges for this type of reactor are in terms of heat of reaction and thermodynamics, heat integration, mass balance and mass transfer. Some of the researchers claimed the use of components along with the reactors like distillation column, heat integration distillation and dividing-wall column (DWC). Out of all these techniques, DWC played an important role at the industrial level as it reduces capital investment (CAPEX) by almost 20–30% and operational cost (OPEX) close to 40%. It is all possible because it could easily separate a mixture of components into individual components in just one distillation unit (DU). These result in an overall reduction of the total cost to build two columns; along with this, one can adjust the cost with the help of a single condenser and reboiler column.

Many research are still going on in this field of reactor modification and design. The microreactor area recently evolved a technique in reactor designing. Here, reactions are carried out in a multiple microreactor system that results in higher selectivity towards the desired product. So, many efforts for reactor designing are going on. A few of the reported reactors and their main applications are listed below.

13.5.1 Conventional Fixed Bed Reactor

A fixed bed reactor is well known for its simple reactor design and ease of operation in the chemical industry. The reaction happens in the gas and solid phases, and the contactors remove the restriction on diffusion. The best feature of the fixed bed reactor is the longitudinal height which helps to gain a broad range of temperatures which ranges from the inlet to the outlet of the reactor. The temperature at the catalyst bed can be maintained with accuracy. This temperature profiling allows the reactor to slowly preheat the methanol from the inlet till it reaches the catalyst zone. The temperature decreases as we reach the outlet of the reactor which helps in the case of the exothermic reaction of methanol to DME. The outlet gases exit at a lower temperature as compared to the estimated exothermic outlet temperature. Formation of coke and catalyst sintering is the major reported problem in fixed bed reactors. During the course of reactions, the creation of hotspots at the catalyst bed may lead to blockage of the active sites of the catalyst and thus reduces the catalyst activity. It can be avoided by maintaining a good contact time. To maintain the contact time, one can vary the WHSV of the feed flow which ensures the smooth operation of the reactor. The out gas from this reactor needs to be sent to the distillation column where the DME is separated from the other undesired gaseous products like CO, H_2 and unreacted methanol which is present in the vapour phase along with the other remaining gases (Yagi et al. 2010) (Fig. 13.3).

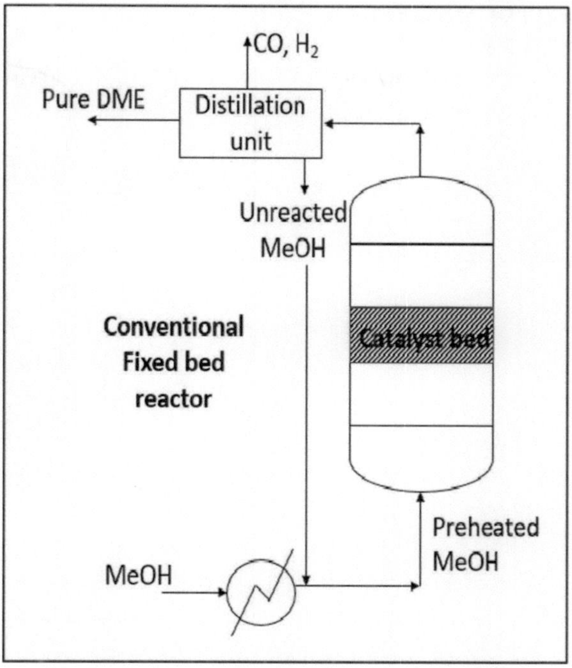

Fig. 13.3 Schematic of the fixed bed reactor used for DME production

13.5.2 Slurry Phase Reactor

In this reactor, the gas is passed into the solvent in which the solid catalyst is suspended. This is also called a three-phase slurry reactor because here, the solid acid catalyst is suspended in a liquid solvent and synthesis gas is passed through the slurry. This reactor can easily function as a lower investment alternative for the single-step synthesis of DME in the slurry phase as it offers good heat transfer than other available reactors. It is easy to maintain the reactor temperature across the reactor as it provides an even temperature distribution in the solvent phase of the reactor using thermic oil or steam. The main limitation of this type of reactor is the lower mass transfer rate in the slurry phase which consequently results in a slower reaction rate. Other limitations include the loss of catalyst in the solvent and difficulty in recycling the catalyst. Due to all such limitations, this technology is not applied at a large scale for such reactions. The reactor can achieve optimal performance when there is an increase in pressure and/or catalyst concentration (Abashar 2017) (Fig. 13.4).

13.5.3 Fluidized Bed Reactor

A fluidized bed reactor is a special type of reactor which is extensively used for gaseous phase catalytic reactions where the catalyst is in a solid state and reactants

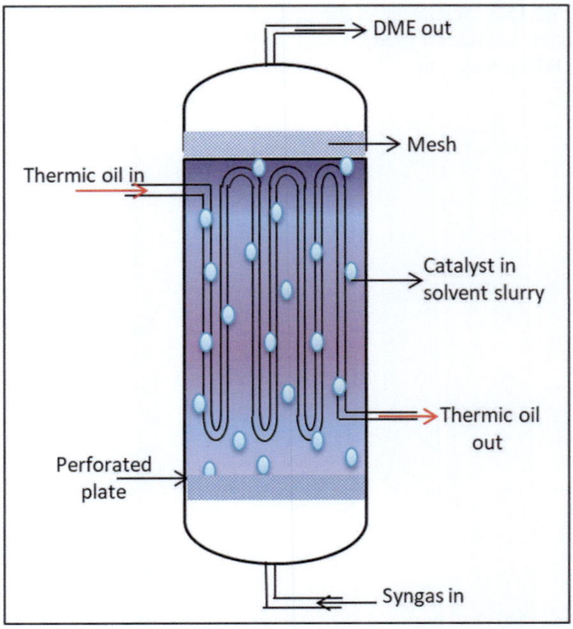

Fig. 13.4 Schematics of the slurry phase reactor

are in a gaseous state. Many researchers suggested this fluidized bed reactor as the best reactor for the setup of DME production reactions. The feasibility of such a type of reactor is yet to be established as it is in the very initial stages of laboratory testing. Once it is scaled up, we may see some interesting facts about these reactors in future. The plus point in this type of reactor is its heat removal characteristic which is better; it allows freely moving catalysts to interact with the feed in the catalytic bed area. Here, the main advantage is the decrement in the resistance of gas–solid mass transfer. This allows intensive mixing and interaction of catalyst particles with that of gaseous feed molecules which indirectly helps to achieve excellent control over temperature. Therefore, in case of reactions where a hotspot is generated which kills the activity of the catalyst, this reactor is a boon. This reactor helps to reduce the extra duty of recirculation and it is easy to operate at moderate conditions; it became of great interest to industries. Due to all the above-mentioned plus points, it gives a close approach to achieve high conversion for the reactions. On the other hand, there is a problem of loss of catalyst due to the collision of catalyst particles and the reactor walls. This reactor can be used in both ways, for direct and indirect routes of synthesis of DME from syngas as shown in the Figs. 13.5 and 13.6 (Falco et al. 2017).

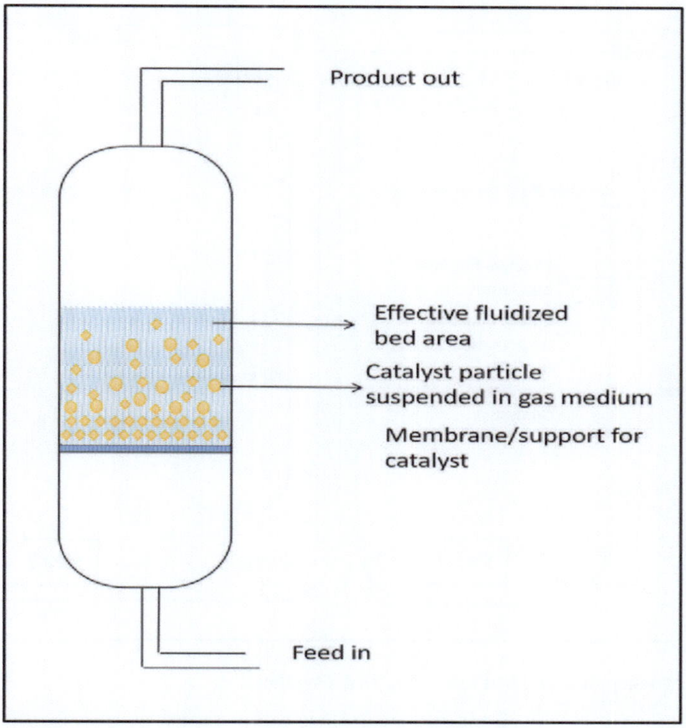

Fig. 13.5 Single-step fluidized bed reactor for DME synthesis

13.5.4 Membrane Reactors

Membranes are highly active due to their perm-selectivity property which provides active surfaces and sites for catalytic activity. For the indirect route of DME synthesis from methanol, one of the side products is water. The adsorbed water primarily deactivates the active sites of the catalyst, thereby reducing overall performance. Selective removal of water vapor formed during the reaction from the reaction zone could mitigate this deactivation and help maintain catalytic activity. Therefore, it can help to achieve a better reaction yields even at the milder temperature condition. Additionally, it can reduce steps of post-processing which involve dimethyl ether separation and purification from water vapours as it is reported to have great miscibility. However, an increment of the water flux through the membrane can produce undesired hydrocarbons also. Therefore, the appropriate volume of water flux through the membrane should be selected (Vega et al. 2021).

At present, ZSM-5, amorphous silica SIL, F-4SF, MOR and polymeric membranes are available and reported membranes for water removal. But they still suffer problems like pore-blockage, and mechanical and thermal stability, and therefore they still have very limited usefulness for methanol dehydration reactions. However,

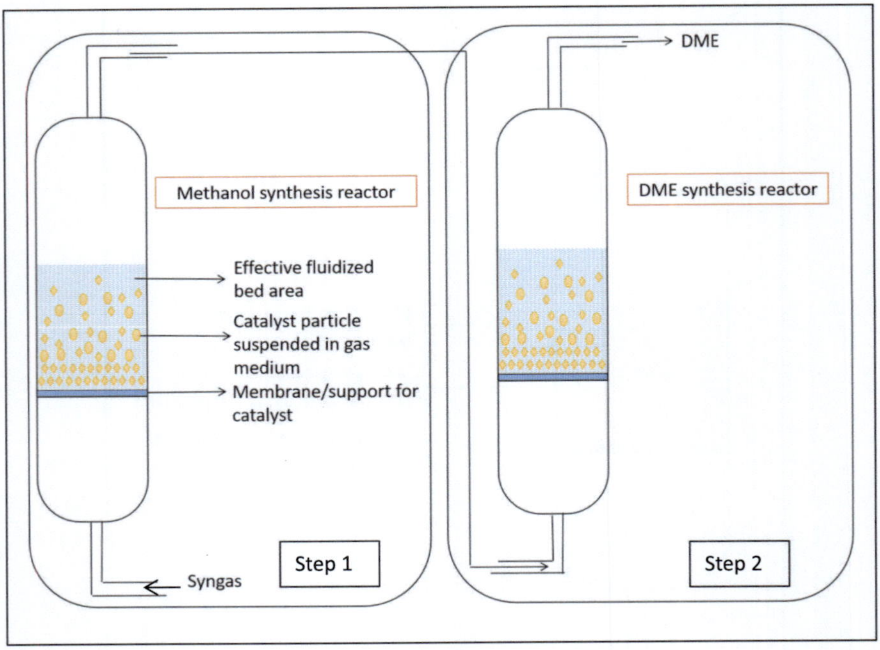

Fig. 13.6 Two-step fluidized bed reactor for DME synthesis

researchers have demonstrated a wide number of experimental and theoretical benefits of the membrane-based reactor systems (Fig. 13.7).

13.5.5 Reactor Coupled with Separation Units

Catalytic distillation (CD), which is also known by the name of reactive distillation (RD), is considered an integrated process. In this process, a single unit is formed which is called a separation unit. This separation unit consists of the distillation column and the reactor. The use of CD unit helps to achieve higher selectivity for DME in methanol dehydration reactions, along with higher conversion and lower operational costs as compared to the traditional single reactor. The CD system needs to be operated at a moderate operational temperature and pressure conditions. The catalysts studied for the methanol dehydration reaction were completely based on solid acid catalysts. All these catalysts were found highly active at higher temperature ranges. Therefore, an extensive study needs to be carried out in order to make this process feasible in milder conditions.

Some other innovative options to overcome the drawback of high-energy consumption for the distillation process are thermally coupled distillation columns, heat-integrated distillation or cyclic distillation, and dividing-wall columns (DWC).

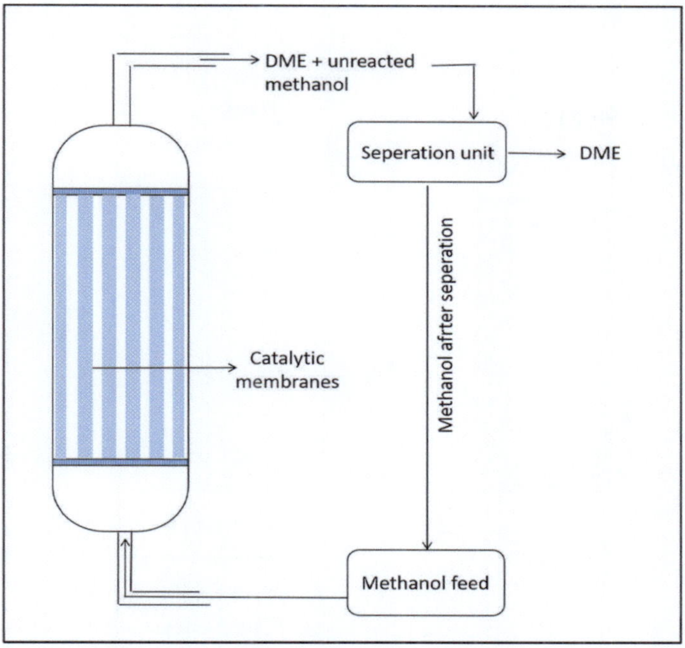

Fig. 13.7 Membrane-based reactor system

The Petlyuk configuration, of techno economics consisting of two thermally coupled distillation columns, is equivalent to DWC which can easily split the middle section of a single tower. This split can be carried out simply by inserting a vertical wall in the vessel at the proper position. DWC has great significance in the chemical process industry due to its capability to separate more components in a single distillation unit. It helps greatly to reduce overall CAPEX and OPEX (Fig. 13.8).

13.6 Conclusions

Recent studies have highlighted the potential of DME as an intermediary in the production of fuel and value-added products, as well as a standalone DME is considered to be an option for diesel engine-based machines and turbines where diesel is used for the generation of power. The present research demonstrates that there is still scope to develop strong, long-lasting and highly stable catalyst systems with which we can increase activity for DME production. The direct approach for DME production from syngas has received attention in the literature and there are numerous studies available. Direct DME synthesis investigations, including the creation of associated bi-functional catalysts, have intensified as a result of the existing state-of-the-art indirect DME from methanol-related reaction's limitations.

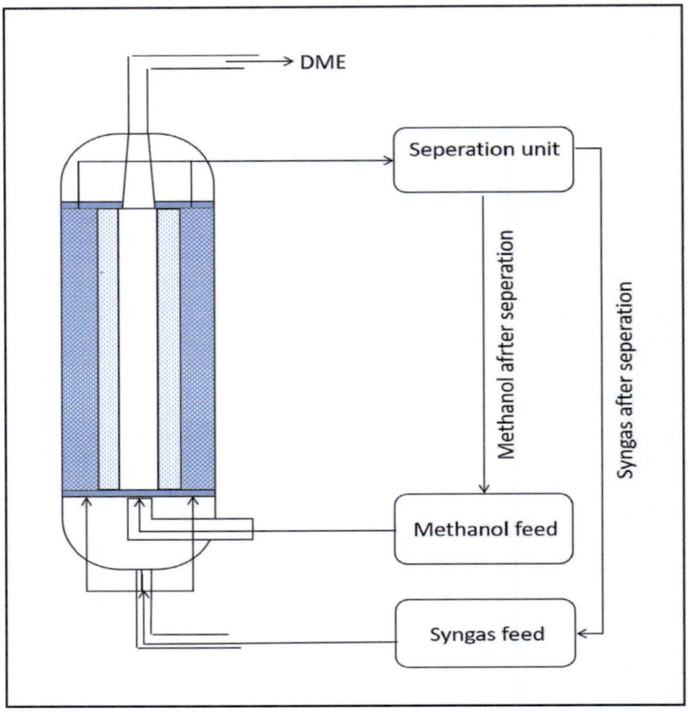

Fig. 13.8 Coupling reactors with separation unit

Investigations are still going on to find catalyst combination and bi-functional catalysts with the maximum conversion of syngas, selectivity towards DME and yield for the direct synthesis of dimethyl ether that can be scaled up for commercial use. For greater synergistic effects with multiple combinations of acidic sites from the metals along with acidic functionalities from metals and supports as well to catalyse reactions for DME production, innovative bi-functional catalysts are still being developed. Since both the production of methanol from syngas and the methanol dehydration reaction to DME processes are carried out in a single reactor, direct synthesis using these catalysts facilitated with active sites is more thermo dynamical than that of the indirect method of DME synthesis. For the production of methanol, primarily Copper-based multi-metallic catalytic systems are employed, but the dehydration of methanol is frequently accomplished using catalysts with solid acidic sites such as alumina, silica, zeolite-based and SAPOs. Physical mixing, co-precipitation, impregnation, sol–gel and other wet chemical techniques are all often utilized in the manufacture of bi-functional catalysts. To determine and design the ideal conditions for the experimental purpose, the aim must be to develop a catalyst and its optimization to co-relate the synergies between various acidic sites of metals that give rise to higher catalytic activity.

Typically, the conventional co-precipitation approach is used to manufacture the CZA oxide-based catalysts. The small concentrations of tri- and tetravalent ions create a cationic structural flaw which enhances and helps to stabilize the active phase of copper on the catalyst's surface. Catalytic activity may be impacted if the Cu surface area varies by Cu strain brought on by the structural faults noted above, improper reductions or changes in the Cu orientation caused due to ZnO. It has also been discovered that zirconium enhances a good dispersion of Cu which can reduce the amount of aluminium on the surface of the catalyst. Under low supersaturation circumstances, zirconium addition significantly increased the activity. The acidic sites quantity and dispersion of metallic copper can greatly affect the activity of a hybrid catalyst towards the DME production from synthesis gas. From this, one can therefore conclude that active acidic sites present in the hybrid catalysts were enough to convert the methanol which was produced during the earliest phase of DME synthesis. It was observed that catalysts with high pH did not directly affect the conversion of CO. Zr-based catalysts showed changes in the textural and structural characteristics as well as with increments in the area of metal and conversion of carbon monoxide. The rise in activity is indirectly associated with the metallic locations; also, it had a synergetic impact along with structural alterations.

Copper-based catalysts were found to be a very promising catalysts for methanol synthesis and direct DME synthesis. An important parameter about copper-based catalysts is the phase of copper which acts as an active component in the reaction. Xiaojing Cui et al. claimed that metallic copper with zinc oxide (Cu^0)-ZnO interface is highly active for the methanol synthesis reactions. The Zn in commercial CZA helps to keep Cu^0 in the active metallic state that leads to form (Cu^0)-ZnO phases. If the Cu^0 from the active phase causes agglomeration, the active interface between Cu and Zn will collapse which may form CO more than methanol. On the other hand (Cu^0)-ZnO if present as an intact phase will form methanol and DME.

Along with the catalyst, there is another parameter that needs to be focused on is feed stream. Numerous catalytic systems and reports are there where co-feed is added in both cases. Mostly inert gas like nitrogen is mixed well with methanol in the vapour phase; on the other hand, for syngas feed nitrogen in various ratios is used. The actual role and impact of co-feed in the form of inert gas must be studied to optimize the proper reaction conditions.

Reactor system design is considered a crucial factor which governs the overall reaction conversion, yield and economics. A simple fixed bed reactors are well reported and used extensively whereas some authors reported the issues like exothermic thermodynamics of reaction, due to which the overall process mainly catalyst collapsing results. To overcome this problem, researchers are actively involved in the development of a specific type of reactor that can control and help to optimize the thermodynamics of the reaction. Many commercial manufacturers are still in search of a promising and long-lasting reactor system.

References

Abashar, M.E.E.: Dimethyl ether synthesis in a multi-stage fluidized bed reactor. Chem. Eng. Process. **122**, 172–180 (2017)

Azizi, Z., Rezaeimanesh, M., Tohidian, T., Rahimpour, M.R.: Dimethyl Ether: A review of technologies and production challenges. Chem. Eng. Process., Chem. Eng. Process.: Process. Intensif. **82**, 150–172 (2014)

Bayat, A., Dogu, T.: Optimization of CO_2/co ratio and temperature for dimethyl ether synthesis from syngas over a new bifunctional catalyst pair containing heteropolyacid impregnated mesoporous alumina. Ind. & Eng. Chem. Res. **55**(44), 11431–11439 (2016)

Bonura, G., Miglion, M., Frusteri, L., Cannilla, C., Catizzone, E., Giordano, G., Frusteri, F.: Acidity control of zeolite functionality on activity and stability of hybrid catalysts during DME production via CO_2 hydrogenation. Journ of CO_2 Utilization **24**, 398–406 (2018)

Cai, M., Palčić, A., Subramanian, V., Moldovan, S., Ersen, O., Valtchev, V., Ordomsky, V.V., Khodakov, A.Y.: Direct dimethyl ether synthesis from syngas on copper–zeolite hybrid catalysts with a wide range of zeolite particle sizes. J. Catal. **338**, 227–238 (2016)

Catizzone, E., Bonura, G., Migliori, M., Frusteri, F., Giordano, G.: CO_2 recycling to dimethyl ether: State-of-the-art and perspectives. Molecules **23**(1), 31 (2017)

Catizzone, E., Bonura, G., Migliori, M., Frusteri, F., Giordano, G.: CO_2 recycling to Dimethyl Ether: State-of-the-art and perspectives. Molecules **23**(1), 31 (2018)

Catizzone, E., Migliori, M., Purita, A., Giordano, G.: Ferrierite vs. γ-Al_2O_3: The superiority of zeolites in terms of water-resistance in vapour-phase dehydration of methanol to dimethyl ether. J. Energy Chem. **30**, 162–169 (2019)

Celik, G., Arinan, A., Bayat, A., Ozbelge, H.O., Dogu, T., Varisli, D.: Performance of Silicotungstic acid incorporated mesoporous catalyst in direct synthesis of dimethyl ether from syngas in the presence and absence of CO_2. Top. Catal. **56**(18–20), 1764–1774 (2013)

Ciftci, A., Varisli, D., Dogu, T.: Dimethyl ether synthesis over novel silico-tungstic acid incorporated nanostructured catalysts. Int. J. Chem. Reactor Eng. **8**(1), (2010)

Din, I.U., Shaharun, M.S., Alotaibi, M.A., Abdulrahman, A.I., Naeem, A.: Recent developments on heterogeneous catalytic CO_2 reduction to methanol. J. CO_2 Util., **34**, 20–33 (2019)

El-Nagar, R.A., Ghanem, A.A.: Syngas production, properties, and its importance, vol. 2. IntechOpen, London, UK (2019)

Ereña, J., Garona, R., Arandes, J.M., Aguayo, A.T., Bilbao, J.: Effect of operating conditions on the synthesis of dimethyl ether over a CuO-ZnO-Al_2O_3/NaHZSM-5 bifunctional catalyst. Catal. Today **107–108**, 467–473 (2005)

Etim, U.J., Song, Y., Zhong, Z.: Improving the Cu/ZnO-based catalysts for carbon dioxide hydrogenation to methanol, and the use of methanol as a renewable energy storage media. Front. Energy Res. **8**, 545431 (2020)

De Falco, Marcello, Mauro Capocelli, and Alberto Giannattasio. "Membrane Reactor for one-step DME synthesis process: Industrial plant simulation and optimization. J. CO_2 Util. **22**, 33–43 (2017)

Flores, J.H., Peixoto, D.P.B., Appel, L.G., Avillez, R.R., Pais, S.M.I.: The influence of different methanol synthesis catalysts on direct synthesis of DME from syngas. Catal. Today **172**, 218–225 (2011)

Guil-López, R., Mota, N., Llorente, J., Millán, E., Pawelec, B., García, R., Fierro, J.L.G., Navarro, R.M.: Structure and activity of Cu/ZnO catalysts co-modified with aluminium and gallium for methanol synthesis. Catal. Today. **355**, 870–881 (2020)

Hassanpour, S., Taghizadeh, M., Yaripour, F.: Preparation, characterization, and activity evaluation of H-ZSM-5 catalysts in vapor-phase methanol dehydration to dimethyl ether. Ind. Eng. Chem. Res. **49**(9), 4063–4069 (2010)

Hosseini, S.Y., Khosravi-Nikou, M.R.: Synthesis and characterization of nano-sized γ-Al2O3 for investigation the effect of temperature on catalytic dehydration of methanol to dimethyl ether. Energy Sources, Part a: Recover., Util., Environ. Eff. **38**(7), 914–920 (2016)

Hua, Y., Guo, X., Mao, D., Lu, G., Rempel, G.L., Ng, F.T.T.: Single-step synthesis of dimethyl ether from biomass-derived syngas over CuO-ZnO-MOx (M=Zr, Al, Cr, Ti)/HZSM 5 hybrid catalyst: Effects of MOx. App.Cat.: A, General, **540**, 68–74 (2017)

Keshavarz, A.R., Rezaei, M., Yaripour, F.: Nanocrystalline gamma-alumina: A highly active catalyst for dimethyl ether synthesis. Powder Tech. **199**(2), 176–179 (2010)

Liu, R., Tian, H., Yang, A., Zha, F., Ding, J., Chang, Y.: Preparation of HZSM-5 membrane packed CuO–ZnO–Al_2O_3 nanoparticles for catalysing carbon dioxide hydrogenation to dimethyl ether. App. Surface Sci. **345**, 1–9 (2015)

Mahmoudi, H., Mahmoudi, M., Doustdar, O., Jahangiri, H., Tsolakis, A., Gu, S., Wyszynski, M.L.: Biofuels Eng. **2**, 11–31 (2017)

Mao, D., Yang, W., Xia, J., Zhang, B., Song, Q., Chen, Q.: Highly effective hybrid catalyst for the direct synthesis of dimethyl ether from syngas with magnesium oxide modified HZSM-5 as a dehydration component. Journ. of Cat. **230**(1), 140–149 (2005)

Merkouri, L.P., Ahmet, H., Reina, T.R., Duyar, M.S.: The direct synthesis of dimethyl ether (DME) from landfill gas: A techno-economic investigation. Fuel. **319**, 123741–123749 (2022)

Millán Ordóñez, E., Mota, N., Guil-López, R., Garcia Pawelec, B., Fierro, J.L.G., Navarro Yerga, R.M.: Direct synthesis of dimethyl ether on bifunctional catalysts based on Cu–ZnO (Al) and supported H3PW12O40: effect of physical mixing on bifunctional interactions and activity. Ind. & Eng. Chem. Res. **60**, 18853–18869 (2021)

Mondal, U., Yadav, G.D.: Perspective of dimethyl ether as fuel: Part I. catalysis. J CO2 Util, **32**, 299–320 (2019)

Park, H., Woo, Y., Jung, H.S., Kim, G., Bae, J.W., Park, M.J.: Development of dimethyl ether synthesis processes using by-product gas from a steel-making plant: Single-vs. two-step processes. J. Clean. Prod. **326**, 129367 (2021)

Peng, J.B., Geng, H.Q. and Wu, X.F.:The chemistry of CO: Carbonylation. Chem, **5**(3), 526–552 (2019)

Phienluphon, R., Pinkaew, K., Yang, G., Li, J., Wei, Q., Yoneyama, Y., Vitidsant, T., Tsubaki, N.: Designing core (Cu/ZnO/Al_2O_3)–shell (SAPO-11) zeolite capsule catalyst with a facile physical way for dimethyl ether direct synthesis from syngas. Chem. Eng. Journal **270**, 605–611 (2015)

Poto, S., Gallucci, F., d'Angelo, M.F.N.: Direct conversion of CO_2 to dimethyl ether in a fixed bed membrane reactor: Influence of membrane properties and process conditions. Fuel 302: 121080–121093 (2021)

Ramos, F.S., Duarte, FA.M., Borges, L.E.P., Monteiro, J.L., Fraga, M.A., Sousa, A.E.F., Appel, L.G.: Role of dehydration catalyst acid properties on one-step DME synthesis over physical mixtures. Cat. Today, **101**: 39–44 (2005)

Ren, H., Xu, C., Zhao, H., Wang, Y., Liu, J.: Methanol synthesis from CO_2 hydrogenation over Cu/γ-Al_2O_3 catalysts modified by ZnO, ZrO_2 and MgO. J. Ind. Eng. Chem. **28**, 261–267 (2015)

Rutkowska, M., Macina, D., Kubień, N.M., Piwowarska, Z., Chmielarz, L.: Hierarchically structured ZSM-5 obtained by desilication as new catalyst for DME synthesis from methanol. App. Cat. b: Environmental **174–175**, 336–343 (2015)

Seddon, D.: Methanol and dimethyl ether (DME) production from synthesis gas. In: Advances in clean hydrocarbon fuel processing, pp. 363–386. Woodhead Publishing, (2011)

Tang, Q., Xu, H., Zheng, Y., Wang, J., Li, H., Zhang, J.: Catalytic dehydration of methanol to dimethyl ether over micro–mesoporous ZSM-5/MCM-41 composite molecular sieves. App. Cat. A: General, **413–414**, 36–42 (2012)

Tokay, K.C., Dogu, T., Dogu, G.: Dimethyl ether synthesis over alumina based catalysts. Chem. Eng. Journ. **184**, 78–285 (2012)

Uriarte, P.P., Gamero, M., Ateka, A., Díaz, M., Aguayo, A.T., Bilbao, J.: Effect of the acidity of HZSM-5 Zeolite and the binder in the DME transformation to Olefins. Ind. Eng. Chem. Res. **55**, 1513−1521 (2016)

Varisli, D., Tokay, K.C., Ciftci, A., Dogu, T., Dogu, G.: Methanol dehydration reaction to produce clean diesel alternative dimethyl ether over mesoporousalumino silicate-based catalysts. Turkish J Chem. **33**, 355–366 (2009)

Vega, P.R., Ateka, A., Kumakiri, I., Vicente, H., Ereña, J., Aguayo, A.T., Bilbao, J.: Experimental implementation of a catalytic membrane reactor for the direct synthesis of DME from H_2+CO/CO_2. Chem. Eng. Sci. **234**, 116396–116407 (2021)

Wang, Y., Wang, W., Chen, Y., Ma, J., Li, R.: Synthesis of dimethyl ether from syngas over core–shell structure catalyst CuO–ZnO–Al_2O_3@SiO_2–Al_2O_3. Chem. Eng. Journal **250**, 248–256 (2014)

Wang, Y., Wang, W.L., Chen, Y.X., Zheng, J.J., Li, R.F.: Synthesis of dimethyl ether from syngas using a hierarchically porous composite zeolite as the methanol dehydration catalyst. J. Fuel Chem. Tech. **41**(7): 873–880 (2013)

Wang, G., Mao, D., Guo, X., Yu, J.: Methanol synthesis from CO2 hydrogenation over CuO-ZnO-ZrO_2-MxOy catalysts (M=Cr, Mo and W). Int. J. Hydrog. Energy, **44**(8), 4197–4207 (2019)

Wang, S., Yang, J., Zhou, H., Xiao, F., Zhao, N.: Performance of Cu-Mn-Zn/ZrO_2 catalysts for methanol synthesis from CO2 hydrogenation: The effect of Zn content. Fuel. **52**(3), 293–303 (2024)

Warmuth, L., Steurer, M., Schild, D., Zimina, A., Grunwaldt, J.D., Pitter, S.: Reversible and irreversible structural changes in Cu/ZnO/ZrO_2 catalysts during methanol synthesis. ACS Appl. Mater. **16**, 8813−8821 (2024)

Xu, Q., Lan, P., Huang, K., Yan, Y.: Effect of different ratios of Si-Al of Zeolite HZSM-5 on the activity of bifunctional catalysts upon dimethyl ether synthesis. Petro. Sci. Tech. **29**, 1080–1092 (2011)

Yagi H, Ohno Y, Inoue N, Okuyama K, Aoki S (2010). Int. J. Chem. React. Eng. 8(1), (2010)

Yang, C., Ma, Z., Zhao, N., Wei, W., Hu, T., Sun, Y.: Methanol synthesis from CO_2 rich syngas over a ZrO_2 doped CuZnO catalyst. Catal. Today **115**, 222–227 (2006)

Yang, G., Thongkam, M., Vitidsant, T., Yoneyama, Y., Tan, Y., Tsubaki, N.: A double shell capsule catalyst with core–shell-like structure for one-step exactly controlled synthesis of dimethyl ether from CO2 containing syngas. Catal. Today **171**, 229–235 (2011)

Chapter 14
A Heterogeneous Catalyst-Driven C–H Bond Activation

Archana Ramakrishnan, Marimuthu Prabu, Siva Prasad Mekala, Samruddhi Mane, Akash Bhatkar, Marimuthu Manikandan, Praveen Dharmalingam, and Raja Thirumalaiswamy

14.1 Activation of Methane Through Reforming Techniques

14.1.1 Introduction

Catalysis is critical to a country's economic growth, notably in the production of chemicals, medicines, and fuel sectors. It contributes to environmental preservation by treating automobile flue gases and eliminating environmentally hazardous substances. Thus, catalysis not only aids in a nation's economic growth but also plays an integral part in people's well-being. In the twenty-first century, more than 80% of the processes involving chemical production relies on catalysts to perform runs in some form. In reality, catalysis has played a significant role in petroleum and petrochemical processes throughout the previous few decades. More recently, various renewable sources like biomass's conversion to value-added platform chemicals or fuels are being practiced through catalysis routes. A catalyst must be strong and multi-functional in order to play a part in all of the aforementioned processes and convert the raw material into products.

To build a catalyst for a specific reaction, one must first understand the properties of the chosen catalyst material as well as its possible function in the selected process. To develop an effective catalyst, it will be helpful to understand the general guidelines of the process through this activity (Kokal et al. 2001). Despite major advancements in innovative catalysts and processes, various constraints still need to be addressed

A. Ramakrishnan · M. Prabu · S. P. Mekala · S. Mane · A. Bhatkar · M. Manikandan · P. Dharmalingam · R. Thirumalaiswamy (✉)
Catalysis and Inorganic Chemistry Division, CSIR-National Chemical Laboratory, Pune, India
e-mail: t.raja@ncl.res.in

Academy of Scientific and Innovative Research (AcSIR), Ghaziabad-201002, India

R. S. Ningthoujam and A. K. Tyagi (eds.), *Handbook of Materials Science, Volume 3*, Indian Institute of Metals Series, https://doi.org/10.1007/978-981-96-8229-4_14

to improve the effectiveness and ecological acceptability of a particular catalytic process.

14.1.1.1 Natural Gas (NG)

Natural gas (NG) is a significant source of energy, and it is used to expand quickly. As per the statistics of International Energy Outlook IEO 2016, its consumption will double from its 2012 level to 203 trillion cubic feet (Tcf) by 2040. Methane and other lower-density hydrocarbons make up the majority of natural gas, with trace amounts of N_2, H_2S, and CO_2, developing under the earth's crust over millions of years when plant and animal stuff decomposes. Due to its high energy content and cleaner burning qualities compared to coal and petroleum products, natural gas plays a crucial role in the industrial sectors.

Compared to others, natural gas (NG) is the cleanest and most environmentally friendly fuel for this use. Despite being a nonrenewable fossil resource, NG is a suitable energy carrier due to its accessibility, ease of storage, and transportation. Natural gas's composition varies depending on where it comes from, although CH_4 is often its main constituent. Although CH_4 could contribute significantly to the production of fuels (between 4 and 9% of GHGs), it is a large GHG that contributes to global warming.

It is challenging to avoid the syngas stage by directly converting CH4 to usable compounds. However, the yields are quite low compared to the conventional procedure and are not cost-effective (Holmen 2009). Methane has a higher stability and resistance to numerous reactants due to the strong C–H bond's dissociation energy (439 kJ/mol).

The thermodynamics do not favor the conversion of methane to higher alkanes without the aid of O_2. The NG's conversion to methanol through syngas has a selectivity of 99.9% and a carbon efficiency of above 95%. But it calls for sizable energy recovery and exchange units. In actuality, the cost of producing syngas to convert natural gas fuels (gas to liquid (GTL)) can account for greater than 60% of the total. As a result, lowering the cost of producing syngas would have a considerable impact on the economics of the GTL technology (Rostrup-Nielsen and Bak Hansen 1993).

14.1.1.2 Global Status of CO_2

China and the US consume the most gas resources, with India following them with a 5.3% worldwide share in 2015. The Government of India began national and regional strategies to reduce and control carbon dioxide (CO_2) emissions in 2004. Natural gas use is being promoted to replace more carbon-intensive coal and liquid fuels.

Earth's CO_2 content increased by 40% due to the industrial revolution. By the end of 2017, it had risen from 280 to approximately 400 ppm. The level of CO_2 has been at its most significant point in the last 20 million years. Deforestation, fossil fuel combustion, and human activity are the major causes of it. The atmosphere's CO_2

concentration has reached a hazardous level and is what is driving global warming. According to recent calculations, the buildup of greenhouse gases (GHGs) since 1880 has increased the earth's surface temperature by 1 °C. Global temperature is predicted to rise by 2 °C if it hits 450 ppm. One hundred ninety-five countries have agreed to work together to cut greenhouse gas emissions through the 2016 Paris Agreement (Hawkins and Jones 2013). Figure 14.1 shows the CO_2 concentration now, and the temperature rises globally from 1960 to 2021.

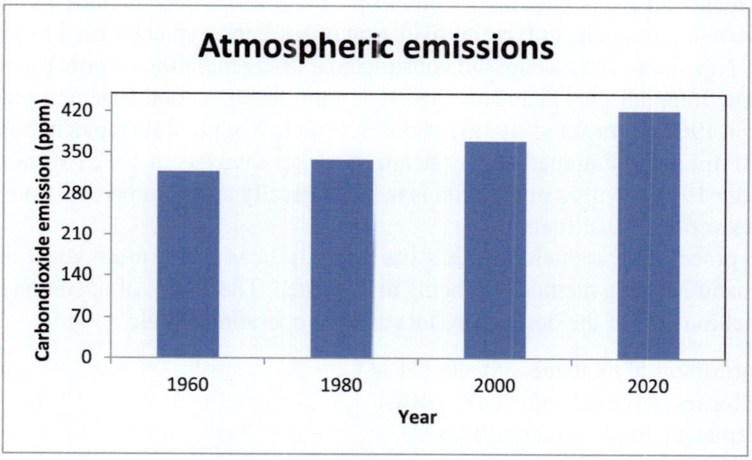

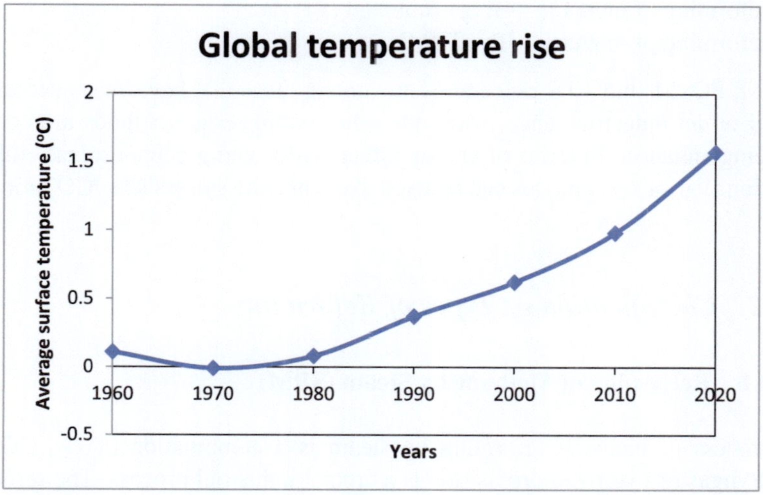

Fig. 14.1 Atmospheric carbon dioxide concentration and average global temperature

14.1.1.3 Synthesis Gas (Syngas)

Syngas is a gas mixture primarily composed of CO and H_2. Syngas is an essential source for the energy conversion process as it is a component of fuels, a source of H_2 and CO, and an input for selective chemical synthesis. Syngas can be produced by reforming natural gas, biomass, oil, and coal. Environmental contaminants include fine airborne coal-related particles, heavy metals, and sulfur compounds that are minimally present during the fuel production, viz. Fischer–Tropsch (FT) synthesis and/or methanol from syngas. Common reforming techniques include steam, dry, auto thermal reforming and partial oxidation of methane, typically used to produce syngas. New Jersey's Standard Oil constructed a first-generation reformer in tubular form and Imperial chemical industry (ICI) introduced second reactor running at 15 atm in 1962 (Byrne et al. 1932). Mittasch et al. first applied for a patent on behalf of BASF for reforming methane by steam on nickel catalysts in 1912 (Mittasch and Schneider 1912). Syngas production is technologically and commercially advanced and uses various fossil fuels.

The process of producing syngas has recently been given more attention, and novel manufacturing methods are being investigated. The choice of operation affects the stoichiometry of the desired product and the operation's scale.

i. Reforming of methane—Steam (SRM)
ii. Reforming of methane—Dry (DRM)
iii. Methane partial oxidation (POxM)
iv. Reforming of methane—Autothermal (ATR)
v. Reforming of methane—Bi or combined (BRM)
vi. Reforming of methane—Tri (TRM).

SRM, POxM, and ATR procedures are among those that have been successfully applied on an industrial scale, while the other cutting-edge methods are continually being enhanced in terms of energy effectiveness and greenhouse gas emission reduction. Various techniques can be used to change the syngas's H_2/CO ratio.

14.1.2 Classification or Types of Reforming

14.1.2.1 Reforming of Methane by Steam (SRM)

The process of methane reforming by steam is a combination of H_2, CO, and CO_2. Syngas or hydrogen production is a crucial industrial process. The reformate produced with a ratio of 3:1 is considered better for the production of H_2. Numerous additional reactions, such as the Water Gas Shift (WGS) reaction (Eq. 14.2), also happen in addition to the SRM (Eq. 14.1) primary reaction.

$$CH_4 + H_2O \leftrightarrow CO + 3H_2 \Delta H = 206\ \text{kJ mol}^{-1} \quad (14.1)$$

$$CO + H_2O \leftrightarrow CO_2 + H_2 \Delta H^\circ (298\ K) = -41\ kJ\ mol^{-1} \tag{14.2}$$

Because this reaction requires a heat (endothermic in nature) process with increasing the products, more conversion can be accomplished at ambient pressure, high temperature and steam to carbon (S/C) ratio, according to the Le Chatelier principle. The procedure is highly expensive as a result of the high capital cost required to meet these parameters.

The standard reforming method for producing syngas or hydrogen consists of a number of reaction units, including a pre-treater unit, a high temperature (HT) reformer, and WGS reactors. In a multi-tubular FBR with an oxide catalyst present, this process is carried out (Gallucci et al. 2004).

Although the SRM is highly efficient and has a high H_2 concentration, start-up times are frequently lengthy. Several factors also constrain it: (a) thermodynamic equilibrium, (b) heat transfer, (c) environment contamination such as emissions of CO_2 and NO_x, and (d) deactivation of the catalyst due to carbon formation and sintering (Ren et al. 2002). Research on the usage of catalytic membrane reactors has also been focused on recently in an effort to overcome thermodynamic constraints. Due to the economic limitations of SRM, alternative processes for converting CH_4 to syngas, such as DRM, POxM, and ATR were developed.

14.1.2.2 Reforming of Methane—Dry (DRM)

Recently, the focus has been placed on GHG like CH_4 and CO_2, which are responsible for global warming. DRM or CO_2 reforming is the term used when CO_2 replaces steam during the reforming cycle. Fischer and Tropsch described this reaction for the first-time using Ni/Co catalysts. Researchers noticed that the catalyst was becoming inactive because of the significant deposition of carbon on the surface. In order to identify this mechanism in the SR or DR of hydrocarbons, Reitmeier et al. examined the link between the composition of the reactant and the deposition or formation of carbon (Reitmeier et al. 1948). In 1964, Rostrup-Nielsen investigated the reformation of methane and other hydrocarbons on Mg and Al catalyst systems involving their spinel oxides. Their understanding explains that Mg is prone to CO_2 activation and works as a promoter that removes carbon from active metal surfaces. According to Gadalla et al., in the DRM reaction, there is an ideal temperature range where carbon creation takes place for every CO_2–CH_4 ratio. For the DRM reaction, they employed a Ni/Al_2O_3 catalyst and discussed how the ratio of CO_2–CH_4 affects the generation of carbon (Gadalla and Sommer 1989). Numerous studies on the use of these two main greenhouse gases that cause global warming were published since 1990. Equation 14.3 shows that the DRM reaction needs a little bit more energy than the SRM reaction. A strong and long-lasting catalyst is needed since the synthesis gas produced in DRM has an H_2/CO ratio of approximately 1. Synthesis gas with a ratio of close to 1 is better suited for the long chain HC synthesis than syngas with a ratio of 2, needed for the FT process.

$$CH_4 + CO_2 \leftrightarrow 2\,CO + 2\,H_2 \Delta H = 247\ \text{kJ mol}^{-1} \tag{14.3}$$

14.1.2.3 Partial Oxidation of Methane

An alternative method for producing syngas for the SRM reaction, which creates the steam needed to make ammonia, is partial oxidation (POX) of NG. In other words, high space velocities are used to supply oxygen to the feed over a solid catalyst in place of steam. In 1929, Liander et al. published the first reports on catalytic methane partial oxidation (Liander 1929). This method is more adaptable and able to handle a variety of hydrocarbons, including waste plastic and asphalt in addition to NG. Non-catalytic and catalytic POX and are the two different kinds of partial oxidation processes. The catalytic process produces 2 ratios of syngas over the catalyst at a lower temperature (Eq. 14.4).

$$CH_4 + {}^1\!/_2\,O_2 \rightarrow CO + 2\,H_2 \quad \Delta H = -38\ \text{kJ mol}^{-1} \tag{14.4}$$

Since the POX of methane is exothermic in nature, it seems more commercially feasible than the SRM technique. However, because of how highly exothermic the reaction of CH_4 with oxygen is, particular care must be taken to prevent explosions. As a result, it is not widely used due to the numerous safety concerns raised.

14.1.2.4 Autothermal Reforming of Methane

Syngas is produced through reforming via an autothermal process where NG reacts with both steam and O_2. It combines exothermic POX and endothermic reactions SR. ATR typically runs with low steam to carbon ratios at 900 and 1150 °C and 1–30 bar of pressure. The reactant oxidant could be oxygen or air, based on the subsequent purifying procedure in the downstream processes. ATR reactor is smaller than a partial oxidation unit but larger than a steam reformer. The reaction benefit is that the production of a syngas ratio of 2 is better suited for various syntheses further down the line, such as FT and methanol synthesis. No further heat is needed at a specific fuel, air, and steam ratio. The exothermic heat generated through the POX reaction is utilized by the endothermic reaction of SRM.

The CH_4 participation in the SR reaction begins at temperatures over 400 °C and continues until the full conversion is achieved. At 700 °C, the concentration of the H_2 is high, where the syngas ratio falls due to the dominance of the reverse water gas shift (RWGS) process. Above 700 °C, carbon production in this process is not anticipated. Therefore, there is less carbon and soot generation during this reaction. For fuel cell applications, a processor using ATR was introduced by Farrauto et al. (2003).

Natural gas transformation into syngas in the industrial processes follows SR, POX, and ATR. Researchers from all over the world have recently focused a lot of attention on bi- and tri-reforming processes. Most of these reactions include the use of flue gases from thermal power plants that burn fossil fuels, biogas produced by biomass anaerobic digestion and landfill gas for the formation of syngas. The process benefit involves non-expensive CO_2 separation equipment, and the composition of syngas may be adjusted for different synthesis processes (such as those for ammonia, methanol, dimethyl ether, and FT products) (Halmann and Steinfeld 2006).

14.1.2.5 Bi-Reforming of Methane (BRM)

NG steam reforming is a proven commercial process for the production of syngas. Due to the extreme coking of the catalysts and the ensuing quick deactivation of those catalysts, the dry reforming process has not advanced much. Adding O_2 and steam to the reactant can solve or prevent carbon formation. Bi-reforming, which combines steam and CO_2 reforming reactions (Eq. 14.5), is an intriguing method for producing syngas with the correct ratio. In this process, CO_2, CH_4, and H_2O are supplied in the appropriate proportion (1:3:2) to produce syngas with a ratio of 2, also termed as "metgas." This gas mixture can be used to create methanol (Eq. 14.6), which can then be utilized to create hydrocarbons (Olah et al. 2013).

$$3CH_4 + 2H_2O + CO_2 \rightarrow 4CO + 8H_2 \quad (14.5)$$

$$4CO + 8H_2 \rightarrow 4CH_3OH \quad (14.6)$$

In bi-reforming, both DRM and SRM work together simultaneously to minimize the deposition of carbon on the catalyst. The nature of the catalyst being utilized and the reaction conditions determine the deposition of carbon contribution due to the decomposition of CH_4 and disproportion of CO. If carbon is created, steam removes it.

14.1.2.6 Tri-Reforming of Methane

Recently, Song et al. published the idea of tri-reforming CH_4 (Song and Pan 2004). It is a synergistic mixture of exothermic methane oxidation and endothermic SR. Therefore, from an industrial and environmental aspect, it is more significant combining SR, DR, and POX to reduce carbon formation, improve the catalyst's life, and enhance the reforming process effectiveness. Typically, carbon production causes a decline in the activity of Ni-based catalysts, ideally prevented by introducing oxygen or carbon dioxide to the reaction ($CH_4 + CO_2$). By removing the coke, oxygen addition reduces the accumulation of coke. However, the oxidation of metallic Ni can take place in the presence of O_2, steam, and CO_2, which can reduce catalytic activity. By introducing rare earth oxides like CeO_2, La_2O_3, etc., to Ni catalysts, this drawback can be partially resolved.

14.1.3 Literature

In the literature, a variety of catalysts for methane reforming have been described. The key characteristics of a reforming catalyst are (i) high activity; (ii) carbon resistance; (iii) tolerance to sulfur; and (iv) least susceptibility to metal sintering. Additionally, a catalyst used in reforming processes needs to be thermally stable and preserve its structural integrity over a long time on stream (thousands of hours). Different transition metals are typically employed for reforming; however, the catalyst most frequently utilized is Ni. It is less expensive and has significant activity in the reforming reaction toward methane cracking. For the reforming reaction, many catalyst supports are also utilized. Al_2O_3 is a popular support among them since it is reasonably priced and has good thermal stability.

It is difficult to overstate the importance of catalytic support in the methane reforming reaction. It helps with excellent carbon removal from the surface of the active metal as well as dispersion of active metal sites and sintering resistance. In the oxidation of carbon or CO, it also contributes labile "oxygen" to the catalytic reaction. As a result, it is a component of a catalytic system for methane reforming. The supports that are employed the most frequently are Al_2O_3, ZrO_2, La_2O_3, MgO, SiO_2, $NiAl_2O_4$, and TiO_2 (Sahli et al. 2006). A significant part in reducing carbon deposition is played by basic catalysts like MgO and CaO. On the other hand, acidic supports promote the polymerization of the formed carbon due to the hydrocarbon thermal cracking that forms over the active surface, which speeds up the decomposition of methane. Either all or a set of metal oxide combinations, as aforementioned, help to strengthen the bond between the support and metal, thus the durability of the catalyst is increased and reduces the susceptible sintering and carbon formation. Similar to reforming catalysts, metal-substituted perovskite ($La_{1-x}Ce_xNiO_3$) (Lima et al. 2012), cobalt-supported catalysts, catalysts with the presence of ceria (CeO_2) with or without doping of active metals like Rh, Ir, and Ru have less inclination to produce carbon.

In comparison to the conventional supported metal catalysts, perovskite-type oxides have received a lot of research attention over the past few years. The active metals should exhibit improved thermal stability as reforming catalysts since they are firmly bound in a perovskite lattice. Furthermore, some perovskites naturally conduct oxygen, which is enhanced by adding an active metal to the perovskite lattice. The active metal will not accumulate as much carbon thanks to this labile oxygen's assistance in oxidizing the coke. The mineral perovskite takes its name from Russian mineralogist L. A. von Perovski. A lanthanide or other alkaline earth metal serves as the larger cation, A in the generic formula for perovskite, ABO_3, whereas transition metals often make up the smaller cation, B.

The perovskite structure's lattice oxygen conduction may speed up the oxidation of carbon species on metallic surfaces. $LaFeO_3$, $SrTiO_3$, $LaAlO_3$, $BaTiO_3$, and $La_{0.4}Ba_{0.6}Co_{0.2}Fe_{0.8}O_3$ are a few of the Ni-supported perovskite catalysts that Urasaki et al. have exploited. They discovered that catalysts such as $LaAlO_3$ and $SrTiO_3$ exhibit greater stability while performing SRM (Urasaki et al. 2005). In Ce-doped perovskite, $La_{1-x}Ce_xFe_{0.7}Ni_{0.3}O_3$ and $La_{0.23}Ce_{0.1}Sr_{0.67}TiO_3$ exhibited substantial resistance toward the deposition of coke with better conversion was noticed. This could be due to the Ce presence in the lattice of the perovskite, which enhanced the reaction's oxygen mobility and Ni dispersion.

14.1.4 *Catalyst Synthesis*

In heterogeneous catalysis, the method used to prepare the catalyst is crucial for achieving the necessary activity and selectivity. Characterizing the prepared materials and correlating the catalyst's activity with its physical–chemical characteristics aid in better understanding the catalyst and enable the creation of tailored catalysts. The procedures for catalyst synthesis and their characterization using various physicochemical methods are discussed in this section.

All the perovskite oxides are prepared through citrate gel methods. Nitrate salts of the desired metals were used as starting materials. The precursors in the form of nitrates are to be weighed and dissolve the same in DI (Deionized) water. An excess amount of citric acid in 1:3 ratio is added to the nitrate solution at ambient conditions. Now the solution is stirred for the formation of a gel at a desired temperature of about 80 °C, then the gel is kept in the oven for 12 h for the evaporation of nitrates at 180 °C. After crushing into a fine powder, the catalyst is calcined at the desired temperature for the formation of perovskite materials (Dama et al. 2018). The obtained final samples were undergone for thorough characterization using Powder X-ray Diffraction (PXRD), N_2-sorption (BET), scanning and transmission electron microscopy (SEM, TEM), thermogravimetry (TGA), temperature programmed reduction (TPR), X-ray photoelectron spectroscopy (XPS), Raman and Fourier Transform Infra-Red (FTIR) spectroscopy, and Inductively Coupled Plasma with Optical Emission Spectroscopy (ICP-OES) for analysis were all used for the detailed structural characterization of these materials.

14.1.5 *Properties*

14.1.5.1 Advantages of SRM

a. More H_2 is produced due to enhanced selectivity.
b. There is no requirement for an oxygen air separation unit.
c. The combustion of low-quality fuel can externally supply of essential heat.
d. Excess steam can easily be recovered and has little effect on the volumetric flow to units downstream.

14.1.5.2 Disadvantages of SRM

a. During the SRM reaction, additional carbon or coke is formed.
b. In addition to its complexity, building an SRM reactor is more expensive.
c. Heat must be transported externally.
d. The reactant should have a high proportion of methane; greater alkanes shorten the life of the catalyst.
e. Active metal sintering is frequently observed since it operates at high temperatures.

14.1.6 Future Scope

The effects of greenhouse gases resulting from CO_2 emissions are currently a major global concern. One solution to this issue might be to use this gas for the creation of syngas by dry reforming, bi-reforming, or tri-reforming methane. These procedures require the use of extremely active and long-lasting catalysts in order to achieve and establish syngas production. This chapter's study has shed some light on several elements of these reactions and the usefulness of different perovskite structures for the production of syngas. These catalysts should be scaled up and perform rigorous testing at various scales of operations. BRM is a helpful method for the reformation of biogas as well, which will reduce the cost involved in the separation of CO_2. Using different catalysts to steam reform and bi-reform biogas, which can be simulated, has a recommendation for future studies.

14.1.6.1 Catalyst Development Biogas SR

Utilizing biogas is crucial because it is created from a variety of sources. However, CO_2 is separated and released to the environment in current procedures, increasing GHG emissions. Therefore, the usage of biogas without removing carbon dioxide in order to obtain the required ratios of syngas is preferred. The production of methanol or other fuels and chemicals can then be done using these syngas. As a result of the study's assistance in identifying a few effective compositions of catalysts and their scale-up studies involving biogas reforming, the performance of the catalyst in this study should also be focused on how various biogas contaminants affect that performance.

14.1.6.2 Developing Sulfur-Tolerance Reforming Catalysts

If the fuel contains sulfur or poisonous chemicals that resemble NH_3 during the reforming reaction, the catalyst will not function for a long time on stream. For instance, the anaerobic decomposition of plant and animal waste results in the production of biogas. The CH_4 and CO_2 feed also contain a little quantity of sulfur. Poisoning of the active catalyst by sulfur and rendered inactive during the reforming of biogas. Therefore, a new sulfur-tolerant catalyst for reforming is needed for this procedure.

14.2 Activation of Methane: A Challenge Through Oxidative Coupling

14.2.1 Introduction

Being aware of the critical energy issues and depleting crude oil reserves with the soaring demands, there is abundant attention given to natural gas as an important feedstock source for chemicals and clean fossil fuel energy. The processes have been developed for converting methane into some higher value-added products, among which oxidative coupling of methane (OCM) is considered challenging work to date. Developing different strategies to produce effective catalysts is being evaluated, including various parameters involving improving metal interaction on the support, appropriate/precise usage of promoters, solid solutions formation, and core-shell structure development. However, it has been unsuccessful in producing the catalyst at the commercial scale.

Ethylene is considered as one of the world's significant chemical commodities for synthesizing most petrochemical products like polyethylene, ethylene oxide, ethylene glycol, polyvinyl chloride, etc. Ethylene is produced from petrochemical refineries where steam and thermal cracking processes are usually adopted, resulting in high reliability for global ethylene production (Galadima and Muraza 2016). From a broader perspective, these processes involve tremendous energy consumption and make it expensive at a large scale, therefore causing environmental un-sustainability. A report from the Organization of Economic Cooperation and Development stated that the demand and ethylene production is more than 100 tons million annually in the global market (Fan et al. 2013). Therefore, there is much attention given to bringing in innovative advancements for the utilization of value-added commodities in various sectors. For making effective usage of methane, one of the possible routes to explore the manufacturing of ethylene, the initial reaction steps of OCM work were carried out in the early 1980s by Keller and Bhasin, involving the non-selective methane oxidation, either occurring in the gaseous phase or heterogeneously catalyzed (Keller and Bhasin 1982). Recent developments in shale gas production have increased interest in methane coupling via oxidation (OCM) to C_2 hydrocarbons (C_2H_4 and C_2H_6). It can be challenging to obtain higher C_2 selectivity using a typical OCM process for facile oxidized products such as C_2H_4 and C_2H_6 for achieving desirable C_2 yields, and different kinds of catalysts have been studied. This chapter gives an overview and summarizes the research on OCM in oxidative reaction conditions, a sturdy reaction system involving heterogeneous reaction steps.

14.2.2 Literature

Several reactions were catalytically carried out by screening different types of catalysts like metal oxides and metal supported on alumina involving normal, sequential or continuous modes of reaction feed. The activity of the catalysts was assessed based on the conversion and selectivity to the C_2 products overcoming the energy barrier and equilibrium conversion. The oxides of Sn, Pb, Sb, Bi, Tl, Cd, and Mn, while Li, Mg, Zn, Ti, Zr, MO, Fe, Cr, W, Cu, Ag, Pt, Ce, V, B, and Al showed meager or no activity. One problem that arises in OCM is the competing gas phase products during the reactions. From the engineering perspective, the reactor design has a crucial role in designing the reactor based on the kinetics and reaction conditions. To minimize these hurdles, the reaction was operated cyclically, where methane and air were sent individually over the catalyst bed, which was followed by interspacing by a short purge of inert gas. In the cyclic operation, the flow rate of feed, varying feed composition and temperature parameters are investigated. Methane gas interacted with the catalyst, which further reduced stoichiometrically and reacted readily with oxygen. All the oxide catalysts required higher temperatures which were above 600 °C to produce ethylene and ethane as the major product, although the yield was around 20%. The considerable number of by-products usually formed in the competing gas phase reaction is likely to be H_2, H_2O, CO_2, CO, C_2H_6 and some quantity of C_3H_8 or C_3H_6, which are preferably lower than 1% of the total yield (Karakaya et al. 2018). The relative importance of gas phase chemistry in this reaction is dependent on the composition of the reactant, contact time, temperature and pressure of the process for carrying out the reaction favoring the thermodynamically forward reaction even in the absence of the catalyst (Schwach et al. 2017). The reactant feed ratio, according to the stoichiometric equation will be 2 (Eq. 14.7) but usually, the preferred ratio of the feed is in between 5 and 10 so as to avoid the runaway of C_2 products to further undesired products. The O_2 concentration plays a crucial role as it has a non-linear trend in selectively forming C_2 products. Although methane conversion increases when O_2 concentration is increased, i.e., the oxygen increases the total oxidation reaction rate, thus creating exothermicity. This temperature rise enhances methane's non-oxidative dehydrogenation to form CH_3^* radicals following rapid recombination of radical reactions to form C–C coupling (Scheme 14.1).

$$CH_{4(g)} + 2O_{2(g)} = CO_{2(g)} + 2H_2O_{(g)} \quad \Delta G^\circ = -801\ \text{kJ mol}^{-1} \tag{14.7}$$

The OCM reaction pathway follows a competency between the homogeneous (gas-phase) being complicated and the heterogeneous (surface-catalyzed) reaction. Oxygen as a reactant, solid oxide heterogeneous catalysts can form active surface oxygen species which selectively abstract one hydrogen from methane to release free methyl radicals $\left(CH_3^*\right)$ in the gas phase, this methyl radical once formed, tries to subsequently couple and form ethane. Then the reaction follows the dehydrogenation of ethane to form ethylene or by the irreversible formation of oxidation products (CO or CO_2). Furthermore, it was reported that electrophilic lattice oxygen species and

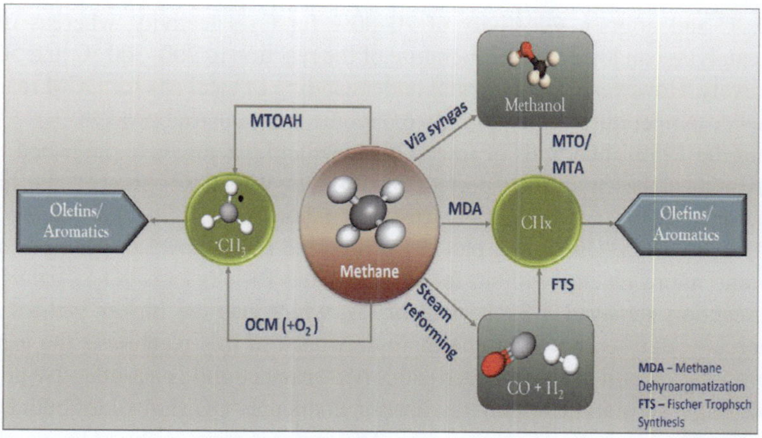

Scheme 14.1 Pathway for C–C coupling reaction to Olefins and Aromatics from methane (Schwach et al. 2017)

the facile filling of surface lattice oxygen vacancies by gas-phase oxygen are key factors in designing efficient catalysts for the oxidative coupling of methane.

For the activation of methane to occur firstly, the abstraction of a hydrogen atom takes place via C–H bond cleavage. However, being a highly stable molecule, methane exhibits strong (first bond dissociation energy = 439.3 kJ mol^{-1}) and weakly polarized (2.84×10^{-40} C^2 m^2 J^{-1}) C–H bonds (Schwach et al. 2017). The reason for this is because of the geometry of the methane molecule, wherein the central carbon atom is tetrahedrally coordinated (point group, T_d) to four hydrogen atoms (sp^3 hybridization) with C–H bond lengths.

It is evident that metal oxides offer a wide range of catalytic activities, but there remain some constraints, the variations are due solely to the intrinsic surface area and hence the density of surface sites of the oxide. The emphasis on the surface area with respect to the yield of C_2 hydrocarbons has been reported by Aika (Iwamatsu et al. 1988), where they concurred that lower surface area materials were mostly preferred; this could be one of the major reasons for the large-scale use of alkali metal compound additives as it results to sintering causing loss of surface area. Thus, comparing different catalysts in terms of their specific activity is highly recommended (Hutchings 1989).

For the temperature around 700–710 °C, the catalysts can be arranged in the order of decreasing specific activity for C_2 production, which is as follows:

$$\begin{aligned} & Li/Sm_2O_3 > Na/CaO > K/CaO > LaAlO_3 > Sm_2O_3 \sim Li/CaO \\ & > PbO > Bi_2O_3 > Ho_2O_3 > Gd_2O_3 \\ & > Li/MgO \sim Yb_2O_3 \sim Y_2O_3 > Na/MgO \sim CaO > MgO \end{aligned}$$

Several catalytic strategies improve C_2 selectivity, yield, and reaction conditions like pressure and temperature. The catalyst $Mn–Na_2WO_4/SiO_2$ activates methane at 800 °C and offers a selectivity of 70–80% for C_2 selectivity, whereas La_2O_3-based catalysts can lower the temperature of the reaction to 200–300 °C and 500 °C, respectively. However, a fundamental understanding of catalysts for OCM reactions with lower temperatures is required to overcome the industrial process.

From the 1980s, hundreds of research papers and patents were published in this area. Amongst the catalysts reported, Mn_2O_3–Na_2WO_4/SiO_2 showed the highest conversion (20–30%) and selectivity (60–80%), but the reaction takes place at 800° to 900 °C. In 2015, Othman et al. (2015) reported that a novel oxygen-permeable membrane micro reactor with an in-situ deposited $Bi_{1.5}Y_{0.3}Sm_{0.2}O_{3-\delta}$ catalyst for OCM reaction achieved a 39% yield of C_2, which has the higher yield than all commercial catalysts, however, the main drawback of this reaction is the usage of the membrane reactor. It has impeded stable performance and a cost-effective process, representing a substantial commercialization challenge. OCM involves both heterogeneous and homogenous catalytic steps. The first step involves heterogeneous activation of CH_4 and O_2 on the catalytic surface to form methyl radials. The second step, the homogeneous gas phase, involves coupling methane followed by dehydration to form ethylene. Researchers focus on finding better catalytic activity with commercialization because of these complex reaction steps and high-temperature reaction conditions.

14.2.3 Synthesis of Materials

Generally, for any catalyst preparation following schematic representation is carried out:

$$\begin{array}{llll}\text{Chemicals} \rightarrow & \text{Catalyst Precursor} \rightarrow & \text{Form} \rightarrow & \text{Final Catalyst} \\ \text{(control impurities)} & \text{(texture)} & \text{(texture, shape)} & \text{(texture, attrition resistance)}\end{array}$$

Most of the catalyst preparation includes the impregnation methods namely the incipient wet and dry impregnations. Metal nitrate precursors are generally preferred to minimize the negative interference by other precursors such as chlorides or sulfates on reactants in the due course of the reaction. The impacts of active ingredients are usually studied by varying the concentration with respect to the support. Usually, lanthanum-based metals have a moderate basic site which helps to minimize the total oxidation products CO_x. Lanthanum-based (La = La, Ce) metal oxide doped on $MnWO_4$–TiO_2 structures were prepared by hydrothermal method with temperature (100–200 °C). Lanthanum-based metal-doped on $MnWO_4$–TiO_2 structure was also prepared by wet impregnation method. Using different surfactants and temperatures in hydrothermal, the formation structure and active phase are different in the reaction atmosphere. The formation of different structural orientations helps to find how the structural and phase transformation induces the activation of methane and oxygen at a lower temperature.

Solution combustion synthesis (SCS) is considered to be helpful technique used for the preparation of simple and complex metal oxides (Ghose et al. 2014). It resembles likely to the sol-gel route and combustion synthesis involving a self-sustained reaction with oxidizers and fuel (e.g., glycine, hydrazine, etc.). When metal nitrates are the precursors for the metals in the final product, the fuel provides energy and also acts as a complexing agent in many catalytic applications.

14.2.3.1 Role of Supports

Finding suitable support has broader applications in catalysis which is essential to the lower the activation of reactant molecules in the active sites with supports. The metal-support interaction at the interface plays a delicate role in enhancing the catalytic activity. The investigation of the influence of support in conventional catalysts like Mn/Na_2WO_4 catalysts (Liu et al. 2017), active sites supported on zeolites are reported frequently for these reactions. OCM reaction requires higher temperature (700–1000 °C) to activate methane and oxygen. In this temperature zeolites structure may collapse easily. Hence, finding new catalytic support with high thermal stability is necessary. Most recently, Wang et al. (2017) observed results with TiO_2-doped Mn_2O_3–Na_2WO_4/SiO_2 catalyst system synthesized via the solution combustion synthesis (SCS) method. Interestingly, they achieved higher ethylene selectivity at a lower temperature (730 °C). This higher selectivity is due to the formation $MnTiO_3$ phase in the Ti-doped $Mn_2O_3/Na_2WO_4/SiO_2$ catalyst; the construction of the $MnTiO_3$ phase is observed only in the Ti-MMW catalysts compared to other supports like silica, TS-1, etc. Thus, it is evident that supports play a role to generate active phase to get higher selectivity.

14.2.3.2 Promoters

In catalysis, a small quantity of promoters plays a massive role in tuning the acid/basic sites, reducibility, oxidation properties, etc. (Liu et al. 2022). An efficient and robust OCM catalysts should be composed of host basic metal oxides (MgO or La_2O_3) promoted with metal oxide dopants that can positively influence C_2 selectivity (e.g., K, Li Cs, Na, Sr, Ba) in the catalytic system. Most reports are based on the Mg and La promoted catalytic system used for the OCM reaction. Notably, I A and II A group metal oxides have a good basic site, which may influence methane conversion and C_2 hydrocarbon selectivity. We can use I A and II A metals as promoters to tune the basic sites to achieve better ethylene selectivity at lower temperatures.

14.2.3.3 Active Sites

The Na_2WO_4–Mn/SiO_2 system is considered as one of the best-reported OCM catalysts with comparatively high C_2 yield with long-standing high stability under applied conditions of higher temperatures. Significant research has been made on various catalyst systems to find better ethylene selectivity. Wang et al. (2017) made report that the $MnTiO_3$ phase triggers at a lower temperature (<750 °C), helping O_2 activation lead to good selectivity. Recent literature says that lanthanum-based catalysts show better activity due to the formation of moderate electrophilic oxygen species and moderate basicity. Also, selective oxygen species are responsible for OCM reaction using modified perovskite ($LaXO_3$, X = Al, Fe, or Ni) with moderate electrophilic oxygen species favoring OCM reaction to get higher C_2 selectivity.

14.2.4 *Properties*

14.2.4.1 OCM Reaction Kinetics

The chemistry of OCM reaction and its understanding of its complex mechanism are still left incomplete. The competing reactions occurring in the gas phase and those consisting of surface reactions both play quite a significant role in CH_4 activation and C_2^+ and CO_X formation. Despite of possessing variations in the active catalytic sites on particular catalysts, the activation of methane takes place through CH_3^* radical formation. The overall rate of reaction is determined and controlled by the preliminary step of the CH_3^* radical. The radical formation is contributed by gas-phase oxygen or oxygen radical (Gambo et al. 2018); the below represented reactions predominantly take place around the 700–850 °C temperature range.

$$CH_4 + O_2 \leftrightarrow CH_3^* + HO_2^* \tag{14.8}$$

$$CH_4 + O_2 \leftrightarrow CH_2O + OH^* \tag{14.9}$$

$$CH_4 + O^* \leftrightarrow CH_3^* + OH^* \tag{14.10}$$

The non-oxidative CH_3^* formation occurs either via thermal cracking or H^* radical attack which preferentially requires very high temperatures (Obradović et al. 2016).

$$CH_4 \leftrightarrow CH_3^* + H^* \tag{14.11}$$

$$CH_4 + H^* \leftrightarrow CH_3^* + H_2 \tag{14.12}$$

In addition to the competing thermal cracking and oxygen attacks, another gas phase activation of methane in the gas phase occurs by OH^* radicals. A study by Takanabe and Iglesia proposed in their study that in addition to the reaction paths (14.9) and (14.10), OH^* radicals are generated through the surface-adsorbed oxygen O(s) and during the gas phase via the reaction between O_2 and H_2O:

$$H_2O + O(s) \leftrightarrow OH^* + OH(s) \quad (14.13)$$

$$2H_2O + O_2 \leftrightarrow 4\,OH^* \quad (14.14)$$

Amongst the above two OH^* formation reactions, preferentially, reaction (14.13) is the dominant pathway. As H_2O is a comparatively stable molecule and O–H bonds in H_2O (497 kJ/mol) are stronger than C–H bonds in CH_4 (439 kJ/mol). The OH^* radical can react with CH_4 in the gas phase to form CH_3^* radicals:

$$CH_4 + OH^* \leftrightarrow CH_3^* + H_2O \quad (14.15)$$

The relative importance of methane activation via gas-phase or surface generation radicals, reactions (14.8)–(14.16) can be determined by analysis of the reaction rates based on experimental data. Following methyl-radical formation, CH_3^* radical undergoes recombination in the gas-phase producing first the C_2 species, C_2H_6, according to the irreversible reaction:

$$CH_3^* + CH_3^* + M \rightarrow C_2H_6 + M \quad (14.16)$$

where M denotes a gas-phase third species. The aforementioned reaction pathways are supposed to be the same for Li/MgO, Sr/CaO, Sr/La_2O_3, Mn/Na_2WO_4/SiO_2 and Mn/Na_2WO_4/MgO catalysts. Figure 14.2 briefly describes the contribution of gas phase and surface reactions to the product distribution. It is also comprehended that gas-phase and surface reactions do contribute to forming C_2^+ and CO_X and H_2O species. The species include O(s), OH(s), H_2O(s), CH_3O(s), CH_2O(s), CHO(s), CO(s), and CO_2(s). The reaction mechanism briefly gives insights into the adsorption of gas-phase species, and the reaction of gas-phase species, either radical or stable molecules, with surface adsorbates or surface–surface reactions.

14.2.5 Emerging Areas

Projecting the soaring reserves of NG, in order to upgrade and develop versatile technologies has been a significant research interest. Meticulous research advancements are being carried out for the reduction and utilization of Greenhouse gases in a sustainable manner. Researchers are provoked over the past few decades in brainstorming to evolve solutions for critical environmental issues. One among some of the establishing advancements is the conversion of methane into value-added products. Due to the formation of syngas at the intermediate step in the indirect route, direct methane conversion pathways are more economical and simpler. OCM possesses a promising route to directly convert methane into C_2 hydrocarbons and has emerging applications in many industrial sectors. Prior to that, at the basic research and development level, there could be a wide exploration in areas of catalysis. Herein, developing

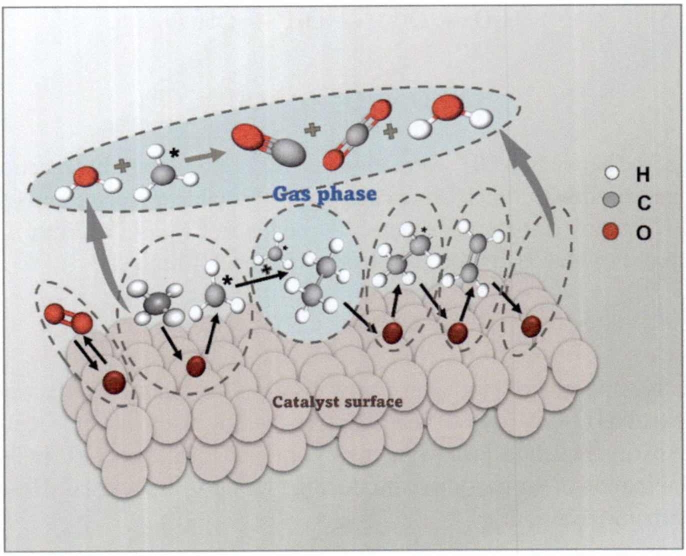

Fig. 14.2 Schematic representation of gas-phase and surface reaction for OCM

catalysts conducive for the long viability and large-scale applications will be tedious yet challenging. Knowing the peculiar thermodynamic and kinetic limitations, the C_2 yields still stand low and have constrained industrial deployment. Despite the challenges, the quest to develop innovative catalysts still remains persistent in all the research communities. Ethylene has commercial value in the fine chemical and petrochemical industries. There exists considerable scope for further improvement on the specific activities, as there and not many detailed studies that have been published yet concerning the optimization of the specific activity by using improved and suitable methods of catalyst preparation. The Dow Chemical plant has constructed a plant that has recently shifted towards green alternatives for manufacturing ethylene to reduce greenhouse gas emissions and dependency on limited fossil fuels.

14.3 Production of Adipic Acid from the Various Feed Stacks

14.3.1 Introduction

In addition to C–H activation in acyclic hydrocarbon molecules, C–H activation in cyclic substrates such as cyclohexane (CYH) is considered a potential strategy from the perspective of producing monomers for polymer synthesis. The chemical richness and potential of adipic acid (AA) make it a valuable intermediate in the top

Fig. 14.3 Structure of AA

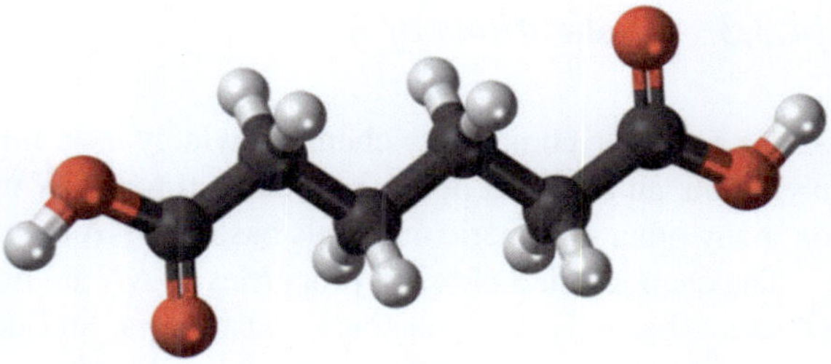

12 value-added platform chemicals. Figure 14.3 describes the structure of AA. It is an imperative feedstock for precursor nylon 66, electronics, automobile, lubricants, thermostat resins, and plasticizers. AA is a crucial organic diacid intermediate used in nylon manufacturing 66. AA produced from industrial processes emits NOx as a by-product and consists of a two-step reaction. It is estimated that 10% of global nitrogen oxide emissions each year result from producing AA, 300 times more potent than carbon dioxide at creating global warming. Single-step conversion of CYH to AA is a grand challenge due to the highly stable C–H bond, its activation, and the formation of over-oxidation products. Herein, we discuss highly efficient and different AA production approaches and their drawbacks.

Recent advances in greener technology with various catalytic routes were discussed in this review article on AA synthesis from various feedstock. It will provide guidelines and enlighten the study of nylon precursors in industry and academia through the catalytic transformation of CYH to green AA. In addition, this chapter discusses the advantages and properties of a novel heterogeneous metal oxide catalyst for the synthesis of AA.

14.3.2 Importance of AA

Dicarboxylic acid-based compounds like AA, glutamic acid, aspartic acid, etc., are globally traded commodity chemicals used for finely synthesizing copolymers such as polyamides and polyesters. One of the broadly used dicarboxylic acids in the chemical industry is AA, the pioneer compound in producing nylon 66, polyesters, polyurethane, low-temperature synthetic lubricant, etc. In the advanced automobile appliances such as electronic and electrical tools, the demand over the AA consumption through the polyamide 66 utilization was required. Global consumption of AA was nearly worth around \$ 6.3 billion every year, and its production rate was anticipated to be 4.1% per annum, leading to 3.3 million per year (Acharyya et al. 2015; Alonso et al. 2015). AA is the most significant and highly demanded dicarboxylic acid with the chemical formula: $COOH{-}(CH_2)_4{-}COOH$.

14.3.3 Applications of AA

AA is a notable commodity chemical widely used for the large-scale production of Nylon 66, an essential polymeric material for daily needs. Figure 14.4 emphasizes on many other polymeric materials based on AA are used in day-to-day life.

The chemical and physical properties of AA are mentioned in Table 14.1. Nylon 66 and other polymers, coatings, plasticizers, and detergents are produced in the chemical industry using AA. Additionally, it serves as a monomer or intermediate in the production of polyurethanes, polyesters, low-temperature synthetic lubrication additives, surface coating, elastomers, and other products, including food acidulants. It is also used in the formulation for many specific applications. AA is generally used as a tanning agent in the leather industry. It is also used as a pH regulator, a detergent, and a cleaning agent. It is additionally utilized as a pelleting agent in pills for disinfecting water intake.

In 1935, Wallace Hume Carothers reported the synthesis of nylon 66 on the laboratory scale for the first time. After three years, DuPont patented the process for the production of nylon 66, which is shown in Scheme 14.2 (Polen et al. 2013).

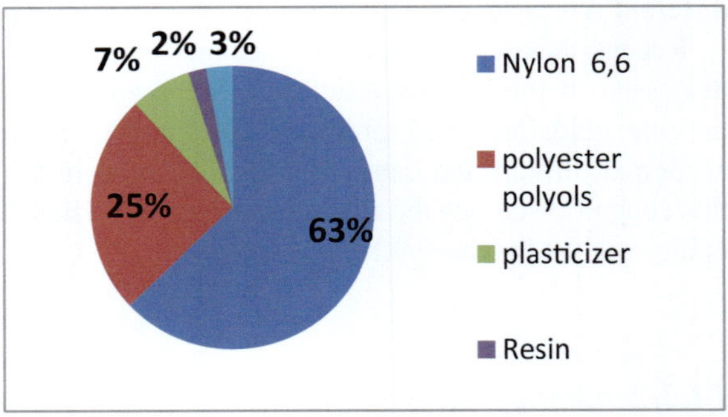

Fig. 14.4 Applications of AA

Table 14.1 Chemical and physical properties of AA

Appearance	Colorless crystalline powder
Odor	Slight odor
Solubility	23 g/l at 25 °C
Molecular weight	146.14 g/mol
Melting point	150.85 °C
Boiling point	337.5 °C
Flashpoint	196 °C
Vapor density	5.04 (1 atm)

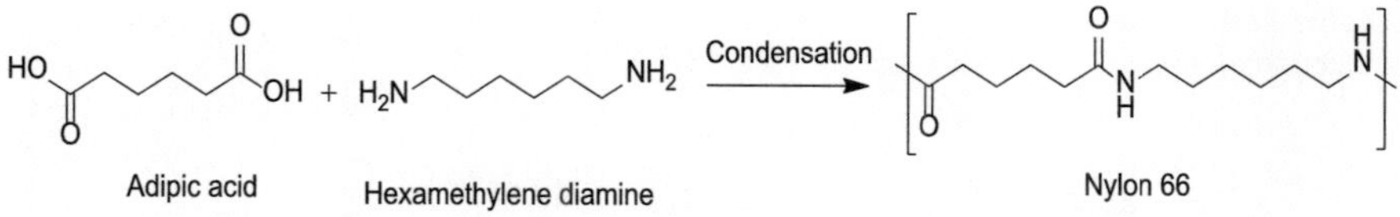

Scheme 14.2 Manufacturing of Nylon 66 from AA and hexamethylenediamine

14.3.4 Consumption of AA

AA consumption worldwide is increasing day by day. China is accounted for roughly 50% of the world's production capacity, with consumption growth predicted to average approximately 6% per year between 2016 and 2021. Furthermore, China contributed more to the overall global exports. Late in 2015, capacity cutbacks in the United States were caused by exports from conventional sources, namely the United States and Western Europe, being replaced. In India, the chemical industry has grown as opportunities have increased worldwide. Investing in nylon and polyurethane industries triggers growth and strengthens the Indian economy. The AA-related market in India was worth USD 186.9 million in 2016. The demand for AA derivatives, such as nylon polymers and polyurethanes, is expected to increase by 2025 (Rios et al. 2021).

14.3.5 Current Industrial Production of AA

AA is commercially manufactured in a two-step process. Step (i) involves the oxidation of CYH to cyclohexanone and cyclohexanol (KA oil) under air (8–15 bar) in the presence of homogenous Co/Mn salt as a catalyst, and the reaction was carried out at very high temperature (125–165 °C). Step (ii) conversion of KA oil to AA using 40–60% conc. Nitric acid in the presence of vanadium and copper-based catalysts. In the first step, selectivity was achieved at around 90%, but the conversion was only <5%. The main disadvantage of the traditional commercial process is that it produces 300 kg of nitrogen oxides (NO_x) per 1000 kg of AA, contributing to global warming, acid rain, and ozone depletion (Scheme 14.3). Nitrous oxide production contributes 7–8% of greenhouse gas emissions despite recycling and recovery attempts (Reimer et al. 1994).

14.3.5.1 First Step of CYH Oxidation with Air

CYH is generated by hydrogenating benzene as a starting material or separating a small amount from the naphtha fraction. In the mid-twentieth century, DuPont revealed preferential oxidation of CYH. The Co- and Mn-based organic salts served as the catalysts for the reaction, which was carried out satisfactorily at 10–20 atm

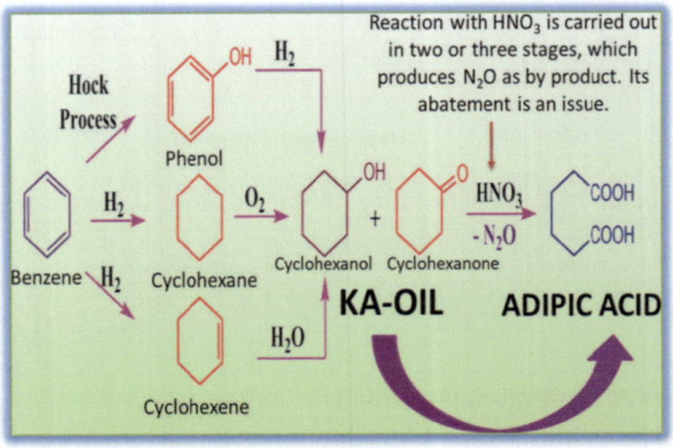

Scheme 14.3 Commercial production of AA

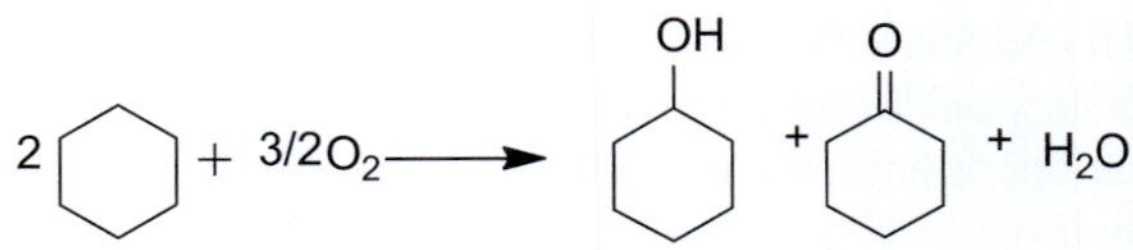

Scheme 14.4 CYH oxidation to KA Oil

pressure and 423–453 K. The procedure is separated into two phases, the first of which is a catalyst-free partial oxidation that produces peroxide. The second phase involves de-peroxidation while a catalyst with conc. Nitric acid. The generation of hydroperoxides, which determines the process's rate, takes place in passivated reactors without any metal complexes to prevent hydroperoxide breakdown. Approximately 75–80% of KA oil is selectivity achieved, and carboxylic acids are its byproducts (Scheme 14.4). The primary disadvantage of the first phase is that only 5% of CYH is converted, and recycling CYH is more expensive.

14.3.5.2 Second Step for the Oxidation of KA Oil with Nitric Acid

The selective oxidation of KA oil to AA in the presence of an excess of 40–60% nitric acid (7/1 molar ratio of HNO_3/KA oil) and a Cu^{II} ammonium metavanadate catalyst is used in the second stage in the production of AA (Scheme 14.5). The reaction was performed in two in-series reactors, one running at 60–80 °C and the other at 90–100 °C with a pressure of 5 atm, which are used to conduct the reaction. Total KA oil conversion has reached 95% molar yield of AA, and the byproducts are succinic acid (selectivity 2%) and glutaric acid (selectivity 3%). The primary disadvantage second step is that it generates more NO_2 as a byproduct.

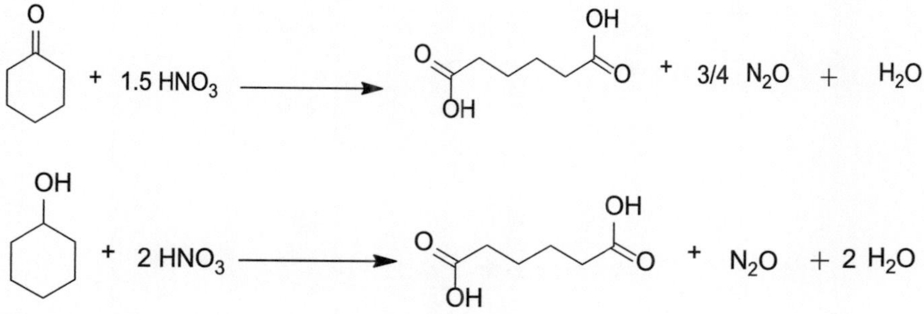

Scheme 14.5 Oxidation of KA oil to AA with nitric acid

14.3.5.3 Environmental Consequence of the Second Step

The HNO_2 serves as a nitric acid oxidation intermediate during the process of converting KA oil to AA, which eventually breaks down into NO and NO_2. In addition to CH_4, CO_2, and various types of aerosols, nitrous oxide is long-lived greenhouse gases (LLGHGS) that affect climate change. Depending on the quantity of catalyst employed and the KA oil makeup, the AA emission factor is predicted to be around one-third of N_2O for the production of AA.

14.3.6 Alternative Pathways for the Production of AA

Recently, in addition to commercial processes, we developed unconventional routes for producing AA from the different substrates shown in Scheme 14.6.

14.3.6.1 Cyclohexene to AA

In 1998, Sato et al. and co-workers reported a synthesis of AA from aqueous 30% H_2O_2 as an oxidant using Na_2WO_2 and $[CH_3(n\text{-}C_8H_{17})_3N]HSO_4$ as a suitable phase transfer catalyst. In the organic phase, Na_2WO_2 undergoes oxidation by H_2O_2 at 75–90 °C, producing an anionic peroxo species. Industrially, this method may be expensive, and using quaternary ammonium salt is not environmentally friendly. Bal et al. reported that an Ag NP on a WO_3 nanorod displays 99% conversion with 94% selectivity of AA in 12 h. Recently, T Shaikh et al. described a CuNP-TiO_2 catalyst with a 95% yield of AA in the presence of H_2O_2 with a low quantity (25 mg) of catalyst (Scheme 14.7) (Shaikh et al. 2022).

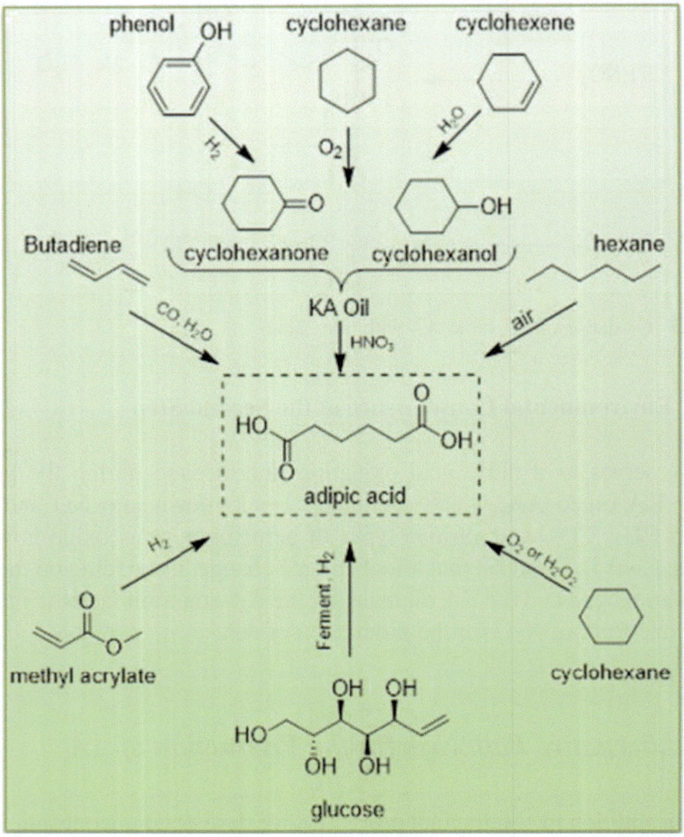

Scheme 14.6 Alternative pathways for the production of AA

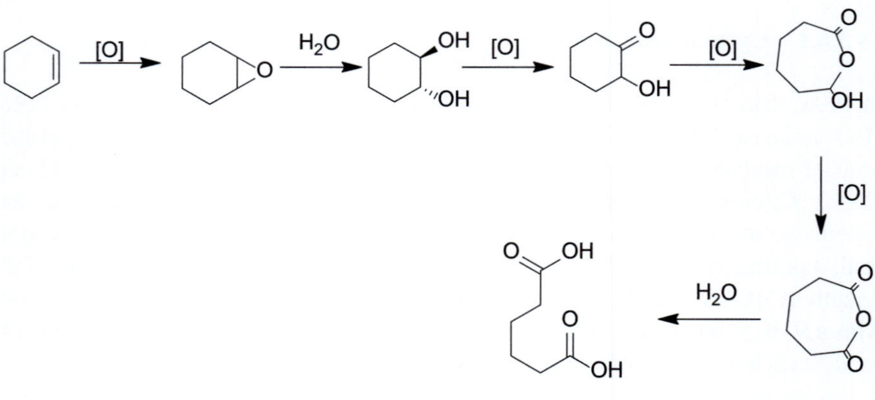

Scheme 14.7 Mechanistic pathways of AA production from cyclohexene with aqueous H_2O_2

14.3.6.2 Oxidation of Cyclohexanone to AA

Zou et al. (2015) reported cyclohexanone to AA conversion over Mn-HTs catalyst with an appropriate amount of Lewis and Bronsted acid site with Mn^{3+}/M^{4+} redox center is responsible for better conversion of cyclohexanone (68%) and selectivity of AA (93%). The recycled catalysts were shown that Mn and Ti content was the same after 15 cycles of reuse. Acetic acid as co-solvent, salient properties have been shown to have greater catalytic activity, which can speed up reactions and increase AA selectivity.

Bhanja et al. (2016) prepared Fe based (FePO-12) an organic–inorganic hybrid material via hydrothermal synthesis. FePO-12 catalyst exhibited superior activity for the oxidation of cyclohexanone to AA because the framework has a high amount of surface acidity and redox center of $Fe^{II/III}$ sites. They also discovered that H_2O has a more substantial impact and better AA selectivity than other solvents due to its higher polarity. The AA yield decreased slightly after 6 successive reactions, and iron leaching was not observed from the catalyst. The same research group later created an oxyfluorinated titanium phosphate (TIPO-1) catalyst that had a cyclohexanone conversion of 92 and 66% selectivity to AA.

Chatterjee et al. described a Mn-containing aluminophosphate material (MnAPO-5) for the complete oxidation of cyclohexanone conversion and AA selectivity. They proposed a mechanism pathway for forming -caprolactone via Baeyer–Villiger oxidation and the ring opening by an oxidative C–C bond to yield AA. At the end of each run, no Mn leaching was detected. Several research reports have been published based on cyclohexanone and cyclohexene to AA conversion over mixed metal oxide. However, catalyst synthesis and methodology still must develop to overcome the industrial process (Yan et al. 2020; Bart and Cavallaro 2015).

14.3.6.3 Butadiene is a Subtract

Methoxy carbonylation of 1, 4 butadiene in the existence of the Co-based catalyst is a unique method that BASF has proposed. Even though this system produced a respectable amount of AA (about 72%), the technology was never industrialized, probably due to the harsh reaction conditions. The first step was performed at a pressure of 300 atm, while the second was conducted at 150–200 atm.

It is also possible to synthesize AA from butane dehydrogenated to yield 1, 3-butadiene and further hydroformylation at both ends to construct dialdehyde. The high reactivity of the aldehyde group goes through aldol before being separated as di-aldehyde (Scheme 14.8). One of the first Rh-catalyzed protocols was patented by Union Carbide in 1994 for in situ acetalizations with a selectivity of 57%. Later, Smith and co-workers reported Rh-based catalysts with a selectivity of 74% for a dialdehyde, which can be oxidized to AA using O_3 as an oxidant. Recently, Liu and co-workers demonstrated economically viable hydroformylation processes that promise to be economically and environmentally beneficial for industrial application.

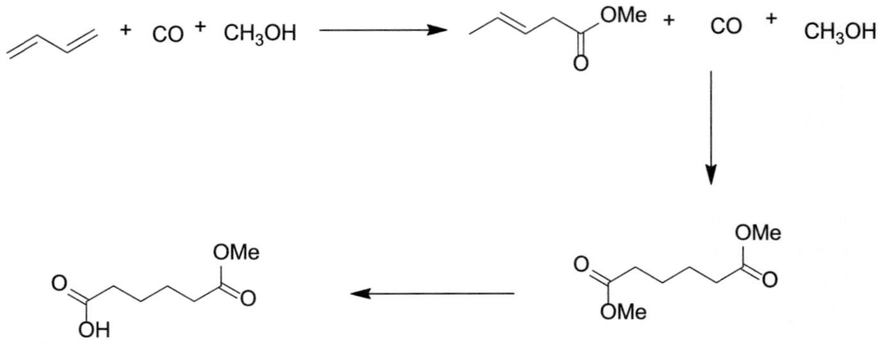

Scheme 14.8 Methoxy carbonylation of butadiene to AA

Compared to traditional CYH routes, butadiene as a starting substrate offers a NOx-free route to produce AA, but its selectivity and efficiency of the catalysts need considerable development. Alternative oxidants are needed for this process to achieve a better yield of AA from butadiene (Capece et al. 2020).

14.3.6.4 Hexane to AA

Oxidizing linear alkanes is extremely difficult, and performing oxy-functionalization of specific terminal positions is grant challenge. A nonheme iron acts as an active center for Enzyme catalytic reaction. Robert Raja et al. established the production of AA from n-hexane as a starting material using cobalt ion substituted aluminophosphate catalyst in the air atmosphere. The reaction was carried out using freshly synthesized catalysts (Co-AlPO-18, Co-AlPO-5) under 1.5 MPa air pressure at 373 K for 24 h. The n-hexane conversion achieved 9.5% with 33.6% selectivity of AA, and 2-hexanol and 2-hexanone were detected as other products. The main drawback of this process is the formation of hexane diol which is more favorable than AA (Capece et al. 2020).

14.3.6.5 Renewable D-Glucose Sources as a Starting Material

According to Frost, AA might be made from glucose as a renewable resource. The use of 3-dehydroshikimic acid, catechol, and protocatechuic acid involves the enzymatic conversion of D-glucose to cis, cis-music. The three enzymes not generally observed in Escherichia coli are the crucial element of these procedures. The cis, cis-music is hydrogenated utilizing a supported Ru, Pt catalyst in the reaction's last stage. After the reaction time at room temperature at 34 bar of H_2 pressure, 97% of cis, cis-muconic acid may be transformed into AA (Scheme 14.9). Green and sustainable technologies are used to produce AA from bio-based renewable glucose and its derivatives. Oxidation of glucose to AA consists two-step process; step 1 is

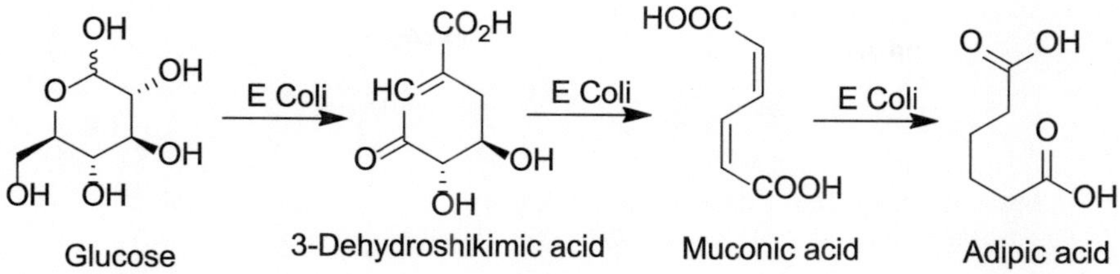

Scheme 14.9 Enzymatic conversion of D-glucose to AA

the oxidation product of glucaric acid obtained from glucose, which is further transformed to AA via the dehydrodeoxygenation (HDO) process. In industry, Boussie et al. disclosed a patent on the HDO process over Pt /Rh bimetallic catalyst system in acidic conditions (acetic acid and HBR as a solvent), yielding AA around 89% (Farrauto et al. 2003). The deoxydehydration of glucaric acid, which is generated from cellulose to an AA ester by using ReOx/ZrO_2-Pd/C catalysts, was reported by Lin et al. in 2019 ($Y = 82\%$). However, scaling up and process implementation issues still need to be resolved in order to make this approach economically viable (Larson 2015).

14.3.6.6 5-Hydroxymythylfurfural to AA

The HMF route can be used to synthesize AA as an alternative to fossil-based synthesis. Step 1, the oxidative conversion of HMF to 2,5-furan dicarboxylic acid (FDCA) and the deoxygenation of FDCA to AA (step 2) are the two most important crucial stages. The system is more expensive when there are two steps (Scheme 14.10). Catalysts based on platinum and palladium are well-studied for the FDCA synthesis from HMF. An exceptionally effective Pt-MoOx/TiO_2 catalyst for FDCA to valeric acid conversion was reported. It can sustain the carboxylic group while cleaving the C–O–C bond of the furan ring. Utilizing 3 MPa H_2, the yield was 20% at 200 °C within 4 h. To overcome this problem, Wei et al. reported platinum-based niobic acid (Pt/Nb_2O_5 cxH_2O) as a catalyst for synthesizing AA from FDCA in 2019. With the complete conversion of FDCA, we observed 38% AA yield under 3 MPa hydrogen in water at 200 °C for 8 h. Even though this catalyst promotes AA synthesis, it is significantly less selective than hydroxymethyl tetrahydrofurans (20–30% compared to 80–90%). FDCA can be used to synthesize AA. However, considerable research is needed to unravel how the C–O bond within a ring can be broken instead of the hydrogenating carboxylic acid group. A catalyst with high acidity and mild hydrogenation needs to be developed to perform a reaction (Lin et al. 2019).

Disadvantages of bio-based derivate pathways

(1) Multiple steps are required to get the desired product
(2) The reaction mechanism is not clear

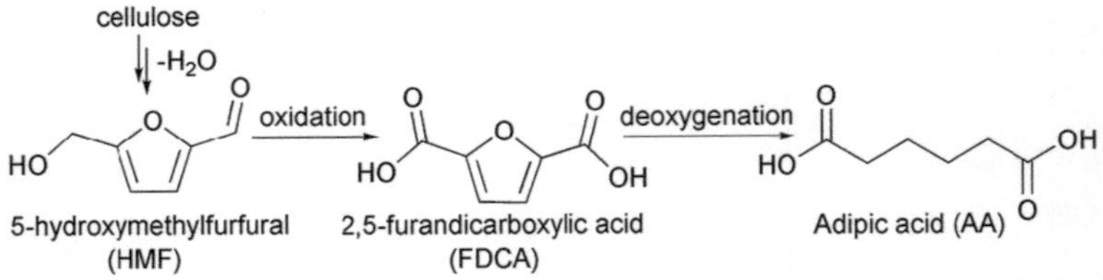

Scheme 14.10 Transformation of HMF to AA

(3) Kinetic and thermodynamic constraints
(4) Reusability and stability of the catalyst.

14.3.6.7 Selective Oxidation of CYH to AA

The chemical industry relies heavily on the selective oxidation of hydrocarbons because it produces several beneficial compounds like alcohols, aldehydes, and carboxylic acids. Since many industrial processes rely on these reactions, creating an effective method for activating hydrocarbons with molecular oxygen can be one of the key technologies for producing oxygenates. AA (AA) is a crucial byproduct of CYH's selective oxidation. By 2017, there is anticipated to be a 6-billion-pound annual increase in AA consumption. Recently, partial oxidation processes are carried out in the gas phase or liquid phases with homogeneous or heterogeneous catalysts. Molecular oxygen is the most feasible oxidizer for partial oxidation processes because it is ubiquitous and inexpensive. Water is produced when molecular oxygen is used as an oxidant. Most heterogeneous catalyst-based procedures deal with very poor selectivity for the anticipated product due to the lack of unique catalyst preparation. Additionally, it results in waste production and ineffective utilization of raw materials. Therefore, obtaining desired product selectivity in the selective oxidation of hydrocarbons utilizing heterogeneous catalyst methods remains a crucial task. Researchers have taken multiple tasks to create a "green" catalytic system for CYH oxidation, such as choosing oxidants, solvents, and catalysts with highly appropriate product selectivity and reduced pollution.

14.3.7 Role of Oxidants

Three oxidants are typically used for CYH oxidation: H_2O_2, tert-butyl hydroperoxide (TBHP), and molecular oxygen (O_2) and air. While molecular oxygen use was supposed to be solvent-free, H_2O_2 and TBHP might both be oxidants and solvents.

14.3.7.1 Hydrogen Peroxide (H_2O_2)

There is much interest in CYH oxidation in green chemistry, which produces fewer pollutants. Since water is the only byproduct of the reaction, oxidation using H_2O_2 as an oxidant is regarded as an environmentally beneficial chemical process. H_2O_2 also takes up 47% more oxygen than air.

In a fundamental contribution, Sato et al. described the oxidation of cyclohexene. They employed $[CH_3(n\text{-}C8H_{17})_3N]\text{–}HSO_4$ as a phase-transfer homogeneous catalyst with 30% H_2O_2 as a green oxidant in the Na_2WO_4 catalyst system. After 8 h of the reaction time, an AA yield of 90% was attained. Freitag et al. improved this system, extending the reaction time to 90 min and increasing the yield to 68%. Cyclohexene was successfully converted to AA by Bohstrom et al. using a mesoporous oxide that contained WO_3.

It has been observed that tungstate-based catalysts are extraordinarily effective in oxidation processes. Examples include the SBA-15 $[BMIm]_2WO_4$ supported on silica sulfamic acid proposed for AA production. Lee et al. found substituted aluminophosphate with a titanium framework for the oxidation of cyclohexene to AA (TAPO-5). In an acetic acid media, cyclohexanone can be converted to AA with an 86% selectivity using a Co/Mn cluster complex catalyst, according to Chavan et al., Rajaram Bal et al. achieved high yields (94%) for converting cyclohexene to AA using silver-supported tungsten oxide nanorod, Cu (II) on WO_3 catalyst using green oxidant of H_2O_2. However, the reaction's safety depends on the oxidation reaction's efficiency, which uses hydrogen peroxide as an oxidant. In cases where the productivity is less than 40%, the oxidant has a greater value than the product. This reaction has several disadvantages, including the high ratio of reactant to oxidant used, which makes it more expensive (Ghosh et al. 2014).

14.3.7.2 Tert-Butyl Hydroperoxide (TBHP)

Polychlorophthalocyanine ($Cl_{16}PcFe^{II}$) and tetrasulfophthalocyanine ($[Fe^{II}TSPc]_4$) complexes catalysts with TBHP, m-chloroperoxybenzoic acid (m-CPBA), and hydrogen peroxide were used as oxidants to convert CYH oxidation to AA. The relative yield of the products is said to depend on the catalyst and the oxidant. In addition, TBHP was used as an oxidant in Fe^{III} and Mn^{III} catalytic systems to oxidize CYH in acetonitrile. The catalyst [Fe(III) (Salen) C_l] produced conversion rates of up to 10.2%. It was discovered that tert-butyl hydroperoxide (TBHP, 70% in water) is a more effective oxidant than H_2O_2 in zeolite covered with complexes [Mn(SALEN)Y] and [Fe(SALEN)Y] systems. The same drawbacks were demonstrated with the low efficiency of the H_2O_2 oxidant. From the commercial point of view, TBHP is an expensive oxidant compared to oxygen and H_2O_2.

14.3.7.3 Oxygen/Air

Using oxygen (or air) instead of H_2O_2 and tert-butyl hydroperoxide in an oxidation reaction is economically and environmentally beneficial due to its inexpensive nature, abundance, and reliability. As oxidants, H_2O_2 and TBHP were widely investigated using homogeneous catalysts. More attention has been paid to heterogeneous catalytic systems in oxygen oxidation systems. Metal complexes, oxides, and metal cations are typically involved in support of these heterogeneous catalysts, which are incorporated in inorganic supports such as SiO_2, zirconia oxides, activated carbon, graphene oxide, alumina, and zeolites. A mild reaction condition was used in these heterogeneous catalytic systems to perform the oxidation reaction, which was well in accordance with green chemistry practices. Consequently, these non-homogenous reaction systems make it easier to separate the catalyst from the reaction medium, save energy, and make it possible for catalyst recycling. The primary advantage of the molecular oxygen system is its ability to protect against the depletion of active metal ions during the process.

Currently, research is centered on the one-pot conversion of CYH to AA using molecular O_2. Mn-TS, FeAlPO-31, $AgWO_3$-SBA-15, N-doped CNTs, and CuO on SAPO-34 have been extensively investigated. Among these, CuO on the SAPO-34 catalyst showed the highest yield (31%), but it required a longer reaction time (20 h) and a higher amount of catalyst (350 mg) (Guo et al. 2020; Goyal et al. 2019; Vernekar et al. 2021).

14.4 Conclusion

Syngas is a crucial raw material for the synthesis of methanol, fuel cell applications and dimethyl ether (DME). SRM or some other shale gases conversion technologies currently produce most syngas. Additionally, significant interest in using CO_2 and CH_4 as main greenhouse gases to produce syngas by dry reforming methane (DRM). However, DME or methanol processes require syngas ratios between 1.5 and 2, whereas DRM produces a syngas ratio of around 1. By switching to BRM or TRM, the desired syngas composition can be achieved. The steam to hydrocarbon and oxygen to hydrocarbon ratios in these processes determine the syngas's H_2/CO ratio. BRM, in addition to dry reforming, is a desirable method for producing syngas since an expensive CO_2 separation phase can be skipped. Methane can be tri-reformed by combining endothermic SR, DR and exothermic POX in a synergistic manner. The tri-reforming concept was proposed as a new processing pathway for the utilization of carbon dioxide in exhaust gases from fossil fuel-powered energy plants without CO_2 removal to produce syngas with adjustable H_2/CO ratios. This adaptability helps the oxo synthetic and long-chain hydrocarbons employing FT synthesis produce syngas having different ratios.

Considering the attractive route of OCM, directly converting methane into ethylene for its major application in the chemical industry is a key building block.

Comparing OCM, the total and partial and oxidation of methane are much more favorable because of the reaction pathway and kinetic stability. Furthermore, the intermediates and products formed in OCM are more highly sensitive than methane to get over-oxidized readily when oxygen is present. The presence of nucleophilic and electrophilic oxygen species importantly determines the product desirably formed. This favors the formation of excess CO_2, resulting in low methane utilization efficiency. However, it is known that OCM is yet to achieve major applications in industries as thermodynamic and kinetic limitations are inherent. Compound like CO_x is usually considered being unselective as they are greatly stable thermodynamically and convert at higher rates than producing C_2 hydrocarbons, resulting in non-economically viable C_2 yields. This can be overcome by carrying out reactions in the suitable presence of oxidants. The selectivity of C_2 can be controlled up to a self-controlling temperature limit, beyond which the CO_x products are formed. For OCM, higher temperature beyond 500 °C is required in order to achieve conversions reasonably. Such parameters for a reaction become energy intensive, increasing the complexity in designing a suitable catalyst that should be metal-based, durable and thermally active. Pertaining to the diversified experimental work being carried out over the decades, it is inevitable to not draw any broad conclusions. Researchers focus on finding better catalytic activity with commercialization, for which we need to determine the oxygen species that are active and favorable for the oxidative coupling reaction followed by dehydrogenation.

Based on the critical review, biomass-based AA can also provide significant benefits that can be used as an alternative source to produce nylon in the future. Multiple reaction steps and noble metal oxide supported on various catalysts are required to achieve a higher yield towards AA. We examined the progress of green oxidants and great promising catalysts for cyclohexane, cyclohexene, cyclohexanol, and cyclohexanone and renewable sources. This study discussed reaction mechanisms and catalyst durability. The catalysts demonstrated superior catalytic activity and selectivity throughout cyclohexane-derived oxidation in the presence of redox and acidity properties. A better understanding of both radical-based and redox-based oxidation pathways could boost research progress in this area. This chapter will look into metal nanocatalysts, metal oxide with oxygen vacancies, which are effective, eco-friendly, safe, stable, environmentally, and financially feasible approaches of AA production and oxidation processes for a variety of other value-added fine chemicals.

References

Acharyya, S.S., Ghosh, S., Bal, R.: Nanoclusters of Cu (II) supported on nanocrystalline W(VI) oxide: a potential catalyst for single-step conversion of cyclohexane to adipic acid. Green Chem. **17**, 3490–3499 (2015)

Alonso, S., Rendueles, M., Diaz, M.: Microbial production of specialty organic acids from renewable and waste materials. Crit Rev. Biotechnol. **35**, 497–513 (2015)

Bart, J.C.J., Cavallaro, S.: Transiting from adipic acid to bioadipic acid, petroleum-based processes. Ind. Eng. Chem. Res. **54**(1), 1–46 (2015)

Bhanja, P., Ghosh, K., Islam, S.S., Patra, A.K., Islam, S.M. and Bhaumik, A.: New hybrid iron phosphonate material as an efficient catalyst for the synthesis of adipic acid in air and water. ACS Sustain. Chem. Eng. **4**(12), 7147-7157 (2016)

Byrne, J.P.J., Gohr, R.J., Haslam, R.T.: Recent progress in hydrogenation of petroleum. Ind. Eng. Chem. **24**, 1129 (1932)

Capece, N., Sadier, A., Ferraz, C.P., Thuriot-Roukos, J., Pietrowski, M., Zielinski, M., Paul, S., Cavani, F., Wojcieszak, R.: Aerobic oxidation of 1,6-hexanediol to adipic acid over Au-based catalysts: the role of basic supports. Catal. Sci. Technol. **10**, 2644–2651 (2020)

Dama, S., Ghodke, S.R., Bobade, R., Gurav, H.R., Chilukuri, S.: Tuning the dimensionality of layered $Sr_{n+1}Ti_{n-x}Ni_xO_{3n+1}$ perovskite structures for improved activity in syngas generation. J. Catal. **360**, 27–39 (2018)

de Lima, S.M., da Silva, A.M., da Costa, L.M.M., Assaf, J.M., Mattos, L.V., Sarkari, R., Venugopal, A., Noronha, F.B.: Hydrogen production through oxidative steam reforming of ethanol over Ni-based catalysts derived from $La_{1-x}Ce_xNiO_3$ perovskite-type oxides. Appl. Catal. B **1**, 121 (2012)

Fan, D., Dai, D.J., Wu, H.S.: Ethylene formation by catalytic dehydration of ethanol with industrial considerations. Materials **6**, 101–115 (2013)

Farrauto, R., Hwang, S., Shore, L., Ruettinger, W., Lampert, J., Giroux, T., Liu, Y., Ilinich, O., Annu.: New material needs for hydrocarbon fuel processing: generating hydrogen for the PEM fuel cell. Rev. Mater. Res. **33**, 1 (2003)

Gadalla, A.M., Sommer, M.E.: Carbon dioxide reforming of methane on nickel catalysts. Chem. Eng. Sci. **44**, 2825 (1989)

Galadima, A., Muraza, O.: Revisiting the oxidative coupling of methane to ethylene in the golden period of shale gas: a review. J. Ind. Eng. Chem. **37**, 1–13 (2016)

Gallucci, F., Paturzo, L., Famà, A., Basile, A.: Experimental study of the methane steam reforming reaction in a dense Pd/Ag membrane reactor. Ind. Eng. Chem. Res. **43**, 928 (2004)

Gambo, Y., Jalil, A.A., Triwahyono, S., Abdulrasheed, A.A.: Recent advances and future prospect in catalysts for oxidative coupling of methane to ethylene: a review. J. Ind. Eng. Chem. **59**, 218–229 (2018)

Ghose, R., Wang, H.T., Varma, A.: Oxidative coupling of methane using catalysts synthesized by solution combustion method: Catalyst optimization and kinetic studies. Appl. Catal. A. Gen. **472**, 39–46 (2014)

Ghosh, S., Acharyya, S.S., Adak, S., Konathala, L.N.S., Sasakib, T., Bal, R.: Selective oxidation of cyclohexene to adipic acid over silver supported tungsten oxide nanostructured catalysts. Green Chem. **16**, 2826–2834 (2014)

Goyal, R., Sarkar, B., Sameer, S., Bag, A., Bordoloi, A.: Ag and WOx nanoparticles embedded in mesoporous SiO_2 for cyclohexane oxidation. ACS App. Nano Mater. **2**(9), 5989–5999 (2019)

Guo, X., Xu, M., She, M., Zhu, Y., Shi, T., Chen, Z., Peng, L., Guo, X., Lin, M., Ding, W.: Morphology-reserved synthesis of discrete nanosheets of CuO@SAPO-34 and pore mouth catalysis for one-pot oxidation of cyclohexane. Angew. Chem. Int. Ed. **59**, 2606–2611 (2020)

Halmann, M., Steinfeld, A.: Thermoneutral tri-reforming of flue gases from coal-and gas-fired power stations. Catal. Today **115**, 170 (2006)

Hawkins, E.D., Jones, P.D.: On increasing global temperatures: 75 years after Callendar. Q. J. Roy. Meteorol. Soc. **139**, 1961 (2013)

Holmen, A.: Direct conversion of methane to fuels and chemicals. Catal. Today **142**, 2 (2009)

Hutchings, G.J.: Oxidative coupling of methane using oxide catalysts. Chem. Soc. Rev. **18**, 251–283 (1989)

Iwamatsu, E., Moriyama, T., Takasaki, N., Aika, K.: Oxidative coupling of methane over promoted magnesium oxide catalysts; Relation between activity and specific surface area. Stud. Surf. Sci. Catal. **36**, 373–382 (1988)

Karakaya, C., Zhu, H., Loebick, C., et al.: A detailed reaction mechanism for oxidative coupling of methane over Mn/Na2WO4/SiO2 catalyst for non-isothermal conditions. Catal. Today **312**, 10–22 (2018)

Keller, G.E., Bhasin, M.M.: Synthesis of ethylene via oxidative coupling of methane: I. Determination of active catalysts. J. Catal. **73**, 9–19 (1982)

Kokal, J.A., Vora, B.V., Imai, T.: Production of linear alkylbenzenes. Appl. Catal. A: General **221**, 295–301 (2001)

Larson, R.T.: Hydrogen gas-mediated deoxydehydration/hydrogenation of sugar acids: catalytic conversion of glucarates to adipates. J. Am. Chem. Soc. **139**(40), 14001–14004 (2015)

Liander, H.: The utilisation of natural gases for the ammonia process. Trans. Faraday Soc. **25**, 462 (1929)

Lin, J., Song, H., Shen, X., Wang, B., Xie, S., Deng, W., Wu, D., Zhang, Q., Wang, Y.: Zirconia-supported rhenium oxide as an efficient catalyst for the synthesis of biomass-based adipic acid ester. Chem. Commun. **55**, 11017–11020 (2019)

Liu, W.C., Ralston, W.T., Melaet, G., Somorjai, G.A.: Oxidative coupling of methane (OCM): effect of noble metal (M = Pt, Ir, Rh) doping on the performance of mesoporous silica MCF-17 supported Mn_xO_y–Na_2WO_4 catalysts. Appl. Catal. A Gen. **545**, 17–23 (2017)

Liu, J., Yue, J., Lv, M.: From fundamentals to chemical engineering on oxidative coupling of methane for ethylene production: a review. Carbon Res. Conv. **5**, 1–14 (2022)

Mittasch, A., Schneider, C.: DRP 296866, BASF (1912)

Obradović, A., Thybaut, J.W., Marin, G.B.: Oxidative coupling of methane: opportunities for microkinetic model-assisted process implementations. Chem. Eng. Techol. **39**, 1996–2010 (2016)

Olah, G.A., Goeppert, A., Czaun, M., Prakash, G.K.S.: Bi-reforming of methane from any source with steam and carbon dioxide exclusively to Metgas (CO–$2H_2$) for methanol and hydrocarbon synthesis. J. Am. Chem. Soc. **135**, 648 (2013)

Othman, N.H., Wu, Z., Li, K.: An oxygen permeable membrane microreactor with an in-situ deposited Bi1.5Y0.3Sm0.2O3−δ catalyst for oxidative coupling of methane. J. Memb. Sci. **488**, 182–193 (2015)

Polen, T., Spelberg, M., Bott, M.: Towards biotechnological production of adipic acid and precursors from biorenewables. J. Biotechnol. **167**(2), 75–84 (2013)

Reimer, R.A., Slaten, C.S., Seapan, M., Lower, M.W., Tomlinson, P.E.: Abatement of N_2O emissions produced in the adipic acid industry. Environ. Prog. **13**(2), 134–137 (1994)

Reitmeier, R.E., Atwood, K., Bennett, H.A., Baugh, H.M.: Production of synthetic gas-reaction of light hydrocarbons with steam and carbon dioxide. Ind. Eng. Chem. **40**, 620 (1948)

Ren, X.H., Bertmer, M., Stapf, S., Demco, D.E., Blümich, B., Kern, C., Jess, A.: Deactivation and regeneration of a naphtha reforming catalyst. Appl. Catal. A Gen. **228**(1), 39 (2002)

Rios, J., Lebeau, J., Yang, T., Li, S., Lynch, M.D.: A critical review on the progress and challenges to a more sustainable, cost-competitive synthesis of adipic acid. Green Chem. **23**, 3172–3190 (2021)

Rostrup-Nielsen, J.R., Bak Hansen, J.H.: CO_2-reforming of methane over transition metals. J. Catal. **144**, 38 (1993)

Sahli, N., Petit, C., Roger, A.C., Kiennemann, A., Libs, S., Bettahar, M.M.: Ni catalysts from $NiAl_2O_4$ spinel for CO_2 reforming of methane. Catal. Today **113**, 187 (2006)

Schwach, P., Pan, X., Bao, X.: Direct conversion of methane to value-added chemicals over heterogeneous catalysts: challenges and prospects. Chem. Rev. **117**, 8497–8520 (2017)

Shaikh, T., Sharma, A.S., Osman, S.M., Luque, R., Kaur, H.: Selective oxidation of cyclohexene to adipic acid over CuNPs supported on PLA/TiO_2. Catal. Comm. **168**, 106460 (2022)

Song, C., Pan, W.: Tri-reforming of methane: a novel concept for catalytic production of industrially useful synthesis gas with desired H_2/CO ratios. Catal. Today **98**, 463 (2004)

Urasaki, K., Sekine, Y., Kawabe, S., Kikuchi, S., Matsukata, M.: Catalytic activities and coking resistance of Ni/perovskites in steam reforming of methane. Appl. Catal. A **183**, 23 (2005)

Vernekar, D., Dayyan, M., Ratha, S., Rode, C.V., Haider, M.A., Khan, T.S., Jagadeesan, D.: Direct oxidation of cyclohexane to adipic acid by a WFeCoO(OH) catalyst: role of bronsted acidity and oxygen vacancies. ACS Catal. **11**(17), 10754–10766 (2021)

Wang, P., Zhao, G., Liu, Y., Lu, Y.: TiO_2-doped Mn_2O_3–Na_2WO_4/SiO_2 catalyst for oxidative coupling of methane: solution combustion synthesis and $MnTiO_3$-dependent low-temperature activity improvement. Appl. Catal. A **544**, 77–83 (2017)

Yan, W., Zhang, G., Wang, J., Liu, M., Sun, Y., Zhou, Z., Zhang, W., Zhang, S., Xu, X., Shen, J., Jin, X.: Recent progress in adipic acid synthesis over heterogeneous catalysts. Front. Chem. Green Sustain. Chem. (2020)

Zou, G., Zhong, W., Xu, Q., Xiao, J., Liu, C., Li, Y., Mao, L., Kirk, S., Yin, D.: Oxidation of cyclohexane to adipic acid catalyzed by Mn-doped titanosilicate with hollow structure. Catal. Commun. **58**, 46–52 (2015)

Chapter 15
Metal–Metal-Oxide Interfaces in Catalysis

Ankita Singh and Balaji Gopalan

15.1 Introduction

15.1.1 Historical Perspective of Catalysis

Catalysis, in various forms, has played a role in human civilization since ancient times. The main ingredient that catalyses milk to form curd consists of lactic acid bacteria, such as *Lactobacillus bulgaricus* and *Streptococcus thermophilus*. These bacteria ferment lactose, a sugar present in milk, into lactic acid, which causes the milk to sour and curdle, resulting in the formation of curd. Another one of the earliest known examples dates back to ancient Egypt, where copper-based compounds were used to accelerate the conversion of sugars into alcohol during the fermentation process for brewing beer. In ancient Rome, the production of lead oxide, known as litharge, was a significant catalytic process. Lead oxide was crucial in extracting silver from its ores through the process of cupellation, where the lead oxide acted as a catalyst for the oxidation of impurities, leaving behind pure silver. In ancient China, the development of porcelain during the Tang Dynasty (618–907 AD) involved the use of catalytic materials such as kaolin clay, which helped lower the firing temperature required for porcelain production, thus enhancing its quality and durability. Throughout history, catalysis has been intertwined with human innovation and technological progress, even if the understanding of the underlying chemical mechanisms was limited compared to modern scientific knowledge. These early examples laid the groundwork for the systematic study and application of catalysis that would later emerge during the Industrial Revolution and continue to evolve into the sophisticated field it is today.

A. Singh · B. Gopalan (✉)
Department of Chemistry, Birla Institute of Technology and Science, Pilani, Hyderabad Campus 500078, Telangana, India
e-mail: gbalaji@hyderabad.bits-pilani.ac.in

R. S. Ningthoujam and A. K. Tyagi (eds.), *Handbook of Materials Science, Volume 3*, Indian Institute of Metals Series, https://doi.org/10.1007/978-981-96-8229-4_15

Catalysis played a significant role in the Industrial Revolution. During the eighteenth and nineteenth centuries, as industries rapidly expanded, there was a growing need for more efficient and cost-effective chemical processes to produce a wide range of goods. The phenomenon of catalysis was not fully understood even at the time due to the lack of understanding regarding the molecular nature of chemicals. One of the most notable examples is the Haber–Bosch process, developed by Fritz Haber and Carl Bosch in the early twentieth century. This process revolutionized agriculture and enabled the large-scale production of ammonia from nitrogen and hydrogen gases. Ammonia is a crucial component of fertilizers, allowing for increased crop yields and helping to sustain the growing global population. However, several chemicals were also universally recognized as being particularly important due to their widespread use and critical role in various industries such as sulphuric acid, chlorine, acetic acid, sodium hydroxide, and nitric acid.

Additionally, catalysis played a vital role in other industrial processes such as petroleum refining, where catalysts were used to convert crude oil into various useful products like gasoline, diesel, and lubricants. The development of catalytic cracking and catalytic reforming processes greatly increased the efficiency and yield of these processes, contributing to the growth of the petroleum industry. Furthermore, catalysis was employed in the production of polymers after the advent of Zeigler–Natta polymerization, and pharmaceuticals, and other materials essential for industrialization. For example, catalytic processes were used in the production of sulphuric acid, nitric acid, and synthetic dyes, amongst many other products.

15.1.2 Role of Imperfections on Catalytic Activity

Imperfections, also known as defects, in solid materials play a major role in diversifying the properties and applications of materials. A few key roles imperfections play are Mechanical, Electrical, Optical, Thermal, and Chemical. By manipulating imperfections, researchers can design materials with enhanced performance and functionality. In chemicals, imperfections can affect the chemical reactivity of solids. Surface defects, such as steps, kinks, and vacancies, can serve as active sites for catalytic reactions by providing sites for adsorption and reaction to occur (Kittel 1957). Grain boundaries can also influence diffusion rates of atoms within materials, affecting processes like oxidation and corrosion. These imperfections can significantly influence the adsorption, diffusion, and activation of reactant molecules on the catalyst surface, thereby affecting the kinetics and selectivity of catalytic reactions (Bond 1974). In detail, imperfections impact catalysis by providing active sites for adsorption and reaction. Defects such as step edges, kinks, and vacancies can serve as high-energy sites where adsorbates preferentially bind, leading to enhanced catalytic activity. For example, in metal catalysts, steps and kinks on the surface can facilitate the dissociation of molecular hydrogen or oxygen, which are key steps in many catalytic reactions, such as hydrogenation and oxidation. Moreover, imperfections can also influence the electronic and geometric structure of catalytic active sites. Point defects and grain boundaries can introduce localized electronic states within the band gap of semiconductors or alter the coordination environment of metal atoms in catalysts, leading to changes in the adsorption energies of reactants and intermediates. This, in turn, can affect the reaction pathways and product selectivity of catalytic reactions. Furthermore, imperfections can modulate the diffusion of reactant molecules and reaction intermediates on the catalyst surface. Grain boundaries and defects can act as barriers to diffusion, limiting the mobility of species and promoting surface reactions. Conversely, defects may also enhance diffusion by providing pathways for atomic or molecular transport across the catalyst surface, facilitating mass transport and increasing reaction rates. Additionally, imperfections can influence the stability and lifetime of catalytic materials. For instance, defects and grain boundaries may promote catalyst deactivation by facilitating the adsorption of poisons or promoting catalyst sintering, where active metal particles coalesce and lose their surface area and activity over time. Understanding and controlling the role of imperfections in catalysis is, therefore, essential for the design of more efficient and durable catalysts for industrial applications. In summary, imperfections in solid catalysts play a multifaceted role in catalysis, affecting adsorption, diffusion, reaction kinetics, selectivity, and catalyst stability. By elucidating the mechanistic insights into the impact of imperfections on catalytic properties, researchers can develop strategies to engineer catalysts with tailored performance for a wide range of catalytic processes.

15.1.3 Importance of Oxygen Vacancies

One such important imperfection is the oxygen vacancies (OVs). OVs are sites in solids (lattice) that are deprived from O^{2-} ions and can dictate the reaction rate, selectivity to reactants, and products that involve oxygen specifically. Amongst the various imperfections, one-dimensional defects like OVs have importance to an extent where the control and manipulations of material properties are possible (Greiner et al. 2013). The historical perspective of OVs in catalysis is a fascinating journey through the evolution of our understanding of solid-state chemistry and heterogeneous catalysis. Not until the mid-twentieth century the role of OVs was appreciated and with the advancement of experimental techniques such as electron microscopy, X-ray diffraction, and surface science tools, researchers began to explore the surface structure of catalysts in greater detail. It became evident that metal oxides, which are commonly used as catalyst supports or active phases, could exhibit a variety of defects, including OVs. The role of OVs in catalysis started to gain more attention in the latter half of the twentieth century, particularly in the context of reactions involving redox chemistry, such as oxidation, reduction, and oxygen activation. OVs were found to be active sites for adsorption, activation, and reaction of gas-phase molecules, particularly oxygen-containing species. One of the landmark studies highlighting the significance of OVs in catalysis was the discovery of the Mars–van Krevelen mechanism in the 1950s

(Mars and Krevelen 1954). This mechanism describes the redox reactions occurring at the surface of metal oxides, where OVs play a crucial role in facilitating the transfer of oxygen atoms between the catalyst surface and gas-phase reactants. In recent decades, with the advent of advanced spectroscopic and computational techniques, our understanding of OVs in catalysis has continued to deepen. Researchers have elucidated the electronic structure, chemical reactivity, and dynamic behaviour of OVs on various catalyst surfaces, paving the way for the design of more efficient and selective catalysts for a wide range of applications. Today, OVs are recognized as active sites for numerous catalytic reactions, including environmental remediation, energy conversion and storage, and chemical synthesis. Understanding and harnessing the catalytic properties of OVs remain an active area of research, with the potential to drive innovations in sustainable and green chemistry (Chen and Goodman 2004).

15.1.4 Metal–Metal-Oxide Interfaces

In that aspect, metal–metal-oxide interface (MMOI) has become a prominent route to explore various catalytic reactions (Jin et al. 2024; Ghosh Chaudhuri and Paria 2012; Zeng et al. 2004; Wang et al. 2005; Nalluri et al. 2019). Metal–support interactions (MSIs) encompass both strong intra-atomic bonding between the metal and support and weaker electronic interactions.

1. Location of Metal Nanoparticles:

The support's role in stabilizing metal nanoparticles is critical, preventing their agglomeration and sintering. Strong interactions, like metal–support bonds, achieve this stabilization, enhancing metal dispersion and resistance to harsh conditions. Importantly, this process maintains the shape and size of both the metal and support.

2. Changing in Shape and Morphology of Metal Nanoparticles:

The equilibrium morphology of metal nanoparticles is influenced by minimizing overall surface free energy. Adhesion energy at the metal–support interface affects nanoparticle morphology. Lattice mismatch induces strain and defects, altering the nanoparticle shape. Strong interactions enhance wetting, leading to shorter, wider forms. Redispersion into atomic species or single atoms on the support is driven by robust metal–support interactions.

3. Reducible Nature of the Supports:

SMSI involves metal nanoparticle encapsulation by suboxides from the support under reducing conditions. This minimizes metal nanoparticle surface energy by mobile support suboxide species. High surface energy metals, especially those activating hydrogen, are prone to encapsulation. Reducible supports generate suboxide

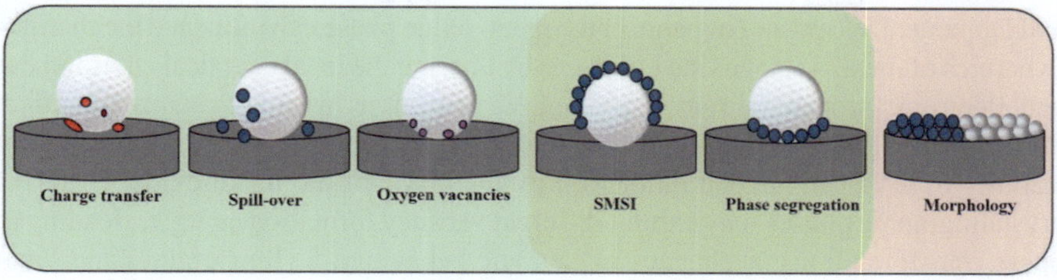

Fig. 15.1 Schematic illustration of the effect of support interaction with metal

species, consisting of a few atomic layers with dynamic structures under different gas atmospheres.

4. Compositional and Structural Rearrangement of Metal Nanoparticles and Support:

RMSI refers to chemical reactions between metal nanoparticles and their support, forming bimetallic structures. This incorporates metals or metalloids challenging to reduce into late transition metal nanoparticles, resulting in novel catalytic systems with unique structures like core–shell formations. It occurs under reducing atmospheres at high temperatures, yielding bimetallic alloys or intermetallic compounds dependent on gas atmospheres.

5. Electronic Interaction between Metal–Support:

Weak interactions involve charge transfer at the metal–support interface, aiming for electron chemical potential equilibration. The extent of electronic interaction depends on metal nanoparticle size. Electronic interaction can alter oxidation states of metal atoms or support metal ions. All the above interactions have been captured in the illustration below (Fig. 15.1).

The thermodynamics of metal–metal oxide interfaces involves understanding the energetics associated with the formation and stability of these interfaces. The factors to be considered are as follows: 1. Formation Energy: The formation enthalpy (ΔH_f) of a metal–metal oxide interface refers to the energy change associated with the formation of the interface from the separate metal and metal oxide phases. This includes the breaking of bonds in the metal and metal oxide and the formation of new bonds at the interface. 2. Interfacial Energy: The interfacial energy (γ) quantifies the excess energy associated with the interface compared to the bulk phases. It represents the energy required to create a unit area of interface. Lower interfacial energies indicate more stable interfaces. 3. Chemical Potential: The chemical potential of the metal and oxygen species at the interface determines the thermodynamic driving force for reactions such as oxidation, reduction, and adsorption. Differences in chemical potential between the metal and metal oxide phases can lead to the migration of species across the interface. 4. Work Function: The work function of the metal and metal oxide surfaces influences charge transfer and electronic structure at the interface. Differences in work function can lead to the formation of built-in electric fields, which can affect catalytic activity and selectivity. 5. Defects and Dopants: Defects and dopants in the metal oxide phase can influence the thermodynamics of metal–metal oxide interfaces by altering the local chemical environment and electronic properties. For example, OVs or surface hydroxyl groups can affect the energetics of adsorption and reaction at the interface. 5. Phase Stability: The stability of OVs can also be influenced by the phase of the metal oxide. Different crystallographic phases may exhibit different vacancy formation energies, leading to phase transformations or changes in catalytic behaviour. 6. Temperature and Pressure Dependence: The thermodynamics of metal–metal oxide interfaces are often temperature- and pressure-dependent. Changes in temperature and pressure can alter the equilibrium concentrations of species at the interface and affect the stability of the interface.

Experimental techniques such as surface science methods (e.g., X-ray photoelectron spectroscopy, scanning tunnelling microscopy) and computational modelling (e.g., density functional theory calculations) are used to study the thermodynamics of metal–metal oxide interfaces and elucidate their properties and behaviour under various conditions. Understanding the thermodynamics of metal–metal oxide interfaces is crucial for the design and optimization of heterogeneous catalysts, electrocatalysts, and other materials with applications in catalysis, energy conversion, and environmental remediation.

In this book chapter, we highlight the importance of MMOIs from their catalytic efficacy point of view. We describe the synthesis, characterization, and their catalytic effect in CO oxidation and hydrogen generation from the hydrolysis of ammonia borane. With regard to synthesis, we give a general account of the strategies and discuss specific work related to the above-mentioned reactions with emphasis on the importance of maximizing the interfaces through synthetic methods. While all the possible characterization techniques are desired, we account for few studies related to the characterization of the MMOIs with specific examples. The role of defects has to be understood in detail due to various pathways that the reaction can proceed in. We emphasize that the role of support need not be simply to provide surface area but actually provide a suitable electronic and structural environment for the pathway of a catalytic chemical reaction.

15.1.5 Thermodynamics of MMOIs

We delve briefly into the thermodynamic considerations concerning the formation and stability of MMOI. Within MOSM, the chemical potential of metal atoms emerges as a pivotal factor. Intuitively, when the chemical potential of atoms is high, they bind effectively to adsorbates, serving as a driving force for the sintering of surface atoms or particles. Moreover, this potential is influenced by particle size and can be impacted by the adhesion energy between the metal and the support.

Expressing the internal energy of the system (Hemmingson and Campbell 2017):

$$U = nU_{m,\text{bulk}} + U_{\text{sup}} + \gamma_m A_m - E_{\text{adh}} A \qquad (15.1)$$

where n is the number of moles of atoms, $U_{m,\text{bulk}}$ is the molar internal energy of the pure bulk metal, U_{sup} is the total internal energy of the pure support (without the metal), γ_m is the surface energy of the metal, A_m is the total surface area, and A is the total area of the metal–support contact.

The equation suggests that when adhesion energy is low, metals tend to aggregate into 3D objects, which is unfavourable for various device applications and catalytic processes. Overcoming this challenge, forming a metallic coating over an oxide surface has been a persistent issue. Typically, surfactants are employed to achieve this, or, alternatively, a surfactant-based metal precursor can be utilized. The presence of a surfactant during the precursor's decomposition reduces surface energy, facilitating the formation of a thin metal film over the metal oxide (Ghosh Chaudhuri and Paria 2012; Zeng et al. 2004; Wang et al. 2005). This method has been successful with metals like Au and Ag due to their low surface energy (Nalluri et al. 2019), whereas the Pd system tends to form nanoislands over iron oxide due to its relatively high surface energy.

Metal vapor deposition on surfaces and subsequent growth occurs via three modes depending on the relative magnitudes of metal surface energy and adhesion energy: the Franck–van der Merwe mode for perfect wetting, the Stranski–Krastanov mode for uniform coating followed by balling up, and the Volmer–Weber growth mode leading to the formation of three-dimensional clusters during initial deposition stages.

The Young–Dupré equation for adhesion energy is given by

$$W_{\text{adh}} = \gamma_m + \gamma_{\text{sup}} - \gamma_{m-\text{sup}} = \gamma_m(1 + \cos\theta) \tag{15.2}$$

where γ_m is the surface energy of metal, γ_{sup} is the energy of the support, and $\gamma_{m-\text{sup}}$ is the energy of the metal–support interface, and θ is the contact angle of metal on the support (Jupille and Lazzari 2006). The conditions dictated for complete uniform coverage, mixture of film and particulate, and full 3D particles structure formation are $W_{\text{adh}} > 2\gamma_m$, $W_{\text{adh}} \sim 2\gamma_m$, and $W_{\text{adh}} < 2\gamma_m$, respectively.

Thus, coordination unsaturation leads to different thermodynamic behaviours, with metal atoms near the surface and interfaces interacting electronically (Hemmingson and Campbell 2017). In its simplest form, this example excludes mixed oxides or intermetallics, and the equilibrium oxidation conditions determine the limiting oxygen partial pressures for the coexistence range. Within this compatibility range, even slight changes in the compositions of bulk phases occur. Additionally, adsorption can modify interfacial energies, necessitating consideration of structural and electronic modifications on the energies in the equation above.

15.2 Synthesis of Catalysts

The rationale behind the synthetic strategy is to maximize the interface between the metal and the support, achieved through a finely dispersed metal. This dispersion is controlled by the metal precursor. Amongst various wet chemical synthesis procedures for supported nanoparticles, the impregnation method stands out as one of the simplest, with many other strategies stemming from it (Fig. 15.2).

15.2.1 A Survey of Synthesis Procedures

There are two well-known types of impregnation methods. Wet impregnation (WI) involves dispersing support materials and metal salts in a highly soluble solvent at very dilute concentrations. The pH of the solution also plays a crucial role. Zhang et al. demonstrated a general synthesis method for Pt group metals and alloy nanoparticles using the wet impregnation approach (Zhang et al. 2016). Dry impregnation (DI) is a method that is used when the adsorption of impregnated materials on the support material is low. Zhu et al. showed Pt nanoparticles can be mounted on carbon using the charge-enhanced dry impregnation method (Zhu et al. 2013).

Another method, pore volume impregnation (PVI), involves filling the pore volume of the support by metal salts followed by reduction to elemental state metal nanoparticles. Proper washing of synthesized materials is necessary if non-gaseous by-products are formed during this reaction (Spieker and Regalbuto 2001). Using this method, Ni nanoparticles was loaded into mesopores of SiO_2 using the

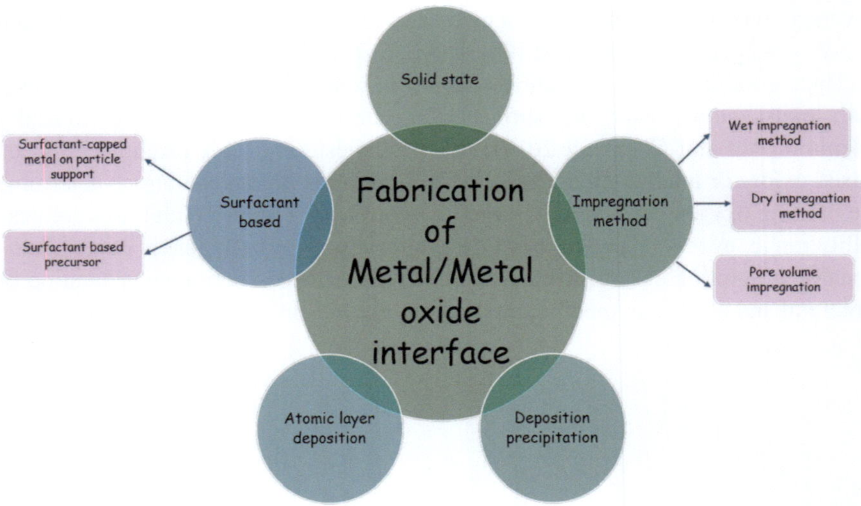

Fig. 15.2 Schematic diagram for the synthetic routes that can be applied to fabricate metal–metal-oxide interfaces

$[Ni(OH)_2](NO_3)_2$ complex as a impregnating metal salt precursor, followed by drying and calcination (Sietsma et al. 2008).

Metal nanoparticles can also be formed on support by mixing the metal salts with support followed the decomposition. However, the nature of interaction between the metal salts and support plays an important role in dispersion of metal nanoparticles on the support material's surface, making the nanoparticles mobile and prone to aggregation during the drying and washing process. A strong electrostatic adsorption between precursor and support was shown to be an important factor (Jiao and Regalbuto 2008). Additionally, controlling the size of metal nanoparticles becomes challenging, resulting in a wide size distribution. Traditionally, MOSM was prepared using the impregnation method, although other available methods can also be employed. The focus remains on increasing the interface area, which enhances catalytic activity significantly.

Manipulating the morphology of MMOI during synthesis is crucial. One approach to increase surface area is by keeping metal nanoislands small, leading to remarkable catalytic activity enhancement. Methods such as coprecipitation and deposition–precipitation can yield higher dispersion compared to impregnation.

15.2.2 Deposition–Precipitation Method

Metal compounds such as hydroxides with low solubility are deposited onto a preformed support material. Various experimental parameter variations like manipulating pH, temperature, and evaporation can be undertaken. Again, the support plays a role in the nucleation process of metal compounds. Care also must be taken while adding alkali solution (for pH) because the concentration of less soluble metal hydroxides does not exceed the critical supersaturation due to the possibility of bulk precipitation. Homogeneous nucleation and precipitation are preferred to achieve uniform dispersion of metal nanoparticles. With the addition of urea, metal complexes become more negatively charged, resulting in increased repulsion and limiting growth. While deposition of metal compounds is possible without urea, the presence of urea plays a crucial role in the adsorption of metal complexes, leading to the formation of narrow-sized metal nanoparticles.

Fang et al. prepared Pt nanoparticles with a loading of 60 wt.% on a carbon black support via the deposition of $PtClx(OH)_y(H_2O)_z$ complex followed by reduction (Fang et al. 2009). Similarly, an initial pH-controlled deposition–precipitation synthesis study was conducted with 40 wt.% PtRu loaded on a carbon support (Kim et al. 2010). Here, controlling the initial pH of the solution with NaOH is crucial as it results in decrease in particle sizes. Supported gold nanoparticles have also been prepared via the formation of $AuCl(OH)_3^-$ and $AuCl_{4-x}(OH)_x^-$ complex species (Zanella et al. 2002, 2005). Furthermore, transition metals such as cobalt and nickel nanoparticles have been synthesized using this method (Lee et al. 2005; Bezemer et al. 2006).

15.2.3 Atomic Layer Deposition Method

Atomic layer deposition (ALD) is a precise two-step method where a thin metal film is deposited using chemical vapors (Cao and Lu 2020). Surface reactions occur at the molecular level, and unreacted reactants and products can be purged by an inert gas. ALD offers advantages such as high-precision deposition and composition control, regardless of the size and shape of the support. Additionally, different metal atoms can be sequentially deposited to obtain AB layers.

15.2.4 Ligand-Based Synthesis Methods

Another method for preparing metal-supported catalysts involves using pre-synthesized gold colloidal particles and depositing them onto pre-formed cores (Fang et al. 2009; Grunwaldt et al. 1999; Tsubota et al. 1998; Chou and McFarland 2004; Lopez-Sanchez et al. 2011; Zheng and Stucky 2006; Zou et al. 2021). This approach has been utilized in the preparation of oxide-supported gold catalysts, where synthesized nanoparticles are deposited on oxide supports through mechanical mixing followed by evaporation. One critical issue in this process is the size and distribution of the metal nanoparticles obtained during synthesis (Tsubota et al. 1998). If ligands are used in the synthesis step, a thermal treatment is often necessary to remove the organic components, allowing for the sintering of nanoparticles or the deposition of carbon or sulphur. Hutchings et al. demonstrated the enhancement of catalytic activity by systematically removing polyvinyl alcohol (PVA) from gold nanoparticles by stirring in a liquid phase solvent at temperatures where nanoparticle aggregation is not observed (Lopez-Sanchez et al. 2011). Exploiting the weak interaction between metal-oxide supports and metal nanoparticles, gold nanoparticles have been successfully deposited on various supports ranging from acidic to basic and from insulating to semi-conducting oxides (Zheng and Stucky 2006). The kinetic equilibrium between the adsorption and desorption of metal nanoparticles in solution results in a homogeneous dispersion of metal nanoparticles on oxide particles.

15.2.5 Other Variations

Zou et al. fabricated nanometre-sized metal/oxide interfaces using a hybrid Pd-Bi_2O_3 cluster ensemble stepwise photochemical method. Through this approach, they tailored Pd_xBi/TiO_2 and Pd_x-Bi_2O_3/TiO_2 systems, showing that Pd-Bi_2O_3 hybrid clusters feature a small Pd–Pd coordination number and intra-cluster electron transfer, enabling weak C_2H_4 adsorption without compromising H_2 activation activity (Zou et al. 2021). In another synthesis method, it was demonstrated that doping Ca(001) films with Mo alters the growth behaviour of Au particles (Shao et al. 2011).

A plethora of innovative strategies were followed to form metals on the oxide support. This small section presents all the possible variations to maximize the metal dispersion. It was suggested that the differently shaped active Au systems reported by Hutchings et al. may result from such processes (Hashmi and Hutchings 2006). They utilized mixed precursors in solid-state decomposition over support oxides, resulting in very small metal and alloy nanoclusters (<2 nm) on TiO_2. By mixing acetates with TiO_2 in solid state at room temperature followed by decomposition at a relatively higher temperature, Hutchings et al. achieved fine control over dispersion and composition of Au on TiO_2 (Herzing et al. 2008). Additionally, mixing mixed metal acetates led to the formation of alloy nanoparticle/clusters on TiO_2. In another study, an ethanolic suspension of TiO_2 was mixed with a known concentration of $RhCl_3$ solution and the mixed solution was stirred at 70 °C till dryness. The resultant solid precursor was heated in a reducing atmosphere at 350 °C, and, similarly, Rh dispersed on TiO_2 was also prepared by reducing the precursor solution by sodium borohydride ($NaBH_4$) (Yuan et al. 2024). TiO_2 was dispersed in the $HAuCl_4$ aqueous solution and after stirring for sufficient time, it was irradiated using Xe lamp. Metals on mixed supports $WO_{2.72}/TiO_2$ were also prepared by in situ reduction of Au^{3+} on the hybrid support (Asim et al. 2022). Pt/MoO_{3-x} by deposition–precipitation of Pt nanosystems on MoO_{3-x}-T nanorods (Zhou et al. 2022). A prepared Au NP solutions were added to the P25 (TiO_2) dispersion in water. Au–CeO_x-C catalyst was also prepared by the same method (Huang et al. 2020). The Au catalysts were prepared by a modified deposition–precipitation method (Tang et al. 2016c). The gold catalysts were positioned at the interface of TiO_2 and hydroxyapatite supports. Pd/ FeO_x NPs were prepared by heating an appropriate amount of $Fe(acac)_3$, oleylamine, and Pd nanoparticles at 300 °C for 1 h (Yan et al. 2022). Apparently, the dispersion of Pd metal precursor was found to depend on the size of the support (Patra et al. 2023). For the same amount of Pd precursor, the particle Pd metal in the final decomposed product is more if the size of the support is higher.

15.2.6 Surfactant-Based Metal Precursor Method

Gopalan and Ganesan et al. used surfactant-based precursor for the synthesis of core–shell nanoparticles (Nalluri et al. 2019). Briefly, Au-tetraoctylammonium bromide and Ag-sarcocinate were mixed with pre-formed oxide core in solid state. The low melting point of the precursors enables them to coat over the oxide uniformly. The thermal decomposition thus leads to the formation of metal coating aided by the surfactant which reduces the surface energy. Whether the metal coating is uniform or not depends on the metal, as Pd tends to ball up in addition to small layer indicating a Stranski–Krastanov type growth (Patra et al. 2021, 2023). There are two ways by which we can create metal–metal-oxide interfaces. One is (1) create a core–shell of metal-oxide@metal nanostructures or nanoislands of metals on the metal-oxide or vice versa. Alternatively, the support could be in the form of brushes and fibres, and metal can be deposited on the so-called brushes or fibres. One more approach is to deposit metal oxide nanoparticles on metal sheets or vice versa (Liu et al. 2015). In all the above cases, the idea is to maximize the interface contact area between metal and metal oxide. Two of the main targets to be achieved through the synthetic methods are the dispersibility of the metal content and the composition (amount of metal content).

15.3 Characterization Techniques

Along with typical characterization techniques for conventional analysis like phase identification and composition, other techniques can be used in tandem for the characterization of MMOIs or OVs including probe techniques X-ray photoelectron spectroscopy (XPS) and surface X-ray absorption near-edge spectroscopy (XANES) (Ahmadi et al. 2016; Naldoni et al. 2012; Kong et al. 2011; Sarkar and Khan 2019; Xu et al. 2016).

15.3.1 Scanning Transmission Electron Microscopy High-Angle Annular Dark-Field (STEM-HAADF)

STEM-HAADF is an advanced microscopy technique used to study the structure and composition of materials at the atomic scale. In STEM-HAADF, a focused electron beam is transmitted through a thin sample, and the transmitted electrons are collected by a detector positioned at a high angle relative to the incident beam. This detector collects electrons that have undergone large-angle scattering, resulting in high-angle annular dark-field (HAADF) images. HAADF imaging is particularly sensitive to variations in atomic number (Z), making it useful for imaging heavy elements and providing high-contrast images of materials with atomic resolution. In HAADF images, bright regions correspond to areas with higher atomic number elements, while dark regions correspond to areas with lower atomic number elements. STEM HAADF is a powerful tool for characterizing nanostructures, interfaces, defects, and individual atomic columns in a wide range of materials, including metals, semiconductors, ceramics, and biological samples. It enables researchers to visualize and analyse the atomic structure of materials with unprecedented detail, providing insights into their properties and behaviour at the nanoscale (Lopez-Sanchez et al. 2011). The lattice distortion due to OVs can be observed by direct images in TEM/HRTEM. Zhuang et al. attributed the disordered structure in HRTEM image of $Fe_1Co_1O_x$ to the OVs (Zhuang et al. 2018).

15.3.2 Diffuse Reflectance Spectroscopy (DRS)

DRS can be a valuable technique for characterizing metal–metal oxide interfaces. 1. Optical Properties: At metal–metal oxide interfaces, changes in the optical properties can provide insights into the interface composition, structure, and electronic interactions. For example, shifts in the absorption edges or changes in the band gap can indicate the presence of interfacial states or charge transfer phenomena. 2. Surface Plasmon Resonance: Metal nanoparticles or thin films at the interface can exhibit surface plasmon resonance (SPR) phenomena, which lead to enhanced absorption and scattering of light at specific wavelengths. DRS can be used to study SPR in metal–metal oxide nanocomposites, providing information about the size, shape, and distribution of metal nanoparticles at the interface. 3. Electronic Transitions: DRS can probe electronic transitions associated with metal and metal oxide components at the interface. For example, transitions between d-orbitals in transition metals or between valence and conduction bands in metal oxides can be observed in the spectra. Changes in these transitions upon interface formation can indicate electronic coupling or charge transfer effects. 4. Surface Modification: DRS can also be employed to study surface modifications or functionalization of metal–metal oxide interfaces. Changes in the reflectance spectra before and after surface treatments or chemical reactions can provide information about the extent and nature of surface functional groups, chemical bonds, or surface coverage at the interface. DRS spectra profile of α-Fe_2O_3-Pd depends on the size of the α-Fe_2O_3 support (Patra et al. 2023). The Kubelka–Monk plot exhibited a cross-over in spectral lines indicating a change in the reflectance due to the MMOIs. It should be noted that Fe_2O_3 is a semiconductor with a band gap in the visible region. One has to be careful in the interpretation of such interfaces using DRS. When methyl orange (MO) was mixed with TiO_2, the spectra apparently showed a change in the reflectivity and the band gap. However, there is no such change as shown in Figs. 15.3 and 15.4.

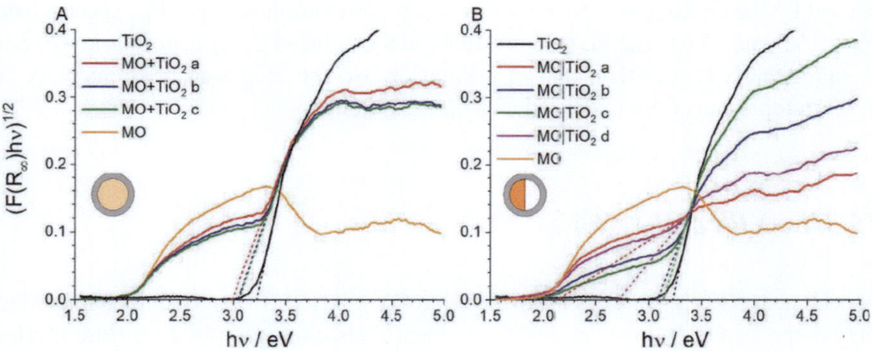

Fig. 15.3 Tauc plots of the bare TiO_2, methyl orange, **a** MO + TiO_2 sample, and (B) MO|TiO_2 system (linear fit for measurement **b** (blue line) and **c** (green line) overlap). Numbered spectra were recorded for the same pellet differently placed in the holder. The insets show schematically the sample in the holder. Adapted with permission from ACS (Makuła et al. 2018)

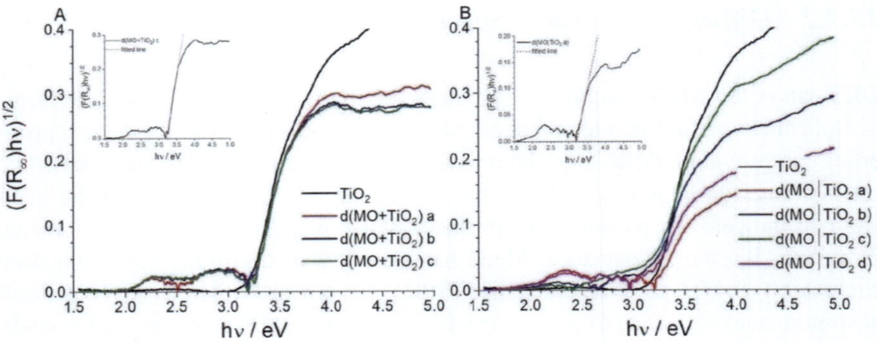

Fig. 15.4 Tauc plots of the TiO_2 components (extracted from the spectra of the MO|TiO_2 system) and bare TiO_2 (**a**). The Tauc plots of the differential spectra of the sample MO + TiO_2 and bare TiO_2 (**b**). The determinations of E_g for measurements 1 (4A) and 3 (4B) are shown as insets. Adapted with permission from ACS (Makuła et al. 2018)

15.3.3 Electron Paramagnetic/Spin Resonance (EPR/ESR) and Photoluminescence (PL)

A combination of EPR and PL can be done, and the data obtained from these two techniques can be correlated. The figures show the EPR and PL data obtained for ZnO, but similar studies can be performed on MMOIs. EPR can determine qualitatively and quantitatively the chemical species with unpaired electrons and the structural characteristics of the environment around such species. The unpaired electrons would exhibit a signal at g-value of 2. Thus, a presence of a peak would confirm the presence of OVs. There could be a limitation in using EPR in the metal that is ferromagnetic/ paramagnetic in nature. The EPR line gets broadened due to the magnetic nature of the samples. There is a correlation in the emission spectral intensity. The formation of OVs can further be confirmed by the increase of the emission peak intensity caused by the photogenerated electrons in the photoluminescence (PL) spectra (see Figs. 15.5 and 15.6). The PL peak intensity of the reduced CeO_2 nanorods at 420 nm is significantly higher than that of normal CeO_2. The increased peak intensity at 420 nm represents OVs in the reduced CeO_2 systems.

15.3.4 XPS and XANES

A peak near 532 eV for O 1s is attributed to OVs. XANES was used to characterize the existence of OVs. In series of WO_3 nanoparticles, a few sample spectra are similar to WO_3-NPs indicating the presence of W in +6 valence, while a specific sample has XANES peak located in between the peaks W foil and WO_3-NPs (Hou et al. 2019). Thus, the valence state of W in that specific sample should be between 0 and

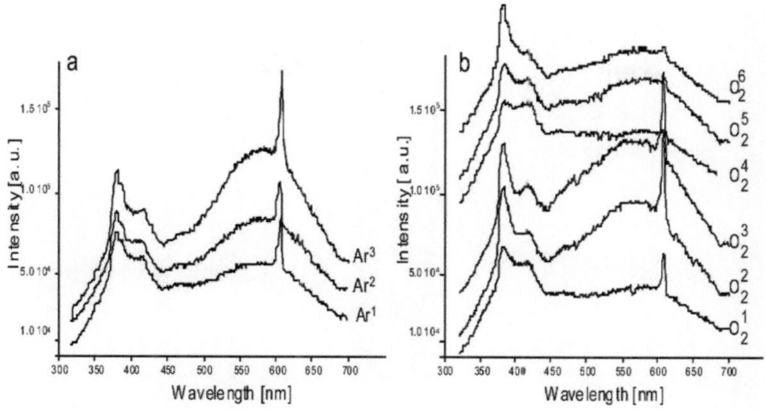

Fig. 15.5 Photoluminescence emission spectra of kadox sample recorded at room temperature during the sequence described in Fig. 15.1. **a** Ar^i series: spectra recorded under argon ($P_{Ar} = 1.2$ bar) and **b** $O_2{}^j$ series: spectra recorded under oxygen ($P_{O2} = 0.25$ bar). Adapted with permission from ACS (Drouilly et al. 2012)

+6. The valence states of metal elements in oxygen-containing compounds could be clearly observed by XANES.

15.3.5 Infrared (IR) and Raman Spectra

The in situ IR or Raman spectra of catalysts may give the adsorption sites and surface chemical reaction or intermediates. In the Raman spectra, the characteristic bands related to OVs appear at 600 cm^{-1}. Hezam et al. characterized the OVs in two CeO_2 samples prepared by modified combustion methods (Hezam et al. 2020). The bands at 465.2 and 602 cm^{-1} were assigned correspond to Ce–O and OVs in the symmetrical stretching F_{2g} mode, respectively. Besides, the IR spectra of the adsorbates like CO molecule give information on the surface binding sites. The chemical bonding between CO and Au particles dispersed on MgO thin films shows that the ad-particles charge up due to an electron transfer from the support, whereas the charge transfer on thicker MgO films was attributed to an electron donation from defects. In both cases, the excess electrons on metals give rise to unusual CO adsorption properties, as deduced from Scanning Tunnelling Microscope (STM) and infrared reflection adsorption spectroscopy (IRAS) measurements. When the MgO support film thickness is less, the CO exclusively binds to the low coordinated Au atoms along the island perimeter. On thicker films, a red shift was observed for the CO stretch mode indicating the filling of the antibonding CO 2π* orbital (Lin et al. 2010). Interestingly, the CO stretching frequency decreased with decreasing Au particle size. The formation of superoxide O_2^- species on thin MgO layers supported

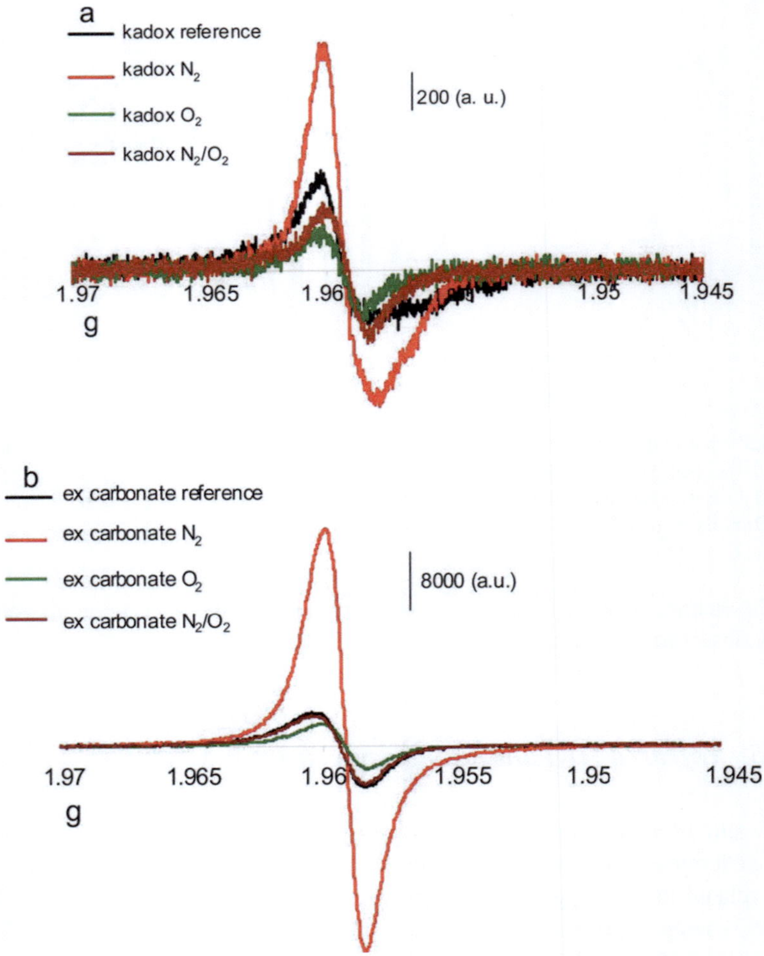

Fig. 15.6 EPR spectra of **a** kadox and **b** ex-carbonate samples. Adapted with permission from ACS (Drouilly et al. 2012)

on Ag(100) has also been proposed (Sun et al. 2010; Hellman et al. 2009). Combining molecular-beam methods and TR-RAIRS, in the study on the kinetics of methanol decomposition on well-ordered Pd crystallites, two competing reaction pathways were proposed: a rapid dehydrogenation to give CO and a slow carbon–oxygen bond breakage leading to formation of carbon and hydrocarbon species (Hoffmann et al. 2003). In fact, the particle edge position was identified as the site of carbon–oxygen bond breakage.

15.4 Role of Structure and Composition on Catalytic Activity

In this section, we highlight the role of MMOIs on the catalytic efficacy of CO oxidation, CO_2 activation, and hydrolysis of ammonia borane hydrolysis.

15.4.1 CO Oxidation

The role of OVs in CO oxidation is significant and has been extensively studied in the field of heterogeneous catalysis. The important parameters under consideration are 1. Active Sites**: OVs on metal oxide surfaces, such as ceria (CeO_2) or titanium dioxide (TiO_2), serve as active sites for CO oxidation. These vacancies provide sites for the adsorption and activation of CO and oxygen molecules, facilitating the reaction. 2. Enhanced Reactivity: OVs can enhance the reactivity of metal oxide catalysts towards CO oxidation by altering the electronic structure of the surface. The presence of vacancies introduces localized electronic states within the band gap, which can lower the activation energy for the reaction, leading to improved catalytic activity. 3. Redox Chemistry: CO oxidation typically involves a redox mechanism, where the adsorbed CO molecules undergo oxidation by oxygen species present on the catalyst surface. OVs play a crucial role in the redox chemistry of CO oxidation by promoting the dissociation and mobility of oxygen species, which are involved in the reaction. 4. Surface Diffusion: OVs can also influence the diffusion of CO and oxygen species on the catalyst surface, affecting the kinetics of the reaction. Vacancies may provide diffusion pathways or act as trapping sites for adsorbed species, influencing their mobility and reactivity. 5. Catalyst Stability: The presence of OVs can impact the stability and durability of metal oxide catalysts for CO oxidation. While vacancies may enhance catalytic activity, excessive vacancy formation can lead to catalyst deactivation through processes such as surface reconstruction or sintering. Overall, OVs play a crucial role in CO oxidation catalysis by serving as active sites for adsorption, promoting redox chemistry, influencing surface diffusion, and impacting catalyst stability. Understanding and controlling the formation and distribution of OVs on metal oxide catalysts are essential for the design of efficient and stable catalysts for CO oxidation and other oxidation reactions.

We present here few specific works: The activation of O_2 can be realized by manipulating interfacial electronic structure by virtue of atomic structures at the Au-TiO_2 interface. The electron transfer from Ti^{3+} ions to Au nanoparticles at the Au-TiO_2 interface forms an Au–O–O–Ti type structure, which enhances CO oxidation activity by approximately 34 times (Huang et al. 2020). This activity is more than the OVs rich Au/TiO_2 interfaces where the electrons from Ti^{3+} species are trapped by interfacial OVs. The interface structure is complicated and can have more than one type of active sites. The sites that can hold more CO at the perimeter of MMOIs prevent the CO poisoning of Pd particles (Suchorski et al. 2018) In Pt-TiO_2, the interface

Pt-TiO_2 was changed by arranging Pt nanowires over TiO_2 and Pt films on TiO_2. It was found that the hot electrons are responsible for the selectivity towards methyl formate from n-butanol. From the DFT, Cr_2O_3 nanoclusters on two polymorphs of TiO_2 surfaces were shown. In TiO_2 rutile (110) and anatase (101) modified by Cr_2O_3, a reduction of the energy band gap due to valence band uplift was shown. It was shown that photoexcited electrons and holes are localized on the Cr_2O_3 nanocluster. It was reported that amongst the three surface oxygen species, namely chemisorbed oxygen, Pt–O like surface oxide, and α-PtO_2 trilayer, the adsorbed oxygen exhibits the highest reactivity towards CO oxidation.

In SMSI, a complete encapsulation of active ingredients takes place. A careful unique configuration can also be designed in such a way that the active metal, i.e., Au nanoparticles in this case partially encapsulated and therefore, partially exposed and still accessible by the reactant molecules. In this way, the sintering of Au nanoparticles can be avoided despite the high-temperature treatment (Tang et al. 2016a). The adsorption and activation of CO_2 can be enhanced at the Au–CeO_x interface. This result is supported by density functional theory calculations as well. A synergy between Au and CeO_x promotes the stability of key carboxyl intermediate (*COOH) and thus facilitates CO_2 reduction reaction (Gao et al. 2017).

The interactions of metals with FeO_x depend on the metal under reducing conditions (Yan et al. 2022). While Pd and Pt interact with FeO_x indicated by the formation of Pd/Pt single atoms on the FeO_x support, Au does not manifest strong bonding with FeO_x. Au_8 bound to the defects on Mg(001) catalyse oxidation of CO–CO_2 (Yoon et al. 2005) and charging of the supported metal cluster enhances catalytic efficacy at low temperatures. The catalytic active structures are created during the synthesis or due to the adsorbates during the reaction experimental conditions (He et al. 2023). In another case, the SMSI was different when the support was changed from TiO_2 to MoO_3, (Xin et al. 2022) while CO formation was dominating when Ru nanoparticles were dispersed on MoO_{3-x}. This SMSI structure was non-porous whereas in Rh-TiO_2 case which is porous, CO_2 was not preferred. Apart from reducible oxides as support for SMSI, other materials act as good supports. For example, Tang et al. showed the presence of a SMSI between Au NPs and hydroxyapatite (HAP), a phosphate instead of an oxide (Tang et al. 2016b). In general, the SMSI formation is normally attributed to reducing conditions, while it has been reported under oxidizing conditions as well as in Pt–BN system where the encapsulating BO_x layer is formed.

15.4.2 AB Hydrolysis

In the presence of strong electronic interaction between Rh nanoparticles and OVs, the Rh surface becomes electron-rich leading to highly efficient hydrogen generation from hydrolysis of AB. This interaction can be maximized in Rh–TiO_2 systems by proper heat treatment during the synthesis. For Ru_1Co_9/TiO_2 system, a turnover frequency based on Ru as high as 1408 $mol_{H2}/(mol_{Ru}\cdot min)$, is obtained at 25 °C (Zhang et al. 2021). The H_2 generation activity depends on the nature of the support.

Amongst anatase and rutile TiO_2 supports, anatase phase exhibited interaction and thus catalysts showed better catalytic activity. In support mixture of $WO_{2.72}$, and TiO_2, it was demonstrated the lifetime of photoinduced charges can be prolonged by the possibility of quick electron transfer from $WO_{2.72}$ to and then to TiO_2 (Asim et al. 2022). This mixed support showed better photocatalytic activity than individual supports for H_2 generation by hydrolysis of AB. In Pt^{2+}–Ov–Mo^{5+} (Ov represents oxygen vacancy), both experimental results and DFT calculations revealed that the Mo^{5+}–Ov sites accelerate the dissociation of H_2O molecules (rate-determining step), while the adjacent Pt^{2+} species facilitates the cleavage of the B–H bond in the AB hydrolysis to produce H_2 (Zhou et al. 2022). One of the ways by which MMOI is designed by using a rational approach to create unique defects concepts. The dual active sites in Cu-$Cu_{0.76}Co_{02.4}O_4$. The dual active sites mean two sites of adsorption of AB and water. While O–H bond is activated by electron-rich Cu defect interaction with transition metal oxides. Such a design can help to attain a TOF of 50.33 min^{-1} which is 45.9 times that of metal Cu (Wang et al. 2022).

Rh nanoparticles in ceria were active due to the charge build up on the support whereas Ru was in active. This is attributed to difference in the morphology and surface crystalline phase. This difference arises because of the metal–support interaction. A weak interaction of Rh metal with TiO_2 led to a unstable catalytic activity of H_2 from AB hydrolysis (Tonbul et al. 2019). The interaction was established by XPS technique. Doping of Pt with Ni was very effective for hydrogen generation upon hydrolysis of AB in terms of catalytic activity. The facile synthesis of ultrafine, highly dispersed PtNi NP nanocatalysts remains challenging. It has been reported that doping noble metals with transition metals in small quantities gives rise to synergy for H_2 generation. The valence states of Pt was modified with unique PtNi/NiO structure leading to a high TOF value of 1240 $mol_{H2}/(mol_{Pt}$ min) (Ge et al. 2017). The hydrogen spillover from Pt reduced the NiO to Ni and in turn this process enhanced the adsorption, activation, and dissociation of H_2O molecules, which was earlier difficult on the single Pt site.

15.5 Future Prospects

Single-atom catalysts (SACs) represent a cutting-edge realm in modern heterogeneous catalysis, offering intriguing distinctions from nanoparticulate counterparts. SACs entail isolated atoms within nanomaterials, and their catalytic efficacy hinges on factors such as atom positioning and metal–support interactions. Wang and colleagues pioneered SAC utilization, employing isolated Rh atoms on VO_2 nanorods to drive efficient hydrogen generation from AB. Their study unveiled a critical link between catalytic activity and metal–support interactions, particularly influenced by the phase transition of the VO_2 support, which modulated the Rh atom's state and heightened AB hydrolysis. Meanwhile, Chen's group showcased the potency of trace Pt (160 ppm) addition to supported Ni catalysts in enhancing hydrogen evolution from AB hydrolysis. Leveraging a unique configuration where atomic Pt was uniformly dispersed on Ni nanoparticle surfaces, they achieved a potent synergistic effect between Pt and Ni, yielding an impressive turnover frequency (TOF) of 12,000 molH_2/(molPt min). While SACs are becoming important, these lack multisite adsorption which is desirable for further quicker catalytic reactions. In that regard, single-cluster catalysts are becoming important.

One other potential area in which the oxygen vacancies would affect the reaction is in preventing the poisoning of the catalyst (Patra et al. 2023). One problem in catalytic hydrolysis of AB is that the catalyst is the deposition of polyborate by-products on the catalyst surface. Due to metal–support electronic interactions between Pt/Co_3O_4 and Pd/Fe_2O_3 poisoning of the catalyst can be avoided due to its water-soluble nature. Such electronic interactions are a direct result of OVs, the regions in the catalysts where congregation of solvent H_2O molecules takes place (Patra et al. 2023; Li et al. 2021).

15.6 Conclusions

We have shown the role of MMOIs on the possible mechanistic pathways of chemical reactions. The chemical reactivity of metal nanoparticles on support is directly linked to the chemical potential of the metal (Hemmingson and Campbell 2017). A higher chemical potential corresponds to a stronger bonding affinity with adsorbates. As the chemical potential increases, the metal tends to be reactive. Moreover, when metal atoms are part of a nanoparticle with a higher chemical potential, they encounter a greater thermodynamic driving force for diffusion resulting in sintering. As the surface-to-volume ratio increases, the chemical potential increases while the adhesion energy between metal and support decreases. All these factors needed to be considered for suitable catalysts design and optimal catalytic performance. The thermodynamic parameters may be temperature-dependent which needs to be accounted for the design of the catalysts. Operando catalytic studies will shine more light on the structural and chemical changes, the information that can be looped. Also, a lot of variation can be envisaged in the current synthetic strategies. While the concept of impregnation and deposition–precipitation techniques can be modified, good attention should also be given to the preparation of the support. As the metal–support interaction depends on the size of the support as well, synthetic design such as supports prepared from pyrolysis of MOF and COF materials can be adopted. With the huge amount of already available reliable data, machine and/or deep learning techniques could be used for the data analysis.

Acknowledgements The authors thank BITS-Pilani for the support and encouragement. AS thanks BITS-Pilani for the financial support.

References

Ahmadi, M., Mistry, H., Roldan Cuenya, B.: Tailoring the catalytic properties of metal nanoparticles via support interactions. J. Phys. Chem. Lett. **7**, 3519–3533 (2016)

Asim, M., et al.: Photohydrolysis of Ammonia Borane for effective H2 evolution via hot electron-assisted energy cascade of Au-WO2.72/TiO2. Ind. Eng. Chem. Res. **61**, 11429–11435 (2022)

Bezemer, G.L., et al.: Preparation of Fischer-Tropsch cobalt catalysts supported on carbon nanofibers and silica using homogeneous deposition-precipitation. J. Catal. **237**, 291–302 (2006)

Bond, G.C.: Heterogeneous Catalysis: principles and Applications. Oxford University Press (1974)

Cao, L., Lu, J.: Atomic-scale engineering of metal-oxide interfaces for advanced catalysis using atomic layer deposition. Catal. Sci. Technol. **10**, 2695–2710 (2020)

Chen, M.S., Goodman, D.W.: The structure of catalytically active gold on titania. Science (80-.) **306**, 252–255 (2004)

Chou, J., McFarland, E.W.: Direct propylene epoxidation on chemically reduced Au nanoparticles supported on titania. Chem. Commun. **4**, 1648–1649 (2004)

Drouilly, C., et al.: ZnO oxygen vacancies formation and filling followed by in situ photoluminescence and in situ EPR. J. Phys. Chem. C **116**, 21297–21307 (2012)

Fang, B., Chaudhari, N.K., Kim, M.S., Jung, H.K., Yu, J.S.: Homogeneous deposition of platinum nanoparticles on carbon black for proton exchange membrane fuel cell. J. Am. Chem. Soc. **131**, 15330–15338 (2009)

Gao, D., et al.: Enhancing CO2 electroreduction with the metal-oxide interface. J. Am. Chem. Soc. **139**, 5652–5655 (2017)

Ge, Y., et al.: PtNi/NiO clusters coated by hollow Sillica: novel design for highly efficient hydrogen production from Ammonia-borane. ACS Appl. Mater. Interfaces **9**, 3749–3756 (2017)

Ghosh Chaudhuri, R., Paria, S.: Core/shell nanoparticles: classes, properties, synthesis mechanisms, characterization, and applications. Chem. Rev. **112**, 2373–2433 (2012)

Greiner, M.T., Chai, L., Helander, M.G., Tang, W.M., Lu, Z.H.: Metal/metal-oxide interfaces: how metal contacts affect the work function and band structure of MoO3. Adv. Funct. Mater. **23**, 215–226 (2013)

Grunwaldt, J.D., Kiener, C., Wögerbauer, C., Baiker, A.: Preparation of supported gold catalysts for low-temperature CO oxidation via 'size-controlled' gold colloids. J. Catal. **181**, 223–232 (1999)

Hashmi, A.S.K., Hutchings, G.J.: Gold catalysis. Angew. Chemie Int. Ed. **45**, 7896–7936 (2006)

He, Y., Zhang, J., Polo-Garzon, F., Wu, Z.: Adsorbate-induced strong metal-support interactions: implications for catalyst design. J. Phys. Chem. Lett. **14**, 524–534 (2023)

Hellman, A., Klacar, S., Grönbeck, H.: Low temperature CO oxidation over supported ultrathin MgO films. J. Am. Chem. Soc. **131**, 16636–16637 (2009)

Hemmingson, S.L., Campbell, C.T.: Trends in adhesion energies of metal nanoparticles on oxide surfaces: understanding support effects in catalysis and nanotechnology. ACS Nano **11**, 1196–1203 (2017)

Herzing, A.A., Kiely, C.J., Carley, A.F., Landon, P., Hutchings, G.J.: Identification of active gold nanoclusters on iron oxide supports for CO oxidation. Science (80-.) **321**, 1331–1335 (2008)

Hezam, A., et al.: CeO2 nanostructures enriched with oxygen vacancies for photocatalytic CO2 reduction. ACS Appl. Nano Mater. **3**, 138–148 (2020)

Hoffmann, J., Schauermann, S., Johánek, V., Hartmann, J., Libuda, J.: The kinetics of methanol oxidation on a supported Pd model catalyst: molecular beam and TR-IRAS experiments. J. Catal. **213**, 176–190 (2003)

Hou, T., et al.: Operando oxygen vacancies for enhanced activity and stability toward nitrogen photofixation. Adv. Energy Mater. **9**, 1–8 (2019)

Huang, J., et al.: Manipulating atomic structures at the Au/TiO2 interface for O2 activation. J. Am. Chem. Soc. **142**, 6456–6460 (2020)

Jiao, L., Regalbuto, J.R.: The synthesis of highly dispersed noble and base metals on silica via strong electrostatic adsorption: I. Amorphous Silica. J. Catal. **260**, 329–341 (2008)

Jin, S., Wang, Z., Li, D., Wang, Y., Zhu, M.: Effect of gas environment on metal-support interactions of supported metal catalysts. ChemCatChem **16**, 1–11 (2024)

Jupille, J., Lazzari, R.: Improving the wetting of oxides by metals. J. Optoelectron. Adv. Mater. **8**, 901–908 (2006)

Kim, M.S., et al.: Effect of pH on electrocatalytic property of supported PtRu catalysts in proton exchange membrane fuel cell. Catal. Today **158**, 354–360 (2010)

Kittel, C.: Introduction to solid state physics. J. Mech. Phys. Solids **6** (1957)

Kong, M., et al.: Tuning the relative concentration ratio of bulk defects to surface defects in TiO2 nanocrystals leads to high photocatalytic efficiency. J. Am. Chem. Soc. **133**, 16414–16417 (2011)

Li, M., Zhang, S., Zhao, J., Wang, H.: Maximizing metal-support interactions in Pt/Co3O4 nanocages to simultaneously boost hydrogen production activity and durability. ACS Appl. Mater. Interfaces **13**, 57362–57371 (2021)

Lin, X., et al.: Charge-mediated adsorption behavior of CO on MgO-supported Au clusters. J. Am. Chem. Soc. **132**, 7745–7749 (2010)

Liu, B.H., Boscoboinik, J.A., Cui, Y., Shaikhutdinov, S., Freund, H.J.: Stabilization of ultrathin zinc oxide films on metals: reconstruction versus hydroxylation. J. Phys. Chem. C **119**, 7842–7847 (2015)

Lopez-Sanchez, J.A., et al.: Facile removal of stabilizer-ligands from supported gold nanoparticles. Nat. Chem. **3**, 551–556 (2011)

Makuła, P., Pacia, M., Macyk, W.: How to correctly determine the band gap energy of modified semiconductor photocatalysts based on UV-Vis spectra. J. Phys. Chem. Lett. **9**, 6814–6817 (2018)

Mars, P., van Krevelen, D.W.: Oxidations carried out by means of vanadium oxide catalysts. Chem. Eng. Sci. **3**, 41–59 (1954)

Naldoni, A., et al.: Effect of nature and location of defects on bandgap narrowing in black TiO2 nanoparticles. J. Am. Chem. Soc. **134**, 7600–7603 (2012)

Nalluri, S.R., Nagarjuna, R., Patra, D., Ganesan, R., Balaji, G.: Large scale solid-state synthesis of catalytically active Fe3O4@M (M = Au, Ag and Au–Ag alloy) core-shell nanostructures. Sci. Rep. **9**, 1–11 (2019)

Patra, D., Ganesan, R., Gopalan, B.: Hydrogen generation rate enhancement by in situ Fe(0) and nitroarene substrates in Fe3O4@Pd catalyzed ammonia borane hydrolysis and nitroarene reduction tandem reaction. Int. J. Hydrogen Energy **46**, 25486–25499 (2021)

Patra, D., Garg, R., Gautam, U.K., Gopalan, B.: Mitigation of polyborate precipitation on Pd/Fe2O3 sites during ammonia borane hydrolysis: an alternate insight into the role of oxygen vacancies. Int. J. Hydrogen Energy **48**, 28333–28342 (2023)

Sarkar, A., Khan, G.G.: The formation and detection techniques of oxygen vacancies in titanium oxide-based nanostructures. Nanoscale **11**, 3414–3444 (2019)

Shao, X., et al.: Tailoring the shape of metal Ad-particles by doping the oxide support. Angew. Chemie Int. Ed. **50**, 11525–11527 (2011)

Sietsma, J.R.A., et al.: Ordered mesoporous silica to study the preparation of Ni/SiO2 ex nitrate catalysts: Impregnation, drying, and thermal treatments. Chem. Mater. **20**, 2921–2931 (2008)

Spieker, W.A., Regalbuto, J.R.: A fundamental model of platinum impregnation onto alumina. Chem. Eng. Sci. **56**, 3491–3504 (2001)

Suchorski, Y., et al.: The role of metal/oxide interfaces for long-range metal particle activation during CO oxidation. Nat. Mater. **17**, 519–522 (2018)

Sun, Y.N., et al.: The interplay between structure and CO oxidation catalysis on metalSupported ultrathin oxide films. Angew. Chemie Int. Ed. **49**, 4418–4421 (2010)

Tang, H., et al.: Ultrastable hydroxyapatite/titanium-dioxide-supported gold nanocatalyst with strong metal-support interaction for carbon monoxide oxidation. Angew. Chemie **128**, 10764–10769 (2016a)

Tang, H., et al.: Strong metal-support interactions between gold nanoparticles and nonoxides. J. Am. Chem. Soc. **138**, 56–59 (2016b)

Tang, H., et al.: Ultrastable hydroxyapatite/titanium-dioxide-supported gold nanocatalyst with strong metal–support interaction for carbon monoxide oxidation. Angew. Chemie Int. Ed. **55**, 10606–10611 (2016c)

Tonbul, Y., Akbayrak, S., Özkar, S.: Group 4 oxides supported Rhodium(0) catalysts in hydrolytic dehydrogenation of ammonia borane. Int. J. Hydrogen Energy **44**, 14164–14174 (2019)

Tsubota, S., Nakamura, T., Tanaka, K., Haruta, M.: Effect of calcination temperature on the catalytic activity of Au colloids mechanically mixed with TiO2 powder for CO oxidation. Catal. Lett. **56**, 131–135 (1998)

Van Der Lee, M.K., Van Jos Dillen, A., Bitter, J.H., De Jong, K.P.: Deposition precipitation for the preparation of carbon nanofiber supported nickel catalysts. J. Am. Chem. Soc. **127**, 13573–13582 (2005)

Wang, L., et al.: Monodispersed core-shell Fe3O4@Au nanoparticles. J. Phys. Chem. B **109**, 21593–21601 (2005)

Wang, C., Ren, Y., Zhao, J., Sun, S., Du, X., Wang, M., Ma, G., Yu, H., Li, L., Yu, X., Zhang, X., Lu, Z., Yang, X.: Oxygen vacancy-attired dual-active-sites Cu/$Cu_{0.76}Co_{2.24}O_4$ drives electron transfer for efficient ammonia borane dehydrogenation. Appl. Catal. B Environ. **314**, 121494 (2022). https://doi.org/10.1016/j.apcatb.2022.121494

Xin, H., et al.: Overturning CO2 hydrogenation selectivity with high activity via reaction-induced strong metal-support interactions. J. Am. Chem. Soc. **144**, 4874–4882 (2022)

Xu, Y., et al.: Pd-catalyzed instant hydrogenation of TiO2 with enhanced photocatalytic performance. Energy Environ. Sci. **9**, 2410–2417 (2016)

Yan, Z., et al.: Metal–metal oxide catalytic interface formation and structural evolution: a discovery of strong metal-support bonding, ordered intermetallics, and single atoms. Nano Lett. **22**, 8122–8129 (2022)

Yoon, B., et al.: Charging effects on bonding and catalyzed oxidation of CO on Au8 clusters on MgO. Science (80-.) **307**, 403–407 (2005)

Yuan, M., Zhong, S., Li, G., Fan, G., Yu, X.: Architecture of urchin-like TiO2 integrated ultrasmall Rh nanoparticles with oxygen vacancy-reinforced electronic metal-support interaction for boosting hydrogen production from ammonia borane hydrolysis. Fuel **358**, 130200 (2024)

Zanella, R., Giorgio, S., Henry, C.R., Louis, C.: Alternative methods for the preparation of gold nanoparticles supported on TiO2. J. Phys. Chem. B **106**, 7634–7642 (2002)

Zanella, R., Delannoy, L., Louis, C.: Mechanism of deposition of gold precursors onto TiO2 during the preparation by cation adsorption and deposition-precipitation with NaOH and urea. Appl. Catal. A Gen. **291**, 62–72 (2005)

Zeng, H., Li, J., Wang, Z.L., J. P. Liu, A., Sun, S.: Magnetic properties of Fe3O4 nanoparticles. **40**, 3015–3017 (2004)

Zhang, C., Oliaee, S.N., Hwang, S.Y., Kong, X., Peng, Z.: A generic wet impregnation method for preparing substrate-supported platinum group metal and alloy nanoparticles with controlled particle morphology. Nano Lett. **16**, 164–169 (2016)

Zhang, J., et al.: Efficient hydrogen production from ammonia borane hydrolysis catalyzed by TiO2-supported RuCo catalysts. Int. J. Hydrogen Energy **46**, 3964–3973 (2021)

Zheng, N., Stucky, G.D.: A general synthetic strategy for oxide-supported metal nanoparticle catalysts. J. Am. Chem. Soc. **128**, 14278–14280 (2006)

Zhou, S., et al.: Metal-support synergistic catalysis in Pt/MoO_{3-x} Nanorods toward Ammonia Borane hydrolysis with efficient hydrogen generation. ACS Appl. Mater. Interfaces **14**, 5275–5286 (2022)

Zhu, X., Cho, H.R., Pasupong, M., Regalbuto, J.R.: Charge-enhanced dry impregnation: a simple way to improve the preparation of supported metal catalysts. ACS Catal. **3**, 625–630 (2013)

Zhuang, L., et al.: Tuning oxygen vacancies in two-dimensional iron-cobalt oxide nanosheets through hydrogenation for enhanced oxygen evolution activity. Nano Res. **11**, 3509–3518 (2018)

Zou, S., et al.: Grafting nanometer metal/oxide interface towards enhanced low-temperature acetylene semi-hydrogenation. Nat. Commun. **12**, 1–11 (2021)

Chapter 16
Group 10 Molecular Materials as Efficient Catalysts for Advanced Coupling Reaction

Atharva Yeshwant Kulkarni, Saili Vikram Lokhande, Suraj Peerappa Yadav, Gourab Karmakar, Adish Tyagi, and Rohit Singh Chauhan

16.1 Introduction

Cross-coupling reactions provide the reliable platform in modern synthetic chemistry. These important pathways generally involve an organometallic reagent and sp^2-hybridized electrophile. The presence of metal catalysts, especially groups 8–10 transition metal catalysts is indispensable for these reactions, which essentially aim to achieve a C–C, C–H, C–N, C–O, C–S, C–P or C–M bond formation in a variety of organic molecules. Since the inception in the early 1970s, eminent scientist like Kochi, Kumada, Murahashi and Corriu have developed organotin, organoboron, organozinc and organosilicon organometallic reagents respectively as effective catalysts for cross-coupling reactions. Various types of electrophiles and metal complexes with numerous combinations have been effectively used in these reactions, resulting in a broad range of synthetic approaches for molecular assemblies

Atharva Yeshwant Kulkarni and Saili Vikram Lokhande—These authors contributed equally.

A. Y. Kulkarni · S. V. Lokhande · S. P. Yadav · R. S. Chauhan (✉)
Department of Chemistry, K. J. Somaiya College of Science and Commerce, Vidyavihar, Mumbai, India
e-mail: rohit.chauhan@somaiya.edu

G. Karmakar (✉) · A. Tyagi (✉)
Chemistry Division, Bhabha Atomic Research Centre, Mumbai, India
e-mail: gourabk@barc.gov.in

A. Tyagi
e-mail: tyagia@barc.gov.in

Homi Bhabha National Institute, Mumbai, India

R. S. Ningthoujam and A. K. Tyagi (eds.), *Handbook of Materials Science, Volume 3*, Indian Institute of Metals Series, https://doi.org/10.1007/978-981-96-8229-4_16

(Nolan and Navarro 2007). Cross-coupling reactions are indispensable in pharmaceutical, natural product and agro-based industries with the need to access specific complexes or target molecules at more efficient and low costs. For instance, the Suzuki–Miyaura cross-coupling reaction, discovered in 1981, has the most vital role in the commercial synesis of the blood pressure drug losartan and the list is not exhaustive. This dictates the supremacy of cross-coupling reactions and, in turn, the corresponding catalysts in organic synthesis.

Among the cross-coupling reactions, Suzuki, Heck and Sonogashira coupling reactions have gained rapid development in the last few decades. It has been considered a boon to the organic chemistry. Suzuki cross-coupling reaction has attracted a notable interest among the researchers because it is one of most preferred methods to constructing C–C bonds. It provides the general methodology by coupling carbon–carbon single bond in the presence of organoborane derivatives, base, aryl/alkyl/vinyl halides (Prashad 2004; Buskes and Blanco 2020), whereas Sonogashira coupling is widely used for the synthesis of organic compounds where C–C bond formation takes place between sp-hybridized carbon of terminal alkyne and sp^2 hybrid carbon centre of aryl halides in the presence or absence of copper as a co-catalyst and a base (Wang and Gao 2014; Tabassum et al. 2022). The Heck (also called as the Mizoroki–Heck) reaction has evolved as one of the most prominent methods for C–C bond formations. It is used to construct carbon–carbon double bonds by connecting unsaturated aryl, vinyl halides with activated olefins. It is important to mention here that all these important coupling reactions demand the presence of catalysts. In recent times, various organometallic and metal–organic complexes have been utilized as catalysts for these reactions, among which group 10 (Ni, Pd)-based complexes are found to be the major contributors (Devendar et al. 2018). In this chapter, we will focus on the developments in C–C bond formation by cross-coupling reactions and the related development of new and more efficient Ni- and Pd-bearing catalysts. Each section is associated with rigorous synthetic methods of the catalysts also.

16.2 Nickel-Based Catalysts

16.2.1 Nickel Thiosemicarbazone Complexes as Catalysts for Cross-Coupling Reactions

Hard acid behaviour of nickel metal resulted in poor tendency to make stronger bond with softer donor sites, i.e., (Se, Te). Despite poor solubility, unstable behaviour and longer reaction time, researchers have made several efforts to synthesize various nickel chalcogenolate complexes.

In this series, derivatives of thiosemicarbazone (TSC) ligand system have been exploited significantly (Scheme 16.1). The reaction of TSCs with $[Ni(H_2O)_6](ClO_4)_2$ in the presence of a base afforded binuclear complexes **1** and **2** of composition $[\{Ni(L)_2\}(N\text{–}N)]$ (L = thiosemicarbazone of salicylaldehyde,

2-hydroxyacetophenone and 2-hydroxynaphthaldehyde (N–N = 2,2′-bipyridine), whereas on performing similar reactions by varying ancillary ligand as 2,2′:6′2′-terpyridine the mononuclear complexes [Ni(L)(N–N)] (N–N = 2,2′:6′2′-terpyridine) (**3**) have been obtained. In the binuclear complexes, nickel pair has been stabilized in two different geometries. One nickel centre is coordinated octahedrally while second nickel adopts square planar geometry (Datta et al. 2011). The complexes of type $[Ni(L)_2]$ having the same ligand moieties were isolated and further reacted with triphenyl phosphine (PPh_3), pyridine (py) and 4,4′-bipyridine (bpy) to obtain mononuclear $[Ni(L)(PPh_3)]$, [Ni(L)(py)] and binuclear $[\{Ni(L)\}_2(bpy)]$ complexes respectively (**4**) (Datta et al. 2012). Another prominent auxiliary phosphine ligand system supported through TSCs nickel complexes $[Ni(L)(PPh_3)]$ (L = 4-diethylamino-salicylaldehyde thiosemicarbazone) has also been explored by reacting $[NiCl_2(PPh_3)_2]$ with 4-N-substituted thiosemicarbazide under refluxing conditions (**5**,**6**) (Priyarega et al. 2011). Furthermore, Anitha and co-workers performed the substitution reaction between $[NiCl_2(PPh_3)_2]$ and N(4)-substituted thiosemicarbazone ligands to isolate four coordinated complexes, but an unexpected octahedral chelate **7** having formula $[Ni(L)_2]$ (L = 9,10-phenanthrenequinone thiosemicarbazone and its derivatives) was obtained (Anitha et al. 2015a). An easily accessible Nickel catalyst **8** derived from naphthalene- and thiophene-substituted TSC was also synthesized by using $Ni(OAc)_2 \cdot 4H_2O$ (Prabhu and Ramesh 2016; Suganthy et al. 2013). All these complexes exhibit nice catalytic performance towards a variety of cross-coupling reactions.

16.2.2 Nickel Pincer Complexes as Catalysts for Cross-Coupling Reactions

Another important class of ligand system is tridentate pincer ligands, which are basically chelating agents that bind tightly to three adjacent coplanar sites of a metal complex.

A series of SPS-Ni(II) pincer complexes (**9**) having formula *trans*-$[NiCl_2(P(C_6H_4\text{-}4\text{-}R)_3)_2]$ (R = H, Cl, F, CH_3, OCH_3, CF_3) (Scheme 16.2) was synthesized by the reaction of $[NiCl_2(PAr_3)_2]$ with $phPS_2H_2$ in the presence of triethylamine in dichloromethane. These complexes exhibited high catalytic activity in C–S couplings of alky- and aryl-disulphides with iodobenzene for the production of non-symmetric sulphides (Gómez-Benítez et al. 2018). Molina and co-workers synthesized SNS novel pincer mononuclear complexes $[NiCl_2\{C_5H_3N\text{-}2,6\text{-}(CH_2SR)_2\}]$ (R = ^tBu, ^iBu, ^sBu) (**10a**) and $[NiCl_2\{C_5H_3N\text{-}2,6\text{-}(CH_2S^tBu)_2\}\text{-}(^iPrOH)]$ (**10a**) (Fig. 16.1). However their isopropyl derivative resulted in binuclear $[NiCl\{C_5H_3N\text{-}2,6\text{-compounds }(CH_2S^iBu)_2\}\mu\text{-}Cl]_2$ in which chloro group acts as bridging nature and thiolate group occupies terminal position. Under heating and microwave irradiation conditions, these compounds appeared to be fine catalysts for C–C and C–S coupling reactions (Scheme 16.3) (Basauri-Molina et al. 2014).

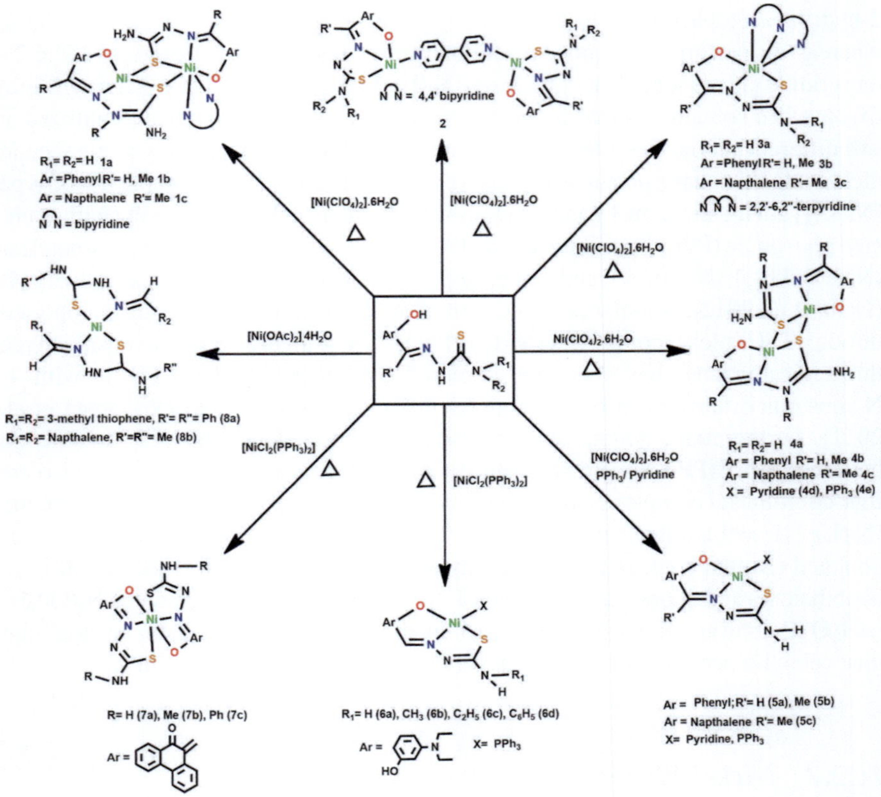

Scheme 16.1 Synthesis of various nickel thiosemicarbazone complexes

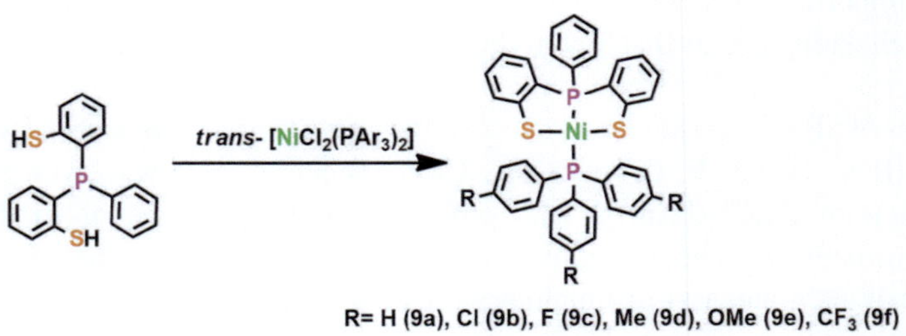

Scheme 16.2 Synthesis of SPS nickel pincer complexes

A bidentate ligand 2-(diphenylphosphanyl) benzenechalcogenol had also been explored to study their reactivity towards [$NiCl_2(PMe_3)_2$] (Scheme 16.4). In this case, available phosphorous as well as chalcogen donor atoms get co-coordinated to metal centre to stabilize square planar geometry. Both these compounds (**11a** and **11b**) have shown efficient activity in the Sonogashira coupling reaction. The sulphur complexes have exhibited superior catalytic activity in comparison to selenium counterpart under

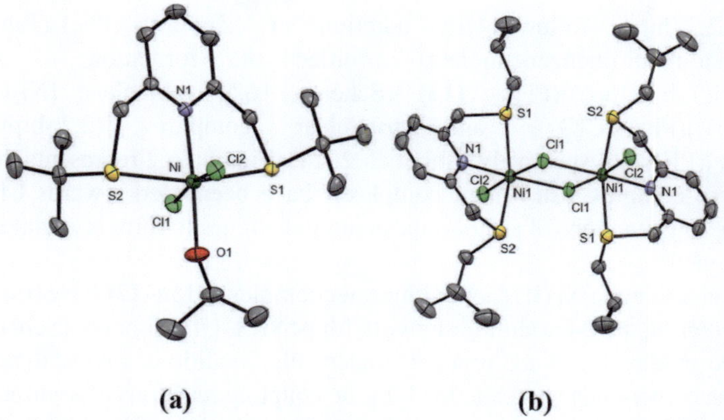

Fig. 16.1 Crystal structure of mononuclear [$NiCl_2\{C_5H_3N$-2,6-$(CH_2S^tBu)_2\}$-(^iPrOH)] (**10a**) and binuclear [$NiCl\{C_5H_3N$-2,6-$(CH_2S^iBu)_2\}\mu$-Cl]$_2$ (**10b**). The crystals are redrawn from CCDC No. 1014500 and 1014501

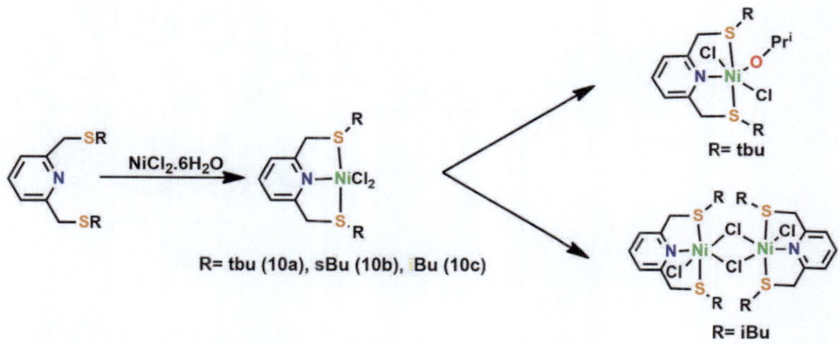

Scheme 16.3 Synthesis scheme of SNS nickel complexes

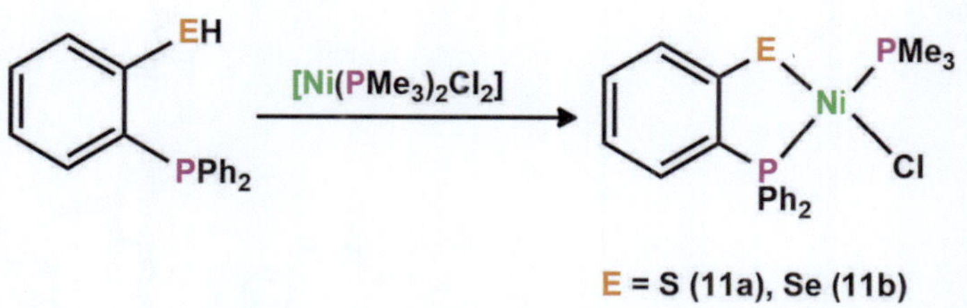

Scheme 16.4 Reaction of 2-(diphenylphosphanyl) benzenechalcogenol with [$NiCl_2(PMe_3)_2$]

mild conditions. The reaction of phenylacetylene with 1-chlorooctane at 50 °C for 12 h was catalysed by catalyst **11** to form octylphenylacetylene (yield = 91%) (Fan et al. 2021).

Transmetallation is another important strategy to lead metal chalcogenolate complexes. Reactions of [$Ni(H_2O)_6]Cl_2$/[$Ni(H_2O)_6](ClO_4)_2$ with (dppe), (dppb), 2,2′-bipy, followed by addition of $[Zn(Tab)_4](PF_6)_2$(Tab = 4-(trimethylammonio)benzenethiolate) afforded the formation of mononuclear $[Ni(Tab)_2(dppe)](PF_6)_2$ (**12**) (Scheme 16.5), binuclear $[Ni_2(dppb)(\mu\text{-}Tab)_2(Tab)_2](PF_6)_2(ClO_4)_2$ and trinuclear complex $[(2,2'\text{-bipy})_4Ni_3(\mu\text{-}Tab)_4]Cl_{0.5}(PF_6)_{5.5}$ respectively (Shi et al. 2014). Figure 16.2 represents the crystal structure of complex **12a**. These complexes have been used towards Chan–Lam cross-coupling reactions of arylboronic acids and amines to form N-arylated amines with high yield.

Novel heteroleptic Ni (II) β-oxodithioester complexes **13a–13f** have been synthesized by treating the β-oxodithioester with $[dppeNiCl_2]$ in methanol (Scheme 16.6). In order to obtain N-arylated products under mild conditions, the catalytic activity of all these compounds was examined for the coupling reactions of arylboronic acid and amines. They exhibit decent catalytic activity with yield of final product varying in the range 65–85% (Kumari et al. 2020).

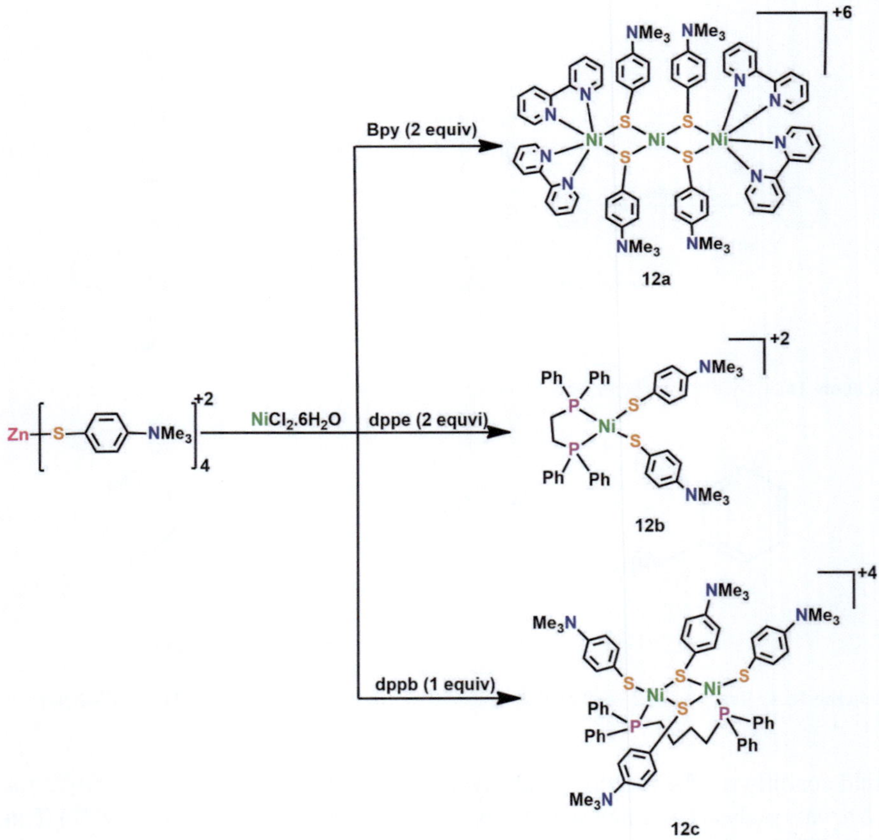

Scheme 16.5 Synthesis of nickel thiolate complexes derived from P and N donors

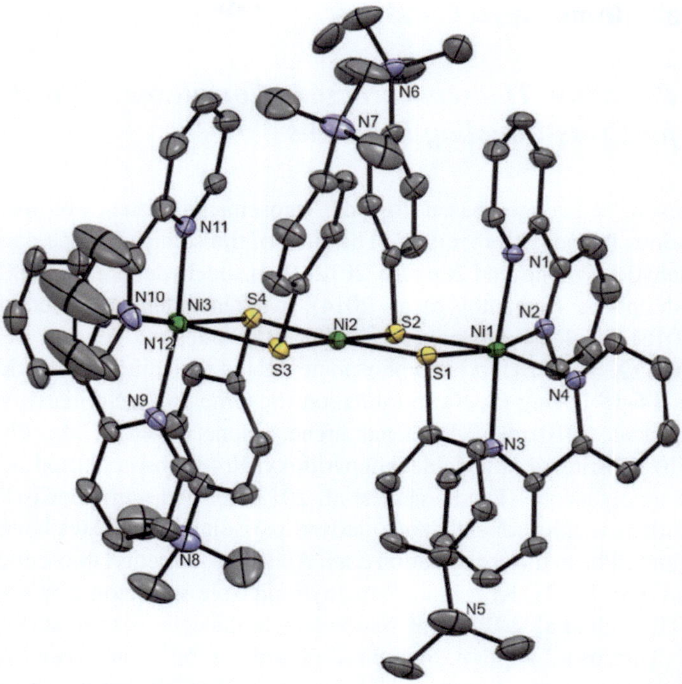

Fig. 16.2 Crystal structure of $[Ni_3(\mu\text{-Tab})_4(2,2'\text{-bipy})_4]Cl_{0.5}(PF_6)_{5.5}$. The crystal is redrawn from CCDC No. 983485

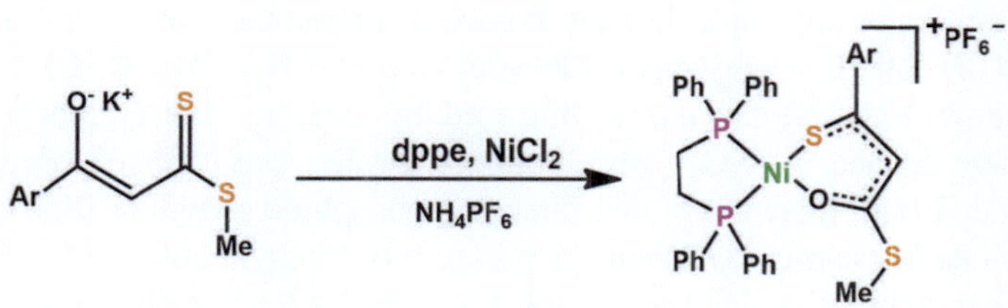

13a-13f

Scheme 16.6 Synthesis of nickel complexes derived from β-oxodithioesters

16.3 Palladium-Based Catalysts

16.3.1 Palladium Thiosemicarbazone Complexes as Catalysts for Cross-Coupling Reactions

Among several chalcogen-based ligands, thiosemicarbazones are well studied against various Palladium precursors. The class of thiosemicarbazone derived from terepthaldehyde (Prabhu and Ramesh 2012), benzaldehyde (Paul et al. 2011), 3-methyl-2-thiophene (Suganthy et al. 2014), 1-pyrenealdehyde-4-methyl (Prabhu and Pal 2015) and 1-napthaldehyde (Prabhu and Ramesh 2017) was treated with $PdX_2(PPh_3)_2$ (X = Cl, Br) to yield phosphine ligated palladium thiosemicarbazone complexes **14–19** having (S, N) coordination (Scheme 16.7). Similarly, salicyladimato (Xie et al. 2010), 9,10-R-phenanthrenequinone (R = H, Me, Ph) (Anitha et al. 2015b), substituted salicylaldehydepyridoxyl N(4)-substituted and sulphonated monothiosemicarbazone (Pandiarajan et al. 2013) reacted with $[PdCl_2(PPh_3)_2]$ to result in tridentate thiosemicarbazone-derived palladium complexes having (O, N, S) donors and PPh_3 in the coordination core. Anisaldin-4-methyl thiosemicarbazone was treated with $PdCl_2$, KPF_6 and N-methylimidazole in acetonitrile under reflux condition (Baruah et al. 2017). 4-R-benzaldehyde thiosemicarbazone (R = OCH_3, NO_2) with 1-nitroso-2-napthol, quinolin-8-ol and Na_2PdCl_4 in ethanol resulted in mixed ligand coordinated complexes (Dutta et al. 2012). The intriguing example β-D-glucopyranosynyl-derived thiosemicarbazone was also applied to stabilize binuclear chloro [4-tert-butyl-benzaldehyde 4-(β-D-glucopyranosyl) thiosemicarbazone] palladium(II) dimer (Tenchiu et al. 2015).

Other similar complexes, having immense importance in catalysis include [(dppe)Pd(LH-R)] [L = (4-R-benzaldehyde; R = (OCH_3, CH_3, H, Cl and NO_2)] (**20**) (Scheme 16.8) which was synthesized by treating $[PdCl_2(dppe)]$ with 4-R-benzaldehyde and $AgNO_3$, were refluxed in the presence of triethylamine (Thapa et al. 2019). Reaction of non-chelating phosphine precursor $[PdCl_2(PPh_3)]$, benzophenone thiosemicarbazone in the presence of triethylamine as a base afforded mononuclear complex which was explored as catalyst for coupling reactions. Paul and his co-workers refluxed benzophenone thiosemicarbazone, triphenyl phosphine and triethylamine with $[PdCl_2(PPh_3)_2]$ in ethanol to afford the complex **21** (Paul et al. 2015; Dutta and Bhattacharya 2013). 1,1′-Bis (diphenylphosphino) ferrocene-bridged palladium complex **22** was synthesized by Yan using 3,4-dichloroacetophenonethiosemicarbazone with K_2PdCl_4 and dppf in ethanol–water mixtures (Yan et al. 2013). High-nuclearity tetranuclear palladium TSC complexes **23a–23c** have also been isolated by treatment of 3-acetyl-7-methoxy-2H-chromen-2-one TSC with potassium tetrachloropalladate in MeOH-DCM mixtures. The tetranuclear palladium complex comprises four monomer fragments which are interlinked through Pd–S bond formation between diagonal/alternate fragments. All the palladium atoms occupied four corners of the square. The palladium–palladium bond distance of Pd···Pd neighbour was 3.186 Å (Dharani et al. n.d.).

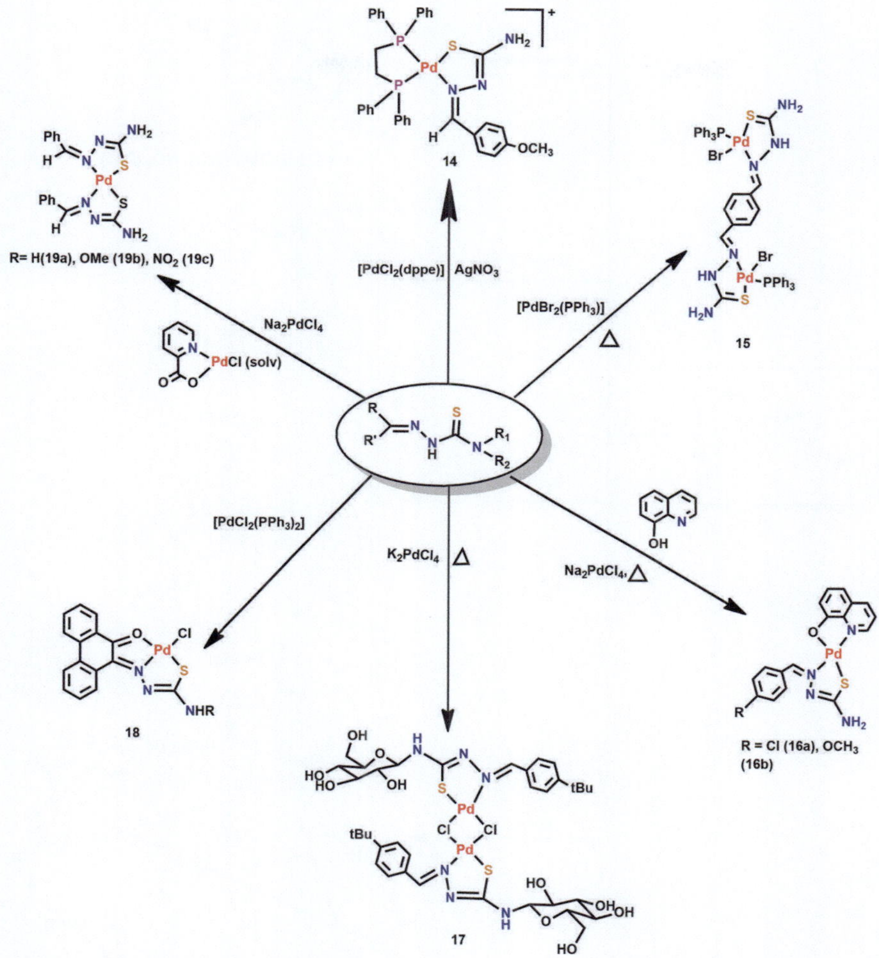

Scheme 16.7 Synthesis of palladium complexes derived from thiosemicarbazone

16.3.2 Palladium Pincer Complexes as Catalysts for Cross-Coupling Reactions

Palladium pincer complexes incorporating substituted 2,6-pyridinedicarbothioate and 2-pyridinyl methyl thiolate were synthesized, displaying (S, N, S) and (N, S) coordination modes, respectively (Gabriel Flores-Rojas et al. 2020; Fiebor et al. 2018; López-Saucedo et al. 2018). Additionally, a series of multidentate ligands derived from functionalized carboxamides was investigated for their coordination behaviour with [$PdCl_2(PhCN)_2$] in chlorinated solvents. The resulting Pd(II) pincer complexes (**24–27**) were stabilized by metallocyclic rings featuring N, N, S and S, N, S donor atoms, which played a crucial role in their catalytic performance, as

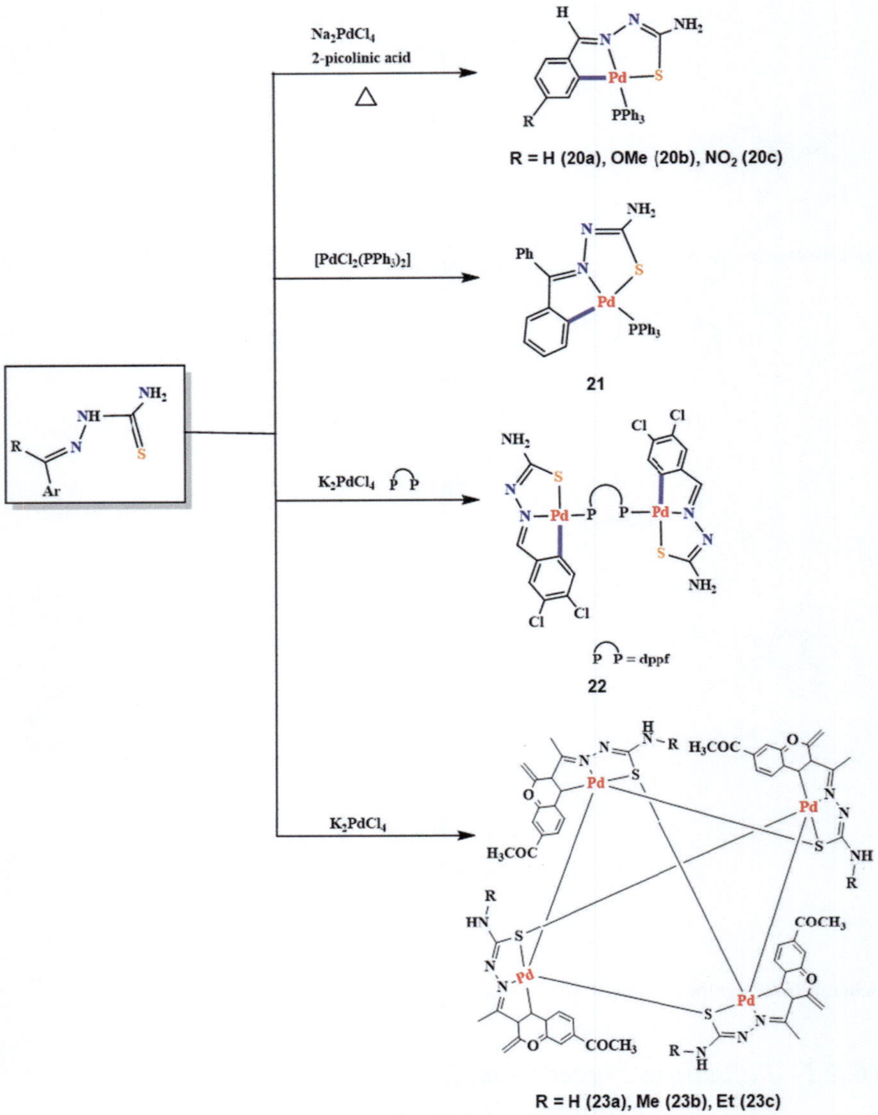

Scheme 16.8 Synthesis of palladacycles derived from thiosemicarbazones

illustrated in Schemes 16.9 and 16.10 (Churusova et al. 2018). Furthermore, unsymmetrical pincer complexes (**28**) incorporating an 8-hydroxyquinoline (oxime) framework, such as 2-(N,N-dimethylthiocarbamoyl)quinolin-8-ol and 2-(pyrrolidin-1-ylthiocarbamoyl)quinolin-8-ol, were examined for their catalytic efficiency in cross-coupling reactions (Kumar et al. 2017). In a separate study, the reaction of phosphine-based ligands with palladium salts led to the formation of complexes with the general formulas $[PdCl_2\{PPh(o\text{-}C_6H_4SMe)_2\}]$, $[Pd(Me)Cl\{PPh(o\text{-}C_6H_4SMe)_2\}]$

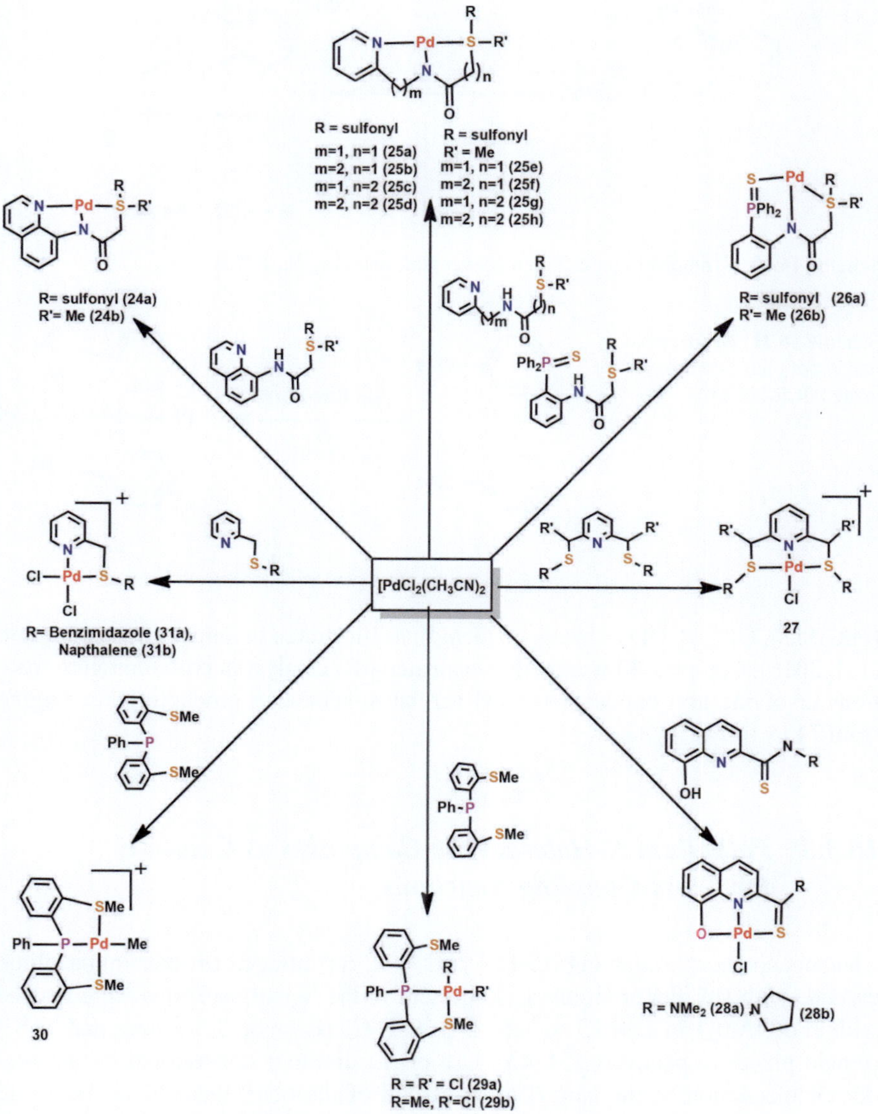

Scheme 16.9 Reaction scheme depicting the synthesis of palladium complexes with various pincer ligands

and [Pd(R){PPh(o-$C_6H_4SMe)_2$}]X, where X = OTf or PF_6 and R = Cl or Me. These complexes (29–32) exhibited S, S, P and S, P coordination modes (Luquin et al. 2016).

A unique asymmetrical pincer ligand, PPh(o-$C_6H_4SMe)_2$, exhibiting an (S, P) coordination mode, was investigated for the synthesis of a distinct Pd(II) complex (**33**). This complex was formed through an addition reaction between the ligand and

E = S (32a), Se (32b)

Scheme 16.10 Synthesis of palladium pincer complexes using Na_2PdCl_4

Scheme 16.11 Synthesis of palladium pincer complex using PdClMe(COD)

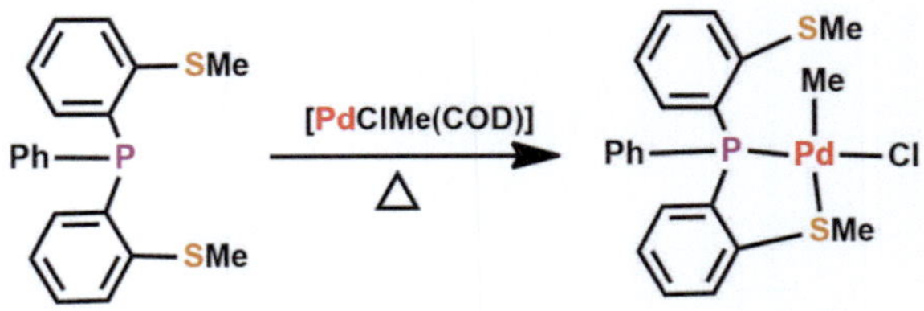

33

[PdClMe(COD)] (COD = cyclooctadienyl), as illustrated in Scheme 16.11 (Luquin et al. 2016). Complex **33** was further evaluated as a catalyst in cross-coupling reactions involving aryl halides and alkyl acrylates, achieving product yields ranging from 74 to 100%.

16.3.3 Palladium N-Heterocyclic Complexes as Catalysts for Cross-Coupling Reactions

Chalcogenoethers containing (C–E–C) (E = S, Se) bridged bis-benzimidazolium carbene with soft donor atoms S, Se present in the ligand backbone were reacted with $[Pd(OAc)_2]$ in DMSO to afford (κ^3C-E-C) (E = S, Se) pincer and (κ^2C) pseudo pincer respectively (**34**–**43**). Use of the dinitrate salt resulted in unprecedented metalation of the benzyl ortho-carbon of the benzimidazolium moiety to form a (C–C–Se–C) pincer complex (Prashanth et al. 2017). N-alkyl-N′-(2-ethyl-1-selenophenyl) containing imidazolium salts and varying alkyl chain length (N = 0, 7, 17) was treated with $[PdCl_2(CH_3CN)_2]$ in acetonitrile at room temperature for 8 h using silver carbene transfer reaction that afforded robust palladium NHC complexes at higher yields. It was found that the complex having a higher alkyl chain length was a more efficient catalyst than 1 and 2 for the Suzuki coupling reaction. Nanoparticles (size ~2–5 nm) formed in the reaction play a role in the catalysis (Sharma et al. 2013). Chalcogenoethers containing benzimidazolium NHC and amido functionalization were treated with palladium salts in DMF at 80 °C that yielded (N, C_{NHC}, S) (Sharma et al. 2019) and (C_{NHC}, Se) (Nayan Sharma et al. 2018) type pincer complexes (Scheme 16.12). P. Dubey et al. prepared Pd-(N, C_{NHC}, E) complexes by reacting (1-benzyl-1H-1,2,3-triazol-4-yl)methyl)-1-(2-phenylthio/seleno)ethyl)-1H-benzo[d]imidazol-3-ium hexafluorophosphate and 4-methyl-1-(2-(phenylthio/seleno)ethyl)-1H-1,2,4-triazol-4-iumiodide with $Pd(OAc)_2$ in acetonitrile for 8 h at 50 °C in the presence of Ag_2O (Dubey and Singh 2020). $[Pd(OAc)_2]$, when reacted with functionalized benzimidazolium salt under refluxing conditions, afforded binuclear NHC species, which on treatment with further excess of Me_3OBF_4 and $AgBF_4$ in dichloromethane resulted in tetranuclear palladium NHC. Similarly, bi and tetra nuclear complexes had been obtained from imidazolium-thioesters salt with $Pd(OAc)_2$ in DMSO at 80 °C. The bulky σ-donor ligand 1,3-bis(2,4,6-trimethylphenyl)imidazol-2-ylidene was treated with 0.5 equiv. of the Pd dimers in toluene at room temperature to yield the corresponding (C_{HNC}, S, N) products at higher yields (Holzer et al. 2018). In a contemporary study, Wang et al. have treated trans-$Pd(IMe)_2Cl_2$ with flexible zwitterionic thiolate $[Zn(Tab)_4]^{2+}$ (Tab = 4-(trimethylammonio)benzenethiolate) to give tetranuclear complex **39**: $[Pd_4Ag_4(IMe)_8(Tab)_{10}](PF_6)_{12}$ as highly efficient catalyst for Suzuki–Miyaura couplings in water (Wang et al. 2017; Dubey et al. 2017).

The reaction of meta-phenylene-bis(1-methyl-1H-imidazole-2(3H)-selone) with $[Pd_2(\mu\text{-}Cl)_2(2\text{-}C_6H_4CH_2NMe_2)_2]$ in benzene and glacial acetic acid resulted in the formation of an unsymmetrical five- and six-membered Pd(II) pincer complex **44**. The reaction proceeds through C–H bond activation which pushes out one selenium (Scheme 16.13). The Pd(II) complex is simultaneously coordinated to a selone and an N-heterocyclic carbene in a tridentate fashion acquiring distorted square planar geometry. The complex was applied as a fine catalyst for coupling of iodobenzene with acrylates and styrene (yield = ~95%) (Ghavale et al. 2015).

16.3.4 Palladacycles as Catalysts for Cross-Coupling Reactions

Palladacycles are among the most extensively studied complexes across various classes of coordination compounds. Various chalcogen-based ligand systems have been investigated in this context. A representative example involves Benzyl-[1-(2-phenylsulfanyl/selanylethyl)-1H-indol-3-ylmethylene]amine, which reacts with Na_2PdCl_4 to form complexes featuring tridentate coordination in an (N, C, E) mode. This coordination pattern promotes the stabilization of both five- and six-membered chelate rings around the central palladium centre (Singh et al. 2016). The iminophosphorane-based ligand presents another fascinating example, leading to the formation of palladacycles with the composition $[PdCl\{C_6H_4(Ph_2P = NC_6H_4S\text{-}R\text{-}\kappa^3\text{-}C,N,S)\}]$ (R = Me, Ph–NO_2, Bn). In these complexes, coordination occurs through the imine moiety, following an (N, C, S) tridentate binding mode (Ramírez-Rave et al. 2017). Various sulphur-functionalized palladacycles based on 1,1′-bis(diphenylphosphino)ferrocene have been synthesized through the

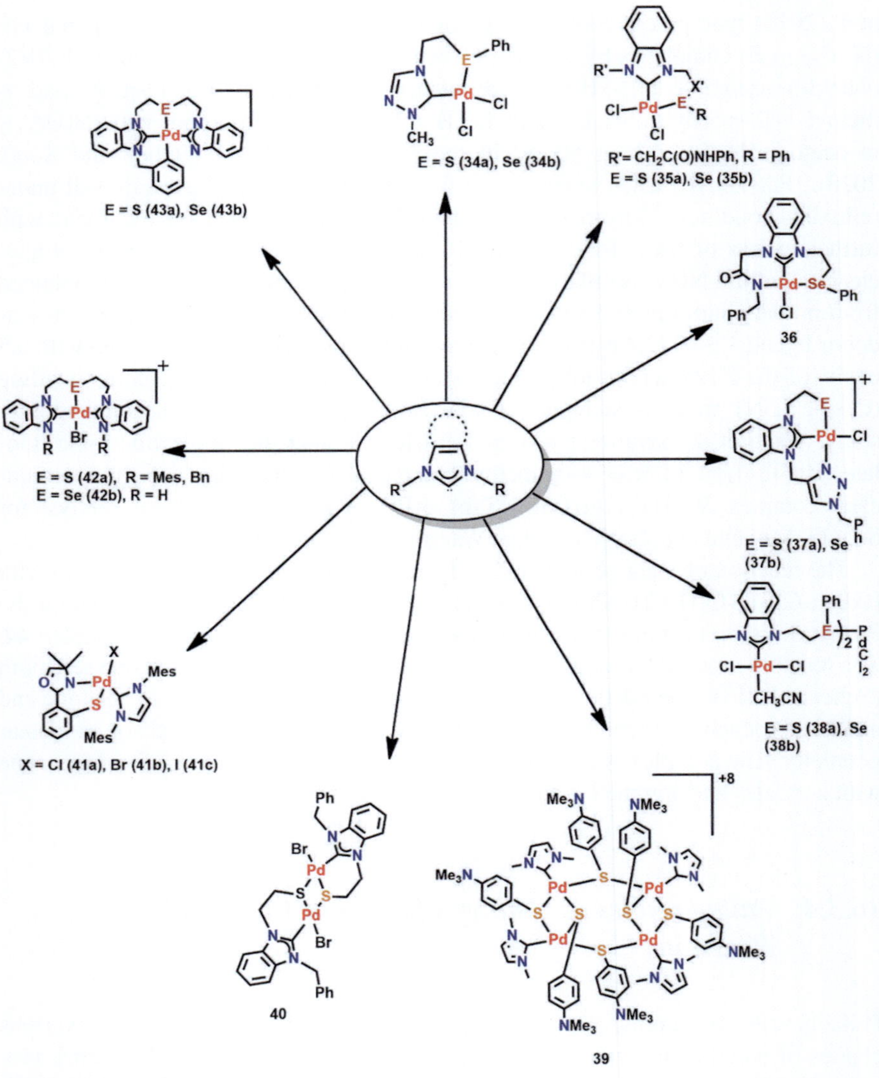

Scheme 16.12 Synthesis of chalcogen containing palladium N-heterocyclic complexes bearing imidazole and benzimidazole moieties

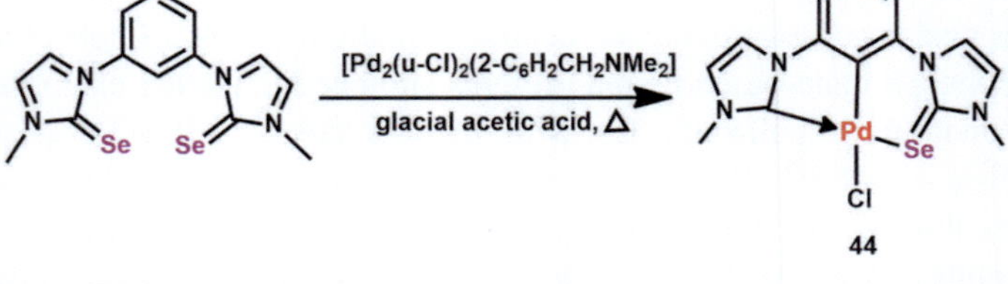

Scheme 16.13 Preparation of palladium NHC complex using binuclear palladium complex

reaction of O-ethyl ferrocenylthionoester with Na_2PdCl_4 and phosphine ligands (PR_3, where R = Ph, O–Tol, $SbPh_3$) (Balam-Villarreal et al. 2016). Additionally, high-nuclearity palladacycles were obtained by reacting 2,3-bis[(phenylthio/seleno)methyl]quinoxaline with Na_2PdCl_4 in an acetone–water mixture. Their crystal structures revealed the unusual stabilization of a seven-membered ring, facilitated by enhanced secondary interactions, particularly Cl···H–C hydrogen bonds, leading to the formation of helical arrays (Saleem et al. 2013). Further, palladacyclic complexes incorporating chalcogen-ligated ferrocene carboxaldehyde amine were prepared by its reaction with Na_2PdCl_4 in the presence of sodium acetate, resulting in phosphine-supported palladacycles (Sharma et al. 2017). Likewise, hydroxy-derived chalcogen ethers of the type 2-OH-4-R-C_6H_3-(C_6H_4)CHNH$(CH_2)_3$EPh (E = S, Se, Te; R = H, OMe) reacted with palladium salts to afford palladacycles featuring a well-defined (C, N, E) coordination core (Rao et al. 2010, 2013, 2014a). Bhaskar et al. reported the synthesis of a palladacyclic complex **(45)** by reacting [N(E)]-N-(1-naphthalenylmethylene)-2-(phenylthio)ethanamine with Li_2PdCl_4 in the presence of sodium acetate, as depicted in Scheme 16.14 (Bhaskar et al. 2017). These complexes demonstrated significant catalytic efficiency in Sonogashira and Suzuki coupling reactions. Interestingly, the catalytic activity was attributed not to the complex itself but to the in situ generation of palladium nanoparticles during the reaction. Under similar conditions, pincer-type palladacycles (46–52) were synthesized, as illustrated in Scheme 16.15 (Bhaskar et al. 2017; Das et al. 2022).

Roffee et al. reported the synthesis of palladacycles by reacting palladium chloride with 2-{3-[(methylsulfanyl)methyl]phenyl}pyridine in the presence of $AgBF_4$ (Roffe et al. 2016). Similarly, Tyson and colleagues developed a palladacycle by treating 1,3-bis(3′-butylimidazolyl-1′-yl)benzene dithione with $PdCl_2(CH_3CN)$ in dichloromethane at room temperature for 16 h. Additionally, a series of related palladium complexes (**53–57**) was synthesized using 1-methyl-3-(2-ethyl-1-(3,5-dimethyl-1H-pyrazol-1-yl)phenylselenyl)-3H-imidazol-1-ium iodide and $PdCl_2$, as illustrated in Scheme 16.16 (Rao et al. 2010). Among these, complex **55** demonstrated outstanding catalytic efficiency in Suzuki reactions, particularly with aryl halides, achieving a turnover frequency (TOF) of 11,880 h^{-1} at 50 °C. Moreover, Suzuki–Miyaura and Heck coupling reactions proceeded more efficiently with compound **56a** than **56b**, which may be attributed to the in situ formation of palladium nanoparticles along with sulphur/selenium-based nanoparticles. Additionally, complexes **57a**

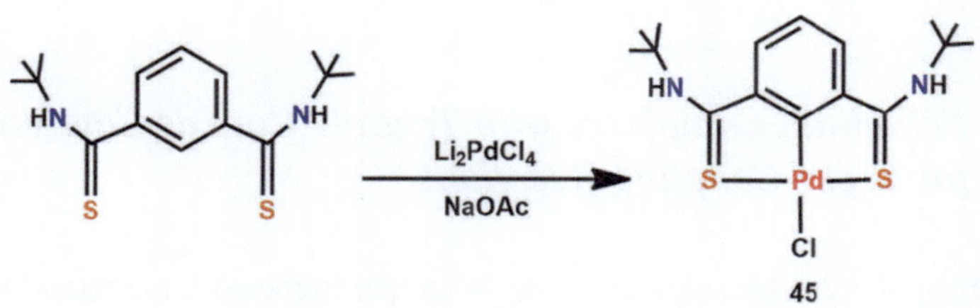

Scheme 16.14 Synthesis of Pd(II) complex of [N(E)]-N-(1-Naphthalenylmethylene)-2-(phenylthio) ethanamine

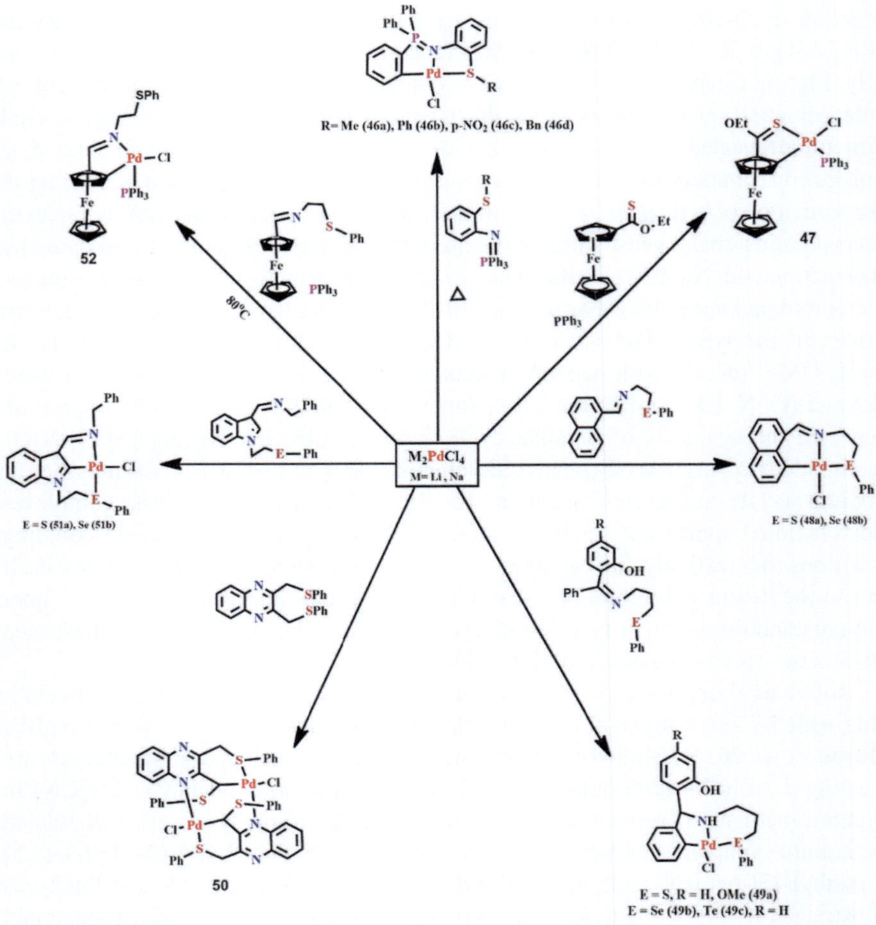

Scheme 16.15 Synthesis of palladacycles using M_2PdCl_4 (M = Li, Na)

and **57b** were explored as catalysts for Suzuki–Miyaura and Sonogashira reactions involving different aryl bromides, with the optimal catalyst loading determined to be between 0.001 and 2 mol% (Tyson et al. 2015; Kumbhare and Dutta 2016).

16.3.5 Palladium Complexes with Triazole Core as Catalysts for Cross-Coupling Reactions

The existence of 1,2,3-triazole core brings a unique framework to render interesting variations with chalcogen units to form palladium complexes. Various palladium

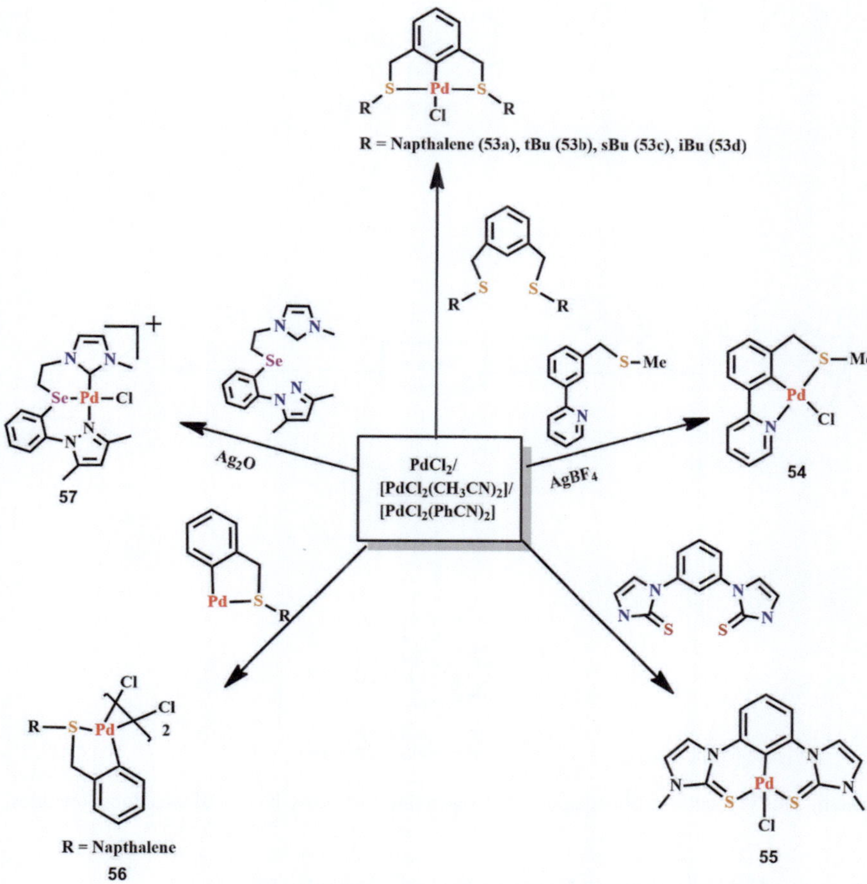

Scheme 16.16 Synthesis of palladacycles using $PdCl_2$ and $PdCl_2$(acetonitrile) complexes

chalcogenolated bearing aryl triazoles (**58–61**) were prepared and found to be competent for Heck and Suzuki coupling with 95000 TON values. It is noteworthy that the complexes of telluroethers are better catalysts selenoether derived complexes (Scheme 16.17) (Das et al. 2010). Similarly, 2-[(Methylthio)methyl]-4-phenyl-2H-1,2,3-triazole (Yan et al. 2019), 1-(Phenylmethyl)-4-[(phenylthio/seleno)methyl]-1H-1,2,3-triazole (Saleem et al. 2014) and 1-[2,6-Bis(1-methylethyl)phenyl]-4-[(phenylthio/seleno)methyl] 1H-1,2,3-triazole (Kumar 2016) on treatment with $[PdCl_2(CH_3CN)_2]$ yielded respective palladium(II) complexes in which coordination takes place through chalcogen (S/Se) and triazolium nitrogen.

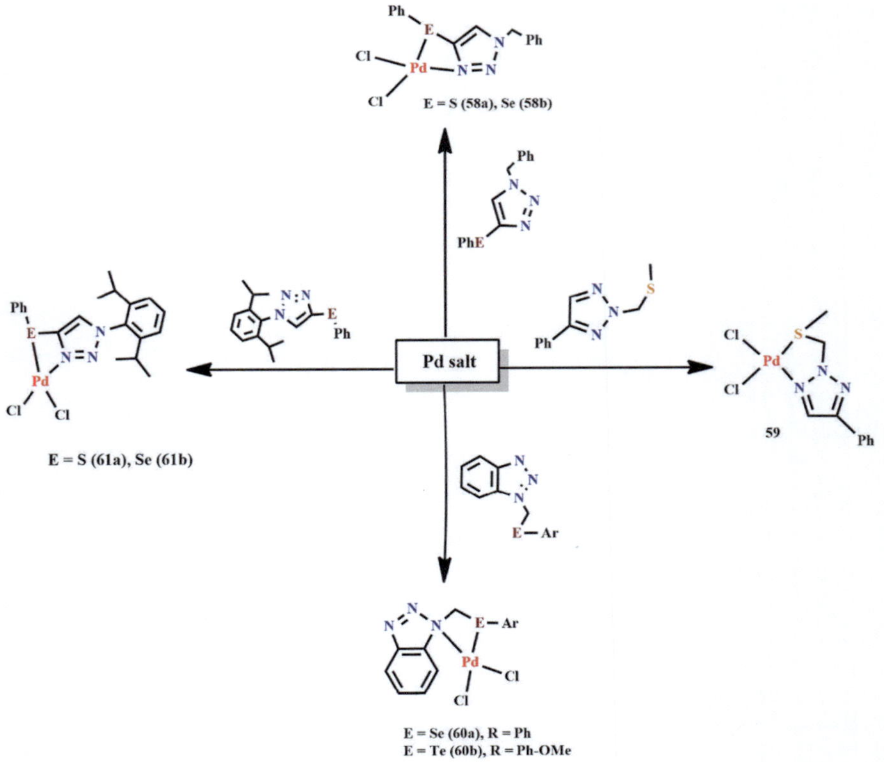

Scheme 16.17 Synthesis of palladium chalcogenolate complexes having triazole functionalities

16.3.6 *Miscellaneous Complexes of Palladium as for Cross-Coupling Reactions*

Apart from the above-mentioned broad classifications of palladium-based catalysts, there are several other complexes of palladium, which utilized diverse ligand systems and offer superior catalytic performance. For instance, Kumar et al. explored the reactivity of 3-Hydroxy-4-{[[3-(phenylseleno)propyl]imino]methyl}phenyl-4-R-benzoate [R = Octadecyloxy, 3,4,5-tris(decyloxy), 4-(decyloxy), 4-methoxy] and 4,6-Bis[1-[[2-[(R)-E]ethyl]imino]ethyl]-1,3-benzenediol (R = 4-methoxyphenyl; E = Te and R = phenyl; E = S, Se) were reacted with M_2PdCl_4 (M = K, Na) in acetone–water mixture (Rao et al. 2012; Kumar et al. 2014a). Chandan and co-workers prepared phosphine-ligated palladium Schiff base complex by applying dichloro-6-{[(2(ethylthio)phenyl)imino]methyl}phenol with Na_2PdCl_4 and PPh_3 in acetonitrile under refluxing condition (Manna et al. 2021). Singh et al. exploited various chalcogen-based Schiff bases, i.e., 2,4-Bis(1,1-dimethylethyl)-6-[[[2-(phenylseleno)ethyl]imino]methyl]phenol,3,5-Bis(1,1-dimethylethyl)-2-(phenylmethoxy)-N-[2-(phenylseleno)ethyl]benzenemethanmine (Kumar et al.

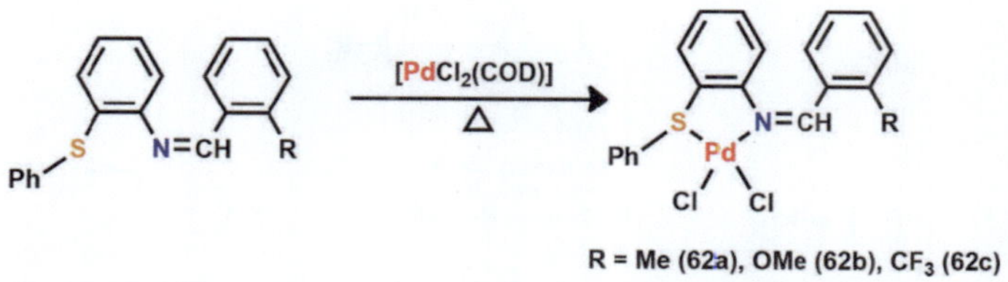

Scheme 16.18 Complexation of $(C_6H_5)SC_6H_4N = CR$ with $[PdCl_2(COD)]$

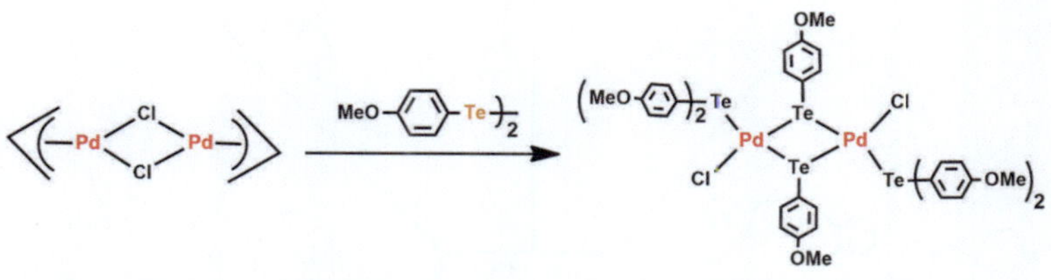

Scheme 16.19 Synthesis of binuclear palladium tellurolate complex

2014b) and N-[2-(Phenylthio)ethyl]-1-naphthalenemethanamine with Na_2PdCl_4 in methanol/acetone–water respectively (Bhaskar et al. 2017).

Schiff bases of 2-(phenylthio)aniline, $(C_6H_5)SC_6H_4N = CR$ [R = $(OCH_3$, CF_3, Ph)] were also studied to explore their coordination with palladium using $[PdCl_2(COD)]$ in DCM under refluxing conditions. The complexes were examined in terms of their potential as Suzuki–Miyaura and Mizoroki–Heck catalysts. The Suzuki–Miyaura reaction at 100 °C in K_2CO_3 media yielded high conversion (Scheme 16.18) (Keleş et al. 2015), while the reaction of bis-(4-methoxyphenyltelluro)methane with allylpalladium(II) chloride resulted in dimericpalladium complex **63** (Scheme 16.19).

Various palladium chalcogenolate complexes (**64–71**) as efficient catalysts were isolated using chalcogenoether ligands of 3-(phenylsulphanyl)propylamine (Singh et al. n.d.), 2-(Phenylseleno)-N-[2-phenylthio/seleno/telluro]ethaneamine (Kumar Rao et al. 2015), N-{2-(aryltelluro)ethyl}morpholine/piperidine) (Singh et al. 2012), [[2-(Phenylseleno)ethyl]thio]benzene,3-(phenylsulphanyl)propylamine,1,1′-thio/Selenobis[docosane] (Kumar et al. 2014b, 2015), bis(phenylsulfanyl) propan-2-ol (Joshi et al. 2014), 2-methylthioaniline (Rao et al. 2014b) and N-(2-bromobenzyl)-2-(phenylthio)ethanamineand $Na_2PdCl_4/PdCl_2(PhCN)_2$ resulting in a mononuclear complex (Scheme 16.20).

Metallo-supramolecular complexes have gained an outgrowing interest in catalysis because of their discrete and thermodynamic stability. This arises through a metal-directed self-assembly process between metal ions and organochalcogenolate ligands. Various thiolate and selenolate-ligated supramolecular complexes were well established. Different aryl-dithiolates/4-pyridyl chalcogenolate were

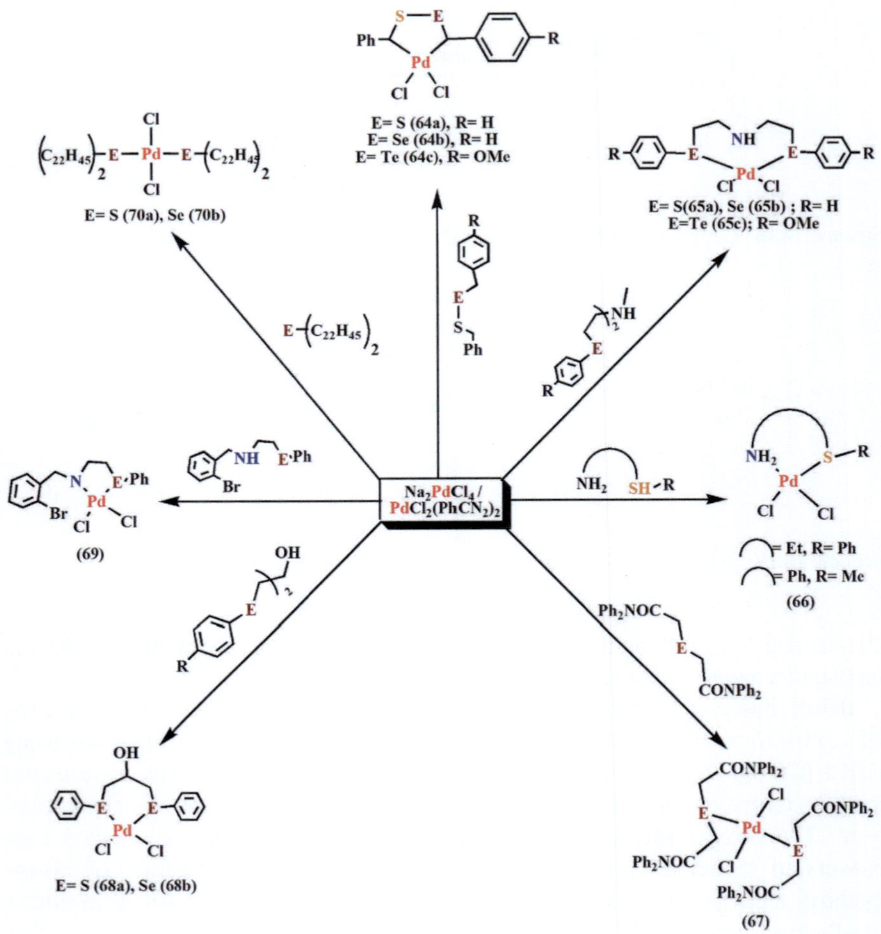

Scheme 16.20 Synthesis of palladium complexes using chalcogenoether ligands

reacted with $[Pd(P\text{–}P)(OTf)_2]$ ((P–P) = dppe, dppf, xantphos) to obtain macromolecules of palladium analogue bridged through aryl chalcogenolate systems (Gaikwad et al. 2020; Mane et al. 2019; Vivekananda et al. 2015) (Scheme 16.21). A different protocol was established by Kumar and co-workers to prepare interesting macrocyclic palladium complexes as catalysts by refluxing 1,8-bis(2-((phenylselanyl)methyl)phenoxy)octane with $[PdCl_2(CH_3CN)]$ in acetonitrile (Scheme 16.22) (Kumar et al. 2021).

Internally functionalized ligands are another interesting class which effectively binds with metal centre, and their palladium complexes are known to act as effective catalysts for a variety of coupling reactions. For example, the ligand $(ECH_2CH_2CH_2NMe_2)_2$ (E = S, Se) on reacting with Na_2PdCl_4 yields bi and trinuclear cyclic palladium complexes. The trimeric structures comprise three palladium atoms bridged through chalcogen atoms of the chelating chalcogenolate ligand

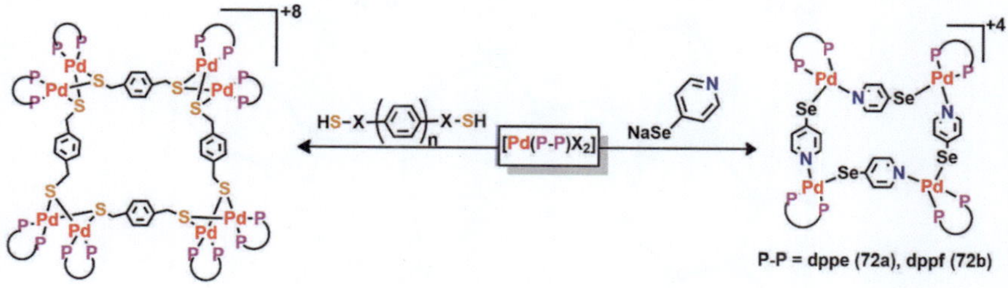

Scheme 16.21 Synthesis of supramolecular palladium complexes

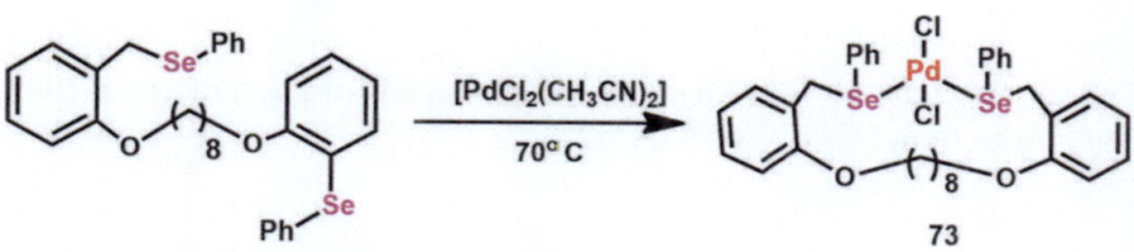

Scheme 16.22 Synthesis of palladium macrocyclic complex

forming a twisted six-membered Pd_3E_3 ring. The coupling reaction of 4-iodotoluene and phenylboronic acid was investigated at 100 °C for 6 h by employing the complexes **74**. Depending upon the nature of catalyst, product yield varied in the range 63–98% (Paluru et al. 2013) (Scheme 16.23).

Another macrocyclic ligand 1,8-bis(2-phenylselenoethoxy)anthracene-9,10-dione was treated with bis(benzonitrile)palladium(II) chloride in ethanol solution to isolate air stable complex **75**. The molecular structure revealed that the ligand chelates with palladium centre to form a convex square planar geometry. The complex possesses 14-membered macrocyclic rings with puckered orientation as confirmed from the crystal structure analysis (Fig. 16.3). The catalytic activity of **75** was studied for Heck reaction for the conversion of styrene to stilbene using 0.001 mol% concentration (Mariappan et al. 2015) (Scheme 16.24).

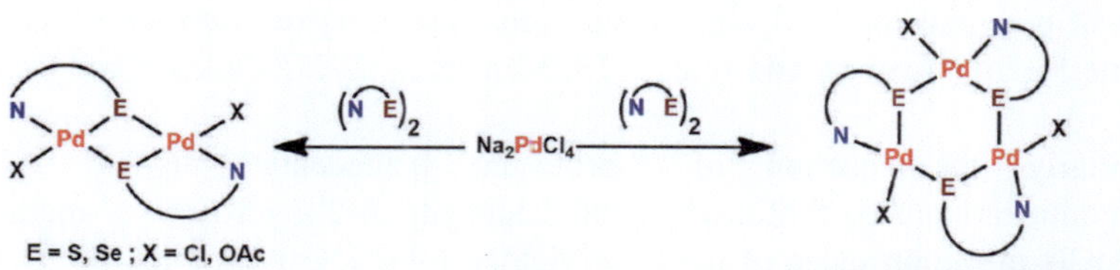

Scheme 16.23 Synthesis of palladium complexes with internally functionalized ligands

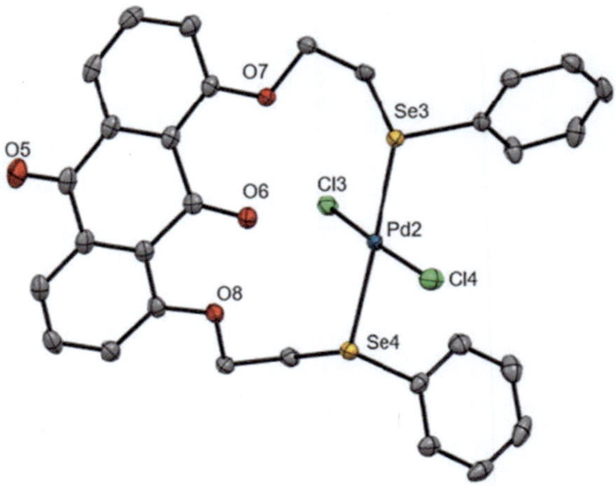

Fig. 16.3 Crystal structure of [Pd{1,8-bis(2-phenylselenoethoxy)anthracene-9,10-dione}$_2$]Cl$_2$. The crystal is redrawn from CCDC No. 1004318

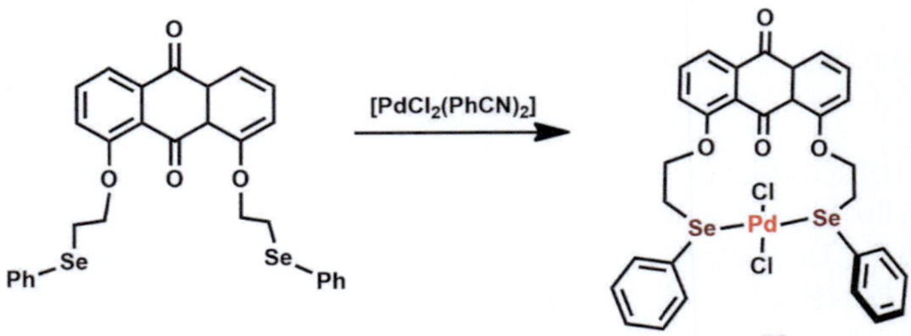

Scheme 16.24 Synthesis of [Pd{1,8-bis(2-phenylselenoethoxy)anthracene-9,10-dione}$_2$]Cl$_2$

A series of palladium complexes with pyridyl chalcogenolate ligands comprise a well-established class of catalysts, which could be isolated by utilizing different palladium phosphine precursors with the sodium salt of 4-pyridyl selenolate ligands (generated in situ via the reaction of 4,4′dipyridyl diselenide using sodium borohydride) giving mono and dinuclear complexes **76–79**. These complexes exhibited great potential for catalysing Suzuki cross-coupling reaction with high TON (Scheme 16.25) (Vivekananda et al. 2013; Mane et al. 2021; Kulkarni and Chauhan 2022).

Similarly, the reaction of S-protected pentadentate ligand: 2-(N,N-dimethylthiocarbamato)-5-Methylisophthalaldehyde-di-2′-hydroxy 5′-methylanil with PdCl$_2$ in the presence of pyrazole yielded a binuclear complex **80** in which both palladium centres were bridged through pyrazolyl moiety (Scheme 16.26). The complex was found to be an active catalyst for Mizoroki–Heck reaction between methyl acrylate and iodobenzene with >99% product conversion in 30 min (Khadir et al. 2016).

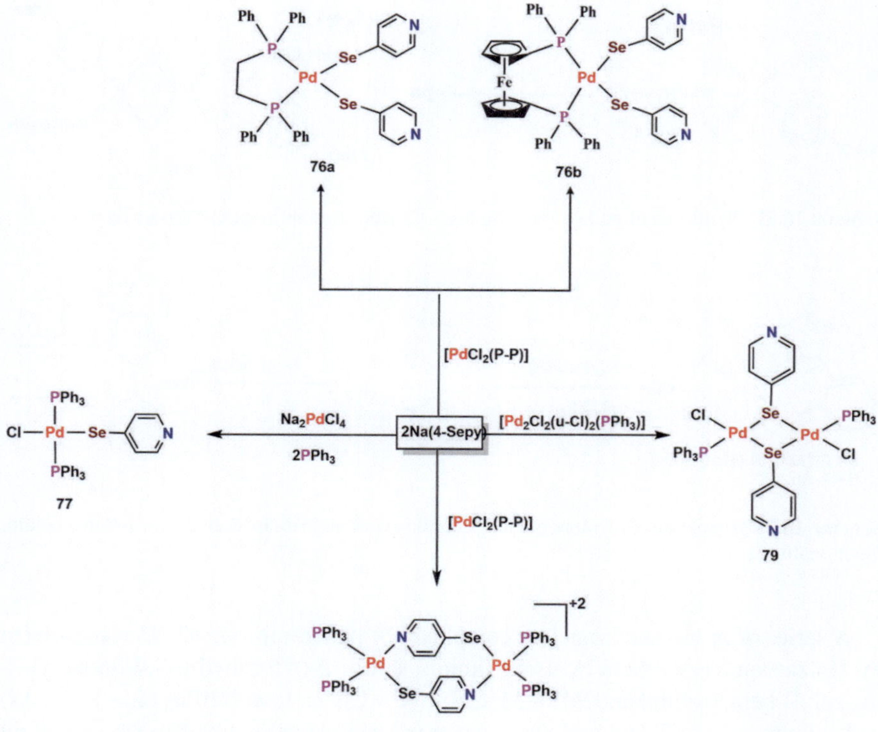

Scheme 16.25 Synthesis of palladium selenolate complexes using 4,4′dipyridyldiselenide. Redrawn from Xie et al. (2010)

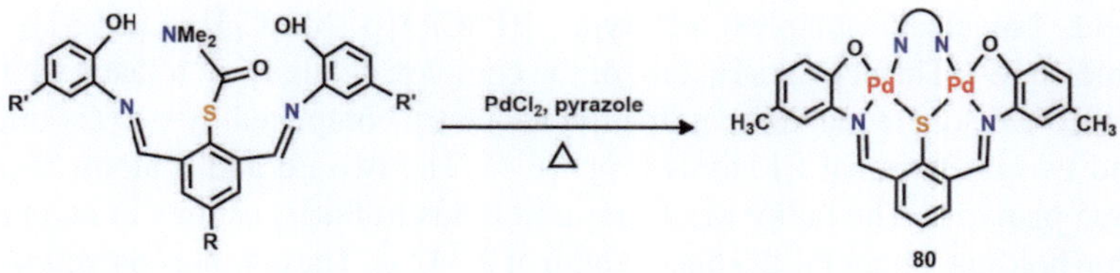

Scheme 16.26 Reaction of 2-(N,N-dimethylthiocarbamato)-5-Methylisophthalaldehyde-di-2′-hydroxy 5′-methylanil with $PdCl_2$ in the presence of pyrazole

Gholivand et al. have utilized a dithiophosphorus ligand $[(PhCH_2NH)(p\text{-}C_6H_4OMe)PS_2]^-[PhCH_2NH_3]^+$ along with $PdCl_2$ for the preparation of the complex **81** in methanol media (Scheme 16.27). This catalyst was used for the coupling of iodobenzene and styrene at very low loading (10^{-5} mol%) to obtain stilbene with 100% yield (Gholivand et al. 2016).

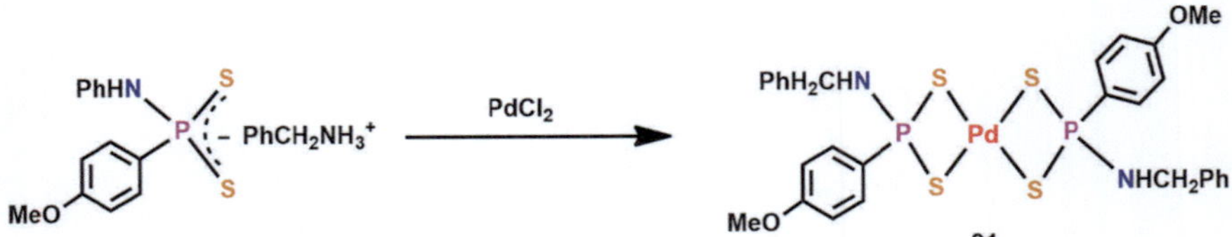

Scheme 16.27 Synthesis of palladium complexes derived from dithiophosphorous ligand

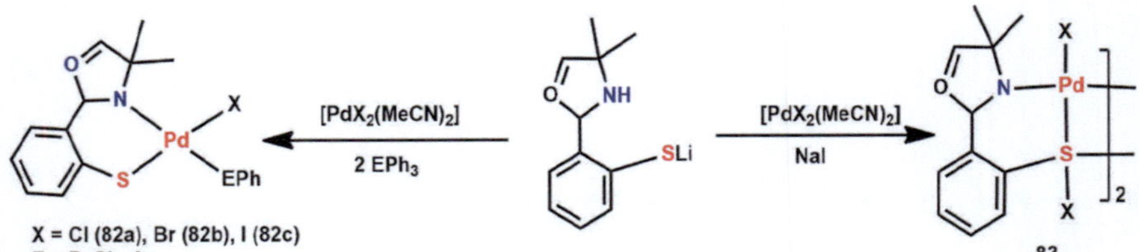

Scheme 16.28 Preparation of mononuclear complexes of palladium derived from benzenethiol functionality

A series of mono and binuclear complexes of palladium viz. **82–83** was isolated by Holzar and co-workers by using lithium salt of 2-(4,5-dihydro-4,4-dimethyl-2-oxazolyl)-benzenethiol and $PdX_2(MeCN)_2$ (X = Cl, Br, I) with EPh_3 (E = P, Sb, As) and sodium iodide. These complexes act as precatalyst in Suzuki–Miyaura coupling reactions (Scheme 16.28). Notably, the sulphur-bridged binuclear complexes have turnover frequencies of up to 16000 h^{-1} for bromoacetophenone and phenylboronic acid coupling reaction (Holzer et al. 2018).

Zang et al. have reacted $PdCl_2$ with N,N′-disubstituted-imidazole-2-thionato to yield binuclear complex of type $[(PdCl_2)\{\mu\text{-}(Me_3C_6H_2)C_3N_2S\}]_2$ (**84**) (Scheme 16.29). The crystal structure of the complex divulges such that each Pd(II) has a distorted square planar coordination geometry completed by two cis bridging μ-S and two *cis* terminal Cl atoms (Fig. 16.4). The two Pd and S atoms reside in the same planarity. The bulky $Me_3C_6H_2$ unit binds to Pd(II) centres in an opposite direction holding longer Pd-μS bond length of 2.342 Å. The complex **84** was worked well in the coupling reaction between 4-bromoacetophenone and pheylboronic acid in water at 60 °C to produce desired product 1-biphenyl-4-yl-ethanone (yield = 99%) (Zhang et al. 2017).

16.4 Conclusions and the Way Forward

In the twenty-first century, significant advancements have been made in developing group 10 metal complexes as efficient catalysts for C–C coupling and related reactions. This chapter provides an in-depth look at the synthetic methodologies of

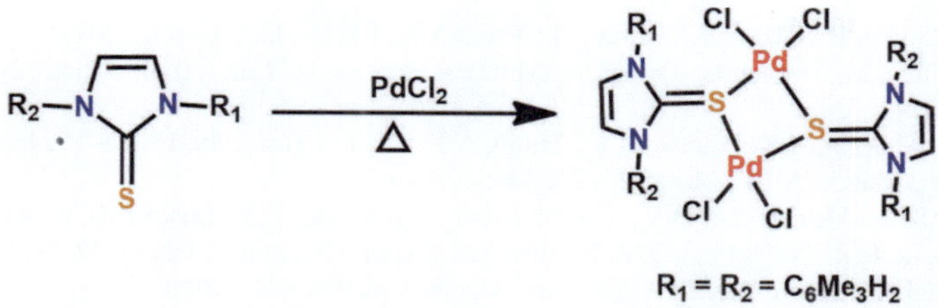

Scheme 16.29 Preparation of binuclear complex of palladium derived from imidazole thione functionality

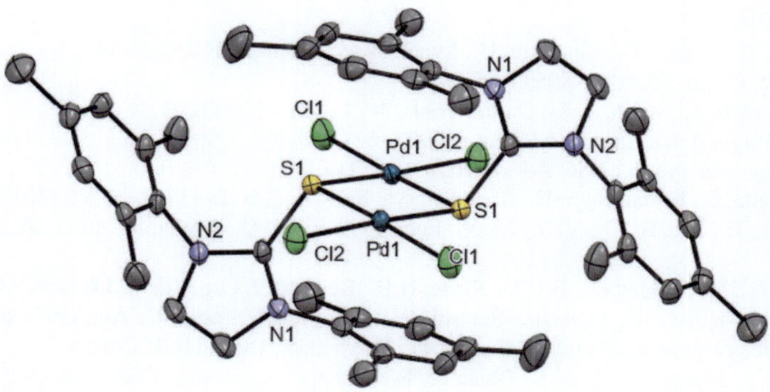

Fig. 16.4 Crystal structure of $[PdCl_2\{\mu\text{-}(MePhC_3N_2S)\}]_2$. The crystal is redrawn from CCDC No. 1494639

various nickel and palladium chalcogen-based systems that have been successfully utilized as homogeneous catalysts. Detailed mechanistic studies have highlighted the role of metal and metal chalcogenide nanoparticles in catalysis. It is believed that further exploration, including kinetic studies and mechanistic analyses of these novel complexes, could lead to the synthesis of previously unknown molecules with significant applications in nanoelectronics, pharmaceuticals and optoelectronic devices. Although considerable effort has been made to provide a comprehensive overview of the subject, it is understandable that summarizing over two decades of research in the field of cross-coupling catalysis is a formidable task.

References

Anitha, P., Manikandan, R., Vijayan, P., Prakash, G., Viswanathamurthi, P., Butcher, R.J.: J. Chem. Sci. **127**, 597–608 (2015a)

Anitha, P., Manikandan, R., Viswanathamurthi, P.: J. Coord. Chem. **68**, 3537–3550 (2015b)

Balam-Villarreal, J.A., Sandoval-Chávez, C.I., Ortega-Jiménez, F., Toscano, R.A., Carreón-Castro, M.P., López-Cortés, J.G., Ortega-Alfaro, M.C.: J. Organomet. Chem. **818**, 7–14 (2016)

Baruah, J., Gogoi, R., Gogoi, N., Borah, G.: Trans. Met. Chem. **42**, 683–692 (2017)
Basauri-Molina, M., Hernández-Ortega, S., Morales-Morales, D.: Eur. J. Inorg. Chem. **2014**, 4619–4625 (2014)
Bhaskar, R., Sharma, A.K., Yadav, M.K., Singh, A.K.: Dalton Trans. **46**, 15235–15248 (2017)
Buskes, M.J., Blanco, M.-J.: Molecules **25**, 3493 (2020)
Churusova, S.G., Aleksanyan, D.V., Vasil'ev, A.A., Rybalkina, E.Y., Susova, O.Y., Klemenkova, Z.S., Aysin, R.R., Nelyubina, Y.V., Kozlov, V.A.: Appl. Organomet. Chem. **32**, e4360 (2018)
Das, D., Singh, P., Singh, A.K.: J. Organomet. Chem. **695**, 955–962 (2010)
Das, D., Kannan, S., Kumar, M., Sadhu, B., Kumbhare, L.B.: Inorg. Chim. Acta **531**, 120704 (2022)
Datta, S., Seth, D.K., Butcher, R.J., Bhattacharya, S.: Inorg. Chim. Acta **377**, 120–128 (2011)
Datta, S., Seth, D.K., Gangopadhyay, S., Karmakar, P., Bhattacharya, S.: Inorg. Chim. Acta **392**, 118–130 (2012)
Devendar, P., Qu, R.-Y., Kang, W.-M., He, B., Yang, G.-F.: Palladium-catalyzed cross-coupling reactions: a powerful tool for the synthesis of agrochemicals. J. Agric. Food Chem. **66**, 8914–8934 (2018)
Dharani, S., Kalaiarasi, G., Sindhuja, D., Lynch, V.M., Shankar, R., Karvembu, R., Prabhakaran, R.: Inorg. Chem. **58**, 8045–8055 (n.d.)
Dubey, P., Gupta, S., Singh, A.K.: Dalton Trans. **46**, 13065–13076 (2017)
Dubey, P., Singh, A.K.: ChemistrySelect **5**, 2925–2934 (2020)
Dutta, J., Bhattacharya, S.: RSC Adv. **3**, 10707–10721 (2013)
Dutta, J., Datta, S., Kumar Seth, D., Bhattacharya, S.: RSC Adv. **2**, 11751–11763 (2012)
Fan, Q., Sun, H., Xie, S., Dong, Y., Li, X., Fuhr, O., Fenske, D.: Organometallics **40**, 2240–2245 (2021)
Fiebor, A., Tia, R., Makhubela, B.C.E., Kinfe, H.H.: Beilstein J. Org. Chem. **14**, 1859–1870 (2018)
Gabriel Flores-Rojas, G., González-Sebastián, L., Reyes-Martínez, R., Aguilar-Castillo, B.A., Hernández-Ortega, S., Morales-Morales, D.: Polyhedron **185**, 114601 (2020)
Gaikwad, V.V., Mane, P.A., Dey, S., Bhanage, B.M.: Appl. Organomet. Chem. **34**, e5255 (2020)
Ghavale, N., Manjare, S.T., Singh, H.B., Butcher, R.J.: Dalton Trans. **44**, 11893–11900 (2015)
Gholivand, K., Salami, R., Farshadfar, K., Butcher, R.J.: Polyhedron **119**, 267–276 (2016)
Gómez-Benítez, V., Valdés, H., Hernández-Ortega, S., Manuel German-Acacio, J., Morales-Morales, D.: Polyhedron **143**, 144–148 (2018)
Holzer, C., Dupé, A., Peschel, L.M., Belaj, F., Mösch-Zanetti, N.C.: Eur. J. Inorg. Chem. **2018**, 568–575
Joshi, H., Sharma, K.N., Sharma, A.K., Singh, A.K.: Nanoscale **6**, 4588–4597 (2014)
Keleş, M., Keleş, H., Emir, D.M.: Appl. Organomet. Chem. **29**, 543–548 (2015)
Khadir, N., Tavakoli, G., Assoud, A., Bagherzadeh, M., Boghaei, D.M.: Inorg. Chim. Acta **440**, 107–117 (2016)
Kulkarni, A.Y., Chauhan, R.S.: SMC Bull. **13**, 60–67 (2022)
Kumar, A., Rao, G.K., Saleem, F., Kumar, R., Singh, A.K.: J. Hazard. Mater. **269**, 9–17 (2014a)
Kumar, U., Dubey, P., Singh, V.V., Prakash, O., Singh, A.K.: RSC Adv. **4**, 41659–41665 (2014b)
Kumar, S., Rao, G.K., Kumar, A., Singh, M.P., Saleem, F., Singh, A.K.: RSC Adv. **5**, 20081–20089 (2015)
Kumar Rao, G., Kumar, A., Saleem, F., Singh, M.P., Kumar, S., Kumar, B., Mukherjee, G., Singh, A.K.: Dalton Trans. **44**, 6600–6612 (2015)
Kumar, S., Saleem, F., Singh, A.K.: Dalton Trans. **45**, 11445–11458 (2016)
Kumar, S., Saleem, F., Mishra, M.K., Singh, A.K.: New J. Chem. **41**, 2745–2755 (2017)
Kumar, S., Singh, S., Gadwal, J., Makar, P., Joshi, H.: J. Organomet. Chem. **946–947**, 121907 (2021)
Kumari, K., Kumar, S., Singh, K.N., Drew, M.G.B., Singh, N.: New J. Chem. **44**, 12143–12153 (2020)
Kumbhare, L.B., Dutta, P.K.: J. Organomet. Chem. **805**, 59–67 (2016)
López-Saucedo, F., Flores-Rojas, G.G., González-Sebastián, L., Reyes-Martínez, R., German-Acacio, J.M., Avila-Sorrosa, A., Hernández-Ortega, S., Morales-Morales, D.: Inorg. Chim. Acta **473**, 83–93 (2018)

Luquin, A., Castillo, N., Cerrada, E., Merchan, F.L., Garrido, J., Laguna, M.: Eur. J. Org. Chem. **2016**, 789–798 (2016)
Mane, P.A., Dey, S., Pathak, A.K., Kumar, M., Bhuvanesh, N.: Inorg. Chem. **58**, 2965–2978 (2019)
Mane, P.A., Pathak, A.K., Bhuvanesh, N., Dey, S.: Inorg. Chem. Front **8**, 3815–3829 (2021)
Manna, C.K., Naskar, R., Bera, B., Das, A., Mondal, T.K.: J. Mol. Struct. **1237**, 130322 (2021)
Mariappan, K., Ragothaman, J.S., Balasubramanian, V., Hoffman, M., Alaparthi, M., Sykes, A.G.: Inorg. Chim. Acta **429**, 46–50 (2015)
Nayan Sharma, K., Satrawala, N., Kumar Joshi, R.: Eur. J. Inorg. Chem. **2018**, 1743–1751 (2018)
Nolan, S.P., Navarro, O.: C–C bond formation by cross-coupling. In: Mingos, D.M.P., Crabtree, R.H. (eds.) Comprehensive Organometallic Chemistry III, pp. 1–38. Elsevier (2007)
Paluru, D.K., Dey, S., Wadawale, A., Jain, V.K.: J. Organomet. Chem. **728**, 52–56 (2013)
Pandiarajan, D., Ramesh, R., Liu, Y., Suresh, R.: Inorg. Chem. Commun. **33**, 33–37 (2013)
Paul, P., Datta, S., Halder, S., Acharyya, R., Basuli, F., Butcher, R.J., Peng, S.-M., Lee, G.-H., Castineiras, A., Drew, M.G.B., Bhattacharya, S.: J. Mol. Catal. A Chem. **344**, 62–73 (2011)
Paul, P., Butcher, R.J., Bhattacharya, S.: Inorg. Chim. Acta **425**, 67–75 (2015)
Prabhu, R.N., Ramesh, R.: Tetrahedron Lett. **53**, 5961–5965 (2012)
Prabhu, R.N., Pal, S.: Tetrahedron Lett. **56**, 5252–5256 (2015)
Prabhu, R.N., Ramesh, R.: Tetrahedron Lett. **57**, 4893–4897 (2016)
Prabhu, R.N., Ramesh, R.: Tetrahedron Lett. **58**, 405–409 (2017)
Prashad, M.: Palladium-Catalyzed Heck Arylations in the synthesis of active pharmaceutical ingredients. In: Larsen, R. (ed.) Organometallics in Process Chemistry, pp. 181–203. Springer (2004)
Prashanth, B., Bawari, D., Mandal, U., Verma, A., Choudhury, A.R., Singh, S.: Dalton Trans. **46**, 6291–6302 (2017)
Priyarega, S., Kalaivani, P., Prabhakaran, R., Hashimoto, T., Endo, A., Natarajan, K.: Mol. Struct. **1002**, 58–62 (2011)
Ramírez-Rave, S., Morales-Morales, D., Grévy, J.-M.: Inorg. Chim. Acta **462**, 249–255 (2017)
Rao, G.K., Kumar, A., Ahmed, J., Singh, A.K.: Chem. Commun. **46**, 5954–5956 (2010)
Rao, G.K., Kumar, A., Kumar, B., Kumar, D., Singh, A.K.: Dalton Trans. **41**, 1931–1937 (2012)
Rao, G.K., Kumar, A., Kumar, S., Dupare, U.B., Singh, A.K.: Organometallics **32**, 2452–2458 (2013)
Rao, G.K., Kumar, A., Singh, M.P., Singh, A.K.: J. Organomet. Chem. **749**, 1–6 (2014a)
Rao, G.K., Kumar, A., Bhunia, M., Singh, M.P., Singh, A.K.: J. Hazard. Mater. **269**, 18–23 (2014b)
Roffe, G.W., Boonseng, S., Baltus, C.B., Coles, S.J., Day, I.J., Jones, R.N., Press, N.J., Ruiz, M., Tizzard, G.J., Cox, H., Spencer, J.: R. Soc. Open Sci. **3**, 150656 (2016)
Saleem, F., Rao, G.K., Singh, P., Singh, A.K.: Organometallics **32**, 387–395 (2013)
Saleem, F., Rao, G.K., Kumar, A., Kumar, S., Singh, M.P., Singh, A.K.: RSC Adv. **4**, 56102–56111 (2014)
Sharma, K.N., Joshi, H., Sharma, A.K., Prakash, O., Singh, A.K.: Organometallics **32**, 2443–2451 (2013)
Sharma, A.K., Joshi, H., Bhaskar, R., Kumar, S., Singh, A.K.: Dalton Trans. **46**, 2485–2496 (2017)
Sharma, K.N., Satrawala, N., Srivastava, A.K., Ali, M., Joshi, R.K.: Org. Biomol. Chem. **17**, 8969–8976 (2019)
Shi, J., Li, F.-L., Li, H.-X., Wang, F., Yu, H., Ren, Z.-G., Zhang, W.-H., Lang, J.-P.: Inorg. Chem. Commun. **46**, 159–162 (2014)
Singh, P., Das, D., Kumar, A., Singh, A.K.: Inorg. Chem. Commun. **15**, 163–166 (2012)
Singh, M.P., Saleem, F., Rao, G.K., Kumar, S., Joshi, H., Singh, A.K.: Dalton Trans. **45**, 6718–6725 (2016)
Singh, V.V., Kumar, U., Tripathi, S.N., Singh, A.K.: Dalton Trans. **43**, 12555–12563 (n.d.)
Suganthy, P.K., Prabhu, R.N., Sridevi, V.S.: Tetrahedron Lett. **54**, 5695–5698 (2013)
Suganthy, P.K., Prabhu, R.N., Sridevi, V.S.: Inorg. Chem. Commun. **44**, 67–69 (2014)
Tabassum, S., Zahoor, A.F., Ahmad, S., Noreen, R., Khan, S.G., Ahmad, H.: Cross-coupling reactions towards the synthesis of natural products. MolDivers **26**, 647–689 (2022)

Tenchiu, A.-C., Ventouri, I.-K., Ntasi, G., Palles, D., Kokotos, G., Kovala-Demertzi, D., Kostas, I.D.: Inorg. Chim. Acta **435**, 142–146 (2015)

Thapa, K., Paul, P., Bhattacharya, S.: Inorg. Chim. Acta **486**, 232–239 (2019)

Tyson, G.E., Tokmic, K., Oian, C.S., Rabinovich, D., Valle, H.U., Hollis, T.K., Kelly, J.T., Cuellar, K.A., McNamara, L.E., Hammer, N.I., Webster, C.E., Oliver, A.G., Zhang, M.: Dalton Trans. **44**, 14475–14482 (2015)

Vivekananda, K.V., Dey, S., Wadawale, A., Bhuvanesh, N., Jain, V.K.: Dalton Trans. **42**, 14158–14167 (2013)

Vivekananda, K.V., Dey, S., Maity, D.K., Bhuvanesh, N., Jain, V.K.: Inorg. Chem. **54**, 10153–10162 (2015)

Wang, D., Gao, S.: Org. Chem. Front. **1**, 556–566 (2014)

Wang, Y.-T., Gao, B.-B., Wang, F., Liu, S.-Y., Yu, H., Zhang, W.-H., Lang, J.-P.: Dalton Trans. **46**, 1832–1839 (2017)

Xie, G., Chellan, P., Mao, J., Chibale, K., Smith, G.S.: Adv. Synth. Catal. **352**, 1641–1647 (2010)

Yan, H., Chellan, P., Li, T., Mao, J., Chibale, K., Smith, G.S.: Tetrahedron Lett. **54**, 154–157 (2013)

Yan, Q., Zheng, L., Li, M., Chen, Y.: J. Catal. **376**, 101–105 (2019)

Zhang, L.-M., Li, H.-Y., Li, H.-X., Young, D.J., Wang, Y., Lang, J.-P.: Inorg. Chem. **56**, 11230–11243 (2017)

Chapter 17
Layered Double Hydroxide (LDH) Derived Versatile Electrocatalysts Towards Water Splitting

Dipti Prava Sahoo, Ritik Mohanty, Lekha Paramanik, Upali Aparajita Mohanty, and Kulamani Parida

17.1 Introduction

With the rapid exhaustion of fossil fuels and slow deterioration of global climate and environment condition owing to the releasing carbon waste, research has been focused towards green and sustainable energy resources. Among all wind and solar energy can act as the most promising alternative energies to fulfil the energy needs giving low carbon footprint. Hydrogen, a well-recognized example of clean and green fuel has bumped up since 1970s and started to replace the extensive uses of fossil fuels as post combustion its by-product is only water (Wang et al. 2018; Cao 2018). Though hydrogen is considered as the plentily available species in the whole world but free hydrogen molecules can't exist. Various preparing techniques have been adopted to generate H_2 fuel, among all steam forming process of hydrocarbons to generate H_2 and CO_2 the most principal route used in industry nowadays. Further, the electrical water splitting represents nearly 4% of contribution towards the world H_2 production. Compared to other methods, as the chemical reaction involved in the water splitting process is indeed a reversible reaction of hydrogen combustion, which allows the complete and close cycle of hydrogen with zero carbon emission, thus it is considered as the most vital and promising method. Because of the zero-carbon emission and high energy density of the hydrogen, it grabbed the increasing research interests in past few years and became the most critical candidate to replace the overuse of fossils fuels. Electrocatalytic water decomposition is measured as one of the most promising and suitable strategies to generate hydrogen which involves the two major half reaction like hydrogen evolution reaction (HER) and oxygen evolution reaction (OER) to achieve benchmark efficiency. Generally, the water splitting reaction is

D. P. Sahoo · R. Mohanty · L. Paramanik · U. A. Mohanty · K. Parida (✉)
Centre for Nano Science and Nano Technology, Siksha O Anusandhan (Deemed to be University), Bhubaneswar, India
e-mail: kulamaniparida@soa.ac.in; paridakulamani@yahoo.com

R. S. Ningthoujam and A. K. Tyagi (eds.), *Handbook of Materials Science, Volume 3*, Indian Institute of Metals Series, https://doi.org/10.1007/978-981-96-8229-4_17

considered as the energy uphill process, where the free energy change ($\Delta G°$) for the conversion of 1 mol of water molecules into respective oxygen and hydrogen gas is +237.2 kJ mol^{-1}. Considering the extra amount of work for 1 mol of H_2 formation the required enthalpy change ($\Delta H°$) is +286 kJ mol^{-1}. Theoretically, for a reversible electrolytic cell thermodynamic electrical potential is 1.23 V and when applied potential is greater than 1.23 V the water reduction and oxidation occur at cathode and anode, respectively, by the release of H_2 and O_2 gases. The practical application of both HER and OER for overall water splitting is hindered by the large overpotential and ultra-slow kinetic (Liu et al. 2021a; Song et al. 2020; Han et al. 2017; Wang et al. 2020). Therefore, fabrication of electrocatalyst with high water splitting activity having lowest overpotential is enormously needed and urgent. Highly valuable metal (Pt) and noble metal oxides such as RuO_2 and IrO_2 are better catalysts for both electrocatalytic water reduction and oxidation, respectively.

Nevertheless, high price, restricted sources and poor functioning stability hamper their extensive practical applications. So, cheap, highly effective and long-standing electrocatalysts for water decomposition are extremely desired. Nowadays, electrocatalysts involving transition metals (Ni, Co, Fe, Mn, Mo) and their respective hydroxides, oxides, phosphides, sulphides, selenides and nitrides have been recognized as noteworthy applicants for the HER and OER. Also, these materials are prepared from the parent layered double hydroxides (LDHs) and used as water splitting electrocatalysts. LDHs exhibit a lamellar ionic structure consisting of brucite-like layers having bivalent or trivalent cations and charge-balancing anionic compound with water molecules present in middle of the layers. LDHs having general formula of $[M^{2+}{}_{1-x}M^{3+}{}_{x}(OH)_2]^{x+}\ [A^{n-})_{x/n} \cdot mH_2O]$, where M represents a transition metal (having +2 and +3 states) and A is the intercalated anion present between the brucite layer; n represents interlayer charge; $m =$ amount of water molecules. The 'x' value lies in between 0.2 and 0.33. In brief, the trivalent metal ions substitute some trivalent metal ion in positive brucite layers which consist of edge-sharing octahedra component ($[M^{2+}, M^{3+}](OH)_6$) of infinite layers in which the −OH bonds perpendicular to the planes. The MO_6 octahedral unit in the brucite layer makes the positively charged metal hydroxide frameworks. The intercalated anion with water molecules boned with metal hydroxide layer through electrostatic interaction and hydrogen bonding. The tuneable ratio of the metal cations and charge compensating anionic species lead to a broad field of host-guest relations and nanoarchitectures. Because of the adjustable physical and chemical properties, changeable composition, LDHs can act as fascinating catalyst in electrochemistry (Sahoo et al. 2022; Hameed et al. 2022).

Substantial research is centred on electrocatalytic water redox reaction catalysed by LDH modified nanomaterials which showed decent electrocatalytic activity especially for HER and OER. Though, LDH-based nanomaterials come across a few drawbacks like low electrical conductivity, limited effective reactive sites, and long-term electrocatalytic stability. Thus, to resolve these issues, numerous LDH derived catalysts have been designed and developed for diverse applications. The derived materials show unbelievable electrocatalytic performance along with fascinating characteristics owing to the synergistic benefits of LDHs themselves (intrinsic

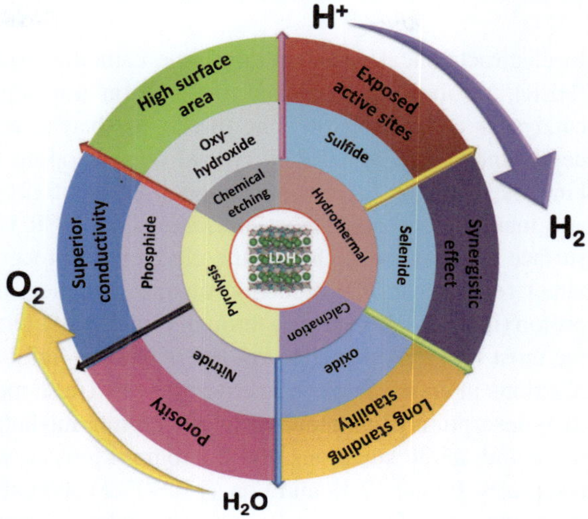

Scheme 17.1 Schematic illustration of strategies for LDH derived electrocatalysts towards water splitting

layered structure and adjustable metal cations composition) and the exceptional characteristic features of their derivatives. Also, the uniform dispersion of multivalent metals present in the laminar structure of LDHs can create a close hetero-interface in the derived compound, and augments the synergistic interaction among the two metal components in the resultant material (Zhou et al. 2017; Karmakar et al. 2021). Thus, the derived material offers an improved surface area, a greater number of electroactive sites, and unified charge-transfer, demonstrating in outstanding activity in HER, OER, and overall water splitting (OWS) performances.

Here, in this book chapter, we summarized mainly the current progress on LDHs derived products for electrocatalytic water splitting (Scheme 17.1). In brief, we have provided the recent advances and efforts in synthetic strategy of LDH derived nanomaterials, discussing the basic principle of electrocatalytic water splitting, characterization techniques, electrocatalytic activity for HER, OER and OWS and perceptions into the improved catalytic performance are given followed by summery and future prospective.

17.1.1 Electrocatalytic Water Splitting

LDH derived materials have been widely investigated towards water electrolysis. It is accomplished by passing a current across corresponding electrodes that had been submerged in an aqueous electrolyte. In a nutshell, electrolysis of universal solvent comprised of two half-reactions: Hydrogen evolution reactions ($HER_{(cathodic)}$: $H^+ + e^- \rightarrow 1/2H_2$) and Oxygen evolution reactions ($OER_{(Anodic)}$: $H_2O \rightarrow 1/2O_2 + 2H^+ + 2e^-$), respectively.

17.1.1.1 Hydrogen Evolution Reaction (HER)

In electrochemical water splitting, the cathodic hydrogen evolution half-reaction (HER), involves conventional two-electron transmission processes in which the catalyst is crucial for lowering the energy barrier at each stage. Additionally, in-depth knowledge of HER is crucial to developing an understanding of reaction kinetics, methods to lower reaction overpotential, and ultimately how to create and manufacture an efficient electrocatalyst. HER is a various-step, pH-influence surface phenomenon. During the electrolytic process, this cathodic half-reaction either results from the reduction of universal electrolytes (in a basic medium) or proton (in an acidic medium) through electrolysis. The following diagram illustrates the most typical Volmer-Heyrovsky or Volmer-Tafel mechanisms for H_2-evolution reactions in an acidic or basic environment. These mechanisms involve the adsorption–desorption phenomena of hydrogen intermediates (H*). Furthermore, in both basic and acidic situations, the hydronium ions or water molecules act as proton reservoirs. Figure 17.1a and Eqs. (17.1–17.6) show the reaction route for hydrogen evolution over LDH derived catalyst in acidic and basic conditions.

In an acidic environment, HER occurs in a two-step mechanism: (I) a proton in electrolyte is adsorbed on the active site of catalyst and trap an electron to create adsorbed hydrogen (H^*_{ads}) (Volmer step); (II) one H^*_{ads} interacts with a proton in electrolyte and an electron from the electrode to generate H_2 (Heyrovsky step) and

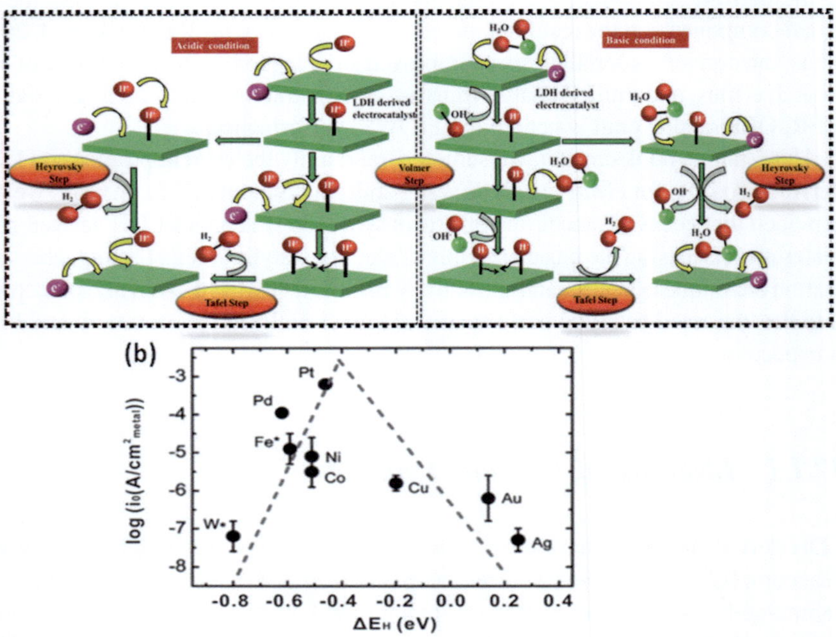

Fig. 17.1 **a** Schematic illustration of HER in acidic and basic mediums. **b** Schematic representation for a HER volcano plot (Hu et al. 2019, Copyright Wiley, adapted with permission)

(III) a second approach is that when the number of H^*_{ads} is excess, two adjacent H^*_{ads} combine to produce molecular hydrogen H_2 (Tafel step). Details of the mechanism mentioned above are as follows:

$$H_3O + e^- + M \rightleftharpoons M - H^*{}_{ads} + H_2O \text{ (Volmer)} \quad (17.1)$$

$$M - H^*{}_{ads} + H^+ + e^- \rightleftharpoons M + H_2 \uparrow \text{ (Heyrovsky)} \quad (17.2)$$

$$M - H^*{}_{ads} + M - H^*{}_{ads} \rightleftharpoons 2M + H_2 \uparrow \text{ (Tafel)} \quad (17.3)$$

However, the reaction moves through three phases in an alkaline and neutral environment (as equated below): In the first step, the water molecule attaches on the reactive site of catalyst and breaks into OH^- and H^+, and then proton adsorbed on the electrode surface (H^*_{ads}) (Volmer step); in the second step, the H^*_{ads} treats with H_2O and an electron to enable the desorption of hydrogen (Heyrovsky) and in the third step, two linked adsorbed hydrogen atoms are finally combined to form molecular hydrogen (Tafel).

$$H_2O + e^- + M \rightleftharpoons M - H^*{}_{ads} + OH^- \text{(Volmer step)} \quad (17.4)$$

$$M - H^*{}_{ads} + H_2O + e^- \rightleftharpoons M + H_2 \uparrow + OH^- \text{(Heyrovsky)} \quad (17.5)$$

$$M - H^*{}_{ads} + M - H^*{}_{ads} \rightleftharpoons 2M + H_2 \uparrow \text{ (Tafel step)} \quad (17.6)$$

(Here, 'M' represents the LDH electrocatalyst).

As different reaction pathways have different energy barriers, the amount of energy needed to complete the overall reactions varies significantly. In contrast to an acidic medium, the catalyst has less performance and current density in a basic medium because the covalent link between the H^+ ion is weaker than the covalent bond in H_2O molecule (Sahoo et al. 2022). For the HER, the experimentally obtained reaction kinetics over Pt in an acidic environment are much superior to those in a basic environment. The slopes of the polarization curve, as described by Tafel in 1905, are obtained by extrapolating the log of the reduced current densities in opposition to the overpotentials, also known as Tafel plots. In addition, it was found that the rate-determining step (RDS) is Tafel reaction, and its slope was 29 mV/dec, just as Heyrovsky and Volmer's RDS has slopes of 39 and 118 mV/dec, respectively. The strong contacts of H^* on the electrode surface deliberate the Volmer phase, generating constrained turn-over rates, as illustrated in the volcano image in Fig. 17.1b. According to Sabatier's principle, the ideal HER electrocatalyst should have a Gibb's free energy of H-adsorption (ΔG_H) of approximately zero, meaning that the binding energy of H-atom on the catalyst surface shouldn't be too strong or loose. In other words, when H-atoms are weakly attached on the catalyst surface, the electrocatalyst is not sufficiently activated. Conversely, if it is strongly bonded to the surface than the majority of the reaction sites are occupied, which results in catalyst poisoning. Because of this, ΔG_H is a key factor in determining how effective the catalyst is at inhibiting HER activity. Additionally, it was assumed that the Pt at the top position of the volcano curve would make the greatest H_2 evolution reaction candidates. Additionally, the W/Fe/Ni/Co/Pd situated on the opposite side had large hydrogen binding energies while the Cu/Ag/Au present on the right side of the volcanic curve had poor hydrogen binding energies (Hu et al. 2019). Pt is still used as a HER catalyst, specifically in an acid environment, despite its inadequacies and astronomical cost. Based on their relative abundances the Co, Fe and Ni are very crucial for modern water-splitting technologies.

17.1.1.2 Oxygen Evolution Reaction (OER)

The water oxidation taking place in anodic half-cell reaction acts as the most important role in defining the total cell potential in water electrolysis. Compared to HER, the combined electrocatalytic process of four electrons and a proton made electrocatalytic OER more difficult. As a result, the OER process that involves the production of O–O bonds has a sluggish kinetic and a significant over-potential to progress the kinetic obstacle. Additionally, Fig. 17.2a illustrates the several electron-transfer steps that were engaged in the OER mechanism on the surface of LDH in both the basic and acidic media, presented in Eqs. 17.7–17.18.

In alkaline condition:

$$M + OH^- \rightarrow MOH \tag{17.7}$$

$$MOH + OH^- \rightarrow MO + H_2O_{(l)} \tag{17.8}$$

$$2MO \rightarrow 2M + O_{2g} \tag{17.9}$$

$$MO + OH^- \rightarrow MOOH + e^- \tag{17.10}$$

$$MOOH + OH^- \rightarrow M + O_{2(g)} + H_2O_{(l)} \tag{17.11}$$

And the overall reaction is,

$$4OH^-_{(aq)} \rightarrow 2H_2O_{(aq)} + 4e^- + O_{2(g)} \tag{17.12}$$

In acidic condition:

$$M + H_2O_{(l)} \rightarrow MOH + H^+ + e^- \tag{17.13}$$

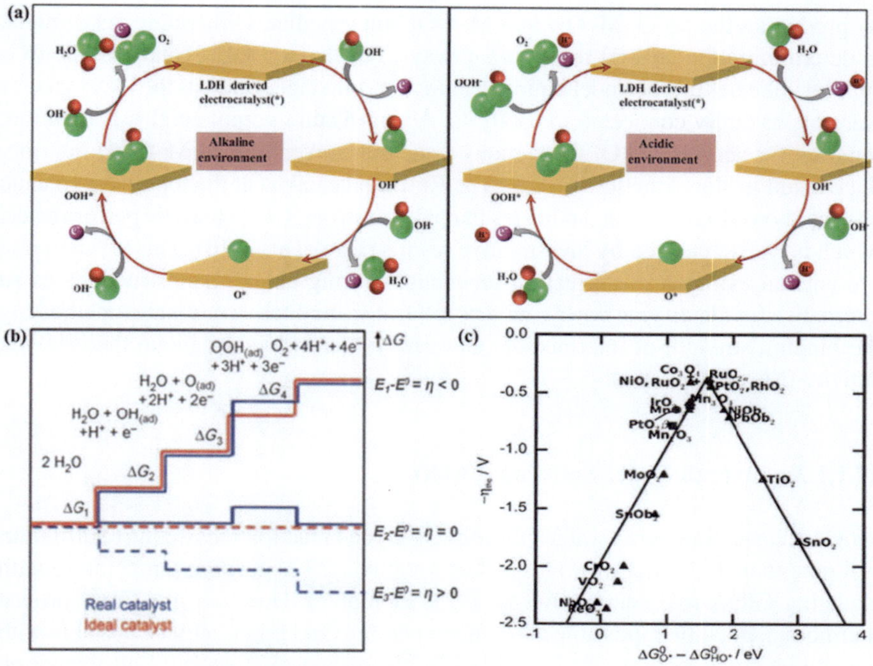

Fig. 17.2 **a** Schematic representation of OER mechanism on LDH derived catalyst. **b** Gibbs free energy depiction over OER vs. the reaction coordination. Dashed lines represent energetics at the electrode potential whereas all the thermochemical obstruct disappeared ("thermochemical overpotential") (Dau et al. 2010, Copyright Wiley, adapted with permission). **c** Plot for the negative values of theoretical overpotential verses the standard free energy difference ($\Delta G_{O*} - \Delta G_{HO*}$) (Man et al. 2011, Copyright Wiley, adapted with permission)

$$MOH + OH^- \rightarrow MO + H_2O_{(l)} + e^- \quad (17.14)$$

$$MO \rightarrow 2M + O_{2(g)} \quad (17.15)$$

$$MO + H_2O_{(l)} \rightarrow MOOH + H^+ + e^- \quad (17.16)$$

$$MOOH + H_2O_{(l)} \rightarrow M + O_{2(g)} + H^+ + e^- \quad (17.17)$$

And overall reaction is,

$$2H_2O_{(aq)} \rightarrow 4H^+{}_{(aq)} + 4e^- + O_{2(g)} \quad (17.18)$$

As seen in Fig. 17.2b, each step must overcome the relative Gibbs free energy blockage (ΔG_n, $n = 1, 2, 3$ and 4). Additionally, the M–O bonding interactions start the adsorption–desorption phenomena at the metal's active sites and later produced by producing the M–O, M–OH and M–OOH intermediates which are very crucial in determining the general reaction pathways. For the real OER catalysts, the ΔG_3 symbol indicates the thermochemically unfavourable stage, whereas the OOH species were haphazardly connected to catalysts. A significant overpotential was therefore required for accelerated OER reactions since the conversions of M–OOH intermediate result in slow kinetics of OER. The Ru-based catalyst at the top of the volcano plot, as shown in Fig. 17.2c, indicates that it has a strong OER catalytic performance, which is well supported by investigative results (Dau et al. 2010). This volcano plot presents an extrapolative direction for manufacturing unique nanostructures as an alternative for identifying novel catalysts, offering a qualitative rationale for adjusting the binding strength of intermediate and catalyst surfaces to regulate the catalytic activity (Man et al. 2011).

17.1.1.3 Overall Water Splitting (OWS)

Moving ahead, the overall water splitting (OWS) cell reaction requires an equilibrium cell potential ($E^0 = E_{anode} - E_{cathode}$) of roughly 1.23 V at room temperature with changing Gibbs free energy ($\Delta G = 237.2$ kJ mol^{-1}). However, the OWS proved thermodynamically impossible at room temperature and might only be possible with enough electrical energy. Additionally, the electrolyte pH may have an impact on the cathode ($E_{cathode}$) and anode (E_{anode}) equilibrium cell voltages. In brief, for OER E_{anode} ($2H_2O \rightarrow O_2 + 4H^+ + 4e^-$) can be 1.23 V vs NHE under acidic conditions (pH = 0), whereas $E_{cathode}$ for HER ($2H^+ + 2e^- \rightarrow H_2$) can be 0 V. The Nernst equation states that every unit increase in pH results in potential shifts of 59 mV. Therefore, E_{anode} for OER ($4OH^- \rightarrow O_2 + 2H_2O + 4e^-$) can be 0.40 V and $E_{cathode}$ for HER ($2H_2O + 2e^- \rightarrow H_2 + 2OH^-$) can be −0.83 V in basic circumstances (pH = 14). In order to identify pH influence on the applied voltage, reversible hydrogen electrode (RHE) was calculated as a reference electrode. This could result for OER the driving theoretical potential is 1.23 V versus RHE at all pH. However, developing catalysts for OER or HER was still crucial at low overpotential to achieve effective, affordable O_2 and H_2 generation (Sahoo et al. 2022). The effectiveness of the cell was unavoidably reduced when different electrolytes (and separation films) were added to the same electrolytes, while the cost of manufacture increased at the same time.

17.1.2 Important Parameters for Evaluating Water Splitting Performance

The basic function of a catalyst in a variety of contexts is to adsorb reactant and create intermediate on its catalyst surfaces, so as to accelerate the transfer of charge between the adsorbate and electrode. Several electrocatalytic kinetic parameters, such as the overpotential (η), Tafel slope (b), exchange current density (j_0), Faradaic efficiency, total electrode activity, turnover frequency (TOF) and long-term stability, need to be properly explained in order to have a thorough understanding of the target catalyst's catalytic behaviour (Wang et al. 2018). The overpotential and Tafel slope were the most crucial characteristics among them since they are helpful in partially developing the electrocatalytic reaction process. Furthermore, the polarization curve was used to calculate overpotential by using the Nernst equation, i.e. $E = E_0 + RT/nf \ln C_0/C_R$, whereas E_0, E, R, T, n, F were the standard potential, applied potential, universal gas constant, absolute temperature, number of transferred electrons and Faraday constant, respectively. C_O and C_R were the concentration of oxidized and reduced reagent, respectively. The overpotential described in the study was calculated at 10 mA cm^{-2} current density. However, exchange current density and Tafel slope were determined using the Tafel equation, i.e. $\eta = a + b \log j_0$, where a is Tafel constant; in the linear portion of a Tafel plot. As the overall current in the process of water splitting typically consists of a non-Faradaic capacitive current, the linear sweep voltammetry (LSV) or cyclic voltammetry (CV) is used to reproduce and quantify the entire electrode activity. The Faradaic efficiency for HER and OER is defined by the molar ratio of experimentally generated hydrogen (nH_2) or oxygen content (nO_2) to theoretically produced hydrogen (nH_2') or oxygen amount (nO_2'), respectively, For OER $nO_2/nO_2' = 4FnO_2/(It)$ and for HER $nH_2/nH_2' = 2FnH_2(It)$, where I and t represent oxidation current constant and active time, respectively. Additionally, to measure the intrinsic catalytic activity of each active site, TOF is measured by using the following equation TOF = (jA)/(4Fn), where j is current density at constant overpotential, A is working electrode area and n is the number of moles of active materials. The individual atoms on the catalyst have different catalytic activity, hence the accurate TOF value cannot calculated. But it is still a good method to compare the electrocatalytic performance of different electrocatalytic materials. The longevity of catalysts was also a crucial factor in determining their quality in practical applications. There are two main methods to assess the longevity of the electrocatalyst; first by recording chronopotentiometry (CP), which measures potential fluctuation over time while maintaining 10 mA cm^{-2} of constant current density, chronoamperometry (CA) measures current variation over time while maintaining a constant voltage, and second by comparing the LSV curves before and after CV measurements (>2000 cycles) (Wang et al. 2018; Sahoo et al. 2022).

17.2 Classification of LDH Derived Nanomaterials

The noble metal-based electrocatalysts such as RuO_2 and IrO_2 are effective for OER and Pt show better HER performances, but their high cost, limited sources, poor stability and low yield restricts its application for water splitting. Therefore, noble metal free compound derived from LDH has been widely explored for water splitting and substituent the noble metal-based electrocatalyst. In this book chapter, the derived LDH products have been categorized into several types, such as:

(1) Mixed metal oxide/oxyhydroxide
(2) Bimetal phosphide
(3) Bimetal sulphides
(4) Bimetal selenides
(5) Bimetal nitrides.

17.3 Synthesis Strategies of LDH Derived Materials

Physico-chemical changeability in LDHs can be regulated with the rational choice of metal cations in brucite layers and exchangeable inter layer anion. The synthesis techniques are the foremost and primary significant requisite to fabricate various types of electrocatalysts. Nevertheless, the derived method unlocks a novel, facial and advanced way to consumption of LDH-based materials as precursor for designing other innovative catalyst. The 2D lamellar structure of LDHs delivers an extra ultra-edge stratagem to plan effective electrocatalysts by investing controlled space for the directed growing of target product. In general, that the stability and integrity were fully dependent on the morphological arrangement. The new transformed mixed oxides, phosphides, nitrides and sulphides are chemically and thermally stable as compared to parent LDH precursor. Since the distinctive modifications of the material delivers a heterogenous interface and huge extent of exposer to the surface area having abundant active sites number which was enormously required for electrocatalytic reaction. Therefore, this section clarifies the fabrication technique for the construction of LDH derived materials and transformation related with it. Based on the reported literature, the preparation method of LDH derived material can be broadly categorized into two types, i.e. heat treatment and chemical treatment.

17.3.1 Heat Treatment Methods

Precisely, the heat treatment technique involves the nanomaterial exposed to high temperature condition under different atmosphere (such as air, N_2, H_2, Ar_2 and NH_3). This fabrication approach can be considered as an extensively used technique for the synthesis of LDH derived materials at a pre-set temperature because of its simplicity and flexibility. Broadly, the heat treatment involves two types of procedure, i.e.: calcination and pyrolysis.

17.3.1.1 Calcination

In the calcination process, the materials are exposed under high temperature in presence of air. The synthesis of mixed metal oxide (MMO) through calcination of LDH is a very important technique in which the divalent and trivalent metal cation in brucite layer of LDH generating the final derived material possessing large surface area. Three key stages are involved in the during thermal breakdown of LDH: (i) releasing of physical adsorbed or interlayer water molecule; (ii) dehydroxylation of metal hydroxide brucite layers and (iii) decomposition of intergallery anion resulting to formation of mesoporous MMO.

Wang et al. designed a 3D hierarchical CoNiFe-LDH nanoflakes and their derived porous $CoNiFeO_x$ spinel Oxides. First α-$Co(OH)_2$ nanoplate precursor was prepared by coprecipitation method then by using anion surfactant assisted hydrothermal technique, multi-metal LDH was synthesized and followed by calcination process mixed oxide was prepared (Wang and Zeng 2018). In the start of Fig. 17.3a the adsorbed sodium dodecyl benzene sulfonate (SDBS) anion surfactant on α-$Co(OH)_2$ nanoplate acts as protector for avoiding hydrolysis and hydroxide during hydrothermal reaction. Next the metal ions (Ni^{2+}, Fe^{3+}) are incorporated into the α-$Co(OH)_2$ surface. When the Fe^{3+} trivalent metal ion was added the SDBS anion surfactant behave as exfoliating agent or intercalating species resulting to exfoliation nanosheets are obtained. In contrast, the exotic divalent metal ions (Co^{2+} and Ni^{2+}) only doped on the surface to create a core-shell architecture. Consequently, the CoNiFe-LDH possesses a thinner flake-like structure than its precursor. After calcination at 350 °C for 1 h, the of derived MMO spinel ($CoNiFeO_x$) retained the morphology same as its parent material. The removal of intercalating species during thermal treatment creates some mesopores in the derived products. For example, a bifunctional heterostructure electrocatalyst composed of $CoFe_2O_4$ nanoflower having macro-porosity with topmost CoNi(oxy)hydroxide mesoporous nanosheets supported on NF substrate (CoNi/$CoFe_2O_4$/NF) was prepared. Here, CoFe LDH was used as parent material for preparing MMO spinel. Using conductive NF substrate having interconnected macroscopic pore structure with robust mechanical strength provides high specific surface area for the growth of LDH. During hydrothermal time, the flower like Co/Fe compound hydrolysis with urea to produce CoFe LDH nanoflower. Next to calcination, the uniformly deposited brown skinny layer of CoFe LDH changes to black colour with formation of metal oxide spinel ($CoFe_2O_4$) portrayed in Fig. 17.3b. Thereafter, the Co^{2+} and Ni^{2+} in electrolyte react with OH− through electrodeposition technique to produce bimetallic oxyhydroxide layers on $CoFe_2O_4$ nanoflower (Li et al. 2018a).

17.3.1.2 Pyrolysis

Apart from calcination, pyrolysis is carried out under inert atmosphere (N_2, Ar_2, NH_3) gas at high temperature. In this process, the transformation of LDH crystal lattice involves the displacement of oxygen atoms. Experimentally, the LDH was subjected to heat-treatments in the presence of phosphorus, nitrogen, or selenium sources to produce bimetal phosphide, bimetal nitride and bimetal selenide, respectively. In this technique, bulk LDH can be utilized to fabricate 2D nanosheets having few layers and a close heterogenous interface permitting for rapid charge transfer, better ion diffusion and enhanced electron migration.

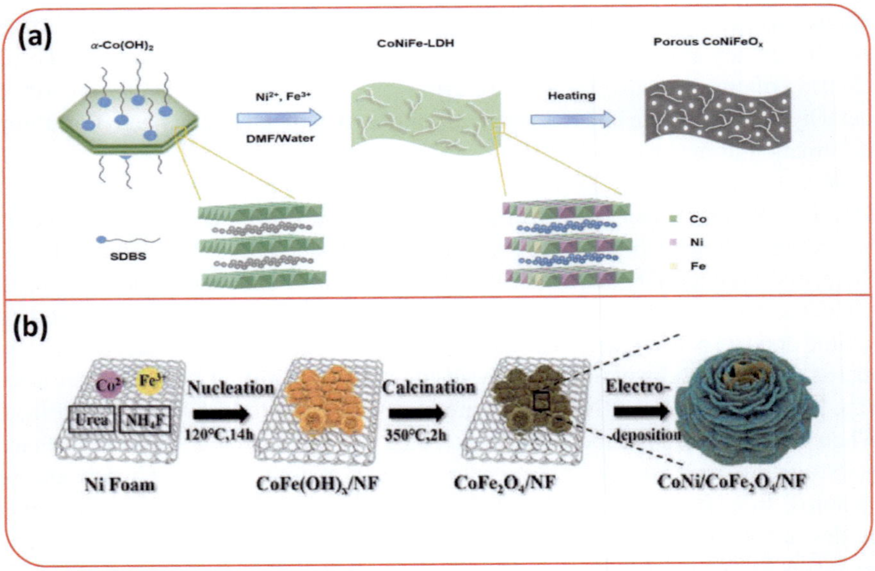

Fig. 17.3 **a** Schematic representation of the synthesis procedure of CoNi/$CoFe_2O_4$/NF porous composite (Wang and Zeng 2018, Copyright ACS, adapted with permission). **b** Schematic outline of fabrication method for CoNi/$CoNiFeO_4$/NF (Li et al. 2018a, Copyright RSC, adapted with permission)

Wang et al. reported a highly topological multi-metal phosphide (CoP@FeNiP) prepared from the heterostructure of ZIF-6 @FeNi-LDH covered with thin carbon networks (ZIF-67/FeNi-LDH) self-supported on NF (Wang et al. 2021a). Figure 17.4a illustrated the preparation mechanism of CoP@FeNiP. At first the FeNi LDH flower like microsphere was growth on NF by simple hydrothermal process. Next, ZIF-67 was in-situ growth and decorated directly onto the FeNi LDH nanosheets and produce the ZIF-67@FeNi-LDH nanohybrid precursor. Subsequently, phosphidation reaction was performed in presence of phosphating agent (NaH_2PO_2). At high temperature, the phosphrous source breaks down by releasing PH_3 gas which transformed the ZIF-67/FeNi-LDH nanohybrid to tri-metallic CoP@FeNiP/NF monoclinic structure. In another example, the holey nanosheet arrays of bimetal nitride ($CoFeN_x$ HNAs) supported on NF was designed by two-step process portrayed in Fig. 17.4b. Foremost, CoFe LDH precursor nanoarrays were established on NF through simple hydrothermal method by using metal source [$Co(NO_3)_2$ and $Fe(NO_3)_3$] structure directing species (NH_4F), and hydrolysis (urea) agent as the precursors. Afterwards, the parent LDH was pyrolyzed under NH_3/Ar_2 atmosphere at high temperature, the CoFe LDH nanosheet was topologically transformed into bimetal $CoFeN_x$ HNAs. This phase conversion can also be replicated by the sample colour change from the yellow to black (Li et al. 2020). In addition, self-templating alkali etching followed by vapour phase selenization method was adopted for constructing the hexagonal porous nickel selenide anchored N-doped

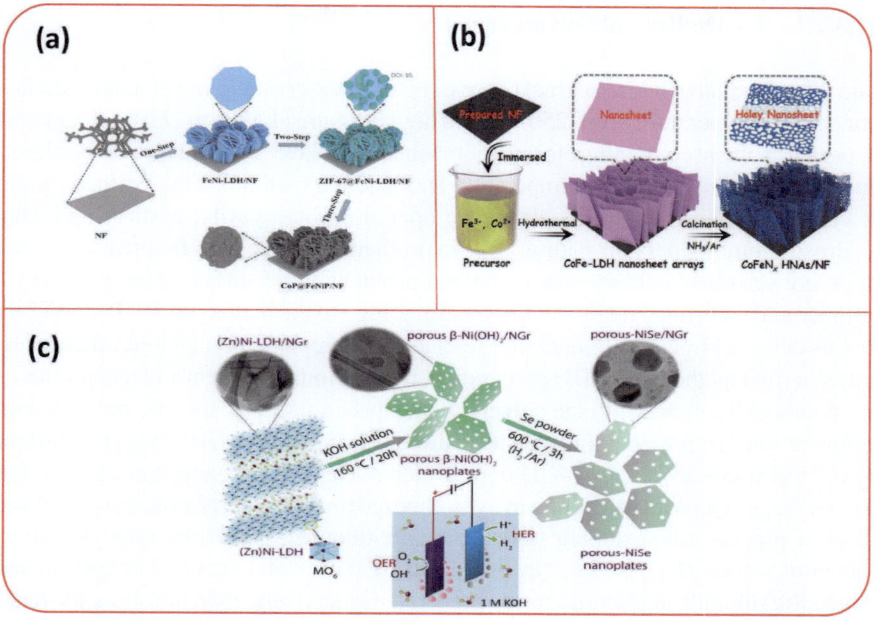

Fig. 17.4 **a** Schematic representation of the synthesis procedure of CoP@NiFeP/NF (Wang et al. 2021a, Copyright RSC, reproduced with permission). **b** Schematic illustration of the preparation process of $CoFeN_xHNAs/NF$ (Li et al. 2020, Copyright ACS, adapted with permission). **c** Schematic representation of preparation procedure for NiSe/NGr nanocomposite (Nadeema et al. 2018, Copyright ACS, adapted with permission)

graphene composite (P-NiSe/NGr) represented in Fig. 17.4c (Nadeema et al. 2018). By taking the advantages of thin layer of LDH nanosheet and amphoteric nature of Zn, a porous NiZn LDH/NGr was obtained. Next, the selective alkali etching of Zn^{2+} in NiZn LDH/NGr composite leads to formation of β-$Ni(OH)_2$/NGr and then it was converted to P-NiSe/NGr though selenization under H_2/Ar_2 gas atmosphere.

17.3.2 Chemical Treatment Methods

This synthetic technique can be behaved as an advanced methodology route towards the better derived electrocatalyst produced from the topotactic transformation of LDH with taking chemical reagent. These can be classified into two sub-techniques such as: (a) Hydrothermal or solvothermal method and (b) chemical etching method, as deliberated below.

17.3.2.1 Hydrothermal/Solvothermal

The solvothermal or hydrothermal process includes the crystallizing of nanomaterials under high temperature (100–250 °C) and high pressure (1 MPa to 1GPa) in aqueous or organic solvent containing closed Teflon lined autoclave. The pressure of the closed autoclave depends upon the temperature and reaction medium. This synthetic route is very challenging due to its simplistic operation, high purity, costless and easy heterostructure dispersion nature. The hydrothermal process is frequently chosen by many scientists because this technique provides large surface area, good crystallinity and controlled nanostructured morphology to the nanomaterial. Parent LDH and its derived bimetal sulphides and selenides are prepared though hydrothermal or solvothermal method. For LDH preparation, the precipitating agent and metal sources are combinedly mixed into the solvent then exposed under at specific pressure and temperature. The precipitators play as main role as it provides OH^- that get attached by the metal ions to create M–OH unity which nucleated and contributes to develop layered structures with increasing in hydrothermal time. Similarly, collectively taking LDH as precursor and sulphur (thioacetamide, thiourea, potassium thiocyanate) or selenium source (Se powder, Na_2SeO_3, SeO_2) with water or organic solvent the bimetallic sulphide or selenide can be prepared, respectively. By controlling the reaction time or temperature, the size, morphology and crystallinity of derived products can be tuned to some extent. Recently Li and his co-workers proposed a heterostructure of Mo doped bimetal NiS–NiS_2 sulphide on NiFe LDH with self-supported NF presented in Fig. 17.5. Because of the solid frame and robust electrical conductivity, NF was used as substrate and the flower like NiFe LDH was developed on it through first step hydrothermal method, this unique 3D nanoflower configuration consists of interlaced nanosheet which provides enough room for subsequent reactions. Thereafter, the simultaneous in-situ Mo doping and vulcanization process was happened in presence of Na_2MoO_4 and thioacetamide (TAA) as Mo and sulphur source, respectively, through second-step hydrothermal method. Finally, NiFe-LDH@Mo–NiS_2–NiS/NF 3D nanoflower having rough surface was obtained (Li et al. 2022). Furthermore, a bimetallic selenide heterostructure consists of $NiFe(OH)_x$ and (Ni, Fe)Se_2 was designed by Liu's group. As represented in Fig. 17.5b, the precursor arrays like NiFe LDH nanosheet were developed on carbon cloth (CC) substrate through hydrothermal technique followed by selenization assisted hydrothermal process to transform the parent LDH into (Ni, Fe)Se_2/CC nanosheet arrays. Thereafter, with the help of electrode deposition technique, an amorphous layer of $NiFe(OH)_x$ was decorated over on the bimetallic selenide surface to construct a NiFe(OH)x/(Ni, Fe)Se_2/CC core-shell heterostructure catalyst (Liu et al. 2021a).

17.3.2.2 Chemical Etching

In this process, a specific reagent was utilized to etching the metal hydroxide brucite layer and intergallery anion of LDH. The etching rate on the crystal surface of can be fluctuated by altering the pH of the etchant solvent. In this respect, different etchant

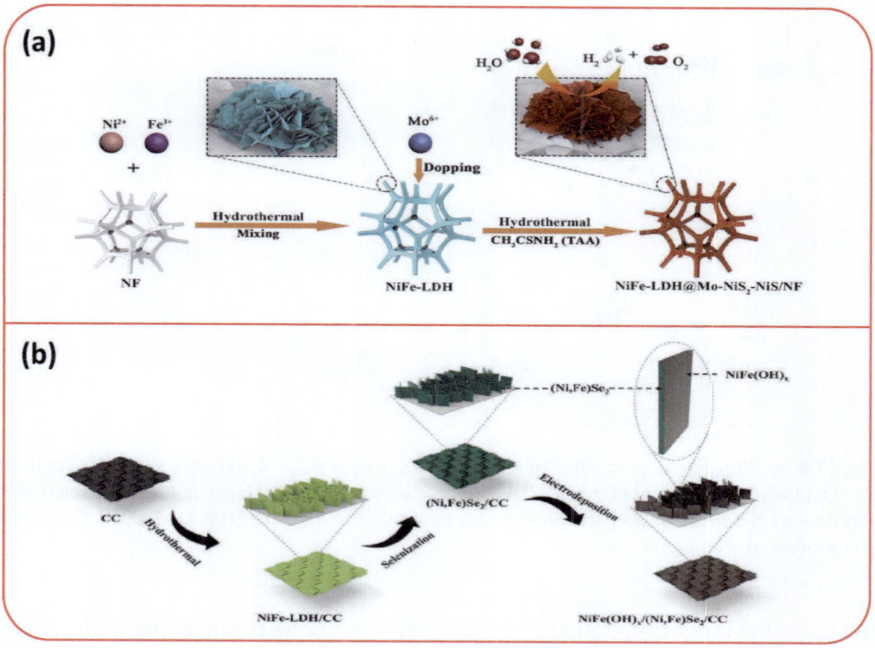

Fig. 17.5 **a** Representation of preparation methods for NiFe LDH@Mo–NiS_2–NiS/NF (Li et al. 2022, Copyright Elsevier, adapted with permission). **b** Synthesis procedure image of NiFe$(OH)_x$/(NiFe)Se_2/CC (Liu et al. 2021a, Copyright Wiley, adapted with permission)

species such as nitric acid, acetic acid, hydrochloric acid, potassium hydroxide, sodium hydroxide and ethanol are usually taken for the synthesis of LDH derived nanomaterials. The selectively etching of LDH from the strongly binding of etchant solution revealed to developed the highly ordered rational designed nanostructure. By taking this consideration, a lateral or vertical aligned CoOOH nanoplate was prepared through from topological transformation of CoAl LDH by alkali etching (Song et al. 2020). As represented in Fig. 17.6a, the CoAl–CO_3^{2-}-LDH microplate was fabricated through hydrothermal method then by adopting ion exchange route. The interlayer CO_3^{2-} ion is replaced with sulphate anions to acquire CoAl–SO_4^{2-} LDH microplates, having large intergallery spacings. Thereafter, these CO_3^{2-} and SO_4^{2-} intercalating LDHs were transformed to distinctive lateral and vertical geometric arrangement of CoOOH hierarchical hexagonal nanoplate through alkali etching with NaOH, respectively. In addition, Han and his research group fabricated a Fe modulated CoOOH nanoparticles assembled graphene matrix 2D nanohybrid (Fe–CoOOH/G) derived from CoFeAl LDH/G nanocomposite. As portrayed in Fig. 17.6b, CoFeAl LDH was developed on the graphene oxide (GO) nanosheet through in-situ urea hydrolysis assist reflux method. Here, the GO acts as substrate, wherever the divalent metal (Co^{2+}, Fe^{2+}) and trivalent metal (Al^{3+}) ions could be easily attached on GO surfaces through electrostatic force. Finally, the as-prepared CoFeAl-LDH/G composite was

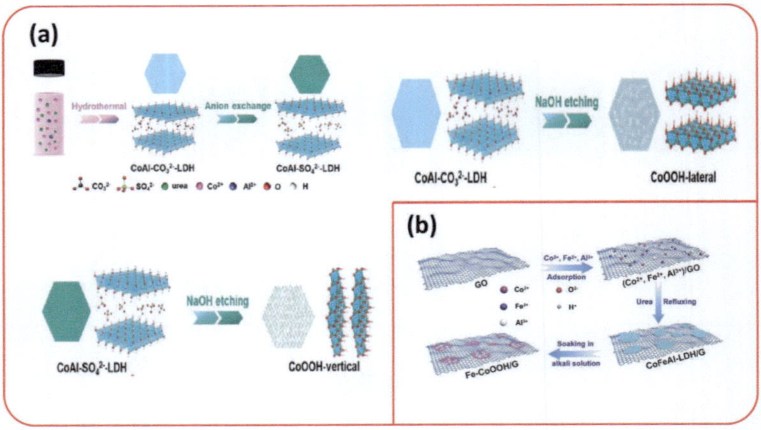

Fig. 17.6 **a** Preparation procedure of CoAl–CO_3^{2-}, CoAl–SO_4^{2-} LDH and synthesis route of CoOOH lateral and vertical hierarchical structures (Song et al. 2020, Copyright RSC, adapted with permission). **b** Schematic preparation route for Fe–CoOOH/Graphene (Han et al. 2017, Copyright Wiley, adapted with permission)

reacted with high concentration of alkaline solution (12 h), resulting the well-defined Fe–CoOOH/G porous 2D nanohybrids structure was produced (Han et al. 2017).

17.4 Physico-Chemical Characterization

The characterizations of LDH derived electrocatalytic materials are very important in order to explore their benefits in structural and physio-chemical alternations as compared to their respective parent LDH. The X-ray diffraction (XRD), X-ray photoelectron spectroscopy (XPS), Brunauer–Emmett–Teller (BET) surface area and Barrett-Joyner-Halenda (BJH) pore size distribution, Field Emission scanning electron microscopy (FESEM) and High-resolution transmission electron microscopy (HRTEM), atomic force microscopy (AFM), frontier transmission infra-red (FTIR), Raman spectra are some of the basic characterization methodologies which have been adopted to explain various properties of LDH derived electrocatalysts.

17.4.1 Metal Oxyhydroxide and Mixed Metal Oxide

Song et al. investigated the CO_3^{2-} or SO_4^{2-} intercalated CoAl LDH microplate and topochemical converted laterally or vertically oriented CoOOH nanoplates (Song et al. 2020). The XRD image in Fig. 17.7a illustrated that the (003) peak of CoAl–CO_3^{2-}-LDH (11.51°) was shifted to lower 2θ of 7.64° after intercalating the thicker layer of SO_4^{2-} ion suggesting the broadening of interlayer spacing. Next, the transformation of LDH to CoOOH superstructure was formed (Fig. 17.7b). The SEM image of CoAl–CO_3^{2-}-LDH displayed the regular hexagonal morphology having lateral size 4 μm shown in Fig. 17.7c. After NaOH etching of the CO_3^{2-} and SO_4^{2-} intercalated CoAl LDH microplate result to formation of lateral (Fig. 17.7d) and vertical (Fig. 17.7e) hierarchical structure of CoOOH nanoplate, respectively. The lateral connected CoOOH nanoplate having pore size of 15–30 nm, whereas the vertical aligned CoOOH nanoplate exhibits 5 nm of thickness with smooth surface in comparison with first one. In another example, Wang et al. investigated the surface elemental composition, internal morphology and surface area analysis of NiCo-LDH, NiCo–CL (without stirring of 12 h) and NiCo mixed oxide (NiCo–H) and NiCo–CH (without stirring of 12 h) through XPS, TEM and BET techniques. The O1s XPS spectrum in Fig. 17.7f fitted into three peaks positioned at 529.2–529.9 (O1), 530.8 (O2), 531.5 (O3) and 532.1(O4) eV attributed for metal–oxygen bond, OH group or CO_3^{2-} anion, surface defect sites and surface adsorbed H_2O molecule, respectively. In comparison with O2 peak area of NiCo–CH, the NiCo–CL exhibits more adsorbed oxygen and OH^- species illustrated the elimination of OH^- and water evaporation during calcination time. Furthermore, the O1 content of NiCo–CH is pointedly higher and O_2 content is comparable lower than that of NiCo–H, signifying that the easier remove OH in case of NiCo–CH sample. Similarly, the C1s spectrum in XPS image Fig. 17.7g displayed two peaks for reference carbon (288.7 eV) and CO_3^{2-} anion (289.0 eV), respectively. The less carbon content in NiCo–CH again proves that decarburization happened throughout the thermal treatment which good agreement with the EDX result. The TEM micrograph of NiCo–CL in Fig. 17.7h revealed the layer-by-layer lamella structure with porous morphology. After calcination, the lamella size of NiCo–CH (Fig. 17.7i) get smaller and clear porous structure was formed in comparison with parent material. The N_2 adsorption–desorption isotherm of NiCo–CL exhibits large surface area (113.3 m^2/g) (Fig. 17.7j) and pore size (13.6 nm) (Fig. 17.7k) in comparison with NiCo-LDH (12.4 m^2/g, 4.3 nm). When the NiCo-LDH and NiCo–CL were converted into NiCo–H and NiCo–CH by thermal treatment, a huge quantity of water vapours and CO_2 gas was appeared, resulting to the reduction of particle size having porous structure with decreased surface area (94.2 m^2/g) and increased pore size (14.9 nm). The decreased surface area suggests the more exposed reactive sides for electrocatalytic reaction (Wang et al. 2020). Moreover, the XRD, FESEM, TEM and colour mapping analysis of self-supported CoMnLDH@ZIF-67/NF and its derived $MnCo_2O_4$@Co_3O_4/NF MMO composite was studied by Zhou and his research team (Zhou et al. 2018). As presented in Fig. 17.7l the XRD image of as prepared $MnCo_2O_4$@Co_3O_4 sample scraped from NF substrate possess both phase of Co_3O_4 and $MnCo_2O_4$ prepared from CoMn-LDH having Mn:Co molar ratio of 0.47. The presented SEM image of CoMn-LDH@ZIF-67 in Fig. 17.7m revealed the rhombic dodecahedron shaped ZIF-67 crystals uniformly penetrated on CoMn LDH nanoneedles having 65 nm diameter and 5 μm length. The vertically oriented CoMn LDH nanoneedles orderly arranged on NF substrates offer a favourable platform for the successive growth of ZIF-67 polyhedron. After annealing treatment, the $MnCo_2O_4$@Co_3O_4 retains same nanoneedles with dodecahedral structure like parent material (Fig. 17.7n). The resulting Co_3O_4 has smaller size as compared to pristine ZIF-67 particle due to contraction through annealing process. The TEM image in Fig. 17.7o also proved the dodecahedron-penetrated needle morphology. Moreover, the colour mapping in scanning transmission electron microscopy (STEM) (Fig. 17.7p) confirmed the elemental distribution of $MnCo_2O_4$@Co_3O_4 in which Co and O were situated in the entire polyhedron and nanoneedle area whereas the Mn was only dispersed among nanoneedle, which authorize the effective integration of Co_3O_4 and $MnCo_2O_4$.

17.4.2 Bimetallic Phosphides

An interfacial Co_xP–Fe_2P bimetallic phosphide heterostructure supported on NF (Co_xP–Fe_2P/NF) was fabricated through phosphidation of CoFe LDH anchored on CoFe-PBA (Prussian blue analogue) nanocubes. Li's group characterized the FESEM, HRTEM and XRD analysis of Co_xP–Fe_2P/NF electrocatalyst (Li et al. 2021). As depicted the FESEM micrograph (Fig. 17.8a) of Co_xP–Fe_2P/NF illustrated the irregular size of PBA nanocubes create closely compact interfaces with CoFe-LDH and this hierarchical heterostructure still well-maintained having copious nano-channels. Figure 17.8b (1, 2, 3, 4) are the magnified HRTEM images of Co_xP–Fe_2P heterostructure from marked area in Fig. 17.8b. The lattice fringes of corresponding planes for CoP [(111), (020)], Co_2P [(130). (211), (121), (201)] and Fe_2P [(300), (201), (111)] phases were observed suggesting a hetero-interface was developed in bimetallic phosphide heterostructure. Furthermore, XRD results support the HRTEM analysis. As enclosed in Fig. 17.8c, the CoFe LDH diffraction peaks are pure and the CoFe-LDH/PBA is composed of CoFe LDH and PBA. Later phosphating of CoFe-LDH/PBA, the XRD image in Fig. 17.8d Co_xP–Fe_2P/NF heterostructure displayed the hexagonal Fe_2P, orthorhombic Co_2P and orthorhombic CoP phases. Liu et al. examined the XPS, FESEM and surface area characterizations of B doped CoP/Ni_2P hollow nanoprism prepared from NiCo LDH. The Ni 2p and Co 2p spectrum (Fig. 17.8e) of all materials showed two spin-orbit doublets (Ni $2p_{3/2}$, Ni $2p_{1/2}$, Co $2p_{3/2}$, Co $2p_{1/2}$) and two shake-up satellites. More in, the Ni 2p3/2 is again deconvoluted into peaks attributed to Ni–P bond (852.8 eV) and Ni^{2+} in nickel oxide (856.7 eV). Similarly, the peaks for Co–P bond and oxidized Co species are observed in Co 2p spectrum. In comparison with Ni–B–P and Co–B–P, the Ni 2p and Co 2p of Ni–Co–B–P displayed negative and positive shifting of binding energy, respectively. This determines that charge transferred from CoP to Ni_2P. the B 1s spectrum exhibits two peaks assigned to amorphous M–B (186.9 eV) and B–O (190.8 eV), respectively, which further validates that incorporation of B-atoms into the Ni–Co precursors through electrons transferred to the adjacent Ni/Co or carrying of O atom. In addition, the P 1s possess three peaks attributed to P $2p_{3/2}$, P $2p_{1/2}$ of Co–P or Ni–P, and P–O, respectively. The XPS analyses, proved that the Ni–Co–B–P are consist of B-doped CoP/Ni_2P prepared by solid-state phosphidation. The SEM image in Fig. 17.8f of $NiCo_2$ precursor (molar ratio of Ni:Co = 2:1) exhibits uniform prisms

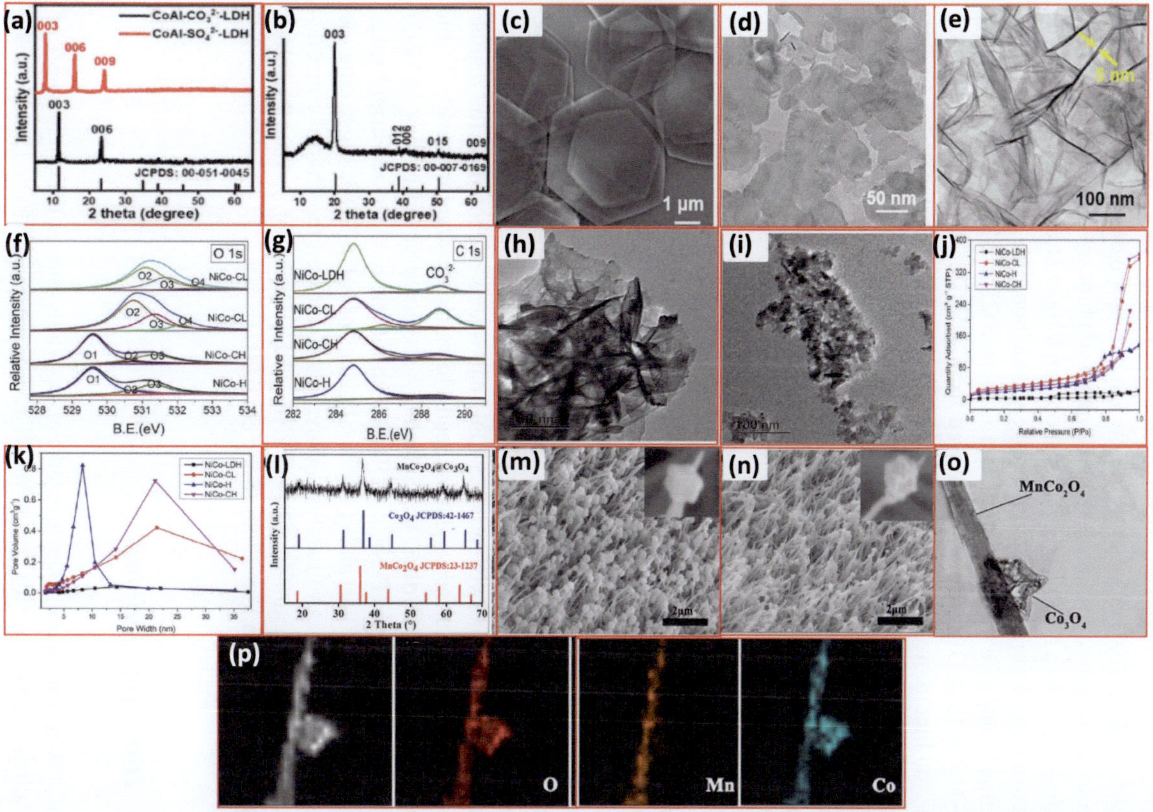

Fig. 17.7 **a** XRD graph of LDH, **b** XRD graph of lateral oriented CoOOH, **c** FESEM micrograph of CoAl–CO_3^{2-} LDH, TEM images of lateral **d** and vertical **e** hierarchical superstructure of CoOOH (Song et al. 2020, Copyright RSC, adapted with permission). XPS image of NiCo-LDH, NiCo–CL, NiCo–H and NiCo–CH, **f** O1s, **g** C1s, TEM Image of **h** NiCo-LDH (i) NiCo CH, **j** surface area and **k** pore size distribution of NiCo-LDH, NiCo–Cl, NiCo–H and NiCo–CH (Wang et al. 2020, Copyright Elsevier, adapted with permission). **l** XRD graph, FESEM image of **m** CoMnLDH@ZIF-67, **n** $MnCo_2O_4$@Co_3O_4 nanocomposite, **o** TEM and **p** colour mapping image of $MnCo_2O_4$@Co_3O_4 nanocomposite (Zhou et al. 2018, Copyright Elsevier, adapted with permission)

like morphology having length of ~500 nm. After treatment with $NaBH_4$ through solvothermal reaction followed by phosphidation process, B doped CoP/Ni_2P (Ni–Co–B–P) well preserved the parental morphology (Fig. 17.8g). The BET surface areas of Ni–B–P, Ni_2Co–B–P (Ni:Co = 1:2), $NiCo_2$–B–P (Ni:Co = 2:1), NiCo–B–P (1:1) and Co–B–P were 126.43, 38.60, 21.64, 27.06, and 90.06 m^{-2} g^{-1}, respectively (Fig. 17.8h). The $NiCo_2$–B–P hollow nanoprism possesses lowest specific surface area as compared to other B-doped bimetal sulphide or single metal sulphide nevertheless more than NiCo LDH. The nanoprism hollow structure exposed abundant reactive sites along with the comparatively wide pore size structure (Fig. 17.8i) which helps to improve the ion migration (Liu et al. 2021b). Moreover, a core-shell $Co(OH)_2$@P-NiCo-LDH nanohybrid derived from ZIF-67@NiCo-LDH-60 having dodecahedral hierarchical structure was constructed by Sing's group and characterized the XRD, FESEM, HRTEM analysis. The XRD image of $Co(OH)_2$@NiCo-LDH-60 [60 mg of $Ni(NO_3)_2$] in Fig. 17.8j displayed the four characteristics diffraction peak corresponding to (003), (006), (009) and (110) lattice planes for NiCo LDH and a strong peak [(001) plane] for $Co(OH)_2$ indicates the construction of core-shell $Co(OH)_2$@-NiCo-LDH nanocomposite. After phosphating method, there was no variations of XRD patterns are observed at the $Co(OH)_2$@P-NiCo-LDH nanohybrid. The FESEM micrograph of ZIF-67 presents the uniform dodecahedron shape having clear edge, smooth surface and sharp corner which suggests that it has the high crystallinity. The $Co(OH)_2$@NiCo-LDH-60 FESEM image in Fig. 17.8k revealed that in-situ development of NiCo-LDH grows on the ZIF-67 surface with no significant changing in core-shell structure. Finally, as-obtained the $Co(OH)_2$@P-NiCo-LDH-60 nanocomposite retains the parental dodecahedral shape after execution of phosphidation process (Fig. 17.8l). The TEM images in Fig. 17.8m exhibit the abundantly interwoven NiCo-LDH nanosheets anchored on the regular $Co(OH)_2$ dodecahedron surface and developed a hierarchical structure having 20 nm thickness of outer shell. Further, the Selected Area Electron Diffraction (SAED) in Fig. 17.8n of $Co(OH)_2$@P-NiCo-LDH-60 exhibits three clear and bright rings corresponding to (006), (009) and (110) basal planes of NiCo-LDH, suggesting the polycrystallinity properties of catalyst. Above results proved the fruitful construction of core-shell dodecahedral $Co(OH)_2$@PNiCo-LDH nanohybrid. Also, the colour mapping is represented in Fig. 17.8o (Song et al. 2021).

17.4.3 Bimetallic Sulphides

Yang and his research team constructed the two-dimensional (2D) nanosheet of bimetal sulphide arrays in-situ growth on NF (Ni_3S_2/Fe_9S_{10}@NF) and analysed the structural and morphological traits via diverse characterization techniques (Yang et al. 2021). The XRD plot displayed in Fig. 17.9a confirms the formation of NiFe LDH on NF by exhibiting the characteristic (003), (006), (012), (015), (018), (0015), (113) and (116) crystalline planes. Post in-situ sulphuration, respective sulphides have been formed by strongly displaying the main peaks for Ni_3S_2 at an 2θ angle of

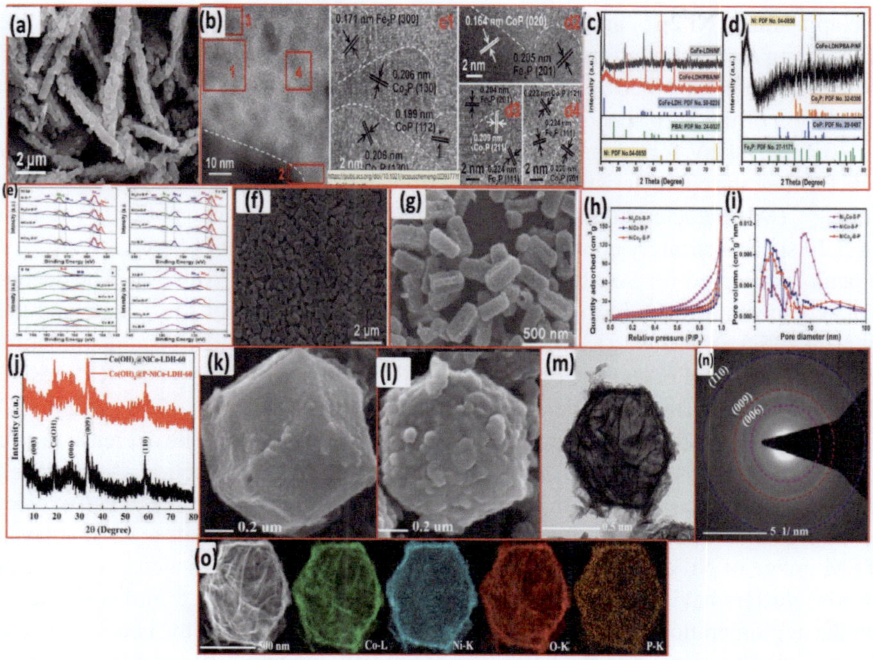

Fig. 17.8 **a** FESEM image, **b** TEM and HRTEM images of Co_xP–Fe_2P/NF, **c**, **d** XRD graph (Li et al. 2021, Copyright ACS, adapted with permission). **e** XPS image of Ni 2p, Co 2p, B 1s, and P 2p spectra at Ni–Co–B–P, Ni–B–P, and Co–B–P. SEM micrographs of **f** Ni_2Co pre, **g** $NiCo_2$–B–P. **h** N_2 sorption isotherms, **i** pore size distributions curve (Liu et al. 2021a, Copyright ACS, adapted with permission). **j** XRD analysis, **k**, **l** FESEM image of $Co(OH)_2$@NiCo-LDH-60 and $Co(OH)_2$@P-NiCo-LDH-60 respectively, **m** TEM image, **n** SAED pattern, **o** elemental mapping of $Co(OH)_2$@P-NiCo-LDH-60 (Song et al. 2021, Copyright Elsevier, adapted with permission)

21.8°, 31.1°,37.8°,44.3°, 49.7°, 50.1 and 55.1° and 11.2°, 12.7°, 12.5°, 21.5°, 33.8°, 43.7° for Fe_9S_{10} (Fig. 17.9b). Vertical anchoring of the NiFe LDHs nanosheets on the surface of NF has been clearly observed from the SEM image of NiFe LDHs@NF displayed in Fig. 17.9c. Porosity in this sample is attributed to the interlacing of the nanosheets which will not only improve the conductivity by the quick migration of electrolyte ions but also let out the generated bubbles in the process of electrolysis. Moreover, in Ni_3S_2/Fe_9S_{10} sample thinner and rougher nanosheet like structure has been observed after the sulphuration process (Fig. 17.9d). The displayed TEM image in Fig. 17.9e with the inserted SAED image confirms formation of numerous nanopores over the nanosheets owing to the ion exchange mechanism in sulphuration process (OH^-–S^{2-}). The formation of Ni_3S_2/Fe_9S_{10} heterostructures again confirmed by the HRTEM image of where lattice fringe of 1.8 nm is attributed to the (211) crystal plane of Ni_3S_2 and 5.8 nm is ascribed to the (322) of Fe_9S_{10}. Besides, Mo doping bimetallic sulphide on NiFe LDH supported with NF was developed by Li's group, where they confirmed the Mo doping of NiS-$NiiS_2$/NiFe LDH through XPS analysis. The +2 oxidation state of Ni 2p spectra (Fig. 17.9f) is confirmed by the appearance Ni $2p_{3/2}$ and $2p_{1/2}$ peaks at binding energy of 856.05 and 873.75 eV. The strong electronic interactions along with Mo doping are confirmed by the 1 eV positive shift in the Ni $2p_{3/2}$ and $2p_{1/2}$ signal of NiFe-LDH@NiS_2–NiS/NF. The high resolution of the S 2p spectrum in Fig. 17.9g shows the higher binding energy shift of ~0.4 and ~1 eV for $2p_{3/2}$ and $2p_{1/2}$ states compared with NiFe-LDH@NiS_2–NiS/NF. Further, the characteristic peaks for S^{2-} are centred at 161.55 and 163.55 eV, and for S_2^{2-} two characteristics peaks at 161.8 and 163.35 eV are also noticed. The shift in S 2p spectra strongly confirms the successful Mo doping in NiS and at 170.2 eV a peak is observed for SeO bond primarily due to the surface oxidation. The deconvulated spectra of Mo 3d in Fig. 17.9h centred at 232.4 eV for Mo^{4+} state and the peak at 235.5 eV is due to Mo^{6+} $3d_{3/2}$ caused by the surface oxidation contamination of sample exposed in air. The FESEM and HRTEM analysis support the XPS result. The SEM micrograph of NiFe-LDH@Mo–NiS_2–NiS/NF in Fig. 17.9i exhibits 3D flower like structure holds the parental NiFe LDH/NF structure. The decrease in size of the nanosheet is primarily attributed to the vulcanization reaction. As obtained 3D flower-like morphology offers more surface sites and contact area of active electrode material with water, thus producing more electroactive sites. As displayed TEM image of NiFe-LDH@Mo–NiS_2–NiS in Fig. 17.9j displays a huge number of flower clusters having sphere like structure having diameter of 1.5 mm constructed by the agglomeration of nano-sheets. The HRTEM image proved the heterostructure formation with displaying the lattice fringes of 0.20 nm and 0.19 nm corresponding to the Ni_2S (220) and NiS(102) plane, respectively. The elemental colour mapping in Fig. 17.9k exhibits obvious uniform signals for Ni, Fe, S and Mo elements, which is also reliable with the XRD data (Li et al. 2022). Furthermore, a free-standing NiCo LDH derived $NiCo_2S_4$ on edge-enriched N, O-codoped 3D rivet graphene (H-3DRG) heterostructure was prepared by Qin et al. The formation of NiCo-LDH is confirmed by Raman spectra displaying two characteristics modes of vibrations at 502 and 651 cm^{-1} (Fig. 17.9l). Further, the presence of Raman spectra peak at 280 cm^{-1} validates the transformation of $NiCo_2S_4$ from NiCo-LDH which is consistent with XPS outcomes. The survey XPS image demonstrated the existence of all elements Ni, Co, S, N, O and C in H-3DRG@$NiCo_2S_4$ heterostructure shown in Fig. 17.9m. The deconvoluted N 1s spectrum clarifies the presence of graphitic, pyridine, pyrrolic and oxides structure in H-3DRG and its interaction with NiCo-LDH has also been proved by the shift towards higher binding energy of N 1s spectrum in composite. Furthermore, in H-3DRG@$NiCo_2S_4$ the N 1s spectrum (Fig. 17.9n) displays the decreased peak intensities of Pyrrolic N, Pyridinic N and oxidized N rather that of H-3DRG@NiCo-LDH indicating the replacement of S atoms in presence of N while anion exchange process. The as displayed TEM image of H-3DRG@NiCo LDH in Fig. 17.9o demonstrated the vertically oriented NiCo-LDH nanoneedles having length of ~1 μm and diameter of 10–20 nm. Post sulphuration the as obtained $NiCo_2S_4$ hybrid consists of numerous nanosheets attached on the nanorod surface of to provide a backbone like architecture (Fig. 17.9p). The HRTEM micrograph of H-3DRG@$NiCo_2S_4$ additionally reveals its polycrystallinity as shown in lattice fringes with different orientations (Qin et al. 2018).

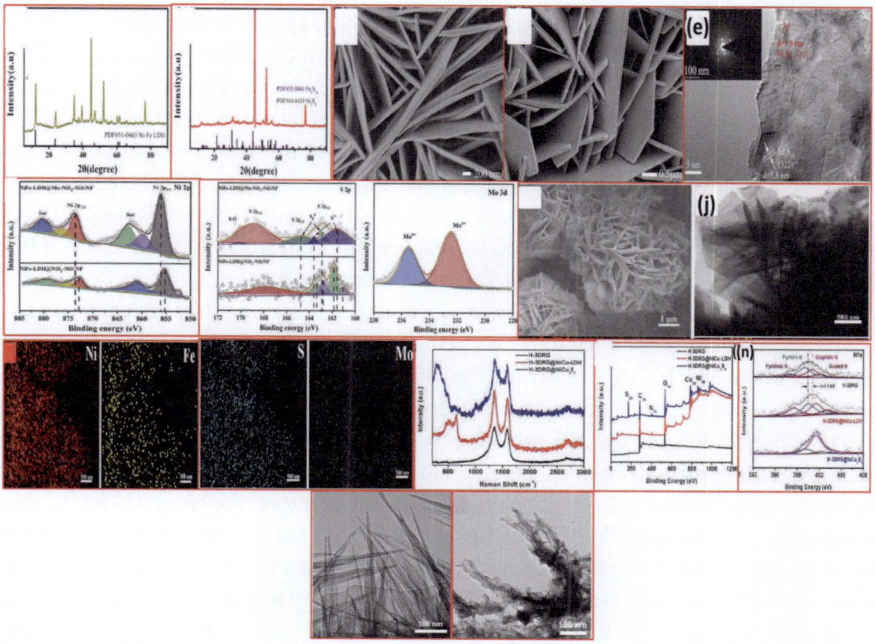

Fig. 17.9 **a**, **b** XRD plot, FESEM images. **c** NiFe LDHs@NF, **d** Ni_3S_2/Fe_9S_{10}@NF, **e** TEM images of Ni_3S_2/Fe_9S_{10}@NF (Yang et al. 2021, Copyright Elsevier, adapted with permission). XPS image of NiFe-LDH@Mo–NiS_2–NiS/NF and NiFe-LDH@NiS_2–NiS/NF of **f** Ni 2P, **g** S 2P. **h** Mo 3d. FESEM, **i** TEM. **j** images of NiFe-LDH@Mo–NiS_2–NiS/NF, **k** elemental mapping (Li et al. 2022, Copyright Elsevier, adapted with permission). **l** Raman spectra, **m** XPS survey scan, TEM images of **o** NiCo-LDH nanoneedles. **p** $NiCo_2S_4$ nanorods (Qin et al. 2018, Copyright RSC, adapted with permission)

17.4.4 Bimetallic Selenides

Wan's group investigated the structural, morphological and compositional characterization of bimetallic selenide (Co_3Se_4 and $FeSe_2$) derived from CoFe LDH (Song et al. 2021). As presented XRD graph in Fig. 17.10a displays the parent diffraction peaks of monoclinic Co_3Se_4 and in contrast weak peaks for $FeSe_2$ noticed due to the low content. Subsequently, the presence of $FeSe_2$ and formation of hybrid can also be verified via HRTEM analysis (Fig. 17.10b) where inter-planar distance of 0.245 nm between lattice fringe corresponds to the (−112) crystal plane of $FeSe_2$, and 0.250 nm matches to the (310) lattice planes of Co_3Se_4. The inductively coupled plasma atomic emission spectroscopy (ICP AES) analysis revels that the compositional ratio of Co, Fe, and Se is 3.76:1:7.2 in Co_3Se_4 and $FeSe_2$. Furthermore, a sheet like structure of $Zn_{1-x}Fe_x$-LDH and $Zn_{1-x}Fe_x$-oxyselenide supported NF was reported and their structural, functional and morphological analysis are characterized. The hydrotalcite-like layered hydroxide structure of the as prepared $Zn_{1-x}Fe_x$-LDH/

NF along with extra small peak corresponds to the oxyhydroxide phase and ultra-high intense peaks for NF substrate is confirmed from the XRD plot (Fig. 17.10c). Post selenization process completely altered diffraction patterns found from neat LDH and Se powder by strongly demonstrating the conversion of LDH to selenide. However, complete selenization is not achieved due to the absence of any reductive agent, hence a few oxide ions remain in the hybrid. Therefore, the observed diffraction patterns correspond as $Zn_{1-x}Fe_x$–oxyselenide/NF. In addition, the FTIR (Fig. 17.10d) spectra of $Zn_{1-x}Fe_x$–LDH displayed the absorption vibrational band at 3370, 3490, 1359 and 1506 cm^{-1} attributed to OH stretching, interlayer water molecule and CO_3^{2-} stretching vibration, respectively. Next the selenium treatment results to completely diminished IR band of $Zn_{1-x}Fe_x$–LDH and transformed to the $Zn_{1-x}Fe_x$-oxyselenide phase. FESEM analysis (Fig. 17.10e) revels nanosheet like morphology of the $Zn_{1-x}Fe_x$–LDH material (having thickness ~18 nm) which is grown by urea assisted condition over the NF substrate. A porous superstructure is noticed by the primary arrangement of nanosheets in a zigzag manner having plenty of mesopores and micro pores. Post selenization reaction, the nanosheet like morphology having rough edges in noticed (Fig. 17.10f). It is anticipated that the 'Se' substitution into the $Zn_{1-x}Fe_x$–LDH changes its layered structure in some extent. The surface roughness and thickness were evaluated via AFM microscopy where Fig. 17.10g(i) shows the surface topograpy and height profiling of the $Zn_{1-x}Fe_x$–LDH. Afore selenization thickness of ≈19 nm and post selenization thickness of ≈30 nm is observed for $Zn_{1-x}Fe_x$–LDH nanosheet (Fig. 17.10g(ii). Also, AFM additionally validates the fact that selenization process improves the surface roughness in the as prepared $Zn_{1-x}Fe_x$–LDH (Rajeshkhanna et al. 2018). Liu and their co-workers fabricated a NiFe LDH derived crystalline pyrite bimetallic (Ni, Fe)Se_2 hybrid heterostructure having amorphous NiFe$(OH)_x$ nanosheet arrays by taking CC as substrate. From the XRD data, the broad peaks for CC are noticed in between 20° and 30° and other diffraction signals of (Ni, Fe)Se_2 are related to pyrite $NiSe_2$. Subsequently after the electrodeposition, the previous diffraction peaks of the remains almost unchanged owing to the amorphous nature of deposited NiFe-based (oxy)hydroxides. For the confirmation, Raman analysis was accomplished as displayed in Fig. 17.10h. The characteristic Raman bands at 132, 145, 206 and 232 cm^{-1} are allocated to (Ni, Fe)Se_2 for (Ni, Fe)Se_2/CC sample, but post electrodeposition results 3 broad peaks at 484, 552 and 678 cm^{-1} strongly agreeing to the characteristic Raman peaks of as deposited amorphous NiFe$(OH)_x$. The SEM image in Fig. 17.10i demonstrated the surface of the (Ni, Fe)Se_2 nanosheets became rough having thickness increased from 26 to 45 nm. A finely flocculent layer of amorphous NiFe$(OH)_x$ is covered on the surface of (Ni, Fe)Se_2 nanosheets after electrodeposition for 300 s. From the HRTEM image, NiFe$(OH)_x$/(Ni, Fe)Se_2/CC displayed at Fig. 17.10j shows the dense nanoplate like morphology having crystalline core and amorphous thin walls of ~5 nm. The displayed lattice fringe having 0.265 nm interplanar spacing (Fig. 17.10j) is attributed to pyrite (Ni, Fe)Se_2, and the equivalent selected area electron diffraction (SAED) pattern represented in Fig. 17.10k also supports this fact. Furthermore, all four elements like Ni, Fe, Se and O are evenly distributed throughout the nanosheet inferred from the elemental mapping analysis in Fig. 17.10l (Liu et al. 2021a).

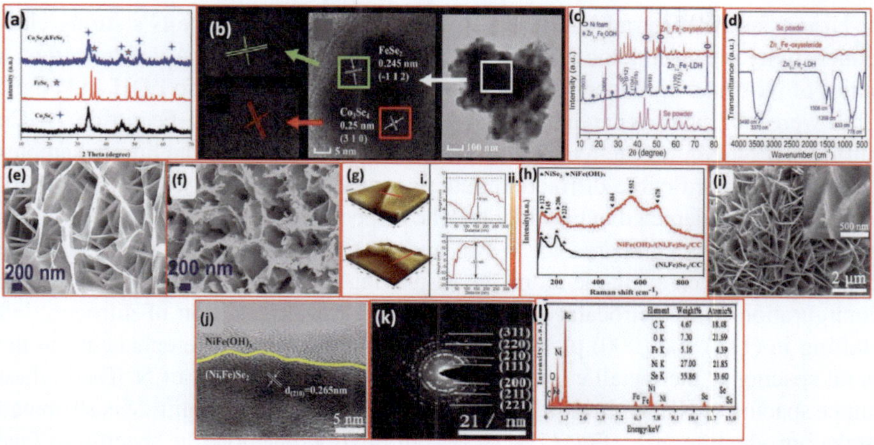

Fig. 17.10 **a** XRD patterns, **b** HRTEM images of Co_3Se_4 & $FeSe_2$ (Wan et al. 2019, Copyright RSC, adapted with permission). **c** XRD image, **d** FT-IR spectra, FESEM image of **e** $Zn_{1-x}Fe_x$–LDH/NF and **f** $Zn_{1-x}Fe_x$-oxyselenide/NF, **g** AFM images $Zn_{1-x}Fe_x$–oxyselenide/NF (Rajeshkhanna et al. 2018, Copyright Wiley, adapted with permission). **h** Raman analysis, **i** SEM micrograph for (Ni, Fe)Se_2/CC, **j** HRTEM images of NiFe$(OH)_x$/(Ni, Fe)Se_2/CC, **k** SAED pattern, **l** EDX analysis of nanocomposite NiFe$(OH)_x$/(Ni, Fe)Se_2/CC (Li et al. 2021, Copyright Wiley, adapted with permission)

17.4.5 Bimetallic Nitrides

Wen et al. reported the unique flower-like morphology of N-NiCo LDH (Ni_3N/Co_2N)/NF derived from pristine NiCo LDH and studied its various physico-chemical properties (Wen et al. 2021). As prepared N–NiCo-LDH shows XRD peak (Fig. 17.11a) positions indexed to Ni_3N, Co_2N and Ni derived from the NF substrate. SEM images in Fig. 17.11b indicates the surface roughness of the NF substrate, including abundant tiny spherical polymer like chains at the range of 3–4 mm. Here, each unit carries numerous layers of NiCo LDH nanosheets having thickness ~10 nm, that were embedded and twisted inside to form a flower like structure. HRTEM image displayed in Fig. 17.11c displays the heterojunction interface by dividing the N–NiCo-LDH into two phases, where lattice fringes of 0.111 nm and 0.122 nm assigned to the (221) and (113) facets in Ni_3N, while lattice fringes of 0.123 nm and 0.135 nm correspond to the (023) and (130) lattice planes of Co_2N, respectively. The XPS analysis displayed in Fig. 17.11d validates the N dopants successfully introduced into the NiCo-LDH structure. The deconvoluted N 1s XPS spectra are assigned to pyridinic-N, pyrrolic-N, graphitic-N and oxidized-N at binding energy of 397.8, 399.8, 401.4 and 403.4 eV, respectively. During catalyst optimization process by varying the pyrolysis time, NiCo-LDH favourably encourages the alteration of pyridine nitrogen rather than graphite and pyrrole nitrogen. Considering the best catalyst (N–NiCo-LDH-6, pyrolyzed 6 h), a slight positive shifting (0.1–0.2 eV) in pyridine and pyrrole nitrogen peaks is noticed which indicates the fully coordination with Co or Ni ions at the material interface. Apart from this owing to the N's strong electronegativity electron delocalization of metal ions is improved. Apart from this, a series of Ni-based LDH nanosheets including Ni_3Mn LDH, Ni_3Co LDH and Ni_3Fe LDH were fabricated as parent material for designing holey 2D Ni_3Fe nitride, Ni_3Co nitride and Ni_3Mn nitride nanosheets. The XRD patterns of all the nitride samples possess a standard cubic diffraction pattern indexed to the (111), (200) and (220) lattice planes as depicted in Fig. 17.11e. The respective interplanar spacing of Ni_3Fe nitride is more as compared to Ni_3Mn and Ni_3Co nitride, which could alter the pathway of surface available atoms by displaying the change in surface electronic configuration and electrocatalytic activity. After the introduction of nitrogen, left shifting in (111) and (200) planes of all the samples validate the enhancing in the basal spacing of the metallic state materials by the incorporation of N. The highest lattice spacing of Ni_3Fe nitride as compared to Ni_3Mn and Ni_3Co nitride is attributed to the introduction of N (Fig. 17.11f). Post nitridation reaction, the specific surface area thin holey 2D structured was improved from 114.9 (Ni_3Fe LDH) to 189.6 m^2 g^{-1} (Ni_3Fe nitride), while abundant pores were caused on lateral surfaces due to the phase transformation and Kirkendall effect (Fig. 17.11g). Further, the AFM analysis in Fig. 17.10h(i) gives more insight into the surface structure and thickness. The nanosheets of Ni_3Fe nitride have the thickness of 0.6–0.8 nm and around 3 atomic layers of (001) lattice planes. The trend found is obtained as displayed in Fig. 17.10h(ii) curves also specify the holey nanostructure on the lateral 2D surfaces present in the materials (Guo et al. 2018).

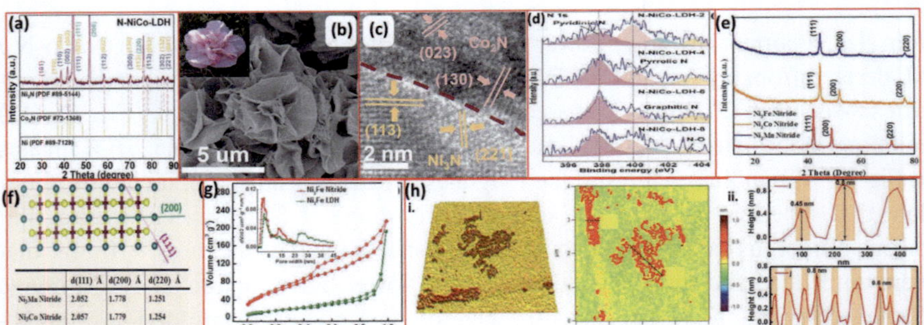

Fig. 17.11 **a** XRD pattern, **b** FESEM and **c** HRTEM image of N–NiCo-LDH. **d** XPS image of N–NiCo-LDH for N1s (Wen et al. 2021, Copyright RSC, adapted with permission). **e** XRD pattern, **f** distance between lattice planes image of 2D holey Ni_3Co nitride, Ni_3Fe nitride, Ni_3Mn nitride, **g** surface area (insert: pore size distributions plot), (**h** (i, ii)) AFM images for the 2D holey Ni_3Fe nitride (Guo et al. 2018, Copyright ACS, adapted with permission)

17.5 Electrocatalytic Water Splitting

17.5.1 Metal Oxyhydroxides and Mixed Metal Oxide

The 3d transitional metal oxide and oxyhydroxide have displayed decent electrochemical activities. However, the OER properties of these electrocatalytic materials are unsatisfactory due to the existence of restricted active site and thus lead to less electrocatalytic activities. The mixed metal oxide (MMO) and oxyhydroxide obtained from LDHs retain the 2D layered structure with high specific area, more approachability to surface-atom, flexible electronic structures and innately high dynamic performances that results in enhanced OER activity. Moreover, the multifaceted chemical composition and mutualism enable the mixed transitional-metal oxide to show high electrochemical activities. These MMOs can overtake the morphological LDH precursors having electro-active encouraged nanoparticle which are extraordinarily dispersed in the matrix of metal oxide which offers abundant active sites for electrocatalytic-based reaction. Through these morphological conversions, the active sites of well-ordered MMOs can be completed.

A vertically and laterally arranged CoOOH hexagonal nanoplate was synthesized from SO_4^{2-} and CO_3^{2-} intercalated CoAl LDH through chemical etching method, respectively, for electrocatalytic OER in 0.1 M KOH solution (Song et al. 2020). After etching through NaOH, the lesser interlayer space was converted to laterally stitched CoOOH and larger interlayer space was transformed to vertically aligned CoOOH nanoplate. As portrayed in OER polarization curve (Fig. 17.12a), the CoOOH-vertical possess a lower onset value of 1.55 V (vs. RHE) in comparison with CoOOH-lateral (1.59 V vs. RHE). These outcomes concur with the Co^{3+}/Co^{4+} redox potential above which the water oxidation happens along with steady rising of current density. In addition, at 1.69 V onset potential (vs. RHE), the CoOOH-vertical superstructure achieved 10 mA cm^{-2} which smaller than the CoOOH-lateral catalyst (1.75 V vs. RHE). Similarly, Fig. 17.12b displays a lowest Tafel slope (80.7 mV dec^{-1}) and smallest charge-transfer resistance ($R_{ct} = 16.6\ \Omega$) was observed in Fig. 17.12c for CoOOH-vertical suggesting more favourable reaction kinetics and better charge transfer than lateral catalyst. Moreover, the effective electrochemically active surface area (ECSA) of all as-prepared materials is strongly associated with the exposed active sites and intrinsic activity. From CV measurement, the ECSA value is estimated through electric double layer capacitance (C_{dl}). The CV curves of all electrodes were measured at different scan rate and by fitting these data to a straight line, the slope value of the line corresponds to the C_{dl}. The C_{dl} is directly proportional to the ECSA and the quantity of reaction active sites of the catalyst. The CoOOH-vertical catalyst displayed larger double layer capacitance (C_{dl}) value of 9.2 mF cm^{-2} with respect to CoOOH-vertical presented in Fig. 17.12d, which implies high electrons are collected on the vertically arranged nanoplates and provide more quantity of electro-active sites. More in, the controlling alignment of building blocks in a hierarchical CoOOH-vertical catalyst improves the OER performance. CA measurement illustrates that the two hierarchical superstructures of CoOOH electrocatalyst exhibits better stability at 1.67 V versus RHE. In addition, a series of Ni/Co-based LDH that NiCo-LDH, NiCo–CL (without stirring 12 h) and their corresponding NiCo oxides such as NiCo–H and NiCo–CH, respectively, electrocatalyst have been effectively synthesized through co-precipitation followed by calcination method. As presented the OER polarization image and overpotential bar diagram in Fig. 17.12e, f, respectively, the NiCo–CH required smallest overpotential of 343 mV for achieving 10 mA cm^{-2} in comparison with other pristine LDH, metal oxide products, even being lesser than the state-of-the-art RuO_2 (~486 mV). Moreover, a ultrasmall Tafel slope of 66 mV/dec^{-1} was observed for NiCo–CH as compared to other catalyst indicating better OER performance. The charge transport between the catalyst and the electrolytic solution at NiCo–CH is more superficial and faster than NiCo-LDH. The NiCo-LDH displayed highest C_{dl} value (Fig. 17.12g) which was nearly 15.5-fold of that NiCo-LDH. Subsequently, the mixed oxide products have higher ECSA value than precursor LDH suggesting good electric conductivity and electron movement ability. Although the NiCo–CL has largest specific surface area than NiCo–CH, but the higher ECSA value and relatively large surface area suitable for increasing its OER behaviour. Moreover, the high performance of OER over NiCo–CH catalyst attributed to (i) the more amount of Co_{2p} and Ni_{3p} being exposed on the NiCo–CH surface, (ii) the introduction and elimination of interlayer $CO_3{}^{2-}$ ion in intergallery of NiCo-LDH resulting to the metal oxide porous nanostructure having small particle sizes (iii) a moderately high BET surface area. In addition, the CA response of NiCo–CH at 1.53 V (vs. RHE) potential displayed (Fig. 17.12h) stable current density from 12.9 mA cm^{-2} to 11.2 mA/cm^{-2} during 9000 s inferring necessary durability (Wang et al. 2020). Zhou et al. fabricated a unique $MnCo_2O_4$@Co_3O_4 nanoneedle supported on NF from ZIF-67 dodecahedron penetrated CoMn LDH nanoneedles array for electrocatalytic OER in 1 M KOH electrolyte. The polarization curve was measured with 90% iR correction at 5 mVs^{-1}. The $MnCo_2O_4$@Co_3O_4 exhibited lower onset potential with high current density due to the strong oxidation peaks changing from M^{2+} to M^{3+}. Figure 17.12i, j illustrated that the $MnCo_2O_4$@Co_3O_4 MMO nanocomposite needed smallest overpotential of 280 mV and low Tafel slope of 109.6 mV dec^{-1} in comparison with $MnCo_2O_4$ and Co_3O_4. Furthermore, a highest C_{dl} value of 232.9 C cm^{-2} was noticed for MMO nanocomposite in Fig. 17.12k signifying improved water oxidation activity. The stability test was measured at a 1.66 V of static potential, illustrated that $MnCo_2O_4$@Co_3O_4 nanoneedle arrays exhibit the 82.38% of the original current density is maintained up to 10 h, whereas 5.1% of Co_3O_4 and 19.1% of $MnCo_2$O4, is reduced (Fig. 17.12l), inferring a better stability. The Faradaic efficiency of 98.76% was calculated from the ratio of the theoretical and experimental quantities of evolved O_2. The enhanced performance over $MnCo_2O_4$@Co_3O_4 MMO nanocomposite may be assigned to (a) the $MnCo_2O_4$@Co_3O_4 nanoneedle on NF substrate offers a straight electrical path from active species of catalyst to the current collector; (b) the ZIF-67 polyhedron derived porous Co_3O_4 can deliver more reactive corner for Faradaic redox reaction; (c) the synergistic effect arise from the different combination of MMO in $MnCo_2O_4$@Co_3O_4 lead to the supercapacitive performance towards electrocatalytic water oxidation (Zhou et al. 2018).

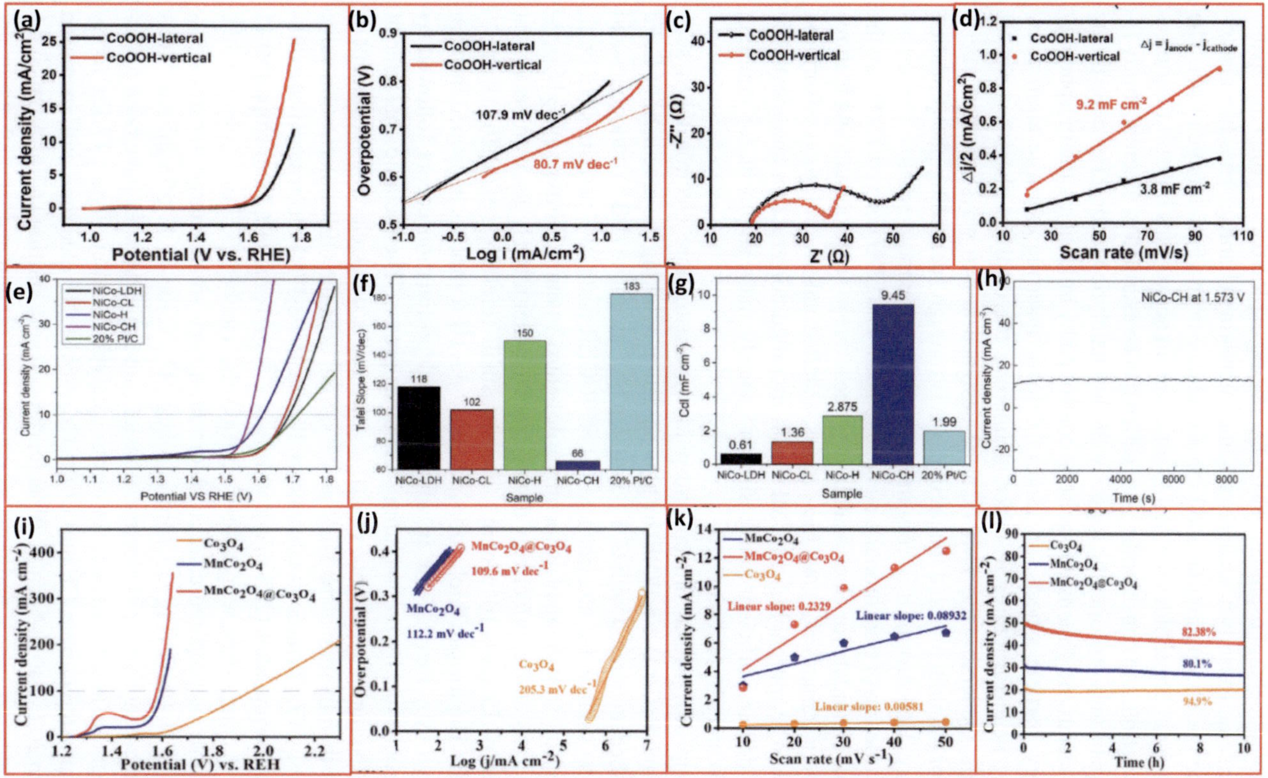

Fig. 17.12 **a** polarization curve, **b** Tafel plots of OER (0.1 M KOH with 20 Vs^{-1} scan rate), **c** Nyquist plot. **d** C_{dl} plot at 1.47 V versus RHE as a function of scan rate (Song et al. 2020, Copyright RSC, adapted with permission). **e** LSV plot, **f** Tafel value image for OER, **g** C_{dl} value of respective samples, **h** CA profile at 1.573 V (Wang et al. 2020, Copyright Elsevier, adapted with permission). **i** polarization curve and **j** Tafel plot with iR correction, **k** C_{dl} value of respective sample with 1.18 V scan rate (vs. RHE). **l** CA curve at 1.66 V scan rate (vs. RHE) with 0.1 M KOH (Zhou et al. 2018, Copyright Elsevier, adapted with permission)

17.5.2 Bimetallic Phosphides

The transitional metal phosphides (TMPs) are being explored as a group of capable bi-functional electrocatalyst because of some interesting characteristics like electrically conductive, cheap, durable and high abundance over a broad range of pH. The transitional metal phosphides are highly abundant in active sites but still their electrocatalytic actions are considerably low which can be ascribed to inherent sturdy H-bonding capabilities, low specific capacity and low catalytic kinetics. Henceforth, the atomic and electronic dimensions must be controlled by influencing the morphology and topology, particle aspects, creation of defects, inserting extrinsic atom or by constructing heterostructures in order to improve the catalytic characteristics of the materials. Generally, the TMPs can be designed in two ways: the first one is via precursor manufacture and the second one is through succeeding phosphidation methodology. Henceforth, it is very much important to choose suitable precursors for the construction of transitional metal phosphides. The LDH derived TMPs are the most suitable candidate for electrocatalytic process due to some outstanding features like thin-layer framework, tuneable cations, inserted anions and good adsorption competence. The morphological TMPs with modified construction and electronic circulation deliver sufficient active site and speedy mass shipment for attaining exceptional catalytic activities.

Li's group designed a bimetal phosphide heterostructure electrocatalyst self-supported on NF (Co_xP–Fe_2P/NF) by phosphidation as taking the precursor of CoFe LDH nanosheet decorated with CoFe-Prussian blue analogue (PBA) nanocubes and examined the catalyst for electrochemical HER OER and OWS activities (Li et al. 2021b). Profiting from bimetal nanocomposite heterostructure, the as prepared Co_xP–Fe_2P/NF electrocatalyst effectively dissociates the water into H_2 gas with the lowest overpotential of 75 mV@10 mA cm^{-2} and ultrasmall Tafel slope of 90.26 mV dec^{-1} in 1 M KOH. The lowest Tafel slope indicating the HER process happened through Volmer-Heyrovsky mechanism, in which rapid deprotonation occurs followed by again electrochemically recombining of another proton. The metallic character of Co_xP–Fe_2P/NF and the presence of Co^0 accelerate the quick electron movement of protons and then facilitates the HER process. Similarly, the OER polarization value depicted in Fig. 17.13a indicates that bimetallic Co_xP–Fe_2P/NF electrocatalyst exhibited lowest overpotential (265 mV@50 mA cm^{-2}) and smallest Tafel slope (50.11 mV dec^{-1}) in comparison with state of art RuO_2/NF catalyst (373 mV), parent CoFe LDH and other as-prepared catalyst. More in, a very small charge transfer resistance (R_{ct}) of 0. 96 Ω (Fig. 17.13b) and large C_{dl} value of 17.505 $mFcm^{-2}$ was noticed suggesting that rapid electron transfer more reactive sites have been generated at bimetallic heterostructure for electrocatalytic reaction. Figure 17.13c represents the multistep current CP plot of the Co_xP–Fe_2P/NF electrode with increased current from 20 to 240 mA cm^{-2}. When current density starting at 20mA cm^{-2}, the potential rise up to 1.44 V and remains constant till 60 min and parallel results are observed for other steps with increasing the current density to 240 mA cm^{-2}, signify that this heterostructure material reveals better mass transport properties and robust performance on electron conductivity. Specially, when $Co_xP–Fe_2P/NF$ acts as both cathode and anode, 1.61 V (vs. RHE) (Fig. 17.13d) of cell voltage is needed for 10 mA cm^{-2} current density with outstanding stability in long-standing operation for 100 h (Fig. 17.13e), which is equivalent to $Pt/C/NF||RuO_2/NF$. The enhanced water splitting performance over bimetal compound assigned to following reasons; (i) the as-prepared $Co_xP–Fe_2P/NF$ heterostructure possesses multi-heterogenous interface which result in rapid charge transfer, enhances the reaction kinetics and able to distribute charge between Co_xP and Fe_2P, accordingly the adsorption ability of reactant was optimized; (ii) the unique 3D cross linked structure fabricated from 3D NF, PBA nanocubes and 2D CoFe LDH increased the reaction rate with mass and charge transport; (iii) the $Co_xP–Fe_2P$ self-supported on NF electrode without using any binder confirming fast charge movement and evading electrocatalyst detaching. In another attempt, a uniformly dispersive Ni_2P implanted in 2D amorphous Al_2O_3 array supported on NF ($NiP–Al_2O_3/NF$) electrocatalyst was derived from reduction of NiAL LDH precursor for efficient water oxidation reaction. The OER LSV plot depicted in Fig. 17.13f was measured in 1 M KOH at 5 mVs^{-1} scan rate. The unique nano/micro structure $NiP–Al_2O_3/NF$ catalyst displayed low overpotential of 193 mV@10 mA cm^{-2} (Fig. 17.13f) and smallest Tafel slope of 63 mV dec^{-1} (Fig. 17.13g) than bench mark RuO_2, $Ni–Al_2O_3/NF$ and bare NF signifying the better catalytic kinetics for O_2 evolution. The as-obtained $NiP–Al_2O_3/NF$ electrocatalyst showed a 96% faradaic efficiency throughout OER process. Furthermore, the electrochemical impedance spectroscopy (EIS) results revealed that $NiP–Al_2O_3/NF$ displays a low radius of Nyquist plot having interfacial R_{ct} value of 0.91 Ω and highest ECSA value having C_{dl} of 4.62 $mFcm^{-2}$ (Fig. 17.13h) reflecting the better electron transfer and improving OER activity. In addition, these outstanding OER performance over $NiP–Al_2O_3/NF$ catalyst could be assigned to as following reasons: at first the exceptional electronic orientation of Ni_2P results to increase the adsorption ability of reactive intermediate and diminished the overpotential for oxidation reaction. Second, the interconnected conductive network, vertically designed nano array and uniformly embedded of Ni_2P nanoparticles on Al_2O_3 bed assurance for fast charge transfer rates and fully disclosing of the active corners during electrocatalysis. Third, the macro/mesopore of the 2D Al_2O_3 nanoarrays and macroscopical pore of NF not only expand the access of the electrode to electrolytic solution but also permit the discharging of the O_2 gas bubble. Lastly, the in-situ development of NiAl LDH parent material supported on the NF skeleton permits the complete interaction and make final $NiP–Al_2O_3/NF$ electrocatalyst with brilliant structural durability (Li et al. 2019). In 2021, Chen et al. designed an interlaced vertically oriented Ni/Fe nanohybrid on MXene surface through a hydrothermal followed by pyrolysis phosphating process and evaluated its electrocatalytic OER performance. The optimized NiFeP/MXene (molar ratio of NiFe LDH = 2:1) exhibited lowest overpotential (286 mV@10 mA cm^{-2}) and also ultrasmall Tafel slope (35 mV dec^{-1}) in comparison with parent NiFe-LDH/MXene and NiFe-LDH/GO nanocomposite, pristine NiFe-LDH and reference IrO_2 (Fig. 17.13i). Furthermore, the NiFeP/MXene nanocomposite presented 92.59% of Faradic efficiency enclosed in Fig. 17.13j due to the close interaction of bimetal phosphide (NiFe-P) and MXene. The Nyquist plot portrayed in Fig. 17.13k reveal that 6.6 Ω of R_{ct} value was noticed for phosphating NiFeP/MXene catalyst which was lower among other catalyst indicating better charge separation from electrolyte to electrode. For stability of catalyst, CP test was measured in 1 M KOH solution at constant 10 mA cm^{-2} current density (Fig. 17.13l). NiFeP/MXene expresses the best stability up to 12 h with little increasing of overpotential only 46 mV. By taking the density function theory (DFT) analysis and experimental results, Chen's group illustrated the following aspects for outstanding OER performance of NiFeP/MXene nanocomposite: (i) The well dispersed NiFeP particle on conductive MXene surface avoid agglomeration. (ii) The synergistic effect between NiFeP and MXene permits the fast electrons transfer between NiFeP and MXene, resulting NiFeP being as efficient electron acceptor. (iii) After phosphidation, the P atoms of NiFeP/MXene nanocomposite have larger electronegativity owing to spontaneous electron transfer from Ni and Fe to P, which effectually endorsing the upshifting d-band centre and valence of Ni atom, allowing the catalyst to energetically adsorb oxygenated intermediates. Moreover, when NiFeP/MXene acts as anode and Pt/C as cathode with loading catalyst on NF in two-electrode configuration, the OWS reaction happened (Fig. 17.13m). As presented in Fig. 17.13n, the cell potential of 1.61 V was needed to gain a current density of 10 mA cm^{-2} which is close to noble metal electrocatalyst $IrO_2(-)$//Pt/C(+). Meanwhile, the CP measurement illustrated that NiFeP/MXene preserves brilliant stability during water splitting reaction for 12 h (Fig. 17.13o) due to its exceptional vertically array interwoven architecture (Chen et al. 2021).

17.5.3 Bimetallic Sulphides

Transitional metal sulphides (TMSs) have been extensively investigated as electrocatalytic material for HER, OER and OWS because of exceptional properties such as exposed edge-sites, inexpensive and high catalytic activities. However, agglomeration and low conductivity along with the instability under operative circumstances are some severe issues which hinder their practical efficacy. The chemical vapour deposition (CVD) and solvothermal are some of the common strategies that have been implemented towards construction of TMS. The tough apparatus requisite, low production, unsatisfactory tuneability and accumulated difficulties hinder the application of CVD strategy for fabricating the TMS. Very recently, by utilizing the benefits of LDH structures and configuration, the TMSs are designed from LDH as sources with H_2S/S_2 which is on the basis of Kirkendall effect. The as-synthesized TMSs maintain their original LDH structure along with identical dispersion of metal cation that provides large specific area and catalytic sites.

A bimetallic NiS/NiS_2 heterostructure composed of NiFe LDH with Mo doping 3D nano-flower structure was studied for electrocatalytic HER, OER and OWS activities in alkaline solution. The bimetal sulphide heterostructure NiFe-LDH@Mo–NiS_2–NiS/NF-2 (0.084 gm of thioacetamide) required the lowest overpotentials of 261 and 120 mV to attain the of 50 and 10 mA cm^{-2} of current density (Fig. 17.14a, b)

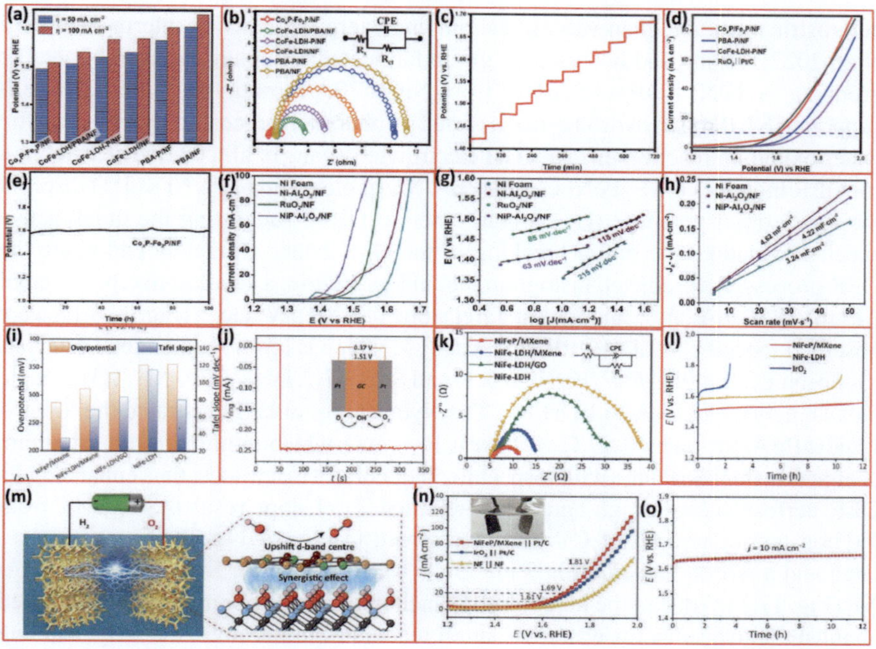

Fig. 17.13 **a** Overpotential comparison at 50 and 100 mA cm^{-2} current density for OER, **b** Nyquist plot, **c** multicurrent process over Co_xP–Fe_2P/NF, **d** polarization curve for OWS at 2 mVs^{-1} for Co_xP-Fe_2P/NF, CoFe LDH-P/NF and PBA-P/NF, **e** CP test at 10 mA cm^{-2} for Co_xP–Fe_2P/NF (Li et al. 2021, Copyright ACS, adapted with permission). **f** LSV curve, **g** Tafel slope for OER, **h** capacitive current with a function scan rate at 1.23 V vs RHE (Li et al. 2019, Copyright RSC, adapted with permission). **i** Overpotentials values of NiFeP/MXene, NiFe-LDH/MXene, NiFe-LDH/GO, NiFe-LDH and IrO_2 at 10 mA cm^{-2}, **j** Faradaic efficiency for NiFeP/MXene, **k** Nyquist plots. **l** CP response at 10 mA cm^{-2}, **m** schematic representation of cell containing two electrode system having NiFeP/MXene//Pt/C, **n** OWS polarization curve, **o** CP measurement for stability test of NiFeP/MXene (Chen et al. 2021, Copyright Elsevier, adapted with permission)

and own the smallest Tafel slope of 86 and 105 mV dec^{-1} for OER and HER, respectively, as compared to NiFe-LDH@NiS_2–NiS/NF, parent NiFe LDH/NF, RuO_2/NF and other catalysts. As depicted from CP curve in Fig. 17.14c the potential did not change up to 20 h at 50 mA cm^{-2} for OER and 10 mA cm^{-2} for HER. These results demonstrated that after Mo doping and sulphuration the bimetallic heterostructure possess long-term stability as compared to the NiFe-LDH. More in, when NiFe-LDH@Mo–NiS_2–NiS/NF acts as the bifunctional catalysts for OWS, it only desires only 1.63 V of cell voltage to gaining 10 mA cm^{-2} current density (Fig. 17.14d). The excellent electrocatalytic behaviour of NiFe-LDH@Mo–NiS_2–NiS/NF nanoflower heterostructure is attributed to: (a) The in-situ growth of Mo doped NiFe LDH on NF and subsequent sulphuration to make the close connection between the NF and active sides of catalyst, resulting to advanced charge transfer kinetics and quick charge transfer rate, (b) there was a robust interaction between Mo element and NiS_2–NiS on NiFe-LDH nanosheets leading to the interfacial charges distribution, constructing heterostructures, and then varying the binding energy of reaction intermediates (Li et al. 2022). Zhang and his research group fabricated a sulphur enriched bimetallic products (S–FeNi LDH@PBA/NF) from NiCo-PBA/Flower like NiFe LDH self-supported NF through hydrothermal assisted sulphuration process for electrocatalytic water oxidation reaction in 1 M KOH electrolyte (Zhang et al. 2022). The LSV plot for OER indicates the S–FeNi LDH@PBA/NF (molar ratio of Fe/Ni = 1:1) displays the lower onset potential (E_{onset}) than others catalyst, suggesting the quick active species oxidation and easier breaking of the O–H bond, ultimately endorsing the OER process. As predicted histogram plot in Fig. 17.14e, a smallest overpotential of 243 mV@50 mA cm^{-2} and lowest Tafel slope of 62.9 mV dec^{-1} (Fig. 17.14f) were observed for S–FeNi LDH@PBA/NF catalyst, which is lower than bimetal catalysts [S–Fe_3Ni LDH@PBA/NF (Fe/Ni = 3:1) and S–$FeNi_3$ LDH@PBA/NF (1:3)], single metal catalyst and S–FeNi LDH/NF. The Nyquist plot of EIS illustrated the S–FeNi LDH@PBA/NF owns 0.89 Ω of lowest R_{ct} value which confirmed the minimum charge transfer resistance with respect to other control electrodes, thus enhancing its OER performance, in high agreement with the Tafel slope results. Similarly, from the non-faradic regions of CV plot represented in Fig. 17.14g, the C_{dl} was calculated and a largest value of 10.33 mF cm^{-2} was obtained for S–FeNi LDH@PBA/NF (Fig. 17.14h) due to the unique cubes anchored nanoflower morphology exposed abundant active sides for water oxidation reaction. Furthermore, the long-standing CP plot of optimized catalyst demonstrated that at 30 mA cm^{-2}, the overpotential almost keep constant up to 40 h operation. The structure of the catalyst did not change after long-standing OER measurement. After 1000 CV run of S–FeNi LDH@PBA/NF catalyst, the LSV curve (Fig. 17.14i) diverges a little bit with increased 6 mV of overpotential to reach the 100 mA cm^{-2}. This group also tested two-electrode overall water splitting by taking S-FeNi LDH@PBA/NF as e anode and marketable Pt/C as the cathode in alkaline medium (Fig. 17.14j). As expected, the S–FeNi LDH@PBA/NF//Pt/C/NF electrolyser needed the much lesser overpotential than the state-of-art RuO_2/NF//Pt/C/NF. In addition, the brilliant OER activity over S-FeNi LDH@PBA/NF with the fact concluded as: (1) the 3D self-supported NiCo-PBA nanocube/flower-like NiFe LDH heterostructure on porous NF provides the catalyst with copious active corners disclosed to the targets. (2) Such exceptional structure successfully avoids the physical damage and integrated by hastening the mass diffusion and discharging of the evolved O_2 gas, thus encouraging its catalytic stability. (3) The synergetic effect between Fe, Ni and S atom in final heterostructure with established NiFe LDH nanoflower and NiCo-PBA nanocube structure significantly reduces the energy barrier, greatly improve the charge migration resulting to better OER performance.

17.5.4 Bimetallic Selenides

Transitional metal selenides (TMSes) have been found to be an important group of active electrocatalyst for water-splitting. During HER, the metal selenide delivers

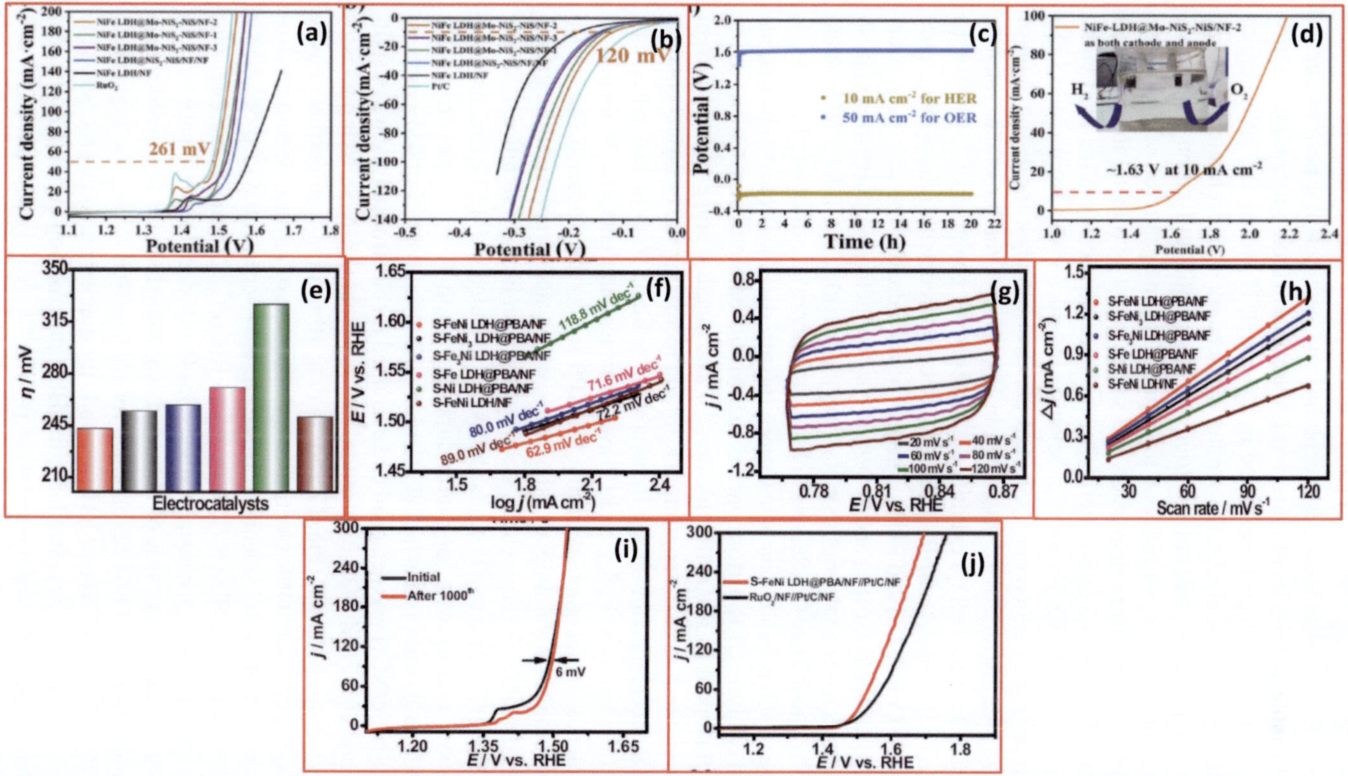

Fig. 17.14 **a**, **b** OER and HER polarization curves, respectively, **c** CP curve of HER and OER at 10 and 50 mA cm^{-2}, **d** OWS polarization curve in 1 M KOH (Li et al. 2022, Copyright Elsevier, adapted with permission). **e** overpotential comparison bar diagram, **f** Tafel plot of respective catalysts, **g** CV curves of S–FeNi LDH@PBA/NF, **h** C_{dl} plot, **i** LSV curve of S–FeNi LDH@PBA/NF before and after 1000 cycles test, **j** polarization curve for OWS (Zhang et al. 2022, Copyright Elsevier, adapted with permission)

very well activity but some restrictions are witnessed such as metal-leaching in acidic environment whereas they can be transformed into hydroxide in alkaline environments. But, in OER the performance is exceptional and metal selenide performs better than those of simple hydroxide, oxy-hydroxides, etc. due to their outstanding electro-conductivity and stable electron occupied in e_g-orbital. However, the designing of such catalyst is quite unwieldy, i.e. the synthesis procedure is problematical and time taking treatment processes that have restricted their potential utilization. Therefore, LDH as originator illustrates a pathway to acquire the amalgamation of diversified metal selenides. The uniform dispersion and scattering of two metal ions in LDH framework that facilitates a close interactions of selenide species in the final yield. Wan et al. prepared the Co/Fe bimetal selenide hybrid derived from CoFe LDH through hydrothermal method and examined the electrocatalyst for dissociation of water into O_2 (Wan et al. 2019). As shown in Fig. 17.15a, the bimetallic selenide composite (Co_3Se_4 and $FeSe_2$) exhibits the lowest over potential of 280 mV@10 mA cm^{-2} towards greatest OER capability as compared with pure $FeSe_2$, pure Co_3Se_4, pristine CoFe LDH and RuO_2. This can be accredited to the core-shell structural arrangement where Co_3Se_4 and $FeSe_2$ play as conductive support to accomplish fast charge transference analogous to metal (oxy)hydroxide. The Tafel plot depicted in Fig. 17.15b revealed that Co_3Se_4 and $FeSe_2$ nanocomposite exhibited 51 mV dec^{-1} of Tafel slope which is minor than bare Co_3Se_4 (66 mV dec^{-1}). These results suggest that the good catalytic OER performance is an intrinsic feature of pure Co_3S_4 having reduced Tafel slope than RuO_2 (86 mV dec^{-1}). The metallic property and inherent electronic configuration near to optimum e_g filling increasing the electrical conductivity of Co_3S_4, accelerate the charge transfer with reducing solution resistance. The synergetic effect between $FeSe_2$ and Co_3Se_4 increased the catalytic behaviour of bimetallic selenide nanocomposite. From the XPS results the oxidation state of Co has been encouraged by coupling with the $FeSe_2$, which again improved the catalytic function. Besides of low overpotential, stability character is also a significant pointer of catalysts for large-scale applications ability. As depicted in Fig. 17.15c, the catalyst holds good stability with long-term durability up to 40 h. Moreover, a NiZn LDH derived hexagonal porous nickel selenide embedded over N-doped graphene composite (p-NiSe/NGr) was designed and evaluated for electrocatalytic water splitting into H_2 and O_2 in 1 M KOH electrolyte. The as-prepared p-NiSe/NGr possesses only 311 mV of overpotential (Fig. 17.15d) to drag 10 mA cm^{-2} of current density signifying highly effectual OER electrocatalysis because of intrinsic in-situ electrochemical oxidation of nickel selenides. However, the overpotential of p-NiSe/NGr is slightly greater than ZnNiNGr (290 mV) and RuO_2/C (240 mV), but Tafel plot revealed the lowest Tafel slope over p-NiSe/NGr (44 mV dec^{-1}) in comparison with ZnNiNGr (49 mV dec^{-1}) because of improved electronic features of selenides than LDH. The low R_{ct} value (11.68 Ω) obtained for p-NiSe/NGr during the impedance analysis measured at 1.55 V versus RHE supports the quicker Faradaic process on the catalysts surface (Fig. 17.15e). Similarly, the HER polarization curve (Fig. 17.15f) in alkaline condition displayed that the main p-NiSe/NGr catalyst required 285 mV negative potential shift to attain the 10 mA cm^{-2} from the benchmark 20% Pt/C, whereas only 122 mV of overpotential was observed in acidic medium (0.5 M

H_2SO_4). The Tafel slope value of 125 mV dec^{-1} for p-NiSe/NGr catalyst reveals that HER mechanism followed Volmer-Heyrovsky pathway. More in, the experimentally measured quantity of gas molecule matches well with the theoretical value showing 98.5% and 96.36% maximum Faradaic efficiency for OER and HER, respectively. Furthermore, the LSV curve of completely water electrolysis for (p-NiSe/NGr)-CC (carbon cloth) and bench mark (Pt/C)-CC‖(RuO_2/C)-CC show that voltage of 1.69 V and 1.70 V, respectively. Figure 17.15g signifies the LSV polarization comparative curves for OWS at 10 mVs^{-1} scan rate in 1 M KOH. The LSV plot for (Pt/C)-CC‖(RuO_2/C)-CC and (p-NiSe/NGr)-CC shows that voltage of 1.62 V and 1.60 V, respectively, producing 5 mA cm^{-2} current density, were enough for complete water splitting reaction using this catalyst and benchmark system. The stability measurement of (p-NiSe/NGr)-CC-based bifunctional electrocatalyst was measured by CP test for 8 h at 1.7 V presenting the 5 mA cm^{-2} current density was retained throughout the OWS analysis (Fig. 17.15h). Th advanced electrocatalytic water splitting over p-NiSe/NGr catalyst assigned to the exceptional structure having $NiSe_2$ crystalline phases, Ni_3Se_4 and amorphous NiO_x, metallic selenium anchored in the NGr skeleton reduced the charge transport path, providing plentiful active canters, enhanced mass transfer phenomena and better structural stability against accumulation, ensuing in good activity for electrochemical OWS reaction (Nadeema et al. 2018).

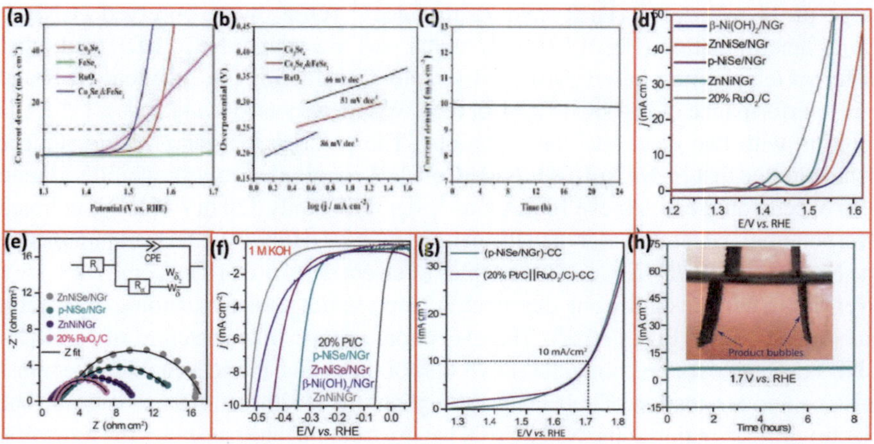

Fig. 17.15 **a** Polarization curves, **b** Tafel slope for OER, **c** stability curves for Co_3Se_4 and $FeSe_2$ (Wan et al. 2019, Copyright RSC, adapted with permission). **d** OER polarization plot at 10 mVs^{-1} in 1 M KOH, **e** EIS plot, **f** HER polarization comparison in 0.5 M H_2SO_4, **g** OWS polarization comparative curves at scanning rate of 10 mV s^{-1}, **h** CP stability test over p-NiSe/NGr (Nadeema et al. 2018, Copyright ACS, adapted with permission)

17.5.5 Bimetallic Nitrides

Transitional metal nitrides (TMNs) parade outstanding activity in water-splitting due to their unique characteristics such as low resistance to exciton shipment, improved active site, exceptional mechanical-strength, great resistant to erosion, ease of transformation to catalytic active classes and metallic behaviour. The developed N-metallic bond in TMNs regulates the inherent characteristics of the D-band which results in an enhanced density of states (DOS) which is subsequent to the Fermi level in respect to the pure metal. Moreover, the bimetallic and ternary nitrides display outstanding activity as compared to that of non-metallic nitrides as the mixing of element can modify the electronic structures and valence states the metal. The nitridation of LDH nanosheet resulted in the nanostructured mixed-metal nitride (MMNs) which outperformed the benchmark catalysts for OER, HER and OWS.

In 2021, Wen's group constructed a bi-functional electrocatalyst by incorporating N atom to unique NiCo LDH nanoflower (N–NiCo-LDH) for enhancing HER, OER and OWS. The electronic relocation at the bimetallic nitride (Ni_3N/Co_2N) heterojunctions interfaces favours the energy-barrier for HER/OER and, thus, increases the electrocatalytic activity. As presented in Fig. 17.16a, c, the N–NiCo-LDH-6 (pyrolyzed for 6 h) shows an ultra-low overpotential of 35 mV @10 mA cm^{-2} and 250 mV@10 mA cm^{-2} for HERs and OER, respectively. Similarly, a smaller Tafel slope of 34 mV dec^{-1} (HER) and 75 mV dec^{-1} (OER) were observed at Ni_3N/Co_2N heterojunction (Fig. 17.16b, d) which was lower than NiCo LDH and other different temperature-based pyrolyzed sample. Considering the excellent HER and OER performance of N–NiCo-LDH-6, the OWS activity also evaluated in IM KOH solution with two electrode configurations. The polarization curve in Fig. 17.16e demonstrated that N–NiCo-LDH//N–NiCo-LDH couple electrolyser requires around 1.55 V cell voltage to achieve 10 mA cm^{-2}, signifying only 320 mV of low combined overpotential, which is lesser than the parent NiCo-LDH//NiCo-LDH couple. Beside the N–NiCo-LDH//N–NiCo-LDH couples a current density of 10 mA cm^{-2} at 1.55 V over 24 h with insignificant deprivation, representing its outstanding long-term catalytic stability (Fig. 17.16f). The exceptional flower like structure of N–NiCo-LDH is valuable for revealing the active sites of the catalysts and hastening the transmission and dissipation of substance throughout the electrocatalytic processes. The electronic distribution at the Ni_3N/Co_2N heterostructure interface helps the reduction in the energy barricade for HERs/OERs and, henceforth, advances the electrocatalytic performance that pyridine nitrogen and metal ion play a crucial role in synergistically promoting water splitting catalytic activities. Herein, the CoNi LDH derived CoNiN composed NiFe LDH hybrid architecture was fabricated as bifunctional electrocatalyst towards OWS through nitridation followed by in-situ electrochemical assembling technique (Wang et al. 2021b). CoNiN@NiFeLDH expresses brilliant OER activities with the 227 and 291 mV of overpotentials to achieve the 10 and 100 mA cm^{-2} of current densities respectively (Fig. 17.16g). The CP study in Fig. 17.16h demonstrated at constant 10, 50, and 100 mA cm^{-2}, the potentials of CoNiN@NiFe LDH are slightly changed. Precisely, the CoNiN@NiFe LDH possesses superior stability than Fe/Co/Ni-based reported electrocatalysts for OER. For water reduction reaction, the best catalyst possesses only 150 mV@10 mA cm^{-2} of overpotential which was lower than CoNiO@NiFe LDH, CoNi@NiFe LDH, NiFe LDH (Fig. 17.16i). Moreover, when CoNiN@NiFe LDH acts as both cathode and anode it needs only 1.63 V of cell voltage (Fig. 17.16j) to attain 10 mA cm^{-2} and exhibits the outstanding stability with 100 h towards OWS. As demonstrated in Fig. 17.16k, the experimental results match with theoretical value of H_2 and O_2 with ratio 2:1, revealing 100% faradic efficiency towards OWS. The outstanding electrocatalytic performance of CoNiN@NiFe LDH bifunctional catalyst might be accredited to the following reasons: (1) the mutualism effect and strong electronic interaction between NiFe LDH and CoNiN in nano-architecture hybrid which improves the OER kinetics; (2) the induced (oxy)hydroxide species during OER and phase transformation from LDH to bimetal nitride provide active sites for promoting water dissociation.

Apart from above literature, other LDH derived heterostructures (such as γ-NiOOH, $NiONiFe_2O_4$/rGO, CoP/Co_2P, NiVP@NiFeV-LDH, CoMnS, $NiCo_2S_4$@NiFe LDH, $(Ni, Co)_{0.85}Se$, Fe-doped Ni_3Se_4, Cu_4N/Ni_3N and V-$FeNi_3N/Ni_3N$) delivered enhanced electrocatalytic water splitting activities (Li et al. 2018a, b; Menezes et al. 2021; Shi et al. 2019; Zhang et al. 2016, 2021; Chen et al. 2018;

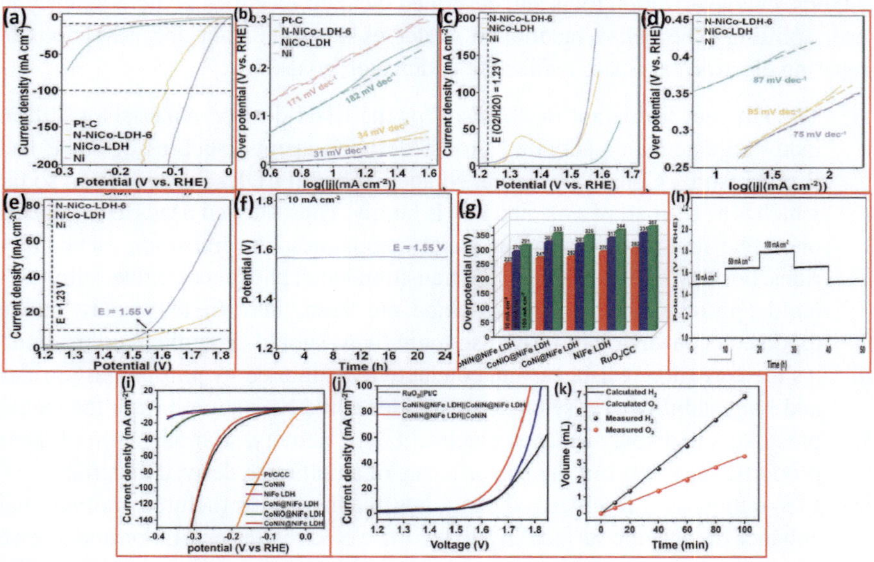

Fig. 17.16 **a** LSV curve, **b** Tafel plots for HER of Pt/C, NiCo-LDH, N–NiCo-LDH-6 and NF, **c** LSV curve, **d** Tafel plot for OER, **e** OWS polarization curve, **f** CP stability test over N–NiCo-LDH-6/ /N–NiCo-LDH-6 (Wen et al. 2021, Copyright RSC, adapted with permission). **g** Overpotential value for OER at current density 10, 50, 100 mA cm^{-2}. **h** CP stability curve at different current density 10, 50, 100 mA cm^{-2}. **i** polarization curve for HER. **j** LSV curve of OWS, **k** theoretical and experimental released O_2 and H_2 gas (Wang et al. 2021b, Copyright Elsevier, adapted with permission)

Xie et al. 2020; Zou et al. 2021; Zeng et al. 2018; Liu et al. 2017; Xiao et al. 2018; Du et al. 2018; Han et al. 2022; Wang et al. 2019, 2021c; Lu et al. 2021).

17.6 Conclusion

Renewable energy such as solar and wind driven electrocatalytic water splitting is an economically achievable approach to produce the green and clean hydrogen energy. HER and OER are considered as the two major half reactions of complete water electrolysis process, that decides the fruitful beginning of one-pot OWS. To optimize and achieve superior activity in these reactions, it's vital to design highly active electrocatalysts. Owing to the intrinsic constraints like poor electrical conductivity and few available exposed active sites of LDHs, its derivatives based on transition metals have recently emerged because of their excellent electrochemical activity towards OER and HER. The design and development of LDH derived materials, including mixed metal oxides, sulphides, selenides, oxyhydroxides, nitrides, and phosphides, are needed urgently for the improved performance due to their impressive electronic and structural traits. This book chapter has summarized the followed synthesis methods, characterization tools and described the their electrocatalytic efficiency in water splitting. The below mentioned strategies are significant and auspicious in preparing electrocatalysts to realize the anticipated goals.

(i) The inherent formation mechanism for the LDH derived materials is still in many arguments, as adopting proper deriving agents (such as S, Se, P and B precursors). Generally, the Fe, Ni and Co-based LDHs have reported so far which is a little part of as-stated LDHs family. Thus, studied needs to be centred on single non-metal transition metal compounds like transition metal fluorides, borides, dual non-metal like transition metal phosphoselenide, sulphoselenide, phosphonitride, borophosphide, etc. Also, there are few reports on the topotactic conversion of LDHs, so more focus should be applied to this field.

(ii) As the corrosion resistance and conductivity exhibited by phosphides/nitrides and long-stability over selenides/sulphides could be inherited from the parent precursors to display superior electrocatalytic activity, so modulation of these properties can improve the performance of an ultimate derived material.

(iii) Preparation of LDH derived materials having nanostructured morphology enhance the specific surface area to facilitate electrolyte penetration and release of gas, amplify charge-transfer, and enhance conductivity as compared to pristine LDHs having multilayer stacked nanosheets. However, scaleup preparation and uniform regulation of size need more crucial observation. Thus, regulation of morphology along with its relationship with activity must be studied more elaborately and should be proved with theoretical analysis.

(iv) Replacing intercalated anion of pristine LDH with huge organic species that can form carbon doped derived catalyst to fastens the diffusion of electrolytes via interface coupling effect. The creation of surface defect, corners and edges in LDH derived catalysts generating dangling bonds and decrease the coordination number of active hotspots. Also, vacancies (like S/P and metal defect) can regulate the electronic structure of active sites neighbouring to the defects, that can augment the electrocatalytic function and adsorption process of intermediates. Hence, both the theoretical and experimental studies need to be centred in this domain to gain the in-depth understanding of defect generation, control and recognition.

(v) Further, heterojunction formation with other conductive material could be an effective approach to modify the LDH derived material's electronic structure. The agglomeration and conductivity issues of LDH derived catalyst hinder the performance and practical applications. Adopting conductive substrates like Ni- and Cu-based foam or C-based substrate materials (graphene, CNTs, carbon fibre, carbon cloth, etc.) will increase the catalytic activity by improving the conductivity, dispersibility, hydrophilicity and exposure of electroactive sites.

According to the summary of the current book chapter, the electrocatalytic activities of LDH derived material have been improved by various methods, still the stability and catalytic activity in large-current density are distance from reasonable and realistic industrial application. Thus, constant efforts should be made in discovering the hierarchical LDH nanohybrid composites which can be quite challenging, not only in encouraging the in-depth depth understanding of catalytic mechanisms but also in creating new electrocatalyst possessing improved activity and stability. Besides the water electrolysis application, LDHs and their derivative heterostructure can also be used in various other applications such as batteries, supercapacitors, electrocatalytic CO_2 reduction and N_2 fixation.

Acknowledgements The authors are very much thankful to SOA (Deemed to be) University management for their support with encouragement. Dr. Dipti Prava Sahoo is thankful to CSIR New Delhi for awarding CSIR-RA (File No.- 09/969(0010)/2020-EMR-I). Ritik Mohanty is grateful to CSIR, New Delhi, for the award of the SRF-Direct fellowship [File No. 09/0969(18687)/2024-EMR-I]. Upali Aparajita Mohanty is grateful to UGC for the award of a SJSGC fellowship (UGCES-22-GE-ORI-F-SJSGC-5637).

Declaration of Competing Interest The authors declare no competing financial interest.

References

Cao, L., Cao, Y., Liu, X., Luo, Q., Liu, W., Zhang, W., Wei, S.: J. Mater. Chem. A **6**, 15684–15689 (2018)

Chen, L., Zhang, Y., Wang, H., Wang, Y., Li, D., Duan, C.: Nanoscale **10**, 21019–21024 (2018)

Chen, J., Long, Q., Xiao, K., Ouyang, T., Li, N., Ye, S., Liu, Z.Q.: Sci. Bull. **66**, 1063–1072 (2021)

Dau, H., Limberg, C., Reier, T., Risch, M., Roggan, S., Strasser, P.: ChemCatChem **2**, 724–761 (2010)

Du, J., Zou, Z., Liu, C., Xu, C.: Nanoscale **10**, 5163–5170 (2018)

Guo, H.P., Ruan, B.Y., Luo, W.B., Deng, J., Wang, J.Z., Liu, H.K., Dou, S.X.: ACS Catal. **8**, 9686–9696 (2018)

Hameed, A., Batool, M., Liu, Z., Nadeem, M.A., Jin, R.: J. Energy Chem. **7**, 3311–3328 (2022)
Han, X., Yu, C., Zhou, S., Zhao, C., Huang, H., Yang, J., Qiu, J.: Adv. Energy Mater. **7**, 1602148 (2017)
Han, B., Du, X., Li, J., Wang, H., Liu, G., Li, J.: Appl. Surf. Sci. Adv. **604**, 154617 (2022)
Hu, C., Zhang, L., Gong, J.: Energy Environ. Sci. **12**, 2620–2645 (2019)
Karmakar, A., Karthick, K., Sankar, S.S., Kumaravel, S., Madhu, R., Kundu, S.: J Mater Chem A **9**, 1314–1352 (2021)
Li, S., Sirisomboonchai, S., Yoshida, A., An, X., Hao, X., Abudula, A., Guan, G.: J. Mater. Chem. A **6**, 19221–19230 (2018a)
Li, J., Xu, W., Luo, J., Zhou, D., Zhang, D., Wei, L., Yuan, D.: Nano-Micro Lett. **10**, 1–10 (2018b)
Li, M., Chen, L., Ye, S., Fan, G., Yang, L., Zhang, X., Li, F.: J. Mater. Chem. A **7**, 13695–13704 (2019)
Li, D., Xing, Y., Yang, R., Wen, T., Jiang, D., Shi, W., Yuan, S.: ACS Appl. Mater. Interfaces **12**, 29253–29263 (2020)
Li, D., Zhou, C., Yang, R., Xing, Y., Xu, S., Jiang, D., Shi, W.: ACS Sustain. Chem. Eng. **9**, 7737–7748 (2021)
Li, Y., Dai, T., Wu, Q., Lang, X., Zhao, L., Jiang, Q.: Mater. Today Energy **23**, 100906 (2022)
Liu, J., Wang, J., Zhang, B., Ruan, Y., Lv, L., Ji, X., Jiang, J.: ACS Appl. Mater. Interfaces **9**, 15364–15372 (2017)
Liu, C., Han, Y., Yao, L., Liang, L., He, J., Hao, Q., Liu, H.: Small **17**, 2007334 (2021a)
Liu, M., He, Q., Huang, S., Zou, W., Cong, J., Xiao, X., Hou, L.: ACS Appl. Mater. Interfaces **13**, 9932–9941 (2021b)
Lu, Y., Li, Z., Xu, Y., Tang, L., Xu, S., Li, D., Jiang, D.: Chem. Eng. J. **411**, 128433 (2021)
Man, I.C., Su, H.Y., Calle-Vallejo, F., Hansen, H.A., Martínez, J.I., Inoglu, N.G., Rossmeisl, J.: ChemCatChem **3**, 1159–1165 (2011)
Menezes, P.W., Yao, S., Beltrán-Suito, R., Hausmann, J.N., Menezes, P.V., Driess, M.: Angew. Chem. Int. Ed. Engl. **60**, 4640–4647 (2021)
Nadeema, A., Walko, P.S., Devi, R.N., Kurungot, S.: ACS Appl. Energy Mater. **1**, 5500–5510 (2018)
Qin, K., Wang, L., Wen, S., Diao, L., Liu, P., Li, J., Zhao, N.: J. Mater. Chem. A **6**, 8109–8119 (2018)
Rajeshkhanna, G., Kandula, S., Shrestha, K.R., Kim, N.H., Lee, J.H.: Small **14**, 1803638 (2018)
Sahoo, D.P., Das, K.K., Mansingh, S., Sultana, S., Parida, K.: Coord. Chem. Rev. **469**, 214666 (2022)
Shi, R., Wang, J., Wang, Z., Li, T., Song, Y.F.: J. Energy Chem. **33**, 74–80 (2019)
Song, Q., Wang, J., Sun, Q., Wang, X., Zhu, Z., Pei, C., Huang, W.: Chem. Commun. **56**, 10285–10288 (2020)
Song, N., Hong, S., Xiao, M., Zuo, Y., Jiang, E., Li, C., Dong, H.: J. Colloid Interface Sci. **582**, 535–542 (2021)
Wan, J., Ye, W., Gao, R., Fang, X., Guo, Z., Lu, Y., Yan, D.: Inorg. Chem. Front. **6**, 1793–1798 (2019)
Wang, J., Zeng, H.C.: ACS Appl. Energy Mater. **1**, 4998–5007 (2018)
Wang, Y., Yan, D., El Hankari, S., Zou, Y., Wang, S.: Adv. Sci. **5**, 1800064 (2018)
Wang, Z., Xu, L., Huang, F., Qu, L., Li, J., Owusu, K.A., Mai, L.: Adv. Energy Mater. **9**, 1900390 (2019)
Wang, S., Wang, T., Wang, X., Deng, Q., Yang, J., Mao, Y., Wang, G.: Int. J. Hydrog. Energy **45**, 12629–12640 (2020)
Wang, J., Chen, C., Cai, N., Wang, M., Li, H., Yu, F.: Nanoscale **13**, 1354–1363 (2021a)
Wang, J., Lv, G., Wang, C.: Appl. Surf. Sci. Adv. **570**, 151182 (2021b)
Wang, J., Sun, Y., Qi, Y., Wang, C.: ACS Appl. Mater. Interfaces **13**, 57392–57402 (2021c)
Wen, Y., Qi, J., Wei, P., Kang, X., Li, X.: J. Mater. Chem. A **9**, 10260–10269 (2021)
Xiao, K., Zhou, L., Shao, M., Wei, M.: J. Mater. Chem. A **6**, 7585–7591 (2018)
Xie, M., Jia, K., Lu, J., Zhao, R.: Cryst. Eng. Commun. **22**, 546–553 (2020)

Yang, Y., Meng, H., Yan, S., Zhu, H., Ma, W., Wang, C., Hu, Z.: J. Alloys Compd. **874**, 159874 (2021)
Zeng, Y., Chen, L., Chen, R., Wang, Y., Xie, C., Tao, L., Wang, S.: J. Mater. Chem. A **6**, 24311–24316 (2018)
Zhang, G., Li, Y., Zhou, Y., Yang, F.: ChemElectroChem **3**, 1927–1936 (2016)
Zhang, X., Xue, Y., Yan, Q., Zhu, K., Ye, K., Yan, J., Wang, G.: Mater. Today Energy **21**, 100741 (2021)
Zhang, L., Ma, Y.T., Duan, J.J., Yao, Y.Q., Feng, J.J., Wang, A.J.: J. Colloid Interface Sci. **611**, 205–214 (2022)
Zhou, L., Shao, M., Wei, M., Duan, X.: J. Energy Chem. **26**, 1094–1106 (2017)
Zhou, J.J., Han, X., Tao, K., Li, Q., Li, Y.L., Chen, C., Han, L.: Chem. Eng. J. **354**, 875–884 (2018)
Zou, X., Wei, X., Bao, W., Zhang, J., Jiang, P., Ai, T.: Int. J. Hydrog. Energy **46**, 32385–32393 (2021)

Chapter 18
Perspective on Catalytic Oxidation of CO on Supported Metal Catalysts

Ansari Palliyarayil, Rohan Keshav Kunkalekar, Karthika Vinayakumar, Premkumar Selvarajan, Sanchita Sil, and Amit Ashok Vernekar

18.1 Introduction

The increased concentration of toxic gaseous pollutants in the atmosphere is one of the major issues that manhood is facing in the twenty-first century. During the last few decades, the number of transport vehicle has increased tremendously which demand an improvement in the technology for removing pollutants, especially those emitted by vehicle exhaust. Carbon monoxide (CO) is one such toxic pollutant gas (kinetic diameter (K.D.) = 0.376 nm) generated by the incomplete burning of fossil fuels. It is a chemical asphyxiant gas that can react with hemoglobin once inhaled, thereby affecting the oxygen transportation ability of blood. It also acts as a neurotoxin as it can penetrate the placenta and can affect the developing fetal brain (Greingor et al. 1998). CO is emitted into the atmosphere by the partial burning of carbonaceous fuels which includes oil, gasoline, coal, wood, tobacco, etc. Exposure to CO is often

A. Palliyarayil (✉) · A. A. Vernekar (✉)
Inorganic and Physical Chemistry Laboratory (IPCL), CSIR-Central Leather Research Institute (CLRI), Chennai 600020, India
e-mail: ansariplyr@gmail.com

A. A. Vernekar
e-mail: amitvernekar@clri.res.in

A. A. Vernekar
Academy of Scientific and Innovative Research (AcSIR), Ghaziabad 201002, India

R. K. Kunkalekar
School of Chemical Sciences, Goa University, Taleigao Plateau 403206, Goa, India

K. Vinayakumar · S. Sil
Defence Bioengineering and Electromedical Laboratory (DEBEL), C V Raman Nagar, Bangalore 560093, India

P. Selvarajan
Department of Physics, School of Advanced Sciences, Vellore Institute of Technology, Vellore 632014, India

R. S. Ningthoujam and A. K. Tyagi (eds.), *Handbook of Materials Science, Volume 3*, Indian Institute of Metals Series, https://doi.org/10.1007/978-981-96-8229-4_18

not identifiable as it is a non-irritating gas. According to Richard J. Levy, there are thousands of emergency hospitalizations, and more than 6000 deaths every year due to non-fire-related CO exposure (Levy 2015).

The greatest sources of CO to the atmospheric air are transport vehicles that burn fossil fuels such as gasoline and diesel. They contribute ~75% of the total CO pollution. The vehicle exhaust contains a maximum of 1,00,000 ppm CO (Raub et al. 2000). The other major sources include steam boilers, industrial activities, solid waste discarding, and other varied sources. On the other hand, in indoor such as houses, offices, and commercial buildings, CO is emitted from various sources which include cigarette smoking, water pipes or hookahs, water heaters, stoves that uses wood or coal as fuels, inappropriate use of generators or charcoal grill, etc. (Iqbal et al. 2010). Depending on the surrounding situations, the indoor CO can raise to 100 ppm. The time average exposure limit for CO is 25–35 ppm for an average of 8–10 h. The low-level exposure (120–520 ppm) to this toxic gas has minor effects such as headache, decreased alertness, nausea, fatigue, rapid breathing, and chest pain. Albeit, high-level exposure to CO can cause disorientation, unconsciousness, neurologic disabilities, coma, cardiorespiratory failure, and sometimes, death. Figure 18.1 represents the major sources and their contribution toward CO production as per the Environmental Protection Agency (EPA), USA.

The low boiling point and critical temperature value of CO (−191.5 °C and −139.5 °C, respectively) make it difficult to remove CO by adsorption, even by using a highly efficient adsorbent. The insolubility of this toxic gas in most of the solvents precludes the plausible removal by physical absorption. The most convenient method for the removal of CO from polluted air is by catalytic

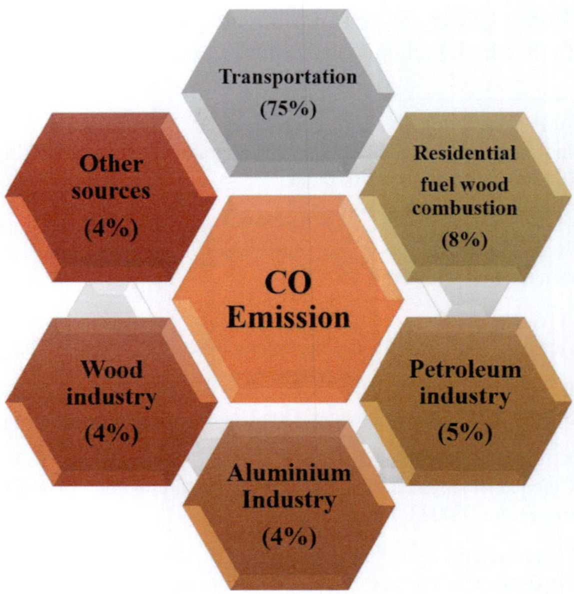

Fig. 18.1 Percentage contribution of different sources toward CO pollution

oxidation. Generally, catalysts made of transition metals, noble metals, and metal oxides are widely used for CO oxidation. The catalyst efficiency in CO oxidation for a catalyst is determined by its ability to oxidize CO–CO_2 completely at a low reaction temperature. The efficiency of a catalyst is highly dependent on structural characteristics such as particle size as well as morphology, its dispersion over the support, the electronic nature of the metal catalysts, metal-support interaction, the surface properties of the support materials like surface area, pore volume, etc.

Oxidants such as hoolamite, silver permanganate, and hopcalite have been reported as catalysts for the catalytic removal of CO (Srivastava et al. 2012). Among these, hopcalite is the most successful one which is a mixture of oxides of manganese and copper (MnO_2 and CuO). However, these catalysts are highly sensitive to moisture and their activity is decreased considerably in the moist condition. Other catalysts such as CuO/CeO_2, CuO_x–CeO_2–ZrO_2, and CuO/Al_2O_3 have also been studied for the oxidative removal of CO, but require a high reaction temperature of 200 °C (Srivastava et al. 2012). So far, the most effective catalyst reported for the abatement of CO uses noble-metal catalysts such as Platinum (Pt), Palladium (Pd), Gold (Au), Ruthenium (Ru), and Rhodium (Rh). However, the high cost and limited abundance of these noble metals reduce their practical application (Liang et al. 2009). During the last several decades, considerable efforts have been made to develop an effective catalyst by replacing these noble metals with other transition metals.

Computational methods are used as a counterpart to reveal invaluable insights into the catalytic mechanism since the computational methods are capable to provide atomic insights into the catalytic reaction. Density functional theory (DFT) is an indispensable tool used to comprehend reaction intermediates, adsorption and desorption characteristics, charge transfer interactions, and the change in electronic properties of the catalyst before and after the CO oxidation reaction. Since the DFT methods are highly accurate and can handle more atoms, it is possible to calculate various parameters for complex catalytic systems (Qin et al. 2018). The present book chapter aims to explore the recent advances of transition metal catalysts including noble metals toward the removal of carbon monoxide. Accordingly, literature published since 2017 has been considered.

18.2 Noble Metal Supported Catalysts for CO Oxidation

CO oxidation, one of the most well-known heterogeneous reactions, can be used as a benchmark system to understand catalysis as it is a simple bimolecular surface reaction. The implementation of catalytic converters in the United States and Europe in the early 1970s and 1990s, respectively, emphasizes the exceptional activity of noble metals toward the conversion of CO–CO_2. Later on, extensive studies have been reported by various research communities toward the use of noble metals such as platinum, palladium, gold, and silver for CO oxidation.

The preparation method has a considerable role on the physical characteristics of the catalyst such as surface area, particle size, dispersion, and interaction between the support and the metal catalysts, etc., which will affect the efficacy of the final catalyst. Wang et al. (2017) explored precipitation-deposition method to synthesize atomically dispersed metal catalyst. They have achieved Au particles possessing a size of ~1.2 nm supported on CeO_2 with exceptional thermal stability. The solid interaction between the atomic Au particles and the CeO_2 support provides better thermal stability as well as enhanced CO activity of >90% continuously even after 1000 min.

The cold plasma method, i.e., the method utilizing high electron energy is another efficient method of synthesizing highly dispersed fine metal nanoparticles (NPs). According to the pressure used, the method is generally named as low pressure (LP) and atmospheric pressure (AP) cold plasma (CP), APCP being the most effective. The catalyst prepared by this method generally shows high atom utilization as well as stability due to the enhanced metal-support interaction. Di et al. (2017) have reported an efficient Pd–Fe_2O_3 catalyst for removing CO synthesized via APCP. The Pd–Fe_2O_3 catalyst, prepared via coprecipitation method, was subjected to cold plasma (generated using a sine-wave of 36 kV peak-to-peak high voltage at 14.1 kHz). Compared to the conventional H_2 reduction method, the reported catalyst, synthesized via cold plasma method was characterized with highly dispersed sub-nanometer clusters of Pd NPs, increased amount of metallic Pd, and high surface Pd-Fe ratio, showing high catalytic activity for low temperature CO oxidation. Burange et al. (Burange et al. 2018) have adopted photo-deposition technique, i.e., irradiation with UV light, for the synthesis of Pd supported $CeZrO_{4-\delta}$ catalyst. To synthesize the catalyst, the support material was added to the methanolic solution of Pd precursor salt, followed by degassing and purging using Ar under stirring. The mixture was then subjected to 400 W UV light for a different duration. By changing the irradiation time, the authors were able to synthesize Pd catalyst with different crystallite sizes (6.6–8.5 nm) with a maximum dispersion of 17%. As the crystallite size decreases, a reduction in the CO conversion efficiency of the catalyst was observed by the authors.

It is a well-known fact that the size of the metal nanocatalyst has a considerable effect on its activity toward different catalysis reactions, especially in the oxidative removal of CO. The size of the metal catalysts has an inverse relationship with its activity, i.e., as the metal NP size decreases, the activity increases. By adopting microwave-assisted alcohol reduction, Rh NPs with controlled particle size have been synthesized and studied for its CO removal efficiency by Nishida et al. (2017). The method helps to carry out the reaction under solvothermal conditions providing rapid uniform heating. Solvothermal methods are known for synthesizing monodispersed highly crystalline NPs. However, the size of the particles will vary depending on the reducibility of the reducing agents, accordingly CO removal efficiency (Nishida et al. 2017). When using reducing agents with high reducing capacity, nucleation will start at a considerably lower temperature. This will prevent further agglomeration of NPs as the nucleation of the particles is a much faster process compared to the particle's nuclear growth at a lower temperature. NPs of size 3.3 ± 0.5 nm, 5.7 ± 1.3 nm, 6.7 ± 0.8 nm, 8.9 ± 1.1 nm, 9.5 ± 1.4 nm, and 10.9 ± 2.1 nm were obtained by Nishida et al. (2017) using ethanol, ethylene glycol, diethylene glycol, 2-ethoxyethanol, 2-(2-ethoxyethoxy)ethanol, or triethyleneglycol as the reducing

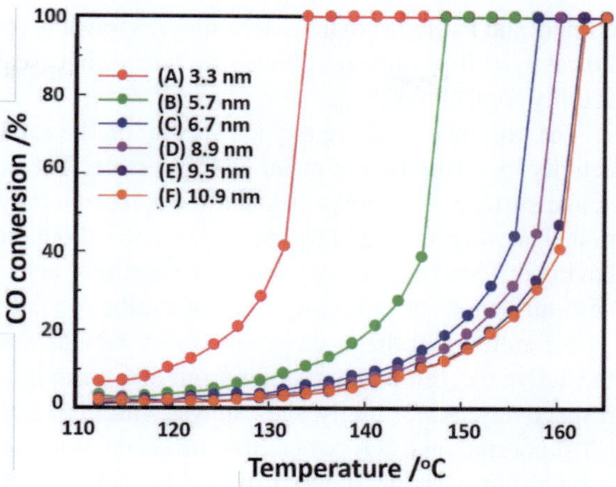

Fig. 18.2 Relationship between the particle size and the CO conversion efficiency of Rh catalyst (Reproduced with permission from Chemistry Letters 46(8) (2017) 1254–1257) (Zhang et al. 2017)

agent and polyvinylpyrrolidone (PVP) as the stabilizing agent. The catalyst having the smallest size of 3.3 ± 0.5 nm shows 100% CO removal at 135 °C whereas the larger particle of size 10.9 ± 2.1 nm at ~165 °C, i.e., dependency of efficiency on particle size (Fig. 18.2).

The presence of heteroatoms such as oxygen and nitrogen on the surface of support materials generally enhances the formation of single atom catalysts (SAC). The single atom catalysts (SAC), i.e., catalysts having isolated single atoms, have shown excellent catalytic activity compared to similar metal NPs. Though, there is an increased probability of agglomeration of these single metal atom catalysts if their interaction with the support is weak. This will affect the activity of the catalyst materials, i.e., the support materials play an important role in the efficiency of the catalyst. Heteropoly acids (HPA) possessing oxygen-enriched Keggin structure enable the incorporation of isolated single metal atoms on its surface. Zhang et al. have used ammonium salt of phosphotungstic acid as support for incorporating Rh metals for removing CO (Zhang et al. 2017). The authors concluded that the use of HPA not only benefits the formation of SAC, but also enhances the oxygen transfer via deoxygenation-oxidation steps and improves the CO oxidation to CO_2. Although it has been reported by many authors that the isolated single atoms show superior catalytic activity, there are contradictory reports by various research groups regarding the size of the catalyst and its effect on activity. Hence, it is necessary to understand the size and structure effect on the activity of a catalyst, which may help to enhance their activity further as well as to design a new efficient catalyst. By considering this, Murata et al. (2019) have synthesized Pd supported Al_2O_3 catalysts; via the conventional impregnation method; with varying Pd particle sizes varying from 1.2 to 19 nm and studied their CO removal efficiency. The activity was increased as the size of the particle increased from 1 to 2 nm (TOF is 0.08–0.18 s^{-1}) and then decreased for 2–4 nm Pd NPs (TOF is decreased to 0.08 s^{-1}). A further increase was observed as the particle size increased to 19 nm. With the help of various characterization techniques such as XAFS, STEM, and IR, they justified this size effect. According to the authors, Pd corners and Pd(111) are the active sites responsible for oxidation of CO and the 2 nm Pd catalyst shows an amorphous structure with increased corner sites, giving highest activity for CO oxidation.

The presence of active oxygen species on the catalyst surface may also affect the activity by changing the metal NP dispersion. The support materials enriched with active surface oxygen species show enhanced catalytic activity toward CO oxidation. This was supported by the studies carried out on Pt incorporated CeO_2 support having different surface oxygen concentrations, achieved by selectively adding suitable surfactants at different stages of synthesis (Zhou et al. 2018). It was observed by the author that the % dispersion of Pt NPs decreases from 76 to 68% to 62% as the active oxygen content varies from 696 mmol/g to 672 mmol/g to 645 mmol/g, respectively. Accordingly, the catalyst with high active oxygen content, i.e., Pt-S-f (Pt supported on CeO_2, where the surfactant 6-aminocaproic acid was added before spray drying), shows superior oxidation activity (T50% is 113 °C) compared to the catalyst with less active oxygen content, i.e., Pt-C-f (Pt supported on CeO_2, where the surfactant PEG-600 was added before crystallization process) (T50% is 121 °C) (Fig. 18.3a). Moreover, the presence of active oxygen species provides a strong interaction between the Pt and the CeO_2 support which further prevent the sintering of NPs even at a higher temperature of 750 °C and the temperature-treated catalyst still shows enhanced activity for CO oxidation (Fig. 18.3b) (Zhou et al. 2018). It is likely that the chemical reduction of the support material with reducing agents generates surface defects which in turn help to atomically disperse the metal catalyst on it. Chen et al. (2018) utilized this strategy to synthesize atomically dispersed Pt catalyst over Ce support. To synthesize, the authors have treated the porous CeO_2 nanorods with ascorbic acid solution before the incorporation of metal NPs. The resulting oxygen vacancies will act as an anchoring site for the incoming metal particles, ensuring the development of atomically dispersed Pt with a Pt-O coordination number of 3.3 and a Pt-O bond length of 1.98 Å. The observed very short Pt-O distance leads to a strong interaction between Pt and CeO_2 showing sinter resistant property even at a higher temperature of 300 °C with superior CO activity. Similar studies have been reported for gold and silver nanoparticles supported on SiO_2 and CeO_x, respectively (Wang et al. 2018).

Palliyarayil et al. (2021a) have reported activated carbon fiber cloth as a support material to incorporate Pd and its use for the oxidative removal of CO. Before Pd incorporation, the carbon supports were treated with 10% HNO_3, making the carbon surface enriched with acidic functional groups. The resultant hydrophilic carbon surface ensures the uniform distribution of Pd over it and shows a continuous 100% CO removal capacity for 300 min (Fig. 18.4). One-dimensional boron nitride nanotube (BNNT), structurally analogous to graphene and carbon nanotube have emerged as a promising catalyst material. The superior mechanical properties, excellent oxidation resistance, high thermal conductivity, and stability make it an important material for various industrial application and catalysis. Kim et al. (2019) have used the same catalyst support for removing CO as CO_2. By the sonochemical method, BNNT has been impregnated with Pd and Pt with an ultrafine Pd and Pt

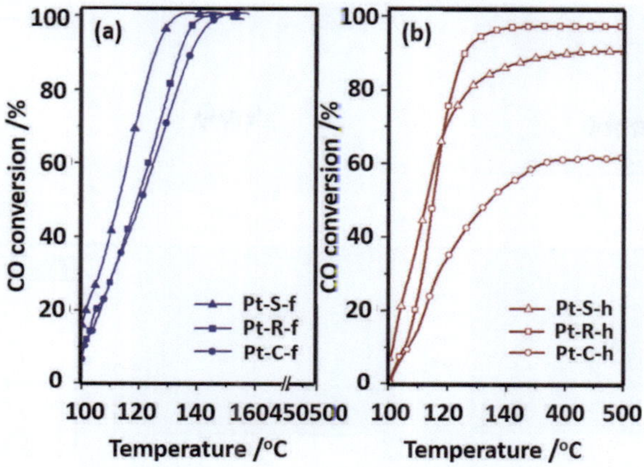

Fig. 18.3 CO activity plot for the **a** Fresh and **b** aged Pt catalyst supported on CeO_2 having different surface oxygen concentrations (Reproduced with the permission from Journal of Rare Earths 36(3) (2018) 257–264) (Zhou et al. 2018)

particle size of 2–3 nm and 1–2 nm, respectively. The boron and nitrogen vacancies present in BNNT lead to a strong metal-BNNT interaction which improves the catalytic activity. The corresponding temperature to get 100% CO conversion with respect to Pt-BNNT and Pd-BNNT is 180 °C and 125 °C, respectively. Unlike Pt, Pd is generally observed in the cluster form on the BNNT defect site. There will be a strong interaction between Pd and the nanotube due to the increased charge transfer between them which stabilizes the CO adsorption, showing enhanced activity compared to Pt-BNNT (Kim et al. 2019).

MOFs are grouped as porous solid organic–inorganic hybrid crystalline materials containing metal ions/clusters and bridging organic linkers. As a support material for metal NPs, MOF has been studied extensively, especially for catalytic applications (Zhang et al. 2018a). Metal NPs can be incorporated into the MOF structure either by post-modification of MOF or by direct incorporation. The latter ensures the uniform distribution of NPs throughout the MOF surfaces. Pt NPs supported on Zr-based UiO-67 MOF have been reported for the catalytic oxidation of CO (Vakili et al. 2018). The linker design method adopted by the research group provides ultrafine Pt NPs of size 1.2 ± 0.4 nm with a T50% (temperature at which the conversion becomes 50%) of 250 °C for CO oxidation. Although the value is slightly on the higher side, the catalyst shows an enhanced stability up to 5 cycles without any sintering/agglomeration of supported Pt NPs. The study ensures the beneficial property of MOF as catalyst support over conventional oxide support materials. Studies were also reported for the use of MOF as a precursor material for synthesizing nanoporous metal oxides possessing high surface area. Zhang et al. (2018b) have developed a catalyst for CO oxidation where Ag is impregnated on Fe_2O_3 support which is derived from MIL-100 (Fe). Hydrothermally synthesized MIL-100 (Fe) was modified with Ag of varying concentrations to obtain catalysts of different Fe/Ag ratios. The catalyst was characterized with a surface area of 49–70 m^2/g, total

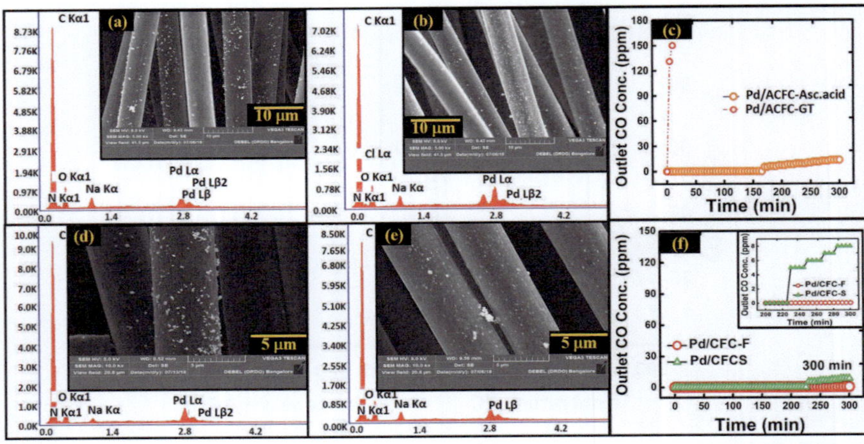

Fig. 18.4 **a–c** SEM-EDS and CO activity plot for the Pd catalysts supported over ACFC reduced with ascorbic acid and green tea (Reproduced with the permission from Diamond and Related Materials 107 (2020) 107884) (Palliyarayil et al. 2020) and **d–f** reduced with formaldehyde and sodium borohydride (Reproduced with the permission from Cleaner Engineering and Technology 4 (2021) 100252) (Palliyarayil et al. 2021)

pore volume of 0.17–0.35 mL/g, and Ag particle size of 3.8–5.7 nm. In general, the incorporation of metal NPs leads to a decrease in the structural properties of the supports. However, in the case of MIL-100 (Fe) derived Ag/Fe_2O_3 catalyst, in addition to the increased metallic Ag%, an increase in the surface characteristics; i.e., surface area as well as pore volume was reported with an increase in Ag %. In the case of MOF-derived support materials, the decomposition of metal salt as well as the organic ligands leads to the production of reductive gases during calcination which promotes the formation of more pores. Further, with the increase in surface area as well as the metallic Ag, the catalyst with an optimum value of Fe/Ag ratio and the pore architecture shows better efficiency toward CO oxidation with T_{100} (temperature at which the conversion becomes 100%) at 160 °C. Very recently, Altass et al. (2022) have used a unique magnesium-based MOF, Mg-BTC, synthesized by using 1,3,5 benzene tricarboxylic acid (BTC) as an organic linker. The authors tried to improve the efficiency of Mg-BTC by making a composite with reduced graphene oxide followed by Au impregnation (Au/rGO@Mg-BTC). They observed that the presence of graphene oxide possessing different functional groups as well as defects helps in the uniform dispersion of metal catalysts. Moreover, it promotes the dissociation of oxygen molecules as well as the transfer of electrons between the reactant molecule and the catalyst (Altass et al. 2022). This helps in increasing the activity of Au/rGO@Mg-BTC compared to Au/Mg-BTC and rGO/Mg-BTC. The performance of a monometallic MOF can be improved by the incorporation of another active metal. The defects produced by the introduction of bimetallic node increases the number of active sites which leads to an enhanced activity (Salama et al. 2021).

Like other parameters, moisture also plays a major role in the oxidative removal of CO. It can act as a promoter as well as an inhibitor depending on the concentration present in the gaseous mixture. At lower concentrations, water has a promoting effect on the oxidative removal of carbon monoxide, whereas an excess amount of water affects the activity negatively by blocking the pores of the support materials, especially in the case of carbon-based supports. Moreover, the amount of water required will vary depending on the structural characteristics, i.e., the porous architecture of the support material as well as the oxidation state of the metal catalysts. For example, in the case of Pd supported granular activated carbon (Pd/GAC), where the support is composed of both micro, meso, and macro pores, the CO oxidation activity increases with increasing the moisture content even up to 15% (Palliyarayil et al. 2021b). At the same time, in Pd-supported activated carbon fiber cloth (Pd/ ACFC), where the support is characterized with well-developed micro porous structure, a moisture content of >5% shows an inhibitory effect toward CO oxidation (Palliyarayil et al. 2021a). The reported Pd/ GAC and Pd/ACFC catalysts had both metallic as well as oxidized Pd (Pd^0 and Pd^{+2}) of varying percentages (the total Pd content in Pd/GAC and Pd/ACFC is 5.98–7% (comprising 38.04–66.46% Pd^0 and 33.54–61.96% Pd^{+2}) and 5.23–5.95% (comprising 3.94–29.43% Pd^0 and 70.57 to 96.06% Pd^{+2}), respectively). Those catalysts with Pd^0 as the major species show catalytic activity independent of the presence/absence of water where the reaction follows the co-adsorption of CO and O_2 molecules followed by the atomic dissociation of O_2 molecules. Thereafter, atomically dissociated O undergoes a reaction with adsorbed CO and forms CO_2. On the other hand, the catalyst with Pd^{+2} as the major form demands water for oxidizing CO, following a different reaction path where OH species derived from water molecules play a major role in converting CO–CO_2 (Eqs. 18.1–18.3) (Palliyarayil et al. 2021b).

$$CO^* + O^* + H_2O_{(g)} \rightarrow COOH^* + OH^* \tag{18.1}$$

$$COOH^* + OH^* \rightarrow CO_2^* + H_2O^* \tag{18.2}$$

$$CO_2^* + H_2O^* \rightarrow CO_2 + H_2O \tag{18.3}$$

In the case of heterogeneous supported metal catalysts, there is a possibility of decreasing the activity of the catalyst via diffusion of metal particles into the bulk support material as well as by agglomeration (Vedyagin et al. 2018). The addition of a second metal atom, i.e., a bimetallic system shows better stability with an enhanced activity, compared to a monometallic catalyst. This was demonstrated by the work reported by Singhania and Gupta (2018). The authors have incorporated a small amount, ~0.95% transition metals such as Co, Ni, and Cu into the Pd-supported activated carbon catalyst. The catalysts were synthesized by impregnating the support carbon material with the aqueous solution of Pd and second metal precursors followed by reduction with liquid media-based reduction using potassium borohydride as the reducing agent. Compared to the monometallic Pd/C catalyst, the bimetallic catalyst

Pd–M/C, shows the increased surface area, smaller particle size of 3–6 nm, and long-term stability of ~50 h with enhanced CO activity (T50% for Pd/C: 70 °C, Pd–Co/C: 55 °C, Pd–Ni/C: 42 °C, Pd–Cu/C: 30 °C). Similarly, Eid et al. (2018) have reported Cu incorporated Pd catalyst supported over carbon nitride nanowires (CN NWs). The CN generally possesses outstanding chemical and thermal stability, electrical conductivity as well as mechanical characteristics. Moreover, the difference in the electron density of C and N causes a strong metal support interaction in CN which in turn make the electron transfer easier during catalysis. The polymerization of melamine carried out in presence of metals to synthesize the Pd–Cu/CN NWs catalyst leads to the formation of Pd and Cu metallic NPs on the surface of CN NWs. The high affinity of Cu toward oxygen and Pd toward CO together leads to an enhanced CO oxidation efficiency of the Pd–Cu/CN NWs catalyst for up to 20 h (Fig. 18.5).

Although the addition of second metal helps in enhancing the activity via generating active centers with a specific geometry and/or changing the electronic structure of the active species, the surface composition of the metals is an important factor that affects the activity of the bimetallic catalyst. Bukhtiyarov et al. (2018) have utilized near-ambient pressure XPS and mass spectroscopy to understand the formation of an active site in the case of a bimetallic catalyst by selecting Pd–Au as a model catalyst supported on highly oriented pyrolytic graphite. The study reveals that at a temperature of ≤150 °C, due to the strong adsorption of CO on Pd, Pd undergoes segregation, preventing alloy formation. At a higher temperature of ≥150 °C, alloy formation occurs as the CO is desorbed from Pd and the surface becomes active toward CO oxidation. The doping of metal nanoclusters with other hetero atoms such as Pt, Pd, and Ag enhances the stability as well as the catalytic property of the metals by changing the physical and chemical properties of the metals, the position, as well as the number of dopant atom, play a major role to get an improved catalytic activity. When Au_{25} clusters are doped with Pd ($PdAu_{24}/TiO_2$), Garcia and co-workers (Garcia et al. 2020) observed more than 90% onset activity compared

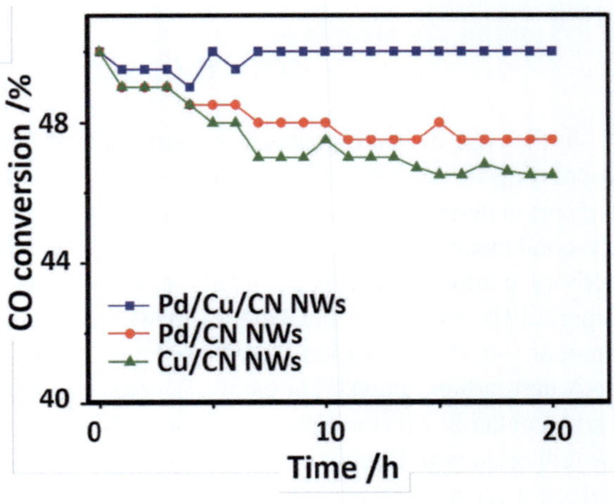

Fig. 18.5 Activity plot for the Pd/Cu/CN NWs, Pd/CN NWs and Cu/CN NWs (Reproduced from Catalysts 8(10) (2018) 411) (Eid et al. 2018)

to Au_{25}/TiO_2 (conversion is only 10%). The combined analysis of XAFS, DRIFT, and XPS helped the authors to conclude that upon applying an oxidative-reductive pretreatment condition, the Pd atom migrated from the Au cluster core to the surface and the resulting available surface single Pd atom enhanced the adsorption of CO and finally the conversion of $CO–CO_2$.

Further details of literature reported on the noble metals supported heterogeneous catalysts for the oxidative removal of CO have been provided in Table 18.1.

18.3 Non-noble Metal Supported Catalysts for CO Oxidation

Transition metal/metal oxide catalysts with different morphological features are also known for the oxidative removal of CO. Although they exhibit lower activity compared to noble metals, researchers are focusing more on the development of base metal catalysts due to several advantages such as low cost, environmentally friendly, and easy availability. Various types of base metal catalysts with specific sizes and shapes, and in different forms such as metal NPs, oxides, mixed oxides, and supported materials are reported for CO oxidation reaction. Supported base metal catalysts were found to show high catalytic CO oxidation activity at a lower temperature due to their unique properties and are excellent materials for this application.

Cr_2O_3 NPs incorporated reduced graphene oxide (rGO) and graphene oxide (GO) hybrids were prepared and tested for CO oxidation activity (Ali et al. 2020). The Cr_2O_3/GO shows higher CO oxidation capacity at low temperatures compared to Cr_2O_3/rGO and pristine Cr_2O_3 NPs. The higher activity of Cr_2O_3/GO is due to its higher surface area and the mixed oxidation state of Cr which makes the mobility of electrons easy, are necessary for CO oxidation. The graphene oxide substrate avoids the aggregation of Cr_2O_3 NPs and the occurrence of carbon vacancy defects and functional groups stabilizes the Cr_2O_3 NPs thereby forming thermally stable material (Ali et al. 2020).

Ni supported on graphene support has been studied for CO oxidation by Wang et al. (2021) wherein binary metal catalysts Ni–Fe were atomically dispersed over nitrogen-doped graphene (NGr) support which serves as a proficient catalyst for CO oxidation reaction. The studies lead to the conclusion of the adsorption of oxygen molecules over electrophilic Fe active sites and the adsorption of CO molecules on nucleophilic Ni active sites. The superior catalytic activity of Ni–Fe@NGr catalyst can be related to the synergistic effect of both Ni and Fe and its amazingly low kinetic energy barriers for CO oxidation reaction via both Langmuir-Hinshelwood and Eley-Rideal mechanisms. Further Ni–Fe@NGr catalyst also exhibits higher activity compared to Ni@NGr and Fe@NGr homonuclear catalysts counterparts. Thus, this novel strategy can be utilized to design new single-atom supported catalysts for low-temperature catalytic abatement of CO.

Table 18.1 Literature reported noble metals supported heterogeneous catalysts for CO oxidation (Lin et al. 2017)

SI No.	Catalyst	Support materials used	Particle size (nm)	CO removal efficiency				References
				CO concentration (%)	Activity (%)	TOF (min^{-1})	Temperature (°C)	
1	Pd	Ce-MOF	4.6 ± 2.2	4	100	–	96	Lin et al. (2017)
2	Ru–Cu alloy	–	9.2 ± 2.5	–	50	–	122	Huang et al. (2017)
3	Pd	CeO_2	0.9	1	100	–	75	Jeong et al. (2017)
4	Pd	α-Fe_2O_3	1.8	1	100	–	40	Wang et al. (2017)
		γ-Fe_2O_3	2.3		100	–	0	Zhang et al. (2017)
5	Ag	α-Fe_2O_3	8–12	–	98	–	157–275	
6	Pt	α-Fe_2O_3	Single atom	1	–	9–53.76	140–180	Lou et al. (2017)
		γ-Al_2O_3				0.18–1.56		
		ZnO				3.18–23.40		
7	Au	α-Fe_2O_3-spindles	2–5	1	90	45.6	120	Zeng et al. (2017)
		α-Fe_2O_3-rod				69	136	
		α-Fe_2O_3-hollows				87	75	
		α-Fe_2O_3-hollow-rods				66.6	95	
8	$Pt_1Fe_{0.025}$	SiO_2	–	0.5	–	17	40	Ro et al. (2018)
	$Pt_1Fe_{0.05}$					46		
	$Pt_1Fe_{0.1}$					48		
	$Pt_1Fe_{0.2}$					68		

(continued)

Table 18.1 (continued)

SI No.	Catalyst	Support materials used	Particle size (nm)	CO removal efficiency				References
				CO concentration (%)	Activity (%)	TOF (min^{-1})	Temperature (°C)	
9	Au	Fe_2O_3	3–10	–	40	–	25	Tanaka et al. (2018)
10	Ag	h-BN	5–10	4	100	–	194	Konopatsky et al. (2018)
11	Ag	SiO_2	3–10	1	98	0.11–0.18	58–66	Dutov et al. (2018)
12	Au	TiO_2	5.6	1	68	–	25	Wei et al. (2018)
13	Au	m-ZrO_2	8–10	0.1	94	–	25	Zakaria et al. (2019)
14	Pt	TiO_2	1.5–2.5	0.9%	72.8	6.54	120	Liu et al. (2018)
15	Au	CeO_2	5.12	–	97.9	–	30	Carltonbird et al. (2018)
16	Au	CeO_2	–	1	100	–	40	Wang et al. (2019)
17	Pt	TiO_2	1.3	1	100		105–220	Zhang et al. (2020)
18	Ag	MTS-9	2–20	0.08	100	–	240	Hamidi et al. (2021)

(continued)

Table 18.1 (continued)

SI No.	Catalyst	Support materials used	Particle size (nm)	CO removal efficiency				References
				CO concentration (%)	Activity (%)	TOF (min^{-1})	Temperature (°C)	
19	Ru	CeO_2	2–3	1	50	–	50	Li et al. (2019)
20	Rh	ZnO	Single atom	1	100	22.2–37.8	210	Han et al. (2019)
	Pt						300	
	Au						400	
21	Pd	CeO_2/Al_2O3	–	1	50	–	15–20	Xie et al. (2021)
22	Rh	TUD-1 meso porous silica	5–10	1	100	–	75	Al-Shehri et al. (2020)
	Pd						55	
	Pt						45	
	Au						30	
23	Co-Pd	Activated carbon	<10	1	100	–	110	Aragão et al. (2022)

Copper-based catalysts such as supported Cu/CuO NPs, supported bimetallic (Cu with other active base metal) and NPs over different active support such as TiO_2, SiO_2, MnO_2, CeO_2, ceria-zirconia are found to be efficient for catalytic CO oxidation. An interesting study was reported for CO oxidation over Cu/CeO_2 catalysts wherein four different types of CeO_2 supports were used with different sizes and morphological features; synthesized using hydrothermal and solvothermal methods (Dong et al. 2019). This comprises CeO_2 as nanoparticles (20 nm), nanospheres (200 nm), nanorods (20–40 nm), and flower-like microspheres (4 μm). 15 wt.% Cu was supported over all these CeO_2 support materials using the deposition precipitation method. Figure 18.6 shows the SEM, TEM, and HRTEM images of different Cu/CeO_2 catalysts which clearly show different morphological features of CeO_2 support. The morphological features of CeO_2 affect the exposed crystal plane and modulate the size of supported CuO NPs which affects the efficiency of these materials. The highest CO conversion activity was obtained over Cu/CeO_2 nanorods followed by Cu/CeO_2 nanoparticles, Cu/CeO_2 nanospheres, and lowest by Cu/CeO_2 microspheres. The high activity of Cu/CeO_2 nanorods is due to the smaller size of Cu and CeO_2 NPs which results in stronger interaction between them and also due to the formation of the increased number of oxygen vacancies. Further, this catalyst also works well in humid conditions and showed complete conversion at a lower temperature. In the case of Cu/CeO_2 microspheres, CeO_2 particles are larger in size and a lesser number of Cu NPs dispersed, resulting in lower activity (Dong et al. 2019).

Mono-metallic (Cu/CNT, Co/CNT, Fe/CNT) and bimetallic (Cu–Fe/CNT, Cu–Co/CNT) catalysts were prepared by a simple impregnation method (Dasireddy et al. 2018). The complete CO conversion was obtained over bimetallic Cu–Fe/CNT catalyst at a much lower temperature followed by Cu–Co/CNT catalyst in presence of CO_2, hydrogen, and water-rich environment indicating that bimetallic catalysts are better for CO oxidation. The increased activity of Cu–Fe/CNT could be credited to the existence of $CuFe_2O_4$ on CNT. On the other hand, the synergistic interaction of Cu and Co is responsible for the good activity of Cu–Co/CNT. Among the monometallic catalysts, Cu/CNT exhibited higher activity.

A series of 7.0 wt.% CuO supported on transition metal (Zr, Mn, Ti, and Ni) doped CeO_2 rod-shaped support was developed by adopting the solvent thermal method (Guo et al. 2018). The presence of dopants in CeO_2 influences CO oxidation activity. The incorporation of Ti and Mn species in the CeO_2 lattice facilitates defect structure formation and helps in the assimilation of Cu ions into the CeO_2 surface which significantly assists the generation of unsaturated Cu^{2+} ions and strong interfacial Cu–Ce interaction. Further due to the presence of Mn species in variable oxidation states and the tunnel-like structure of todorokite, the surface oxygen mobility of CuO/Mn-doped CeO_2 catalyst is high at low temperatures. The incorporation of Zr species does not improve the catalytic activity since the interaction between Cu and CeO_2 support in the interface of CuO/Zr-doped CeO_2 catalyst remains the same with CuO/CeO_2 catalyst. CO oxidation activity decreases over CuO/Ni-doped CeO_2 catalyst, this is due to a decrease in CuO NPs dispersion over CeO_2 support and weakens CuO–CeO_2 interaction due to the addition of Ni ions in CeO_2 lattice (Guo et al. 2018). The Cu–CeO_2/SiO_2 and Cu–CeO_2/Al_2O_3 catalysts were prepared by the

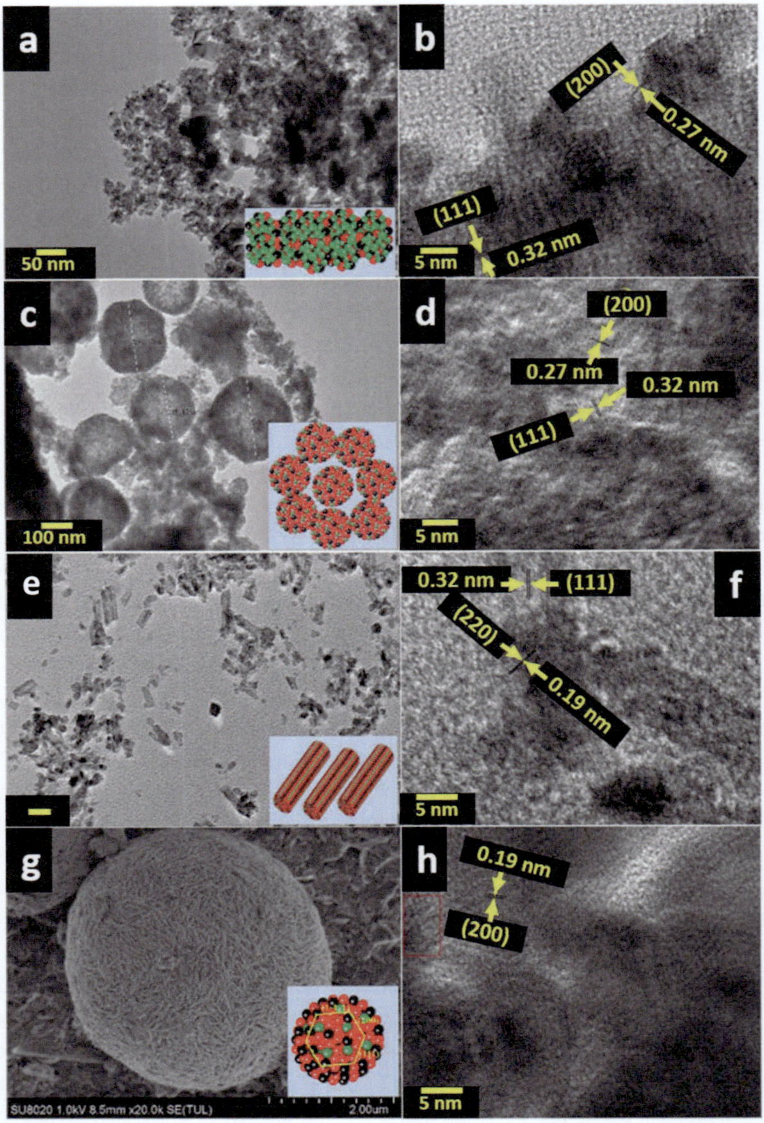

Fig. 18.6 **a**, **c**, **e** TEM. **b**, **d**, **f**, **h** HRTEM and **g** SEM images of different Cu/CeO_2 catalyst (Reproduced with the permission from Scientific reports 9(1) (2019) 1–14) (Dong et al. 2019)

co-impregnation method to study the role of support on $Cu–CeO_2$ interface and its thermal stability at a temperature of 500, 700, and 900 °C for catalytic CO oxidation reaction (Alvarez et al. 2018). $Cu–CeO_2$ interface is stable over the SiO_2 support at a temperature of 500 °C, but interfacial sites decrease when calcined at 700 °C, and separation of the Cu and Ce species occurs on the surface of the SiO_2 at 900 °C resulting in complete loss of catalytic activity. On the other hand, $Cu–CeO_2$ interface is much more stable on Al_2O_3 and exhibits considerable catalytic activity even when calcined at 900 °C. Due to superior interaction between Cu and CeO_2 species with Al_2O_3, the sintering process of Cu and CeO_2 is much lower on Al_2O_3 support thus exhibiting higher activity.

One study reported nanostructured films of Cu oxide and Zn oxide supported on a brass substrate prepared via a simple vapor oxidation process. These prepared materials were utilized as microreactors for catalytic CO oxidation reactions and were found to show high activity. The increased activity of these materials is due to the synergistic interaction between nano-sized Cu and Zn oxides and higher exposed active sites (Cabello et al. 2022). 10 atomic wt.% Cu supported on tungsten oxide (with varying loadings 0, 0.4, 2.0, and 3.2 at%) modified ceria nanorods were synthesized by a two-step sequential deposition-precipitation method (Gan and Si 2021). The role of tungsten oxide on Cu supported on ceria catalyst for the catalytic oxidation reaction was investigated. The modification that occurred in the physical characteristics (pore size, surface area, and pore volume) of ceria support significantly changes the interaction between Cu and the ceria support which reduces the catalytic performance of copper-ceria catalysts toward the oxidation of CO–CO_2, thus showing the negative effect of incorporation of tungsten oxide in ceria.

In the case of transition metal oxides of cobalt, nickel, and iron, Co_3O_4 behaves as a strong catalyst for CO oxidation under dry atmospheric conditions along with its good stability. The iron oxide (Fe_2O_3) catalyst shows good CO oxidation abilities at room temperature depending on the structure and oxidation state of iron. The smaller particle size, increased outer surface area, as well as closely occupied unsaturated surface coordination sites are its unique features that aid CO catalytic activity. On the other hand, nickel oxide shows bimodal porosity which provides more space for chemisorption, along with high Ni metal particle dispersion on the surface of the catalyst which is responsible for its high efficiency. Alshehri and Narasimharao (2017) prepared Fe-based catalyst using Fe_2O_3 as support and gold particles were dispersed over it via the deposition-precipitation method. The Fe_2O_3 support was prepared by hydrothermal method followed by Au deposition. In the deposition-precipitation method, the support material is heated in a water bath followed by adjusting pH and loading of the Au solution. A very less amount of Au (i.e., 0.5%) is loaded onto the support providing a good CO conversion of 100% at 323 K. The catalyst possesses a mesoporous structure with a large specific surface area which helps to disperse the active metal NPs uniformly over the catalyst surfaces. The interaction between Au NPs and the mesoporous Fe_2O_3 supports with high specific surface area helps in CO oxidation (Alshehri and Narasimharao 2017). Recently, Fedorov and co-workers (Fedorov et al. 2018) synthesized Fe–Al and Cu–Fe–Al nanocomposite catalysts by the fusion of mixing of aluminum, iron, and copper salts for CO oxidation. The catalysts were prepared by adopting the co-precipitation method by mixing and heating the salts of the precursors [Cu $(NO_3)_2 \cdot 3H_2O$, Fe $(NO_3)_3 \cdot 9H_2O$, and Al $(NO_3)_3 \cdot 9H_2O$] in the required ratios with a high concentration of Fe_2O_3 at a temperature of 200 °C followed by calcination at 400 °C for 1 h and then at 700 °C for 1 h in presence of air. It was observed that an increase in Fe content causes an increase in CO activity. The catalysts demonstrate moderate activity in the oxidation of CO.

The addition of secondary metals also enhances the redox properties of catalysts thereby providing a positive effect on CO oxidation. The chemical interaction caused by the addition of metals to ceria enhances the mobility of surface oxygen. The synergic effect of cobalt oxide with ceria changes the structural as well as the redox characteristics of composite oxides, thereby increasing the efficiency of the Co^{3+}/Co^{2+} redox couple. Gomez and Boix prepared cobalt catalysts supported on ZrO_2, CeO_2, and CoCeMnOx for preferential CO oxidation (Gómez and Boix 2019). The catalysts were evaluated by their CO oxidation efficiency as well as selectivity. At lower temperatures, CoCeMnOx showed superior CO oxidation capacity whereas the cobalt oxide supported on ceria exhibited the best selectivity for CO_2. Co_3O_4 spinel was the active phase in all the catalysts. The presence of ceria dopants enhanced the efficiency of the catalyst by increasing its redox properties. Similarly, Mn addition also increases the catalytic efficiency at low temperatures. Catalysts have also been prepared by wet impregnation method using zirconia support with an aqueous $Co(NO_3)_2$ solution followed by evaporation and drying at 120 °C overnight. CoO_x–CeO_2 catalysts are obtained by wet impregnation or coprecipitation method. In the wet impregnation method, CoO_x–CeO_2 catalyst with 10 wt.% of cobalt was received by evaporating a mixture of an aqueous solution of $Co(NO_3)_2$ and CeO_2 powder (the support CeO_2 was obtained by precipitating $Ce(NO_3)_3$ solution with $NH_4(OH)$). This was followed by the overnight drying of the resulting precipitate at 110 °C. In the coprecipitation method, precipitation of the aqueous solution of $Co(NO_3)_2$ and $Ce(NO_3)_3$ is done by adding $NH_4(OH)$ dropwise under vigorous stirring. The precipitate thus obtained was filtered, washed, and dried at 110 °C. Similarly, 10 wt.% of Co supported on CoCeMnOx oxide mixture with Mn/Co molar ratio of 1/4 has been synthesized via the coprecipitation method, by adding $NH_4(OH)$ to an aqueous $Co(NO_3)_2$, $Ce(NO_3)_3$, and $Mn(NO_3)_2$ solution.

Table 18.2 discusses the properties of non-noble metals supported catalysts reported in the literature for CO oxidation.

18.4 Theoretical Aspects of CO Oxidation

It is experimentally proven that the CO oxidation catalytic properties of the metal/metal oxides can be evaluated by the number of active sites as well as the activity of each active site. However, the identification of the activity of each active site is extremely challenging because it is challenging to precisely determine the atomic and electronic structure of active sites using experimental techniques. Besides, it is not easy to comprehend the intermediate states of the reaction and the associated electron transfer from the experimental techniques during the CO oxidation catalytic processes. Understanding the catalytic mechanism through experimental methods is a nightmare, which obstructs the rational design of novel catalysts for CO oxidation.

Table 18.2 Literature reported non-noble metals supported heterogeneous catalysts for CO oxidation (Ali et al. 2020; Dong et al. 2019; Dasireddy et al. 2018; Alvarez et al. 2018; Cabello et al. 2022; Gan and Si 2021; Alshehri and Narasimharao 2017; Li et al. 2019)

Sl No.	Catalyst	Support materials used	Particle size (nm)	CO removal efficiency				References
				CO concentration (%)	Activity (%)	TOF (min^{-1})	Temperature °C	
1	Cr_2O_3	rGO, GO	47	–	100	–	240 120	Ali et al. (2020)
2	Cu	CeO_2 nanorods	6	1	90	–	58	Dong et al. (2019)
3	Cu–Fe	CNT	–	1	100	–	180	Dasireddy et al. (2018)
	Cu–Co						220	
4	Cu–CeO_2	SiO_2 Al_2O_3	–	2	100	–	140	Alvarez et al. (2018)
5	CuO ZnO	Brass	–	1	100	–	~190	Cabello et al. (2022)
6	Cu	CeO_2, TiO_2 (2at%) modified CeO_2	–	2	~68 ~62	–	100	Gan and Si (2021)
7	Au–Fe_2O_3	Fe_2O_3	20–40	0.01	100	2.28	50	Alshehri and Narasimharao (2017)
8	Cu–Mn	Al_2O_3	–	1	100	–	120	Li et al. (2017)
9	Fe, Cr	Co_3O_4	18–33	1	100	–	<–60	Baidya et al. (2017)
10	Cu–Mn oxide	–	5–10	1	100	–	45	Li et al. (2017)

(continued)

Table 18.2 (continued)

Sl No.	Catalyst	Support materials used	Particle size (nm)	CO removal efficiency				References
				CO concentration (%)	Activity (%)	TOF (min^{-1})	Temperature °C	
11	Fe Oxide	Mesoporous alumina	<1 nm	1	100	4.44×10^{-3}	100	Kim et al. (2017)
12	Co oxide	Ceria	10–20	1	100	–	−60	Huan et al. (2019)

The energy and the intrinsic properties of a system are calculated using the density functional theory (DFT) calculations, where the energy is a function of electron density (r) and the electron density is a function of three positional variables ((r) $= f(x, y, z)$). The Hohenberg-Kohn theorem is used by the DFT to predict a molecular system's ground state energy from its electron density, which demonstrated that the ground state energy is a singular functional of electron density. The Kohn-Sham method provides the form for functional of the Hohenberg-Kohn theorems. The precise exchange-correlation functional must be known for the DFT method to produce the ground state energy and electron density with accuracy. The exact functional is unknown in practice, but one may attempt an approximation, which prompts a thorough search for functional with new variations. Local density approximation (LDA) and generalized gradient approximation (GGA) are well-known exchange correlation functions used in DFT calculations. In comparison with LDA, the GGA method provides geometrical structures with experimental-level accuracy, bond energies, transition, and adsorption energies with a few percent errors. As a result, the GGA is frequently used in DFT calculations to predict the descriptors of the molecule–surface interactions in CO oxidation catalytic reaction (Wang et al. 2013).

The descriptors to describe the rate of CO oxidation reactions are the reaction pathways, formation of intermediates, energy barrier, and charge transfer interactions. The reaction pathway of CO oxidation depends on several factors like moisture, oxygen vacancy, functionalization, active sites, etc. The CO oxidation mechanism typically follows three main reaction pathways, namely Eley-Rideal (E-R), Langmuir-Hinshelwood (L-H), and Mars-van Krevelen (MvK) mechanisms. In all the mechanisms, the reactant CO molecule is in the gas phase. The O_2 molecule is required to activate the CO oxidation in E-R and L-H mechanisms, whereas in MvK mechanism, the oxygen from metal oxides is used to activate the CO oxidation reaction (Chen et al. 2019).

In the E-R mechanism, the gas-phase reactant (O_2) directly reacts with the adsorbent (metal/metal oxide surface). In the initial step, the O_2 molecule is pre-adsorbed on the adsorbate and dissociates into two O atoms. The dissociation of the O_2 molecule is a reaction rate determination intermediate step. The O_2 dissociation energy depends on the intrinsic properties of the adsorbent. Subsequently, the CO molecule adsorbs at the bridge of two O atoms via physisorption. Depending upon the barrier energy, the physisorbed CO molecule directly interacts with one of the O atoms formed via dissociation and produces CO_2 molecule or the CO molecule reacts with the two dissociated O atoms and forms a carbonate as an intermediate. If the carbonate forms in the CO oxidation reaction, then the subsequent reaction is the cleavage of C–O bond of carbonate to generate CO_2. The CO oxidation activity will be higher for the adsorbates which directly produce the CO_2 molecule (Allian et al. 2011). In the L-H mechanism, both the reactants such as CO and O_2 are co-adsorbed on the adsorbent surface, which is an initial step for CO oxidation. The second step is O_2 molecule dissociation. As mentioned in E-R mechanism, the O_2 decomposition reaction determines the CO oxidation reaction activity in the L-H mechanism also. The bond length of O_2 elongates first and the adsorbed O_2 molecule dissociates into two O atoms. Following this, the dissociated O atom interacts with the adsorbed CO molecule and forms an OOCO intermediate through the C–O coupling. The C–O coupling intermediate reaction also needs to overcome a large activation barrier. The next step is the formation of a bent CO_2 molecule adsorbed on the surface. In the final step, the CO_2 molecule dissociates from the metal surface and forms a linear CO_2 molecule (Liu et al. 2012). The MvK mechanism reveals the CO oxidation reaction pathway, where the CO reacts with the lattice oxygen of the adsorbent. In the MvK mechanism, the initial step is CO molecule adsorption on the lattice oxygen of the adsorbent surface (metal oxides surface), where the metal oxides surface acts as an oxidant to interact with the CO molecule. The CO molecule adsorbed on the O atom of the adsorbent surface through the carbon atom. Then COO intermediates form through the C–O coupling. The next step is the formation of a bent CO_2 molecule. The bent CO_2 molecule dissociates from the adsorbent surface and forms a linear CO_2 molecule (Zhang et al. 2014).

The experimental conditions used to enhance the CO oxidation activity of metal oxides are still being debated. In most of the cases, the DFT calculated reaction barriers are very consistent with the experimental findings. Hence, significant computational research work has been carried out to understand the effect of different experimental conditions on CO oxidation. Some experimental studies suggested that the oxygen-rich characteristics of metal oxide surfaces will promote CO catalytic oxidation reactions. Some of them argue that the CO oxidation rate will be dependent on the ratio of O_2 and CO gas as well as the surface temperature. Chen-Hao Yeh et al. (2021) used DFT calculations to address the factors which determine the CO oxidation rate. They considered rutile IrO_2 (110) surface for the study and used Climbing Image Nudged Elastic Band (CI-NEB) method and Dimer method to predict the probable transition states and minimum energy path of CO oxidation reactions. The computational results revealed that the O-rich IrO_2 surface follows the E-R mechanism in the CO oxidation reaction. The O_2 can be easily dissociated from IrO_2, leading to an oxygen-rich environment. Direct CO_2 production and CO_3 formation are the two possible reaction routes on the surface. The calculation concluded that the desorption energy for CO_2 is still large on the IrO_2 surface and this reaction will be the rate-determining step in the E-R mechanism.

Lin et al. (2021) attempted to reveal the CO oxidation mechanism of polyoxometalates-supported Rh single-atom via DFT calculations. Single Rh atom doped phosphomolybdic acid (PMA) and phosphotungstic acid (PTA) have been considered for the study. The DFT calculations revealed that both the Rh doped PMA and PTA follow the modified MvK mechanism. The difference between the proposed and common MvK mechanism is the formation of a second CO_2 molecule through the central O atom (Fig. 18.7). They proposed that the Rh single atom doping is responsible for the three CO_2 molecule production. The computational study also indicates that the experimental condition not only influences the reaction mechanism but also the reaction path and formation of intermediates. Song et al. (2018) investigated the CO oxidation mechanism of Mn-doped CeO_2 catalysts with the aid of DFT and the microkinetic study. The study was an attempt to reveal the influence of Mn concentration on CeO_2 towards CO oxidation. The calculated CO adsorption energy on the surface (-2.25 eV) was found to be significantly higher than the O_2

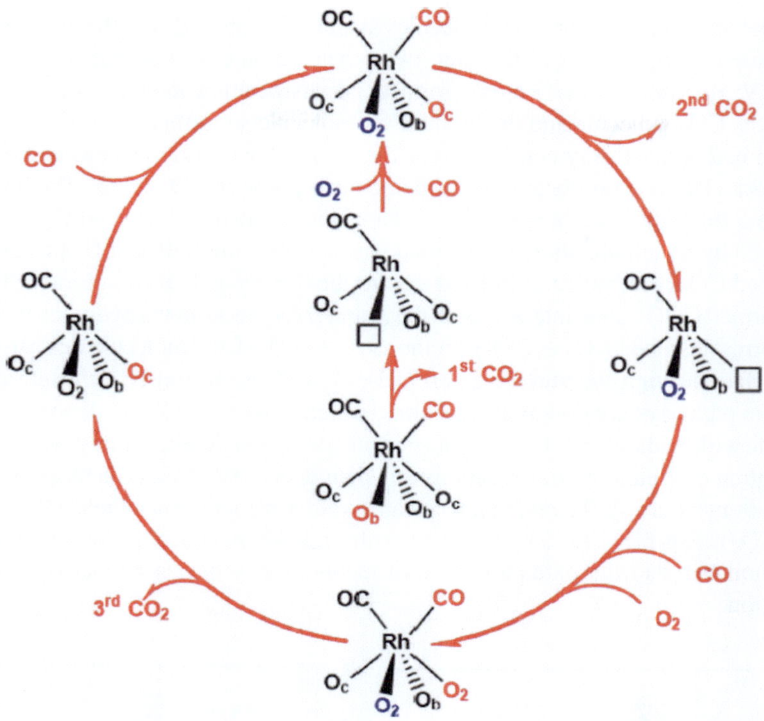

Fig. 18.7 Rh-assisted Mar-van Krevelen mechanism for CO oxidation (Reproduced with Molecular Catalysis 512 (2021) 111761) (Lin et al. 2021)

adsorption energy (−0.08 eV). Before the O_2 adsorption, a carbonite-like species was already formed. The results indicate that CO oxidation follows the MvK catalytic mechanism. The microkinetics simulations predict that the optimal temperature for CO oxidation depends on oxygen adsorption energies. The oxygen adsorption energy is controlled by the Mn concentration. Hence, the Mn doping significantly influence the CO oxidation rates and optimal temperatures.

In another study, Palliyarayil et al. (2021b) demonstrated the influence of moisture on CO oxidation of granular activated carbon (GAC) adsorbed by two different species of Pd such as Pd^0 and Pd^{2+}. The experimental findings show that Pd^0 adsorbed on GAC exhibit a significant CO oxidation reaction rate both with and without moisture. However, the Pd^{2+} adsorbed GAC shows the CO oxidation reaction only in presence of moisture, without moisture CO oxidation reaction is forbidden. A computational investigation was undertaken to predict the origin of the peculiar properties of Pd adsorbed GAC. In the computational approach, the Pd^{2+} species was represented by the PdO cluster as per the experimental findings. The L-H mechanism-based reaction pathway was considered for the study since most of the Pd-based catalysis follows the same. In the L-H mechanism reaction pathway considered for both the Pd and Pd^{2+} without moisture, it is emerged that (i) the reaction is initiated with the adsorption of reactant molecules such as CO and O_2; (ii) the adsorbed O_2 molecule undergoes dissociation into two oxygen atoms with an energy barrier of ~1.5 eV; (iii) one of the O atoms formed via dissociation interacts with CO and produces CO_2 molecule and finally the CO_2 molecule so formed gets desorbed.

The reaction pathway considered for the Pd and Pd^{2+} with moisture is shown in Fig. 18.8. The reaction begins with the co-adsorption of CO and O. The gaseous moisture (H_2O) was adsorbed at the center of the co-adsorbed CO and O. This was followed by the dissociation of adsorbed gaseous H_2O into OH and H species. The adsorbed CO molecule gets attracted toward the dissociated OH and forms COOH. The formed COOH then interacts with the nearest oxygen atom via hydrogen bonding and forms adsorbed CO_2 and OH. In the later step, the CO_2 molecule desorbs from the surface leaving the surface as an OH-rich surface. In the Pd^0 cluster, all the reaction pathways are down-hill reactions, which makes the CO oxidation reaction feasible with/without moisture. For the Pd^{2+} cluster system without moisture, the CO_2 desorption is an endothermic reaction, which hinders its CO oxidation reaction rates. But with moisture all the reaction paths are exothermic and hence the conversion of CO–CO_2 is feasible. The computational results indicate that not only the experimental condition but also the oxidation state of metal species play a crucial role in CO oxidation.

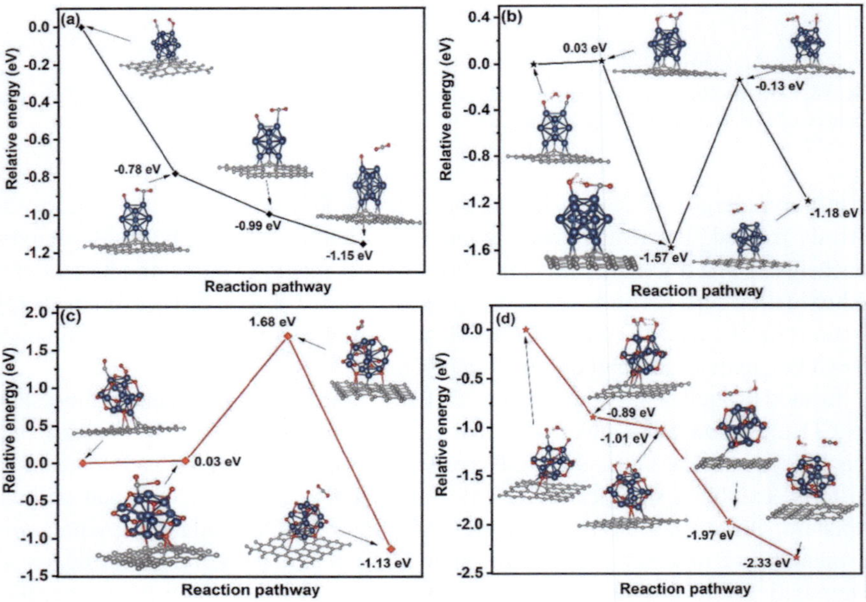

Fig. 18.8 Energy profile plot of CO oxidation, on **a–b** supported Pd^0 clusters in the absence and presence of moisture. **c–d** Pd^{+2}clusters in the absence and presence of moisture (Reproduced with the permission from ChemCatChem, 13(5), (2021) 1326–1339) (Singhania and Gupta 2018)

18.5 Conclusions

The abatement of carbon monoxide is becoming increasingly important as the change in human lifestyle and the growth of transport vehicles continues to increase the emission of this toxic gas. Among the different methods available, oxidative removal by using a suitable catalyst is considered the most efficient one. The studies reported in the literature lead to the conclusion that noble metals-based catalysts are more active toward CO oxidation and generally require mild operation conditions. On the other hand, non-noble metals demand drastic experimental conditions, especially high temperatures for better CO activity.

Different strategies have been developed for controlling the catalyst structures precisely by reducing metal sizes from nano to sub-nano or even single metal atoms. The recent advances in characterization techniques have helped in the development of novel catalysts and reaction routes for CO oxidation. The preparation method plays a major role in the structural characteristics of the metal nanocatalysts. Depending on the methods adopted, such as deposition-precipitation, cold plasma, and photo-deposition, the structural characteristics of the catalysts will be varied. Accordingly, the prepared catalyst shows varying activity which is related mainly to the particle size, particle dispersion, and its electronic state. It is also observed from the reported studies that the catalytic property of the catalyst material is influenced by the metal-support interaction as well as the presence of moisture in the feed gas. The incorporation of a secondary active metal, i.e., metal alloy formation is also reported for enhancing the CO removal capacity of the metal nanocatalysts. The molecular level understanding of the CO oxidation carried out via DFT studies proposes that the reaction follows three reaction pathways namely Eley-Rideal (E-R), Langmuir-Hinshelwood (L-H), and Mars-van Krevelen (MvK) mechanisms. However, research is still going on in this field to investigate the exact mechanism of how the different support materials affect catalytic activity. The future scope of the study involves the development of catalysts with low-loading noble metals possessing high CO oxidation efficiency under complex reaction conditions.

Acknowledgements We acknowledge MLP-10 CSIR-CLRI for funding. AP thanks the Science and Engineering Research Board (SERB), New Delhi, for SERB-National Postdoctoral Fellowship Award. CSIR-CLRI communication number 1799.

Authors Contribution The literatures on noble metal-supported catalysts have been reviewed by AP. KV and RKK carried out the literature survey on non-noble metal catalysts. Literatures on theoretical studies have been collected and reviewed by PS. SS has supervised, investigated, analyzed the discussions, reviewed, and edited the manuscript; AAV has conceptualized the chapter; supervised, investigated, and analyzed the discussions; coordinated with the team; and reviewed and edited the manuscript.

References

Ali, A.A., Madkour, M., Sagheer, F., Al, Zaki, M.I., Abdel Nazeer, A.: Catalysts **10**(1), 105 (2020)
(a) Allian, A.D., Takanabe, K., Fujdala, K.L., Hao, X., Truex, T.J., Cai, J., Buda, C., Neurock, M., Iglesia, E.: J. Am. Chem. Soc. **133**(12), 4498–4517 (2011); (b) Freund, H., Meijer, G., Scheffler, M., Schlögl, R., Wolf, M.: Angew. Chemie Int. Ed. **50**(43), 10064–10094 (2011)
Alshehri, A.A., Narasimharao, K.: J. Nanomater. **2017** (2017)
Altass, H.M., Ahmed, S.A., Salama, R.S., Moussa, Z., Jassas, R.S., Alsantali, R.I., Al-Rooqi, M.M., Ibrahim, A.A., Khder, M.A., Morad, M.: Catal. Lett. 1–11 (2022)
Alvarez, P., Aguila, G., Guerrero, S., Araya, P.: J. Chil. Chem. Soc. **63**(3), 4102–4108 (2018)
Bukhtiyarov, A.V., Prosvirin, I.P., Saraev, A.A., Klyushin, A.Y., Knop-Gericke, A., Bukhtiyarov, V.I.: Faraday Discuss. **208**, 255–268 (2018)
Burange, A.S., Reddy, K.P., Gopinath, C.S., Shukla, R., Tyagi, A.K.: Mol. Catal. **455**, 1–5 (2018)
Cabello, A.P., Ulla, M.A., Zamaro, J.M.: J. Mater. Sci. **57**(27), 12797–12809 (2022)
Chen, J., Wanyan, Y., Zeng, J., Fang, H., Li, Z., Dong, Y., Qin, R., Wu, C., Liu, D., Wang, M.: ACS Sustain. Chem. Eng. **6**(11), 14054–14062 (2018)
(a) Chen, Z.W., Chen, L.X., Wen, Z., Jiang, Q.: Phys. Chem. Chem. Phys. **21**(43), 23782–23802 (2019); (b)Tang, Y., Wang, Y.-G., Li, J.: J. Phys. Chem. C **121**(21), 11281–11289 (2017)
Dasireddy, V.D.B.C., Likozar, B., Valand, J.: Appl. Catal. B Environ. **237**, 1044–1058 (2018)
Di, L., Li, Z., Park, D.-W., Lee, B., Zhang, X.: Jpn. J. Appl. Phys. **56**(6), 60301 (2017)
Dong, F., Meng, Y., Han, W., Zhao, H., Tang, Z.: Sci. Rep. **9**(1), 1–14 (2019)
Eid, K., Ahmad, Y.H., Mohamed, A.T., Elsafy, A.G., Al-Qaradawi, S.Y.: Catalysts **8**(10), 411 (2018)
Fedorov, A.V., Tsapina, A.M., Bulavchenko, O.A., Saraev, A.A., Odegova, G.V., Ermakov, D.Y., Zubavichus, Y.V., Yakovlev, V.A., Kaichev, V.V.: Catal. Lett. **148**(12), 3715–3722 (2018)
Gan, L., Si, R.: J. Rare Earths **39**(1), 43–50 (2021)
Garcia, C., Truttmann, V., Lopez, I., Haunold, T., Marini, C., Rameshan, C., Pittenauer, E., Kregsamer, P., Dobrezberger, K., Stöger-Pollach, M.: J. Phys. Chem. C **124**(43), 23626–23636 (2020)
Gómez, L.E., Boix, A.V.: Cobalt Compounds and Applications. IntechOpen (2019)
(a) Greingor, J.L., Tosi, J.M., Ruhlmann, S., Aussedat, M.: Emerg. Med. J. **18**(5), 399–401 (2001); (b) McGregor, H.P., Westcott, K., Walker, D.W.: Pediatr. Res. **43**(1), 126–131 (1998)
Guo, X., Qiu, Z., Mao, J., Zhou, R.: Phys. Chem. Chem. Phys. **20**(40), 25983–25994 (2018)
(a) Iqbal, S., Clower, J.H., Hernandez, S.A., Damon, S.A., and Yip, F.Y.: Am. J. Public Health **102**(10), 1957–1963 (2012); (b) Daher, N., Saleh, R., Jaroudi, E., Sheheitli, H., Badr, T., Sepetdjian, E., Al Rashidi, M., Saliba, N., Shihadeh, A.: Atmos. Environ. **44**(1), 8–14 (2010)
Kim, J.K., Jin, C., Park, J., Iloska, M., Kim, M., Seo, D., Jung, J., Kim, T., Kim, J.: Ind. Eng. Chem. Res. **58**(43), 20154–20161 (2019)
Levy, R.J.: Neurotoxicol. Teratol. **49**, 31–40 (2015)
(a) Li, L., Han, W., Dong, F., Zong, L., Tang, Z., Zhang, J.: Microporous Mesoporous Mater. **249**, 1–9 (2017); (b) Baidya, T., Murayama, T., Bera, P., Safonova, O.V., Steiger, P., Katiyar, N.K., Biswas, K., Haruta, M.: J. Phys. Chem. C **121**(28), 15256–15265 (2017); (c) Li, Z., Wang, H., Wu, X., Ye, Q., Xu, X., Li, B., Wang, F.: Appl. Surf. Sci. **403**, 335–341 (2017); (d) Kim, I.H., Seo, H.O., Park, E.J., Han, S.W., Kim, Y.D.: Sci. Rep. **7**(1), 1–11 (2017); (e) Huan, W., Li, J., Ji, J., Xing, M.: Chinese J. Catal. **40**(5), 656–663 (2019)
Liang, F., Zhu, H., Qin, Z., Wang, G., Wang, J.: Catal. Commun. **10**(5), 737–740 (2009); (b) Wang, F., Xu, Y., Zhao, K., He, D.: Nano-Micro Lett. **6**(3), 233–241 (2014)
Lin, C.-H., Sun, Z.-Y., Liu, C.-G.: Mol. Catal. **512**, 111761 (2021)
Lin, A., Ibrahim, A.A., Arab, P., El-Kaderi, H.M., El-Shall, M.S.: ACS Appl. Mater. Interfaces **9**(21), 17961–17968 (2017); (b) Huang, B., Kobayashi, H., Yamamoto, T., Matsumura, S., Nishida, Y., Sato, K., Nagaoka, K., Kawaguchi, S., Kubota, Y., Kitagawa, H.: J. Am. Chem. Soc. **139**(13), 4643–4646 (2017); (c) Jeong, H., Bae, J., Han, J.W., and Lee, H.: ACS Catal. **7**(10), 7097–7105 (2017); (d) Wang, L., Pu, C., Xu, L., Cai, Y., Guo, Y., Guo, Y., Lu, G.: Fuel Process. Technol. **160**, 152–157 (2017); (e) Zhang, X., Yang, Y., Lv, X., Wang, Y., Cui, L.: Catalysts

7(12), 382 (2017); (f) Lou, Y., Liu, J.: Ind. Eng. Chem. Res. **56**(24), 6916–6925 (2017); (g) Zeng, L., Li, K., Wang, H., Yu, H., Zhu, X., Wei, Y., Ning, P., Shi, C., Luo, Y.: J. Phys. Chem. C **121**(23), 12696–12710 (2017); (h) Ro, I., Aragao, I.B., Chada, J.P., Liu, Y., Rivera-Dones, K.R., Ball, M.R., Zanchet, D., Dumesic, J.A., Huber, G.W.: J. Catal. 358: 19–26 (2018); (i) Tanaka, S., Lin, J., Kaneti, Y.V., Yusa, S., Jikihara, Y., Nakayama, T., Zakaria, M.B., Alshehri, A.A., You, J., Hossain, M.S.A.: Nanoscale **10**(10), 4779–4785 (2018); (j) Konopatsky, A.S., Leybo, D.V, Firestein, K.L., Popov, Z.I., Bondarev, A.V., Manakhov, A.M., Permyakova, E.S., Shtansky, D.V., Golberg, D.V.: J. Catal. **368**, 217–227 (2018); (k) Dutov, V.V., Mamontov, G.V., Zaikovskii, V.I., Liotta, L.F., Vodyankina, O.V.: Appl. Catal. B Environ. **221**, 598–609 (2018); (l) Wei, S., Fu, X.-P., Wang, W.-W., Jin, Z., Song, Q.-S., Jia, C.-J.: J. Phys. Chem. C **122**(9), 4928–4936 (2018); (m) Zakaria, M.B., Malgras, V., Nagata, T., Kim, J., Bando, Y., Fatehmulla, A., Aldhafiri, A.M., Farooq, W.A., Jikihara, Y., Nakayama, T.: Microporous Mesoporous Mater. **288**, 109530 (2019); (n) Liu, J., Ding, T., Zhang, H., Li, G., Cai, J., Zhao, D., Tian, Y., Xian, H., Bai, X., Li, X.: Catal. Sci. Technol. **8**(19), 4934–4944 (2018); (o) Carltonbird, M., Eaimsumang, S., Pongstabodee, S., Boonyuen, S., Smith, S.M., Luengnaruemitchai, A.: Chem. Eng. J. **344**, 545–555 (2018); (p) Wang, X.-L., Fu, X.-P., Wang, W.-W., Ma, C., Si, R., Jia, C.-J.: J. Phys. Chem. C **123**(14), 9001–9012 (2019); (q) Zhang, L., Cheng, X., Zhang, G., Qiu, W., He, H., Chen, G.: Appl. Catal. B Environ. **266**, 118629 (2020); (r) Hamidi, F., Chaker, H., Gil, S., Cherif, L., Saidi, K., Vernoux, P.: Res. Chem. Intermed. **47**(11), 4443–4456 (2021); (s) Li, J., Liu, Z., Cullen, D.A., Hu, W., Huang, J., Yao, L., Peng, Z., Liao, P., Wang, R.: ACS Catal. **9**(12), 11088–11103 (2019); (t) Han, B., Lang, R., Tang, H., Xu, J., Gu, X.-K., Qiao, B., Liu, J.J.: Chinese J. Catal. **40**(12), 1847–1853 (2019); (u) Xie, S., Wang, Z., Tan, W., Zhu, Y., Collier, S., Ma, L., Ehrlich, S.N., Xu, P., Yan, Y., Xu, T.: Environ. Sci. Technol. **55**(11), 7624–7633 (2021); (v) Al-Shehri, B.M., Shkir, M., Khder, A.S., Kaushik, A., Hamdy, M.S.: Nanomaterials **10**(6), 1067 (2020); (w) Aragão, I.B., Estrada, F.R., Barrett, D.H., Rodella, C.B.: Mol. Catal. **526**, 112377 (2022)

(a) Liu, S., Jin, P., Hao, C., Zhang, D., Yang, X., Chen, S.: Appl. Surf. Sci. **258**(8), 3980–3985 (2012); (b) Günay, M.E., Yildirim, R.: ChemCatChem **5**(6), 1395–1406 (2013)

Murata, K., Eleeda, E., Ohyama, J., Yamamoto, Y., Arai, S., Satsuma, A.: Phys. Chem. Chem. Phys. **21**(33), 18128–18137 (2019)

Nishida, Y., Sato, K., Yamamoto, T., Wu, D., Kusada, K., Kobayashi, H., Matsumura, S., Kitagawa, H., Nagaoka, K.: Chem. Lett. **46**(8), 1254–1257 (2017)

Palliyarayil, A., Selvarajan, P., Prakash, P.S., Sathish, C.I., Dasireddy, V.D.B.C., Vinu, A., Kumar, N.S., Sil, S.: ChemCatChem **13**(5), 1326–1339 (2021b)

(a) Palliyarayil, A., Prakash, P.S., Nandakumar, S., Kumar, N.S., Sil, S.: Diam. Relat. Mater. **107**, 107884 (2020); (b) Palliyarayil, A., Kumar, N.S., Sil, S.: Clean. Eng. Technol. **4**, 100252 (2021)

Qin, L., Cui, Y., Deng, T., Wei, F., Zhang, X.: ChemPhysChem **19**(24), 3346–3349 (2018)

Raub, J.A., Mathieu-Nolf, M., Hampson, N.B., Thom, S.R.: Carbon monoxide poisoning—A public health perspective. Toxicology **145**(1), 1–14 (2000)

Salama, R.S., Mannaa, M.A., Altass, H.M., Ibrahim, A.A., Khder, A.E.-R.S.: RSC Adv. **11**(8), 4318–4326 (2021)

Singhania, A., Gupta, S.M.: Catal. Lett. **148**(3), 946–952 (2018)

Song, W., Chen, L., Deng, J., Jing, M., Zheng, H., Liu, J., Zhao, Z.: J. Phys. Chem. C **122**(44), 25290–25300 (2018)

Srivastava, A.K., Saxena, A., Shah, D., Mahato, T.H., Singh, B., Shrivastava, A.R., Gutch, P.K., Shinde, C.P.: J. Hazard. Mater. **241**, 463–471 (2012)

Vakili, R., Gibson, E.K., Chansai, S., Xu, S., Al-Janabi, N., Wells, P.P., Hardacre, C., Walton, A., Fan, X.: ChemCatChem **10**(19), 4238–4242 (2018)

Vedyagin, A.A., Plyusnin, P.E., Rybinskaya, A.A., Shubin, Y.V., Mishakov, I.V., Korenev, S.V.: Mater. Res. Bull. **102**, 196–202 (2018)

Wang, Y.-G., Mei, D., Li, J., Rousseau, R.: J. Phys. Chem. C **117**(44), 23082–23089 (2013)

Wang, H., Shen, J., Huang, J., Xu, T., Zhu, J., Zhu, Y., Li, C.: Nanoscale **9**(43), 16817–16825 (2017)

Wang, Q., Jin, B., Hu, M., Jia, C., Li, X., Sharman, E., Jiang, J.: J. Phys. Chem. C **125**(10), 5616–5622 (2021)
(a) Wang, L., Wang, L., Zhang, J., Wang, H., Xiao, F.-S.: Chinese J. Catal. **39**(10), 1608–1614 (2018); (b) Wang, H., Luo, S., Zhang, M., Liu, W., Wu, X., Liu, S.: Roles of oxygen vacancy and Ox− in oxidation reactions over CeO2 and Ag/CeO2 nanorod model catalysts. J. Catal. **368**, 365–378 (2018)
Yeh, C.-H., Ji, B.-C., Nachimuthu, S., Jiang, J.-C.: Appl. Surf. Sci. **539**, 148244 (2021)
Zhang, B., Asakura, H., Yan, N.: Ind. Eng. Chem. Res. **56**(13), 3578–3587 (2017)
Zhang, X., Yang, Y., Song, L., Wang, Y., He, C., Wang, Z., Cui, L.: Mol. Catal. **447**, 80–89 (2018b)
(a) Zhang, S.-T., Li, C.-M., Yan, H., Wei, M., Evans, D.G., Duan, X.: J. Phys. Chem. C **118** (7), 3514–3522 (2014); (b) Xue, P., Fu, Z., Chu, X., Zhang, Y., Yang, Z.: Appl. Surf. Sci. **317**, 752–759 (2014)
(a) Zhang, X., Hou, F., Li, H., Yang, Y., Wang, Y., Liu, N., Yang, Y.: Microporous Mesoporous Mater. **259**, 211–219 (2018); (b) Yang, Y., Dong, H., Wang, Y., Wang, Y., Liu, N., Wang, D., Zhang, X.: Inorg. Chem. Commun. **86**, 74–77 (2017)
Zhou, A., Wang, J., Wang, H., Li, H., Wang, J., Shen, M.: J. Rare Earths **36**(3), 257–264 (2018)

Chapter 19
An Overview on MOF-Based Nanocatalysts for Dehydrogenation of Chemical Hydrogen Storage Sources

Ansari Palliyarayil, Rajani Kumar Borah, Rohan Keshav Kunkalekar, and Amit Ashok Vernekar

19.1 Introduction

The exponential growth of the human population as well as industries escalates the energy demand worldwide. Using fossil fuels as energy sources may lead to many environmental problems including air pollution, global warming, extreme climate change, and harmful health issues for humans and animals (Alpaydın et al. 2020). To overcome this increased demand for energy and the problems associated with using fossil fuels as energy resources, the research community is constantly searching for an alternative energy source that is clean, sustainable, and environmentally benign. Due to its high calorific value (~282 kJ mol^{-1}), zero pollutant emissions, and renewable nature, hydrogen is recognized as a promising alternative and sustainable energy resource (Furukawa et al. 2019; Feng et al. 2020; Su and Li 2021; Hou et al. 2019). Figure 19.1 represents the methods generally employed to generate hydrogen from various sources.

Albeit being a suitable alternative, safe storage and release, as well as transportation, are significant challenges to be addressed for the upcoming hydrogen economy. A more expedient option to solve this problem is to use chemical hydrogen storage sources in which hydrogen is chemically bonded with other stable elements (Liu et al. 2020; Navlani-García et al. 2019). They offer excellent safety, great convenience,

A. Palliyarayil · R. K. Borah · A. A. Vernekar (✉)
Inorganic and Physical Chemistry Laboratory (IPCL), CSIR-Central Leather Research Institute (CLRI), Chennai 600020, Tamil Nadu, India
e-mail: amitvernekar.clri@csir.res.in

R. K. Borah · A. A. Vernekar
Academy of Scientific and Innovative Research (AcSIR), Ghaziabad 201002, India

R. K. Kunkalekar
School of Chemical Sciences, Goa University, Taleigao Plateau 403206, Goa, India

R. S. Ningthoujam and A. K. Tyagi (eds.), *Handbook of Materials Science, Volume 3*, Indian Institute of Metals Series, https://doi.org/10.1007/978-981-96-8229-4_19

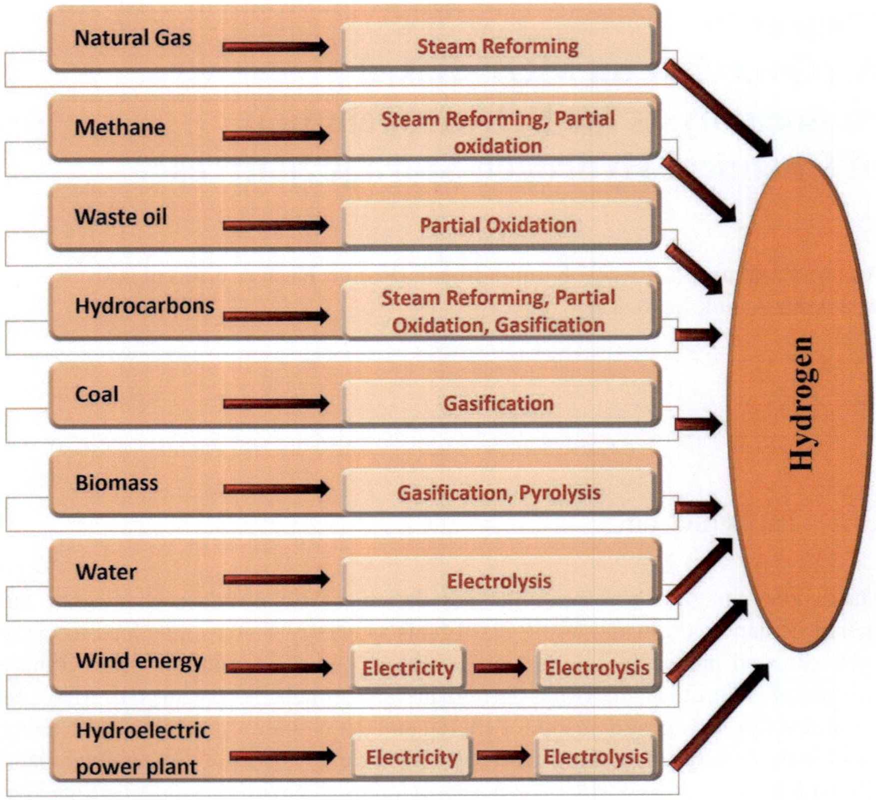

Fig. 19.1 Different methods of generation of hydrogen from various sources

and superior efficacy and can be suitable for large-scale practical applications. In the presence of a suitable catalyst, easy hydrogen release can be achieved by solvolysis or decomposition of various hydrogen storage mediums such as formic acid (HCOOH) and ammonia borane (NH_3BH_3), as compared to other sources like nitrides and imides. The latter demands high temperatures to release hydrogen, characterized by slow reaction kinetics and poor recyclability (Sun et al. 2020). The most common solid-state boron compounds reported for hydrogen generation include NH_3BH_3, hydrazine borane ($N_2H_4BH_3$), and sodium borohydride ($NaBH_4$). The advantage of using these boron compounds lies in the fact that the byproducts formed such as boric acid (H_3BO_3) and sodium metaborate ($NaBO_2$) are environmentally benign.

In 1953, Schlesinger et al. (1953) initially reported the use of $NaBH_4$ as a hydrogen storage material, and it has been considered the most favorable hydrogen fuel source. Nevertheless, its hydrogen content by weight is relatively modest at 10.6%, in contrast to other hydrogen carriers like NH_3BH_3 and $N_2H_4BH_3$. Furthermore, the formation of borates as byproducts and subsequent clogging of the piping system, along with the imperative need for efficient temperature management due to the exothermic nature of the process, represent notable drawbacks associated with the use of $NaBH_4$ (Schlesinger et al. 1953). Later on, in 2006, Chandra and Xu put forward NH_3BH_3 as a replacement for $NaBH_4$ (Schlesinger et al. 1953). This low molecular weight NH_3BH_3 (30.7 g mol^{-1}) attains much attention toward generating hydrogen because of its non-toxic nature and the high solubility and stability in an aqueous solution. It possesses a high hydrogen content of 19.6 wt.%. A maximum of 3 mol of hydrogen can be released per mole of NH_3BH_3. Albeit, the cost of the system is considerably high as the synthesis of NH_3BH_3 uses $NaBH_4$ along with other materials such as ammonium sulfate and tetrahydrofuran. Moreover, the system demands an additional filtration unit for removing the formed byproduct NH_3 (due to high pH and temperature) which may further increase the total cost of the system (Schlesinger et al. 1953). On the other hand, the hydrogen content of $N_2H_4BH_3$ is relatively high compared to $NaBH_4$, i.e., 15.4%, but the release of hydrazine byproduct and unwanted residue creation are some disadvantages regarding their hydrolysis (Schlesinger et al. 1953). The hydrogen production from NH_3BH_3, $N_2H_4BH_3$, and $NaBH_4$ via solvolysis/decomposition has been represented stoichiometrically by the following Eqs. 19.1–19.6 (Sun et al. 2020):

Ammonia Borane:

$$\text{Hydrolysis: } NH_3BH_3 + 2H_2O \rightarrow NH_4BO_2 + 3H_2 \tag{19.1}$$

$$\text{Methanolysis: } NH_3BH_3 + 4CH_3OH \rightarrow NH_4B(OCH_3)_4 + 3H_2 \tag{19.2}$$

Hydrazine Borane:

$$\text{Hydrolysis: } N_2H_4BH_3 + 3H_2O \rightarrow N_2H_4 + H_3BO_3 + 3H_2 \tag{19.3}$$

$$\text{Decomposition: } N_2H_4 \rightarrow N_2 + 2H_2 \tag{19.4}$$

Sodium Borohydride:

$$\text{Hydrolysis: } NaBH_4 + 2H_2O \rightarrow NaBO_2 + 4H_2 \tag{19.5}$$

$$\text{Methanolysis: } NaBH_4 + 4CH_3OH \rightarrow NaB(OCH_3)_4 + 4H_2 \tag{19.6}$$

Although the homogenous and heterogenous catalytic systems have been reported to release hydrogen from the different chemical hydrogen storage materials, the former is less efficient mainly due to fast deactivation and loss of catalyst during separation and recycling. The heterogeneous catalytic systems where metal nanocatalysts are supported on porous solid support attract much attention. Using porous materials as support will help to create specific reaction sites and inhibit the agglomeration of particles, thereby affecting the activity of the catalysts (White et al. 2009). Generally, materials like zeolites (Sun et al. 2019), metal–organic frameworks (MOFs) (Zhu et al. 2013), porous silica (Mori et al. 2019), activated carbon (Cai et al. 2013), and porous organic cages (POC) (Fang et al. 2018) as well as porous organic polymers (POP) (Cui et al. 2018) have been used as a support to incorporate ultrafine metal nanocatalysts.

MOFs are categorized as porous organic–inorganic hybrid crystalline materials containing metal ions/clusters and bridging organic linkers. They possess a high surface area which ensures easy access to active sites, thereby facilitating the reactions (Reddy et al. 2021). The presence of functionalizable organic linkers is another advantage of MOFs, which guarantees the possibility of tuning the property of MOFs thereby making them specific for particular applications. Moreover, these linkers will act as a skeleton to link the active metal node, necessary to synthesize a highly porous MOF structure. Owing to these special features, MOFs have been studied for various applications by the research community, especially for catalysis. The present chapter discusses the recent advances in MOFs-based catalysts for the generation of hydrogen from various hydrogen storage materials such as NH_3BH_3, $NaBH_4$, and $N_2H_4BH_3$. In this regard, literature published from 2017 onwards has been covered and discussed.

Prior to 2017, there are a lot of MOF-based catalysts (both noble and non-noble metal catalysts) that have been reported for hydrogen production specifically from NH_3BH_3. Yaoqi Li and co-workers, in 2009, developed metallic Ni catalyst via the in situ reduction of a MOF, [(Ni(4,4′-bipy) (HBTC)] (HBTC = 1,3,5-benzenetricarboxylic acid; 4,4′-bipy = 4,4′-bipyridine) for NH_3BH_3 hydrolysis (Li et al. 2009). The MOF featured a honeycomb channel with dimension 8 Å at the wide and 5 Å at the narrow spacing, along with a rectangular channel measuring 7 × 6 Å^2. To prepare the catalyst, the aqueous solution of the MOF was reduced with methanolic solution of NH_3BH_3 in a glove box under ambient temperature. The microporous nature of the MOF facilitated the creation of tiny clusters of active metallic site, resulting in an enhanced activity for hydrogen generation with an activation energy value of 26 kJ mol^{-1}.

In the following year, Qimimg Li et al. and Pei-Zhou Li et al. developed Co-ZIF-9 (solvo thermal method) and Ni immobilized ZIF-8 (CVD-reduction method) catalysts for the hydrolytic dehydrogenation of $NaBH_4$ and NH_3BH_3, respectively (Li and Kim 2012). Later In 2016, a bimetallic (NiRu) and trimetallic (RuCuCo) catalysts supported on MIL-101 was reported by Sandipan et al. and Kanzhou et al., respectively. While Sandipan et al. used a double-solvent method (followed by the reduction of metal ions) to prepare NiRu-MIL-101 catalyst, Kanzhou et al. followed a simple liquid impregnation method. The synergetic effect between the metals enhances the activity of the catalysts (TOF_{NiRu} = 272.72 mol H_2 min^{-1}(mol cat)$^{-1}$, TOF_{RuCuCo} = 241.2 mol H_2 min^{-1} (mol Ru)$^{-1}$) (Yang et al. 2016). The reported noble metal catalysts include Pt@MIL (MatérialInstitut Lavoisier)-101, Au_7Ni_{93}@MIL-101, Pd@MIL-101, Ru@MIL-101, Au_6Co_{94}@MIL-101, AgNi/MIL-101, $Ni_{70}Ru_{30}$@MIL-101, $Rh_{15}Ni_{85}$@ZIF (Zeolitic Imidazolate Framework)-8, and Ag_1Pd_4@UiO (University of Oslo)-66-NH_2 (Aijaz et al. 2012). On the other hand, the reports on MOF-based catalysts for hydrogen production from $NaBH_4$, and $N_2H_4BH_3$, prior to 2017 are limited.

19.2 MOFs for Hydrogen Generation

The history of MOFs started in the early 1990s. Robsons group (Yu et al. 2017) has developed a copper-based 3-D network polymer $\{CuI[C(C_6H_4 \cdot CN_4)]\}_n$, by using Cu(I) and the nitrogen-containing organic ligands as the metal node and bridging linker, respectively, which pave a path for the development of MOFs. Later on, from the widespread attention received from the research communities, there have been several research articles published on MOFs/MOF-based catalysts for the hydrogen production from different hydrogen storage sources. The performance of MOFs can be controlled by the proper selection of metal nodes and bridging organic linkers. Furthermore, the activity can be tuned by regulating the pore size as well as the electronic nature of the active center and post-synthetic modification with other metal catalysts. Albeit, their stability toward temperature and chemical steadiness should be considered while using them for catalytic application.

Another factor affecting the catalytic efficiency is the property of the solvents used. To understand the effect of solvent type and property on catalyst efficiency, Li et al. (2019) studied different solvents such as water, alcohol, acid, alkali, and rich NH_4^+ solution for hydrogen production using RuCu/β-SiC NWs as a catalyst. As per their observation, both acid and alkali can enhance the activity. However, the mechanism is different in both acid and base mediated hydrogen generation reaction. In the case of acids, it will react directly with NH_3BH_3 and enhance the activity, whereas in the case of base, it is not the case. The reason can either be (i) the OH^- promotes the activation of B–H bond in BH_3 or the O–H bond of H_2O or (ii) the OH^- helps to remove the surface NH_4^+ ions generated from the byproduct NH_4BO_2. Similarly, the catalyst shows difference in activity when water and alcohol were used as solvents. In the presence of water, 3 equiv. of H_2 were generated from 1 equiv. of NH_3BH_3, whereas in alcohols, the generated hydrogen is 2.79, 2.76, and 2.76 equiv., when CH_3OH, CH_3CH_2OH, and $(CH_2OH)_2$ were used as solvents, respectively. The reason could be due to the difference in the energy barrier for the hydrolysis and methanolysis of NH_3BH_3. The value is smaller for hydrolysis reaction (1.2962 eV) compared to the methanolysis (1.66 eV). The case is different for the NH_4^+ rich solvents. The activity shows a decrease as the active catalyst surface was covered by the excess NH_4^+ ions.

19.3 Transition Metal-Supported MOF for Hydrogen Generation

Unlike noble metals such as Silver (Ag), Gold (Au), Palladium (Pd), and Platinum (Pt), non-noble metals like Cobalt (Co), Iron (Fe), Nickel (Ni), Copper (Cu), etc. received significant attention among researchers due to their abundance and low cost. As the activity of these transition metal catalysts was low as compared to the noble metal catalysts, researchers took efforts to find a new synthesis route to develop noble metal-free catalysts with increased catalytic activity. Amorphous materials are characterized by a large number of unsaturated coordination sites, which arise due to the increased structural defects present in them. Liu et al. (2017) utilized this property of amorphous materials to synthesize highly active catalysts for hydrogen generation, assuming that small-sized amorphous nanoparticles (NPs) will show high hydrogen generation efficiency. To avoid the complex experimental procedure, the authors have adopted an ultrasound-assisted rapid in situ reduction protocol to develop a noble metal-free catalyst supported on MIL-101MOF ($Cr_3F(H_2O)_2O$–$[C_6H_4(CO_2)_2]_3 \cdot nH_2O(n \approx 25)$). In general, the pore size of the support materials plays a key role in the development of small-sized NPs with uniform distribution. MIL-101, a chromium-based MOF, possesses a large pore size of 2.9–3.4 nm. Four different monometallic, i.e., Co, Cu, Ni, and Fe as well as bimetallic catalysts, i.e., Co alloyed with Cu, Ni, and Fe were synthesized accordingly with a metal particle size of 1.6–2.6 nm. The MIL-101 is retaining its crystalline structure even after the incorporation of the metal NPs as evident from the XRD analysis. Among the four monometallic catalysts, the activity toward hydrogen generation by NH_3BH_3 hydrolysis was reported to be highest for Co/MIL-101 with a total TOF of 51.4 mol H_2 (mol cat)$^{-1}$ min^{-1}. The incorporation of the second metal doesn't make much difference as the activity is more or less similar with a TOF value of 51.7, 50.8, and 44.3 mol H_2 (mol cat)$^{-1}$ min^{-1}, respectively, for CuCo-MIL-101, FeCo/MIL-101, and NiCo/MIL-101, respectively. The authors have observed extended stability of the catalysts up to 30 cycles without any change in the activity as well as the size of the NPs, ensuring the possibility of developing highly stable and active catalysts via the proposed ultrasound-assisted rapid in situ reduction protocol.

To develop an economically feasible catalyst with enhanced activity, Xiaopeng et al. (2017) introduced a small percentage of noble metals into Co-based catalysts and synthesized RhCo bimetallic catalyst supported on ZIF-67 (Zeolite Imidazole Framework-$C_8H_{10}N_4Co$), a cobalt-based MOF. To synthesize the catalyst with varying Co and Rhodium (Rh) ratios, Xiaopeng and co-workers followed a simple wet chemical reduction protocol where ZIF-67 MOF was treated with the required amount of Co and Rh salt solution under vigorous stirring for 24 h. Following this, the mixture was reduced by using $NaBH_4$ as the reducing agent to result in the final bimetallic catalyst RhCo@ZIF-67. The uniform dispersion of ultrafine RhCo NPs of size 0.52 ± 0.3 nm over the ZIF-67 support causes a drastic decrease in the surface area from 1371.3318 to 195.6174 m^2 g^{-1}. Among the different series of catalysts studied (prepared by changing the mol ratio of Rh and Co), the $Rh_{25}Co_{75}$@ZIF-67 shows the best NH_3BH_3 dehydrogenation efficiency. The studies of NH_3BH_3 dehydrogenation by RhCo@ZIF-67 catalyst have shown that the combined effect of both RhCo alloy and ZIF-67 increased the performance of the catalysts considerably compared to the respective monometallic catalysts with a TOF of 100.21 mol H_2 (mol cat)$^{-1}$ min^{-1}. Similar studies have been reported by Lu et al. (2017), where Co was alloyed with Rh, which is immobilized on an aluminum-based MOF, i.e., MIL-96 ($Al_{12}O(OH)_{18}(H_2O)_3(Al_2(OH)_4)[btc]_6 \cdot 24H_2O$), toward the hydrolytic hydrogen generation from NH_3BH_3. The $AlO_2(OH)_4$ and $AlO_4(OH)_2$ octahedra present in the structure of MIL-96 were interconnected to form an eighteen-membered hexagonal ring structure. This linkage formed via trimesate (btc) ligand gives a 3D network structure to the MOF. The catalyst was prepared by the direct mixing of the respective metal salt solution with MIL-96 dispersed in water under stirring followed by reduction with $NaBH_4$. The catalyst was characterized with a high surface area of 205.5 m^2 g^{-1}. The MIL-96 possesses a nanofibrous morphology with width 5–10 nm and length 100–200 nm (Fig. 19.2a) with uniformly dispersed CoRu alloy particles on it (Fig. 19.2b) comprising Ruthenium (Ru) in its metallic state and Co in both oxidized and reduced states. The particle size distribution curve as given in Fig. 19.2c ensures an average CoRu particle size of 1.7 nm. The lattice contraction arises due to the incorporation of Co into the lattice of Ru, i.e., CoRu alloy formation leads to a shift in the XRD pattern of Ru as reported by the authors; a positive shift in 2 theta values from 44.1° to 44.7° is observed. The synergetic effect which arises due to the difference in the electronegativity of Co and Ru combined with the bi-functional effect between the alloy NPs and the MOF considerably increases the hydrogen generation activity of the catalysts (TOF = 320.7 mol H_2 (mol Ru)$^{-1}$ min^{-1}, TON = 30400, and activation energy $E_a = 36$ kJ mol^{-1} for NH_3BH_3). Albeit the catalyst retains 90.5% of its initial activity during the second cycle, the activity is reduced during the consecutive 3rd, 4th, and 5th cycles and shows only 69.7% of its initial activity caused by the leaching out of active metal nanocatalyst during each run.

In another study, a multi-layered core-shell structure type transition metal-based catalyst has also been reported for the generation of hydrogen from NH_3BH_3 by hydrolysis. A copper-cobalt-nickel-based triple-layered catalyst (Cu@Co@Ni/MIL-101) with a core-shell structure (having a particle size of ~3.3 nm) deposited on the surface of a porous MOF (MIL-101) has been synthesized via double-solvent approach (Sun et al. 2018), where the aqueous solution of the metal salts was added into a suspension of MIL-101 in dry n-hexane under stirring followed by filtration and reduction with NH_3BH_3. The produced copper hydride Cu–H (Cu has a high reduction potential of +0.34 eV) during the reduction will ensure the continuous reduction of other Co and Ni ions. The authors confirmed metal hydride formation by observing a broad peak at ~6.5 ppm in the proton NMR spectra of Cu@Co@Ni/MIL-101. The three-layered core-shell catalyst, Cu@Co@Ni/MIL-101, possessing a high surface area of 2148 m^{2}g^{-1} with all the three active metals in its reduced form, shows dehydrogenation efficiency superior to the respective mono/bi metallic and alloy catalysts. It ceases the hydrogen release in 10 min at 25 °C (Fig. 19.3) with a TOF of 31 mol H_2 (mol cat)$^{-1}$ min^{-1} with excellent stability ensuring the

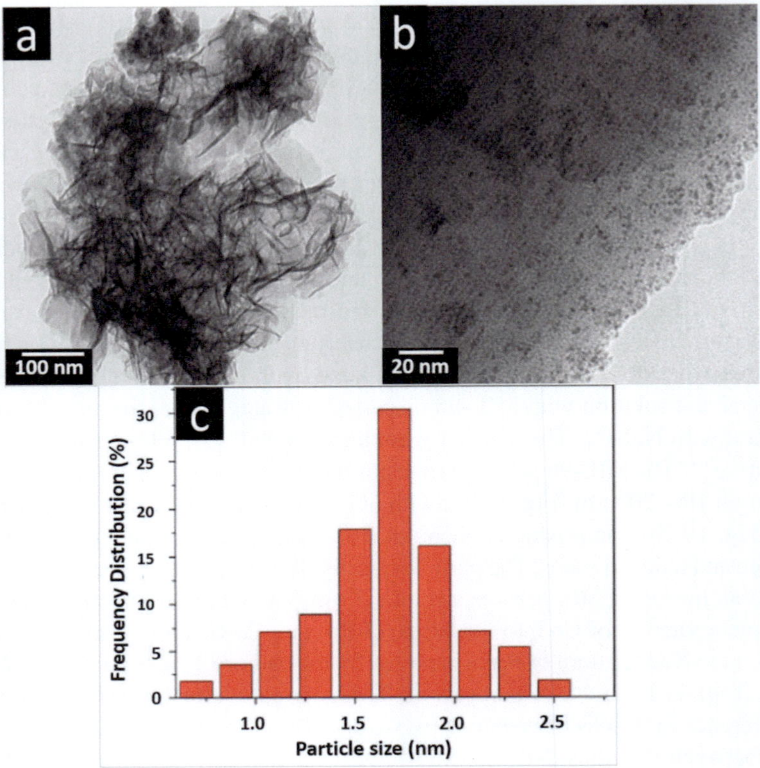

Fig. 19.2 **a** TEM, **b** HRTEM images, and **c** the particle size distribution of Ru_1Co_1/MIL-96 catalyst. Reproduced with the permission from Journal of alloys and compounds 694 (2017) 662–671 (Lu et al. 2017)

reusability of the catalyst up to 5 consecutive cycles. Similar studies of NH_3BH_3 dehydrogenation by using trimetallic Cu–Ni–Co and Co–Fe–Cu have been reported by other research groups (Liang et al. 2018; Li and Li 2020).

To understand the mechanism involved in the generation of hydrogen by non-noble metal catalysts, isotopic mass spectrometry analysis was carried out by Liang et al. (2018). They proposed that upon treating NH_3BH_3 with $Cu_{0.8}Ni_{0.1}Co_{0.1}$/MIL-101 catalyst, an active metal-H complex forms on the surface of metals. Simultaneously, the B–N bond of the intermediate NH_3BH_3-M undergoes water-associated dissociation leading to the generation of borate ions followed by the release of hydrogen molecules by the interaction between water and metal-H complex (Fig. 19.4) (Liang et al. 2018).

Although MOFs have been reported as support for various nanocatalysts, the study reported by Xu et al. (2021) ensures that other than the incorporated metal NPs, the metal nodes of MOF can also act as active sites toward the reactants. They have synthesized Cu NPs-Co-MOF (Cu_x@Co_{1-x}-MOF) interfacial ensemble by following an in situ solvothermal method where MOF forms by coordinating Co^{2+}

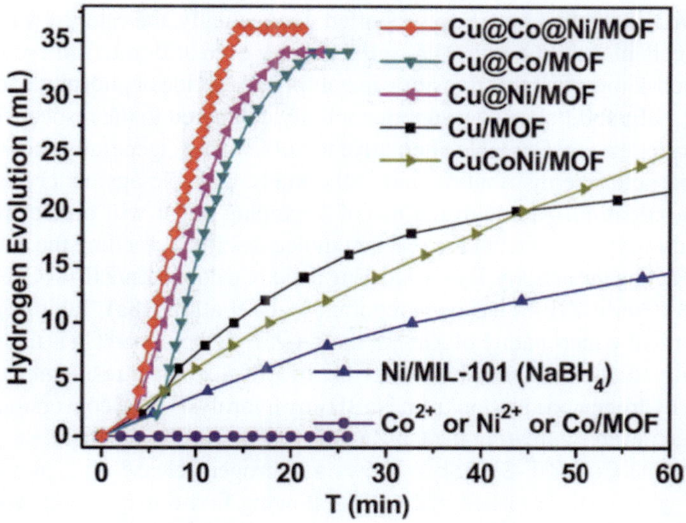

Fig. 19.3 Comparison plot of time-dependent volume of hydrogen generated from NH_3BH_3 using the Cu@Co@Ni/MOF and the other control samples. Reproduced with the permission from ACS applied materials & interfaces 11(1) (2019) 940–947 (Sun et al. 2018)

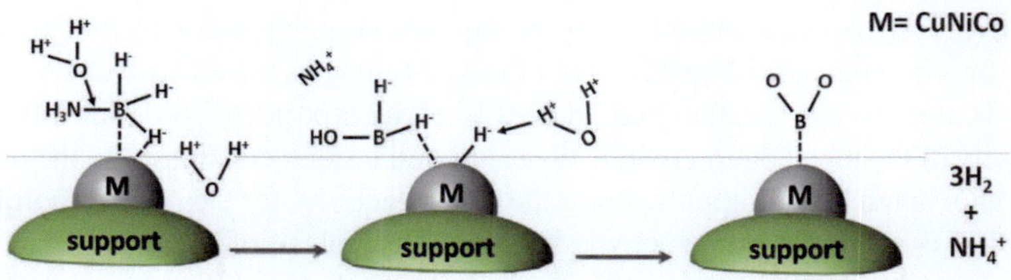

Fig. 19.4 Schematic representation of NH_3BH_3 hydrolysis using $Cu_{0.8}Ni_{0.1}Co_{0.1}$/MIL-101 Catalyst. Reproduced with the permission from Journal of Alloys and Compounds 741 (2018) 501–508 (Liang et al. 2018)

with 1,3,5-benzene tricarboxylic acid and Cu NPs by the methanolic reduction of Cu^{2+} under acidic pH at 150 °C. The alloying effect between the metals has been exempted by the authors as the B.E. value w.r.t. Co^{2+} and Cu^0 obtained from XPS do not show any shift indicating that there is no electronic interaction occurring between Co and Cu. The prepared catalyst has been studied for hydrogen production from NH_3BH_3 hydrolysis. A substantial increase in activity of the catalyst $Cu_{0.5}$@$Co_{0.5}$-MOF has been observed by Xu et al. (2021) upon the incorporation of Cu. The incorporation of Cu causes the TOF value to jump from 16.8 min^{-1} to 36.6 min^{-1}, approximately a 118% increase. Further increase in the activity (255%) by subjecting the catalyst to activation via heating at 200 °C under air has also been observed. The reason could be attributed to the abundant defects formed at the Cu–Co^{2+} interface, which acts as the active catalytic centers; a maximum TOF of 129.8 min^{-1} for the

Cu@Co-MOF activated for 5 h was reported. Interestingly, the catalyst was reported to have excellent stability with 95% initial activity even after the fifth cycle.

It is a well-known fact that transition metals in their boride or phosphide form (e.g., CoB, CoP, NiB, and FeB) show superior activity compared to the respective metals. Cobalt boride, the most suitable alternative to noble metals, generally shows a broad range of particle size distribution and is thermally unstable against crystallization as well as oxidation. Although the use of a capping agent will protect NPs from agglomeration, it may prevent the reactant molecules from reaching the active sites, thereby affecting the activity. By considering these, Co doped Zn/ZIF-8 ($C_8H_{10}N_4Zn$) framework (Co/Zn-ZIF-8) has been reported by Li et al. (2018a). Taking advantage of the chemically inert nature of Zn^{2+} in the Zn-ZIF-8, the doped Co is then reduced to CoB without affecting the phase structure of ZIF-8, and the resultant CoB/ZIF-8 is used for hydrogen production from $NaBH_4$ by hydrolysis. The conversion of Co to CoB was achieved by treating the Co/Zn-ZIF-8 with a reducing agent, i.e., $NaBH_4$, for ~96 h. The CoB/ZIF-8 catalyst shows a hydrogen release efficiency of 453.6 mL min^{-1} g^{-1} in the presence of NaOH following first-order kinetics with an E_a of 57.72 kJ mol^{-1}. The authors also observed a correlation between the hydrogen release efficiency of the CoB/ZIF-8 catalyst and the concentration of the base. An optimum NaOH concentration of 5% is suitable to achieve maximum hydrogen release efficiency. Likely, Luo et al. (2019) in their study have prepared a series of Co, Cu, Fe, and Ni NPs (average particle sizes are 1.5 nm, 3.5 nm, 1.4 nm, and 2.1 nm, respectively) encapsulated on ZIF-8 by the deposition-precipitation method followed by in situ reduction using $NaBH_4$. From the TEM images, the authors observed that although most of the Co, Cu, and Ni NPs were incorporated inside the MOF, the selectivity of the metal counterpart is the result of the steric constraints in the interior of the MOF cavity. The support shows the best selectivity for Co NPs. Accordingly, the Co@ZIF-8 shows superior activity toward the hydrolysis of $NaBH_4$ (TOF = 2935 mL H_2 (gcat)$^{-1}$ min^{-1}, E_a = 62.9 kJ mol^{-1}). They proposed that during the initial stage, a hydrogen bonding is formed between the hydridic borohydride hydrogen atom and the acidic hydrogen atom from water. The presence of the catalysts causes the oxidative addition of B–H and O–H bonds. During this, the hydridic as well as protonic hydrogen are shifted to the surface of NPs generating metal NPs-dihydride. This is followed by reductive elimination that produces H_2 and BH_3OH^- (Fig. 19.5).

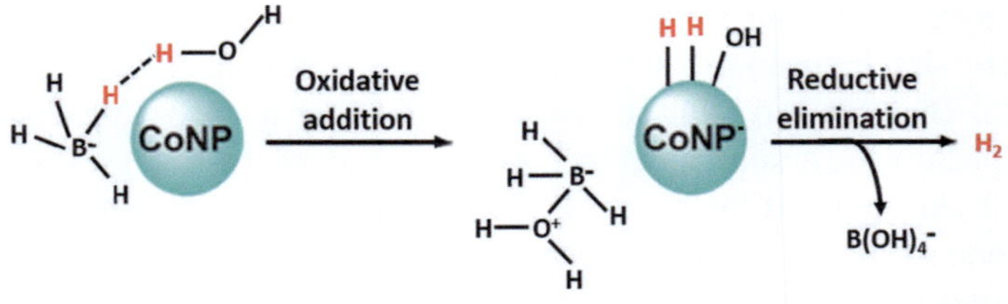

Fig. 19.5 Proposed mechanism of hydrolysis of $NaBH_4$ by Co@ZIF-8 catalyst. Reproduced with the permission from Chem Cat Chem 11(06) (2019) 1643–1649 (Luo et al. 2019)

Very recently, an amorphousless crystalline NiPt alloy catalyst supported on MIL-101 was synthesized by Bai and co-workers (Bai et al. 2022). A combined effect of the NPs dispersion, electron synergy by metal alloying, and the amorphous-poor crystalline nature toward hydrogen generation has been reported by the authors. The space constraint effect of MIL-101 ensures the uniform dispersion of the NPs over the surface of MIL-101. By modifying the NiPt NPs with CeO_x, the authors put efforts to reduce the crystallinity of NiPt NPs, thereby introducing unsaturated coordination sites in the system. Moreover, the electron donor characteristics of CeOx promote the electron transfer from CeOx to Ni and Pt. The reported amorphous-poor crystalline NiPt-CeOx/MIL-101 catalyst shows superior activity toward hydrogen generation from $N_2H_4BH_3$, 100% activity in 1.32 min with a TOF of 2951.1 h^{-1}. The recyclability studies guarantee the reusability of the catalyst as it shows a 100% H_2 generation rate up to 6th cycle. Mechanistically, the B-N bond in $N_2H_4BH_3$ molecule adsorbed on the surface of NiPt-CeOx catalyst undergoes dissociation to generate BH_3 and N_2H_4. The BH_3 molecules then undergo a replacement reaction with water to form H_2. Moreover, the electron-rich NiPt catalyst selectively breaks the N–H bond of N_2H_4, leading to 100% generation of H_2 from NH_3BH_3. Li et al. have also studied CeOx-modified NiFe system toward hydrogen generation from $N_2H_4BH_3$ (Li et al. 2018b).

As is known, the dehydrogenation of $N_2H_4BH_3$ includes the hydrolysis as well as the decomposition of BH_3 and N_2H_4 groups, respectively. However, most of the reported catalysts were active toward the hydrolysis of the BH_3 group. To achieve the selective decomposition of N_2H_4 part, it is mandatory to use noble metals as catalysts. However, the high cost, as well as the resource scarcity, pulls the catalyst interest among researchers down. Albeit, alloying the noble metals with other non-noble metals helps in tuning the electronic nature as well as the surface state of the noble metals, which can lead to an economically feasible catalyst with enhanced activity. A chromium-based MOF possessing a high surface area and large pore size (~3.5 nm) was used by Zhang et al. (2017) to encapsulate Pt catalyst alloyed with Ni. By controlling the reduction rate as well as by changing the metal (Ni + Pt): reducing agent ($NaBH_4$) ratio, the authors were able to achieve a well-dispersed PtNi alloy catalyst of size ~2.5 nm showing a 100% hydrogen generation activity from $N_2H_4BH_3$ in 1.92 min with a total TOF of 1515 mol H_2 (mol(Ni + Pt))$^{-1}$ h^{-1} (Fig. 19.6). The corresponding E_a reported for H_2 generation from both BH_3 and N_2H_4 groups are 18.6 and 44.6 kJ mol^{-1}. Table 19.1 briefs out the properties of non-noble metal-supported MOF-based catalysts reported in the literature toward hydrogen generation from solid hydrogen storage materials.

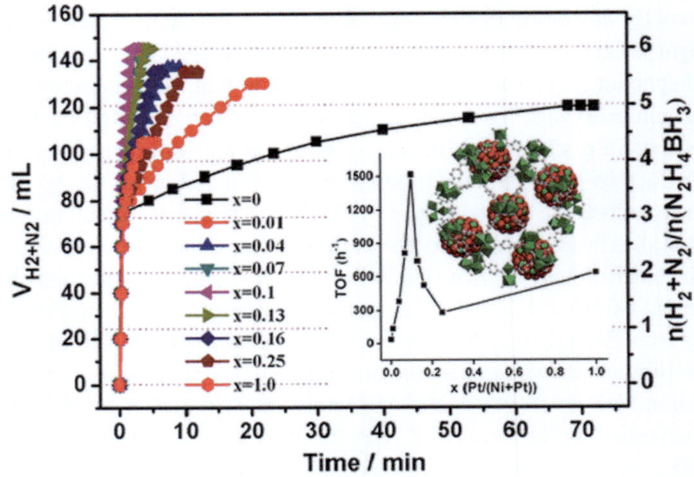

Fig. 19.6 H_2 production from $N_2H_4BH_3$ solution using $Ni_{1-x}Pt_x$/MIL-101. Inset shows the respective TOF values of the catalysts. Reproduced with the permission from Inorganic Chemistry 56(19) (2017) 11938–11945 (Zhang et al. 2017)

19.4 Noble Metal-Supported MOF for Hydrogen Generation

From the literature survey, it is proven that noble metals like Rh and Ru have also shown great catalytic activity toward hydrogen generation by dehydrogenating various hydrogen storage materials. The confinement of the noble metal NPs in the pores of MOF having big pore sizes inhibits the aggregation and stabilizes the NPs. The metal NPs can be encapsulated within the framework of UiO-66, and their nucleation and growth can also be controlled. Zhang et al. (2019) reported the synthesis of Rh@UiO-66, with Rh NPs having a mean size of 2.7 nm confined in UiO-66. They followed a double-solvent host-guest strategy as shown in Fig. 19.7, to prevent the aggregation of NPs on the outer surface of the MOF. For comparison, they also synthesized another catalyst, Rh/UiO-66, by supporting Rh NPs on the outer surface of the MOF via a simple impregnation strategy (Fig. 19.7). From the BET analysis, the decreased pore volume and surface area of Rh@UiO-66, compared to UiO-66, indicate the encapsulation of Rh NPs in the MOF pores. The activity of Rh@UiO-66 for hydrolysis of NH_3BH_3 is 3.5 times greater than that of Rh NPs dispersed on the outer surface of UiO-66 (Rh/UiO-66) due to the confinement effect. At room temperature, Rh@UiO-66 showed a TOF of 219.8 mol H_2 (mol cat)$^{-1}$ min^{-1}, and TOF for Rh/UiO-66 catalyst was 62.9 mol H_2 (mol cat)$^{-1}$ min^{-1}. The catalyst showed high reusability with nearly unchanged activity even after four cycles. Under alkaline conditions (0.2 M of NaOH), the hydrogen generation rate was significantly higher with TOF and E_a of 490.6 mol H_2 (mol cat)$^{-1}$ min^{-1} and 38.4 kJ mol^{-1}, respectively.

Table 19.1 The properties of non-noble metal-supported MOF-based catalysts reported in literature toward the generation of hydrogen from solid hydrogen storage materials

SI No.	Catalyst used	Particle size nm	Hydrogen evolution activity						References
			Rate mL min^{-1} g^{-1}	TOF min^{-1}	Activation energy kJ mol^{-1}	Temperature °C	Reusability		
							No. of cycles	Percentage of retention	
Ammonia borane									
1	Co/MIL-101 Cu/MIL-101 Ni/MIL-101 Fe/MIL-101 CuCo/MIL-101 FeCo/MIL-101 NiCo/MIL-101	1.6–24.5	–	51.4 0.5 12.2 0.8 51.7 50.8 44.3	31.3 (Co/MIL-101) 30.5 (CuCo/MIL/101)	25	30	100 (Co/MIL-101)	Liu et al. (2017)
2	$Rh_{25}Co_{75}$/ZIF-67	0.52 ± 0.3	–	100.21	–	25	5	–	Xiaopeng et al. (2017)
3	Ru_1Co_1/MIL-96	1.7	–	320.7	36	25	5	67.9	Lu et al. (2017)
4	$Cu_{0.33}$@$Co_{0.33}$@$Ni_{0.33}$/MIL-101	3.3	–	31	–	25	5	100	Sun et al. (2018)
5	$Cu_{0.8}Ni_{0.1}Co_{0.1}$/MIL-101	2.8 ± 0.2	–	70.1	29.1	25	8	73	Liang et al. (2018)
6	$Cu_6Fe_{0.8}Co_{3.2}$/MIL-101	2.6 ± 0.8	–	23.2	37.1	25	7	60	Li and Li (2020)
7	$Cu_{0.5}$@$Co_{0.5}$-MOF	5.5	–	129.8	26.5	25	10	95	Xu et al. (2021)
Sodium borohydride									
8	CoB/ZIF-8	–	453.6	–	57.72	25	–	–	Li et al. (2018a)

(continued)

Table 19.1 (continued)

SI No.	Catalyst used	Particle size nm	Hydrogen evolution activity						References
			Rate mL min^{-1} g^{-1}	TOF min^{-1}	Activation energy kJ mol^{-1}	Temperature °C	Reusability		
							No. of cycles	Percentage of retention	
9	Co/ZIF-8 Cu/ZIF-8 Ni/ZIF-8 Fe/ZIF-8	1.5 3.5 2.1 1.4	–	2935 137 1044 155	62.9	30	–	–	Luo et al. (2019)
Hydrazine borane									
10	$Ni_{0.9}Pt_{0.1}$/MIL-101	3.35	–	49.185	10.5 (for BH_3) 43.9 (for N_2H_4)	50	6	100	Bai et al. (2022)
11	$Ni_{0.5}Fe_{0.5}$–CeO_x/ MIL-101	10.3	–	5.855	33.36 45.14	70	5	100	Li et al. (2018b)
12	$Ni_{0.9}Pt_{0.1}$/MIL-101	2.5	–	25.25	18.6 44.6	50	20	100	Zhang et al. (2017)

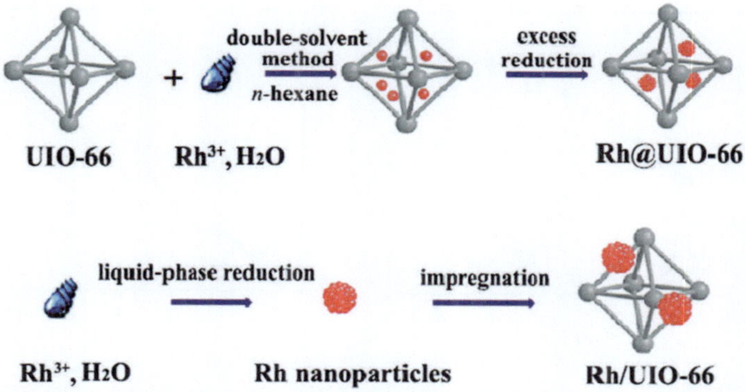

Fig. 19.7 Schematic representation of double-solvent method for the synthesis of Rh@UIO-66 and simple impregnation strategy for Rh/UiO-66. Reproduced with the permission from Chemical Communications 55 (32) (2019) 4699–4702 (Zhang et al. 2019)

Rh-based catalysts generally show excellent hydrolysis of NH_3BH_3. Therefore, Rh NPs supported on UiO-66-NH_2 is also investigated for NH_3BH_3 hydrolysis by Cheng et al. (2021). They reported the in situ formation of Rh NPs on UiO-66-NH_2. The doping of NH_2 in the UiO-66 matrix helps to stabilize Rh NPs and impedes their sintering during preparation. The XPS analysis indicates a strong interaction between the metal and support in the catalyst, Rh/UiO-66-NH_2, because of electron transfer to NH_2 groups. The Rh/UiO-66-NH_2 catalyst with 4.38 wt.% Rh loading resulted in excellent catalytic activity, showing a maximum TOF of 876.6 min^{-1} at 25 °C. Low loading of Rh in the catalyst results in the less availability of catalytic active sites; on the other hand, high Rh loading results in the aggregation of Rh NPs. The kinetic studies of the reaction showed that it follows first- and zero-order kinetics with respect to Rh and NH_3BH_3 concentration, respectively. The E_a of the reaction is 22.3 kJ mol^{-1}, which is comparatively low with respect to other catalysts. The Rh/UiO-66-NH_2 catalyst showed good reusability and maintained good activity after five consecutive cycles. Slight agglomeration of Rh NPs and the adsorption of by-product (BO_2^-) reduces the catalytic activity.

MOF-functionalized flow-through wood-based catalytic microreactors have been studied to continuously release hydrogen from NH_3BH_3 by Keplinger and his co-workers (Tu et al. 2022). They functionalized beech wood scaffolds with UiO-66 MOF to uniformly support AgPd bimetallic NPs. The low-cost, renewable wood microreactors provide high-efficiency natural microchannels for liquid transportation. Beech wood species were selected for functionalization because of their higher elastic modulus than spruce, basswood, and poplar wood species. In the synthesized catalyst, AgPd/UiO-66@beech, MOF particles containing Ag and Pd were uniformly distributed over the wood microchannel surfaces. From argon sorption, the reduced pore volume of UiO-66 suggests that metal NPs are present on the outer surface and/or inside the pores of UiO-66. The bimetallic AgPd/UiO-66@beech catalysts exhibit a high rate of 45.4 mol H_2 (mol metal)$^{-1}$ min^{-1} for NH_3BH_3 hydrolysis compared to

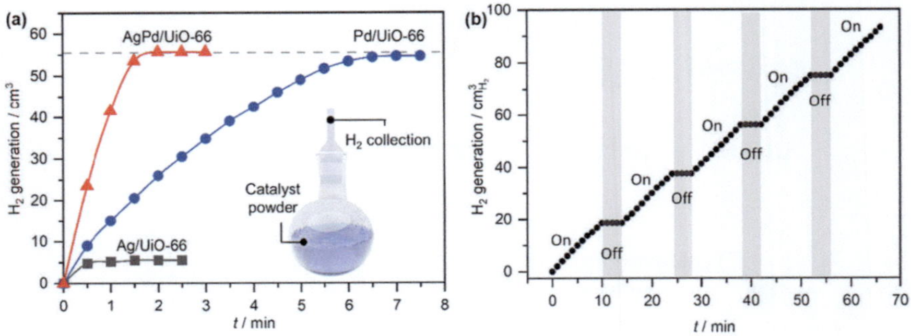

Fig. 19.8 **a** Hydrogen generation performance of AgPd/UiO-66, Pd/Uio-66, and Ag/UiO-66. The inset is set up for hydrogen production using the powder catalysts in the batch mode. **b** Controllable switch on/off hydrogen release using AgPd/UiO-66@beech. Flow rate is 4- and 0–cm^3 min^{-1} for on and off, respectively. Reproduced with the permission from ACS Applied Materials and Interfaces 14 (6) (2022) 8417–8426 (Tu et al. 2022)

its monometallic counterparts, Ag/UiO-66 and Pd/UiO-66, as shown in Fig. 19.8a. This is because Ag behaves as a promoter of palladium, resulting in a higher rate of bimetallic catalyst. One of the major advantages of this structured catalyst over powder catalysts is its controllable switch on/off hydrogen release (Fig. 19.8b). The wood microstructure strongly influences the performance of hydrogen generation. The catalyst shows good reusability and great potential for large-scale applications.

MOF MIL-53(Al) has also been studied for the dehydrogenation of NH_3BH_3. Zhang et al. (2018a) used amine-functionalized MIL-53(Al) to immobilize Ru NPs within the pores via an in situ impregnation-reduction approach. The presence of the amine group in the catalyst function as an anchoring site to capture Ru (III) ions. From the N_2 sorption isotherms, the reduced surface area, pore size, and pore volume of Ru/MIL-53(Al)-NH_2 compared to MIL-53(Al)-NH_2 indicates that the fine Ru NPs are located in the pores of the MOF. The 18.10 wt.% Ru/MIL-53(Al)-NH_2 performed higher activity than the other wt.% Ru loaded catalysts toward NH_3BH_3 dehydrogenation with a TOF of 287 mol H_2 (mol Ru)$^{-1}$ min^{-1} and E_a of 30.5 kJ mol^{-1} at 25 °C. With the increase in the Ru loading, the rate of NH_3BH_3 dehydrogenation follows a volcano-shaped trend due to the agglomeration of Ru NPs on higher Ru loading. The ICP-AES analysis of Ru/MIL-53(Al) and Ru/MIL-53(Al)–NH_2 revealed that Ru loading is higher in Ru/MIL-53(Al)–NH_2, although an equal amount of $RuCl_3$ was added. This is because the amine group of the linker helped in anchoring the Ru (III) ions and resulting in high Ru loaded Ru/MIL-53(Al)–NH_2 that performed better NH_3BH_3 hydrolysis activity compared to Ru/MIL-53(Al). Reusability studies of the catalyst show that the catalyst retained around 72.4% of the initial activity in the fifth cycle. After multiple cycles, the viscosity of the solution increases, impeding the diffusion of NH_3BH_3 toward Ru NPs and thus reducing the activity. Adsorption of increasing metaborate on the active sites may be another reason for its reduced activity. In another study, Menghuan et al. (2018) synthesized Ru_1Co_1@MIL-53(Al) catalyst by immobilizing RuCo bimetallic NPs on MIL-53(Al) and checked its activity for NH_3BH_3 hydrolysis. Ru_1Co_1@MIL-53(Al) performed higher activity than Co@MIL-53(Al), Ru@MIL-53(Al), RuCo NPs, and RuCo@MIL-53(Al) catalysts with different Ru/Co molar ratios because of the synergistic effect. The synergistic effect is generated from the geometric effect and electronic interactions between Co and Ru atoms, and the interaction between metal NPs and the MOF support. The Ru_1Co_1@MIL-53(Al) catalyst performed the NH_3BH_3 hydrolysis reaction with a TOF and E_a of 87.24 mol H_2 (mol Ru)$^{-1}$ min^{-1} and 34.32 kJ mol^{-1}, respectively, at 25 °C. It showed good reusability without any notable change in activity after five cycles. Pd@MIL-101, containing 0.30 wt.% Pd, was synthesized using the double-solvent method, as shown in Fig. 19.9, and investigated for NH_3BH_3 methanolysis by Zahmakiran and co-workers (Caner et al. 2020). The nitrogen adsorption-desorption isotherm and pore-size distributions of MIL-101 and Pd@MIL-101 indicate the deposition of Pd NPs in the MOF cavities because the surface area and micropore volume of Pd@MIL-101 reduced compared to MIL-101. The Pd@MIL-101 catalyst, with *ca.* 3.5 nm Pd NPs, exhibited good catalytic activity at 25 °C with TOF and E_a of 1080 min^{-1} and 50 kJ mol^{-1}, respectively, showing remarkable stability against sintering and leaching.

Platinum NPs have also been reported to be excellent catalytic materials for hydrogen production from solid chemicals such as NH_3BH_3. MOF-supported bimetallic (Pt NPs with some other active metal NPs) catalysts show higher dehydrogenation activity due to the synergistic effect. One such study reported Ni-Pt NPs supported on ZIF-8 support catalysts with varying compositions of mono-dispersed Ni-Pt NPs prepared by the co-reduction technique (Fu et al. 2018). The prepared catalysts were evaluated for hydrogen evolution from NH_3BH_3 hydrolysis at 20 °C. The Ni_2Pt@ZIF-8 catalyst shows much superior activity due to very significant volcano-type positive synergistic effect of Ni_2Pt compared to other prepared catalysts with varying Ni-Pt amounts (Ni@ZIF-8, Ni_3Pt@ZIF-8, NiPt@ZIF-8, $NiPt_2$@ZIF-8, $NiPt_3$@ZIF-8, and Pt@ZIF-8). It also shows high activity than other types of catalysts when Ni and Pt NPs are replaced by other transition metal NPs and noble metal NPs (Co_2Pt@ZIF-8, Cu_2Pt@ZIF-8, Ni_2Rh@ZIF-8, and Ni_2Ru@ZIF-8). Further, the

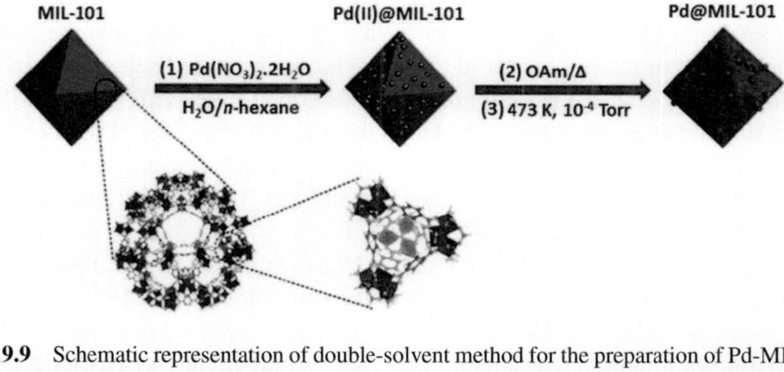

Fig. 19.9 Schematic representation of double-solvent method for the preparation of Pd-MIL-101. Reproduced with the permission from New Journal of Chemistry 44(29) (2020) 12435–12439 (Caner et al. 2020)

ZIF-8 support was found to be excellent support than other types such as SBA-15, graphene oxide, and active carbon with these Ni-Pt NPs. In another study, Au-Ni alloy NPs incorporated in MIL-101 support without aggregation of Au-Ni NPs on the external surface were synthesized by double-solvents method (to control the size and location of the MIL-101 immobilized Au–Ni NPs) along with a liquid-phase concentration-controlled reduction approach (Cai et al. 2013). Prepared catalysts were tested for hydrogen evolution from NH_3BH_3 hydrolysis at ambient temperature. The $Au_{0.07}Ni_{0.93}$/MIL-101 catalyst exhibits higher activity compared to other compositions, Au/MIL-101 and Ni/MIL-101. The very small size of Au and Ni NPs, the synergistic effect between Au and Ni, and the much better dispersion of Au–Ni NPs into the porous structure of MIL-101 are responsible for the better activity of this catalyst for NH_3BH_3 hydrolysis.

Ru metal is widely used for the catalytic hydrolysis of $NaBH_4$ due to its excellent catalytic behavior and comparatively lower price than other noble metals. Metal NPs supported on ZIF-67 showed good activity toward $NaBH_4$ hydrolysis due to its high stability in water even at high temperatures and hierarchically consisted of Co, which is also an active catalytic component for $NaBH_4$ hydrolysis in an alkaline medium. Tuan and Lin (2018) synthesized Ru/ZIF-67 composites by immobilizing Ru particles of average size 41.6 ± 7.1 nm on ZIF-67 nanocrystals, as shown in Fig. 19.10. The presence of Ru in the catalyst has been confirmed by TEM and EDS analyses. The performance of the catalyst depends on the amount of Ru loading. Higher loading of Ru in Ru/ZIF tremendously boosted the hydrogen production rate and reduced the E_a of the reaction. With 9 wt.% of Ru, the catalyst Ru-9.0/ZIF-67 exhibits a rate of 5520 mL min^{-1} g^{-1} and 22,400 mL min^{-1} g^{-1} for hydrolysis of $NaBH_4$ at 30 °C and 60 °C, respectively. But the reduction in E_a with higher loading of Ru is not significant after 4.0 wt.% Ru loading. The hydrogen generation rate increases with the increase in Ru/ZIF-67 concentration, and the concentration of $NaBH_4$ does not affect the rate. Turn et al. investigated the role of NaOH on the rate of hydrolysis of $NaBH_4$. In the presence of 7.5 wt.% of NaOH, the E_a of the reaction is low, 36.4 kJ mol^{-1}, but a higher concentration of NaOH ($\geq$10 wt.%) negatively impacts the rate. The catalyst is quite stable and shows good reusability as the rate of hydrogen generation and H_2 yield remains almost the same for over five subsequent cycles. In another report, Wu et al. (2019) investigated the effect of Pd nanoclusters supported on ZIF-67 for $NaBH_4$ hydrolysis. They synthesized Pd nanoclusters supported on partially decomposed cobalt-based ZIF (Pd/PD-ZIF-67) through solution deposition. Because of the synergetic effect between Pd nanoclusters and improved conductivity of partially decomposed ZIF-67, the catalyst showed good activity with a TOF and E_a of 495.0 mol H_2 (mol Pd)$^{-1}$ min^{-1} and 58.5 kJ mol^{-1}, respectively, toward $NaBH_4$ hydrolysis at 25 °C.

NH_2-UiO-66 is widely used as support for NPs owing to its large surface area and the amine groups of the linker, which provide binding sites for the metal ions. Farrag (2022) incorporated Ru-Co nanocluster having a mean size of 1.5 nm inside the NH_2-UiO-66 framework through an anionic exchange technique followed by reduction of metal ions using hydrazine hydrate. The Ru–Co/NH_2-UiO-66, with 2 wt.%, exhibits an excellent rate of 10526 mL H_2 g^{-1} min^{-1} for hydrolysis of $NaBH_4$. Ultra-small

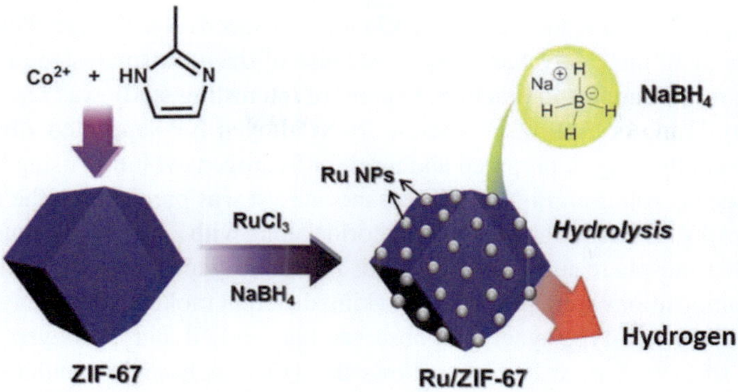

Fig. 19.10 A schematic representation of the synthesis of Ru/ZIF-67 and dehydrogenation of $NaBH_4$ over Ru/ZIF-67 surface. Reproduced with the permission from Chemical Engineering Journal 351 (2018) 48–55 (Tuan and Lin 2018)

particle size, the synergistic effect of Ru and Co, and the higher surface area of MOF support collectively contributed to the excellent performance of the catalyst. The E_a of the reaction is 42.55 kJ mol^{-1}, and the catalyst shows good reusability for several cycles as there are no significant changes in their activity.

Alloy NPs supported on MOF have been shown to exhibit excellent catalytic performance for the dehydrogenation of $N_2H_4BH_3$. Yang et al. (2018) synthesized NiFePd/MIL-101 catalyst using trimetallic NiFePd NPs anchored on MIL-101 through impregnation, followed by the co-reduction method. The TEM and EDX analysis confirmed that the trimetallic NPs having a size of 2–3 nm are dispersed on MIL-101. From BET surface area analysis, the reduced surface area of NiFePd/ MIL-101, compared to MIL-101, attributes to the occupied pores and the surface area of MIL-101 by highly dispersed NPs. The XPS analysis indicates the electronic communication between Pd, Ni, and Fe, which may lead to their improved catalytic activity. The $Ni_{0.36}Fe_{0.24}Pd_{0.4}$/MIL-101 catalyst shows better performance for complete dehydrogenation of aqueous alkaline $N_2H_4BH_3$ solution with 100% H_2 selectivity compared to $Ni_{0.36}Fe_{0.24}Pd_{0.4}$NPs, and its mono- and bimetallic counterparts. The catalyst shows a TOF of 60 h^{-1} at 50 °C, and E_{a1} for the hydrolysis of BH_3 and E_{a2} for N_2H_4 decomposition are 30.3 and 58.1 kJ mol^{-1}, respectively. The catalyst is stable and reusable as there is only a slight decrease in activity after five cycles.

In another report, Zhang et al. (2018b) synthesized Rh-Ni alloy NPs having an average size of 2.8 nm immobilized on MIL-101 through impregnation and then reduction of metal precursors. From the N_2 physisorption measurement, the reduced surface area and pore size of RhNi/MIL-101 compared to MIL-101 implies that tiny RhNi NPs occupy the pores of MIL-101 in RhNi/MIL-101. The catalyst, $Rh_{0.8}Ni_{0.2}$/ MIL-101, performs an excellent activity for hydrogen production from the aqueous solution of $N_2H_4BH_3$ with a TOF of 1200 mol H_2 (mol Rh Ni)$^{-1}$ h^{-1} at 50 °C. The E_{a1} for the hydrolysis of BH_3 and E_{a2} for N_2H_4 decomposition over $Rh_{0.8}Ni_{0.2}$/

MIL-101 are 17.5 ± 2.0 and 47.1 ± 1.2 kJ mol^{-1}, respectively. The excellent activity and durability of the catalyst are mainly because of the ultra-small size of NPs, and the strong metal-support interaction. One more interesting study was reported very recently by Wang et al. (2022) wherein Pt-Ni alloyed NP-supported ZIF-8 nanocomposite catalysts were prepared and used for hydrogen evolution using hydrazine borane aqueous solution. Pt-Ni@ZIF-8 nanocatalyst was prepared by the reduction of platinum(IV) chloride and nickel(II) chloride along with ZIF-8 in aqueous solution under inert atmosphere and using $NaBH_4$ as reducing agent (Fig. 19.11 (scheme 1)). Various compositions were prepared by taking different molar ratios of the Ni and Pt precursors. Volcano-type synergy is observed between Ni and Pt alloyed NPs with the ZIF-8 support. Figure 19.11a–g shows the TEM, high-angle annular dark-field (HAADF)-STEM, and SE-STEM images for Pt-Ni@ZIF-8 catalyst which reveals the ultrafine nano-size particle nature as well as the uniform dispersion of the metal particles over the ZIF-8 support having an average particle size in the range of 1.8–0.3 nm. The Pt and Ni NP-supported ZIF-8 with varying amounts of Ni and Pt were tested for hydrogen generation from hydrazine borane. Among all, the Pt-Ni@ZIF-8 (with equimolar amounts of Ni and Pt) nanocatalyst shows excellent catalytic performance in the dehydrogenation of $N_2H_4BH_3$ (100% hydrogen selectivity) at room temperature which is found to be one of the best catalysts reported for this reaction. The high proficiency of PtNi@ZIF-8 catalyst is mainly due to homogeneous dispersion of Ni and Pt nanoalloys, ultrafine nano-size of the particles, and significant triple volcano-type synergy among ZIF-8, Pt, and Ni.

The catalysts for dehydrogenation of $N_2H_4BH_3$ perform better in alkaline conditions because OH^- converts undesirable $N_2H_5^+$–N_2H_4 (Eq. 19.1) and accelerates the rate-limiting deprotonation step (Eq. 19.8), and also prevents the generation of NH_3 byproduct (Yang et al. 2018):

$$N_2H_5^+ + OH^- \rightarrow N_2H_4 + H_2 \qquad (19.7)$$

$$N_2H_4 \rightarrow N_2H_3^* + H^* \qquad (19.8)$$

Table 19.2 briefs out the properties of non-noble metals supported MOF-based catalysts reported in the literature toward generation of hydrogen from the solid hydrogen storage materials.

19.5 Conclusions

The increased demand for energy can be addressed by utilizing hydrogen as an alternative. To avoid the practical difficulties associated with the storage, release, and transportation of hydrogen, the research community has utilized solid hydrogen storage materials as effective hydrogen precursor materials. The utilization of porous MOFs enhances the uniform dispersion of metal NPs over the MOF. Moreover,

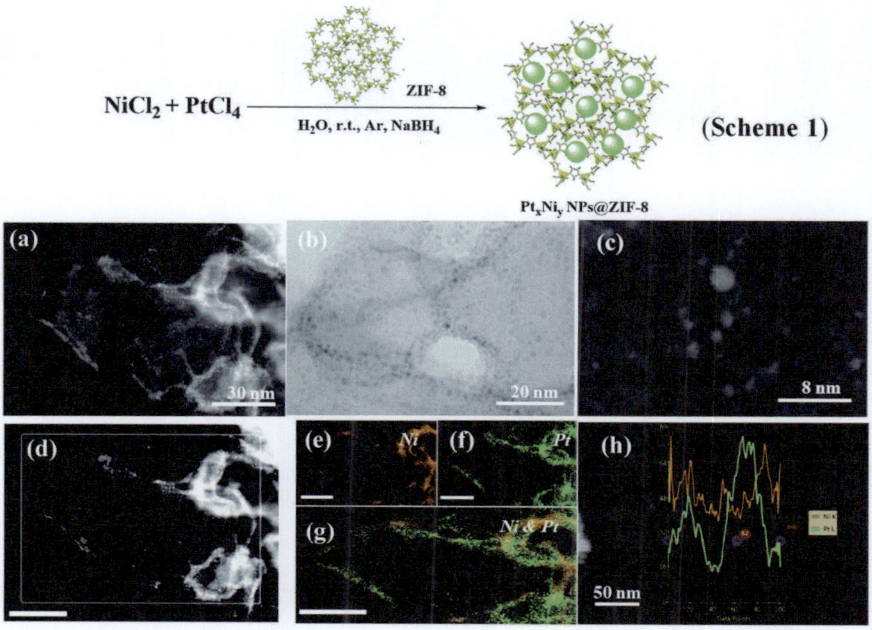

Fig. 19.11 A scheme showing the synthesis of the Pt_xNi_y@ZIF-8 nanocatalyst. **a** HAADF-STEM image. **b** TEM image **c** SE-STEM image. **d–g** HAADF-STEM image and element mapping images (scale bar 50 nm). **h** The line scan reveals the PtNi nanoalloys. Reproduced with the permission from Journal of Material Chemistry A 10 (2022) 17614 (Wang et al. 2022)

the constraint effect of MOFs ensures the construction of small-sized NPs without agglomeration which in turn enhances the catalytic activity of the supported catalysts. Different metal-based MOFs such as MIL-101 (Cr as a metal node), ZIF-8 (Zn as a metal node), ZIF-67 (Co as a metal node), and MIL-96 (Al as a metal node) have been studied and reported for hydrogen generation studies. The synthesis protocol, such as ultrasound-assisted rapid in situ reduction and wet chemical reduction, for developing heterogenous catalysts produces metal nanocatalysts with an ultrafine size of ~1.6–2.7 nm. The semicrystalline or amorphous nature of the NPs is also vital for the activity of the catalysts as such NPs enhance the activity due to the presence of structural defects which act as the activation sites for the reactants.

The noble metals such as Ru, Rh, Ag, and Pd have been reported for the hydrolysis of NH_3BH_3, $NaBH_4$, and $N_2H_4BH_3$, with the highest TOF of 1080 min^{-1} and activation energy of 50 kJ mol^{-1} for the Pd-supported MIL-101 catalyst. It has also been shown by the reported studies that the activity of the catalysts can be enhanced by functionalizing the support MOF with groups such as $-NH_2$ as it increases the anchoring sites for the NPs and thereby the uniform dispersion of the same. The noble metal-free catalysts such as NPs of Cu, Co, Ni, and Fe also show considerable activity for NH_3BH_3, $NaBH_4$, and $N_2H_4BH_3$ hydrolysis. These non-noble metal catalysts supported on different MOFs such as MIL-101, MIL-96, ZIF-8, and ZIF-67

Table 19.2 The properties of noble metal-supported MOF-based catalysts reported in literature toward generation of hydrogen from the solid hydrogen storage materials

SI No.	Catalyst used	Particle size nm	Hydrogen evolution activity						References
			Rate mL min^{-1} g^{-1}	TOF min^{-1}	Activation energy kJ mol^{-1}	Temperature °C	Reusability		
							No. of cycles	Percentage of retention	
Ammonia borane									
1	Rh@UiO-66	2.7	–	490.6	38.4	Room temp	4	100	Zhang et al. (2019)
2	Rh/UiO-66	–	–	62.9	–	Room temp	4	–	Zhang et al. (2019)
3	Rh/UiO-66-NH_2	3.38	–	876.7	22.3	25	5	62.5	Cheng et al. (2021)
4	AgPd/UiO-66@beech	–	–	45.4	–	Room temp	5	>70	Tu et al. (2022)
5	Ru/MIL-53(Al)-NH_2	1.22 ± 0.34	–	287	30.5	25	5	86.3	Zhang et al. (2018a)
6	Ru_1Co_1@MIL-53(Al)	1.10 ± 0.20	–	87.24	34.32	25	5	77	Menghuan et al. (2018)
7	Pd@MIL-101	3.5	–	1080	50	25	5	>85	Caner et al. (2020)
Sodium borohydride									
8	Ru-9.0/ZIF-67	41.6 ± 7.1	5520	–	39	30	5	100	Tuan and Lin (2018)
9	Pd/PD-ZIF-67	3.4	–	495	58.5	25	5	90	Wu et al. (2019)
10	Ru-Co/NH_2-UiO-66	1.5	10526	–	42.55	30	5	–	Farrag (2022)

(continued)

Table 19.2 (continued)

SI No.	Catalyst used	Particle size nm	Hydrogen evolution activity						References
			Rate mL min^{-1} g^{-1}	TOF min^{-1}	Activation energy kJ mol^{-1}	Temperature °C	Reusability		
							No. of cycles	Percentage of retention	
Hydrazine borane									
11	$Ni_{0.36}Fe_{0.24}Pd_{0.4}$/MIL-101	2–3	–	1	30.3 (for BH_3) 58.1 (for N_2H_4)	50	5	100	Yang et al. (2018)
12	$Rh_{0.8}Ni_{0.2}$/MIL-101	2.8	–	20	17.5 ± 2.0 (for BH_3) 47.1 ± 1.2 (for N_2H_4)	50	6	100	Zhang et al. (2018b)
13	Pt/Ni@ZIF-8	1.8 ± 0.3	–	9.54	19.94 (for BH_3) 30.89 (for N_2H_4)	Room temp	5	–	Wang et al. (2022)

show a TOF varying from 0.8 to 2935 min^{-1} with activation energy ranging from 29 to 62 kJ mol^{-1}, ensuring the efficiency of the catalyst.

The alloying of metals enhances the catalytic activity via the electron synergy between the metals. This in turn increases the efficiency of the catalyst. Moreover, alloying the transition metals with other noble metals helps in the development of an economically feasible catalyst with superior catalytic activity 320.7 min^{-1} for RuCo supported over MIL-96. The recyclability studies reported for hydrogen generation using the MOF-supported metal nanocatalysts indicate that they are highly stable and can be re-used effectively even after 30 cycles. Thus, it is clear from the reported literature that MOFs can be a suitable support for the nanocatalyst materials and superior hydrogen generation activity can be achieved by tuning the property of the support MOF as well as by alloying the nanometal catalyst with other metal ions.

Although mono/multi metallic nanocatalysts show activity toward hydrogen generation from the chemical hydrogen storage sources, the durability of the catalyst has to be treated as a major issue from the practical viewpoint. Moreover, the generation of byproducts such as borates is another important issue. As they are highly stable compounds, the formation of the same and its deposition on the catalysts' surface may affect the reusability of the catalysts. Future studies have to be focused on exploring the possibility of reusing/recycling the catalyst as well as the regeneration of precursor hydrogen storage materials like NH_3BH_3. Aerial oxidation is another factor that will affect the activity of catalysts, especially in the case of non-noble metal catalysts. Subsequent research has to be carried out to prevent the activity loss of first raw transition metals via aerial oxidation. Like metal borides, metal phosphides are gaining much attention among researchers due to their structural stability and catalytic ability. These unique properties of metal phosphides can be exploited for the upcoming hydrogen economy. Atomic level understanding is very crucial to developing a robust catalyst and the catalytic reaction has to be critically investigated to achieve the optimum experimental conditions. The design and development of simulation calculations are very advantageous in this regard. Additionally, the utilization of in situ characterization techniques will help to understand the structure as well as the electronic nature of the catalyst material during the reaction process.

Acknowledgements We acknowledge MLP-10 CSIR-CLRI for funding. AP thanks the Anusandhan National Research Foundation (ANRF), New Delhi, for SERB-National Postdoctoral Fellowship Award. RKB thanks the Council of Scientific and Industrial Research (CSIR), New Delhi, for the research fellowship. CSIR-CLRI communication number 1792.

Authors Contribution The literature on non-noble metal-supported catalysts have been reviewed by AP. RKB and RKK carried out the literature survey on noble metal catalysts. AAV has conceptualized the chapter; supervised, investigated, and analyzed the discussions; coordinated with the team; and reviewed and edited the manuscript.

References

(a) Aijaz, A., Karkamkar, A., Choi, Y.J., Tsumori, N., Ronnebro, E., Autrey, T., Shioyama, H., Xu, Q.: J. Am. Chem. Soc. **134**, 13926–13929 (2012); (b) Zhu, Q.L., Li, J., Xu, Q.: J. Am. Chem. Soc. **135**, 10210–10213 (2013); (c) Dai, H., Su, J., Hu, K., Luo, W., Cheng, G.: Int. J. Hydrog. Energy **39**, 4947–4953 (2014); (d) Cao, N., Liu, T., Su, J., Wu, X., Luo, W., Cheng, G.: New J. Chem. **38**, 4032 (2014); (e) Li, J., Zhu, Q.L., Xu, Q.: Chem. Commun. **50**, 5899–5901 (2014); (f) Chen, Y.Z., Liang, L., Yang, Q., Hong, M., Xu, Q., Yu, S.H., Jiang, H.L.: Mater. Horiz. **2**, 606–612 (2015); (g) Roy, S., Pachfule, P., Xu, Q.: Eur. J. Inorg. Chem. **27**, 4353–4357 (2016); (h) Xia, B., Liu, C., Wu, H., Luo, W., Cheng, G.: Int. J. Hydrog. Energy **40**, 16391–16397 (2015); (i) Shang, N.Z., Feng, C., Gao, S.T., Wang, C.: Int. J. Hydrog. Energy **41**, 944–950 (2016)

Alpaydın, C.Y., Gülbay, S.K., Colpan, C.O.: Int. J. Hydrog. Energy **45**(5), 3414–3434 (2020)

Bai, Y., Liu, Y., Shang, H., Li, S., Liang, J.: New J. Chem. **46**(29), 13971–13980 (2022)

(a) Cai, Y., Li, X., Zhang, Y., Wei, X., Wang, K., Chen, J.: Angew. Chemie **125**(45), 12038–12041 (2013); (b) Song, F., Zhu, Q., Yang, X., Zhan, W., Pachfule, P., Tsumori, N., Xu, Q.: Adv. Energy Mater. **8**(1), 1701416 (2018)

Caner, N., Yurderi, M., Bulut, A., SaydanáKanberoglu, G., Kaya, M., Zahmakiran, M.: New J. Chem. **44**(29), 12435–12439 (2020)

Cheng, W., Peng, Y., Wang, Y., Jiang, W., Long, Y., Fan, G.: Int. J. Hydrog. Energy **46**(2), 2204–2212 (2021)

Cui, C., Tang, Y., Ziaee, M.A., Tian, D., Wang, R.: ChemCatChem **10**(6), 1431–1437 (2018)

(a) Fang, Y., Li, J., Togo, T., Jin, F., Xiao, Z., Liu, L., Drake, H., Lian, X., Zhou, H.-C.: Chem **4**(3), 555–563 (2018); (b) Fang, Y., Xiao, Z., Li, J., Lollar, C., Liu, L., Lian, X., Yuan, S., Banerjee, S., Zhang, P., Zhou, H.: Angew. Chemie Int. Ed. **57**(19), 5283–5287 (2018)

Farrag, M.: J. Alloys Compd. **920**, 165893 (2022)

Feng, Y., Wang, H., Chen, X., Lv, F., Li, Y., Zhu, Y., Xu, C., Zhang, X., Liu, H.-R., Li, H.: Int. J. Hydrog. Energy **45**(35), 17164–17173 (2020)

Fu, F., Wang, C., Wang, Q., Martinez-Villacorta, A.M., Escobar, A., Chong, H., Wang, X., Moya, S., Salmon, L., Fouquet, E.: J. Am. Chem. Soc. **140**(31), 10034–10042 (2018)

Furukawa, S., Nishimura, G., Takayama, T., Komatsu, T.: Front. Chem. **7**, 138 (2019)

Hou, C.-C., Chen, Q.-Q., Li, K., Wang, C.-J., Peng, C.-Y., Shi, R., Chen, Y.: J. Mater. Chem. A **7**(14), 8277–8283 (2019)

Li, Y., Li, S.: Int. J. Hydrog. Energy **45**(17), 10433–10441 (2020)

Li, Y., Xie, L., Li, Y., Zheng, J., Li, X.: Chem. Eur. J. **15**(36), 8951–8954 (2009)

Li, Q., Yang, W., Li, F., Cui, A., Hong, J.: Int. J. Hydrog. Energy **43**(1), 271–282 (2018a)

Li, S., Wang, H., Wulan, B., Zhang, X., Yan, J., Jiang, Q.: Adv. Energy Mater. **8**(21), 1800625 (2018b)

Li, H., Yan, Y., Feng, S., Zhu, Y., Chen, Y., Fan, H., Zhang, L.: Fuel **255**, 115771 (2019)

(a) Li, Q., Kim, H.: Fuel Process. Technol. **100**, 43–48 (2012); (b) Li, P. Z., Aranishi, K., Xu, Q.: ChemComm **48**(26), 3173–3175 (2012)

Liang, Z., Xiao, X., Yu, X., Huang, X., Jiang, Y., Fan, X., Chen, L.: J. Alloys Compd. **741**, 501–508 (2018)

Liu, P., Gu, X., Kang, K., Zhang, H., Cheng, J., Su, H.: ACS Appl. Mater. Interfaces **9**(12), 10759–10767 (2017)

Liu, M., Zhou, L., Luo, X., Wan, C., Xu, L.: Catalysts **10**(7), 788 (2020)

Lu, D., Yu, G., Li, Y., Chen, M., Pan, Y., Zhou, L., Yang, K., Xiong, X., Wu, P., Xia, Q.: J. Alloys Compd. **694**, 662–671 (2017)

Luo, C., Fu, F., Yang, X., Wei, J., Wang, C., Zhu, J., Huang, D., Astruc, D., Zhao, P.: ChemCatChem **11**(6), 1643–1649 (2019)

Menghuan, C., Zhou, L., Di, L., Yue, L., Honghui, N., Yaxi, P., Hongkun, X., Weiwei, P., Shuren, Z.: Int. J. Hydrog. Energy **43**(3), 1439–1450 (2018)

(a) Mori, K., Futamura, Y., Masuda, S., Kobayashi, H., Yamashita, H.: Nat. Commun. **10**(1), 1–10 (2019); (b) Verma, P., Yuan, K., Kuwahara, Y., Mori, K., Yamashita, H.: Appl. Catal. B Environ. **223**, 10–15 (2018)

Navlani-García, M., Mori, K., Salinas-Torres, D., Kuwahara, Y., Yamashita, H.: Front. Mater. **6**, 44 (2019)

Reddy, D.A., Kim, Y., Gopannagari, M., Kumar, D.P., Kim, T.K.: Sustain. Energy Fuels **5**(6), 1597–1618 (2021)

(a) Schlesinger, H.I., Brown, H.C., Finholt, A.E., Gilbreath, J.R., Hoekstra, H.R., Hyde, E.K.: J. Am. Chem. Soc. **75**(1), 215–219 (1953); (b) Chandra, M., Xu, Q.: J. Power Sour. **156**(2), 190–194 (2006); (c) Lai, Q., Alligier, D., Aguey-Zinsou, K.F., Demirci, U.B.: Nanoscale Adv. **1**(7), 2707–2717 (2019); (d) Alpaydın, C.Y., Gülbay, S.K., Colpan, C.O.: Int. J. Hydrog. Energy **45**(5), 3414–3434 (2020)

Su, X., Li, S.: Int. J. Hydrog. Energy **46**(27), 14384–14394 (2021)

Sun, J.-L., Chen, Y.-Z., Ge, B.-D., Li, J.-H., Wang, G.-M.: ACS Appl. Mater. Interfaces **11**(1), 940–947 (2018)

Sun, Q., Wang, N., Xu, Q., Yu, J.: Adv. Mater. **32**(44), 2001818 (2020)

(a) Sun, Q., Wang, N., Zhang, T., Bai, R., Mayoral, A., Zhang, P., Zhang, Q., Terasaki, O., Yu, J.: Angew. Chemie Int. Ed. **58**(51), 18570–18576 (2019); (b) Sun, Q., Wang, N., Bing, Q., Si, R., Liu, J., Bai, R., Zhang, P., Jia, M., Yu, J.: Chem **3**(3), 477–493 (2017)

Tu, K., Büchele, S., Mitchell, S., Stricker, L., Liu, C., Goldhahn, C., Allaz, J., Ding, Y., Günther, R., Zhang, Z.: ACS Appl. Mater. Interfaces **14**(6), 8417–8426 (2022)

Tuan, D.D., Lin, K.-Y.A.: Chem. Eng. J. **351**, 48–55 (2018)

Wang, C., Liu, X., Wu, Y., Astruc, D.: J. Mater. Chem. A **10**, 17614 (2022)

White, R.J., Luque, R., Budarin, V.L., Clark, J.H., Macquarrie, D.J.: Chem. Soc. Rev. **38**(2), 481–494 (2009)

Wu, C., Guo, J., Zhang, J., Zhao, Y., Tian, J., Isimjan, T.T., Yang, X.: Renew. Energy **136**, 1064–1070 (2019)

Xiaopeng, Q., Zhiqiang, Y., Zeping, L., Xiangyu, G., Rui, J., Zhimou, X., Cunhua, C., Jing, P.: Int. J. Hydrog. Energy **42**(51), 30037–30043 (2017)

Xu, W., Li, W., Wen, H., Ding, J., Liu, Y., Li, B.: Appl. Catal. B Environ. **286**, 119946 (2021)

Yang, K., Yang, K., Zhang, S., Luo, Y., Yao, Q., Lu, Z.-H.: J. Alloys Compd. **732**, 363–371 (2018)

(a) Yang, K., Zhou, L., Xiong, X., Ye, M., Li, L., Xia, Q.: Microporous Mesoporous Mater. **225**, 1–8 (2016); (b) Roy, S., Pachfule, P., Xu, Q.: Eur. J. Inorg. Chem. **2016**(27), 4353–4357 (2016)

Yu, J., Mu, C., Yan, B., Qin, X., Shen, C., Xue, H., Pang, H.: Mater. Horizons **4**(4), 557–569 (2017)

Zhang, Z., Zhang, S., Yao, Q., Chen, X., Lu, Z.-H.: Inorg. Chem. **56**(19), 11938–11945 (2017)

Zhang, S., Zhou, L., Chen, M.: RSC Adv. **8**(22), 12282–12291 (2018a)

Zhang, Z., Zhang, S., Yao, Q., Feng, G., Zhu, M., Lu, Z.-H.: Inorg. Chem. Front. **5**(2), 370–377 (2018b)

Zhang, H., Huang, M., Wen, J., Li, Y., Li, A., Zhang, L., Ali, A.M., Li, Y.: Chem. Commun. **55**(32), 4699–4702 (2019)

(a) Zhu, Q.-L., Li, J., Xu, Q.: J. Am. Chem. Soc. **135**(28), 10210–10213 (2013); (b) Wang, C., Tuninetti, J., Wang, Z., Zhang, C., Ciganda, R., Salmon, L., Moya, S., Ruiz, J., Astruc, D.: J. Am. Chem. Soc. **139**(33), 11610–11615 (2017)

MIX
Papier aus verantwortungsvollen Quellen
Paper from responsible sources
FSC® C105338

If you have any concerns about our products,
you can contact us on
ProductSafety@springernature.com

In case Publisher is established outside the EU,
the EU authorized representative is:
Springer Nature Customer Service Center GmbH
Europaplatz 3, 69115 Heidelberg, Germany

Printed by Libri Plureos GmbH
in Hamburg, Germany